Contemporary Physicists

Gian-Carlo Rota
David Sharp

Editors

Birkhäuser
Boston · Basel · Stuttgart

James Glimm
Arthur Jaffe

Collected Papers
Volume 2

Constructive Quantum Field Theory

Selected Papers

Birkhäuser
Boston · Basel · Stuttgart
1985

James Glimm
Courant Institute
New York, N.Y.
USA

Arthur Jaffe
Harvard University
Cambridge, Mass.
USA

Library of Congress Cataloging in Publication Data

Glimm, James.
Constructive quantum field theory.

(Contemporary physicists)
Reprint of articles originally published 1968–1980.
Includes bibliographies.
1. Quantum field theory—Addresses, essays, lectures.
2. Statistical physics—Addresses, essays, lectures.
I. Jaffe, Arthur, 1937– . II. Title. III. Series.
QC174.46.G584 1985 530.1'43 84–21525
ISBN-13: 978-1-4612-5420-1 e-ISBN-13: 978-1-4612-5418-8
DOI: 10.1007/ 978-1-4612-5418-8

CIP-Kurztitelaufnahme der Deutschen Bibliothek

Glimm, James:
Collected papers / James Glimm ; Arthur Jaffe. –
Boston ; Basel ; Stuttgart : Birkhäuser

(Contemporary physicists)

NE: Jaffe, Arthur:; Glimm, James: [Sammlung];
Jaffe, Arthur: [Sammlung]

Vol. 2. Constructive quantum field theory :
selected papers. – 1985.

Softcover reprint of the hardcover 1st edition 1985

James Glimm/Arthur Jaffe

Collected Papers – Volume 2

I Construction of $P(\varphi)_2$

II Phase Cell Localization and $\varphi_3{}^4$ Stability

III Phase Transitions Exist

IV Phase Transitions and Critical Behavior

V Particle Structure

VI Bounds on Coupling Constants

VII Confinement and Instantons

VIII Reflection Positivity

Collected Papers – Volume 1

Part II. The Solution of Two-Dimensional Boson Models 153

IV Boson Quantum Field Models:
Part III. Further Developments 183

V The Particle Structure of the Weakly Coupled $P(\varphi)_2$ Model and Other Applications of High Temperature Expansions:
Part I. Physics of Quantum Field Models 203

VI The Particle Structure of the Weakly Coupled $P(\varphi)_2$ Model and Other Applications of High Temperature Expansions:
Part II. The Cluster Expansion 273

I Construction of $P(\varphi)_2$

Introduction

The series of papers constructing the $\lambda\phi_2^4$ quantum field model without cutoffs gave the first example of a quantum field model with a nontrivial scattering matrix. In the first paper, a self adjoint and locally correct Hamiltonian H(B) is constructed for each bounded region B of space-time. The quantum field $\phi(x,t) = \exp(itH(B))\phi(x)\exp(-itH(B))$ can be defined in terms of the time zero value $\phi(x)$. This field is studied in the second paper and is shown to be a self adjoint operator which is independent of B as long as B contains the causal shadow of the space-time point (x,t). The intuitive idea is to define H(B) on the time-zero surface as the correct Hamiltonian for the intersection of this surface with the interior of the light cones which eminate from points in B. Since in a relativistic theory influence propagates at a finite speed (at most the speed of light), local field expressions such as $\phi(x,t)$ can be constructed from a locally correct generator H(B) of time translations. In addition the ground state $\Omega(B)$ for H(B) is shown to exist. The eigenstates of H(B), however, depend on B. Thus expectations of ϕ in $\Omega(B)$ do depend on B.

The Wightman axioms require the existence of a global Hamiltonian H, not just the locally correct H(B)'s, as well as a ground state Ω for H. The construction of these objects is carried out in paper III, using C*-algebra methods. The basic estimate shows the finiteness of the vacuum energy shift per unit volume due to the interaction (after renormalization). As a consequence it follows that local functions $A(\phi)$ of the space-time field $\phi(x,t)$ have convergent power series expansions in the time zero fields $\phi(x,0)$ and their canonical conjugate fields $\pi(x,0)$. Using this expansion, the vacuum states $\{\Omega(B)\}$ are shown to have a subsequence which converges when B increases to R^2. For this subsequence, expectations $< \Omega(B), A\Omega(B) >$ converge to limits $< \Omega, A\Omega >$ which define the vaccum Ω and Hamiltonian H. The question of uniqueness is quite subtle and depends on the value of λ. While first developed for the ϕ^4 interaction, the methods also were applied to the $P(\phi)_2$ and Yukawa$_2$ models. Here P denotes a polynomial, bounded from below to ensure classical stability of the energy.

The paper on the Wightman axioms uses Euclidean functional integrals to establish all the Wightman axioms for a subset of the $\lambda P(\phi) + \phi^2$models constructed by the Hamiltonian approach above, namely those with weak coupling $0 < \lambda \ll 1$. The method relies on a convergent expansion to decouple different space-time regions in expectations of products of fields. For this reason the expansion has become known as the cluster expansion. Uniqueness of the ground state Ω follows for λ in its domain of convergence.

The cluster expansion is sufficiently powerful to investigate the spectrum of $\exp(-tH)$, and hence the spectrum of the Hamiltonian H. To study the spectrum of H on the interval $[0,\varepsilon]$ one proves $< A,(I - P_\varepsilon)\exp(-tH)A > \ \leq O(1)\exp(-\varepsilon t)$ as $t \to \infty$. Here P_ε are the spectral projections of H, which are constructed by an inductive procedure using the cluster expansion.

Boson quantum field Hamiltonians are equivalent to Schrödinger operators acting on an infinite dimensional space. They have the form $H = -\Delta + V$, where Δ is an infinite dimensional Laplacian, where V is a multiplication operator, and where H is bounded from below. Let $E_0 \leq E_1 \leq E_2 \leq \cdots$ denote the eigenvalues of a finite dimensional approximation to H. The cluster expansion method gives a tool to study the gaps $E_{i+1} - E_i$ between the smallest eigenvalues of such operators, uniformly in the dimension of the space on which they act. This is a difficult mathematical problem. Independent of the significance for physics, mathematicians may well wish to study spectral properties of such Laplacians further by these methods. In fact the heat kernel $\exp(-tH)$ can be considered as generating a random process labelled by time. Such random processes arise both in pure mathematics (geometry and topology as well as analysis) and in many fields of application, such as electrical engineering, stochastic control theory and presumably econometrics and population biology. The random field (labeled by space as well as time) has the infinite dimensional Laplacian H as its generator. It would be reasonable to guess that its abstract theory and its applications would be as rich as those of the random process indexed by time alone.

Bibliography

A $\lambda(\phi^4)_2$ quantum field theory without cutoffs I, Phys. Rev. 176, 1945–1951 (1968).

The $\lambda(\phi^4)_2$ quantum field theory without cutoffs II. The field operators and the approximate vacuum, Ann. Math. 91, 362–401 (1970).

The $\lambda(\phi^4)_2$ quantum field theory without cutoffs III. The physical vacuum, Acta Math. 125, 204–267 (1970).

The Wightman axioms and particle structure in the $P(\phi)_2$ quantum field model, with T. Spencer, Ann. Math. 100, 585–632 (1974).

A $\lambda\varphi^4$ Quantum Field Theory without Cutoffs. I

James Glimm*
Department of Mathematics, Massachusetts Institute of Technology, Cambridge, Massachusetts 02139

AND

Arthur Jaffe†
Lyman Laboratory of Physics, Harvard University, Cambridge, Massachusetts 02138
(Received 20 May 1968)

We construct the dynamics for a local quantum field theory in two-dimensional space-time. Our model describes a scalar field φ with a $\lambda\varphi^4$ self-interaction. We rely upon the fact that, in the Heisenberg picture, influence propagates at the speed of light. The resulting dynamics is independent of any cutoff, and hence the theory is formally Lorentz-covariant.

I. INTRODUCTION

In this paper we obtain the dynamics for a local quantum field theory in two-dimensional space-time. We study a scalar $\lambda\varphi^4$ self-interaction, and no cutoffs appear in the solution. Formally, our model is Lorentz-covariant, and a covariant local quantum field theory should result.

This $\lambda\varphi^4$ model is simple because the only momentum-space divergences are removed by Wick-ordering the Hamiltonian density. The vacuum energy has a divergence linear in the volume in each order of perturbation theory, as is expected in any theory with vacuum polarization. In perturbation theory the scattering matrix for this interaction is nontrivial. However, the perturbation series is known to diverge when summed to all orders,[1] so we cannot solve the theory in this manner.

We construct the dynamics in the Heisenberg picture which means that we have a map σ_t which takes the field operators from time zero to time t. The main advantage of the Heisenberg picture, is that the volume divergence does not enter in this formulation.

In order to prove that the theory exists, we introduce a spatial cutoff in the interaction Hamiltonian. Thus the formal Hamiltonian expression

$$H(g)=H_0+\lambda\int_{t=0} :\varphi^4(x):dx$$

is replaced by the operator on Fock space,

$$\begin{aligned} H(g)&=H_0+\int_{t=0} :\varphi^4(x):g(x)dx \\ &=H_0+H_I(g), \end{aligned} \tag{1.1}$$

where the space-dependent coupling constant $g(x)$ is a smooth, non-negative function which vanishes outside a large interval. We derive an estimate for fixed $g(x)$,

$$-[H_0^{1/2},[H_0^{1/2},H_I(g)]]\leq \epsilon H_0^2+b, \tag{1.2}$$

where ϵ is any positive number and $b(\epsilon)$ a suitably large number. This estimate, along with the previously known estimate[2,3]

$$-H_I(g)\leq \epsilon H_0+b, \tag{1.3}$$

and a knowledge of the self adjointness of H_0 and $H_I(g)$, allows us to prove that the total Hamiltonian $H(g)$ is self adjoint. The connection between estimates (1.2) and (1.3) and self adjointness leads to a general theory of singular perturbations developed elsewhere.[4]

We then discuss the Heisenberg picture dynamics

$$\sigma_t(A)=e^{iH(g)t}Ae^{-iH(g)t},$$

of an observable A associated with a bounded region of space at $t=0$. The locality of the Hamiltonian density is essential to our method. As physicists have long known, it ensures that influence propagates in the Heisenberg picture at the speed of light. A local Hamiltonian density, therefore, assures a local theory on the formal level. Guenin pointed out that in perturbation theory the finite propagation speed is not destroyed by a space cutoff.[5] Segal states this result as a rigorous theorem, under the hypothesis that the total and the interaction Hamiltonians both are self adjoint.[6] In other words $\sigma_t(A)$ does not depend on $g(x)$, provided that $g(x)$ equals a constant, i.e., the desired coupling constant λ, for x in some suitably large region. By choosing an appropriate $g(x)$ for each bounded region of space-time, we can patch together a time translation for any operator. A similar argument shows that the theory is local. Since we have proved that the Hamiltonian $H(g)$ is self adjoint, we have constructed a local quantum field theory without a cutoff.

* Supported in part by the National Science Foundation under Grant No. GP-7477.

† Supported in part by the Air Force Office of Scientific Research.

1 A. M. Jaffe, Commun. Math. Phys. **1**, 127 (1965).

2 E. Nelson, in *Mathematical Theory of Elementary Particles*, edited by R. Goodman and I. Segal (The M.I.T. Press, Cambridge, Mass., 1966).

3 J. Glimm, Commun. Math. Phys. **8**, 12 (1968); P. Federbush (to be published).

4 J. Glimm and A. M. Jaffe, Commun. Pure Appl. Math. (to be published).

5 M. Guenin, Commun. Math. Phys. **3**, 120 (1966).

6 I. Segal, Proc. Natl. Acad. Sci. U.S. **57**, 1178 (1967).

II. ESTIMATES ON THE HAMILTONIAN

Let $\mathfrak{F}$ be the Fock space for a massive, neutral scalar field in two-dimensional space-time. The elements of $\mathfrak{F}$ are sequences of functions on momentum space. Let the annihilation and creation operators be normalized by the relation

$$[a(k),a^*(k')]=\delta(k-k'), \tag{2.1}$$

so that the free-field Hamiltonian is

$$H_0=\int a^*(k)a(k)\omega(k)dk,$$

where

$$\omega(k)=(k^2+m^2)^{1/2}.$$

The $t=0$ field

$$\varphi(x)=\frac{1}{(2\pi)^{1/2}\sqrt{2}}\int e^{-ikx}[a^*(k)+a(-k)]\omega(k)^{-1/2}dk \tag{2.3}$$

is used to construct the spatially cut-off interaction Hamiltonian

$$\begin{aligned}H_I(g)&=\int :\varphi^4(x):g(x)dx\\&=\sum_{j=0}^{4}\binom{4}{j}\int a^*(k_1)\cdots a^*(k_j)a(-k_{j+1})\cdots\\&\qquad\times a(-k_4)\tilde{g}\Big(\sum_{i=1}^{4}k_i\Big)\prod_{i=1}^{4}\omega(k_i)^{-1/2}dk_i.\end{aligned} \tag{2.4}$$

The total Hamiltonian for the spatially cut-off interaction is

$$H(g)=H_0+H_I(g). \tag{2.5}$$

We also need the number operator

$$N=\int a^*(k)a(k)dk, \tag{2.6}$$

and the domain

$$D_0=\bigcap_{n=0}^{\infty}D(H_0^n). \tag{2.7}$$

We let $\mathfrak{S}(R^1)$ denote the Schwartz space of infinitely differentiable functions, which, along with all their derivatives, vanish faster than any inverse polynomial at infinity.

Theorem 2.1: For any $\epsilon>0$ and for fixed $g\in\mathfrak{S}(R^1)$ there is a constant b such that as bilinear forms on $D_0\times D_0$

$$-[H_0^{1/2},[H_0^{1/2},H_I(g)]]\leqq\epsilon H_0^2+b \tag{2.8}$$

and

$$-[N,[N,H_I(g)]]\leqq\epsilon N^3+b. \tag{2.9}$$

Let us state two lemmas, which then yield the theorem.

Lemma 2.2: Let W be an operator of the form

$$W=\int w(k)a^*(k_1)\cdots a(-k_m)dk. \tag{2.10}$$

Then

$$\|(N+I)^{-j/2}W(N+I)^{-(m-j)/2}\|\leqq\text{const}\,\|w\|_{L^2},\quad |j|\leqq m \tag{2.11}$$

$$\|(H_0+I)^{-1}[H_0^{1/2},[H_0^{1/2},W]](H_0+I)^{-1}(N+I)^{-(m-4)/2}\|\leqq\text{const}\,\|\omega^{1/2}\Big(\sum_{i=1}^{m}k_i\Big)w\|_{L^2}, \tag{2.12}$$

and

$$\|[H_0^{1/2},[H_0^{1/2},W]](N+I)^{-m/2}\|\leqq\text{const}\,\|\sum_{i=1}^{m}\omega(k_i)w\|_{L^2}. \tag{2.13}$$

Lemma 2.3: Let W be as above, and suppose that $w(k)=0$ whenever $|k_i|>\kappa$ for any i. Then

$$\|[H_0^{1/2},[H_0^{1/2},W]](N+I)^{-(m-2)/2}\|\leqq\text{const}\,\kappa^2\|w\|_{L^2}. \tag{2.14}$$

Proof of Theorem 2.1: Introduce the $t=0$ field φ_κ with an ultraviolet cutoff:

$$\varphi_\kappa(x)=\frac{1}{(2\pi)^{1/2}\sqrt{2}}\int_{-\kappa}^{\kappa}e^{-ikx}[a^*(k)+a(-k)][\omega(k)]^{-1/2}dk.$$

The interaction $H_{I,\kappa}(g)$ is defined by

$$H_{I,\kappa}=\int :\varphi_\kappa^4(x):g(x)dx, \tag{2.15}$$

and formally

$$H_I=\lim_{\kappa\to\infty}H_{I,\kappa}. \tag{2.16}$$

If we write H_I as a sum of five operators of the form W in (2.10), then by Lemma 2.2 taken for the case $m=4$

$$\|(H_0+1)^{-1}[H_0^{1/2},[H_0^{1/2},W]](H_0+1)^{-1}\|\leqq\text{const}\,\|\omega^{1/2}\Big(\sum_{i=1}^{4}k_i\Big)w\|_{L^2}. \tag{2.17}$$

Since the kernel $w(k)$ has an over-all factor $\tilde{g}(\sum_{i=1}^{4}k_i)$, where $\tilde{g}$ is the Fourier transform of the spatial cutoff $g(x)$, the fast decrease of $\tilde{g}(k)$ ensures that $\omega^{1/2}(\sum_{i=1}^{4}k_i)w$ is in L^2. Thus the kernel for the corresponding cutoff interaction term w_κ approximates w in the sense that

$$\|\omega^{1/2}\Big(\sum_{i=1}^{4}k_i\Big)(w-w_\kappa)\|_{L^2}\to 0, \tag{2.18}$$

as $\kappa\to\infty$. This is true for each W making up $H_I(g)$, so we infer that there exists a κ_0 such that for $\kappa>\kappa_0$,

$$\|(H_0+I)^{-1}[H_0^{1/2},[H_0^{1/2},H_I-H_{I,\kappa}]]\times(H_0+I)^{-1}\|\leq\tfrac{1}{2}\epsilon. \quad (2.19)$$

By Lemma 2.3, we have the relations

$$-[H_0^{1/2},[H_0^{1/2},H_{I,\kappa}]]\leq\text{const }\kappa_0^2(N+I)\leq\tfrac{1}{2}\epsilon H_0^2+\text{const}, \quad (2.20)$$

on the domain $D_0\times D_0$. Thus combining (2.20) with the estimate on the tail (2.19),

$$-[H_0^{1/2},[H_0^{1/2},H_I]]=-[H_0^{1/2},[H_0^{1/2},H_I-H_{I,\kappa}]]-[H_0^{1/2},[H_0^{1/2},H_{I,\kappa}]]\leq\epsilon H_0^2+b,$$

which is (2.8). In order to prove (2.9), we expand $H_I(g)$ once more as a sum of terms W of the form (2.10). Since $[N,W]=\text{const }W$, and W is a fourth-order operator with an L^2 kernel, we infer from (2.11) that

$$|(\psi[N,[N,W]]\psi)|\leq\text{const }(\psi,(N+I)^2\psi)\leq\tfrac{1}{5}\epsilon(\psi,N^3\psi)+\tfrac{1}{5}b(\psi,\psi). \quad (2.21)$$

As $H_I(g)$ has 5 such terms,

$$[N,[N,H_I]]\leq\epsilon N^3+b,$$

which completes the proof of the theorem.

Proof of Lemmas 2.2 and 2.3: Let θ be an n-particle wave function. Then $(a(k)\theta)(k_1,\cdots,k_{n-1})$ is an $(n-1)$-particle state defined by

$$(a(k)\theta)(k_1,\cdots,k_{n-1})=(\sqrt{n})\theta(k,k_1,\cdots,k_{n-1}). \quad (2.22)$$

Using (2.22) and the Schwarz inequality, it is easy to establish the bound (2.11).

Note that

$$([H_0^{1/2},a(k)]\theta)(k_1,\cdots,k_{n-1})=\lambda_1(k;k_1,\cdots,k_{n-1})(a(k)\theta)(k_1,\cdots,k_{n-1})\equiv(a_1(k)\theta)(k_1,\cdots,k_{n-1}), \quad (2.23)$$

where

$$\lambda_1(k;k_1,\cdots,k_{n-1})=[\omega(k)+\sum_{i=1}^{n-1}\omega(k_i)]^{1/2}-[\sum_{i=1}^{n-1}\omega(k_i)]^{1/2}. \quad (2.24)$$

For convenience, we have defined a modified annihilation operator $a_1(k)$, which decreases the particle number by one.

For $x\geq 0$, we see that

$$(1+x)^{1/2}-1\leq x^{1/2},$$

so

$$|\lambda_1(k;k_1,\cdots,k_{n-1})|\leq[\omega(k)]^{1/2}. \quad (2.25)$$

Furthermore, for $x\geq 0$,

$$(1+x)^{1/2}-1\leq\tfrac{1}{2}x,$$

so

$$|\lambda_1(k;k_1,\cdots,k_{n-1})|\leq\tfrac{1}{2}\omega(k)[\sum_{i=1}^{n-1}\omega(k)]^{-1/2}\leq\tfrac{1}{2}\omega(k)\frac{1}{m^{1/2}(n-1)^{1/2}}. \quad (2.26)$$

In the same way

$$([H_0^{1/2},[H_0^{1/2},a(k)]]\theta)(k_1,\cdots,k_{n-1})=\lambda_2(k;k_1,\cdots,k_{n-1})(a(k)\theta)(k_1,\cdots,k_{n-1})\equiv(a_2(k)\theta)(k_1,\cdots,k_{n-1}). \quad (2.27)$$

Here

$$|\lambda_2(k;k_1,\cdots,k_{n-1})|\leq\begin{cases}\omega(k)\\ \text{const }\omega^2(k)/n\end{cases}. \quad (2.28)$$

Taking adjoints,

$$[H_0^{1/2},a^*(k)]=-a_1^*(k), \quad (2.29)$$

and

$$[H_0^{1/2},[H_0^{1/2},a^*(k)]]=a_2^*(k). \quad (2.30)$$

Note that the double commutation $[H_0^{1/2},[H_0^{1/2},W]]$ can be expressed as a sum of two kinds of terms. There will be m contributions arising from double commutators of $H_0^{1/2}$ with one of the m creation or annihilation operators, and these will be of the form of the original W in (2.10), but they will have one $a(k)$ or one $a^*(k)$, replaced by an $a_2(k)$ or $a_2^*(k)$. There will also be $\frac{1}{2}m(m-1)$ terms arising from the commutator $H_0^{1/2}$ with two distinct creators or annihilators of W. These terms will be of the form of the right-hand side of (2.10) but will involve two of the modified operators $a_1(k)$ or $-a_1^*(k)$.

In order to prove (2.12), we use the definitions (2.22)–(2.24) and (2.27). In order to estimate the matrix element $|(\psi,[H_0^{1/2},[H_0^{1/2},W]]\theta)|$, we use the bound (2.25) and the first of the inequalities (2.28). As in the proof of (2.11), the Schwarz inequality then yields

$$\|[H_0^{1/2},[H_0^{1/2},W]](N+I)^{-m/2}\|\leq\text{const }\|\sum_{i=1}^{m}\omega(k_i)w\|_{L^2}, \quad (2.31)$$

which is (2.13). The proof of (2.14) proceeds in a similar fashion, using the second set of bounds for the $|\lambda_j|$.

In order to prove (2.12), we investigate the operator

$$X=(H_0+I)^{-1}W(H_0+I)^{-1}(N+I)^{-(m-4)/2}. \quad (2.32)$$

By following the proof of Lemma 2.41 and Theorem 2.43 of Ref. 7, we find that

$$\|X\|\leq\text{const }\|E^{-1/2}w\|_{L^2},$$

where

$$E=\begin{cases}\sup_{i\neq j}\omega(k_i)\omega(k_j) & \text{if } m>1\\ \omega(k_i) & \text{if } m=1.\end{cases}$$

[7] J. Glimm, Commun. Math. Phys. 5, 343 (1967).

Replacing the operator W by $\hat{W}=[H_0^{1/2},[H_0^{1/2},W]]$, we find that for the corresponding $\hat{X}$,

$$\|\hat{X}\|\leq\text{const}\ \|E^{-1/2}\sum_{i=1}^{m}\omega(k_i)w\|_{L^2}.$$

Each $\omega(k_i)$ is dominated by a constant multiple of $E^{1/2}$, unless k_i is very much larger than the magnitude of all the other momenta, or unless $m=1$. The former can occur only when $\sum_{i=1}^m k_i$ is large in magnitude, and in either case

$$\omega(k_i)\leq\text{const}\ E^{1/2}\omega^{1/2}(\sum_{i=1}^{m}k_i).$$

Thus

$$\|\hat{X}\|\leq\text{const}\ \|\omega^{1/2}(\sum_{i=1}^{m}k_i)w\|_{L^2},$$

which was to be proved.

III. SELF ADJOINTNESS OF THE INTERACTION HAMILTONIAN

For a real spatial cutoff $g(x)$ in the Schwartz space $\mathfrak{S}(R^1)$, the interaction part of the Hamiltonian $H_I(g)$ is self adjoint. This result was proved by Lanford, and also by Doplicher and Jaffe. Statement of these theorems are given in Wightman.[8] With a slightly different formulation, the result has been announced by Segal.[6]

We initially define $H_I(g)$ on the domain

$$D_0=\bigcap_{n=0}^{\infty}D(H_0^n),$$

and we will prove that its closure is self adjoint. It follows from (2.11) that H_I is well defined on D_0.

Theorem 3.1: If $g\in\mathfrak{S}(R^1)$ is real, then

$$H_I(g)=\int_{t=0}:\varphi^4(x):g(x)dx \tag{3.1}$$

is essentially self adjoint on D_0.

Let us introduce a domain D_1 obtained by applying any polynomial of the $t=0$ fields $\varphi(f_i)$, $f_i\in\mathfrak{S}(R^1)$, to the no-particle state Ω_0. Clearly $D_1\subset D_0$, and any vector Ω in D_1 is an entire vector for $\varphi(f)$, which means that the power series

$$\sum_{n=0}^{\infty}\frac{\|\varphi(f)^n\Omega\|}{n!}z^n \tag{3.2}$$

defines an entire function of z. Since D_1 is dense in Fock space, a result of Nelson,[9] shows that for real f, $\varphi(f)$ is essentially self adjoint on D_1. A similar argument can be made for the canonically conjugate $t=0$ fields $\pi(f)$.

Let $\mathfrak{M}$ denote the von Neumann algebra of operators generated by the spectral projections of all the $t=0$ fields $\varphi(f)$, $f\in\mathfrak{S}(R^1)$. The algebra $\mathfrak{M}$ is maximal Abelian. In other words, a bounded operator which commutes with all operators in $\mathfrak{M}$ is itself in $\mathfrak{M}$.

Let us consider $\varphi(f)$ for supp $f\subset O\subset R^1$, where O is an open region of space. (The support of a function is the smallest closed set outside of which the function vanishes identically.) Define $\mathfrak{A}(O)$ as the von Neumann algebra of operators generated by the spectral projections of all the fields $\varphi(f)$ and $\pi(f)$, supp $f\subset O$.

Since

$$\varphi(x,t)=\int\left(\Delta(x-y,t)\pi(y)-\frac{\partial\Delta}{\partial t}(x-y,t)\varphi(y)\right)dy, \tag{3.3}$$

and $\Delta(x,t)$ vanishes outside the light cone, we infer that

$$e^{iH_0t}\mathfrak{A}(O)e^{-iH_0t}\subset\mathfrak{A}(O_t), \tag{3.4}$$

where O_t is the region O expanded by t.

Theorem 3.2: If $g(x)\in\mathfrak{S}(R^1)$ is real and has its support in an open interval O, then for the H_I of (3.1)

$$\exp(iH_I(g)t)\in\mathfrak{A}(O)\cap\mathfrak{M}.$$

We now prove two lemmas.

Lemma 3.3: Let T be any operator with domain D_1 such that

$$TD_1\subset D(\varphi(f)^n), \tag{3.5}$$

$$TD_1\subset D(T|_{D_1}{}^*), \tag{3.6}$$

and

$$[T,\varphi(f)^n]D_1=0. \tag{3.7}$$

Then

$$\mathfrak{M}D_1\subset D(T|_{D_1}{}^-) \tag{3.8}$$

and

$$[T^-,\mathfrak{M}]D_1=0. \tag{3.9}$$

Proof: For $\Omega\in D_1$, from (3.5) and (3.7) we have

$$T\varphi(f)^n\Omega=\varphi(f)^nT\Omega.$$

But by (3.6), for real f

$$\|T\varphi(f)^n\Omega\|^2=(T\Omega,\varphi(f)^{2n}T\Omega)=(T^*T\Omega,\varphi(f)^{2n}\Omega)$$
$$\leq\|T^*T\Omega\|\ \|\varphi(f)^{2n}\Omega\|.$$

Thus the convergent power series (3.2) shows that for Ω in D_1,

$$T^-e^{i\varphi(f)}\Omega=e^{i\varphi(f)}T\Omega. \tag{3.10}$$

It is clear that (3.10) is still valid with $\exp(i\varphi(f))$ replaced by strong limits of sums of such exponentials, and hence (3.8) and (3.9).

Lemma 3.4: Let $\mathfrak{M}$ be a maximal Abelian algebra of bounded operators on a Hilbert space $\mathfrak{H}$ with a cyclic vector Ω_0. Let T be a symmetric operator with domain

---

[8] A. S. Wightman, in *1964 Cargèse Summer School Lectures*, edited by M. Lévy (Gordon and Breach, Science Publishers, Inc., New York, 1967).

[9] E. Nelson, Ann. Math. **70**, 572 (1959); H. J. Borchers and W. Zimmermann, Nuovo Cimento **31**, 1047 (1963).

$\mathfrak{M}\Omega_0$, and let T commute with $\mathfrak{M}$. Then T is essentially self adjoint.

Proof: Without loss of generality, $\mathfrak{M}=L_\infty(X)$ and $\mathfrak{H}=L_2(X)$ for some measure space X, and Ω_0 is the function 1. Let $t=T\Omega_0$. Then $t\in L_2$ and T is multiplication by t, with domain L_∞. Let $f\in L_2$ and suppose $tf\in L_2$ also and let

$$f_n(x)=\begin{cases} f(x) & \text{if } |f(x)|\leq n \\ 0 & \text{otherwise.}\end{cases}$$

Then $f_n\in L_\infty=D(T)$ and $f_n\to f$, $tf_n\to tf$ in L_2 norm by the bounded convergence theorem. Thus $\{f,tf\}$ is in the graph of the closure of T. Thus the closure of T is self adjoint, and T is essentially self adjoint.

Remark: Let T_n be a sequence of operators with the property of T in the lemma. Then $T_n\to T$ strongly on the domain $\mathfrak{M}\Omega_0$ if and only if $T_n\Omega_0\to T\Omega_0$.

Proof of Theorems 3.1 and 3.2: We apply the Lemmas 3.3 and 3.4 with the case $T=H_I(g)$, $\mathfrak{M}$ in Lemma 3.4 as in Lemma 3.3, the Hilbert space Fock space, and Ω_0 the Fock no-particle state. The hypotheses (3.5) and (3.6) can be verified by a direct computation.[10] Thus $H_I(g)$ is essentially self adjoint on $D_1\subset D_0$, and hence $H_I(g)$ is essentially self adjoint on D_0.

If we assume that supp $g\subset O$, then as O is an open interval, supp $g\subset O_1$ where O_1 is O contracted by some small amount $\epsilon>0$. Since $H_I(g)$ commutes with $\mathfrak{M}$, and $\mathfrak{M}$ is maximal Abelian, $\exp(iH_I(g)t)\in\mathfrak{M}$. Furthermore the argument in the proof of Lemma 3.3, can be repeated to show that $H_I(g)$ commutes with $\mathfrak{A}(O_1')$, where O_1' is the complement of the closure of O_1.[11] Since $\mathfrak{A}(R^1)$ is irreducible and $H_I(g)$ commutes with $\mathfrak{A}(O_1')$, $\exp(iH_I(g)t)\in\mathfrak{A}(O_2)$ where O_2 is O_1 expanded by any amount $\epsilon'>0$.[12] Taking $\epsilon'<\epsilon$, we have $\exp(iH_I(g)t)\in\mathfrak{A}(O)$, which completes the proof.

IV. SELF ADJOINTNESS OF THE TOTAL HAMILTONIAN

Theorem 4.1: (a) For real $g(x)\in\mathfrak{S}(R^1)$, the total Hamiltonian $H(g)=H_0+H_I(g)$ is self adjoint with the domain $D(H(g))=D(H_0)\cap D(H_I(g))$.

(b) The total Hamiltonian $H(g)$ is essentially self adjoint on the domain

$$D_0=\bigcap_{n=0}^{\infty} D(H_0^n).$$

In order to prove the self adjointness of H, we combine the estimates of Sec. II, the self adjointness of $H_I(g)$ proved in Sec. III, and a singular perturbation theory developed elsewhere.[4] We need the following result which is a special case of Theorem 8 of Ref. 4.

Theorem 4.2: Under the hypotheses (i)–(iii) below, the operator $H=H_0+H_I$ is self adjoint.

(i) Both H_0 and H_I are self adjoint. The domain

$$D_0=\bigcap_{n=0}^{\infty} D(H_0^n)$$

is contained in the domain of H_I, and H_I is essentially self adjoint on D_0.

(ii) Let $N\geq 0$ be a positive self adjoint operator, commuting with H_0, and such that $N\leq\text{const } H_0$. Suppose that the operators $(I+N)^{-1}H_I(I+N)^{-1}$ and $(I+N)^{+1}H_I(I+N)^{-3}$ are bounded.

(iii) Suppose that for any $\epsilon>0$, there exists a number b such that as bilinear forms on $D_0\times D_0$,

$$-H_I\leq\epsilon N+bI, \tag{4.1}$$

$$-[H_0^{1/2},[H_0^{1/2},H_I]]\leq\epsilon H_0^2+bI, \tag{4.2}$$

and

$$-[N,[N,H_I]]\leq\epsilon N^3+bI. \tag{4.3}$$

Proof of Theorem 4.1: In order to prove that H is self adjoint, we apply Theorem 4.2 in the case that H_0 is the free Hamiltonian, N is the number operator, and H_I is the interaction Hamiltonian $H_I(g)$. Thus we need to verify (i–iii). Condition (i) was dealt with in Theorem 3.1, while condition (ii) is a consequence of (2.11). In Refs. 2, and 3, it is shown that for any $\epsilon>0$, there is a number b such that

$$-H_I(g)\leq\epsilon H_0+bI.$$

By following that proof, but using the smoothing operator $\exp(-tN)$, in place of $\exp(-tH_0)$, one arrives at the estimate (4.1) required in (iii). The remaining estimates (4.2) and (4.3) were established in Theorem 2.1. Thus we conclude from Theorem 4.2 that $H(g)$ is self adjoint on the domain $D(H_0)\cap D(H_I(g))$.

We now show that $H(g)$ is essentially self adjoint on D_0. We first show that $H(g)$ is essentially self adjoint on $D_2=D(H_0)\cap D(N^2)$. By (2.11) it is clear that the domain of $H(g)$ contains D_2. For $\psi\in D(H(g))=D(H_0)\cap D(H_I(g))$, consider $\psi_n\in D_2$ defined by

$$\psi_n=n(nI+N)^{-1}\psi. \tag{4.4}$$

Thus $\|\psi_n-\psi\|+\|H_0\psi_n-H_0\psi\|\to 0$ as $n\to\infty$. We study

$$H_I\psi_n-H_I\psi=-N(nI+N)^{-1}H_I\psi+n[H_I,\ (nI+N)^{-1}]\psi.$$

Since $N(nI+N)^{-1}$ is a uniformly bounded sequence converging to zero on the dense set $D(N)$, it converges to zero and $\|N(nI+N)^{-1}H_I\psi\|\to 0$ as $n\to\infty$. But for the other term

$$\begin{aligned} n[H_I,\ (nI+N)^{-1}]\psi &=[H_I,\ (nI+N)^{-1}](nI+N)n(nI+N)^{-1}\psi \\ &=(nI+N)^{-1}[N,H_I]n(nI+N)^{-1}\psi \\ &=(nI+N)^{-1}(I+N)(I+N)^{-1}[N,H_I] \\ &\quad\times(I+N)^{-1}n(nI+N)^{-1}(I+N)\psi. \end{aligned}$$

[10] A. M. Jaffe, J. Math. Phys. 7, 1250 (1966).
[11] A. M. Jaffe, Ann. Phys. (N. Y.) 32, 127 (1965).
[12] H. Araki, J. Math. Phys. 5, 1 (1964).

Note that as $n\to\infty$, $n(nI+N)^{-1}(I+N)$ converges strongly to $(I+N)\psi$, that by (2.11), $(I+N)^{-1}[N,H_I]\times(I+N)^{-1}$ is bounded, and that $(nI+N)^{-1}(I+N)$ converges strongly to zero. Thus $\|[H_I,(nI+N)^{-1}]\psi\|\to 0$ as $n\to\infty$, and so $\|H_I\psi_n-H_I\psi\|\psi\to 0$. Thus we see that $H(g)$ is the closure of $H(g)$ restricted to D_2, so $H(g)$ is essentially self adjoint on D_2.

Let D_2 be a Hilbert space with the norm

$$(\|\psi\|_{D_2})^2=\|\psi\|^2+\|H_0\psi\|^2+\|N^2\psi\|^2.$$

From (2.11) we infer that

$$\|H\psi\|\leq\text{const}\|\psi\|_{D_2},$$

so that $H(g)$ is essentially self adjoint on any subset of D_2 which is dense in the Hilbert space D_2. For $\psi\in D_2$, $\psi_\lambda=\exp(-\lambda H_0)\psi\in D_0=\bigcap_{n=0}^\infty D(H_0^n)$, and $\|\psi_\lambda-\psi\|_{D_2}\to 0$ as $\lambda\to 0$. Thus $H(g)$ is essentially self adjoint on D_0.

V. REMOVING THE SPATIAL CUTOFF AND LOCALITY

For the reader's convenience, we sketch a proof of Segal's theorem[6] that the self adjointness of $H(g)$ allows the removal of the spatial cutoff. In fact, if A is a bounded function of the free fields localized in a bounded region of space at $t=0$, then

$$\sigma_t(A)=e^{itH(g)}Ae^{-itH(g)}$$

is independent of $g(x)$ provided that $g(x)=\lambda$, the desired coupling constant, on a sufficiently large interval, depending on t. Furthermore, if A is localized in the region of space O, then $\sigma_t(A)$ is localized in the region O_t, where O_t is the region O expanded by t. (We have taken the velocity of light to be one.) In other words, the time translation σ_t gives rise to a local theory. If one chooses for the operator A a spectral projection of the $t=0$ field $\varphi(f)$, one can piece together the time translation operator for the fields themselves.

In section IV, we showed that $H=H_0+H_I$, which is sum of two self adjoint operators, is itself self adjoint. As a consequence of this fact, the Trotter product formula[13] says that for all ψ

$$e^{itH}\psi=\lim_{n\to\infty}(e^{itH_0/n}e^{itH_I/n})^n\psi.$$

Thus

$$\sigma_t(A)\psi=\lim_{n\to\infty}(e^{iH_0t/n}e^{iH_It/n})^nA\,(e^{-iH_It/n}e^{-iH_0t/n})^n\psi.$$

Let O be the region defined by $|x|<M$, $t=0$, and let $A\in\mathfrak{A}(O)$, where $\mathfrak{A}(O)$ is defined in Sec. III.

Given an arbitrary, positive ϵ, split $g(x)$ into two infinitely differentiable parts

$$g(x)=g_1(x)+g_2(x),$$

where supp $g_1(x)\subset O_\epsilon$, and supp $g_2\cap O_{\epsilon/2}$ is empty. Write

$$H_I(g)=H_I(g_1)+H_I(g_2),$$

so that as a consequence of theorems 3.1 and 3.2, $H_I(g_1)$ and $H_I(g_2)$ commute, and

$$\exp[iH_I(g)t/n]=\exp[iH_I(g_1)t/n]\exp[iH_I(g_2)t/n].$$

Furthermore,

$$\exp[iH_I(g_1)t/n]\in\mathfrak{A}(O_\epsilon),$$

and $\exp[iH_I(g_2)t/n]$ commutes with $\mathfrak{A}(O_{\epsilon/4})$. Therefore,

$$A_1(t)=\exp(iH_0t/n)\exp[iH_I(g)t/n]A \times\exp[-iH_I(g)t/n]\exp(-iH_0t/n)$$

depends on $g(x)$ only in the region O_ϵ, and by the free propagation property (3.4),

$$A_1\in\mathfrak{A}(O_{(t/n)+\epsilon}).$$

We continue step by step, and after n steps we conclude that

$$A_n(t)=[\exp(iH_0t/n)\exp(iH_I(g)t/n)]^nA \times[\exp(-iH_I(g)t/n)\exp(-H_0t/n)]^n$$

depends on $g(x)$ only in the region $O_{t+n\epsilon}$, and

$$A_n(t)\in\mathfrak{A}(O_{t+n\epsilon}).$$

Since ϵ can be chosen arbitrarily, $A_n(t)$ depends on $g(x)$ only in the region $\bar{O}_t$, the closure of O_t, and

$$A_n(t)\in\bigcap_{\epsilon>0}\mathfrak{A}(O_{t+\epsilon}).$$

Thus $A_n(t)$ commutes with any local observable B localized in open region of space O' such that O' and O_t are disjoint. As this is true for each n, it is true for

$$\sigma_t(A)=\text{strong}\lim_{n\to\infty}A_n(t).$$

Hence $\sigma_t(A)$ is *local*, and it depends on $g(x)$ only in the region $\bar{O}_t$, where we choose $g(x)=\lambda$. We therefore conclude that the spatial cutoff has been removed and the resulting theory is local.

VI. THEORY IN A BOX

We can consider a somewhat different cutoff theory, namely the $g\varphi^4$ theory in a periodic box. This gives a cut-off interaction which is translation invariant, and therefore it is useful for the study of the vacuum state.[14] In a finite interval, but with no ultraviolet cutoff, we prove that the total Hamiltonian is self adjoint and has a complete set of normalizable eigenstates.

The theory in volume V is constructed by taking a Fock space $\mathfrak{F}_V$ of functions defined on the momentum

[13] H. F. Trotter, Proc. Am. Math. Soc. 10, 545 (1959); E. Nelson, J. Math. Phys. 5, 332 (1964).

[14] A. M. Jaffe and R. T. Powers, Commun. Math. Phys. 7, 218 (1968); J. Glimm and A. M. Jaffe (to be published).

space lattices Γ_V, $\Gamma_V\times\Gamma_V$, etc., where

$$\Gamma_V=\{k:\ k=2\pi n/V,\ n=0,+1,\pm 2,\cdots\}. \quad (6.1)$$

Thus the free Hamiltonian is

$$H_{0,V}=\sum_{k\in\Gamma V} a_V^*(k)a_V(k)\omega(k), \quad (6.2)$$

and the $t=0$ field is

$$\varphi_V(x)=\frac{1}{(2V)^{1/2}}\times\sum_{k\in\Gamma_V} e^{-ikx}[a_V^*(k)+a_V(-k)]\omega(k)^{-1/2}. \quad (6.3)$$

The interaction Hamiltonian $H_{I,V}$ is defined by

$$H_{I,V}=g\int_V :\varphi_V(x)^4:dx, \quad g\geq 0 \quad (6.4)$$

and $H_{I,V}$ is self adjoint as was shown in Sec. III for $H_I(g)$. The total Hamiltonian in a box is

$$H_V=H_{0,V}+H_{I,V}.$$

It is possible to regard $\mathfrak{F}_V$ as a subspace of $\mathfrak{F}$ defined by functions on R, $R\times R$, etc., which are piecewise constant between lattice sites. The correspondence between a_V and a is

$$a_V(k)=\left(\frac{V}{2\pi}\right)^{1/2}\int_0^{2\pi/V} a(k+l)dl, \quad k\in\Gamma_V. \quad (6.5)$$

Therefore the estimates of Sec. II are valid for H_V, and so H_V is self adjoint on the domain $D(H_{0,V})\cap D(H_{I,V})$, and H_V is essentially self adjoint on the domain

$$D_0=\bigcap_{n=0}^{\infty} D(H_{0,V}^n).$$

Theorem 6.1: The spectrum of H_V is discrete with finite multiplicity, so H_V has a complete set of normalizable eigenstates.

Proof: Let $\Gamma_{\kappa,V}=\Gamma_V\cap\{k:|k|\leq\kappa\}$, so that an ultraviolet cut-off field $\varphi_{\kappa,V}(x)$ is obtained from φ_V by summing (6.4) over $\Gamma_{\kappa,V}$. Then a sharply cut-off Hamiltonian $H_{\kappa,V}=H_{0,V}+H_{I,\kappa,V}$ comes from

$$H_{I,\kappa,V}=g\int_V :\varphi_{\kappa,V}(x)^4:dx.$$

The operator $H_{\kappa,V}$ has pure discrete spectrum with finite multiplicity.[15] Furthermore, $H_{I,\kappa,V}$ is the sum of five expressions of the form $W_{\kappa,V}$ of (2.10), such that $\|w_{\kappa,V}-w_{\kappa',V}\|_{L^2}\rightarrow 0$ as $\kappa,\kappa'\rightarrow\infty$. Therefore the bounded operators $(I+N_V)^{-1}H_{I,\kappa,V}(I+N_V)^{-1}$ converge in norm as $\kappa\rightarrow\infty$ to $(I+N_V)^{-1}H_{I,V}(I+N_V)^{-1}$. We now appeal to Corollary 10 of Ref. 4 to infer that the resolvents of $H_{\kappa,V}$ converge in norm as $\kappa\rightarrow\infty$ to the resolvent of H_V. Since each $R_{\kappa,V}=(H_{\kappa,V}+c)^{-1}$ is compact, so is their uniform limit R_V, and the theorem is proved.

We further see that the projections onto each eigenvalue $\lambda_n(\kappa,V)$ of $H_{\kappa,V}$ converge as $\kappa\rightarrow\infty$. This is a consequence of Theorem IV. 3.16 of Kato.[16] Thus, in particular the vacuum vectors for $H_{\kappa,V}$ converge in Fock space to a vacuum vector for H_V.

VII. CONCLUSIONS

We have shown the existence of a local time translation for a two-dimensional $\lambda\varphi^4$ theory without cutoffs. The method relies on basic inequalities satisfied by the first and second power of the Hamiltonian. Similar estimates of all the higher powers of the Hamiltonian proved useful to investigate the existence of vacuum-expectation values in a $\lambda\varphi^4$ theory with a sharp momentum-space cutoff.[15] It does not seem likely, in our case, that H_0^3 is dominated by higher powers of H, since the ground state of H must be in the domain of H^n, for all n, and the first-order perturbation correction to the Fock vacuum is in $D(H_0)$, but not in $D(H_0^{3/2})$.

[15] A. M. Jaffe, Ph.D. thesis, Princeton University (to be published).

[16] T. Kato, *Perturbation Theory for Linear Operators* (Springer-Verlag, Berlin, 1966).

The $\lambda(\varphi^4)_2$ quantum field theory without cutoffs: II. The field operators and the approximate vacuum

JAMES GLIMM[1] and ARTHUR JAFFE[2]

TABLE OF CONTENTS

1. Introduction

1.1. *Discussion of the results.* We study a simple model for quantum field theory and we solve the equations of motion. Our model describes bosons with a non-linear φ^4 self-interaction in two dimensional space time (a $(\varphi^4)_2$ interaction for short). The essential simplification in this model is the reduction in the number of dimensions, which renders finite many (but not all) of the standard divergent integrals of quantum field theory. We verify most of the Haag-Kastler axioms for quantum field theory for this model. The remaining axiom is valid on the level of formal perturbation theory[3], as are the Wightman axioms.

The equation of motion in the Schrödinger picture is

$$i\frac{\partial\theta}{\partial t} = H\theta$$

[1] Supported by the New York State Science and Technology Foundation, Grant SSF-(8)-8.

[2] Alfred P. Sloan Foundation Fellow. Supported in part by the Air Force Office of Scientific Research and by the Schweizerische Nationalfonds.

[3] *Added in proof.* The missing axiom, Lorentz covariance, is estabilished in **[30]**.

where the state θ of the quantum field is a vector of unit length in a space $\mathcal{F}$, and H, the hamiltonian operator, is a self-adjoint operator on $\mathcal{F}$. The solution of this Schrödinger equation is

$$\theta_S(t) = e^{-itH}\theta_S(0)$$

and this solution describes the dynamics in the Schrödinger picture. The observables are represented by certain self-adjoint operators A_S on $\mathcal{F}$ and the inner product $(\theta_S(t), A_S\theta_S(t))$ is the expected value for the measurement of the quantity A at the time t if the quantum field is in the state $\theta_S(0)$ at the time zero. The Schrödinger picture observable A is independent of time.

In the Heisenberg picture, the state θ_H in $\mathcal{F}$ is independent of time (and equals $\theta_S(0)$) while the observable $A_H(t) = e^{itH}A_H(0)e^{-itH}$ depends on time. We take $A_H(0) = A_S$ and then

$$(\theta_S(t), A_S\theta_S(t)) = (\theta_H, A_H(t)\theta_H) ,$$

so that the expected values of measurements in the Heisenberg and Schrödinger pictures agree.

As is typical for quantum field theory problems, the hamiltonian H for the $(\varphi^4)_2$ interaction is not given as a self-adjoint operator, but rather as a singular formal expression (some of the terms having infinite coefficients) with no vectors obviously in its domain, other than $\theta = 0$. For this model it is easier to construct the dynamics in the Heisenberg picture, and this was accomplished in [8, 9]. The bounded observables $A_H(0)$ generate a C^* algebra $\mathfrak{A}$. (See § 4.) The observable $A_H(t)$ belongs to $\mathfrak{A}$ for all t and so $\mathfrak{A}$ is invariant under t. The map σ_t: $A_H(0) \longrightarrow A_H(t)$ defines an automorphism of $\mathfrak{A}$ and σ is a one parameter group $\sigma_{t+s} = \sigma_t\sigma_s$.

In § 3 of this paper we establish the existence of self adjoint field operatos $\varphi(f)$,

$$\varphi(f) = \int\varphi(x, t)f(x, t)dxdt$$

acting on the Fock-Hilbert space $\mathcal{F}$. They are local fields, so that $\varphi(f)$ and $\varphi(h)$ commute when the test functions f and h have supports which are space like separated. In other words, the field in one space time region can affect measurements in a second space time region only if a light signal can be sent from one region to the other. We construct a domain in $\mathcal{F}$ which is invariant under the field operator $\varphi(f)$ and prove that the field has a unique self adjoint extension when defined on this domain. The Heisenberg field equation

$$(\Box + m_0^2)\varphi(x, t) = -4\lambda : \varphi(x, t)^3 :$$

is satisfied, see § 3.2.

The hamiltonian operator $H(g)$ is self adjoint [8, 9] and semi-bounded [21, 7.]. We show that the bottom of the spectrum of $H(g)$ is an isolated eigenvalue (§ 2.2) and that it has multiplicity one (§ 2.3). Thus if

$$E_g = \inf\{\text{spectrum}\,(H(g))\} \tag{1.1.1}$$

there is a vector $\Omega_g \in \mathcal{F}$ such that

$$H(g)\Omega_g = E_g\Omega_g\,, \qquad \|\Omega_g\| = 1\,, \tag{1.1.2}$$

and Ω_g is unique up to a scalar multiple.

In the next paper in this series [29] we establish the existence of a physical vacuum, of the renormalized hamiltonian operator H and of the Schrödinger picture dynamics. None of these three will exist on the Hilbert space $\mathcal{F}$; in fact there is a general theorem due to Haag [10, 12] which says roughly that in order for these objects to exist, the Hilbert space $\mathcal{F}$ must be replaced by a new Hilbert space $\mathcal{F}_{\text{ren}}$.

1.2. *Notation.* The Fock space $\mathcal{F}$ is the Hilbert space completion of the symmetric tensor algebra over $L_2(R)$,

$$\mathcal{F} = \mathfrak{S}(L_2(R)) = \bigoplus_{n=0}^{\infty} \mathcal{F}_n\,, \tag{1.2.1}$$

where $\mathcal{F}_n$ is the space of n non-interacting particles.

$$\mathcal{F}_n = L_2(R) \otimes_s L_2(R) \otimes_s \cdots \otimes_s L_2(R) \qquad (n \text{ factors}) \tag{1.2.2}$$

The variable $k \in R$ denotes momentum. For $\psi = \{\psi_0, \psi_2, \cdots\} \in \mathcal{F} = \mathcal{F}_0 \oplus \mathcal{F}_1 \oplus \cdots$, we have $\|\psi\|^2 = \sum_{n=0}^{\infty} \|\psi_n\|^2$. The no particle space $\mathcal{F}_0$ is the complex numbers, and

$$\Omega_0 = \{1, 0, 0, \cdots\} \in \mathcal{F}\,, \tag{1.2.3}$$

is the (bare) vacuum or (bare) no-particle state vector. We define operators N and H_0 by

$$(N\psi)_n = n\psi_n\,, \tag{1.2.4}$$

and

$$(H_0\psi)_n(k_1, \cdots, k_n) = \sum_{j=1}^{n} \mu(k_j)\psi_n(k_1, \cdots, k_n)\,, \tag{1.2.5}$$

where

$$\mu(k) = (k^2 + m_0^2)^{1/2}\,. \tag{1.2.6}$$

Here N is the number of particles operator, and H_0 is the free energy operator (the free hamiltonian). The rest mass of the non-interacting particles is m_0, and $\mu(k)$ is the energy of a free particle with momentum k.

We use the standard annihilation and creation operators $a(k)$ and $a^*(k)$,

$$(a(k)\psi)_{n-1}(k_1, \cdots, k_{n-1}) = n^{1/2}\psi_n(k, k_1, \cdots, k_{n-1})\,.$$

As a convenient minimal domain for $a(k)$, we use the set $\mathcal{S}$ of vectors ψ in $\mathcal{F}$ with $\psi_n = 0$ for large n and $\psi_n \in \mathcal{S}(R^n)$ for all n.

$$(1.2.7)\quad \begin{aligned} &(a^*(k)\psi)_{n+1}(k_1, \cdots, k_{n+1}) \\ &\quad = (n+1)^{-1/2} \textstyle\sum_{j=1}^{n+1} \delta(k - k_j)\psi_n(k_1, \cdots, \hat{k}_j, \cdots, k_{n+1}) \end{aligned}$$

Here the variable k_j is omitted. While $a^*(k)$ is not an operator, it is a densely defined bilinear form on $\mathcal{S}\times\mathcal{S}$. For a function or distribution b we can define a bilinear form

$$(1.2.8)\quad \begin{aligned} B = \int &b(k_1, \cdots, k_m; k'_1, \cdots, k'_n)a^*(k_1) \cdots a^*(k_m) \\ &\times a(-k'_1) \cdots a(-k'_n)dk_1 \cdots dk'_n . \end{aligned}$$

The integration helps in (1.2.8) and B is not only a bilinear form, but often an operator. This is the case if, for example, b is the kernel of a bounded operator B_0 from $\mathcal{F}_n$ to $\mathcal{F}_m$. In this case

$$(1.2.9)\quad \|(N+I)^{-\alpha/2}B(N+I)^{-\beta/2}\| \leqq \text{const.}\, \|B_0\| ,$$

provided that $m + n \leqq \alpha + \beta$. The constant depends only on α, β, m and n. Intuitively we think of B as being dominated by $N^{(n+m)/2}$; in particular B is an operator on $D(N^{(m+n)/2})$, the domain of $N^{(n+m)/2}$. The inequality (1.2.9) is one of our basic estimates and in using it we will often dominate $\|B_0\|$ by the Hilbert Schmidt norm $\|B_0\| \leqq \|b\|_2$.

By definition the field $\varphi(x)$ is

$$(1.2.10)\quad \varphi(x) = (4\pi)^{-1/2}\int e^{-ikx}\mu(k)^{-1/2}\{a^*(k) + a(-k)\}dk .$$

Since the kernels $b(k) = e^{\pm ikx}\mu(k)^{-1/2}$ are not in L_2, the B_0 are not bounded operators. In fact $\varphi(x)$ is a bilinear form and not a densely defined operator. We also define the bilinear form

$$(1.2.11)\quad \pi(x) = i(4\pi)^{-1/2}\int e^{-ikx}\mu(k)^{1/2}\{a^*(k) - a(-k)\}dk ,$$

the conjugate momentum to $\varphi(x)$. For real $f(x)$, $g(x)$ such that $\mu(k)^{-1/2}\tilde{f}(k) \in L_2$ and $\mu(k)^{1/2}\tilde{g}(k) \in L_2$, the bilinear forms $\varphi(f)$ and $\pi(g)$ define operators whose closures on $\mathcal{D}(N^{1/2})$ are self-adjoint. They satisfy the canonical commutation relations

$$(1.2.12)\quad e^{i\pi(g)}e^{i\varphi(f)} = e^{-i(f,g)}e^{i\varphi(f)}e^{i\pi(g)} .$$

It is furthermore possible to define polynomial functions of the field $\varphi(x)$, the Wick polynomials $:\varphi(x)^n:$ (see [28] and also [5, 3] for a definition of the Wick dots : :). Explicitly, as a bilinear form on $D(N^{n/2}) \times D(N^{n/2})$,

$$(1.2.13)\quad :\varphi(x)^n: = \sum_{j=0}^{n} \binom{n}{j} b_x(k_1, \cdots, k_n) a^*(k_1) \cdots a^*(k_j) a(-k_{j+1}) \cdots a(-k_n)$$

where

$$b_x(k_1, \cdots, k_n) = (4\pi)^{-n/2}\{\mu(k_1) \cdots \mu(k_n)\}^{-1/2} e^{-i(k_1+\cdots+k_n)x} .$$

Thus for real $f(x) \in \mathfrak{S}$, the bilinear form $:\varphi^n:(f) = \int :\varphi^n(x): f(x)dx$ has a kernel proportional to $\{\mu(k_1) \cdots \mu(k_n)\}^{-1/2} \tilde{f}(k_1 + \cdots + k_n) \in L_2$. Thus from (1.2.9) we conclude that $:\varphi^n:(f)$ defines a symmetric operator on the domain $\mathfrak{D}(N^{n/2})$. It was shown in [8] that $:\varphi^n:(f)$ is essentially self-adjoint on this domain. This result was proved independently by Lanford and by Segal. (See for example [25, p. 82, lines 4–15].)

2. The cutoff hamiltonian $H(g)$

2.1. *Introduction.* The hamiltonian $H(g)$ acts on $\mathcal{F}$ and can be written in terms of the field operator $\varphi(x)$ as

$$(2.1.1)\quad \begin{aligned} H(g) &= H_0 + \int :\varphi(x)^4: g(x)dx \\ &= H_0 + H_{I,g} , \end{aligned}$$

where $H_0 = H(0)$ is the free hamiltonian, and $0 \leqq g$. Let

$$C^\infty(H_0) = \{\psi : \psi \in D(H_0^n),\ n = 1, 2, \cdots\}$$

be the set of C^∞ vectors for H_0. It was shown in [8–9] that $H(g)$ and $H_{I,g}$ are essentially self-adjoint on $C^\infty(H_0)$, that

$$(2.1.2)\quad \mathfrak{D}(H(g)) = \mathfrak{D}(H_0) \cap \mathfrak{D}(H_{I,g}) ,$$

and that there are finite a, $b = b(g)$ such that

$$(2.1.3)\quad \| H_0\psi \| + \| H_{I,g}\psi \| \leqq a \| (H(g) + b)\psi \|$$

for all $\psi \in \mathfrak{D}(H(g))$.

We find that it is convenient to introduce a periodic approximation in configuration space. Under this approximation, the Fourier transform, or momentum space variable $k \in R$ is replaced by a discrete variable $k \in \Gamma_V$,

$$\Gamma_V = \left\{k : k = \frac{2\pi n}{V},\ n = 0, \pm 1, \pm 2, \cdots\right\} .$$

Thus we define $\mathcal{F}_V$, the Fock space for volume V, as

$$\begin{aligned} \mathcal{F}_V &= \mathfrak{S}(l_2(\Gamma_V)) \\ &= C \oplus l_2(\Gamma_V) \oplus [l_2(\Gamma_V) \otimes_s l_2(\Gamma_V)] \cdots . \end{aligned}$$

We identify $\mathcal{F}_V$ with the subspace of $\mathcal{F}$ consisting of piecewise constant functions which are constant on each cube of volume $(2\pi/V)^j$ centered about a lattice point

$$\{k_1, \cdots, k_j\} \in \Gamma_V \times \Gamma_V \cdots \times \Gamma_V = \Gamma_V^j .$$

The periodic annihilation and creation operators $a_V(k)$ and $a_V^*(k)$ can be extended from $\mathcal{F}_V$ to $\mathcal{F}$ through the formulas

$$a_V(k) = \left(\frac{V}{2\pi}\right)^{1/2} \int_{-\pi/V}^{\pi/V} a(k+l)dl \tag{2.1.4}$$

and

$$a_V^*(k) = \left(\frac{V}{2\pi}\right)^{1/2} \int_{-\pi/V}^{\pi/V} a^*(k+l)dl . \tag{2.1.5}$$

Similarly the periodic field $\varphi_V(x)$ and the periodic hamiltonian H_V can be extended to act on $\mathcal{F}$ through the formulas

$$\varphi_V(x) = (2V)^{-1/2} \sum_{k \in \Gamma_V} e^{-ikx}\big(a_V^*(k) + a_V(-k)\big)\mu(k)^{-1/2} , \tag{2.1.6}$$

and

$$H_V = H_{0,V} + H_{I,V} . \tag{2.1.7}$$

Here

$$H_{I,V} = \lambda \int_{-V/2}^{V/2} :\varphi_V(x)^4: dx , \tag{2.1.8}$$

and in (2.1.7) we take

$$H_{0,V} = \int a^*(k)a(k)\mu(k_V)dk , \tag{2.1.9}$$

with k_V a lattice point close to k,

$$k_V \in \Gamma_V, \ |k - k_V| \leqq \pi/V . \tag{2.1.10}$$

Note the absence of a V in the a^* and a in (2.1.9). On $\mathcal{F}_V$, this definition of $H_{0,V}$ agrees with the standard definition $\sum_{k \in \Gamma_V} a_V^*(k)a_V(k)\mu(k)$.

The operators $H_{I,V}$ and H_V are essentially self adjoint on $C^\infty(H_{0,V})$, and

$$\mathcal{D}(H_V) = \mathcal{D}(H_{0,V}) \cap \mathcal{D}(H_{I,V}) . \tag{2.1.11}$$

For all $\psi \in D(H_V)$,

$$\| H_{0,V}\psi \| + \| H_{I,V}\psi \| \leqq a \| (H_V + b)\psi \| \tag{2.1.12}$$

where b depends on V [8–9]. On $\mathcal{F}_V$, the operator H_V has a compact resolvent.

We want to approximate $H(g)$ by operators with compact resolvents on $\mathcal{F}_V$, so we define

$$\begin{aligned} H(g)_V &= H_{0,V} + \int :\varphi_V(x)^4: g(x)dx \\ &= H_{0,V} + H_{I,g,V} . \end{aligned} \tag{2.1.13}$$

As in [8, 9] we can show that $H(g)_V$ and $H_{I,g,V}$ are essentially self-adjoint on $C^\infty(H_{0,V})$, and that

$$\mathfrak{D}(H(g)_V) = \mathfrak{D}(H_{0,V}) \cap \mathfrak{D}(H_{I,g,V}) . \tag{2.1.14}$$

Furthermore, for all $\psi \in \mathfrak{D}(H(g)_V)$,

$$\| H_{0,V}\psi \| + \| H_{I,g,V}\psi \| \leqq a \| (H(g)_V + b)\psi \| . \tag{2.1.15}$$

In this case both g and V serve as volume cutoffs, and the constant $b = b(g, V)$ can be chosen independently of V for fixed g. On the space $\mathcal{F}_V$, the operator $H(g)_V$ has a compact resolvent.

Our hamiltonians are semi-bounded and for each $\varepsilon > 0$, there is a constant b such that

$$0 \leqq \varepsilon H_0 + H_{I,g} + b , \tag{2.1.16}$$

$$0 \leqq \varepsilon H_{0,V} + H_{I,V} + b , \tag{2.1.17}$$

and

$$0 \leqq \varepsilon H_{0,V} + H_{I,g,V} + b , \tag{2.1.18}$$

see [21, 7, 2]. In (2.1.18), the b can be chosen to be independent of V. Taking $\varepsilon = \frac{1}{2}$, we have $\frac{1}{2}H_0 \leqq H(g) + b$, which implies that for all $\psi \in \mathfrak{D}(H(g)^{1/2})$,

$$\| H_0^{1/2}\psi \| \leqq \sqrt{2} \| (H(g) + b)^{1/2}\psi \| . \tag{2.1.19}$$

Here we must choose b at least $\frac{1}{2} | E_{2g} |$, where E_{2g} is the vacuum energy for the cutoff $2g$.

2.2. *The existence of a vacuum.* We prove the existence of a vacuum vector Ω_g for $H(g)$, and in the following section we prove that the vacuum is unique. Since the hamiltonian $H(g)$ is bounded from below, we can define the vacuum energy E_g to be the infimum of the spectrum of $H(g)$. We also refer to E_g as the lower bound of $H(g)$. We show that E_g is an isolated point in the spectrum. In a relativistic theory, the gap between the ground state and the first excited state is the mass of the interacting particle. For this reason we say that $H(g)$ has a mass gap. A vacuum vector $\Omega_g \in \mathcal{F}$ is defined as a normalized eigenvector of $H(g)$ corresponding to the eigenvalue E_g.

$$H(g)\Omega_g = E_g\Omega_g, \quad \| \Omega_g \| = 1 . \tag{2.2.1}$$

Our result is

THEOREM 2.2.1. *There exists a vacuum vector Ω_g for $H(g)$. For any $\varepsilon > 0$, the operator $H(g)$, restricted to the spectral interval $[E_g, E_g + m_0 - \varepsilon]$ is compact.*

In this section we also prove

THEOREM 2.2.2. *The approximate hamiltonian $H(g)_V$ has a vacuum vector $\Omega_{g,V}$. Any sequence of volumes V_j tending to infinity has a subsequence V_l such that*

$$\Omega_g = \lim_l \Omega_{g,V_l} \tag{2.2.2}$$

exists and satisfies (2.2.1).

Let $E_{g,V}$ be the lower bound of $H(g)_V$ on $\mathcal{F}_V$. Since $H(g)_V$ has a compact resolvent on $\mathcal{F}_V$, there is a vacuum vector $\Omega_{g,V} \in \mathcal{F}_V$ for $H(g)_V \mid \mathcal{F}_V$. We now see that $E_{g,V}$ is the lower bound for $H(g)_V$ on $\mathcal{F}$, so that $\Omega_{g,V}$ is a vacuum vector for $H(g)_V$.

Let $\mathcal{F}_V^{\perp}$ be the orthogonal complement of $\mathcal{F}_V$. Since $H(g)_V$ leaves $\mathcal{F}_V$ invariant and is self-adjoint, $H(g)_V$ also leaves $\mathcal{F}_V^{\perp}$ invariant.

LEMMA 2.2.3. *The lower bound of $H(g)_V$ on $\mathcal{F}_V^{\perp}$ is $E_{g,V} + m_0$, where m_0 is the rest mass of the Fock space bosons.*

This lemma shows that $\Omega_{g,V}$ is a vacuum for $H(g)_V$.

PROOF. We have an orthogonal decomposition in the single particle space

$$\mathcal{F}_1 = L_2(R) = \mathcal{F}_{1V} \oplus \mathcal{F}_{1V}^{\perp} . \tag{2.2.3}$$

Here $\mathcal{F}_{1V} = \mathcal{F}_1 \cap \mathcal{F}_V$ consists of functions piecewise constant on intervals centered at lattice points. Thus we may write

$$\mathcal{F} = \bigoplus_{j=0}^{\infty} \mathcal{F}^{(j)} , \qquad \mathcal{F}_V^{\perp} = \bigoplus_{j=1}^{\infty} \mathcal{F}^{(j)} , \tag{2.2.4}$$

where $\mathcal{F}^{(j)}$ consists of vectors with exactly j particles from $\mathcal{F}_{1V}^{\perp}$ and

$$\mathcal{F}^{(j)} = \mathcal{F}_{1V}^{\perp} \otimes_s \cdots \otimes_s \mathcal{F}_{1V}^{\perp} \otimes_s \mathcal{F}_V . \tag{2.2.5}$$

In this tensor product decomposition there are j factors $\mathcal{F}_{1V}^{\perp}$. The hamiltonian $H(g)_V$ leaves each subspace $\mathcal{F}^{(j)}$ invariant, and on $\mathcal{F}^{(j)}$ we have $H(g)_V = I \otimes A + B \otimes I$, where $A = H(g) \mid \mathcal{F}_V$ and B is a sum of j copies of $H_{0,V}$, each acting on a single factor $\mathcal{F}_{1V}^{\perp}$. Since

$$j m_0 \leqq B , \tag{2.2.6}$$

the lemma follows from this decomposition.

LEMMA 2.2.4. *For $V \leqq \infty$, and for b sufficiently large we have*

$$\mathfrak{D}(H_0) \subset \mathfrak{D}(H_0^{1/2}) \cap \mathfrak{D}(N) \subset \mathfrak{D}\big(H(g)_V + b)^{1/2}\big) , \tag{2.2.7}$$

and

$$\mathfrak{D}(H_0) \subset \mathfrak{D}\big([(N + I)^{-1}(H(g)_V + b)]^{-}\big) . \tag{2.2.8}$$

PROOF. We take b large enough so that $H(g)_V + b$ is positive, see (2.1.18). By (1.2.9) and (2.1.14)

$$\mathfrak{D}(H_0) \cap \mathfrak{D}(N^2) \subset \mathfrak{D}(H_0) \cap \mathfrak{D}(H_{I,g,V}) = \mathfrak{D}\big(H(g)_V\big) \subset \mathfrak{D}\big((H(g)_V + b)^{1/2}\big) .$$

Thus for $\psi \in \mathfrak{D}(H_0) \cap \mathfrak{D}(N^2)$,

$$\begin{aligned} \| (H(g)_V + b)^{1/2}\psi \|^2 &= (\psi, (H(g)_V + b)\psi) \\ &\leq (\psi, (H_{0,V} + b)\psi) \\ &\quad + \| (N + I)^{-1}H_{I,g,V}(N + I)^{-1} \| \, \| (N + I)\psi \|^2 . \end{aligned}$$

Since $(H(g)_V + b)^{1/2}$ is a closed operator, we can extend this inequality by continuity. As N and $H_{0,V}$ commute, the inequality extends by continuity to all $\psi \in \mathfrak{D}(H_0^{1/2}) \cap \mathfrak{D}(N) \supset \mathfrak{D}(H_0)$. The proof of (2.2.8) is similar.

LEMMA 2.2.5. *Let z be non-real or real and sufficiently negative. Then as V tends to infinity,*

$$\| (H(g)_V - zI)^{-1} - (H(g) - zI)^{-1} \| = O(V^{-1}) , \tag{2.2.9}$$

and $E_{g,V} \rightarrow E_g$.

PROOF. (See also [9, Cor. 10(b)].) Let us fix g and z and suppress g when possible. In [8] we have shown that $H(g)$ is essentially self-adjoint on $C^\infty(H_0)$. Thus vectors of the form $\chi = (H - z)\psi$, $\psi \in C^\infty(H_0)$, are dense in $\mathcal{F}$. On such vectors

$$\begin{aligned} \{(H_V - z)^{-1} - (H - z)^{-1}\}\chi &= (H_V - z)^{-1}\{(H - z)\psi - (H_V - z)\psi\} \\ &= (H_V - z)^{-1}(H - H_V)(H - z)^{-1}\chi \\ &= (H_V - z)^{-1}(N + I)(N + I)^{-1}(H - H_V)(N + I)^{-1} \\ &\quad \times (N + I)(H - z)^{-1}\chi . \end{aligned}$$

For $\theta \in \mathcal{F}$,

$$\begin{aligned} &| (\theta, \{(H_V - z)^{-1} - (H - z)^{-1}\}\chi) | \\ &\leq \| (N + I)(H_V - \bar{z})^{-1} \| \, \| \theta \| \, \| (N + I)^{-1}(H - H_V)(N + I)^{-1} \| \\ &\quad \| (N + I)(H - z)^{-1} \| \, \| \chi \| . \end{aligned} \tag{2.2.10}$$

Using (2.1.15), we find that $\| (N + I)(H_V - \bar{z})^{-1} \|$ is bounded uniformly in V, since

$$\begin{aligned} \| (N + I)(H_V - \bar{z})^{-1}\psi \| &\leq \text{const.} \| (H_{0,V} + I)(H_V - \bar{z})^{-1}\psi \| \\ &\leq \text{const.} \| H_V(H_V - \bar{z})^{-1}\psi \| \\ &\quad + \text{const.} \| (H_V - \bar{z})^{-1}\psi \| , \end{aligned}$$

where the constants can be chosen independently of V. In a similar fashion, (2.2.3) shows that $(N + I)(H - z)^{-1}$ is a bounded operator. Thus from (2.2.10), and the fact that the χ are dense, we infer

$$\| (H_V - z)^{-1} - (H - z)^{-1} \| \leq \text{const.} \| (N + I)^{-1}(H - H_V)(N + I)^{-1} \| , \tag{2.2.11}$$

with a constant independent of V.

The difference $H - H_V = (H_0 - H_{0,V}) + (H_{I,g} - H_{I,g,V})$, and for large V,

$$\| (N + I)^{-1/2}(H_0 - H_{0,V})(N + I)^{-1/2} \| = O(V^{-1}) . \tag{2.2.12}$$

This is a simple direct computation, using $|\mu(k_V) - \mu(k)| = O(V^{-1})$. For the interaction terms, we use (1.2.10) to estimate

$$\|(N+I)^{-1}(H_{I,g} - H_{I,g,V})(N+I)^{-1}\| . \tag{2.2.13}$$

The kernel $b(k_1, \cdots, k_4)$ corresponding to a monomial in $H_{I,g}$ is

$$b(k_1, \cdots, k_4) = \binom{4}{j} \prod_{i=1}^{4} \mu(k_i)^{-1/2} \tilde{g}(k_1 + k_2 + \cdots + k_4) , \qquad 0 \leqq j \leqq 4 .$$

The kernel $b_V(k_1, \cdots, k_4)$ for the corresponding monomial in $H_{I,g,V}$ is obtained by replacing the factor $\prod_{i=1}^{4} \mu(k_i)^{-1/2}$ with $\prod_{i=1}^{4} \mu(k_{iV})^{-1/2}$. Inspection of $b - b_V$ shows that as V tends to infinity, $\|b - b_V\|_{L^2} = O(V^{-1})$, from which we conclude that (2.2.13) is $O(V^{-1})$. The convergence of the resolvents follows from (2.2.11–13). The limit $E_{g,V} \to E_g$ follows from the convergence of the resolvents, since for large positive b,

$$(E_{g,V} + b)^{-1} = \|(H(g)_V + b)^{-1}\| .$$

PROOF OF THEOREMS 2.2.1 AND 2.2.2. Let $f(x)$ be a smooth positive function with support in the interval $[-\varepsilon, m_0 - \varepsilon]$. Then $f(H(g)_V - E_{g,V}) \,|\, \mathcal{F}_V$ is compact, since the resolvent of $H(g)_V \,|\, \mathcal{F}_V$ is compact on $\mathcal{F}_V$. By Lemma 2.2.3, $f(H(g)_V - E_{g,V}) \,|\, \mathcal{F}_V^{\perp} = 0$. Thus $f(H(g) - E_{g,V})$ is compact on the full Fock space $\mathcal{F}$.

By Lemma 2.2.5, the resolvents $(H(g)_V - E_{g,V} - z)^{-1}$ converge in norm as $V \to \infty$. Therefore

$$\|f(H(g)_V - E_{g,V}) - f(H(g) - E_g)\| \to 0 ,$$

since $f(H(g) - E_g)$ is a bounded function of $(H(g) - E_g - z)^{-1}$ which vanishes at infinity. Since the uniform limit of compact operators is compact, $H(g)$ restricted to the spectral interval $[E_g, E_g + m_0 - \varepsilon]$ is compact. This means furthermore that only a finite number of eigenvalues of $H(g)_V$ converge to E_g. Theorem 3.16 of [18, p. 212] shows that the projection onto the corresponding set of eigenvectors of $H(g)_V$ converge as $V \to \infty$. Since $\Omega_{g,V}$ is an eigenvector of $f(H(g)_V - E_{g,V})$, a subsequence of the $\Omega_{g,V}$ converge to a limit as V tends to ∞. For this limit

$$\begin{aligned}(E_g + b)^{-1}\Omega_g &= \lim_{l\to\infty} (E_{g,V_l} + b)^{-1}\Omega_{g,V_l} \\ &= \lim_{l\to\infty} (H(g)_{V_l} + b)^{-1}\Omega_{g,V_l} \\ &= (H(g) + b)^{-1}\Omega_g ,\end{aligned}$$

by Lemma 2.2.5. Hence $\Omega_g \in \mathcal{D}(H(g))$, $H(g)\Omega_g = E_g\Omega_g$, and Ω_g is a vacuum vector for $H(g)$.

In the following section we will see that Ω_g and $\Omega_{g,V}$ are unique except for an arbitrary phase multiple $e^{i\theta}$, and that there is a natural choice for this arbitrary phase. With this choice, we then will prove that the $\Omega_{g,V}$ converge to Ω_g as $V \to \infty$.

2.3. *Uniqueness of the vacuum.*

THEOREM 2.3.1. *The vacuum vector Ω_g for $H(g)$ is unique.*

In other words E_g, the lower bound of $H(g)$ is a simple eigenvalue. Using the Feynman-Kac formula we show that the bounded operator $e^{-H(g)}$ has a positive, ergodic kernel. We also show that a positive ergodic kernel ensures that the eigenvalue $e^{-E_g} = \| e^{-H(g)} \|$ is simple.

Let $\mathcal{H} = L^2(Q, df)$ be a Hilbert space. We say that a bounded operator A has a *strictly positive kernel* provided that

$$0 < (\psi, A\chi) \tag{2.3.1}$$

whenever ψ and χ are non-negative L_2 functions with non-zero norms. Such an operator transforms a function χ, $\chi \geqq 0$, $\| \chi \| \neq 0$, into a function $A\chi$ which is strictly positive almost everywhere.

A weaker requirement is that A have a *positive, ergodic kernel.* We say that A has a positive, ergodic kernel if for each ψ, χ as above $0 \leqq (\psi, A\chi)$ and

$$0 < (\psi, A^j\chi) \tag{2.3.2}$$

for some j, depending on ψ and χ. Clearly every A with a strictly positive kernel has a positive, ergodic kernel.

THEOREM 2.3.2. *Let A have a positive ergodic kernel, and suppose that $\|A\|$ is an eigenvalue of A. Then $\|A\|$ is a simple eigenvalue and the corresponding eigenvector can be chosen to be a strictly positive function.*

Remark. Classical theorems of this sort make assumptions about A which are strong enough to prove both the existence of an eigenvalue and its uniqueness. In 1912, Jentzsch [16] showed that on a compact interval, any symmetric integral operator with a strictly positive, continuous kernel has a maximum eigenvalue, and it is simple. For positive matrices see Perron [22] and Frobenius [4].

PROOF. Since A maps positive functions into positive functions, it also maps real functions into real functions. If $\psi \in \mathcal{H}$ satisfies $A\psi = \|A\| \psi$, then so do $\operatorname{Re}\psi$ and $\operatorname{Im}\psi$. Therefore without loss of generality we may assume that ψ is real. Since $\| A^j \| = \| A \|^j$, and $A^j\psi = \| A \|^j \psi$, we infer that

$$\| A^j \| \, \| \psi \|^2 = (\psi, A^j\psi) \leqq (| \psi |, A^j \, | \psi |) \leqq \| A^j \| \, \| \psi \|^2 ,$$

and

$$(\psi, A^j\psi) = (| \psi |, A^j \, | \psi |) .$$

Writing $\psi = \psi^+ - \psi^-$, where ψ^+ and ψ^- are the positive and negative parts of ψ,

$$(\psi^+, A^j\psi^+) - (\psi^+, A^j\psi^-) - (\psi^-, A^j\psi^+) + (\psi^-, A^j\psi^-)$$
$$= (\psi^+, A^j\psi^+) + (\psi^+, A^j\psi^-) + (\psi^-, A^j\psi^+) + (\psi^-, A^j\psi^-) ,$$

or

$$(\psi^+, A^j\psi^-) + (\psi^-, A^j\psi^+) = 0 . \tag{2.3.3}$$

Unless $\psi^+ = 0$ or $\psi^- = 0$, each term of (2.3.3) could be made strictly positive by choosing an appropriate j. Thus either ψ^+ or ψ^- must vanish, and we may choose the eigenvector ψ to be non-negative.

If $\chi \geqq 0$, $\|\chi\| \neq 0$, then for some j, $0 < (\chi, A^j\psi) = \|A\|^j (\chi, \psi)$. This proves that $\chi\psi$ is not zero almost everywhere, and that ψ is strictly positive almost everywhere.

Finally, if ψ and χ were linearly independent eigenvectors of A with the eigenvalue $\|A\|$, then we could repeat the above argument with the component of χ orthogonal to ψ. This would yield two positive, orthogonal eigenvectors, which would be impossible, and the proof is complete.

Let $\varphi(h) = \int \varphi(x)h(x)dx$ denote the smeared, time zero free field operators. The spectral projections of the $\varphi(h)$, or the functions $\exp(i\varphi(h))$ generate a maximal abelian algebra $\mathfrak{M}$ of bounded operators on $\mathcal{F}$. Let Q be the spectrum of $\mathfrak{M}$. The no particle vector $\Omega_0 \in \mathcal{F}$ is a cyclic vector for $\mathfrak{M}$, namely $\mathcal{F} = (\mathfrak{M}\Omega_0)^-$. Therefore we may introduce a measure df on Q so that $\mathcal{F}$ is unitarily equivalent to $L^2(Q, df)$ and so that the equivalence carries $\mathfrak{M}$ into $L_\infty(Q)$ and takes Ω_0 into the function 1.

LEMMA 2.3.3. *With $\mathcal{F}$ represented as $L^2(Q, df)$, e^{-H_0} has a positive, ergodic kernel.*

PROOF. Let ψ and χ be non-negative. Write $\psi = \psi_1 + \psi_2$, where ψ_1 is the component of ψ along Ω_0. Thus the L_1 norm of ψ is given by $\|\psi\|_1 = (\psi, \Omega_0) = (\psi_1, \Omega_0)$. Note $\|\psi\|_1 \neq 0$ whenever ψ is non-zero, and $\|e^{-tH_0}\psi_2\| \leqq e^{-tm_0}\|\psi_2\|$, where m_0 is the boson mass. Thus

$$(\psi, e^{-tH_0}\chi) \geqq \|\psi\|_1 \|\chi\|_1 - \|\psi_2\| \|\chi_2\| e^{-tm_0} . \tag{2.3.4}$$

By choosing t sufficiently large, (2.3.4) is positive, which proves (2.3.2). If

$$e^{-tm_0} < \frac{1}{2}\frac{\|\psi\|_1 \|\chi\|_1}{\|\psi_2\| \|\chi_2\|} = \frac{1}{2}\frac{\|\psi\|_1 \|\chi\|_1}{(\|\psi\|^2 - \|\psi\|_1^2)^{1/2}(\|\chi\|^2 - \|\chi\|_1^2)^{1/2}} \tag{2.3.5}$$

then

$$(\psi, e^{-tH_0}\chi) \geqq \frac{1}{2}\|\psi\|_1 \|\chi\|_1 . \tag{2.3.6}$$

We need to show that $0 \leq (\psi, e^{-tH_0}\chi)$ for all t. In fact, it is sufficient to prove this for a dense set of non-negative ψ and χ.

Let us consider an approximate free energy operator

$$H_{0,K,V} = \int_{-K}^{K} a^*(k)a(k)\mu(k_V)dk \ .$$

For vectors $\psi \in C^\infty(H_0)$, as K, $V \to \infty$, $\| H_{0,K,V}\psi - H_0\psi \| \to 0$. Since H_0 is essentially self-adjoint on $C^\infty(H_0)$, the resolvents of $H_{0,K,V}$ converge strongly [18, p. 429]. Thus the semigroup convergence theorem [18, p. 502] ensures that for all $\psi \in \mathcal{F}$, $\| e^{-tH_{0,K,V}}\psi - e^{-tH_0}\psi \| \to 0$ as K, $V \to \infty$, and the convergence is uniform on compact sets of t. Therefore we need only show that for a dense set of non-negative ψ and χ, $0 \leq (\psi, e^{-tH_{0,K,V}}\chi)$. Let $F(x_1, \cdots, x_n)$ be a non-negative, infinitely differentiable function with compact support, and let

(2.3.8) $$\psi = F(\varphi(f_1), \cdots, \varphi(f_n))\Omega_0 \ ,$$

where $f_1, \cdots, f_n$ are real. The set of all such vectors are dense in $\mathcal{F}^+$, the non-negative vectors in $\mathcal{F}$. Furthermore, we define

(2.3.9) $$\psi_{K,V} = F(\varphi_{K,V}(f_1) \cdots \varphi_{K,V}(f_n))\Omega_0 \ ,$$

where $\varphi_{K,V}(x)$ is defined by restricting the sum in (2.1.6) to those $k \in \Gamma_{K,V} = \Gamma_V \cap \{k: |k| \leq K\}$. Then $\psi_{K,V} \in \mathcal{F}^+_{K,V} \subset \mathcal{F}^+$ where $\mathcal{F}_{K,V}$ is the Fock space corresponding to the modes $k \in \Gamma_{K,V}$. For any vector $\chi \in C^\infty(H_0)$,

$$\| \varphi_{K,V}(f)\chi - \varphi(f)\chi \| \to 0 \ , \qquad \text{as K, } V \to \infty \ ,$$

and as $C^\infty(H_0)$ is a core for $\varphi(f)$, the resolvents of $\varphi_{K,V}(f)$ converge strongly to the resolvent of $\varphi(f)$[18, p. 429]. Thus the semigroup convergence theorem [18, p. 502] ensures that for each $\chi \in \mathcal{F}$, s real,

$$\| e^{is\varphi_{K,V}(f)}\chi - e^{is\varphi(f)}\chi \| \to 0 \ , \qquad \text{as K, } V \to \infty \ ,$$

and the convergence is uniform for compact sets of s. By (2.3.9)

$$\psi_{K,V} = \int \tilde{F}(s_1, \cdots, s_n) \exp\left(i \textstyle\sum_{j=1}^n s_j\varphi_{K,V}(f_j)\right)ds_1 \cdots ds_n \ ,$$

and $\tilde{F}(s_1, \cdots, s_n)$ vanishes rapidly at infinity, so we conclude that

$$\| \psi_{K,V} - \psi \| \to 0 \ , \qquad \text{as K, } V \to \infty \ .$$

Thus for such vectors ψ, χ,

$$(\psi, e^{-tH_0}\chi) = \lim_{K,V\to\infty} (\psi_{K,V}, e^{-tH_{0,K,V}}\chi_{K,V})$$

and we need only show that

(2.3.10) $$0 \leq (\psi_{K,V}, e^{-tH_{0,K,V}}\chi_{K,V}) \ .$$

However on $\mathcal{F}_{K,V}$, $H_{0,K,V} = \sum_{k\in\Gamma_{K,V}} a_V^*(k)a_V(k)\mu(k) = \sum_{k\in\Gamma_{K,V}} H_{0,K,V}$, so $e^{-tH_{0,K,V}} =$

$\prod_{k \in \Gamma_{K,V}} e^{-tH_{0,K,V}}$. Each operator $e^{-tH_{0,K,V}}$ is known by explicit computation to have a strictly positive kernel [21], so (2.3.10) holds and the proof is complete.

LEMMA 2.3.4. *With $\mathcal{F}$ represented as $L^2(Q,df)$, the operator $e^{-H(g)}$ has a positive, ergodic kernel.*

Remark. We expect that e^{-H_0} and $e^{-H(g)}$ have strictly positive kernels.

PROOF. As in Lemma 2.3.3, formula (2.3.7), we consider $H(g)_{K,V} = H_{0,K,V} + H_{I,g,K,V}$. The approximate interaction $H_{I,g,K,V}$ is constructed with $\varphi_{K,V}$ in place of φ_V. Since $C^\infty(H_0)$ is a core for $H(g)$, we can argue as in the previous lemma that for all $\psi \in \mathcal{F}$,

$$\exp\left(-tH(g)_{K,V}\right)\psi \to \exp\left(-tH(g)\right)\psi , \qquad \text{as K}, V \to \infty .$$

Thus we need only prove that for ψ, χ as in Lemma 2.3.2,

$$0 < \varepsilon < (\psi_{K,V}, e^{-tH(g)_{K,V}}\chi_{K,V}) , \tag{2.3.11}$$

and that for sufficiently large t, the constant $\varepsilon = \varepsilon(\psi, \chi, t, \mathrm{K}, V)$ can be chosen independently of K and V.

On $\mathcal{F}_{K,V}$ we have an explicit representation of $\exp\left(-tH(g)_{K,V}\right)$ given by the Feynman-Kac integral formula [17].

$$\begin{aligned}&\left(\psi_{K,V}, \exp\left(-tH(g)_{K,V}\right)\chi_{K,V}\right)\\ &\quad = \int_{C_{K,V}} \exp\left(-\int_0^t H_{I,g,K,V}(q(s))ds\right)\psi_{K,V}(q(0))\chi_{K,V}(q(t))dq(\cdot) .\end{aligned} \tag{2.3.12}$$

Here q denotes a point in the spectrum of the modes

$$q_V(k) = a_V(k) + a_V(-k) + a_V^*(k) + a_V^*(-k) ,$$

and

$$q'_V(k) = a_V(k) - a_V(-k) + a_V^*(k) - a_V^*(-k) ,$$

for $k \in \Gamma_{K,V}$, and $C_{K,V}$ is the path space for these modes. Since $\exp\left(-tH_{0,K,V}\right)$ has a strictly positive kernel, (2.3.12) exhibits $\exp\left(-tH(g)_{K,V}\right)$ explicitly as an operator with a strictly positive kernel. Thus (2.3.11) is valid, and taking the limit K, $V \to \infty$ shows that

$$0 \leqq (\psi, e^{-tH(g)}\chi) . \tag{2.3.13}$$

We now establish a uniform lower bound on ε in (2.3.11) to prove that for t sufficiently large (2.3.13) is strictly positive. Given any positive M we can split the integral (2.3.13) into two parts. Let $C_{K,V}^{(1)}$ be those paths such that the exponent in the Feynman-Kac formula satisfies $-\int_0^t H_{I,g,K,V}(q(s))ds \geqq -M$, and let $C_{K,V}^{(2)}$ be the complementary set of paths. Thus

$$
\begin{aligned}
(\psi_{K,V}, e^{-tH(g)_{K,V}}\chi_{K,V}) &\geqq e^{-M}\int_{C^{(1)}_{K,V}} \psi_{K,V}(q(0))\chi_{K,V}(q(t))dq(\cdot) \\
&= e^{-M}\Big\{(\psi_{K,V}, e^{-tH_{0,K,V}}\chi_{K,V}) \\
&\quad - \int_{C^{(2)}_{K,V}} \psi_{K,V}(q(0))\chi_{K,V}(q(t))dq(\cdot)\Big\}\,.
\end{aligned} \tag{2.3.14}
$$

First we choose t by (2.3.5) so that (2.3.6) holds. Then for sufficiently large K and V (depending on t),

$$
\begin{aligned}
(\psi_{K,V} e^{-tH_{0,K,V}}\psi_{K,V}) &\geqq \frac{1}{2}(\psi, e^{-tH_0}\chi) \\
&\geqq \frac{1}{4}\|\psi\|_1 \|\chi\|_1\,.
\end{aligned}
$$

Thus (2.3.14) becomes

$$
\begin{aligned}
&(\psi_{K,V}, e^{-tH(g)_{K,V}}\chi_{K,V}) \\
&\quad \geqq e^{-M}\Big\{\frac{1}{4}\|\psi\|_1\|\chi\|_1 - \int_{C^{(2)}_{K,V}} \psi_{K,V}(q(0))\chi_{K,V}(q(t))dq(\cdot)\Big\}\,.
\end{aligned}
$$

Let Pr denote the measure on path space, so that by the Hölder inequality

$$
\begin{aligned}
&\left|\int_{C^{(2)}_{K,V}} \psi_{K,V}(q(0))\chi_{K,V}(q(t))dq(\cdot)\right| \\
&\qquad \leqq (\Pr\{C^{(2)}_{K,V}\})^{(r-1)/r}\, \|\psi_{K,V}(q(0))\chi_{K,V}(q(t))\|_r\,,
\end{aligned}
$$

where $1 < r < 2$. By the smoothing property of $\exp(-tH_{0,K,V})$ for sufficiently large t [7],

$$
\|\psi_{K,V}(q(0))\chi_{K,V}(q(t))\|_r \leqq \|\psi_{K,V}\|_2 \|\chi_{K,V}\|_2\,,
$$

and for K and V sufficiently large, this is dominated by $2\|\psi\|_2\|\chi\|_2$. Thus with the choices so far made for K, V, t, M,

$$
\begin{aligned}
(\psi_{K,V}, e^{-tH(g)_{K,V}}\chi_{K,V}) &\geqq e^{-M}\Big\{\frac{1}{4}\|\psi\|_1\|\chi\|_1 \\
&\quad -2\|\psi\|_2\|\chi\|_2(\Pr\{C^{(2)}_{K,V}\})^{(r-1)/r}\Big\}\,, \\
&\geqq \frac{1}{8}e^{-M}\|\psi\|_1\|\chi\|_1 > \varepsilon(\psi, \chi, t) > 0\,,
\end{aligned}
$$

provided in addition that

$$
\Pr\{C^{(2)}_{K,V}\} \leq \left(\frac{\|\psi\|_1\|\chi\|_1}{16\|\psi\|_2\|\chi\|_2}\right)^{r/(r-1)}\,. \tag{2.3.15}
$$

We now show that for M sufficiently large, (2.3.15) is satisfied and therefore the lemma is proved.

$$\begin{aligned}\Pr\{C^{(2)}_{K,V}\} &= \Pr\left\{M \leq \int_0^t H_{I,g,K,V}(q(s))ds\right\} \\ &= \Pr\left\{1 \leq M^{-1}\int_0^t H_{I,g,K,V}(q(s))ds\right\} \\ &\leq M^{-2}\int_{C_{K,V}}\left|\int_0^t H_{I,g,K,V}(q(s))ds\right|^2 dq(\cdot)\ .\end{aligned}$$

Replacing the integral over s by a limit of Riemann sums, we obtain a bound in terms of Wiener integrals depending on a finite number of times.

$$\begin{aligned}\Pr\{C^{(2)}_{K,V}\} \leq \lim_{n\to\infty} (t/nM)^2 \textstyle\sum_{i,j=1}^n \displaystyle\int_{C_{K,V}} & H_{I,g,K,V}(q(ti/n)) \\ & \times H_{I,g,K,V}(q(tj/n))dq(\cdot)\ .\end{aligned}$$

By the definition of the Wiener integral, this expression can be evaluated in terms of no-particle expectation values, and it equals

$$\lim_{n\to\infty} (t/nM)^2 \textstyle\sum_{i,j=1}^n (\Omega_0, H_{I,g,K,V} \exp(-\,|i-j|\,tH_0/n)\cdot H_{I,g,K,V}\Omega_0)\ .$$

By the Schwarz inequality

$$\begin{aligned}\Pr\{C^{(2)}_{K,V}\} &\leq (t/M)^2\,\|H_{I,g,K,V}\Omega_0\|^2 \\ &\leq (tD/M)^2\ ,\end{aligned}$$

for some constant D independent of K and V. Thus we choose

$$M \geq Dt\left(\frac{16\,\|\psi\|_2\,\|\chi\|_2}{\|\psi\|_1\,\|\chi\|_1}\right)^{r/2(r-1)}\ .$$

Combining Lemma 2.3.4 with Theorem 2.3.2 yields a proof of Theorem 2.3.1. Clearly the same proof applies to $H(g)_V$, to show that its vacuum is unique.

COROLLARY 2.3.5. *Let $\Omega_{g,V}$ be the vacuum for $H(g)_V$ with its phase determined by the requirement*

(2.3.16) $$(\Omega_0, \Omega_{g,V}) > 0\ .$$

Then $\lim_{V\to\infty}\Omega_{g,V} = \Omega_g$ exists, Ω_g is the vacuum for $H(g)$, and

(2.3.17) $$(\Omega_0, \Omega_g) > 0\ .$$

PROOF. A sequence Ω_{g,V_j} with $V_j \to \infty$ has a convergent subsequence by Theorem 2.2.2, converging to a vacuum for $H(g)$. The phase (2.3.16) fixes the phase (2.3.17) so every convergent subsequence has the same limit Ω_g. Thus the $\Omega_{g,V}$ converge to Ω_g, as required.

COROLLARY 2.3.6. *The vacuum Ω_g is a cyclic vector for $\mathfrak{M}$.*

PROOF. The function Ω_g is positive for almost all Q, and $\mathfrak{M} = L_\infty(Q)$ in the $L_2(Q, df)$ representation of $\mathcal{F}$.

3. The Heisenberg picture field operators

3.1. *Introduction.* In the Heisenberg picture operators have the time dependence

$$A(t) = e^{itH(g)}A(0)e^{-itH(g)} . \tag{3.1.1}$$

This definition of the dynamics contains the cutoff function $g(x)$ explicitly. For an important class of operators $A(0)$, however, $A(t)$ is independent of $g(x)$ provided that $g(x) = \lambda$, the coupling constant, on a suitably large set. For example, we take $A(0)$ to be an observable representing a measurement performed in some region B of space (at time $t = 0$). Then $A(t)$ represents the same measurement performed at time t. A hamiltonian with no ultraviolet cutoff, such as $H(g)$, propagates information with at most the speed of light. Therefore if $g(x) = \lambda$ on an interval containing B, and t is sufficiently small, the fact that $g(x)$ does not equal λ everywhere will never be recorded by a measurement $A(t)$. See Figures 1-2.

For each localized observable $A(0)$ and each t, we make an appropriate choice for $g(x)$. Therefore (3.1.1) provides the correct dynamics for the $(\varphi^4)_2$

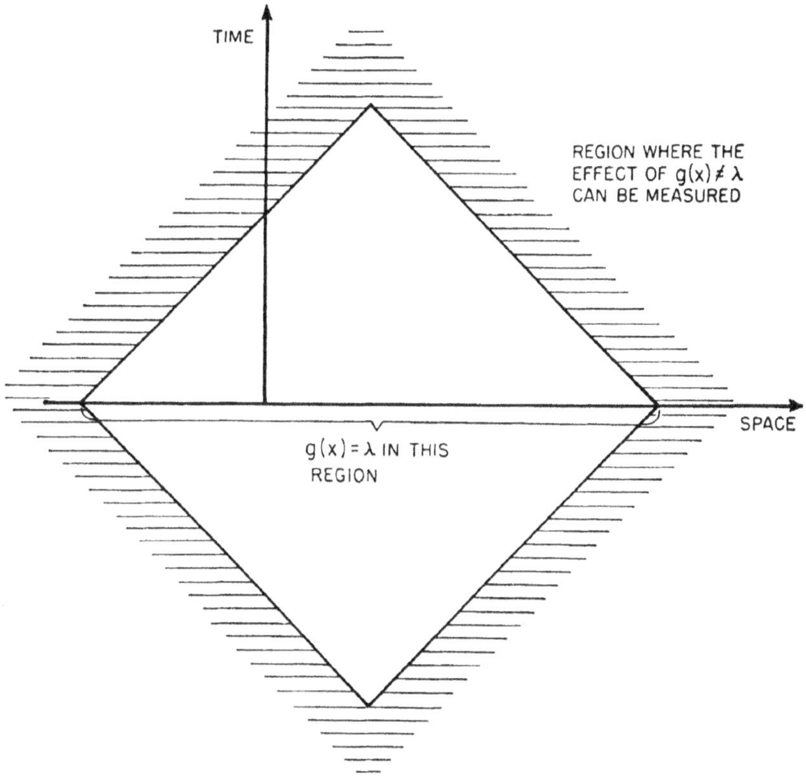

FIG. 1 Information propagates at the speed of light for a local hamiltonian

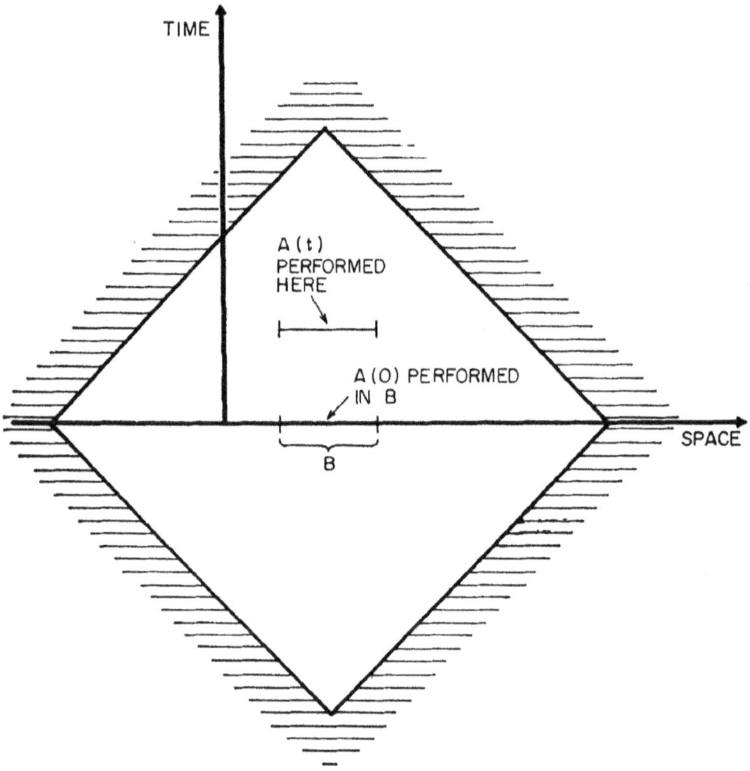

FIG. 2 Measurements performed inside the square obey laws of physics independent of any cutoff

quantum field theory with all the cutoffs removed. The mathematical proof of these statements is the main result of [8–9]. This method of removing the spatial cutoff was first proposed by Guenin. In [8] we used a theorem of Segal [26] to the effect that the above finite propagation speed follows from the Trotter product formula. In order to apply the Trotter product formula, we proved in [8] that $H(g)$ is self-adjoint. See [8] for further discussion.

In this section we discuss the field operators $\varphi(x, t)$ or

$$\varphi(f) = \int \varphi(x, t) f(x, t) dx dt \, . \tag{3.1.2}$$

The operator $\varphi(f)$ is unbounded but for real f it is self-adjoint. The field $\varphi(x, t)$ is not an operator, but it is a bilinear form. It is possible to regard the field at a given time t as an amplitude for a collection of oscillators, one at each point of space x. Then the field strength $\varphi(x, t)$(field deformation) measures the amplitude of the oscillator at x at time t, and $\pi(x, t) = \dot{\varphi}(x, t)$ represents the momentum of the corresponding oscillator. The kinetic energy density for the oscillators is $\frac{1}{2}\colon \pi(x, t)^2 \colon$, while the total energy density is

$$\frac{1}{2}(:\pi(x, t)^2: + :(\nabla\varphi(x, t))^2: + m_0^2:\varphi(x, t)^2:) + \lambda:\varphi(x, t)^4: ,$$

where $\lambda \geqq 0$. For both the classical and the quantum field theory, the positivity of the energy ensures that in states of finite energy, the field strength $\varphi(x, t)$ has one L_2 derivative as a function of x. For the quantum case, the precise mathematical meaning is that for $\psi \in \mathfrak{D}((H(g) + b)^{1/2})$,

$$\left|\int(\psi, \varphi(x, t)\psi)(D_x f)(x, t)dxdt\right| \leqq \text{const.}\ \| f(\cdot, t) \|_2 (\psi, (H(g) + b)\psi) .$$

In particular, in one dimensional space a Sobolev inequality ensures that a function with one L_2 derivative is continuous. Hence the expectation value of the field strength $(\psi, \varphi(x, t)\psi)$ is a continuous function of x, for vectors ψ with finite energy $H(g)$. See Lemma 3.2.1.

We see that integration helps in (3.1.2) because $\varphi(f)$ is an operator while $\varphi(x, t)$ is a bilinear form. Actually the time integration is not required and for real f,

$$A(t) = \int\varphi(x, t)f(x)dx \tag{3.1.3}$$

is also a self-adjoint operator depending continuously on t. We expect that this is a special feature of the two dimensional model we are considering and that sharp time fields will not be operators in four dimensions. For this reason, basic physical concepts have been formulated in terms of the time averaged fields (3.1.2) rather than the sharp time fields (3.1.3). For example, Wightman's axioms for a quantum field theory are expressed in terms of the operators (3.1.2), and we will show that many of his axioms are satisfied for our model.

The time integration in (3.1.2) presents some new difficulties (for example in the proof of self-adjointness or of locality) which would not occur if we considered only the sharp time operators (3.1.3). An advantage of the time averaged field is that products $\varphi(f_1) \cdots \varphi(f_n)$ can be defined on vectors with finite energy (Corollary 3.2.5). In fact we will construct a dense domain which the localized field operator $\varphi(f)$ leaves invariant, and on which $\varphi(f)$ is essentially self-adjoint for real f.

3.2. *An invariant domain for localized fields.* In this section we study the Heisenberg picture field localized in a region of space time $\mathfrak{B}$. We find that $\varphi(x, t)$ is a bilinear form and that for real f, $\varphi(f)$ is a densely defined symmetric operator. We start with the region $\mathfrak{B}$, a bounded open subset of space time. We require that $H(g)$ be a hamiltonian for $\mathfrak{B}$. This means that the spatial cutoff $g(x)$ equals the coupling constant λ on a sufficiently large interval to contain the domain of dependence of $\mathcal{B}$. In other words, assuming that the velocity of light is one, for every point (y, t) in $\mathcal{B}$

$$g(x) = \lambda , \qquad \text{if } |x - y| \leq |t| . \tag{3.2.1}$$

It is convenient to deal with the field $\varphi_g(x, t) = e^{itH(g)}\varphi(x)e^{-itH(g)}$, and its time derivative

$$\pi_g(x, t) = e^{itH(g)}\pi(x)e^{-itH(g)} = \dot{\varphi}_g(x, t) .$$

The time zero field $\varphi(x)$ and its conjugate momentum $\pi(x)$ were defined in (1.2.10–11). We shall see that for $(x, t) \in \mathcal{B}$, $\varphi_g(x, t)$ is independent of g, and equals the field $\varphi(x, t)$. Thus all the cutoffs have been removed in the definition of $\varphi(x, t)$. For each C^∞ function $f(x, t)$ with support in $\mathcal{B}$, we show that

$$\varphi(f) = \int \varphi(x, t)f(x, t)dxdt \tag{3.2.2}$$

is an operator whose domain contains

$$\mathcal{D}_g = C^\infty(H(g)) . \tag{3.2.3}$$

In fact $\mathcal{D}_g$ is an invariant domain,

$$\varphi(f)\mathcal{D}_g \subset \mathcal{D}_g , \tag{3.2.4}$$

so that $\mathcal{D}_g \subset C^\infty(\varphi(f))$. We note that this invariant domain may depend on the region $\mathcal{B}$ in which the field is localized. For $\psi \in \mathcal{D}_g$ the expectation value

$$(\psi, \varphi(x_1, t_1) \cdots \varphi(x_n, t_n)\psi) \tag{3.2.5}$$

is a Schwartz distribution in $\mathcal{D}'(\mathcal{B} \times \cdots \times \mathcal{B})$.

If $f(x, t)$ is a function in $\mathcal{S}$, then $\varphi_g(f)$ still is defined on $\mathcal{D}_g$ and leaves it invariant. The expectation values (3.2.5) of $\varphi_g(x, t)$ are tempered distributions. However, the fields $\varphi_g(f)$ may depend on g.

LEMMA 3.2.1. *The field* $\varphi_g(x, t)$ *is a bilinear form on* $\mathcal{D}((H(g) + b)^{1/2}) \times \mathcal{D}((H(g) + b)^{1/2})$ *continuous in* x *and* t. *Namely for* $\psi \in \mathcal{D}((H(g) + b)^{1/2})$, $(\psi, \varphi(x, t)\psi)$ *is a continuous function. Furthermore*

$$\left|\int (\psi, \varphi_g(x, t)\psi)D_x f(x)dx\right| \leq \text{const.}\, \|f\|_2 (\psi, (H(g) + b)\psi) . \tag{3.2.6}$$

PROOF. The free field $\varphi(x, 0)$ is the sum of two expressions of the form (1.2.8). The kernels $b(k)$ are not in L_2 and so (1.2.9) does not help. We have however, $b(k)\mu(k)^{-1/2} = e^{\pm ikx}\mu(k)^{-1} \in L_2$. The estimate (1.2.9) has been generalized [6] to cover such kernels, giving us

$$\| (H_0 + I)^{-1/2}\varphi(x, 0)(H_0 + I)^{-1/2} \| \leq \text{const.}\, \| b\mu^{-1/2} \|_2 < \infty . \tag{3.2.7}$$

Thus for $\psi \in \mathfrak{D}((H(g)+b)^{1/2})$, $e^{-itH(g)}\psi \in \mathfrak{D}((H(g)+b)^{1/2}) \subset \mathfrak{D}(H_0^{1/2})$, by (2.1.19) and so $(\psi, \varphi_g(x,t)\psi) = (e^{-itH(g)}\psi, \varphi(x,0)e^{-itH(g)}\psi)$ is defined and

$$|(\psi, \varphi_g(x,t)\psi)| \leq \text{const. } \|b\mu^{-1/2}\|_2 (\psi, (H(g)+b)\psi) .$$

Since $\|b(k)\mu(k)^{-1/2}k\tilde{f}(k)\|_2 \leq \|b\mu^{-1/2}k\|_\infty \|f\|_2 \leq \text{const. } \|f\|_2$, the inequality (3.2.6) holds.

Let us write b_x for b to denote the dependence of b on x. Then $\|(b_x - b_y)\mu^{-1/2}\|_2$ is a function of $(x-y)$ only and it tends to zero as $|x-y| \to 0$. Since

$$\begin{aligned} |(\psi, (\varphi_g(x,t) - \varphi_g(y,t))\psi| &\leq \text{const. } \|(1+H_0)^{1/2}\psi\|^2 \|(b_x - b_y)\mu^{-1/2}\|_2 \\ &\leq \text{const. } \|(H(g)+b)^{1/2}\psi\|^2 \|(b_x - b_y)\mu^{-1/2}\|_2 \end{aligned} \tag{3.2.8}$$

we have continuity with respect to x. Also $\|(H(g)+b)^{1/2}(e^{itH(g)} - e^{isH(g)})\psi\| \to 0$, as $|t-s| \to 0$. Thus

$$\begin{aligned} &|(\psi, (\varphi_g(x,t) - \varphi_g(x,s))\psi)| \\ &\quad \leq \text{const. } \|(H_0+I)^{1/2}(e^{itH(g)} - e^{isH(g)})\psi\| \, \|b_x\mu^{-1/2}\|_2 \\ &\qquad \times \{\|(H_0+I)^{1/2}e^{itH(g)}\psi\| + \|(H_0+I)^{1/2}e^{isH(g)}\psi\|\} \\ &\quad \leq \text{const. } \|(H(g)+b)^{1/2}(e^{itH(g)} - e^{isH(g)})\psi\| \, \|b_x\mu^{-1/2}\|_2 \\ &\qquad \times \|(H(g)+b)^{1/2}\psi\| \to 0 \qquad \text{as } |t-s| \to 0. \end{aligned} \tag{3.2.9}$$

From (3.2.8–9) we see that $\varphi(x,t)$ is jointly continuous in x and t.

Probably $\pi(x,t)$ is a bilinear form on $\mathfrak{D}((H(g)+b)^{3/2}) \times \mathfrak{D}((H(g)+b)^{3/2})$ continuous in x and t, but our estimates are not strong enough to prove this.

The functions $f(x,t)$ in $\mathcal{S}(R^2)$ determine bounded measures $d\nu = f(x,t)dxdt$, so $\varphi_g(f) = \int \varphi_g(x,t)d\nu$ is a bilinear form. If $d\nu_n \to d\nu$ in the w^* topology for measures, then $\int \varphi_g d\nu_n \to \int \varphi_g d\nu$ in the weak sense that for $\psi \in \mathfrak{D}((H(g)+b)^{1/2})$,

$$\left(\psi, \int \varphi_g d\nu_n \psi\right) \to \left(\int \psi, \int \varphi_g d\nu \psi\right) . \tag{3.2.10}$$

We define the sharp time field

$$A_g(t) = \int \varphi_g(x,t)f(x,t)dx \tag{3.2.11}$$

and

$$B_g(t) = \int \pi_g(x,t)f(x,t)dx \tag{3.2.12}$$

LEMMA 3.2.2. *Let $f \in \mathcal{S}(R^2)$ be real. Then $A_g(t)$ and $B_g(t)$ define self-adjoint operators, and their domains include $\mathfrak{D}((H(g)+b)^{1/2})$. With a constant c independent of g,*

(3.2.13) $\|A_g(t)\psi\| + \|B_g(t)\psi\| \leq c\{\|f(\cdot, t)\|_2 + \|D_x f(\cdot, t)\|_2\}\|(H(g) + b)^{1/2}\psi\|$,
for all $\psi \in \mathfrak{D}((H(g) + b)^{1/2})$.

PROOF. It is sufficient to consider $\varphi(f_t) = \int \varphi(x)f(x, t)dx$ in place of $A_g(t)$, as they are unitarily equivalent *via* the unitary operator $e^{-itH(g)}$, and this unitary leaves $\mathfrak{D}((H(g) + b)^{1/2})$ invariant. The same is true for $\pi(f_t)$. By (2.1.19)

$$\|(N + I)^{1/2}\psi\|^2 \leq m_0^{-1}\|(H_0 + I)^{1/2}\psi\|^2 \leq 2m_0^{-1}\|(H(g) + b)^{1/2}\psi\|^2 ,$$

so we need only prove that

(3.2.14) $\|\varphi(f_t)\psi\| + \|\pi(f_t)\psi\| \leq c\{\|f(\cdot, t)\|_2 + \|D_x f(\cdot, t)\|_2\}\|(N + I)^{1/2}\psi\|$.

The lemma now follows from (1.2.9). For example, $\pi(f_t)$ is the sum of two operators of the form (1.2.8) with kernels $b(k) = \mu(k)^{1/2}\int f(x, t)e^{\pm ikx}dx$, satisfying

$$\begin{aligned}\|b\|_2 &= \|(-D_x^2 + m_0^2)^{1/4}f(\cdot, t)\|_2 \\ &\leq \text{const.}\left(\|f(\cdot)\|_2 + \|D_x f(\cdot, t)\|_2\right) .\end{aligned}$$

The kernel for $\varphi(f_t)$ can be bounded by the $\|f(\cdot, t)\|_2$ norm alone. The estimate (3.2.13) now follows from (3.2.14). The self-adjointness of $A_g(t)$ and $B_g(t)$ can be proven by showing that $\varphi(f_t)$ and $\pi(f_t)$ are self-adjoint. But (3.2.14) ensures that every vector in $\mathcal{F}$ with a finite number of particles is analytic for $\varphi(f_t)$ and for $\pi(f_t)$, so these operators are essentially self-adjoint on the domain of vectors with a finite number of particles. Hence they are uniquely determined by their definition on that domain.

We now explain the sense in which the integral (3.2.12) converges, since we did not show that $\pi_g(x, t)$ was a bilinear form. If $\psi \in \mathfrak{D}((H(g) + b)^{1/2})$, then $\psi \in \mathfrak{D}((N + I)^{1/2})$ and

(3.2.15) $$(\psi, \tilde{\pi}(k)\psi) = \mu(k)^{1/2}\{(a(k)\psi, \psi) + (\psi, a(-k)\psi)\}$$

is a slowly increasing, locally summable function, and hence a tempered distribution. Thus $(\psi, \pi(x)\psi)$ is by definition the distribution Fourier transform of (3.2.15), and hence a tempered distribution. Finally (3.2.12) is the weak integral

$$(\psi, B_g(t)\psi) = \int f(x, t)(e^{-itH(g)}\psi, \pi(x)e^{-itH(g)}\psi)dx .$$

THEOREM 3.2.3. *Let* $\psi \in \mathfrak{D}((H(g) + b)^{1/2})$, *and let* $f(x, t)$ *be a real function in* $\mathcal{S}(R^2)$. *Then the vectors* $A(t)\psi$ *and* $B(t)\psi$ *are strongly continuous and are rapidly decreasing functions of t. The integrals* $\int A_g(t)\psi dt = \varphi_g(f)\psi$, *and* $\int B_g(t)\psi dt = \pi_g(f)\psi$ *exist and define* $\varphi_g(f)$ *and* $\pi_g(f)$ *as closed symmetric opera-*

tors with domains containing $\mathfrak{D}((H(g)+b)^{1/2})$. *We have*

$$\|\varphi_g(f)\psi\| + \|\pi_g(f)\psi\| \leqq c\int\{\|f(\cdot,t)\|_2 + \|D_x f(\cdot,t)\|_2\}dt \times \|(H(g)+b)^{1/2}\psi\|\,, \tag{3.2.16}$$

with a constant c independent of f and g.

PROOF. We write

$$|f(\cdot,t)|_1 = c\|f(\cdot,t)\|_2 + c\|D_x f(\cdot,t)\|_2 \tag{3.2.17}$$

and

$$\begin{aligned} A_g(s)\psi - A_g(t)\psi &= (I - e^{i(t-s)H(g)})A_g(s)\psi \\ &\quad + e^{itH(g)}\int\varphi(x)(f(x,s)-f(x,t))dx e^{-isH(g)}\psi \\ &\quad + A_g(t)(e^{i(t-s)H(g)} - I)\psi\,. \end{aligned}$$

Thus by (3.2.13),

$$\begin{aligned} \|A_g(s)\psi - A_g(t)\psi\| &\leqq \|(I - e^{i(t-s)H(g)})A_g(s)\psi\| \\ &\quad + |f(\cdot,s) - f(\cdot,t)|_1\,\|(H(g)+b)^{1/2}\psi\| \\ &\quad + |f(\cdot,t)|_1\,\|(e^{i(t-s)H(g)} - I)(H(g)+b)^{1/2}\psi\| \to 0 \end{aligned}$$

as $t \to s$. This proves the continuity. The rapid decrease is ensured by (3.2.13) and the fact that $f \in \mathfrak{S}$. A similar argument works for $B_g(t)\psi$. The integrals defining $\varphi_g(f)\psi$ and $\pi_g(f)\psi$ now exist, (3.2.16) follows from integrating (3.2.13). Since $A_g(t)$ and $B_g(t)$ are self adjoint, for $\psi \in \mathfrak{D}((H(g)+b)^{1/2}) \subset \mathfrak{D}(A_g(t)) \cap \mathfrak{D}(B_g(t))$,

$$(\psi, \varphi_g(f)\psi) = \int(\psi, A_g(t)\psi)dt = \int(A_g(t)\psi, \psi)dt$$

is real, and similarly for π. Symmetric operators are closable and we now define $\varphi_g(f)$ and $\pi_g(f)$ as the closure of the above operators on the domain

$$\mathfrak{D}((H(g)+b)^{1/2})\,. \tag{3.2.18}$$

Remarks. (a). The integrals defining $\varphi_g(f)\psi$ and $\pi_g(f)\psi$ are strong Riemann integrals, $\varphi_g(f)$ is a strong limit of operators of the form

$$A_g(t_1)\Delta t_1 + \cdots + A_g(t_n)\Delta t_n\,. \tag{3.2.19}$$

Conversely using the continuity of $A_g(t)\psi$, we see that an operator of the form (3.2.19) is a strong limit of a sequence of $\varphi_g(f_j)$, and the f_j can be chosen with the norm

$$|f|_1 = c\int\{\|f(\cdot,t)\|_2 + \|D_x f(\cdot,t)\|_2\}dt \tag{3.2.20}$$

uniformly bounded. For both limits the convergence occurs on the domain

(3.2.18) and similar considerations apply to π_g. Furthermore $\varphi_g(f)$ and $\pi_g(f)$ can be defined whenever $|f|_1 < \infty$.

(b). Using (2.1.19) to estimate H_0, we have from (3.2.16),

$$\|\varphi_g(f)\Omega_g\| + \|\pi_g(f)\Omega_g\| \leqq |f|_1\left(\left|\frac{1}{2}E_{2g} - E_g\right| + 1\right)^{1/2}, \tag{3.2.21}$$

but the bound on the right grows in the diameter of the support of g. We derive a better estimate in [29, § 3.3].

THEOREM 3.2.4. *Let $|f|_1 + |D_t f|_1$ defined in* (3.2.20) *be finite. On the domain $\mathfrak{D}((H(g) + b)^{3/2})$, the field φ_g satisfies the differential equation*

$$\begin{aligned}(D_t\varphi_g)(f) &= -\varphi_g(D_t f)\\ &= \pi_g(f) = [iH(g), \varphi_g(f)]\,.\end{aligned} \tag{3.2.22}$$

PROOF. The first equality in (3.2.22) is the definition of a distribution derivative, and is not to be proved. Part of the theorem is to show that the commutator is defined on $\mathfrak{D}((H(g)+b)^{3/2})$. We write out the difference quotient for $D_t f$, $(\Delta_\varepsilon f)(x, t) = \{f(x, t + \varepsilon) - f(x, t)\}/\varepsilon$ which converges to $D_t f$ in the norm $|f|_1$. Thus by (3.2.16), for $\psi \in \mathfrak{D}((H(g) + b)^{1/2})$,

$$\|\varphi_g(\Delta_\varepsilon f)\psi - \varphi_g(D_t f)\psi\| \to 0\,.$$

We have for $\psi \in \mathfrak{D}((H(g) + b)^{3/2})$,

$$\begin{aligned}\varphi_g(\Delta_\varepsilon f)\psi &= \varepsilon^{-1}(I - e^{i\varepsilon H(g)})\int\varphi_g(x, t - \varepsilon)f(x, t)dx\psi dt\\ &\quad + \varepsilon^{-1}\int A(t)(e^{i\varepsilon H(g)} - I)\psi dt\,.\end{aligned} \tag{3.2.23}$$

By Lemma (3.2.2) the last term in (3.2.23) converges as $\varepsilon \to 0$ and its limit is

$$i\int A(t)H(g)\psi dt = i\varphi(f)H(g)\psi\,.$$

Since $\varphi_g(\Delta_\varepsilon f)\psi$ also converges, the remaining term in (3.2.23) necessarily converges also to a limit ψ_1.

For $\chi \in \mathfrak{D}(H(g))$,

$$\begin{aligned}(\chi, \psi_1) &= \lim_{\varepsilon\to 0}\left(\chi, \varepsilon^{-1}(I - e^{i\varepsilon H(g)})\int\varphi_g(x, t - \varepsilon)f(x, t)dx\psi\right)dt\\ &= \left(iH(g)\chi, \varphi_g(f)\psi\right).\end{aligned}$$

Since $H(g) = H(g)^*$, we conclude that $\varphi_g(f)\psi \in \mathfrak{D}(H(g))$ and $\psi_1 = -iH(g)\varphi_g(f)\psi$. Therefore $-\varphi_g(D_t f)\psi = i[H(g), \varphi_g(f)]\psi$.

We still need to prove that $-\varphi_g(D_t f) = \pi_g(f)$, and it is sufficient to prove this as a bilinear form on $\mathfrak{D}_g \times \mathfrak{D}_g$. From the above equation we have

$$i(\psi, \varphi_g(D_t f)\psi) = \int\left(H(g)\psi(t), \int\varphi(x, 0)f(x, t)dx\psi(t)\right)dt \tag{3.2.24}$$
$$- \int dt\left(\int\varphi(x, 0)f(x, t)dx\psi(t), H(g)\psi(t)\right),$$

where $\psi(t) = e^{-itH(g)}\psi$. Since $\psi(t) \in \mathfrak{D}(H(g)) = \mathfrak{D}(H_0) \cap \mathfrak{D}(H_{I,g})$, and

$$\| H_{I,g}(\psi(t) - \psi(s)) \| \leqq a \| (H(g) + b)(\psi(t) - \psi(s)) \| \to 0 ,$$

as $|t - s| \to 0$, we may substitute $H_0 + H_{I,g}$ for $H(g)$ and consider each term separately. It is known that $H_{I,g}$ and $\int\varphi(x, 0)f(x, t)dx$ commute; see [8, Ths. 3.1–3.2] for example. Thus $H_{I,g}$ contributes zero to (3.2.24). Furthermore the identity

$$\left(H_0\psi, \int\varphi(x, 0)f(x, t)dx\psi\right) - \left(\int\varphi(x, 0)f(x, t)dx\psi, H_0\psi\right)$$
$$= \left(\psi, - i\int\pi(x, 0)f(x, t)dx\psi\right)$$

is a standard computation valid for any $\psi \in \mathfrak{D}(H_0)$, for example $\psi = \psi(t)$. Thus (3.2.24) yields

$$i(\psi, \varphi_g(D_t f)\psi) = \int\left(\psi(t), -i\int\pi(x, 0)f(x, t)dx\psi(t)\right)dt$$
$$= (\psi, - i\pi_g(f)\psi) ,$$

and the theorem is proved.

Remarks. (a). In exactly the same fashion one proves that

$$(D_{tt}\varphi_g)(f) = (D_t\pi_g)(f) = -\pi_g(D_t f)$$
$$= [iH(g), \pi_g(f)]$$

if $|D_{tt}f|_1$ is also finite. The commutator is a bilinear form on $\mathfrak{D}_g \times \mathfrak{D}_g$, namely

$$(D_{tt}\varphi_g)(f) = (D_{xx}\varphi_g)(f) - m_0^2\varphi(f) - 4\int \,\vdots\, \varphi_g^3(x, t \,\vdots\, f(x, t)g(x)dxdt . \tag{3.2.25}$$

Here we define the $\vdots\ \vdots$ product by

$$\vdots\, \varphi_g^3(x, t) \,\vdots = e^{iH(g)t} : \varphi^3(x) : e^{-iH(g)t} , \tag{3.2.26}$$

which we now prove is an operator valued distribution. First we note that $\int : \varphi^3(x) : f(x, t)dx$ is a sum of monomials in creation and annihilation operators with kernels in L_2, and their L_2 norms are continuous in t. Thus by (1.2.9) $\int : \varphi^3(x) : f(x, t)g(x)dx$ is a bilinear form on $\mathfrak{D}(N) \times \mathfrak{D}(N)$. By (2.1.3), $(H(g) + b)^{-1}\int : \varphi^3(x) : f(x, t)g(x)dx(H(g) + b)^{-1}$ is a bounded operator, norm continuous in t. Thus on $\mathfrak{D}(H(g)) \times \mathfrak{D}(H(g))$, $\vdots\, \varphi_g^3 \,\vdots\, (f) = \int \,\vdots\, \varphi_g^3(x, t) \,\vdots\, f(x, t)dxdt$ is defined as a bilinear form. Hence (3.2.25) holds as an equation for bilinear forms on $\mathfrak{D}(H(g)) \times \mathfrak{D}(H(g))$. But each term except the last is an operator defined on $\mathfrak{D}((H(g) + b)^{1/2})$. Thus $\vdots\, \varphi_g^3 \,\vdots\, (f)$ is actually an operator on

$\mathfrak{D}((H(g)+b)^{1/2})$, and in fact for real $f(x, t)$ it is essentially self adjoint, as we shall see in § 3.3.

Furthermore, on $\mathfrak{D}((H(g)+b)^{1/2}) \times \mathfrak{D}((H(g)+b)^{1/2})$ each term in (3.2.25) except the last is a bilinear form which is a distribution of order two. Thus the same is true for $(\psi, :\varphi^3(x, t): \psi)$.

We have used (3.2.26) to define the cube of the interacting field. It would be interesting to determine whether this definition agrees with conventional notions involving the separation of points.

We shall see in this section and the following one that the $g(x)$ in equation (3.2.25) can be removed if $f(x, t)$ has compact support and $g(x) = \lambda$ on a sufficiently large set. Then (3.2.25) becomes $(\square + m_0^2)\varphi(f) = -4\lambda :\varphi^3: (f)$, which is a non-linear equation for a self-adjoint operator valued distribution.

(b). The identity $\pi_g(f) = [iH(g), \varphi_g(f)]$ implies

$$B_g(t) = i[H(g), A_g(t)] \tag{3.2.27}$$

provided that the right and left sides of (3.2.27) make sense and are continuous in t. They are certainly defined and are continuous as bilinear forms on $\mathfrak{D}((H(g)) \times \mathfrak{D}(H(g))$. To see that (3.2.27) makes sense as operators on $\mathfrak{D}((H(g)+b)^{3/2})$ we need only show that $A_g(t)$ maps $\mathfrak{D}((H(g)+b)^{3/2})$ into $\mathfrak{D}(H(g))$. We choose a sequence $f_j(x, t) = f(x, t_0)\delta_j(t - t_0)$ where δ_j is a sequence of smooth functions converging to δ in the w^* topology on measures and with the norms $|f_j|_1$ uniformly bounded. Then the bilinear forms converge, which means that the inner products

$$(\theta, iH(g)\varphi_g(f_j)\psi) = (\theta, \varphi_g(f_j)iH(g)\psi) + (\theta, \pi_g(f_j)\psi)$$

converge for $\theta \in \mathfrak{D}(H(g))$. However the norms

$$\begin{aligned} \| H(g)\varphi_g(f_j)\psi \| &\leqq \| \varphi(f_j)H(g)\psi \| + \| \pi(f_j)\psi \| \\ &\leqq |f|_1 \, \| (H(g)+b)^{3/2}\psi \| \end{aligned}$$

are uniformly bounded, and so the inner products converge for all $\theta \in \mathcal{F}$. Thus the limit $A(t)\psi = \text{weak lim}\, \varphi(f_j)\psi$ is in $\mathfrak{D}(H(g)^*) = \mathfrak{D}(H(g))$ which proves (3.2.27) on the domain $\mathfrak{D}((H(g)+b)^{3/2})$.

COROLLARY 3.2.5. *Let $f \in \mathcal{S}$. Then $\mathfrak{D}_g = C^\infty(H(g)) \subset C^\infty(\varphi_g(f))$, and $\varphi_g(f)\mathfrak{D}_g \subset \mathfrak{D}_g$.*

PROOF. Using Theorem 3.2.4, we prove by induction on m that $\varphi_g(f)\mathfrak{D}_g \subset \mathfrak{D}(H(g)^m)$ and that for $\psi \in \mathfrak{D}_g$,

$$H(g)^m\varphi_g(f)\psi = \varphi_g(f)H(g)^m\psi + \sum_{j=1}^m \binom{m}{j} i^j \varphi_g(D_t^j f)H(g)^{m-j}\psi .$$

This formula is a special case of the standard identity

$A^m B \supset \sum_{j=1}^{m} \binom{m}{j} [(adA)^j B] A^{m-j}$. Thus

$$\| H(g)^m \varphi_g(f)\psi \| \leqq \sum_{j=1}^{m} \binom{m}{j} | D_t^j f |_1 \, \| (H(g) + b)^{m-j+(1/2)} \psi \| , \tag{3.2.28}$$

where $| f |_1$ is the norm (3.2.20).

THEOREM 3.2.6. *Let $f \in C_0^\infty(\mathfrak{B})$, that is f is C^∞ with support in the open region of space time $\mathfrak{B}$. Let $H(g)$ be a hamiltonian for $\mathfrak{B}$, so that $g = \lambda$ on a large set. Then*

$$\varphi_g(f) = \varphi(f) \qquad \text{and} \qquad \pi_g(f) = \pi(f)$$

are independent of g/λ.

PROOF. The spectral projections $E^t(\lambda)$ of the sharp time field

$$\begin{aligned} A_0(t) &= \int \varphi(x, t) f(x, t_0) dx \\ &= \int \lambda dE^t(\lambda) \end{aligned}$$

are given by the formula $E^t(\lambda) = e^{itH(g)} E^0(\lambda) e^{-itH(g)}$ and are independent of g by [8]. Thus $A_0(t)$ is independent of g and so is $A(t)$. By (2.2.7), for all g,

$$\mathfrak{D}_0 = \mathfrak{D}(H_0^{1/2}) \cap \mathfrak{D}(N) \subset \mathfrak{D}((H(g) + b)^{1/2}) \subset \mathfrak{D}(\varphi_g(f)) ,$$

so that $\varphi_g(f) \mid \mathfrak{D}_0 = \varphi(f) \mid \mathfrak{D}_0$ is independent of g. Thus to complete the proof, we only need to show that the domain of $\varphi_g(f) = (\varphi_g(f) \mid \mathfrak{D}((H(g) + b)^{1/2}))^-$ is independent of g. Since $H(g)$ is essentially self-adjoint on $C^\infty(H_0) \subset \mathfrak{D}_0$, [8], so is $(H(g) + b)^{1/2}$. Thus by (3.2.16) $(\varphi(f) \mid \mathfrak{D}_0)^- \supset \varphi_g(f) \mid \mathfrak{D}((H(g) + b)^{1/2})$. Therefore $(\varphi(f) \mid \mathfrak{D}_0)^- = \varphi_g(f)$, so $\varphi_g(f) = \varphi(f)$ is independent of g. Similarly $\pi(f)$ is independent of g.

THEOREM 3.2.7. *Let $\psi \in \mathfrak{D}_g$, with $H(g)$ a hamiltonian for $\mathfrak{B}$. Then*

$$(\psi, \varphi(x_1, t_1) \cdots \varphi(x_n, t_n)\psi)$$

is a distribution in $\mathfrak{D}'(\mathfrak{B} \times \cdots \times \mathfrak{B})$.

PROOF. This follows from our previous estimates (3.2.16) and (3.2.28).

3.3. *Essential self-adjointness.* The main result of this section is the proof that for real test functions $f = f(x, t)$ with compact support, the field $\varphi(f)$ is self-adjoint, and essentially self-adjoint on $\mathfrak{D}_0 = \mathfrak{D}(H_0^{1/2}) \cap \mathfrak{D}(N)$, or on any $\mathfrak{D}_g$ where $H(g)$ is a hamiltonian for the support of f. We see furthermore that if f is real and $| f |_1$ defined by (3.2.20) is finite, then $\varphi_g(f)$ is self-adjoint and essentially self-adjoint on $\mathfrak{D}_g$.

The proof has three main steps. First, we assume that f is a regular function of t; in that case we use an analytic vector argument to show that

$\varphi_g(f)$ is essentially self-adjoint on $\mathfrak{D}_g$. As a second step, we take limits in the resolvents $(\varphi_g(f) - z)^{-1}$ as f tends to a more general function. In this way, we obtain a self-adjoint operator $\varphi_g(f)$. As a third step, we show that $\varphi_g(f)$ is essentially self-adjoint on $\mathfrak{D}(H(g))$.

The regularity we impose on f is the requirement that its Fourier transform be a smooth function with compact support, or more generally that for the norm $|f|_1$ of (3.2.20) there exist constants $\alpha = \alpha(f)$, $\beta = \beta(f)$ such that

$$|D_t^r f| \leqq \alpha\beta^r , \qquad r = 0, 1, 2 \cdots . \tag{3.3.1}$$

For a vector ψ, we consider the condition

$$\|(H(g) + b)^r \psi\| \leqq ad^r , \qquad r = 0, 1, \cdots . \tag{3.3.2}$$

LEMMA 3.3.1. *Assume* (3.3.1) *and* (3.3.2). *Then*

$$\|(H(g) + b)^r \varphi_g(f)\psi\| \leqq \alpha adc(b + \beta)^r \tag{3.3.3}$$

for some constant c independent of r, ψ and f.

PROOF. By (3.2.28) and (3.3.1–2)

$$\begin{aligned}\|(H(g) + b)^r \varphi_g(f)\psi\| &\leqq \textstyle\sum_{j=0}^{r} \binom{r}{j} c\alpha\beta^j \|(H(g) + b)^{r-j+1}\psi\| \\ &\leqq \textstyle\sum_{j=0}^{r} \binom{r}{j} c\alpha\beta^j a d^{r-j+1} \\ &= \alpha adc(d + \beta)^r .\end{aligned}$$

LEMMA 3.3.2. *Assume* (3.3.1) *and* (3.3.2). *Then ψ is an analytic vector for $\varphi_g(f)$. In particular for real $f(x, t)$, $\varphi_g(f)$ is essentially self-adjoint on $\mathfrak{D}_g$.*

PROOF. We apply the preceding lemma successively. We see that multiplication by $\varphi_g(f)$ changes the constants a and d of (3.3.2) as follows: $a \to \alpha adc$, $d \to d + \beta$. Thus

$$\begin{aligned}&\|(H(g) + b)^r \varphi_g(f)^k \psi\| \\ &\qquad \leqq a(\alpha c)^k d(d + \beta)(d + 2\beta) \cdots (d + (k-1)\beta)(d + k\beta)^r ,\end{aligned}$$

and

$$\|\varphi_g(f)^k \psi\| \leqq aB^k k!$$

for some constant B, which proves that ψ is analytic for $\varphi_g(f)$. The essential self-adjointness of $\varphi_g(f)$ follows from a theorem of Nelson [20].

We can draw more information from (3.3.3). If we write $H(g) + b = \int \lambda dE(\lambda)$, then (3.3.2) is equivalent to $\psi \in \text{Range } E(d)$ and (3.3.3) ensures that

$$\varphi_g(f) \text{ Range } E(d) \subset \text{Range } E(d + \beta) . \tag{3.3.4}$$

Because $\varphi_g(f)$ is self-adjoint, we also have

$$\varphi_g(f) \text{ Range } (I - E(d)) \subset \text{Range } (I - E(d - \beta)) . \tag{3.3.5}$$

These two inclusions have simple physical interpretations. We imagine that $\varphi_g(f)$ is written as a sum of two operators, one creating physical wave packets associated with $H(g)$, and the other annihilating them. Because of (3.3.1) the wave packets have energy at most β, and so $\varphi_g(f)$ can increase or decrease the total energy, $H(g)$, by at most β.

We note that $\varphi_g(f)$ is essentially self-adjoint on the domain $\bigcup_l \text{Range } E(l)$, by the proof of Lemma 3.3.2 and the remarks above.

Our next step is to take limits with respect to f in the resolvents $R = R(f, z) = (\varphi_g(f) - z)^{-1}$. As preparation, we now prove that R preserves regularity, which means

$$|| (H(g) + b)^{1/2} R(f, z)\psi || \leqq M || (H(g) + b)^{1/2}\psi || . \tag{3.3.6}$$

LEMMA 3.3.3. *Let f be real and satisfy* (3.3.1). *Then the estimate* (3.3.6) *holds for* $\text{Im}\, z \neq 0$. *The constants M and b depend only on z, g, and $|f|_1$.*

PROOF. To prove this lemma, we obtain uniform estimates on approximating operators R_n. If (3.3.6) holds for R_n, with M independent of n, and

$$R = \text{strong lim } R_n , \tag{3.3.7}$$

then (3.3.6) also holds for R. In fact $|| (H(g) + b)^{1/2}\psi ||$ defines a norm on the domain $\mathcal{D}((H(g) + b)^{1/2}) = \mathcal{H}_1$ which makes it into a Hilbert space. The inequality (3.3.6) in equivalent to R_n being a bounded operator on $\mathcal{H}_1$, and the norm $|| R_n ||_{1,1}$ of R_n as an operator from $\mathcal{H}_1$ to itself is defined by

$$|| R_n ||_{1,1} = || (H(g) + b)^{1/2} R_n (H(g) + b)^{-1/2} || \leqq M . \tag{3.3.8}$$

From the strong convergence (3.3.7) on $\mathcal{F}$, we conclude that on a dense set of vectors in $\mathcal{H}_1$, R_n converges weakly to R. Since the R_n are uniformly bounded on $\mathcal{H}_1$, $R_n \rightarrow R$ in weak operator convergence on $\mathcal{H}_1$. Thus the norm $|| R ||_{1,1}$ is bounded by the lim sup of the $|| R_n ||_{1,1}$ and (3.3.6) holds for R.

Let

$$H(g) + b = \int \lambda dE(\lambda) . \tag{3.3.9}$$

We approximate $\varphi_g(f)$ by the bounded self-adjoint operator $C_n = E(n)\varphi_g(f)E(n)$. From (3.3.4) it is clear that $C_n \rightarrow \varphi_g(f)$ on vectors with compact support in the energy. Since $\varphi_g(f)$ is essentially self-adjoint on this domain, the resolvents also converge strongly [18, p. 429] $\lim_{n\rightarrow\infty} R_n(z) = \lim_{n\rightarrow\infty} (C_n - z)^{-1} \rightarrow R(z)$, proving (3.3.7).

We now show that (3.3.8) holds, and it is sufficient to prove

(3.3.10) $$\|(H(g)+b)^{1/2}R_n(H(g)+b)^{-1/2}\psi\| \leqq M\|\psi\|$$

for ψ in the dense set $\mathfrak{D}_g$. Since $(H(g)+b)^{-1/2}$ and R_n both map $\mathfrak{D}_g$ onto $\mathfrak{D}_g$, we need only prove that on $\mathfrak{D}_g \times \mathfrak{D}_g$,

(3.3.11) $$\begin{aligned}(H(g)+b) &\leqq M^2(C_n-\bar{z})(H(g)+b)(C_n-z)\\ &= M^2(C_n-x)(H(g)+b)(C_n-x)+(My)^2(H(g)+b)\\ &\qquad + iM^2y[H(g), C_n]\,,\end{aligned}$$

where $z = x + iy$. As the first term is positive, it is sufficient to show that

(3.3.12) $$0 \leqq ((My)^2-1)(H(g)+b) + M^2y[iH(g), C_n]\,.$$

But

$$\begin{aligned}[iH(g), C_n] &= E(n)[iH(g), \varphi_g(f)]E(n)\\ &= E(n)\pi_g(f)E(n)\,.\end{aligned}$$

By Theorem 3.2.3,

$$\begin{aligned}|(\psi, [iH(g), C_n]\psi)| &= |(E(n)\psi, \pi_g(f)E(n)\psi)| \leqq \|\psi\|\,\|\pi_g(f)E(n)\psi\|\\ &\leqq \|\psi\|\,|f|_1\,\|(H(g)+b)^{1/2}E(n)\psi\|\\ &\leqq \frac{1}{2}|f|_1\Big\{\varepsilon(\psi, (H(g)+b)\psi) + \frac{1}{\varepsilon}(\psi, \psi)\Big\}\,,\end{aligned}$$

for any $\varepsilon > 0$. Furthermore, the norm $|f|_1$ of (3.2.20) can be chosen independent of b for large b, since $H(g) + b_1 \leqq H(b) + b_2$ for $b_1 \leqq b_2$. Therefore (3.3.12) is valid as long as

$$0 \leqq \Big\{(My)^2 - 1 - \frac{1}{2}M^2y\varepsilon|f|_1\Big\}(H(g)+b) - \frac{M^2y}{2\varepsilon}|f|_1\,.$$

For each $|f|_1, |y| \neq 0$, we can pick M large so that $(My)^2 > 3$, ε small so that $M^2y\varepsilon|f|_1/2 < 1$, and b large so that the inequality is valid. This completes the proof.

We now show that the resolvents of approximate field operators converge. We use the spectral projections $E(n)$ defined by (3.3.9) to cut off the field. If $\varphi_g(f)$ is a closed symmetric field operator, then $E(n)\varphi_g(f)E(n)$ is a bounded, self-adjoint approximation to $\varphi_g(f)$.

LEMMA 3.3.4. *Let $\{f_n\}$ be a sequence of real functions satisfying* (3.3.1) *with β depending on n. If the graphs $\mathfrak{G}(\varphi_g(f_n))$ converge to the graph of a densely defined operator, if*

(3.3.13) $$\|(H(g)+b)^{-1/2}\{\varphi_g(f_n)-\varphi_g(f_m)\}(H(g)+b)^{-1/2}\| \to 0$$

and if the norms $|f_n|_1$ are uniformly bounded, then the resolvents

(3.3.14) $$R_n(z) = (C_n - z)^{-1}\,,$$

of

$$C_n = E(n)\varphi_g(f_n)E(n) \tag{3.3.15}$$

converge strongly to the resolvent of a self-adjoint operator C.

PROOF. This result is a special case of [9, Th. 5 and Cor. 6]. See that paper for notation. Recall that $E(n)$ is defined by (3.3.9). Note that

$$\begin{aligned} &\| (H(g) + b)^{-1/2}(C_n - \varphi_g(f_n))(H(g) + b)^{-1/2} \| \\ &\quad \leq \| (E(n) - I)(H(g) + b)^{-1/2} \| \{\| \varphi_g(f_n)(H(g) + b)^{-1/2} \| \\ &\qquad + \| (H(g) + b)^{-1/2}\varphi_g(f_n) \|\} \\ &\quad \leq n^{-1/2}\{\| \varphi_g(f_n)(H(g) + b)^{-1/2} \| + \| (H(g) + b)^{-1/2}\varphi_g(f_n) \|\} . \end{aligned} \tag{3.3.16}$$

By Theorem 3.2.3, $\varphi_g(f_n)(H(g) + b)^{-1/2}$ and its adjoint are bounded with $\| \varphi_g(f_n)(H(g) + b)^{-1/2} \| \leq c \, |f_n|_1$, which is bounded uniformly in n. Then

$$\| (H(g) + b)^{-1/2}(C_n - \varphi_g(f_n))(H(g) + b)^{-1/2} \| = O(n^{-1/2}) \tag{3.3.17}$$

and so by (3.3.13), $\| (H(g) + b)^{-1/2}(C_n - C_m)(H(g) + b)^{-1/2} \| \to 0$. The required uniform boundedness of the resolvents $\| (C_n - z)^{-1} \|_{1,1} \leq$ const. follows from Lemma 3.3.3.

We now discuss when the hypotheses of Lemma 3.3.4 are satisfied. If the $\varphi_g(f_n)$ converge strongly on a dense domain, then the graphs converge. The $\varphi_g(f_n)$ will converge on $\mathfrak{D}_g$ if $f_n \to f$ in the norm $|\cdot|_1$; they will also converge for some sequence f_n

$$f_n \to f(\cdot, t_1)\delta(t - t_1) + \cdots + f(\cdot, t_j)\delta(t - t_j) \tag{3.3.18}$$

with $f(\cdot, t_i) \in \mathcal{S}(R)$. We can choose $f_n(x, t)$ to have the form $f_n(x, t) = \sum_{r=1}^{j} f(x, t_r)\delta_n(t - t_r)$ where $\delta_n(t) \geq 0$ has support in $|t| \leq n^{-1}$, and $\int \delta_n(t)dt = 1$. For such a sequence $|f_n|_1$ is uniformly bounded in n. From (3.2.9) we see that w^* convergence of the δ_n as bounded measures implies (3.3.13). Thus the hypotheses are satisfied for the sequence (3.3.18). They are also satisfied if the f_n converge in the norm $|\cdot|_1$, and every f with finite $|f|_1$ is the limit of such a sequence.

THEOREM 3.3.5. *Let f be real and $|f|_1$ finite. Then $\varphi_g(f)$ is self-adjoint and essentially self-adjoint on $\mathfrak{D}_g$. A real linear combination of sharp time fields with real test functions in $\mathcal{S}$, $A_g(t_1) + \cdots + A_g(t_r)$, is also essentially self-adjoint on $\mathfrak{D}_g$.*

PROOF. The two cases are similar and we only consider $\varphi_g(f)$. We first prove that the C of Lemma 3.3.4 extends $\varphi_g(f)$

$$C \supset \varphi_g(f) . \tag{3.3.19}$$

This is a consequence of [9], but we give the following explicit proof. As in the proof of Theorem 3.2.6, we have

$$\varphi_g(f) = (\varphi_g(f) \mid \mathfrak{D}_g)^- = (\varphi_g(f) \mid \mathfrak{D}_0)^- , \tag{3.2.20}$$

where $\mathfrak{D}_0 = \mathfrak{D}(H_0^{1/2}) \cap \mathfrak{D}(N)$. Let $R_n(z)$ be defined by (3.3.14), where f_n approximates f and satisfies the hypotheses of Lemma 3.3.4. Thus $R(z) = \lim_{n\to\infty} R_n(z)$ exists and is the resolvent of a self-adjoint operator C. For $\psi \in \mathfrak{D}_g$, $C_n\psi = E(n)\varphi_g(f_n)E(n)\psi \to \varphi_g(f)\psi$, and convergence can be shown on $\mathfrak{D}((H(g) + b)^{3/2})$. For $\chi = (\varphi_g(f) - z)\psi$, $R(z)\chi = \lim_{n\to\infty} R_n(z)\chi = \lim_{n\to\infty} R_n(z)(C_n - z)\psi = \psi$. Thus

$$(C - z)^{-1}(\varphi_g(f) - z) \mid \mathfrak{D}_g = I \mid \mathfrak{D}_g ,$$

and by (3.3.20), (3.3.19) is valid.

We now show that $\varphi_g(f)$ is equal to C, which completes the proof. We need only show that if $\psi \in \mathfrak{D}(C)$, then $\psi \in \mathfrak{D}(\varphi_g(f))$. We first notice that

$$R(z)\mathfrak{D}((H(g) + b)^{1/2}) \subset \mathfrak{D}((H(g) + b)^{1/2}) \tag{3.3.21}$$

and that (3.3.8) is valid for $R(z)$. The argument for this is the same as the proof of Lemma 3.3.3, but the approximate operator $C_n = E(n)\varphi_g(f_n)E(n)$ replaces the C_n of the former proof. The remaining calculation is the same since the $|f_n|_1$ are assumed uniformly bounded.

We now introduce the smoothing operator

$$P_j = \left(1 + \frac{1}{j}(H(g) + b)^2\right)^{-1} \tag{3.3.22}$$

with the properties $\|P_j\| \leqq 1$,

$$\text{strong } \lim_{j\to\infty} P_j = I , \tag{3.3.23}$$

and for $r \leqq 2$,

$$\|P_j(H(g) + b)^r\| = \|(H(g) + b)^r P_j\| \leqq j^{r/2} . \tag{3.3.24}$$

Let $\psi \in \mathfrak{D}(C)$ and $\psi = R(z)\chi$. Then $P_j\psi = \psi_j \to \psi$ and $\psi_j \in \mathfrak{D}(H(g)) \subset \mathfrak{D}(\varphi_g(f))$. If $\varphi_g(f)\psi_j$ converges, then ψ is in the domain of the closed operator $\varphi_g(f)$, so we prove this.

$$\begin{aligned}(\varphi_g(f) - z)\psi_j &= (\varphi_g(f) - z)P_j\psi \\ &= (\varphi_g(f) - z)P_jR(z)\chi \\ &= (\varphi_g(f) - z)R(z)P_j\chi + (\varphi_g(f) - z)[P_j, R(z)]\chi .\end{aligned} \tag{3.3.25}$$

The last equality is valid since $P_j\chi \in \mathfrak{D}(H(g)^2) \subset \mathfrak{D}((H(g) + b)^{1/2})$, and by (3.3.21) $R(z)P_j\chi \in \mathfrak{D}((H(g) + b)^{1/2}) \subset \mathfrak{D}(\varphi_g(f))$. Since C extends $\varphi_g(f)$, $(\varphi_g(f) - z)\psi_j = P_j\chi + (\varphi_g - z)[P_j, R(z)]\chi$. As $P_j\chi \to \chi = (C - z)\psi$, to conclude that $\psi \in \mathfrak{D}(\varphi_g(f))$, we need to show that

(3.3.26) $$\Lambda_j = (\varphi_g - z)[P_j, R(z)]\chi \to 0 .$$

We now argue that

(3.3.27) $$\Lambda_j = \lim_{n\to\infty} (C_n - z)[P_j, R_n(z)]\chi ,$$

where C_n and R_n are defined in (3.3.14–15). Since $(C_n - z)R_n(z) = (C - z)R(z) = I$, we need only prove the existence of the limit (3.3.27) with the commutator removed. As observed in the first paragraph of the proof, for $\psi \in \mathfrak{D}((H(g) + b)^{3/2})$, $\| C_n\psi - C\psi \| = \| C_n\psi - \varphi_g(f)\psi \| \to 0$. Since $P_jR(z)\chi \in \mathfrak{D}(H(g)^2) \subset \mathfrak{D}((H(g) + b)^{3/2})$, as $n \to \infty$, $(C_n - z)P_jR(z)\chi \to (\varphi_g(f) - z)P_jR(z)\chi$. Also $R_n(z)\chi \to R(z)\chi$, and by Theorem 3.2.3 and (3.3.24), $\| (C_n - z)P_j \| \leq \text{const.}\, | f_n |_1 j^{1/4}$, which is bounded uniformly in n. Therefore

$$(C_n - z)P_jR_n(z)\chi \to (\varphi_g(f) - z)P_jR(z)\chi ,$$

and (3.3.27) is established. Thus $\Lambda_j = \lim_{n\to\infty} \Lambda_{j,n}$, where

$$\begin{aligned}
\Lambda_{j,n} &= (C_n - z)[P_j, R_n(z)]\chi \\
&= (C_n - z)R_n(z)P_j[P_j^{-1}, (C_n - z)]P_jR_n(z)\chi \\
&= \frac{1}{j}P_j[(H(g) + b)^2, C_n - z]P_jR_n(z)\chi \\
&= \frac{1}{j}P_j\{(H(g) + b)[H(g), C_n] + [H(g), C_n](H(g) + b)\}P_jR_n(z)\chi \\
&= \frac{-i}{j}P_j\{(H(g) + b)E(n)\pi_g(f_n)E(n) \\
&\qquad + E(n)\pi_g(f_n)E(n)(H(g) + b)\}P_jR_n(z)\chi .
\end{aligned}$$

Now by (3.3.24) $\| P_j(H(g) + b) \| = \| (H(g) + b)P_j \| \leq j^{1/2}$ and

$$\begin{aligned}
\| E(n)\pi_g(f_n)E(n)P_j \| + \| P_jE(n)\pi_g(f_n)E(n) \| &\leq \text{const.}\, | f_n |_1 j^{1/4} \\
&\leq \text{const.}\, j^{1/4} ,
\end{aligned}$$

as the $| f_n |_1$ are assumed uniformly bounded. The constant is independent of j and n. Therefore $\| \Lambda_{jn} \| \leq \text{const.}\, j^{-1/4}$, and

$$\lim_{j\to\infty} \| \Lambda_j \| \leq \lim_{j\to\infty} \lim_{n\to\infty} \| \Lambda_{jn} \| = 0 .$$

Thus (3.3.26) is established and the proof is complete.

3.4. *The field as a tempered distribution.* In the previous sections we studied the field $\varphi_g(f)$ corresponding to the hamiltonian $H(g)$. We found that if $\varphi_g(f)$ is localized, namely if f has compact support in $\mathcal{B}$ and $H(g)$ is a hamiltonian for $\mathcal{B}$, then $\varphi_g(f) = \varphi(f)$ is independent of the spatial cutoff g. In this section we show that there is a cutoff independent field $\varphi(f)$ defined for all

$f \in \mathcal{S}$, and $\varphi(f)$ agrees with the previous one when f has compact support. The domain of $\varphi(f)$ includes $\mathfrak{D}_0 = \mathfrak{D}(H_0^{1/2}) \cap \mathfrak{D}(N)$, and on this domain $\varphi(f)$ is a tempered distribution.

LEMMA 3.4.1. *Let* $\mathfrak{D}_0 = \mathfrak{D}(H_0^{1/2}) \cap \mathfrak{D}(N)$. *For* $\psi \in \mathfrak{D}_0$, $(\psi, \varphi(x, t)\psi)$ *is a continuous, polynomially bounded function and*

$$\left|\int(\psi, \varphi(x, t)\psi)D_x f(x, t)dx\right| \leqq O(|t|)\,\|f(\cdot, t)\|_2\,(\psi, (H_0 + N^2 + I)\psi)\,. \tag{3.4.1}$$

PROOF. We divide space time into a number of similar regions with a partition of unity. Let $\zeta(x, t)$ be a C^∞ function satisfying

$$0 \leqq \zeta(x, t) \leqq 1\,, \tag{3.4.2}$$

$$\operatorname{supp}\zeta \subset \{|x| \leqq 1, |t| \leqq 1\}\,. \tag{3.4.3}$$

and such that

$$\sum_{i,j}\zeta_{ij}(x, t) = \sum_{i,j}\zeta(x - i, t - j) = 1\,. \tag{3.4.4}$$

Thus if $f(x, t) \in \mathcal{S}$, $f = \sum_{i,j} f_{ij} = \sum_{i,j} f\zeta_{ij}$ with $f_{ij}(x, t)$ a C^∞ function with support in the cube

$$\mathcal{B}_{ij} = \{(x, t) : |x - i| \leqq 1, |t - j| \leqq 1\}\,. \tag{3.4.6}$$

We also pick a C^∞ function $g_0(x)$ such that

$$g_0(x) = \lambda, \text{ if } |x| \leqq 2\,, \tag{3.4.7}$$

and

$$g_0(x) = 0, \text{ if } |x| \geqq 3\,. \tag{3.4.8}$$

Thus $H(g_{ij})$ is a hamiltonian for $\mathcal{B}_{ij}$ when

$$g_{ij}(x) = g_0\big((x - i)/(|j| + 1)\big)\,. \tag{3.4.9}$$

Furthermore

$$\|(I + N)^{-1}H_{I,g_{ij}}(I + N)^{-1}\| = O(j)\,, \tag{3.4.10}$$

as the kernels of operators contributing to $H_{I,g_{ij}}$ have L_2 norms with are $O(j)$. For $(x, t) \in \mathcal{B}_{ij}$ and $\psi \in \mathfrak{D}_0$, we have by Lemma 3.2.1, that $(\psi, \varphi(x, t)\psi)$ is continuous and

$$|(\psi, \varphi(x, t)\psi)| \leqq \text{const.}\,(\psi, (H(g_{ij}) + b)\psi) \tag{3.4.11}$$

where the constant is independent of x, t, i, and j. Here b is proportional to the lower bound of $H(g_{ij})$, see (2.1.19). In [29, § 5] we prove that the lower bound of $H(g_{ij})$ is of order the diameter of the support of g_{ij}, namely $O(j)$.

Thus (3.4.11) gives the bound for $(x, t) \in \mathcal{B}_{ij}$, $\psi \in \mathfrak{D}_0$

$$\begin{aligned}
&|(\psi, \varphi(x, t)\psi)| \\
(3.4.12)\qquad &\leqq \text{const.}\,\{\|H_0^{1/2}\psi\|^2 + \|(N+I)\psi\|^2\,\|(I+N)^{-1}H_{I,g_{ij}}(I+N)^{-1}\| \\
&\qquad + b\,\|\psi\|^2\} \\
&\leqq \text{const.}\,(\psi, (H_0 + N^2 + I)\psi)O(j)\,,
\end{aligned}$$

by (3.4.10) and the above discussion of b. Since $O(j) = O(|t|)$, we have proved polynomial boundedness. Thus, as in Lemma 3.2.1,

$$\left|\int(\psi, \varphi(x, t)\psi)D_x f(x, t)dx\right| \leqq O(|t|)\,\|f(\cdot, t)\|_2\,(\psi, (H_0 + N^2 + I)\psi)\,,$$

which yields (3.4.1).

We now define the sharp time fields

$$A(t) = \int\varphi(x, t)f(x, t)dx \tag{3.4.13}$$

and

$$B(t) = \int\pi(x, t)f(x, t)dx\,. \tag{3.4.14}$$

LEMMA 3.4.2. *Let* $f \in \mathcal{S}(R^2)$ *be real. Then* $A(t)$ *and* $B(t)$ *define self-adjoint operators, and their domain includes* $\mathfrak{D}_0$. *For* $\psi \in \mathfrak{D}_0$, $\|A(t)\psi\| + \|B(t)\psi\| \leqq |f(\cdot, t)|_2\,\|F\psi\|$ *where* $F = (H_0 + N^2 + I)^{1/2}$, *and* $|f(\cdot, t)|_2 = c(1 + |t|)\{\|f(\cdot, t)\|_2 + \|D_x f(\cdot, t)\|_2\}$.

PROOF. The proof is similar to that of Lemmas 3.2.2 aud 3.4.1.

THEOREM 3.4.3. *Let* $f(x, t) \in \mathcal{S}$ *be a real function in* $\mathcal{S}$. *The vectors* $A(t)\psi$ *and* $B(t)\psi$, $\psi \in \mathfrak{D}_0$, *are continuous and rapidly decreasing in* t. *Their integrals over* t *exist and define closed symmetric operators* $\varphi(f)$ *and* $\pi(f)$ *with domains containing* $\mathfrak{D}_0$. *The fields* $\varphi(f)$, $\pi(f)$, $A(t)$ *and* $B(t)$ *are all independent of* $g(x)$. *For* $\psi \in \mathfrak{D}_0$ *we have* $\|\varphi(f)\psi\| + \|\pi(f)\psi\| \leqq |f|_2\,\|F\psi\|$, *where* $|f|_2 = \int|f(\cdot, t)|_2\,dt$ *and* $F = (H_0 + N^2 + I)^{1/2}$.

PROOF. This proof is based on the proofs of Lemma 3.4.1, Theorem 3.2.3, and Theorem 3.2.6. The fields $\varphi(f)$ and $\pi(f)$ are defined as their closures on $\mathfrak{D}_0$.

3.5. *Locality*. In this section we derive locality of the field operators. Locality means that two field operators $\varphi(f)$ and $\varphi(h)$ commute provided the supports of f and h are spacelike separated. In other words, whenever

$$(x, t) \in \operatorname{supp} f\,,$$
$$(y, s) \in \operatorname{supp} h\,,$$

we have

$$|x - y| > |t - s|\,.$$

Under this hypothesis a signal originating in supp f (caused, for example, by the process of performing the measurement of $\varphi(f)$) cannot be recorded by the measurement of $\varphi(h)$. Thus one expects that the measurement of $\varphi(f)$ does not interfere with the measurement of $\varphi(h)$, and that the joint measurement of $\varphi(f)$ and $\varphi(h)$ can be performed in either order. The precise mathematical statement that the measurements can be performed in either order is that $\varphi(f)$ and $\varphi(h)$ commute.

For any closed operator A, a core $\mathfrak{D}$ of A is defined to be a dense domain contained in $\mathfrak{D}(A)$ such that $A = (A \mid \mathfrak{D})^{-}$. Self-adjoint operators A and B commute if and only if for any spectral projection E of B, and core $\mathfrak{D}$ of A, $E\mathfrak{D} \subset \mathfrak{D}(A)$ and for $\psi \in \mathfrak{D}$, $EA\psi = AE\psi$.

THEOREM 3.5.1. *If* supp f *and* supp h *are spacelike separated,* $\varphi_g(f)$ *and* $\varphi_g(h)$ *commute.*

PROOF. Let $A_g(t)$ and $B_g(t)$ be the sharp time fields obtained from the test functions f and h. First we prove that $A_g(s)$ and $B_g(t)$ commute. For any open set $\mathcal{I}$ in space, we define the algebra $\mathfrak{A}(\mathcal{I})$ as the weak closure of the bounded functions of the $t = 0$ fields

$$\int \varphi(x, 0)f(x)dx \qquad \text{and} \qquad \int \pi(x, 0)f(x)dx$$

as f runs over the C^∞ functions with support in $\mathcal{I}$. If $\mathcal{I}$ and $\mathcal{J}$ are disjoint open sets, then elements of $\mathfrak{A}(\mathcal{I})$ and $\mathfrak{A}(\mathcal{J})$ commute, and it was shown in [8] that

$$e^{i\sigma H(g)}\mathfrak{A}(\mathcal{I})e^{-i\sigma H(g)} \subset \mathfrak{A}(\mathcal{I}_\sigma)\ , \tag{3.5.1}$$

where $\mathcal{I}_\sigma$ is the set of all points in space with distance less than $|\sigma|$ from $\mathcal{I}$. The proof in [8] is valid whether or not $g(x) =$ const. on the set $\mathcal{I}$.

If $\mathcal{I}$ is a small neighborhood of supp $f \cap \{\text{time} = s\}$ and $\mathcal{J}$ is similarly defined with respect to h at time t, then $\mathcal{I}$ and $(\mathcal{J})_{t-s}$ are disjoint. Since the bounded functions of $A_g(s)$ belong to $e^{isH(g)}\mathfrak{A}(\mathcal{I})e^{-isH(g)}$ and the bounded functions of $B_g(t)$ belong to

$$e^{itH(g)}\mathfrak{A}(\mathcal{J})e^{-itH(g)} \subset e^{isH(g)}\mathfrak{A}\big((\mathcal{J})_{s-t}\big)e^{-isH(g)}\ ,$$

$A_g(s)$ and $B_g(t)$ commute.

Let E be a spectral projection of $A_g(s)$ and let $\psi \in \mathfrak{D}_g$, a core for $\varphi_g(h)$. Then $E\psi \in \mathfrak{D}(B_g(t))$ for all t and

$$(\varphi_g(h)\theta, E\psi) = \int(\theta, B_g(t)E\psi)dt = \left(\theta, E\int B_g(t)\psi dt\right) = (\theta, E\varphi_g(h)\psi)$$

for all $\theta \in \mathfrak{D}_g$. Thus

$$E\psi \in D(\varphi_g(h)^*) = \mathfrak{D}(\varphi_g(h)) ,$$
$$\varphi_g(h)E\psi = E\varphi_g(h)\psi ,$$

and $A_g(s)$ commutes with $\varphi_g(h)$. Now let F be a spectral projection for $\varphi(h)$. Then $F\psi \in \mathfrak{D}(A(s))$ for all s and

$$(\varphi_g(f)\theta, F\psi) = \left(\theta, \int A_g(s)F\psi ds\right) = \left(\theta, F\int A_g(s)\psi ds\right)$$

as before, so that $F\psi \in \mathfrak{D}(\varphi_g(f)^*) = \mathfrak{D}(\varphi(f))$ and $\varphi_g(f)F\psi = F\varphi_g(f)\psi$. Therefore, $\varphi_g(f)$ and $\varphi_g(h)$ commute.

3.6. *Space time covariance.* Space time covariance means that the field transforms in the expected fashion under the space time translation α, τ,

(3.6.1) $$\varphi(x, t) \longrightarrow \varphi(x + \alpha, t + \tau) .$$

By its very definition (3.1.1) and § 3.4, the field transforms correctly under time translation. Let $U(\alpha)$ be the unitary operator on $\mathcal{F}$ which implements the free field space translation $x \longrightarrow x + \alpha$. By definition, $U(\alpha)$ acts on each vector θ_j in the j particle subspace $\mathcal{F}_j$ by

(3.6.2) $$U(\alpha)\theta(k_1, \cdots, k_j) = e^{+i\alpha(k_1+\cdots+k_j)}\theta(k_1, \cdots, k_j) .$$

We use the convention that $U(\alpha)$ is the Schrödinger picture operator. The active transformation $V(\alpha) = U(-\alpha)$ commonly occurs in the literature. On a suitable domain,

(3.6.3) $$U(-\alpha)a^*(k)U(+\alpha) = e^{-ik\alpha}a^*(k) ,$$

(3.6.4) $$U(-\alpha)a(k)U(+\alpha) = e^{ik\alpha}a(k) ,$$

and from the definition (1.2.11) we have $U(-\alpha)\varphi(x)U(+\alpha) = \varphi(x + \alpha)$. Now $U(\alpha)$ does not commute with $H(g)$, and in fact $U(-\alpha)H(g)U(+\alpha) = H(g')$, where $g'(x) = g(x - \alpha)$. But $\varphi(x, t)$ is independent of the space cutoff function $g(y)$ provided that $g(y) = \lambda$ for $|x - y| > |t|$. Thus if $g(y) = \lambda$ for $|x + \alpha - y| > |t|$,

$$\begin{aligned} U(-\alpha)\varphi(x, t)U(+\alpha) &= U(-\alpha)e^{itH(g)}\varphi(x)e^{-itH(g)}U(\alpha) \\ &= e^{itH(g')}\varphi(x + \alpha)e^{-itH(g')} \\ &= \varphi(x + \alpha, t) \end{aligned}$$

on a suitable domain, for example on $\mathfrak{D}_0 \times \mathfrak{D}_0$. Thus $U(\alpha)$ implements the space translation for fields $\varphi(f)$.

4. The algebra of local observables

To each open region $\mathcal{B}$ of space time, we associate a C^* algebra $\mathfrak{A}(\mathcal{B})$ in such a way that the self-adjoint elements of $\mathfrak{A}(\mathcal{B})$ are exactly the operators corresponding to experiments which may be performed in $\mathcal{B}$. A number of authors have formulated postulates for quantum field theory in terms of such algebras, see Segal [23], Haag [10], Araki [1], or Haag and Kastler [11]. The requirements of [11] for a local quantum theory are

(a) To each bounded open region $\mathfrak{B}$ of space time, there is an associated C^* algebra $\mathfrak{A}(\mathfrak{B})$ containing the identity.

(b) Isotony: If $\mathfrak{B}_1 \supset \mathfrak{B}_2$, then $\mathfrak{A}(\mathfrak{B}_1) \supset \mathfrak{A}(\mathfrak{B}_2)$.

(c) Locality: If $\mathfrak{B}_1$ and $\mathfrak{B}_2$ are space like separated, then $\mathfrak{A}(\mathfrak{B}_1)$ commutes with $\mathfrak{A}(\mathfrak{B}_2)$.

(d) The algebra of local observables $\mathfrak{A}$ is defined as the norm closure of the union of the $\mathfrak{A}(\mathfrak{B})$. The algebra is *primitive*; in other words, it has a faithful, irreducible representation.

(e) Lorentz covariance: Let $\{a, \Lambda\}$ be an element of the inhomogeneous Lorentz group $\mathfrak{G}$. Then there is a representation $\sigma_{\{a,V\}}$ of $\mathfrak{G}$ by a group of $*$ automorphisms of $\mathfrak{A}$, such that for a bounded region $\mathfrak{B}$

$$\sigma_{\{a,\Lambda\}}\big(\mathfrak{A}(\mathfrak{B})\big) = \mathfrak{A}\big(\{a, \Lambda\}\mathfrak{B}\big) \, . \tag{4.1}$$

In this section we consider several possible definitions for the algebra $\mathfrak{A}(\mathfrak{B})$. The different definitions undoubtedly lead to different C^* algebras. In order to arrive at a natural and aesthetic definition, we prove that all reasonable candidates for $\mathfrak{A}(\mathfrak{B})$ have the same weak closure; we take this weakly closed algebra as the definition of $\mathfrak{A}(\mathfrak{B})$. A further justification of the use of the weak closure in defining $\mathfrak{A}(\mathfrak{B})$ is the fact proved in [29] that the physical representation π of $\mathfrak{A}$ is weakly continuous on each local subalgebra $\mathfrak{A}(\mathfrak{B})$.

In general terms, the $\mathfrak{A}(\mathfrak{B})$ is formed from bounded functions of the $\varphi(f)$, where $\operatorname{supp} f \subset \mathfrak{B}$. We make this notation precise as follows.

Definition. $\mathfrak{A}(\mathfrak{B})$ is the weakly closed operator algebra generated by the operators

$$\{F\big(\varphi(f)\big)\colon F \in L_\infty,\ \operatorname{supp} f \subset \mathfrak{B},\ |f|_1 < \infty\} \, .$$

The definition is unchanged if we replace L_∞ by some $*$ algebra which is dense in the weak operator topology. It is also unchanged if we replace the class of test functions by another (for example $\mathfrak{D}(\mathfrak{B})$) having the same closure in the $|\cdot|_1$ norm. In fact, if $|f_n - f|_1 \to 0$, then $\big(\varphi(f_n) - z\big)^{-1} \to \big(\varphi(f) - z\big)^{-1}$ in the strong operator topology by Lemma 3.3.4 and by the semigroup convergence theorem [18, p. 502] $e^{i\varphi(f_n)} \to e^{i\varphi(f)}$. Thus $e^{i\varphi(f)}$ and $F\big(\varphi(f)\big)$ belong to the weak closure if each f_n is admitted as a test function in defining $\mathfrak{A}(\mathfrak{B})$.

The same algebra $\mathfrak{A}(\mathfrak{B})$ is generated by the bounded functions of sharp time fields $A(t) = \int \varphi(x, t) f(x, t) dx$, $f \in \mathfrak{D}(\mathfrak{B})$. In fact, using a sequence f_n such as (3.3.18), we have the resolvents converging $\big(\varphi(f_n) - z\big)^{-1} \to \big(A(t) - z\big)^{-1}$, and so $F\big(A(t)\big) \in \mathfrak{A}(\mathfrak{B})$. Thus the sharp time fields generate a smaller algebra.

However, if $f \in \mathfrak{D}(\mathcal{B})$, we can approximate $\varphi(f)$ by $A(t_1)\Delta t_1 + \cdots + A(t_n)\Delta t_n$ with strong convergence on $\mathfrak{D}_g$. By [18, pp. 429, 502] or by Lemma 3.3.4 the resolvents converge, so $F(\varphi(f))$ belongs to the weakly closed algebra generated by the bounded functions of finite linear combinations of the sharp time fields. We now see that all such $F(\varphi(f))$ belong to the algebra generated by the bounded functions of the sharp time fields themselves. Let F be a bounded operator commuting with $A(t_1), A(t_2), \cdots$, and $A(t_n)$. Then by the spectral theorem, F commutes with $A(t_1)\Delta t_1 + \cdots + A(t_n)\Delta t_n$ on the domain $\mathfrak{D}_g$, which by Theorem 3.3.5 is a core for $A(t_1)\Delta t_1 + \cdots + A(t_n)\Delta t_n$. Thus F commutes with $A(t_1)\Delta t_1 + \cdots + A(t_n)\Delta t_n$. Thus the commutant of $A(t_1)\Delta t_1 + \cdots + A(t_n)\Delta t_n$ is larger than that of $A(t_1), \cdots, A(t_n)$, and the double commutant smaller. Therefore, the sharp time fields generate $\mathfrak{A}(\mathcal{B})$ as asserted.

THEOREM 4.1. *With our definition of* $\mathfrak{A}(\mathcal{B})$, *the axioms* (a)-(e) *are satisfied with the possible exception*[4] *of covariance under Lorentz rotations.*

PROOF. The properties (a) and (b) are obvious, and (c) is a consequence of Theorem 3.5.1. To prove (d) we observe that $F(A(0)) \in \mathfrak{A}$ if $F \in L_\infty$ and $A(0)$ is a time zero field operator. Furthermore $(A(t) - A(0))/t$ converges strongly on $\mathfrak{D}_g$ to $B(0) = \int \pi(x, 0)f(x, 0)dx$.

Since $B(0)$ is essentially self-adjoint on $\mathfrak{D}_g$, the resolvent of the difference quotients converge. Thus $F(B(0)) \in \mathfrak{A}$ also. It is known that any bounded operator commuting with all the $F(A(0))$ and the $F(B(0))$ is a multiple of the identity, which proves (d).

Let $(\alpha, t) = a$ be a space time translation, and let $A \in \mathfrak{A}(\mathcal{B})$. We define

$$\sigma_a(A) = U(-\alpha)e^{itH(g)}Ae^{-itH(g)}U(+\alpha) ,$$

where $g(x) = \lambda$ on a sufficiently large interval. It is clear that this σ_a preserves operator norm, and thus extends to $\mathfrak{A}$ by continuity. The verification of (4.1) for σ_a follows from the space time covariance of the field φ.

COURANT INSTITUTE, NEW YORK UNIVERSITY
HARVARD UNIVERSITY

REFERENCES

[1] H. ARAKI, *A lattice of von Neumann algebras associated with the quantum theory of a free Bose field,* J. Math. Phys. **4** (1963), 1343-1362.

[2] P. FEDERBUSH, *A partially alternative derivation of a result of Nelson,* J. Math. Phys. **10** (1969), 50-52.

[3] K. FRIEDRICHS, Perturbation of Spectra in Hilbert Space, American Mathematical Society, Providence, 1965.

[4] *Added in proof.* This possible exception is removed in [30].

[4] G. FROBENIUS, *Über Matrizen mit positiven Elementen*, Sitzungsber. d.k. Preuss. Akad. d. Wissen., Berlin, (1908), 471-476.

[5] L. GÅRDING and A. WIGHTMAN, *Fields as operator-valued distributions in relativistic quantum theory*, Arkiv för Fysik, **28** (1964), 129-184.

[6] J. GLIMM, *Yukawa coupling of quantum fields in two dimensions*, Comm. Math. Phys. **5** (1967), 343-386.

[7] ———, *Boson fields with non-linear self-interaction in two dimensions*, Comm. Math. Phys. **8** (1968), 12-25.

[8] ——— and A. JAFFE, *A $\lambda\varphi^4$ quantum field theory without cutoffs*: I, Phys. Rev. **176** (1968), 1945-1951.

[9] ———, *Singular perturbations of self adjoint operators*, Comm. Pure Appl. Math. **22** (1969), 401-414.

[10] R. HAAG, *On quantum field theories*, Mat.-Fys. Medd. Kong. Danske Videns. Selskab, **29** (1955), No. 12.

[11] ——— and D. KASTLER, *An algebraic approach to quantum field theory*, J. Math. Phys. **5** (1964), 848-861.

[12] D. HALL and A. WIGHTMAN, *A theorem on invariant functions with applications to relativistic quantum field theory*, Mat.-Fys. Medd. Kong. Danske Videns. Selskab, **31** (1957), No. 5.

[13] A. JAFFE, *Divergence of perturbation theory for Bosons*, Comm. Math. Phys. **1** (1965), 127-149.

[14] ———, Constructing a $\lambda(\varphi^4)_2$ quantum field theory, Varenna lectures, 1968.

[15] ——— and R. POWERS, *Infinite volume limit of a $\lambda\varphi^4$ field theory*, Comm. Math. Phys. **7** (1968), 218-221.

[16] R. JENTZSCH, *Über Integralgleichungen mit positiven Kern*, J. für d. reine u. angew. Math. **141** (1912), 235-244.

[17] M. KAC, Probability and Related Topics in Physical Sciences, Interscience Publishers, New York, 1959.

[18] T. KATO, Perturbation Theory for Linear Operators, Springer Verlag, New York, 1966.

[19] M. NAIMARK, Normed Rings, Noordhoff, Groningen, 1964.

[20] E. NELSON, *Analytic vectors*, Ann. of Math. **70** (1959), 572-615.

[21] ———, "A quartic interaction in two dimensions", in Mathemtical Theory of Elementary Particles, R. Goodman and I. Segal, Editors, M.I.T. Press, Cambridge, 1966.

[22] O. PERRON, *Zür Theorie der Matrices*, Math. Ann. **64** (1907), 248-263.

[23] I. SEGAL, *Postulates for general quantum mechanics*, Ann. of Math. **48** (1947), 930-948.

[24] ———, *Foundations of the theory of dynamical systems of infinitely many degrees of freedom*, Mat.-Fys. Medd. Kong. Danske Videns. Selskab, **31** (1959), No. 12.

[25] ———, "Nonlinear functions of random distributions and generalized normal products", in Proceedings of the Conference on Functional Integration and Constructive Quantum Field Theory, edited by I. Segal, April 1966.

[26] ———, *Notes towards the construction of non-linear relativistic quantum fields*: I. *The hamiltonian in two dimensions as the generator of a C* automorphism group*, Proc. Nat. Acad. Sci. U.S.A. **57** (1967), 1178-1183.

[27] S. WEINBERG, *High energy behavior in quantum field theory*, Phys. Rev. **118** (1960), 838-849.

[28] G. C. WICK, *The evaluation of the collision matrix*, Phys. Rev. **80** (1950), 268-272.

[29] J. GLIMM and A. JAFFE, *The $\lambda(\varphi^4)_2$ quantum field theory without cutoffs*: III. *The physical vacuum*, to appear.

[30] J. CANNON and A. JAFFE, *Lorentz covariance of the $\lambda(\varphi^4)_2$ quantum field theory*, to appear.

(Received March 3, 1969)
(Revised June 30, 1969)

THE $\lambda(\phi^4)_2$ QUANTUM FIELD THEORY WITHOUT CUTOFFS

III. *The physical vacuum*

BY

JAMES GLIMM and ARTHUR JAFFE

Courant Institute of Mathematical Sciences, New York University, New York, New York (1) *and Lyman Laboratory of Physics, Harvard University, Cambridge, Massachusetts* (2)

Contents

1. Introduction

In this series of papers we construct a quantum field theory model. This model describes a spin-zero boson field ϕ with a nonlinear ϕ^4 selfinteraction in two dimensional space time. The corresponding classical field equation is

$$\frac{\partial^2 \phi}{\partial t^2} - \frac{\partial^2 \phi}{\partial x^2} + m_0^2 \phi + 4\lambda\phi^3 = 0. \tag{1.1}$$

The classical field ϕ is by definition a real valued function of x and t which is a solution to (1.1). The quantum field ϕ is also a function of x and t, but its values $\phi(x, t)$ are densely defined bilinear forms on some Hilbert space. The quantum field ϕ is a solution to (1.1),

(1) Supported in part by the New York State Science and Technology Foundation, Grant SSF-(8)-8.
(2) Alfred P. Sloan Foundation Fellow. Supported in part by the Air Force Office of Scientific Research, AF 49(638)-1380.

14 – 702903 *Acta mathematica*. 125. Imprimé le 22 Octobre 1970.

provided that ϕ^3 is suitably interpreted. The quantum field $\phi(x, t)$ depends continuously on x and t and the derivatives in (1.1) are taken in the sense of distributions.

Certain averages of the quantum field

$$\phi(f) = \int \phi(x, t)\, f(x, t)\, dx\, dt \tag{1.2}$$

and

$$\phi(f, t) = \int \phi(x, t)\, f(x)\, dx \tag{1.3}$$

are self adjoint operators on the Hilbert space. In addition, the quantum field ϕ satisfies the canonical equal time commutation relations

$$\left[\phi(f, t), \frac{\partial}{\partial t}\phi(f, t)\right] = i \int f(x)^2 dx. \tag{1.4}$$

The construction of a quantum field ϕ satisfying (1.1)–(1.4) was one of the main results of the first papers in this series [10, 12] (denoted hereafter as I, II). General background material can be found in [9, 17, 19].

The field ϕ constructed in I and II is a bilinear form on the Fock space $\mathcal{F}$, the Hilbert space for noninteracting bosons. The Cauchy data for the quantum field

$$\phi(x, 0) \quad \text{and} \quad \left(\frac{\partial}{\partial t}\phi\right)(x,0) \tag{1.5}$$

are the same Cauchy data that give a solution to the free field equation

$$\frac{\partial^2}{\partial t^2}\phi_{\text{free}} - \frac{\partial^2}{\partial x^2}\phi_{\text{free}} + m_0^2\phi_{\text{free}} = 0, \tag{1.6}$$

and that satisfy (1.4). In other words,

$$\phi(x, 0) = \phi_{\text{free}}(x, 0) \tag{1.7}$$

and

$$\left(\frac{\partial\phi}{\partial t}\right)(x, 0) = \left(\frac{\partial}{\partial t}\phi_{\text{free}}\right)(x, 0). \tag{1.8}$$

It is expected that this quantum field theory of Iand II, with ϕ realized on the Fock space $\mathcal{F}$, will not possess all the properties conventionally assumed in quantum field theory. For example, it seems that there will not exist a Hamiltonian operator H to generate time translations

$$\phi(x, t + \tau) = e^{iH\tau}\phi(x, t)\, e^{-iH\tau}.$$

Also there will not exist a vacuum vector (a translation invariant ground state of H). The cause of these difficulties is made clear by the circle of ideas and results known as Haag's theorem [16, 24]. Haag's theorem states that if the field $\phi \neq \phi_{\text{free}}$ satisfies (1.7)–(1.8) and if there is a Hamiltonian H for ϕ, then H will not possess a unique Euclidean invariant vacuum vector.

In order to understand and deal with these difficulties—the lack of a Hamiltonian operator H on $\mathfrak{F}$ and of a vacuum vector Ω in $\mathfrak{F}$—we first consider the cutoff field equation

$$\frac{\partial^2}{\partial t^2}\phi - \frac{\partial^2}{\partial x^2}\phi + m_0^2\phi + 4\lambda g\phi^3 = 0. \tag{1.9}$$

Here the spatial cutoff $g(x)$ is a smooth, positive function that equals one on some bounded set and that vanishes off some larger bounded set. Corresponding to the cutoff field equation (1.9), there is a Hamiltonian operator

$$H(g) = H_0 + \lambda\int :\phi^4(x): g(x)\,dx - E_g = H_0 + H_{I,g} - E_g. \tag{1.10}$$

The operator H_0 is the free field Hamiltonian and it corresponds to the free field equation (1.6). The interaction energy operator $H_{I,g}$ contains the spatial cutoff g. The operator $H_0 + H_{I,g}$ is a self adjoint operator [10, 11] and is bounded from below [22, 8]. The constant E_g is chosen so that

$$0 = \inf\{\text{spectrum } H(g)\}.$$

The constant E_g is one of the standard renormalization counterterms of quantum field theory. E_g is called the self energy of the vacuum. E_g is finite because of the spatial cutoff and because of the limitation to only one space dimension [13]. Both g and the restriction to a single space variable serve to reduce the singularity of the perturbing operator $H_{I,g}$.

In I, II we constructed the Heisenberg picture dynamics for (1.1). We started from the locally correct equation (1.9) for which the solution is

$$\phi_g(x,t) = e^{iH(g)t}\phi(x,0)\,e^{-iH(g)t}. \tag{1.11}$$

The solution ϕ_g is independent of $g(\cdot)$ provided $g(y) = 1$ for $|y| \leq |x| + |t|$. Thus by patching together different local solutions (1.11), corresponding to $g = 1$ on different sets, we constructed a solution ϕ to (1.1). This solution is associated with a time translation automorphism σ_τ implemented locally by $H(g)$ as in (1.11).

$$\sigma_\tau(\phi(x,t)) = \phi(x, t+\tau). \tag{1.12}$$

While a vacuum vector Ω does not exist in $\mathfrak{F}$, we showed in II that the approximate

Hamiltonian $H(g)$ does have a vacuum. In fact, zero is an isolated eigenvalue of $H(g)$ with multiplicity one. Thus there is a vector Ω_g in $\mathfrak{F}$ such that

$$H(g)\Omega_g = 0, \quad \|\Omega_g\| = 1, \tag{1.13}$$

and Ω_g is unique up to a scalar multiple.

We wish to obtain the Hamiltonian H and the vacuum Ω corresponding to (1.1) by taking the limit $g(\cdot)\to 1$. We cannot use the most conventional types of limits because as $g(\cdot)\to 1$,

$$E_g \to -\infty \tag{1.14}$$

[13]. Furthermore, perturbation theory predicts that

$$\Omega_g \to 0, \tag{1.15}$$

in the sense of weak convergence on $\mathfrak{F}$. In fact (1.15) can be compared to the simpler weak limit in $L_2(R)$,

$$g(\cdot)/\|g\|_2 \to 0.$$

Instead of studying these quantities, we define the expectation values

$$\omega_g(A) = (\Omega_g, A\Omega_g), \tag{1.16}$$

where A is a bounded function of the field operators (1.2) or (1.3). The set of all such A form a C^* algebra $\mathfrak{A}$ of bounded operators on $\mathfrak{F}$, and $\omega_g \in \mathfrak{A}^*$, the dual space of $\mathfrak{A}$. Furthermore, ω_g is positive and has norm one, and thus ω_g is a *state* in the sense of C^* algebras. We use a limiting process as $g(\cdot)\to 1$ to obtain a limiting expectation value $\omega(A)$, and then $\omega \in \mathfrak{A}^*$ is necessarily a state also.

According to the Gelfand–Segal construction, ω comes from a vector Ω in some new Hilbert space $\mathfrak{F}_{\rm ren}$ in the sense that

$$\omega(A) = (\Omega, \pi(A)\Omega), \quad \|\Omega\| = 1, \tag{1.17}$$

where

$$\pi\colon A \to \pi(A) = A_{\rm ren}$$

is a representation of $\mathfrak{A}$ as operators on $\mathfrak{F}_{\rm ren}$. The Schrödinger picture dynamics exists on $\mathfrak{F}_{\rm ren}$. In the Schrödinger picture, the time translation is given by a one parameter family of unitary operators

$$U(t) = e^{-iHt}. \tag{1.18}$$

If

$$A \to A(t) = \sigma_t(A)$$

is the Heisenberg picture dynamics constructed in I, II, then

$$\pi(\sigma_t(A)) = e^{iHt}\pi(A)\,e^{-iHt}. \tag{1.19}$$

Here H is positive and Ω is a vacuum for H,

$$H\Omega = 0. \tag{1.20}$$

The operator H is the renormalized Hamiltonian, and it is the limit of the $H(g)$ in the sense that

$$(\pi(B)\,\Omega,\, e^{iHt}\pi(A)\,\Omega)$$

is obtained through a limit as $g(\cdot)\to 1$ of

$$(B\Omega_g,\, e^{iH(g)t}A\Omega_g).$$

We call Ω the physical vacuum and π the physical representation. The vectors in $\mathcal{F}_{\text{ren}}$ are called physical vectors.

The space translations are also given by a one parameter group, commuting with the $U(t)$ and leaving Ω fixed.

$$\pi(\sigma_{x,t}(A)) = e^{iHt-iPx}\pi(A)\,e^{-iHt+iPx}, \qquad e^{iPx}\Omega = \Omega. \tag{1.21}$$

It is well known that certain representations of $\mathfrak{A}$ (coming from states of $\mathfrak{A}$ via the Gelfand–Segal construction) cannot be extended to the unbounded field operators $\phi(f)$ of (1.2), (1.3). Such states and representations seem to be totally unsuited for use in physics. The deepest result of this paper is the fact that the physical representation π is not one of these pathological representations. We now formulate this result in a stronger and more precise form.

The C^* algebra $\mathfrak{A}$ is an inductive limit. Let $\mathcal{B}$ be a bounded region of space time (or a bounded region of space at time zero). Let $\mathfrak{A}(\mathcal{B})$ be the weakly closed (von Neumann) algebra generated by bounded functions of the operators (1.2) or (1.3), but with f restricted to have support in $\mathcal{B}$. Then $\mathfrak{A}$ is defined as the norm closure of

$$\bigcup_{\mathcal{B}}\mathfrak{A}(\mathcal{B}).$$

We prove that π is locally equivalent to the representation of $\mathfrak{A}$ as operators on the Fock space $\mathcal{F}$; in brief π is *locally Fock*. For each bounded region $\mathcal{B}$, there is a unitary operator $U_{\mathcal{B}}$,

$$U_{\mathcal{B}}\colon \mathcal{F}\to\mathcal{F}_{\text{ren}} \tag{1.22}$$

such that for all A in $\mathfrak{A}(\mathcal{B})$,

$$\pi(A) = A_{\text{ren}} = U_{\mathcal{B}}AU_{\mathcal{B}}^*. \tag{1.23}$$

We have remarked that there is no vacuum vector Ω in $\mathcal{F}$. It immediately follows from the locally Fock property that for each bounded region $\mathcal{B}$ there is a vector $\Omega_{\mathcal{B}}$ in $\mathcal{F}$ which serves as a vacuum vector for the algebra $\mathfrak{A}(\mathcal{B})$. In other words, for all $A \in \mathfrak{A}(\mathcal{B})$,

$$\omega(A) = (\Omega, A_{\text{ren}}\Omega) = (\Omega_{\mathcal{B}}, A\Omega_{\mathcal{B}}). \tag{1.24}$$

The vector $\Omega_{\mathcal{B}}$ could be called a local vacuum. The vector $\Omega_{\mathcal{B}}$ is not unique, but one choice would be $U_{\mathcal{B}}^* \Omega$. The unitary $U_{\mathcal{B}}$ is also not unique.

An important consequence of the locally Fock property of π is the fact that the field ϕ exists on the physical Hilbert space $\mathcal{F}_{\text{ren}}$. Let $\phi(f)$ be self adjoint on $\mathcal{F}$. This is the case if, for example, f is a real, twice differentiable function with compact support [II]. Then for supp. $f \subset \mathcal{B}$ and for real s,

$$s \to \pi(e^{is\phi(f)}) = U_{\mathcal{B}}\, e^{is\phi(f)}\, U_{\mathcal{B}}^*$$

is a weakly continuous one parameter group of unitary operators on $\mathcal{F}_{\text{ren}}$. We let

$$\pi(\phi(f)) = \phi_{\text{ren}}(f) = U_{\mathcal{B}}\, \phi(f)\, U_{\mathcal{B}}^*$$

be the self adjoint operator that is the infinitesimal generator of the group.(1) Furthermore,

$$\phi_{\text{ren}}(x,t) = U_{\mathcal{B}}\phi(x,t)\, U_{\mathcal{B}}^*, \quad (x,t) \in \mathcal{B},$$

is a densely defined bilinear form depending continuously on x and t, and ϕ_{ren} is a solution to (1.1)–(1.4). All the local properties of ϕ established in II go over immediately to ϕ_{ren} For example, ϕ_{ren} is local, so that $\phi_{\text{ren}}(f_1)$ and $\phi_{\text{ren}}(f_2)$ commute if the supports of f_1 and f_2 are spacelike separated. Also, $\phi_{\text{ren}}(x,t)$ transforms correctly under the space-time translation group.

$$\phi_{\text{ren}}(x+\alpha, t+\tau) = e^{iH\tau - iP\alpha}\phi_{\text{ren}}(x,t)\, e^{-iH\tau + iP\alpha}.$$

The Haag–Kastler axioms [15] are valid for $\mathfrak{A}$ [12, 27] and carry over to $\pi(\mathfrak{A})$ on $\mathcal{F}_{\text{ren}}$.

The main technical step in proving the locally Fock property of π is an estimate on E_g,

$$-MV \leqslant E_g.$$

Here M is a positive constant and V is essentially the length of the support of the spatial cutoff $g(\cdot)$. From this estimate it easily follows that

$$\omega_g\left(\frac{H_0}{V}\right) \leqslant \text{const.}$$

(1) Our definition of ϕ_{ren} is $Z_3^{\frac{1}{2}}$ times the usual renormalized field. The field strength renormalization constant Z_3 is chosen to normalize the one particle states produced by the field ϕ. Perturbation theory predicts that Z_3 is strictly positive in our model. Hence $Z_3^{-\frac{1}{2}}$ should be finite and neglecting Z_3 is merely a matter of convenience.

We use space translation averaging in the limiting process which defines ω, and therefore we are able to conclude that

$$\omega(H_0^{\mathrm{loc}}) \leqslant \text{const.},$$

where H_0^{loc} resembles a local energy operator for free particles. An estimate of this nature, valid for a sequence ω_n of Fock states implies that

$$\omega_n \restriction \mathfrak{A}(\mathcal{B})$$

lies in a *norm* compact subset of the dual $\mathfrak{A}(\mathcal{B})^*$. We note that the set of all states is w^* compact, and that any state $\varrho \in \mathfrak{A}^*$ can be reached as a w^* limit point of Fock states [5, 7]. In particular, the pathological states referred to above are w^* limits of Fock states. However, a norm convergent limit of Fock states is normal on $\mathfrak{A}(\mathcal{B})$, and this excludes the unwanted states. In order to complete the construction of $U_{\mathcal{B}}$, we use the result of Araki [1] that $\mathfrak{A}(\mathcal{B})$ is a factor of type III, for any bounded open set $\mathcal{B}$ at time zero. We also use a result of Griffin [14] that isomorphisms between separable factors of type III are unitarily implementable.

There are several properties of $\Omega \in \mathcal{F}_{\mathrm{ren}}$ which we have not established. For example, the uniqueness of Ω is one of the Wightman axioms. Also Ω should be in the domain of any polynomial in the field operators $\phi_{\mathrm{ren}}(f)$.

The main phenomenon that we encounter in this paper is the necessity of changing Hilbert spaces. This is a common feature of quantum field theory. In all cases this phenomenon can be traced to the fact that the fundamental objects, such as the Hamiltonian or energy operator, are given in the form

$$\text{(Self adjoint operator)} + \text{(Perturbation)}$$

with the perturbation exceedingly singular. (In the $(\phi^4)_2$ theory, the perturbation is a bilinear form, but not an operator on $\mathcal{F}$. In fact the perturbation is so singular that the total Hamiltonian, as a bilinear form on Fock space, is unbounded below. It does not yield an operator on Fock space, but on $\mathcal{F}_{\mathrm{ren}}$ the renormalized Hamiltonian is a positive self adjoint operator.) Two important reasons for this phenomenon are the translation invariance of the Hamiltonian and the fact that the interaction involves an infinite number of degrees of freedom.

The results of this paper extend to the interaction $P(\phi)_2$, where P is a semibounded polynomial. L. Rosen [28] has proved for this model that the cutoff Hamiltonian $H(g) = H_0 + H_{I,g}$ is essentially self adjoint and has a vacuum vector. We expect that the conclusions of section 2 of this work remain valid for certain models, even when the estimates on the vacuum self energy derived in section 5 fail to hold. In particular, we expect that if the vacuum energy in a finite volume has a logarithmic (but not a linear) ultraviolet divergence, the corresponding physical representation is still locally Fock. In two dimensional space-time, the interaction $P(\phi)+Q(\phi)\bar{\psi}\psi$ has this property. Here $P(\phi)$ is a polynomial bounded from below and $Q(\phi)$ is a polynomial of lower degree than P. The two dimensional Yukawa interaction $\bar{\psi}\psi\phi$ also has a logarithmic vacuum energy divergence so we expect that the theory is locally Fock.

The logical order of the following sections is 5, 3, 4, 2. We have rearranged this order to present results, first, followed by details of increasing complexity.

2. The physical representation and renormalization

In this section we construct the physical vacuum vector Ω and the corresponding Hilbert space $\mathfrak{F}_{\text{ren}}$. On $\mathfrak{F}_{\text{ren}}$ we construct the renormalized self adjoint Hamiltonian H, and the self adjoint momentum operator P. The space time translations are implemented by the unitary group

$$U(\alpha, \tau) = e^{iH\tau - iP\alpha} \tag{2.1}$$

and the vacuum satisfies

$$H\Omega = P\Omega = 0. \tag{2.2}$$

Furthermore, H and P commute and $H \geqslant 0$.

We construct Ω as a limit of the vacuum vectors Ω_g of the cutoff Hamiltonians $H(g)$. For concreteness, we take $g(x)$ to be a nonnegative C_0^∞ function equal to one on the interval $[-3, 3]$. Let

$$g_n(x) = g(x/n) \tag{2.3}$$

and for $A \in \mathfrak{A}$, let

$$\omega_n(A) = \frac{1}{n}\int (\Omega_{g_n}, \sigma_\alpha(A)\,\Omega_{g_n})\, h(\alpha/n)\, d\alpha, \tag{2.4}$$

where σ_α is the space translation automorphism (see II, sections 3.6 and 4). The function $h(\cdot) \geqslant 0$ is C^∞, has support in $[-1, 1]$ and

$$\int h(x)\, dx = 1 = \frac{1}{n}\int h(\alpha/n)\, d\alpha. \tag{2.5}$$

By a general compactness principle for states on a C^* algebra, there is a w^* convergent subnet ω_{n_j} of the states $\{\omega_n\}$.

Thus for each $A \in \mathfrak{A}$,

$$\omega_{n_j}(A) \to \omega(A).$$

The limiting state ω can be used to construct the inner product of a new Hilbert space $\mathfrak{F}_{\text{ren}}$, and the operators $A \in \mathfrak{A}$ can be realized as operators on $\mathfrak{F}_{\text{ren}}$. The advantage of working on the Hilbert space $\mathfrak{F}_{\text{ren}}$ is that the space–time automorphisms σ_a are given by unitary operators (2.1). We state some results.

THEOREM 2.1. *Let ω be a w^* limit point of the sequence $\omega_n \in \mathfrak{A}^*$. There is a subsequence of the ω_n which converges w^* to ω. There is a separable Hilbert space $\mathcal{F}_{\text{ren}}$, a * isomorphism π of $\mathfrak{A}$, a continuous unitary representation U of R^2, and a vector $\Omega \in \mathcal{F}_{\text{ren}}$ such that*

$$\omega(A) = (\Omega, \pi(A)\Omega) = (\Omega, A_{\text{ren}}\Omega) \tag{2.6}$$

$$U(a)\pi(A)U(-a) = \pi(\sigma_a(A)) \tag{2.7}$$

and

$$U(a)\Omega = \Omega. \tag{2.8}$$

The spectrum of the generator H of time translations is contained in the interval $[0, \infty)$.

The existence of $\mathcal{F}_{\text{ren}}$, π and Ω satisfying (2.6) is the Gelfand–Segal theorem [21]. We will prove Theorem 2.1 using the locally Fock property of π, which also leads to the properties of ϕ_{ren} explained in section 1.

THEOREM 2.2. *Let ω be a w^* limit point of the sequence $\{\omega_n\}$ and let π be the corresponding representation of $\mathfrak{A}$. Let $\mathcal{B}$ be a bounded region of space time or of space at $t=0$. Then there is a unitary operator $U_{\mathcal{B}}$: $\mathcal{F} \to \mathcal{F}_{\text{ren}}$ such that for $A \in \mathfrak{A}(\mathcal{B})$*

$$\pi(A) = U_{\mathcal{B}} A U_{\mathcal{B}}^*. \tag{2.9}$$

In short, we say that ω and π are *locally Fock*.

The locally Fock property of ω rests on the following theorem.

THEOREM 2.3. *Let $\mathcal{B}$ be a bounded region of space time or of space at $t=0$. Then the sequence*

$$\omega_n \restriction \mathfrak{A}(\mathcal{B}) \in \mathfrak{A}(\mathcal{B})^* \tag{2.10}$$

lies in a norm compact subset of $\mathfrak{A}(\mathcal{B})^*$. Any limit point $\omega \restriction \mathfrak{A}(\mathcal{B})$ is normal.

The proof of Theorem 2.3 will be given in the following sections. We introduce some von Neumann algebra terminology [2]. A state ϱ of a von Neumann algebra $\mathfrak{M}$ is called *normal* if

$$\sup \varrho(A_\alpha) = \varrho\,(\sup A_\alpha) \tag{2.11}$$

for each monotone increasing generalized sequence $\{A_\alpha\}$, $A_\alpha \in \mathfrak{M}$, provided the sequence A_α is bounded from above, so that $\sup A_\alpha$ exists.

Let us consider linear functionals on $\mathfrak{M}$ of the form $A \to l(A)$,

$$l(A) = \sum_{i=1}^{\infty} (A\theta_i, \psi_i), \tag{2.12}$$

where
$$\sum_{i=1}^{\infty} \|\theta_i\|^2 < \infty, \quad \sum_{i=1}^{\infty} \|\psi_i\|^2 < \infty. \tag{2.13}$$

Ultraweak convergence in $\mathfrak{M}$ is defined by convergence of all linear functionals of the form (2.12)–(2.13), and conversely each ultraweakly continuous linear functional on $\mathfrak{M}$ can be represented in the form (2.12)–(2.13) [2, page 40]. A state ϱ is ultraweakly continuous if and only if it is normal, and every ultraweakly continuous linear functional can be written as a linear combination of four normal states [2, page 54]. The norm of $l \in \mathfrak{M}^*$ defined by
$$\|l\| = \sup \{|l(A)|: A \in \mathfrak{M}, \|A\| \leqslant 1\}. \tag{2.14}$$

The A in (2.14) can be restricted to a subalgebra $\mathfrak{M}_0$ of $\mathfrak{M}$ which is weakly or ultraweakly dense in $\mathfrak{M}$, by Kaplansky's density theorem [2, page 46, page 43].

We apply these concepts to the von Neumann algebra $\mathfrak{A}(\mathcal{B})$ and the weakly dense subalgebra $\mathfrak{A}_0(\mathcal{B})$ generated (algebraically) by the operators
$$\{e^{i\phi(f)}, e^{i\dot{\phi}(f)}: \text{supp } f \subset \mathcal{B}, f \in C^\infty\}, \tag{2.15}$$

where $\mathcal{B}$ is a region of space time or of space at $t=0$.

Proof of Theorem 2.2. We first prove that $\pi \restriction \mathfrak{A}(\mathcal{B})$ is ultraweakly continuous. The right and left multiplications
$$A \to AC \quad \text{and} \quad A \to CA$$

are ultraweakly continuous on $\mathcal{B}(\mathcal{J})$. For C_1 and C_2 in some $\mathfrak{A}(\mathcal{B}_1)$, with $\mathcal{B}_1 \supset \mathcal{B}$,
$$A \to C_1^* A C_2 \in \mathfrak{A}(\mathcal{B}_1)$$

is ultraweakly continuous. By Theorem 2.3, $\omega \restriction \mathfrak{A}(\mathcal{B}_1)$ is normal, and hence ultraweakly continuous. Thus
$$A \to \omega(C_1^* A C_2) = (\pi(C_1)\,\Omega, \pi(A)\,\pi(C_2)\,\Omega) \tag{2.16}$$

is ultraweakly continuous. Since the vectors $\pi(C_i)\Omega$ run over a dense subset of $\mathcal{J}_{\text{ren}}$, and since $\pi(A)$ is bounded, the map
$$A \to \pi(A), \quad A \in \mathfrak{A}(\mathcal{B})$$

is ultraweakly–weakly continuous. (This map is also ultraweakly continuous.)

Since $\mathfrak{A}(\mathcal{B})$ is ultraweakly separable, $\pi(\mathfrak{A}(\mathcal{B}))$ has a countable, weakly dense subset.

Since the weak and strong closures of a linear subspace of $\mathcal{F}_{\text{ren}}$ coincide, $\pi(\mathfrak{A}(\mathcal{B}))$ has a countable, strongly dense subset. But

$$\bigcup_{\mathcal{B}} \pi(\mathfrak{A}(\mathcal{B}))\,\Omega$$

is dense in $\mathcal{F}_{\text{ren}}$, so $\mathcal{F}_{\text{ren}}$ is separable also.

In proving the theorem, we replace $\mathcal{B}$ by a bounded open set at time zero (the domain of dependence of $\mathcal{B}$), since this change merely enlarges $\mathfrak{A}(\mathcal{B})$. For such a $\mathcal{B}$ at time zero, Araki has shown that $\mathfrak{A}(\mathcal{B})$ is a factor of type III [1]. His proof requires a computation that a certain operator is not of trace class, and he shows that the operator in question is closed but not everywhere defined due to the conditions at the boundary of $\mathcal{B}$. Thus his proof which is given in three spatial dimensions is valid in one spatial dimension, and is somewhat easier in the latter case.

The kernel of $\pi \restriction \mathfrak{A}(\mathcal{B})$ is a weakly closed two sided ideal. Since there are no proper weakly closed two sided ideals in any factor, we see that ker $(\pi \restriction \mathfrak{A}(\mathcal{B}))$ equals zero and $\pi \restriction \mathfrak{A}(\mathcal{B})$ is an isomorphism. According to a theorem of E. Griffin [14], $\pi \restriction \mathfrak{A}(\mathcal{B})$, as an isomorphism between separable type III factors, is unitarily implemented. This completes the proof.

We remark that the argument of the above paragraph combined with the fact that there are no norm closed two sided ideals in a factor of type III shows that $\mathfrak{A}$ is simple, and thus π is an isomorphism of $\mathfrak{A}$.

Proof of Theorem 2.1. On the metric space $\mathfrak{A}^*(\mathcal{B})$, compactness implies sequential compactness. Thus using Theorem 2.3, we can find a subsequence ω_{n_j} with

$$\|(\omega - \omega_{n_j}) \restriction \mathfrak{A}(\mathcal{B})\| \to 0.$$

Using the diagonal process, the subsequence can be chosen independently of $\mathcal{B}$.

The existence of the unitary $U(a)$ satisfying (2.7)–(2.8) follows from the Gelfand–Segal construction if for each space–time translation σ_a,

$$\omega(\sigma_a(A)) = \omega(A). \tag{2.17}$$

Thus we must show that ω is a fixed point of σ_a^*. Since ω is norm continuous, it is sufficient to take A in the dense subalgebra

$$\bigcup_{\mathcal{B}} \mathfrak{A}(\mathcal{B}).$$

Thus we let A belong to $\mathfrak{A}(\mathcal{B})$ for some region $\mathcal{B}$, and we choose n sufficiently large so that for all $(x, t) \in \mathcal{B}$,

$$n > |x| + |t|.$$

We define $\mathcal{B}_{|\tau|}$ as the set of points within the distance $|\tau|$ of $\mathcal{B}$. With the above restrictions, $H(g_n)$ is a Hamiltonian for $\mathcal{B}_{|\tau|+|\alpha|}$, provided that $|\alpha| < n$ and $|\tau| < n$. The restriction on α is satisfied if

$$\alpha \in \operatorname{supp} h(\cdot/n).$$

Under these restrictions
$$\sigma_\alpha(\sigma_\tau(A)) \in \mathfrak{A}(\mathcal{B}_{|\tau|+|\alpha|})$$

and
$$\sigma_\alpha(\sigma_\tau(A)) = \sigma_\tau(\sigma_\alpha(A)) = e^{iH_n t}\sigma_\alpha(A)\, e^{-iH_n\tau},$$

see II. We have set
$$H_n = H(g_n) = H_0 + H_{I,g_n} - E_{g_n}. \tag{2.18}$$

Since
$$e^{-iH_n\tau}\Omega_{g_n} = \Omega_{g_n},$$

we see that for α in supp $h(\cdot/n)$ and $|\tau| < n$,

$$(\Omega_{g_n}, \sigma_\alpha(\sigma_\tau(A))\,\Omega_{g_n}) = (\Omega_{g_n}, \sigma_\alpha(A)\,\Omega_{g_n})$$

and
$$\omega_n(\sigma_\tau(A)) = \omega_n(A).$$

The formula (2.17) for time translations follows immediately.

In order to prove (2.17) for space translations, we substitute $\sigma_\beta(A)$ for A in (2.4) and perform a change of variables $\alpha \to \alpha - \beta$ in the integration. This leads to the estimate

$$|\omega_n(\sigma_\beta(A) - A)|$$
$$= \frac{1}{n}\left|\int (\Omega_{g_n}, \sigma_\alpha(A)\,\Omega_{g_n})\,\{h((\alpha-\beta)/n) - h(\alpha/n)\}\,d\alpha\right| \leq \|A\| \int \left| h\left(\alpha - \frac{\beta}{n}\right) - h(\alpha)\right| d\alpha.$$

This vanishes as $n \to \infty$, so that for $A \in \mathfrak{A}(\mathcal{B})$,

$$\omega(\sigma_\beta(A)) = \omega(A),$$

(2.17) is valid, and $U(a)$ does exist.

We prove now the strong continuity of $U(a)$. The vectors $\pi(A)\Omega$ are dense in $\mathcal{J}_{\text{ren}}$ for A in $\bigcup_{\mathcal{B}} \mathfrak{A}(\mathcal{B})$ for bounded space–time regions $\mathcal{B}$. If $\sigma_a(A)$ is in some $\mathfrak{A}(\mathcal{B})$, we have

$$U(a)\,\pi(A)\,\Omega = \pi(\sigma_a(A))\,\Omega$$

by (2.7) and (2.8). By (2.9),

$$U(a)\,\pi(A)\,\Omega = U_{\mathcal{B}}\,\sigma_a(A)\,U_{\mathcal{B}}^*\,\Omega. \tag{2.19}$$

Since σ_a is implemented locally on $\mathcal{J}$ by unitary operators of the form

$$\exp(iH(g)\,\tau)\,\exp(-iP\alpha),$$

the right side of (2.19) is strongly continuous in a. This proves the strong continuity of $U(a)$ on a dense set of vectors, which is sufficient.

The remaining assertion of the theorem is that $H \geqslant 0$. Let

$$H = \int \mu dE(\mu)$$

be the spectral resolution of H. Then we define

$$F(\tau) = \omega(A^* \sigma_\tau(A)) = (\pi(A)\,\Omega,\, e^{iH\tau}\pi(A)\,\Omega) = \int e^{i\mu\tau} d\,\|E(\mu)\,\pi(A)\Omega\|^2, \tag{2.20}$$

In order to prove that H has nonnegative spectrum, we must show that only the half line $[0, \infty)$ contributes to the integral (2.20). For A in some local algebra $\mathfrak{A}(\mathcal{B})$, we infer from Theorem 2.3 that $F(\tau)$ is the limit through a subsequence of the similar functions

$$F_n(\tau) = \omega_n(A^* \sigma_\tau(A)) = \frac{1}{n}\int (\sigma_\alpha(A^*)\,\Omega_{g_n},\, e^{iH_n\tau}\sigma_\alpha(A)\,\Omega_{g_n}) \times h(\alpha/n)\, d\alpha,$$

where H_n is defined in (2.18). Since $H_n \geqslant 0$, $F_n(\tau)$ has a Fourier transform $\tilde{F}_n(\mu)$ with support in the half line $[0, \infty)$. The subsequence $F_{n_j}(\tau)$ converges pointwise to F, and is uniformly bounded,

$$|F_{n_j}(\tau)| \leqslant \|A\|^2.$$

Hence the F_{n_j} converge to F as tempered distributions, and consequently the Fourier transforms $\tilde{F}_{n_j}$ converge as tempered distributions also. Their limit $\tilde{F}$ must therefore have support in $[0, \infty)$, and so only the interval $[0, \infty)$ contributes to (2.20). As the vectors $\pi(A)\Omega$, $A \in \mathfrak{A}(\mathcal{B})$, are dense in $\mathcal{H}_{\text{ren}}$, this shows that $H \geqslant 0$.

3. Local number operators

In this section we define local number operators N_B, $N_{\tau,B}$ and $N_{\tau,\zeta}$ and we investigate their properties. Each of these operators is the biquantization of an operator on the one particle space. We reduce the study of such quantized operators to the study of the corresponding one particle operators. We analyze the one particle operators in detail, and thereby arrive at results for the quantized operators. In Section 3.3 we assume the estimates derived in Section 5, and we show that expectation values of the above local number operators in the approximate vacuum states ω_n are bounded uniformly in n. We shall use such results in Section 4.

3.1. One particle operators. The one particle space consists of Lebesque square integrable functions $L_2(R^1)$. We study operators c on $L_2(R^1)$ that are defined on the dense domain

$S(R^1)$, the Schwartz space of rapidly decreasing C^∞ functions with rapidly decreasing derivatives. We require that c be a bounded transformation from $S(R^1)$ to $L_2(R^1)$,

$$\|cf\|_2 \leq \|f\|_s, \tag{3.1.1}$$

where $\|\cdot\|_s$ denotes some Schwartz space norm. Such an operator c can be represented by a tempered distribution kernel. The Fourier transform is an operator of this type.

$$f(x) = (2\pi)^{-\frac{1}{2}} \int \tilde{f}(p)\, e^{-ipx}\, dp. \tag{3.1.2}$$

Corresponding to the physical interpretation of position and momentum, we say that Fourier transformation connects the momentum representation with the configuration space (position) representation. This physical identification corresponds to the introduction of Newton–Wigner position coordinates.

To each bounded operator $c_p: S(R^1) \to L_2(R^1)$ in the momentum representation, there is a bounded operator c_x: $S(R^1) \to L_2(R^1)$ in the configuration space representation, where

$$(c_x f)(x) = (2\pi)^{-\frac{1}{2}} \int (c_p \tilde{f})(p)\, e^{-ipx}\, dp. \tag{3.1.3}$$

Conversely given c_x there is a corresponding c_p.

We now mention a few other examples. Let $\mu(p)$ be the energy of a particle of mass m_0. The operator $c_p = \mu(p)^\tau$ of multiplication by

$$\mu(p)^\tau = (p^2 + m_0^2)^{\tau/2} \tag{3.1.4}$$

maps $S(R^1)$ onto $S(R^1)$.

The configuration space operator μ_x^τ corresponding to $\mu(p)^\tau$ is convolution by a kernel k_τ. Here $k_\tau(x)$ is C^∞ except for $x=0$. If $\tau/2$ is a nonnegative integer, k_τ has support at $x=0$. Otherwise $k_\tau(x)$ decreases exponentially at infinity. Explicitly [25, page 185],

$$k_\tau(x) = \frac{2^{\tau/2+1}}{\Gamma\left(-\frac{\tau}{2}\right)} \left(\frac{m_0}{|x|}\right)^{(\tau+1)/2} \int_0^\infty e^{-m_0|x|\cosh t} \cosh\left(\frac{t}{2} + \tau t\right) dt, \tag{3.1.5}$$

from which we see that for $\tau \geq -1$,

$$\left|\frac{d^n}{dx^n} k_\tau(x)\right| \leq O(e^{-m_0|x|}), \quad \text{as } |x| \to \infty, \tag{3.1.6}$$

for $n=0, 1, 2, \ldots$ For $\tau < -1$, estimates of the form (3.1.6) hold if m_0 is replaced by $m_0 - \varepsilon$, for any $\varepsilon > 0$.

A third example of an operator C_x which we will need is multiplication by the characteristic function $E(x)$ of an interval $\mathcal{B}=[a, b]$. The Fourier transformed operator c_p is convolution by

$$\left(\frac{2}{\pi}\right)^{\frac{1}{2}}(e^{i(a+b)/2})\frac{\sin((a-b)p/2)}{p}=(2\pi)^{-\frac{1}{2}}\tilde{E}(p). \tag{3.1.7}$$

A fourth example of an operator c_x is multiplication by a C_0^∞ function $\zeta(x)$. The corresponding c_p is convolution by $(2\pi)^{-\frac{1}{2}}\tilde{\zeta}(p)\in S(R^1)$. Both c_x and c_p map $S(R^1)$ into itself.

Thus the operator

$$c_x=\zeta(x)\,\mu_x^\tau\zeta(x) \tag{3.1.8}$$

is a bounded transformation of $S(R^1)$ into itself, and it is a localization of the operator μ_x^τ. We also want a localized version of μ_x^τ with a sharp boundary. Thus we wish to replace $\zeta(x)$ in (3.1.8) by $E(x)$.

THEOREM 3.1.1. *Let $\tau<\frac{1}{4}$. Then $E(x)\mu_x^\tau E(x)$ is a bounded operator from $S(R^1)$ to $L_2(R^1)$. Let $\tau<\frac{1}{2}$. If $\zeta(x)$ is a positive C_0^∞ function equal to one in a neighborhood of the support of $E(x)$, and if $\varepsilon>0$, then as bilinear forms on $S(R^1)\times S(R^1)$,*

$$E(x)\,\mu_x^\tau E(x)\leqslant \text{const. }\zeta(x)\,\mu_x^{2\tau+\varepsilon}\zeta(x), \tag{3.1.9}$$

where the constant depends only on τ, ε and supp E.

Proof. It is sufficient to prove that for $\tau<\frac{1}{4}$, $\varepsilon>0$, and $f\in S(R^1)$,

$$(f, E\mu_x^{2\tau}Ef)\leqslant \text{const. }(f, \mu_x^{4\tau+\varepsilon}f). \tag{3.1.10}$$

In that case

$$\|E\mu_x^\tau Ef\|^2\leqslant\|\mu_x^\tau Ef\|^2=(f, E\mu_x^{2\tau}Ef)\leqslant \text{const. }\|\mu_x^{2\tau+\varepsilon/2}f\|^2, \tag{3.1.11}$$

which shows that $E(x)\,\mu_x^\tau E(x)$ is an operator and a bounded transformation from $S(R^1)$ to $L_2(R^1)$. Furthermore, $f=\zeta f+(1-\zeta)f$ and

$$Ef=E\zeta f.$$

Hence

$$(f, E\mu_x^\tau Ef)=(\zeta f, E\mu_x^\tau E\zeta f)\leqslant \text{const. }(f, \zeta\mu_x^{2\tau+\varepsilon}\zeta f),$$

which is (3.1.9).

In order to prove (3.1.10), we write

$$(f, E\mu_x^\tau Ef)=(f, E\mu_x^\tau f)+(f, E[\mu_x^\tau, E]\,f).$$

Since

$$|(f, E\mu_x^\tau f)|\leqslant\|f\|\,\|\mu_x^\tau f\|\leqslant \text{const. }\|\mu_x^\tau f\|^2,$$

we infer that

$$(f, E\mu_x^\tau Ef)\leqslant \text{const. }(f, \mu_x^{2\tau}f)+|(f, E[\mu_x^\tau, E]\,f)|. \tag{3.1.12}$$

We now estimate
$$(f, E[\mu_x^\tau, E]f) = (\tilde{f}, g), \tag{3.1.13}$$

where
$$g(p) = \frac{1}{(2\pi)} \int \tilde{E}(p-q)\,(\mu(q)^\tau - \mu(s)^\tau)\,\tilde{E}(q-s)\,\tilde{f}(s)\,dq\,ds.$$

Since for $\tau \leqslant 1$,
$$|\mu(q)^\tau - \mu(s)^\tau| \leqslant \text{const. } \mu(q-s)^\tau,$$

and since by (3.1.4),
$$|\tilde{E}(p)| \leqslant \text{const. } \mu(p)^{-1},$$

we have
$$|g(p)| \leqslant (h * |\tilde{f}|)(p), \tag{3.1.14}$$

where
$$h(p) = \text{const.} \int \mu(p-q)^{-1} \mu(q)^{-1+\tau} dq. \tag{3.1.15}$$

We now show that for any $\varepsilon > 0$, there is a constant such that $h(p) \leqslant \text{const. } \mu(p)^{-1+\tau+\varepsilon}$. We break up the integral in (3.1.15) into two regions:

$$\text{I: } |q| \leqslant \tfrac{1}{2}|p|$$

and
$$\text{II: } |q| > \tfrac{1}{2}|p|.$$

In region I, $\mu(p-q) \geqslant \text{const. } \mu(p)$, so that

$$\int_{\text{I}} \mu(p-q)^{-1} \mu(q)^{-1+\tau} dq \leqslant \text{const. } \mu(p)^{-1} \int_{\text{I}} \mu(q)^{-1+\tau} dq \leqslant \text{const. } \mu(p)^{-1+\tau},$$

since
$$\int_{\text{I}} \mu(q)^{-1+\tau} dq \leqslant \text{const.} \int_1^{|p|} |q|^{-1+\tau} dq \leqslant \text{const. } \mu(p)^\tau.$$

On the other hand, for $\varepsilon > 0$,

$$\mu(q)^{-\varepsilon/2} \leqslant \text{const. } \mu(p-q)^{-\varepsilon/2} \mu(p)^{\varepsilon/2},$$

so that
$$\int_{\text{II}} \mu(p-q)^{-1} \mu(q)^{-1+\tau} dq \leqslant \text{const. } \mu(p)^{\varepsilon/2} \int_{\text{II}} \mu(p-q)^{-1-\varepsilon/2} \mu(q)^{-1+\tau+\varepsilon/2}\, dq.$$

But in the region II, $\mu(q)^{-1} \leqslant 2\mu(p)^{-1}$, so

$$\int_{\text{II}} \mu(p-q)^{-1} \mu(q)^{-1+\tau} dq \leqslant \text{const. } \mu(p)^{(\varepsilon/2)-1+\tau+\varepsilon/2},$$

from which we conclude that

$$h(p) \leqslant \text{const. } \mu(p)^{-1+\tau+\varepsilon}. \tag{3.1.16}$$

We now bound (3.1.13) by writing for $\nu \geq 0$,

$$|(\tilde{f}, g)| \leq \|\tilde{f}\|_{2-\nu/(1+\nu)} \|g\|_{2+\nu} \leq \|\tilde{f}\|_{2-\nu/(1+\nu)} \|h * |\tilde{f}|\|_{2+\nu}, \tag{3.1.17}$$

where subscripts denote L_p norms. By the Hausdorff–Young inequality

$$\|h * |\tilde{f}|\|_{2+\nu} \leq \|\tilde{h}(x)|\tilde{f}|^{\sim}(x)\|_{2-\nu/(1+\nu)}$$

and by the Hölder inequality

$$\leq \|\tilde{h}(x)\|_{(4+2\nu)/\nu} \|f\|_2,$$

since

$$\||\tilde{f}|^{\sim}\|_2 = \||\tilde{f}|\|_2 = \|f\|_2.$$

Again applying the Hausdorff–Young inequality

$$\|h * |\tilde{f}|\|_{2+\nu} \leq \|h\|_{1+\nu/(4+\nu)} \|f\|_2, \tag{3.1.18}$$

and by (3.1.16), $h \in L_{1+\nu/(4+\nu)}$ if

$$(1-\tau)\left(1+\frac{\nu}{4+\nu}\right) > 1.$$

This can be satisfied if

$$\tau < \tfrac{1}{2} \tag{3.1.19}$$

by choosing ν sufficiently large so that $\nu > (1-2\tau)/4\tau$, or in other words

$$\tau < \frac{\nu}{4+2\nu}. \tag{3.1.20}$$

For such a value of ν, we have by (3.1.17)

$$|(\tilde{f}, g)| \leq \text{const. } \|\tilde{f}\|_{2-\nu/(1+\nu)} \|f\|_2.$$

On the other hand, we shall prove that for certain ν satisfying (3.1.20) and $\varepsilon > 0$,

$$\|\tilde{f}\|_{2-\nu/(1+\nu)} \leq \text{const. } \|\mu_x^{\tau+\varepsilon/2} f\|_2, \tag{3.1.21}$$

so that

$$|(f, E[\mu_x^{\tau}, E]f)| = |(\tilde{f}, g)| \leq \text{const. } \|\mu_x^{\tau+\varepsilon/2} f\|_2 \|f\|_2 \leq \text{const. } (f, \mu_x^{2\tau+\varepsilon} f). \tag{3.1.22}$$

Since this τ is restricted to $\tau < \frac{1}{2}$, the inequalities (3.1.22) and (3.1.12) complete the proof of (3.1.10). In order to establish (3.1.21), we write

$$\|\tilde{f}\|_{2-\nu/(1+\nu)}^{2-\nu/(1+\nu)} = \int |\tilde{f}(p)|^{2-\nu/(1+\nu)} dp = \int |\mu(p)^{\sigma} \tilde{f}(p)|^{2-\nu/(1+\nu)} \mu(p)^{-\sigma(2-\nu/(1+\nu))} dp$$

$$\leq \|\mu(p)^{\sigma} \tilde{f}\|_2^{2-\nu/(1+\nu)} \|\mu(p)^{-2\sigma+\sigma\nu/(1+\nu)}\|_{2(1+\nu)/\nu}.$$

15 – 702903 *Acta mathematica*. 125. Imprimé le 23 Octobre 1970.

The function $\mu(p)^{-2\sigma+\sigma\nu/(1+\nu)}$ is in $L_{2(1+\nu)/\nu}$ if

$$\sigma\left(2-\frac{\nu}{1+\nu}\right)\left(2\cdot\frac{1+\nu}{\nu}\right)>1.$$

This is true for any σ satisfying

$$\sigma>\frac{\nu}{4+2\nu}.$$

Since ν may be any positive number satisfying (3.1.20), we can choose ν so that

$$\tau+\varepsilon/2>\frac{\nu}{4+2\nu}>\tau,$$

and then choose $\sigma=\tau+\varepsilon/2$. Hence

$$\|f\|_{2-\nu/(1+\nu)}\leqslant \text{const. } \|\mu(p)^{\tau+\varepsilon/2}f\|_2=\text{const. } \|\mu_x^{\tau+\varepsilon/2}f\|_2,$$

which is (3.1.21). This completes the proof of the theorem.

For E, ζ, τ as above, the operators

$$E\mu_x^\tau E \text{ and } \zeta\mu_x^\tau\zeta$$

are positive operators with domain $S(R^1)$. Let c_τ and s_τ denote their respective Friedrichs extensions. We note that for

$$f\in\mathcal{D}(s^{\frac{1}{2}}_{2\tau+\varepsilon}),$$

we have

$$f\in\mathcal{D}(c_\tau^{\frac{1}{2}})$$

and for a constant depending only on supp E, τ and ε,

$$\|c_\tau^{\frac{1}{2}}f\|^2\leqslant\text{const. } \|s^{\frac{1}{2}}_{2\tau+\varepsilon}f\|. \tag{3.1.23}$$

This is a consequence of the fact that inequality (3.1.9) extends by continuity to the closure of $S(R^1)$ in the norm

$$|||f|||_1=\|\mu_x^{\tau+\varepsilon/2}\zeta f\|,$$

which is the domain of $s^{\frac{1}{2}}_{2\tau+\varepsilon}$. Furthermore,

$$|||f|||_2=\|\mu_x^{\tau/2}Ef\|\leqslant\text{const. } |||f|||_1$$

so that the domain of $c_\tau^{\frac{1}{2}}$ includes the domain of $s^{\frac{1}{2}}_{2\tau+\varepsilon}$.

THEOREM 3.1.2. *Let $0<\tau<\frac{1}{2}$ and let c_τ be the Friedrichs extension of*

$$E(x)\,\mu_x^\tau E(x),$$

where $E(x)$ is the characteristic function for a bounded interval $\mathcal{B}$. Then c_τ commutes with the orthogonal projection of $L_2(R^1)$ onto $L_2(\mathcal{B})$, and

$$[c_\tau \restriction L_2(\mathcal{B})]^{-1}$$

is a compact operator.

Proof. The projection of $L_2(R^1)$ onto $L_2(\mathcal{B})$ is given by multiplication by $E(x)$. Clearly $E(x)\mu_x^\tau E(x)$ commutes with multiplication by $E(x)$ on $\mathcal{S}(R^1)$, since $E(x)^2=E(x)$. By continuity, this extends as a bilinear form to the domain $\mathcal{D}(c_\tau^{\frac{1}{2}})\times\mathcal{D}(c_\tau^{\frac{1}{2}})$, and so holds on $\mathcal{D}(c_\tau)$ as an operator equality.

In order to show that c_τ has a compact resolvent on $L_2(\mathcal{B})$, it is sufficient to show that the resolvent of $c_\tau^{\frac{1}{2}}$ is compact on $L_2(\mathcal{B})$. The latter is true if whenever $\mathcal{D}$ is a set of vectors such that

$$\mathcal{D}\subset\mathcal{D}(c_\tau^{\frac{1}{2}}\restriction L_2(\mathcal{B})) \tag{3.1.24}$$

and

$$\sup_{f\in\mathcal{D}}\|c_\tau^{\frac{1}{2}}f\|<\infty, \tag{3.1.25}$$

then $\mathcal{D}$ has a compact closure. It is sufficient to replace $\mathcal{D}(c_\tau^{\frac{1}{2}}\restriction L_2(\mathcal{B}))$ in (3.1.24) by any core of $c_\tau^{\frac{1}{2}}\restriction L_2(\mathcal{B})$. By definition of the Friedrichs extension, $\mathcal{S}(R^1)$ is a core for $c_\tau^{\frac{1}{2}}$ and $E(x)\,\mathcal{S}(R^1)$ is a core for $c_\tau^{\frac{1}{2}}\restriction L_2(\mathcal{B})$.

It is convenient to imbed $L_2(\mathcal{B})$ in $L_2(R^1)$ and to write a nonlocal expression for $c_\tau^{\frac{1}{2}}$. For $f\in E(x)\,\mathcal{S}(R^1)$, we have

$$\|c_\tau^{\frac{1}{2}}f\|^2=\|\mu_x^{\tau/2}E(x)f\|^2=\|\mu_x^{\tau/2}f\|^2, \tag{3.1.26}$$

where we call this nonlocal since $\mu_x^{\tau/2}f\in L_2(R^1)$. We remark that (3.1.26) implies a continuity condition on f. Let $f_a(x)=f(x+a)$, so that

$$\|f_a-f\|^2=\int|\tilde{f}(p)\,(e^{-ipa}-1)|^2\,dp.$$

Since for $0<\tau\leqslant 1$,
$$|(e^{-ipa}-1)|\leqslant 2\,|pa|^\tau\leqslant 2\,|a|^\tau\mu(p)^\tau,$$

we infer that
$$\|f_a-f\|\leqslant 2\,|a|^\tau\left(\int|\tilde{f}(p)|^2\mu(p)^{2\tau}dp\right)^{\frac{1}{2}}=2\,|a|^\tau\,\|\mu_x^\tau f\|.$$

Thus a uniform bound (3.1.25) for $f\in\mathcal{D}$, gives

$$\sup_{f\in\mathcal{D}}\|f_a-f\|\leqslant\text{const.}\,|a|^\tau,$$

or equicontinuity of the functions in $\mathcal{D}$. It is a classical result that an equicontinuous set $\mathcal{D}$ of L_2 functions with support in a bounded interval has a compact closure $\mathcal{D}^-$. This completes the proof of the theorem.

3.2. Quantized operators on Fock space. We first review [1] some standard notation. The Fock space $\mathcal{F}$ for noninteracting bosons is the Hilbert space completion of the direct sum of n-particle Fock spaces,

$$\mathcal{F} = \bigoplus_{n=0}^{\infty} \mathcal{F}_n. \tag{3.2.1}$$

Here $\mathcal{F}_0$ is the complex numbers and $\mathcal{F}_n$ is the completion in $L_2(R^n)$ of the n-fold symmetric tensor product of $\mathcal{F}_1 = L_2(R_1)$. Thus a vector $f = \{f_n\} \in \mathcal{F}$ is a sequence of n particle vectors in $\mathcal{F}_n$, $\|f\|^2 = \sum_{n=0}^{\infty} \|f_n\|^2$. As for the one particle space, we represent n particle vectors in the momentum representation $f_n(k_1, \ldots, k_n)$ or in the configuration representation $f_n(x_1, \ldots, x_n)$.

$$f_n(x_1, \ldots, x_n) = (2\pi)^{-n/2} \int e^{-ikx} \tilde{f}(k_1, \ldots, k_n)\, dk, \tag{3.2.2}$$

where $kx = \sum_{i=1}^{n} k_i x_i$ and $dk = \prod_{i=1}^{n} dk_i$. Fourier transformation is an isomorphism of $\mathcal{F}_n$ and of $\mathcal{F}$. We use a domain $\mathcal{D}_0 \subset \mathcal{F}$ of well behaved vectors, The vectors in $\mathcal{D}_0$ have a finite number of particles with wave functions in the Schwartz space $\mathcal{S}(R^n)$,

$$\mathcal{D}_0 = \{f : f \in \mathcal{F},\ f_n = 0 \text{ for large } n, \text{ and } f_n \in \mathcal{S}(R_n)\} \tag{3.2.3}$$

Fourier transformation maps $\mathcal{D}_0$ onto itself.

We define the annihilation operator $a(k)$ on $\mathcal{D}_0$, which maps $\mathcal{F}_n$ onto $\mathcal{F}_{n-1}$. For $f \in \mathcal{D}_0$,

$$(a(k) f)_{n-1}(k_1, \ldots, k_{n-1}) = n^{\frac{1}{2}} f_n(k, k_1, \ldots, k_{n-1}), \tag{3.2.4}$$

so that $a(k)\mathcal{D}_0 \subset \mathcal{D}_0$. Clearly $\quad [a(k), a(k')]\,\mathcal{D}_0 = 0$

and any product $a(k_1) \ldots a(k_\beta)$ is defined on the domain $\mathcal{D}_0$.

The adjoint $a(k)^*$ of $a(k)$ has domain zero as an operator, but it is a densely defined bilinear form on $\mathcal{D}_0 \times \mathcal{D}_0$. Even though $a(k)^*$ is not a densely defined operator, we follow convention and call it the creation operator for a particle with momentum k.

Any monomial of creation and annihilation operators

---

(1) Formulations of Fock space and operators on it may be found in many places. Convenient for our point of view is the systematic treatment of Kristensen, Mejlbo and Poulsen [20]. Creation and annihilation operators as bilinear forms were also considered by Galindo [6]. Our use in [10] of weak integrals of bilinear forms has been questioned [23], and for this reason we present the elementary details leading to formula (3.2.7). We note that every Wightman field $\phi(x)$ is a densely defined bilinear form with C^∞ dependence on x [3].

$$a^*(k_1) \ldots a^*(k_\alpha)a(k_1) \ldots a(k_\beta), \tag{3.2.5}$$

with the creation operators to the left of the annihilators, is a bilinear form on $\mathcal{D}_0 \times \mathcal{D}_0$ with values

$$(f, a^*(k_1) \ldots a^*(k_\alpha)\, a(p_1) \ldots a(p_\beta)\, g) \in \mathcal{S}(R^{\alpha+\beta}). \tag{3.2.6}$$

Thus if $c_{\alpha\beta}(k;p) \in \mathcal{S}'(R^{\alpha+\beta})$ is a tempered distribution,

$$C_{\alpha\beta} = \int c_{\alpha\beta}(k;p)\, a^*(k_1) \ldots a^*(k_\alpha)\, a(p_1) \ldots a(p_\beta)\, dk\, dp \tag{3.2.7}$$

is a bilinear form on $\mathcal{D}_0 \times \mathcal{D}_0$ that exists as a weak integral of bilinear forms. The $C_{\alpha\beta}$ of (3.2.7) is a *Wick ordered monomial* and is the quantization of the tempered distribution $c_{\alpha\beta}(k;p)$.

If $c_{\alpha\beta}(k;p)$ is the kernel of a bounded operator $c_{\alpha\beta}$ from $\mathcal{S}(R^\beta)$ to $L_2(R^\alpha)$, then the bilinear form $C_{\alpha\beta}$ on $\mathcal{D}_0 \times \mathcal{D}_0$ determines an unbounded operator on $\mathcal{F}$ with domain $\mathcal{D}_0$. In order to establish this, we note that if $i_{\gamma\gamma}$ is the identity operator on $L_2(R^\gamma)$, then $c_{\alpha\beta} \otimes i_{\gamma\gamma}$ is a bounded operator from $\mathcal{S}(R^\beta) \otimes \mathcal{S}(R^\gamma)$ into $L_2(R^{\alpha+\gamma})$, and it extends uniquely to a bounded operator from $\mathcal{S}(R^{\beta+\gamma})$ into $L_2(R^{\alpha+\gamma})$.

Symmetrization is a projection on L_2 and on $\mathcal{S}$. Thus the symmetrization of $c_{\alpha\beta} \otimes i_{\gamma\gamma}$ in the $\beta+\gamma$ initial variables and in the $\alpha+\gamma$ final variables yields a bounded operator from $\mathcal{S}(R^{\beta+\gamma})$ to $L_2(R^{\alpha+\gamma})$, and an unbounded operator from $\mathcal{F}_{\beta+\gamma}$ to $\mathcal{F}_{\alpha+\gamma}$ with domain $\mathcal{S}(R^{\beta+\gamma}) \cap \mathcal{F}_{\beta+\gamma}$. Finally, $C_{\alpha\beta}$ is the sum over γ of the symmetrizations of $((\alpha+\gamma)!/\gamma!)\, ((\beta+\gamma)!/\gamma!)^{\frac{1}{2}}$ $c_{\alpha\beta} \otimes i_{\gamma\gamma}$ with domain $\mathcal{D}_0$.

If c_{11} is a bounded operator from $\mathcal{S}(R^1)$ to $L_2(R^1)$, then the operator C_{11} has a closure on $\mathcal{F}$ if the operator c_{11} has a closure as an unbounded operator from $L_2(R^1)$ to $L_2(R^1)$. If $c_{\alpha\beta}$ is bounded from $L_2(R^\beta)$ to $L_2(R^\alpha)$, then $C_{\alpha\beta}$ has a closure with domain containing $\mathcal{D}(N^{(\alpha+\beta)/2})$, where N is the total number of particles operator defined by

$$(Nf)_n = n f_n. \tag{3.2.8}$$

If $\|c_{\alpha\beta}\|$ is the norm of $c_{\alpha\beta}$ as a transformation from $L_2(R^\beta)$ to $L_2(R^\alpha)$ then

$$\|(N+I)^{-\alpha/2} C_{\alpha\beta} (N+I)^{-\beta/2}\| \leq \|c_{\alpha\beta}\|. \tag{3.2.9}$$

This we derive from the bound

$$\begin{aligned} |(f_{\alpha+\gamma}, C_{\alpha\beta}\, g_{\beta+\gamma})| &\leq \left(\frac{(\alpha+\gamma)!}{\gamma!}\,\frac{(\beta+\gamma)!}{\gamma!}\right)^{\frac{1}{2}} \|c_{\alpha\beta}\|\, \|f_{\alpha+\gamma}\|\, \|g_{\beta+\gamma}\| \\ &\leq \|c_{\alpha\beta}\|\, (\alpha+\gamma)^{\alpha/2}\, \|f_{\alpha+\gamma}\|\, (\beta+\gamma)^{\beta/2}\, \|g_{\beta+\gamma}\|. \end{aligned}$$

If $c_{\alpha\beta}$ maps $\mathcal{S}(R^\beta)$ continuously into $\mathcal{S}(R^\alpha)$, then $C_{\alpha\beta}$ maps $\mathcal{D}_0$ into $\mathcal{D}_0$.

By Fourier transformation we define the annihilation operator $A(x)$ for a particle at the space point x,

$$A(x) = (2\pi)^{-\frac{1}{2}} \int a(p)\, e^{-ipx}\, dp, \tag{3.2.10}$$

corresponding to the quantization of $c_{01}(p) = (2\pi)^{-1/2} \exp(-ipx)$. In the configuration space representation, for $f \in \mathcal{D}_0$,

$$(A(x) f)_{n-1}(x_1, \ldots, x_{n-1}) = n^{\frac{1}{2}} f_n(x, x_1, \ldots, x_{n-1}). \tag{3.2.11}$$

As above, we can quantize a tempered distribution $c_{\alpha\beta}(x; y)$ yielding

$$C_{\alpha\beta} = \int c_{\alpha\beta}(x; y)\, A^*(x_1) \ldots A^*(x_\alpha)\, A(y_1) \ldots A(y_\beta)\, dx\, dy, \tag{3.2.12}$$

a bilinear form on $\mathcal{D}_0 \times \mathcal{D}_0$. These $C_{\alpha\beta}$ have the same properties as the $C_{\alpha\beta}$ in the momentum representation discussed above.

We also use the Fock space $\mathcal{F}(\mathcal{B})$ defined over the one particle space $L_2(\mathcal{B})$ where $\mathcal{B}$ is an interval in R^1.

$$\mathcal{F}(\mathcal{B}) = \bigoplus_{n=0}^{\infty} \mathcal{F}_n(\mathcal{B}), \tag{3.2.13}$$

where $\mathcal{F}_n(\mathcal{B})$ is the Hilbert space completion of the n-fold symmetric tensor product of $L_2(\mathcal{B}) = \mathcal{F}_1(\mathcal{B})$. If $\mathcal{B} \subset \mathcal{B}_1$, then $\mathcal{F}(\mathcal{B}) \subset \mathcal{F}(\mathcal{B}_1)$. If $\mathcal{B} = \mathcal{B}_1 \cup \mathcal{B}_2$ and $\mathcal{B}_1 \cap \mathcal{B}_2 = \emptyset$, then

$$\mathcal{F}(\mathcal{B}) = \mathcal{F}(\mathcal{B}_1) \otimes_s \mathcal{F}(\mathcal{B}_2), \tag{3.2.14}$$

where $\otimes_s$ denotes the completion of the symmetric tensor product.

We now specialize to study the biquantization C of operators c on the one particle space $\mathcal{F}_1$. For simplicity, we do not write c_{11} or C_{11}. We consider operators c studied in Section 3.1; they are defined on the domain $\mathcal{S}(R^1)$ and are continuous transformations of $\mathcal{S}(R^1)$ into $L_2(R^1)$. Then on the domain $\mathcal{D}_0$,

$$C = \int c(x, y)\, A^*(x)\, A(y)\, dx\, dy, \tag{3.2.15}$$

where $c(x, y)$ is the kernel of c.

We give some examples which we call local number operators:

$N_{\mathcal{B}}$ is the quantization of multiplication by $E(x)$. (3.2.16)

N_τ is the quantization of μ_x^τ. (3.2.17)

$N_{\tau,\zeta}$ is the quantization of $\zeta(x)\mu_x^\tau\zeta(x)$. (3.2.18)

$N_{\tau,\mathcal{B}}$ is the quantization of $E(x)\mu_x^\tau E(x)$, for $\tau < \frac{1}{4}$. (3.2.19)

In the case $\tau = 1$, the operator $N_{1,\zeta}$ provides a local energy operator H_0^{loc}, discussed in Section 1.

THEOREM 3.2.1. *The operator* $C \restriction \mathcal{D}_0$ *of* (3.2.15) *has any of the following properties if and only if the operator* $c \restriction \mathcal{S}(R^1)$ *has the same property.*

(a) *The operator is essentially self adjoint.*

(b) *The operator is positive.*

(c) *The operator is positive and the Friedrichs extension (or closure) has a compact resolvent.*

Proof. (a) If $C \restriction \mathcal{D}_0$ has a self adjoint closure, the domain of the self adjoint closure contains a dense set of analytic vectors. Since C commutes with the projection of $\mathcal{F}$ onto $\mathcal{F}_1$, there is a dense set of one particle analytic vectors, obtained as limits of one particle states in $\mathcal{S}(R^1)$. Thus they are in the domain of the closure of c, and c is essentially self adjoint. Conversely, if $c \restriction \mathcal{S}(R^1)$ has a self adjoint closure, the domain of the closure c^- contains a total set of analytic unit vectors $e_i(x)$, $i=1,2,\ldots$ The vectors $e_{i_1 \ldots i_n} = e_{i_1} \otimes_s \ldots \otimes_s e_{i_n}$ are n-particle vectors in the domain of the closure of $C \restriction \mathcal{D}_0$. Clearly they are C^∞ vectors for C^-, since

$$C^- e_{i_1 \ldots i_n} = c^- e_{i_1} \otimes_s \ldots \otimes_s e_{i_n} + \ldots + e_{i_1} \otimes_s \ldots \otimes_s c^- e_{i_n}.$$

Furthermore if, $$\|(c^-)^m e_i\| \leqslant \beta_i^m m!,$$

then $$\|(C^-)^m e_{i_1 \ldots i_n}\| \leqslant n^m \beta^m m!, \qquad \beta = \max\{\beta_{i_1}, \ldots, \beta_{i_n}\}.$$

Thus the $e_{i_1 \ldots i_n}$ are analytic vectors for C^- and $C \restriction \mathcal{D}_0$ is essentially self adjoint.

(b) If C is positive, c is clearly positive. Conversely, suppose that c is positive and $f_n(x_1, \ldots, x_n) \in \mathcal{D}_0 \cap \mathcal{F}_n$. Then for each $x_2, \ldots, x_n$, $f_n(x_1, x_2, \ldots, x_n) \in \mathcal{S}(R^1)$ and

$$\int \bar{f}_n(x_1, x_2, \ldots, x_n)\, c(x,y)\, f_n(y, x_2, \ldots, x_n)\, dx\, dy \geqslant 0$$

is a function in $\mathcal{S}(R^{n-1})$. Thus integrating over $x_2, \ldots, x_n$ we have shown that c is a positive operator on each n particle coordinate and hence C is positive on $\mathcal{D}_0 \cap \mathcal{F}_n$, and so on $\mathcal{D}_0$.

(c) Let C_F and c_F be the Friedrichs extensions of $C \restriction \mathcal{D}_0$ and $c \restriction \mathcal{S}(R^1)$ respectively. It is sufficient to show that $c_F^{\frac{1}{2}}$ has a compact resolvent if and only if $C_F^{\frac{1}{2}}$ does. Note that $\mathcal{D}_0$ is a core for $C_F^{\frac{1}{2}}$ and $\mathcal{S}(R^1)$ is a core for $c_F^{\frac{1}{2}}$. The Friedrichs extension C_F commutes with the projection P_1 onto $\mathcal{F}_1$ as a bilinear form on $\mathcal{D}(C_F^{\frac{1}{2}}) \times \mathcal{D}(C_F^{\frac{1}{2}})$, and hence as an operator on $\mathcal{D}(C_F)$. Thus $C_F^{\frac{1}{2}}$ commutes with P_1, and if $C_F^{\frac{1}{2}}$ has a compact resolvent, so does $c_F^{\frac{1}{2}} = P_1 C_F^{\frac{1}{2}} P_1$. Conversely, if $c_F^{\frac{1}{2}}$ has a compact resolvent, we choose orthonormal eigenvectors e_i of $c_F^{\frac{1}{2}}$ corresponding to eigenvalues $0 \leqslant \lambda_i$, $\lambda_i \to \infty$. Since $e_i = \lim_{j\to\infty} e_{ij}$ where $e_{ij}(x) \in \mathcal{S}(R^1)$ and

$$\lim_{j,k\to\infty} (e_{ij}-e_{ik}, ce_{ij}-e_{ik})=0, \tag{3.2.20}$$

it follows that $e_{i_1\ldots i_n}=e_{i_1}\otimes_s \cdots \otimes_s e_{i_n}$ is in the domain of $C_F^{\frac{1}{2}}$, and hence is an eigenvector with eigenvalue $\lambda_{i_1}+\ldots+\lambda_{i_n}$. These eigenvectors are total in $\mathcal{J}$, so that $C_F^{\frac{1}{2}}$ has been diagonalized. It has discrete spectrum and eigenvalues

$$\sum_{i=0}^{\infty} n_i \lambda_i \tag{3.2.21}$$

for nonnegative integers n_i, only a finite number being non-zero. Since $\lambda_i\to\infty$, the eigenvalues have no point of accumulation except ∞ and hence $C_F^{\frac{1}{2}}$ has a compact resolvent. This completes the proof.

Corollary 3.2.2. (a) *The operators N_B and N are essentially self adjoint on $\mathcal{D}_0$. The operators $N_{\tau,\zeta}$ for $\zeta(x)\geqslant 0$ and $N_{\tau,B}$ for $\tau<\frac{1}{4}$ are positive.*

(b) *If $0<\tau<\frac{1}{4}$, the Friedrichs extension of $N_{\tau,B}$ has a compact resolvent on $\mathcal{J}(\mathcal{B})$.*

(c) *Let $\mathcal{B}$ be a bounded interval and*

$$\mathcal{B}=\bigcup_{j=1}^{J}\mathcal{B}_j.$$

Then for $0<\tau<\frac{1}{4}$ the Friedrichs extension of

$$\sum_{j=1}^{J} N_{\tau,B_j}$$

has a compact resolvent on $\mathcal{J}(\mathcal{B})$.

Proof. We need only prove (b) and (c). Since the Friedrichs extension of $E\mu_x^\tau E$ commutes with the projection of $L_2(R^1)$ onto $L_2(\mathcal{B})$ (see the proof of Theorem 3.1.2) a similar argument shows that the Friedrichs extension of $N_{\tau,B}$ commutes with the orthogonal projection P_B of $\mathcal{J}$ onto $\mathcal{J}(\mathcal{B})$. The corollary now follows from part (c) of the theorem, which can be proved for operators $C\restriction P_B\mathcal{D}_0$ and $c\restriction ES(R^1)$ on $\mathcal{J}(\mathcal{B})$ and $L_2(\mathcal{B})$. The compactness of the resolvent of $(E\mu_x^\tau E)_F$ is proved in Theorem 3.1.2. This completes the proof of (b). The proof of (c) follows from the compactness on $L_2(\mathcal{B})$ of the resolvent of

$$\left(\sum_{j=1}^{J} E_j\mu_x^\tau E_j\right)_F, \tag{3.2.22}$$

where E_j is the characteristic function of $\mathcal{B}_j$. Noting that $\|f\|^2\leqslant\sum_{j=1}^{J}\|E_j f\|^2$, we follow the proof of Theorem 3.1.2 in order to establish the compactness of the resolvent of the operator (3.2.22).

3.3. Vacuum expectation values of local number operators. In this section we assume the results of Section 5, and give estimates on the vacuum expectation values of local number operators. We first recall some definitions. In terms of the $A(x)$ and $A(x)^*$, the local fields $\phi(x)$ and $\pi(x)$ are defined by

$$\phi(x) = \frac{1}{\sqrt{2}}\, \mu_x^{-\frac{1}{2}}(A^*(x) + A(x)) \tag{3.3.1}$$

$$\pi(x) = \frac{i}{\sqrt{2}}\, \mu_x^{\frac{1}{2}}(A^*(x) - A(x)), \tag{3.3.2}$$

and are $S(R^1)$ valued bilinear forms on $\mathcal{D}_0 \times \mathcal{D}_0$.

In Section 2 we defined the approximate vacuum ω_n by

$$\omega_n(A) = \frac{1}{n}\int (\Omega_{g_n}, \sigma_\alpha(A)\,\Omega_{g_n})\, h(\alpha/n)\, d\alpha. \tag{3.3.3}$$

THEOREM 3.3.1. (a) *Let $\mathcal{B}$ be a bounded interval in R_1. Then*

$$\omega_n(N_{\mathcal{B}}) \leqslant M, \tag{3.3.4}$$

where M is a translation invariant constant, independent of n.

(b) *If $0 \leqslant \zeta(x) \in C_0^\infty$, and $\tau \leqslant 1$, then*

$$\omega_n(N_{\tau,\zeta}) \leqslant M, \tag{3.3.5}$$

where M is a translation invariant constant, independent of n.

(c) *If $f = \bar{f} \in S(R^1)$, then*

$$\omega_n(\phi(f)^2 + \pi(f)^2) \leqslant |f|^2,$$

where $|f|$ is a Schwartz space norm of f, independent of n.

(d) *If $\tau < \frac{1}{4}$*

$$\omega_n(N_{\tau,\mathcal{B}}) \leqslant M, \tag{3.3.6}$$

where the constant M is translation invariant and independent of n.

Proof. (a) Let $\mathcal{B} = [a, b]$ and

$$c_n(x, y) = \frac{1}{n}\int E(x-\alpha)\, E(y-\alpha)\, h(\alpha/n)\, d\alpha\,\delta(x-y),$$

so that

$$\omega_n(N_{\mathcal{B}}) = (\Omega_{g_n}, C_n \Omega_{g_n}). \tag{3.3.7}$$

Since $$\left|\int E(x-\alpha)\,E(y-\alpha)\,h(\alpha/n)\,d\alpha\right| \leqslant \|h\|_\infty\,\|E\|_1 = \|h\|_\infty\,|b-a|,$$

$c_n(x, y)$ is the kernel of an operator on $\mathfrak{J}_1$ with norm bounded by

$$\text{const.}\ \frac{1}{n}\,|b-a|.$$

Thus by (3.2.9), $$\|(N+I)^{-\frac{1}{2}}C_n(N+I)^{-\frac{1}{2}}\| \leqslant \text{const.}\ \frac{1}{n}\,|b-a|$$

and by (3.3.7) $$\omega_n(N_B) \leqslant \text{const.}\ \frac{1}{n}\,|b-a|\,\omega_n(N) \leqslant \text{const.}\ |b-a|$$

by Theorem 5.1.

(b) We follow the above proof with

$$c_n(x, y) = \frac{1}{n}\int \zeta(x-\alpha)\,k_\tau(x-y)\,\zeta(y-\alpha)\,h(\alpha/n)\,d\alpha.$$

Let $\zeta_\alpha = \zeta(\cdot - \alpha)$. We note that for $f \in S(R^1)$,

$$(f, c_n f) = \frac{1}{n}\int (f, \zeta_\alpha \mu_x^\tau \zeta_\alpha f)\,h(\alpha/n)\,d\alpha \leqslant \frac{1}{n}\,\|h\|_\infty \left\{\int (f, \mu_x^{\tau/2}\zeta_\alpha^2\mu_x^{\tau/2} f)\,d\alpha\right.$$

$$\left. + \int |(f, [\zeta_\alpha, \mu_x^{\tau/2}]\,[\mu_x^{\tau/2}, \zeta_\alpha]\,f)|\,d\alpha + 2\int |(f, \mu_x^{\tau/2}\zeta_\alpha[\zeta_\alpha, \mu_x^{\tau/2}]\,f)|\,d\alpha\right\}$$

$$\leqslant \frac{1}{n}\,\|h\|_\infty \left\{\int \|\zeta_\alpha \mu_x^{\tau/2} f\|^2\,d\alpha + \int \|[\mu_x^{\tau/2}, \zeta_\alpha]\,f\|^2\,d\alpha + 2\int \|[\mu_x^{\tau/2}, \zeta_\alpha]\,f\|\,\|\zeta_\alpha \mu_x^{\tau/2} f\|\,d\alpha\right\}$$

$$\leqslant \frac{2}{n}\,\|h\|_\infty \int (\|\zeta_\alpha \mu_x^{\tau/2} f\|^2 + \|[\mu_x^{\tau/2}, \zeta_\alpha]\,f\|^2)\,d\alpha$$

$$= \frac{2}{n}\,\|h\|_\infty \left(\|\zeta\|_2^2\,\|\mu_x^{\tau/2} f\|^2 + \int \|[\mu_x^{\tau/2}, \zeta_\alpha]\,f\|^2\,d\alpha\right).$$

The commutator is a bounded operator, since

$$|([\mu_x^{\tau/2}, \zeta]\,f)^\sim(p)| \leqslant (2\pi)^{-\frac{1}{2}}\int |(\mu(p)^{\tau/2} - \mu(q)^{\tau/2})\,\tilde{\zeta}(p-q)\,\tilde{f}(q)|\,dq$$

$$\leqslant \text{const.}\ \int (p-q)^{\tau/2}\,|\tilde{\zeta}(p-q)\,\tilde{f}(q)|\,dq$$

and therefore $$\|[\mu_x^{\tau/2}, \zeta]\,f\|_2 \leqslant \text{const.}\ \|(\mu^{\tau/2}\,|\tilde{\zeta}|)^\sim\|_\infty\,\|f\|_2 \leqslant M\|f\|_2,$$

as $\tilde{\zeta}$ is rapidly decreasing.

Let $B=[a,b]$ be a neighborhood of the support of ζ, let E be the characteristic function of B, and let $E_\alpha = E(\cdot - \alpha)$. Then

$$\int \|[\mu_x^{\tau/2}, \zeta_\alpha] E_\alpha f\|^2 d\alpha \leqslant \text{const.} \int \|E_\alpha f\|^2 d\alpha \leqslant \text{const.}\ (b-a)\,\|f\|^2.$$

Furthermore, the operator

$$[\mu_x^{\tau/2}, \zeta_\alpha](I - E_\alpha) = \zeta_\alpha \mu_x^{\tau/2}(I-E_\alpha)$$

is an integral operator with a kernel dominated by const. $e^{-m_0(|x-\alpha|+|y-\alpha|)}$. (See (3.1.6).) Thus the operator

$$K_\alpha = [\mu_x^{\tau/2}, \zeta_\alpha](I-E_\alpha)(1+(x-\alpha)^2)$$

has an L_2 kernel and is consequently bounded with a bound M_1 independent of α. It follows that

$$\begin{aligned}\int \|[\mu_x^{\tau/2}, \zeta_\alpha](I-E_\alpha) f\|^2 d\alpha &= \int \|K_\alpha (1+(x-\alpha)^2)^{-1} f\|^2 d\alpha \\ &\leqslant M_1^2 \int \|(1+(x-\alpha)^2)^{-1} f\|^2 d\alpha = M_1^2 \int (1+t^2)^{-2} dt\, \|f\|^2.\end{aligned}$$

Combining these results

$$(f, c_n f) \leqslant \frac{1}{n}\ \text{const.}\ (\|\mu_x^{\tau/2} f\|^2 + \|f\|^2) \leqslant \frac{1}{n}\ \text{const.}\ \|\mu_x^{\tau/2} f\|^2.$$

Therefore there is a constant M such that

$$\frac{M}{n}\mu_x^\tau - c_n$$

is a positive operator, and hence by Theorem 3.2.1 b we have after biquantization

$$\frac{M}{n} N_\tau - C_n \geqslant 0.$$

Thus for $\tau \leqslant 1$, $$\omega_n(C_n) \leqslant \frac{M}{n}\,\omega_n(N_\tau) \leqslant \frac{M}{n}\ \text{const.}\ \omega_n(H_0).$$

By Theorem 5.1 and Remark 2, equation (5.10),

$$\omega_n(C_n) \leqslant \text{const.}$$

(c) Since

$$\phi(f)^2 + \pi(f)^2 = :\phi(f)^2: + :\pi(f)^2: + \|\mu_x^{-\frac{1}{2}} f\|^2 + \|\mu_x^{\frac{1}{2}} f\|^2, \tag{3.3.8}$$

we need only consider the Wick ordered terms in (3.3.8). Each such term has j annihilators, $j=0, 1, 2,$ and $2-j$ creators. We now show that it is sufficient to consider the terms with $j=1$. Consider, for example, the pure creation term $(A^*(\mu_x^{\frac{1}{2}} f))^2$. We have

$$|\omega_n((A^*(\mu_x^{\frac{1}{2}}f))^2)| \leqslant \omega_n(A^*(\mu_x^{\frac{1}{2}}f)\, A(\mu_x^{\frac{1}{2}}f)) + \tfrac{1}{2}\,\|\mu_x^{\frac{1}{2}}f\|^2.$$

Similarly
$$\omega_n(\phi(f)^2 + \pi(f)^2) \leqslant \text{const. } \omega_n(A^*(\mu_x^{\frac{1}{2}}f)\, A(\mu_x^{\frac{1}{2}}f)) + |f|^2 \tag{3.3.9}$$

for a suitable norm $|f|$. We estimate (3.3.9) as in the proof of parts (a) and (b).

(d) From Theorem 3.1.1, and Theorem 3.2.1 b, we conclude that on $\mathcal{D}_0 \times \mathcal{D}_0$,

$$N_{\tau, B} \leqslant \text{const. } N_{2\tau+\varepsilon, \zeta}, \tag{3.3.10}$$

where $\varepsilon > 0$. Hence (3.3.10) extends to the domain $\mathcal{D} \times \mathcal{D}$, where $\mathcal{D}$ is the domain of the square root of the Friedrichs extension of $N_{2\tau+\varepsilon, \zeta}$. The desired inequality now follows from part (b) of this theorem, by choosing ε sufficiently small so that $2\tau + \varepsilon \leqslant 1$.

4. Norm compactness of the approximate vacuums

In this section we use the bounds on the local number operators derived in Section 3. We give a proof of Theorem 2.3, and hence complete the proof of Theorems 2.1 and 2.2. We restate our result as Theorem 4.1, in a somewhat more general form.

THEOREM 4.1. *Let ω_n be a sequence of normal states defined on the algebra $\mathfrak{A}$ and suppose that for some $\tau > 0$ and for each $\zeta(x) \geqslant 0$ in C_0^∞,*

$$\omega_n(N_{\tau, \zeta}) \leqslant M, \tag{4.1}$$

where $M = M(\zeta)$ a translation invariant constant independent of n. Then the sequence $\omega_n \restriction \mathfrak{A}(\mathcal{B})$ lies in a norm compact subset of the dual $\mathfrak{A}(\mathcal{B})^*$, and any limit point $\omega \restriction \mathfrak{A}(\mathcal{B})$ is normal.

Remarks. 1. The state ω_n defined in (2.4) is a normal state on $\mathfrak{A}$, since the vector state $(\Omega_{g_n}, \cdot\, \Omega_{g_n})$ is normal and the space translation automorphism σ_α in (2.4) is implemented by a one parameter continuous unitary group of operators on $\mathcal{F}$. Hence σ_α extends to a strongly continuous automorphism of all bounded operators on $\mathcal{F}$.

2. In Theorem 3.3.1 we established the bound (4.1) for all $\tau \leqslant 1$ in the case that the ω_n are the approximate vacuum states defined in (2.4) for the $(\phi^4)_2$ interaction. Thus a proof of Theorem 4.1 provides a proof of Theorem 2.3.

3. If the inequality (4.1) holds for some $\tau > 0$, then for $\sigma \leqslant \tau$, a similar inequality holds.

$$\omega_n (N_{\sigma,\zeta}) \leqslant m_0^{\tau-\sigma} \omega_n (N_{\tau,\zeta}).$$

If $\sigma < \frac{1}{2}\tau$, and $\sigma < \frac{1}{4}$, then for a bounded interval $\mathcal{B}$ we conclude from (4.1) and the proof of Theorem 3.3.1 (d) that

$$\omega_n(N_{\sigma,\mathcal{B}}) \leqslant M, \tag{3.3.6a}$$

for a translation invariant constant $M = M(\mathcal{B})$ that is independent of n. The estimate (3.3.6 (a)), for some $\sigma > 0$, is the crucial estimate in the proof of Theorem 4.1.

The method of proof is to use (3.3.6 (a)) to replace the sequence ω_n by a regularized sequence ω_n^ε that is close to ω_n in norm.

$$\|\omega_n - \omega_n^\varepsilon\| \leqslant \varepsilon. \tag{4.2}$$

In the regularized state ω_n^ε, there will be only a finite number of particles in each bounded region of space, and that number will grow at most as a power of the diameter of the region. These bounds on ω_n^ε will be uniform in n.

The algebra $\mathfrak{A}_0(\mathcal{B})$ defined in (2.15) is weakly dense in $\mathfrak{A}(\mathcal{B})$. Since

$$\|(\omega_n^\varepsilon - \omega_m^\varepsilon) \restriction \mathfrak{A}(\mathcal{B})\| = \|(\omega_n^\varepsilon - \omega_m^\varepsilon) \restriction \mathfrak{A}_0(\mathcal{B})\|,$$

it is sufficient to prove that any sequence $\omega_n^\varepsilon \restriction \mathfrak{A}_0(\beta)$ has a norm convergent subsequence.

The next step in the proof is to approximate $C \in \mathfrak{A}_0(\mathcal{B})$ by an operator C^ε such that

$$|\omega_n^\varepsilon (C - C^\varepsilon)| \leqslant \varepsilon \|C\|, \tag{4.3}$$

where C^ε is independent of n. The approximation C^ε of C is obtained by expanding the exponentials in C in power series, each term of which is a product of unbounded field operators. Because of the regularity of ω_n^ε, the expectation value $\omega_n^\varepsilon(C)$ has a corresponding infinite series expansion which converges uniformly in n. To obtain C^ε, we neglect the tail of this expansion and we also localize each term in a finite interval of space. The localization also leads to a small error, uniform in n, and satisfying (4.3).

Finally it will be possible to apply the compactness of the resolvent of the localized number operator $N_{\tau,\mathcal{B}}$ proved in Corollary 3.2.2 (c), in order to prove that the twice regularized expectation values $\omega_n^\varepsilon(C^\varepsilon)$ have a subsequence such that

$$|(\omega_{n_i}^\varepsilon - \omega_{n_j}^\varepsilon)\,(C^\varepsilon)| \leqslant o(1)\,\|C\|. \tag{4.4}$$

From (4.2)–(4.4) we will conclude that Theorem 4.1 is valid.

We prove Theorem 4.1 in the case where $\mathcal{B}$ is an open interval of space whose closure is contained in (0, 1). This restriction on $\mathcal{B}$ is for convenience only. Let

$$X_j = [j, j+1).$$

The local number operators N_{X_j}, $j=0, \pm 1, \ldots$, are commuting self adjoint operators. Let $P_{l,j}$ denote the spectral projection of the self adjoint operator N_{X_j} for the spectral interval $[0, l]$.

If $0 \leqslant \zeta(x)$ is a C_0^∞ function equal to one on X_0, then $\zeta_j(x) = \zeta(x+j)$ equals one on X_j and

$$N_{X_j} \leqslant N_{0,\zeta_j}.$$

By (3.3.6 (a)) in the case $\tau = 0$,

$$\omega_n(N_{X_j}) \leqslant M_1. \tag{4.5}$$

The constant M_1 is independent of n and of j. In the following, we introduce constants $M_2, \ldots$, independent of n and of C in $\mathfrak{A}$.

From (4.5) we have

$$\omega_n(I - P_{l,j}) \leqslant M_1 l^{-1}. \tag{4.6}$$

We use this fact to construct the approximate vacuum ω_n^ε, by keeping a finite number of particles in each interval X_j. For some constant M_2 to be chosen later, we define

$$l = l(j) = M_2 \varepsilon^{-2}(1+j^2). \tag{4.7}$$

We define the projection

$$P = \prod_{j=-\infty}^{\infty} P_{l(j),j}, \tag{4.8}$$

and we define the linear functional $\omega_n^\varepsilon(\cdot)$ on $\mathfrak{A}$ by

$$\omega_n^\varepsilon(C) = \omega_n(PCP). \tag{4.9}$$

By (4.8) and the fact that the $P_{l,j}$ are commuting projections, we have

$$(I-P) \leqslant \sum_{j=-\infty}^{\infty} (I - P_{l(j),j}). \tag{4.10}$$

Thus

$$\omega_n(I-P) \leqslant \sum_{j=-\infty}^{\infty} \omega_n(I - P_{l(j),j})$$

and by (4.6) – (4.7)

$$\leqslant \varepsilon^2 \left(\frac{M_1}{M_2}\right) \sum_{j=-\infty}^{\infty} (1+j^2)^{-1}.$$

We now choose M_2 large, so that

$$\omega_n(I-P) \leqslant (\varepsilon/3)^2. \tag{4.11}$$

In order to verify that ω_n and ω_n^ε are close in norm, we write

$$\omega_n(C) - \omega_n^\varepsilon(C) = \omega_n((I-P)\,C(I-P)) + \omega_n((I-P)\,CP) + \omega_n(PC(I-P)). \tag{4.12}$$

Note that
$$|\omega_n(A^*B)|^2 \leqslant \omega_n(A^*A)\,\omega_n(B^*B). \tag{4.13}$$

Since
$$\|P\| = 1 = \|I-P\|,$$

and
$$(I-P)^2 = (I-P),$$

we have
$$|\omega_n(C) - \omega_n^\varepsilon(C)| \leqslant 3\,\|C\|\,[\omega_n(I-P)]^{\frac{1}{2}}.$$

Thus from (4.11) we conclude that

$$|\omega_n(C) - \omega_n^\varepsilon(C)| \leqslant \varepsilon\,\|C\|.$$

In terms of the states,
$$\|\omega_n - \omega_n^\varepsilon\| \leqslant \varepsilon. \tag{4.14}$$

We now study an operator $C \in \mathfrak{A}_0(\mathcal{B})$. Thus

$$C = \sum_{j=1}^{J} \alpha_j e^{i\phi(f_j) + i\pi(h_j)}. \tag{4.15}$$

The test functions f_j and h_j are smooth and their support is contained in $\mathcal{B}$. For such a C, it is obvious what an expansion in terms of field operators means. First we expand C in terms of annihilation and creation operators A and A^*. We write

$$C = \sum_{\alpha,\beta=0}^{\infty} C_{\alpha\beta}, \tag{4.16}$$

where $C_{\alpha\beta}$ is a Wick ordered monomial of degree α in creation operators A^* and of degree β in annihilation operators A. Thus

$$C_{\alpha\beta} = \int c_{\alpha\beta}(x_1, \dots, x_\alpha, x_{\alpha+1}, \dots, x_{\alpha+\beta})\, A^*(x_1) \dots A^*(x_\alpha)\, A(x_{\alpha+1}) \dots A(x_{\alpha+\beta})\, dx. \tag{4.17}$$

The kernel $c_{\alpha\beta}(x)$ is symmetric in the α creation variables $x_1, \dots, x_\alpha$ and in the β annihilation variables $x_{\alpha+1}, \dots, x_{\alpha+\beta}$. We note several properties of the operators $C_{\alpha\beta}$, the kernels $c_{\alpha\beta}(x)$, and the corresponding operator $c_{\alpha\beta}$ from $\mathcal{F}_\beta$ to $\mathcal{F}_\alpha$.

Suppose that the kernel $c_{\alpha\beta}(x)$ has support in the region

$$x_i \in \mathcal{B}_i, \qquad i = 1, \dots, \alpha+\beta,$$

for intervals $\mathcal{B}_i \subset R^1$. Then if

$$N_+ = \prod_{i=1}^{\alpha} N_{B_i}, \qquad N_- = \prod_{i=\alpha+1}^{\alpha+\beta} N_{B_i}, \tag{4.18}$$

we have
$$\|(N_+ + I)^{-\frac{1}{2}} C_{\alpha\beta} (N_- + I)^{-\frac{1}{2}}\| \leq \|c_{\alpha\beta}\|. \tag{4.19}$$

The bound (4.19) is a localization of (3.2.9), and is a consequence of the local action of $A^*(x)$ and $A(x)$ on the Newton–Wigner wave functions $f_n(x_1, \ldots, x_n) \in \mathfrak{F}_n$. (See Section 3.2.) We also note that for $\psi_\alpha \in \mathfrak{F}_\alpha$, $\psi_\beta \in \mathfrak{F}_\beta$,

$$\|c_{\alpha\beta}\| = \sup_{\substack{\|\psi_\alpha\|=1 \\ \|\psi_\beta\|=1}} |(\alpha!\,\beta!)^{-\frac{1}{2}} (\psi_\alpha, C_{\alpha\beta}\psi_\beta)| \tag{4.20}$$

where $C_{\alpha\beta}$ is the quantization (4.17) of $c_{\alpha\beta}$ on Fock space. Thus for

$$0 \leq \gamma \leq \min(\alpha, \beta),$$

we infer from (4.20) that

$$|(\psi_\alpha, C_{\alpha-\gamma,\beta-\gamma}\psi_\beta)| \leq \|\psi_\alpha\|\,\|\psi_\beta\| \left\{\frac{\alpha!\,\beta!}{\gamma!\,\gamma!}\right\}^{\frac{1}{2}} \|c_{\alpha-\gamma,\beta-\gamma}\| \tag{4.21}$$

Lemma 4.2. *For a bounded operator C of the form* (4.16) – (4.17),

$$\|c_{\alpha\beta}\| \leq 2^{\alpha+\beta}\|C\|. \tag{4.22}$$

Proof. Note that

$$(\psi_\alpha, C_{\alpha\beta}\psi_\beta) = (\psi_\alpha, \{C - \sum_{\gamma=1}^{\min(\alpha,\beta)} C_{\alpha-\gamma,\beta-\gamma}\}\psi_\beta)$$

By (4.20) – (4.21)

$$\|c_{\alpha\beta}\| \leq (\alpha!\,\beta!)^{-\frac{1}{2}} \left\{\|C\| + \sum_{\gamma=1}^{\min(\alpha,\beta)} \frac{(\alpha!\,\beta!)^{\frac{1}{2}}}{\gamma!} \|c_{\alpha-\gamma,\beta-\gamma}\|\right\}$$

$$= \|C\| \left\{(\alpha!\,\beta!)^{-\frac{1}{2}} + \sum_{\gamma=1}^{\min(\alpha,\beta)} \frac{1}{\gamma!} \frac{\|c_{\alpha-\gamma,\beta-\gamma}\|}{\|C\|}\right\}. \tag{4.23}$$

We reason by induction on $\min(\alpha, \beta)$. If $\min(\alpha, \beta) = 0$, then the sum in (4.23) does not occur and

$$\|c_{\alpha\beta}\| \leq \|C\|.$$

If $\min(\alpha, \beta) \neq 0$, then by the induction hypothesis (4.22) is valid for each $\|c_{\alpha-\gamma,\beta-\gamma}\|$ in the sum on the right side of (4.23). Thus

$$\|c_{\alpha\beta}\| \leq \|C\| \left\{(\alpha!\,\beta!)^{-\frac{1}{2}} + \sum_{\gamma=1}^{n} \frac{1}{\gamma!} 2^{\alpha+\beta-2\gamma}\right\} \leq \|C\| \left\{(a!\,\beta!)^{-\frac{1}{2}} + \sum_{\gamma=1}^{\infty} 2^{\alpha+\beta} \frac{4^{-\gamma}}{\gamma!}\right\}$$

$$= \|C\| \{(\alpha!\,\beta!)^{-\frac{1}{2}} + 2^{\alpha+\beta}(e^{\frac{1}{4}} - 1)\} \leq 2^{\alpha+\beta}\|C\|,$$

and this completes the proof of Lemma 4.2.

We want to show that $c_{\alpha\beta}(x)$ is exponentially small at infinity in each of the variables x_i. For this purpose we define a localization multi-index $L=\{L_1, L_2, ..., L_{\alpha+\beta}\}$. We let L_k be an integer corresponding to the localization of the coordinate x_k in the interval X_{L_k}. We use the localization indices L to localize the operator $C_{\alpha\beta}$. We define

$$c_{\alpha\beta}^L(x) = \text{Sym.}\, c_{\alpha\beta}(x) \prod_{k=1}^{\alpha+\beta} E(X_{L_k}), \quad L=\{L_k\}, \tag{4.24}$$

where $E(X_j)$ is the characteristic function for the interval X_j. The symmetrization in (4.24) symmetrizes the kernel in the α creation variables $(x_1, ..., x_\alpha)$ and in the β annihilation variables $(x_{\alpha+1}, ..., x_{\alpha+\beta})$. The corresponding localized operator $c_{\alpha\beta}^L$ from $\mathfrak{F}_\beta$ to $\mathfrak{F}_\alpha$ has the kernel $c_{\alpha\beta}^L(x)$. We denote by $C_{\alpha\beta}^L$ the localized, quantized operator arising from $c_{\alpha\beta}^L$.

$$C_{\alpha\beta}^L = \int c_{\alpha\beta}^L(x)\, A^*(x_1) \dots A^*(x_\alpha)\, A(x_{\alpha+1}) \dots A(x_{\alpha+\beta})\, dx. \tag{4.25}$$

We define

$$N_+^L = \prod_{k=1}^{\alpha} N_{X_{L_k}} \tag{4.26}$$

and

$$N_-^L = \prod_{k=\alpha+1}^{\alpha+\beta} N_{X_{L_k}}. \tag{4.27}$$

Then by (4.19)

$$\|(N_+^L+I)^{-\frac{1}{2}} C_{\alpha\beta}^L (N_-^L+I)^{-\frac{1}{2}}\| \leqslant \|c_{\alpha\beta}^L\|. \tag{4.28}$$

We note that the kernels $c_{\alpha\beta}(x)$ from operators $C \in \mathfrak{A}_0(\mathcal{B})$ are functions in $\mathcal{S}(R^{\alpha+\beta})$. Thus each term in (4.24) before symmetrization is a bounded operator from $L_2(R^\beta)$ to $L_2(R^\alpha)$ with a norm bounded by $\|c_{\alpha\beta}\|$. Hence the same holds for $c_{\alpha\beta}^L(x)$ and

$$\|c_{\alpha\beta}^L\| \leqslant \|c_{\alpha\beta}\|.$$

Thus by Lemma 4.2,

$$\|c_{\alpha\beta}^L\| \leqslant 2^{\alpha+\beta}\|C\|. \tag{4.29}$$

We now define a measure $D=D(L)$ of the total distance from the origin of the localization L in $C_{\alpha\beta}^L$.

$$D = D(L) = \sum_{k=1}^{\alpha+\beta} |L_k|. \tag{4.30}$$

LEMMA 4.3. *There is a constant M_4 independent of C such that for $C \in \mathfrak{A}_0(\mathcal{B})$,*

$$\|c_{\alpha\beta}^L\| \leqslant M_4^{\alpha+\beta} e^{-m_0 D(L)} \|C\|. \tag{4.31}$$

This lemma improves the bound (4.29). We postpone the proof to the end of this section. We remark, however, that the basis for the localization (4.31) is the fact that when

16 – 702903 *Acta mathematica*. 125. Imprimé le 23 Octobre 1970.

C is expanded in terms of the fields $\phi(y)$ and $\pi(y)$, only points $y \in \mathcal{B}$ enter the expansion. The expansion (4.17) approximately preserves this localization, and (4.31) is a quantitative estimate of the loss of locality that results from using the A^*, A expansion.

We now study the operator $PC^L_{\alpha\beta}P$ where P is the projection defined in (4.8). By definition P projects onto a subspace of $\mathcal{F}$ with a finite number $l(j)$ of particles in each interval X_j. Since the localization L restricts the kernel $c^L_{\alpha\beta}(x)$ to have compact support, the operator $C^L_{\alpha\beta}$ creates or annihilates particles in a bounded region. Hence by (4.25),

$$PC^L_{\alpha\beta}P \tag{4.32}$$

is a bounded operator.

Lemma 4.4. *Given $\varepsilon_1 > 0$, there exists a constant M_5 such that for all $C \in \mathfrak{A}_0(\mathcal{B})$,*

$$\|PC^L_{\alpha\beta}P\| \leqslant M_5^{\alpha+\beta} e^{-(m_0-\varepsilon_1)D(L)} \|C\|.$$

Proof. We first prove that given $\varepsilon_1 > 0$, there is a constant M_6 such that

$$\|P(N^L_+ + I)^{\frac{1}{2}} P\| \, \|P(N^L_- + I)^{\frac{1}{2}} P\| \leqslant M_6^{\alpha+\beta} e^{\varepsilon_1 D(L)}.$$

By the definition of P and the fact that each N_{X_j} commutes with P, we have

$$\|P(N^L_+ + I)^{\frac{1}{2}} P\| \, \|P(N^L_- + I)^{\frac{1}{2}} P\| \leqslant \prod_{k=1}^{\alpha+\beta} (l(L_k)+1)^{\frac{1}{2}}. \tag{4.33}$$

For any $\varepsilon_1 > 0$, there is a constant M_6 such that

$$(l(L_k)+1)^{\frac{1}{2}} = \{M_2 \varepsilon^{-2}(1+L_k^2)+1\}^{\frac{1}{2}} \leqslant M_6 e^{\varepsilon_1 |L_k|}.$$

Thus
$$\prod_{k=1}^{\alpha+\beta} (l(L_k)+1)^{\frac{1}{2}} \leqslant M_6^{\alpha+\beta} e^{\varepsilon_1 D(L)}.$$

Using (4.33),
$$\|P(N^L_+ + I)^{\frac{1}{2}} P\| \, \|P(N^L_- + I)^{\frac{1}{2}} P\| \leqslant M_6^{\alpha+\beta} e^{\varepsilon_1 D(L)}.$$

Using estimate (4.28), we find that

$$\begin{aligned}
\|PC^L_{\alpha\beta}P\| &= \|P(N^L_+ + I)^{\frac{1}{2}} (N^L_+ + I)^{-\frac{1}{2}} C^L_{\alpha\beta} (N^L_- + I)^{-\frac{1}{2}} (N^L_- + I)^{\frac{1}{2}} P\| \\
&\leqslant \|P(N^L_+ + I) P\| \, \|P(N^L_- + I)^{\frac{1}{2}} P\| \, \|(N^L_+ + I)^{-\frac{1}{2}} C^L_{\alpha\beta} (N^L_- + I)^{-\frac{1}{2}}\| \\
&\leqslant M_6^{\alpha+\beta} e^{\varepsilon_1 D(L)} \|c^L_{\alpha\beta}\|
\end{aligned}$$

and by Lemma 4.3,
$$\leqslant (M_4 M_6)^{\alpha+\beta} e^{-(m_0-\varepsilon_1)D(L)} \|C\|,$$

to complete the proof.

Our next estimate bounds $D(L)$ from below.

LEMMA 4.5. *If P and $C^L_{\alpha\beta}$ are defined as in* (4.8) *and* (4.25), *and if*

$$PC^L_{\alpha\beta}P \neq 0, \tag{4.34}$$

then there is a constant M_7 such that

$$D(L) + 1 \geqslant M_7(\alpha^{\frac{4}{3}} + \beta^{\frac{4}{3}}). \tag{4.35}$$

Proof. For fixed values of α and β, we can minimize (4.30) by choosing the $|L_k|$ as small as possible. However, (4.34) will hold only if $C^L_{\alpha\beta}$ both creates and annihilates no more than $l(j)$ particles in each region X_j. For this reason it is convenient to count the number of particles created and annihilated in the region X_j for the localization L. We define $L_+(j)$ and $L_-(j)$ to be these numbers:

$$L_+(j) = \text{the number of } L_k = j, \quad \text{for } k = 1, 2, \ldots, \alpha.$$

$$L_-(j) = \text{the number of } L_k = j, \quad \text{for } k = \alpha+1, \ldots, \alpha+\beta.$$

These numbers must satisfy

$$\sum_{j=-\infty}^{\infty} L_+(j) = \alpha, \quad \text{and} \quad \sum_{j=-\infty}^{\infty} L_-(j) = \beta. \tag{4.36}$$

In addition, (4.34) will be valid only if

$$L_+(j) \leqslant l(j) = M_2\varepsilon^{-2}(1+j^2), \tag{4.37}$$

and

$$L_-(j) \leqslant l(j) = M_2\varepsilon^{-2}(1+j^2). \tag{4.38}$$

In order to minimize $D(L)$ consistent with (4.34), we choose

$$L_\pm(j) = \begin{cases} l(j) = M_2\varepsilon^{-2}(1+j^2), & \text{if } |j| < J_\pm \\ 0 & , \text{ if } |j| > J_\pm. \end{cases} \tag{4.39}$$

The maximum value J_+ or J_- of any localization index $|L_k|$ in L, and the number of indices with this maximum value, are fixed by the requirements (4.36). Using (4.39) we conclude that there is a constant M_8 (approximately equal to $M_2\varepsilon^{-2}/3$) such that (excluding the trivial case $J_+ = 0$)

$$\alpha = \sum_{|j|<J_+} M_2\varepsilon^{-2}(1+j^2) + L_+(J_+) \leqslant M_8 J_+^3, \tag{4.40}$$

and similarly (excluding the trivial case $J_- = 0$),

$$\beta \leqslant M_8 J_-^3. \tag{4.41}$$

For our choice of L, there is a constant M_9, independent of α and β, such that

$$D(L) = \sum_{k=1}^{\alpha+\beta} |L_k| = \sum_{|j| \leqslant J_+} |j| L_+(j) + \sum_{|j| \leqslant J_-} |j| L_-(j) \geqslant M_9 (J_+^4 + J_-^4),$$

where we have used (4.39). Thus by (4.40) – (4.41) which bound J_+ and J_-,

$$D(L) \geqslant M_9 M_8^{-\frac{4}{3}} (\alpha^{\frac{4}{3}} + \beta^{\frac{4}{3}}).$$

If $J_+ = 0$, then $\alpha \leqslant M_2 \varepsilon^{-2}$ and if $J_- = 0$, then $\beta \leqslant M_2 \varepsilon^{-2}$. The proof is thus complete.

These estimates are useful for constructing the operator C^ε and for analyzing the error $C - C^\varepsilon$. Let us define the set of localization indices

$$\mathfrak{L}_M = \{L\colon\ |L_k| \leqslant M,\ k = 1, 2, \ldots, \alpha + \beta\}. \tag{4.42}$$

If $L \in \mathfrak{L}_M$, then all the particles that are created or annihilated by $C_{\alpha\beta}^L$ are in the region of space

$$Z = Z_M = \bigcup_{|j| \leqslant M} X_j. \tag{4.43}$$

We define for $C \in \mathfrak{A}_0(\mathcal{B})$,

$$C^\varepsilon = \sum_{\alpha,\beta=0}^{M} \sum_{L \in \mathfrak{L}_M} C_{\alpha\beta}^L, \tag{4.44}$$

where $M = M(\varepsilon)$ will be chosen later, independently of C.

Lemma 4.6 (a). *There exists a constant M_{10} independent of M such that for $C \in \mathfrak{A}_0(\mathcal{B})$ and for C^ε defined in* (4.44),

$$\|PC^\varepsilon P\| \leqslant M_{10} \|C\|. \tag{4.45}$$

(b) *Given $\varepsilon > 0$, there exists $M = M(\varepsilon)$ sufficiently large such that for all $C \in \mathfrak{A}_0(\mathcal{B})$*

$$\|P(C - C^\varepsilon) P\| \leqslant \varepsilon \|C\|, \tag{4.46}$$

and

$$|\omega_n^\varepsilon (C - C^\varepsilon)| \leqslant \varepsilon \|C\|. \tag{4.47}$$

Proof. By Lemma 4.4 with $\varepsilon_1 = \varepsilon$,

$$\|PC^\varepsilon P\| \leqslant \sum_{\alpha,\beta=0}^{M} \sum_{L \in \mathfrak{L}_M}{}' M_5^{\alpha+\beta} e^{-(m_0-\varepsilon) D(L)} \|C\|,$$

where $\sum_L'$ denotes the sum over those L for which $PC_{\alpha\beta}^L P \neq 0$, namely the L which satisfy the hypotheses of Lemma 4.5. Thus

$$\|PC^{\varepsilon}P\| \leqslant M_3 \|C\| \sum_{\alpha,\beta=0}^{M} \sum_{L\in \mathfrak{C}_M} M_5^{\alpha+\beta} \exp\{-\tfrac{1}{2}m_0 D(L) - M_{11}(\alpha^{\frac{1}{2}}+\beta^{\frac{1}{2}})\}, \tag{4.48}$$

where $M_{11} = M_7(\frac{1}{2}m_0 - \varepsilon)$ and $M_3 = e^{m_0-\varepsilon}$. The sum over all L converges exponentially fast, for

$$\sum_L \exp\{-\tfrac{1}{2}m_0 D(L)\} = \left[\sum_{l=-\infty}^{\infty} e^{-m_0|l|/2}\right]^{\alpha+\beta} = (M_{12})^{\alpha+\beta}. \tag{4.49}$$

The sum over the tail, $L \notin \mathfrak{C}_M$, is characterized by the fact that for some k, $|L_k| > M$. Thus

$$\sum_{L\notin \mathfrak{C}_M} \exp\{-\tfrac{1}{2}m_0 D(L)\} \leqslant (\alpha+\beta)\, e^{-Mm_0/2}(M_{12})^{\alpha+\beta}. \tag{4.50}$$

Using the estimate (4.49) in (4.48) shows that

$$\|PC^{\varepsilon}P\| \leqslant M_3 \|C\| \sum_{\alpha,\beta=0}^{M} (M_5 M_{12})^{\alpha+\beta} \exp\{-M_{11}(\alpha^{\frac{1}{2}}+\beta^{\frac{1}{2}})\}, \tag{4.51}$$

which converges faster than exponentially as $M\to\infty$. Hence (4.48) is bounded independently of $M = M(\varepsilon)$.

To establish (4.46), we write

$$\|P(C-C^{\varepsilon})P\| \leqslant \sum_{\alpha,\beta=0}^{M} \sum_{L\notin \mathfrak{C}_M} + \sum_{\substack{\alpha>M\\ \beta\geqslant 0}} \sum_L + \sum_{\substack{0\leqslant\alpha\leqslant M\\ \beta>M}} \sum_L \|PC_{\alpha\beta}^L P\|. \tag{4.52}$$

From (4.50) we conclude that the first sum in (4.52) is bounded by const. $\exp(-m_0 M/2)$, which converges to zero exponentially fast as $M\to\infty$. The remaining two terms in (4.52) converge to zero faster than exponentially as $M\to\infty$, as a consequence of the estimate on the tail of (4.51). Hence by choosing $M = M(\varepsilon)$ sufficiently large, we can assure that

$$\|P(C-C^{\varepsilon})P\| \leqslant \varepsilon \|C\|.$$

Lastly, we remark that

$$|\omega_n^{\varepsilon}(C-C^{\varepsilon})| = |\omega_n(P(C-C^{\varepsilon})P)| \leqslant \|P(C-C^{\varepsilon})P\| \leqslant \varepsilon\|C\|, \tag{4.53}$$

to complete the proof.

We next study the convergence of the sequence $\{\omega_n^{\varepsilon}\}$ of approximate states applied to the approximate operators C^{ε}. In other words, we study the convergence of the states ω_n on the bounded operators $PC^{\varepsilon}P$.

As a first reduction, we restrict our attention to the Fock space of a bounded region. Let Z be the interval defined in (4.43). Then if Z_1 is the complement of Z in R^1,

$$R^1 = Z \cup Z_1 \tag{4.54}$$

and by (3.2.14)
$$\mathcal{F} = \mathcal{F}(Z) \otimes_s \mathcal{F}(Z_1) \tag{4.55}$$

With this decomposition we write

$$P = (P_0 \otimes I)(I \otimes P_1). \tag{4.56}$$

We now study the local number operator N_{τ, X_j} defined in (3.2.19).

Here we choose $\mathcal{B} = X_j$ and we restrict τ to be less than $\frac{1}{4}$. Since the Friedrichs extension $(N_{\tau, X_j})_F$ of $N_{\tau, X_j} \restriction \mathcal{D}_0$ commutes with the projection onto $\mathcal{F}(X_i)$ for all i, it follows that the various $(N_{\tau, X_j})_F$ commute. Also $(N_{\tau, X_j})_F$ commutes with each N_{X_i} and with the spectral projections $P_{l,i}$. Thus $(N_{\tau, X_j})_F$ commutes with $P_0 \otimes I$, with $I \otimes P_1$, and with P. We now let N_{τ, X_j} denote the Friedrichs extension of $N_{\tau, X_j} \restriction \mathcal{D}_0$.

As a consequence of the commutativity with P,

$$\omega_n^\varepsilon (N_{\tau, X_j}) = \omega_n (P N_{\tau, X_j} P) = \omega_n (N_{\tau, X_j}^{\frac{1}{2}} P N_{\tau, X_j}^{\frac{1}{2}}) \leqslant \omega_n (N_{\tau, X_j}).$$

We choose $\tau > 0$ sufficiently small so that (3.3.6 a)' is valid. (See Remark 3 following Theorem 4.1.) Hence

$$\omega_n^\varepsilon (N_{\tau, X_j}) \leqslant \text{const.}$$

We now choose M as in (4.42) – (4.44) and define

$$N_{\tau, M} = (\sum_{|j| \leqslant M} N_{\tau, X_j})_F. \tag{4.57}$$

Thus
$$\omega_n^\varepsilon (N_{\tau, M}) \leqslant \text{const.} \tag{4.58}$$

We note that $N_{\tau, M}$ commutes with each $P_0 \otimes I$ and with $I \otimes P_1$. Also $N_{\tau, M}$ leaves $\mathcal{F}(Z)$ invariant, and by Corollary 3.2.2 c,

$$N_{\tau, M} \restriction \mathcal{F}(Z)$$

has a compact resolvent.

We now study the sequence $\{\omega_n^\varepsilon\}$ of functionals restricted to $\mathfrak{B}(Z)$, the algebra of all bounded operators on the Hilbert space $\mathcal{F}(Z)$. Each functional ω_n^ε is normal, since ω_n is normal by assumption in Theorem 4.1, and P is a projection. The normal functional

$$\omega_n^\varepsilon \restriction \mathfrak{B}(Z)$$

has the form
$$\omega_n^\varepsilon (A) = \operatorname{Tr}(\Lambda_n A), \quad A \in \mathfrak{B}(Z), \tag{4.59}$$

where Λ_n is a positive trace class operator in $\mathfrak{B}(Z)$ with

$$\|\omega_n^\varepsilon \restriction \mathfrak{B}(Z)\| = \operatorname{Tr}(\Lambda_n) \leqslant 1. \tag{4.60}$$

(See [2].) The operator $N_{\tau, M}$ in (4.57) can be restricted to $\mathcal{F}(Z)$, and thus by (4.58)

$$\omega_n^\varepsilon(N_{T,M} \restriction \mathcal{J}(Z)) \leqslant M. \tag{4.61}$$

We wish to study the convergence of

$$\omega_n^\varepsilon(C^\varepsilon) = \omega_n(PC^\varepsilon P).$$

We note that $PC^\varepsilon P = P_0 C^\varepsilon P_0 \otimes P_1$ where we regard C^ε on the right side as an operator on $\mathcal{J}(Z)$. Thus

$$\omega_n^\varepsilon(C^\varepsilon) = \operatorname{Tr}(\Lambda_n P_0 C^\varepsilon P_0).$$

The desired convergence of the $\omega_n^\varepsilon(C^\varepsilon)$ now follows from

LEMMA 4.7. *Let $\mathfrak{B}$ be the algebra of all bounded operators on some Hilbert space $\mathcal{H}$, and let N be a positive (unbounded) operator with N^{-1} a compact element of $\mathfrak{B}$. Then the set*

$$\{\Lambda: \Lambda \in \mathfrak{B}, 0 \leqslant \Lambda \leqslant I, \operatorname{Tr}(\Lambda N) \leqslant 1\} = \mathfrak{J}_N$$

is compact in the trace norm.

Proof. Choose an orthonormal basis $\{e_i\}$ for $\mathcal{H}$ consisting of eigenvectors of N and let b_i be the corresponding eigenvalues. For Λ in $\mathfrak{J}_N$, we have

$$\operatorname{Tr}(\Lambda N) = \operatorname{Tr}(N^{\frac{1}{2}} \Lambda N^{\frac{1}{2}}) = \sum_i \lambda_{ii} b_i$$

if (λ_{ij}) is the matrix representing the operator Λ. If $P = \Lambda^{\frac{1}{2}}$ and if ϱ_{ij} is the corresponding matrix, then

$$\operatorname{Tr}(\Lambda N) = \sum_{i<j} (b_i + b_j) |\varrho_{ij}|^2 + \sum_i b_i |\varrho_{ii}|^2.$$

Let Λ_n be a sequence in $\mathfrak{J}_N$ and let $P_n = \Lambda_n^{\frac{1}{2}}$. Since $b_i \to \infty$ as $i \to \infty$, we see that a subsequence P_n of the P's converge to a limit P in the Hilbert–Schmidt norm. For a trace class operator A, the trace norm $\|A\|_1$ is given by

$$\|A\|_1 = \sup_U \operatorname{Tr}(UA),$$

where the supremum runs over all unitary operators. Thus if $\Lambda = P^2$,

$$\|\Lambda_n - \Lambda\|_1 = \sup_U \operatorname{Tr}(U(\Lambda_n - \Lambda)) = \sup_U \{\operatorname{Tr}(U(P_n - P)P_n) + \operatorname{Tr}(UP(P_n - P))\}$$

$$\leqslant \sup_U \|U(P_n - P)\|_2 \|P_n\|_2 + \sup_U \|UP\|_2 \|P_n - P\|_2 = (\|P_n\|_2 + \|P\|_2) \|P_n - P\|_2 \to 0,$$

with
$$\|P\|_2^2 = \sum_{i,j} |\varrho_{ij}|^2$$

denoting the Hilbert–Schmidt norm, and n belonging to the subsequence.

We apply this lemma to the case $\mathcal{J}=\mathcal{J}(Z)$ and $\mathfrak{B}=\mathfrak{B}(Z)$. We use the operator $N_{\tau,M}$ of (4.57) to give an

$$N=(N_{\tau,M}+I)$$

for Lemma 4.7. The states $\omega_n^\varepsilon \restriction \mathfrak{B}(Z)$ satisfy the hypotheses of the lemma by (4.59)–(4.61). By (4.58) and Corollary 3.3.2 (c) the operator N has the desired properties. Thus there is a norm convergent subsequence ω_{n_j} satisfying

$$|(\omega_{n_i}-\omega_{n_j})(PC^\varepsilon P)| \leqslant o(1)\,\|PC^\varepsilon P\| \leqslant o(1)\,\|C\|$$

by (4.45). For this subsequence

$$|(\omega_{n_i}^\varepsilon-\omega_{n_j}^\varepsilon)(C^\varepsilon)| \leqslant o(1)\,\|C\|, \tag{4.62}$$

which is the announced bound (4.4).

In order to complete the proof of Theorem 4.1, we write for $C\in\mathfrak{A}_0(\boldsymbol{B})$.

$$|(\omega_{n_i}-\omega_{n_j})(C)| \leqslant |(\omega_{n_i}-\omega_{n_i}^\varepsilon)(C)| + |\omega_{n_i}^\varepsilon(C-C^\varepsilon)| + |(\omega_{n_i}^\varepsilon-\omega_{n_j}^\varepsilon)(C^\varepsilon)| + |\omega_{n_j}^\varepsilon(C^\varepsilon-C)|$$
$$+|(\omega_{n_j}^\varepsilon-\omega_{n_j})(C)| \leqslant 4\,\varepsilon\,\|C\|+o(1)\,\|C\|. \tag{4.63}$$

Here we have chosen n_i, n_j to belong to the convergent subsequence in (4.62), and we used (4.14), (4.46) and (4.62) to dominate (4.63). Since

$$\|(\omega_{n_i}-\omega_{n_j})\restriction \mathfrak{A}_0(\boldsymbol{B})\| = \|(\omega_{n_i}-\omega_{n_j})\restriction \mathfrak{A}(\boldsymbol{B})\|,$$

we conclude that the subsequence in (4.63) is norm convergent on $\mathfrak{A}(\boldsymbol{B})$.

It is known and easy to prove that the norm limit of a sequence of normal states is normal. Thus any norm limit ω of the ω_n is normal on $\mathfrak{A}(\boldsymbol{B})$, and Theorem 4.1 is established.

We now return to prove Lemma 4.3. We analyze the expansion of $C\in\mathfrak{A}_0(\boldsymbol{B})$ in terms of the local, time zero fields. We start by proving a useful lemma.

Let α, β, δ, ε, μ, ν be nonnegative integers satisfying $\alpha+\beta=\delta+\varepsilon$, $\mu\leqslant\delta$, $\nu\leqslant\varepsilon$ and $\alpha-\delta=\nu-\mu$.

LEMMA 4.8. *Let $c_{\delta\varepsilon}$ be a bounded operator from $L_2(R^\varepsilon)$ to $L_2(R^\delta)$ with bound $\|c_{\delta\varepsilon}\|$. Let k_C and k_A be Hilbert–Schmidt operators on $L_2(R^\mu)$ and on $L_2(R^\nu)$ respectively, with Hilbert–Schmidt norms*

$$\|k_C\|_2 \text{ and } \|k_A\|_2.$$

Then

$$\int k_A(x_1, \ldots, x_\nu; w_1, \ldots, w_\nu)\, c_{\delta\varepsilon}(x_{\nu+1}, \ldots, x_\alpha, z_1, \ldots, z_\mu; w_1, \ldots, w_\nu, y_{\mu+1}, \ldots, y_\beta) \times k_C(z_1, \ldots, z_\mu; y_1, \ldots, y_\mu)\, dw\, dz$$

is the kernel of a bounded operator $c_{\alpha\beta}$ *from* $L_2(R^\beta)$ *to* $L_2(R^\alpha)$ *with norm* $\|c_{\alpha\beta}\|$ *satisfying*

$$\|c_{\alpha\beta}\| \leqslant \|k_A\|_2 \|c_{\delta\varepsilon}\| \|k_C\|_2.$$

Proof. We introduce the variables

$$x_a = (x_1, \ldots, x_\nu), \quad x_b = (x_{\nu+1}, \ldots, x_\alpha), \quad x = (x_a, x_b),$$
$$y_a = (y_1, \ldots, y_\mu), \quad y_b = (y_{\mu+1}, \ldots, y_\beta), \quad y = (y_a, y_b)$$
$$w = (w_1, \ldots, w_\nu), \quad z = (z_1, \ldots, z_\mu).$$

Then for $f_\alpha \in \mathfrak{F}_\alpha$, $g_\beta \in \mathfrak{F}_\beta$,

$$|(f_\alpha, c_{\alpha\beta} g_\beta)| = \left| \int f_\alpha(x)\, k_A(x_a; w)\, c_{\delta\varepsilon}(x_b, z; w, y_b)\, k_C(z; y_a)\, g_\beta(y)\, dx\, dy\, dw\, dz \right|.$$

By the definition of $\|c_{\delta\varepsilon}\|$,

$$|(f_\alpha, c_{\alpha\beta} g_\beta)| \leqslant \|c_{\delta\varepsilon}\| \int \|f_\alpha(x_a, \cdot\,)\|_2 \|k_C(\,\cdot\,; y_a)\|_2 \|k_A(x_a; \cdot\,)\|_2 \|g_\beta(y_a, \cdot\,)\|_2\, dx_a\, dy_a,$$

and by the Schwartz inequality

$$|(f_\alpha, c_{\alpha\beta} g_\beta)| \leqslant \|c_{\alpha\beta}\| \|k_C\|_2 \|k_A\|_2 \|f_\alpha\| \|g_\beta\|,$$

to complete the proof.

We now expand $C \in \mathfrak{A}_0(\mathcal{B})$ in terms of ϕ and π fields. Let

$$C_\gamma = \sum_{\alpha+\beta=\gamma} C_{\alpha\beta}.$$

Then
$$C_\gamma = \sum_{\mu+\nu=\gamma} B_{\mu\nu},$$

where
$$B_{\mu\nu} = \int b_{\mu\nu}(y){:}\,\phi(y_1) \ldots \phi(y_\mu)\, \pi(y_{\mu+1}) \ldots \pi(y_{\mu+\nu}){:}\, dy.$$

The locality of $C \in \mathfrak{A}_0(\mathcal{B})$ means that

$$\operatorname{supp} b_{\mu\nu}(y_1, \ldots, y_{\mu+\nu}) \in \mathcal{B} \times \mathcal{B} \times \ldots \times \mathcal{B}.$$

The kernels $b_{\mu\nu}(y)$ are symmetric in the μ variables $y_1, \ldots, y_\mu$ corresponding to the ϕ's, and also in the ν variables $y_{\mu+1}, \ldots, y_{\mu+\nu}$ corresponding to the π's. For fixed $\mu+\nu=\gamma$, there are exactly $\gamma+1$ different $B_{\mu\nu}$'s and for fixed $\alpha+\beta=\gamma$, there are exactly $\gamma+1$ different $C_{\alpha\beta}$'s. We now specify the relation between them. First we inspect the simple case $\alpha+\beta=\gamma=\mu+\nu=1$. From (3.9) we conclude that

$$c_{10}(x)=\frac{1}{\sqrt{2}}\mu_x^{-\frac{1}{2}}b_{10}(x)+\frac{i}{\sqrt{2}}\mu_x^{\frac{1}{2}}b_{01}(x) \tag{4.64}$$

and

$$c_{01}(x)=\frac{1}{\sqrt{2}}\mu_x^{-\frac{1}{2}}b_{10}(x)-\frac{i}{\sqrt{2}}\mu_x^{\frac{1}{2}}b_{01}(x). \tag{4.65}$$

We also have the inverse relation

$$b_{10}(x)=\frac{1}{\sqrt{2}}\mu_x^{\frac{1}{2}}c_{10}(x)+\frac{1}{\sqrt{2}}\mu_x^{\frac{1}{2}}c_{01}(x) \tag{4.66}$$

and

$$b_{01}(x)=\frac{-i}{\sqrt{2}}\mu_x^{-\frac{1}{2}}c_{10}(x)+\frac{i}{\sqrt{2}}\mu_x^{-\frac{1}{2}}c_{01}(x). \tag{4.67}$$

The transformations (4.64)–(4.67) by themselves are valid for a wide variety of $b_{01}(x)$, $b_{10}(x)$; they are defined, for instance on tempered distributions. These relations do not require that the $b_{01}(x)$ and $b_{10}(x)$ be local. We now make the locality explicit by means of a C_0^∞ function $\zeta(x)$ which has its support in X_0 and which equals one on a neighborhood of $\mathcal{B}$. We rewrite (4.66) as

$$b_{10}(x)=\frac{1}{\sqrt{2}}\zeta(x)\mu_x^{\frac{1}{2}}c_{10}(x)+\frac{1}{\sqrt{2}}\zeta(x)\mu_x^{\frac{1}{2}}c_{01}(x), \tag{4.68}$$

and similarly for $b_{01}(x)$. We note that the operators μ_x^τ, which were introduced in Section 3.1 have integral kernels $k_\tau(x)$ satisfying (3.1.5)–(3.1.6).

$$(\mu_x^\tau f)(x)=\int k_\tau(x-z)\,f(z)\,dz.$$

By substituting (4.68) into (4.64) – (4.65), we obtain the equations

$$c_{10}(x)=(K_+c_{10})(x)+(K_-c_{01})(x) \tag{4.69}$$

and

$$c_{01}(x)=(K_-c_{10})(x)+(K_+c_{01})(x), \tag{4.70}$$

where we define

$$K_\pm=\tfrac{1}{2}\mu_x^{-\frac{1}{2}}\zeta\mu_x^{\frac{1}{2}}\pm\tfrac{1}{2}\mu_x^{\frac{1}{2}}\zeta\mu_x^{-\frac{1}{2}}. \tag{4.71}$$

Let $k_\pm(x,z)$ be the (tempered distribution) kernel of $K_\pm$. For $x\notin X_0=[0,1)$, the property (3.1.6) of k_τ implies that $k_\pm(x,z)$ is a C^∞ function of x and z and that for a constant M_{14},

$$|k_\pm(x,z)| \leqslant M_{14} \exp(-m_0|x| - m_0|z|). \tag{4.72}$$

We note that the locality requirement

$$\operatorname{supp} b_{10}(x) \subset \mathcal{B}, \qquad \operatorname{supp} b_{01}(x) \subset \mathcal{B},$$

imposes implicit restrictions on—and relations between—$c_{01}(x)$ and $c_{10}(x)$; for instance, if either one is nonzero, they must both be nonzero. Thus we cannot expect, and we do not find, that the identity transformation (4.69)–(4.70) yields the identity on all functions as in (4.64)–(4.67). In fact for the kernels of (4.71), we have

$$k_+(x,z) \neq \delta(x-z), \quad \text{and} \quad k_-(x,z) \neq 0.$$

We now define localized kernels and estimate their norms. Let

$$k^i_\pm(x,z) = \begin{cases} k_\pm(x,z), & x \in X_i \\ 0 \quad , & \text{otherwise}, \end{cases} \tag{4.73}$$

and let $K^i_\pm$ be the operator with kernel $k^i_\pm$.

LEMMA. 4.9. *For some constant M_{15}, the operator norm of K^i_+ is bounded by*

$$\|K^i_+\| \leqslant M_{15} e^{-m_0|i|}, \tag{4.74}$$

and the Hilbert–Schmidt norm of k^i_- is bounded by

$$\|k^i_-\|_2 \leqslant M_{15} e^{-m_0|i|}. \tag{4.75}$$

Proof. If $i \neq 0$, the bound (4.72) immediately yields

$$\|K^i_\pm\| \leqslant \|k^i_\pm\|_2 \leqslant M_{15} \exp(-m_0|i|),$$

so we need only consider the case $i = 0$. Since

$$\mu_x^{-\frac{1}{2}} \zeta \mu_x^{\frac{1}{2}} = \zeta + \mu_x^{-\frac{1}{2}} [\zeta, \mu_x^{\frac{1}{2}}],$$

the lemma follows from (4.71) and an estimate on the Hilbert–Schmidt norm of the operator

$$A = \mu_x^{-\frac{1}{2}} [\zeta, \mu_x^{\frac{1}{2}}].$$

Let $a(x,z)$ be the kernel of A. The Fourier transform of $a(x,z)$ is

$$\tilde{a}(p,q) = \mu(p)^{-\frac{1}{2}} \tilde{\zeta}(p-q) \{\mu(q)^{\frac{1}{2}} - \mu(p)^{\frac{1}{2}}\},$$

where $\tilde{\zeta}$ is the Fourier transform of ζ. Since

$$|\mu(p)^{\frac{1}{2}} - \mu(q)^{\frac{1}{2}}| = \frac{|\mu(p) - \mu(q)|}{\mu(p)^{\frac{1}{2}} + \mu(q)^{\frac{1}{2}}} \leqslant \frac{\mu(p-q)}{\mu(p)^{\frac{1}{2}} + \mu(q)^{\frac{1}{2}}}$$

we have
$$|\tilde{a}(p,q)| \leqslant \frac{\mu(p-q)\,|\tilde{\zeta}(p-q)|}{\mu(p)^{\frac{1}{2}}\{\mu(p)^{\frac{1}{2}} + \mu(q)^{\frac{1}{2}}\}} \in L_2.$$

Thus
$$\|a\|_2 = \|\tilde{a}\|_2 < \infty,$$

and the proof is complete.

Proof of Lemma 4.3. Let $L = \{L_1\}$ be a localization index in the case $\gamma = 1$, which localizes the single particle in X_{L_1}. Then

$$c_{10}^L(x) = (K_+^{L_1} c_{10})(x) + (K_-^{L_1} c_{01})(x) \tag{4.76}$$

and
$$c_{01}^L(x) = (K_-^{L_1} c_{10})(x) + (K_+^{L_1} c_{01})(x). \tag{4.77}$$

The bound of Lemma 4.3 for c_{10}^L follows from

$$\|c_{10}^L\| \leqslant \|K_+^{L_1} c_{10}\| + \|K_-^{L_1} c_{01}\| \leqslant \|K_+^{L_1}\|\,\|c_{10}\| + \|K_-^{L_1} c_{01}\| \tag{4.78}$$

$$\leqslant \|K_+^{L_1}\|\,\|c_{10}\| + \|k_-^{L_1}\|_2\,\|c_{01}\|$$

by Lemma 4.8

$$\leqslant M_{15}(\|c_{10}\| + \|c_{01}\|) \exp(-m_0|L_1|)$$

by Lemma 4.9

$$\leqslant 2M_{15} \exp(-m_0|L_1|)\,\|C\| = 2M_{15}e^{-m_0 D(L)}\|C\|$$

by Lemma 4.2. A similar result holds for c_{01}^L.

We now generalize this method to deal with any $c_{\alpha\beta}^L$, and we need an expansion similar to (4.76) − (4.77). We first obtain such an expansion for operators C_γ which have the form

$$C_\gamma = \sum_{\alpha+\beta=\gamma} C_{\alpha\beta} = :C_1^\gamma:, \tag{4.79}$$

where C_1 is a degree one expression

$$C_1 = A^*(c_{10}) + A(c_{01}) = \phi(b_{10}) + \pi(b_{01}), \tag{4.80}$$

and supp $b_{10} \subset \mathcal{B}$, supp $b_{01} \subset \mathcal{B}$. For our special case

$$C_\gamma = \sum_{\alpha+\beta=\gamma} \int A^*(x_1)\dots A^*(x_\alpha) A(x_{\alpha+1})\dots A(x_{\alpha+\beta}) \binom{\alpha+\beta}{\alpha}$$

$$\times\, c_{10}(x_1)\dots c_{10}(x_\alpha)\, c_{01}(x_{\alpha+1})\dots c_{01}(x_{\alpha+\beta})\, dx \tag{4.81}$$

or
$$C_\gamma = \sum_{\mu+\nu=\gamma} \int :\phi(y_1)\dots\phi(y_\mu)\,\pi(y_{\mu+1})\dots\pi(y_{\mu+\nu}): \binom{\mu+\nu}{\nu} \times b_{10}(y_1)\dots b_{10}(y_\mu)\,b_{01}(y_{\mu+1})\dots b_{01}(y_{\mu+\nu})\,dy. \tag{4.82}$$

We now apply (4.69) to the creation variables, expanding $c_{10}(x_1), \dots, c_{10}(x_\alpha)$. We apply (4.70) to the annihilation variables, expanding $c_{01}(x_{\alpha+1}) \dots c_{01}(x_{\alpha+\beta})$. The kernels $c_{\alpha\beta}(x)$ for (4.79)
$$c_{\alpha\beta}(x) = \binom{\alpha+\beta}{\alpha} \prod_{j=1}^{\alpha} c_{10}(x_j) \prod_{j=\alpha+1}^{\beta} c_{01}(x_j) \tag{4.83}$$
can be written
$$c_{\alpha\beta}(x) = \binom{\alpha+\beta}{\alpha} \prod_{j=1}^{\alpha} \{(K_+c_{10})(x_j) + (K_-c_{01})(x_j)\} \prod_{j=\alpha+1}^{\alpha+\beta} \{(K_+c_{01})(x_j) + (K_-c_{10})(x_j)\}. \tag{4.84}$$

In a more compact form
$$c_{\alpha\beta}(x) = \sum_{\delta+\varepsilon=\alpha+\beta} \int l_{\alpha\beta,\,\delta\varepsilon}(x, z)\, c_{\delta\varepsilon}(z)\, dz, \tag{4.85}$$

where the kernel $l_{\alpha\beta,\,\delta\varepsilon}(x, z)$ is a sum of tensor products of the kernels $k_\pm(x, z)$ times some numerical factors. We can write
$$l_{\alpha\beta,\,\delta\varepsilon}(x, z) = \text{Sym.} \sum_{\mu=0}^{\delta} \binom{\delta}{\mu} \binom{\varepsilon}{\alpha-\mu} \prod_{j=1}^{\mu} k_+(x_j, z_j) \times \prod_{j=\mu+1}^{\alpha} k_-(x_j, z_{j+\delta-\mu}) \prod_{j=\alpha+1}^{\delta+\alpha-\mu} k_-(x_j, z_{j+\mu-\alpha}) \prod_{j=\delta+\alpha-\mu+1}^{\delta+\varepsilon} k_+(x_j, z_j) \tag{4.86}$$

where the symmetrization operation symmetrizes the kernel in the α creation variables $x_1, \dots, x_\alpha$ and in the β annihilation variables $x_{\alpha+1}, \dots, x_{\alpha+\beta}$.

We note some properties of the monomials contributing to the sum (4.86). In each kernel $k_+(x_j, z_j)$ that occurs, either z_j is a creation variable for $c_{\delta\varepsilon}$ and x_j is a creation variable for $c_{\alpha\beta}$, or else z_j is an annihilation variable for $c_{\delta\varepsilon}$ and x_j is an annihilation variable for $c_{\alpha\beta}$. In other words, the K_+ operators in (4.86) do not connect creation variables to annihilation variables. On the other hand, the kernels $k_-(x_j, z_{j'})$ connect an annihilation (creation) variable in $c_{\delta\varepsilon}$ to a creation (annihilation) variable in $c_{\alpha\beta}$. In majorizing (4.86) we will later use the operator norm to dominate k_+ kernels, and Lemma 4.8 to deal with the k_- kernels.

The next step in our argument is to prove that the transformation (4.85)–(4.86) which we derived for a class of $c_{\alpha\beta}(x)$ defined in (4.83), is correct for any $c_{\alpha\beta}(x)$ that might occur in the expansion of an operator $C \in \mathfrak{A}_0(\mathcal{B})$. We note that by a generalized polarization identity, (4.85)–(4.86) is valid for kernels arising from operators

$$C_\gamma = :C_1^{(1)} C_1^{(2)} \dots C_1^{(\gamma)}:, \tag{4.87}$$

where
$$C_1^{(j)} = A^*(c_{10}^{(j)}) + A(c_{01}^{(j)}) = \phi(b_{10}^{(j)}) + \pi(b_{01}^{(j)}), \quad j = 1, 2, \dots, \gamma, \tag{4.88}$$

and supp $b_{10}^{(j)} \subset \mathcal{B}$, supp $b_{01}^{(j)} \subset \mathcal{B}$.

Here we use the identity (for real $a_1, a_2 \dots, a_n$),

$$2^{n-1} n!\, a_1 a_2 \dots a_n = \sum_{\varepsilon_j = \pm 1} \varepsilon_2 \dots \varepsilon_n (a_1 + \varepsilon_2 a_2 + \dots + \varepsilon_n a_n)^n. \tag{4.89}$$

Kernels that are finite sums of kernels of the operators (4.87) are the only kernels $c_{\alpha\beta}(x)$ that occur in the expansion of operators $C \in \mathfrak{A}_0(\mathcal{B})$. Hence the identity transformation (4.85)–(4.86) holds for all our kernels.

We now return to the localized kernels $c_{\alpha\beta}(x)$ and the proof of Lemma 4.3. Using (4.85) we can write

$$c_{\alpha\beta}^L(x) = \sum_{\delta+\varepsilon=\alpha+\beta} \int l_{\alpha\beta,\delta\varepsilon}^L(x, z)\, c_{\delta\varepsilon}(z)\, dz, \tag{4.90}$$

where the localized kernel $l_{\alpha\beta,\delta\varepsilon}^L(x, z)$ is defined by

$$l_{\alpha\beta,\delta\varepsilon}^L(x, z) = \operatorname{Sym} \sum_{\mu=0}^{\delta} \binom{\delta}{\mu} \binom{\varepsilon}{\alpha-\mu} \prod_{j=1}^{\mu} k_+^{L_j}(x_j, z_j) \prod_{j=\mu+1}^{\delta+\alpha-\mu} k_-^{L_j}(x_j, z_{j'}) \prod_{j=\delta+\alpha-\mu+1}^{\delta+\varepsilon} k_+^{L_j}(x_j, z_j) \tag{4.91}$$

and $L = \{L_1, \dots, L_{\alpha+\beta}\}$ is the localization multi-index. Each kernel (4.91), before symmetrization, is the sum of $\delta+1$ terms with a numerical coefficient dominated by $2^{\delta+\varepsilon} = 2^{\alpha+\beta}$. The kernels from one term, obtained by fixing μ, are a tensor product of k_+ kernels and of k_- kernels. The k_- factor

$$\prod_{j=\mu+1}^{\delta+\alpha-\mu} k_-^{L_j}(x_j, z_{j'}) = \prod_{j=\mu+1}^{\alpha} k_-^{L_j}(x_j, z_{j+\delta-\mu}) \prod_{j=\alpha+1}^{\delta+\alpha-\mu} k_-^{L_j}(x_j, z_{j+\mu-\alpha})$$

is the tensor product of Hilbert–Schmidt kernels and has the form of a tensor product of kernels $k_C k_A$ of Lemma 4.8. The first product in (4.91) of k_+ kernels is a bounded operator O_α on $\mathcal{F}_\alpha$ and the second product in (4.91) of k_+ kernels is a bounded operator O_β on $\mathcal{F}_\beta$. Thus applying Lemma 4.8 to the $k_C k_A$ kernels and using the operator norm on the k_+ kernels, we find that each term $l_{\alpha\beta,\delta\varepsilon}^L(x, z)_\mu$ in (4.91) with fixed μ is a kernel such that

$$\int l_{\alpha\beta,\delta\varepsilon}^L(x, z)_\mu\, c_{\delta\varepsilon}(z)\, dz$$

is the kernel of an operator from $L_2(R^\beta)$ to $L_2(R^\alpha)$ with a norm dominated by

$$2^{\alpha+\beta} \|O_\alpha\|\, \|k_C\|_2 \|k_A\|_2 \|O_\beta\|\, \|c_{\delta\varepsilon}\| = 2^{\alpha+\beta} \prod_{j=1}^{\alpha_1} \|K_+^{L_j}\| \prod_{j=\alpha_1+1}^{\delta+\alpha+\alpha_1} \|k_-^{L_j}\|_2 \prod_{j=\delta-\alpha-\alpha_1+1}^{\delta+\varepsilon} \|K_+^{L_j}\|\, \|c_{\delta\varepsilon}\|.$$

By Lemma 4.9, this is less than

$$2^{\alpha+\beta} M_{15}^{\alpha+\beta} \prod_{j=1}^{\alpha+\beta} e^{-m_0 |L_j|} \|c_{\delta\varepsilon}\| = (2\, M_{15})^{\alpha+\beta} e^{-m_0 D(L)} \|c_{\delta\varepsilon}\|,$$

and by Lemma 4.2, $\qquad \leqslant (4\, M_{15})^{\alpha+\beta} e^{-m_0 D(L)} \|C\|.$

Summing over μ in (4.91) and over $\delta+\varepsilon=\alpha+\beta$ in (4.90) yields fewer than

$$(\delta+1)\,(\alpha+\beta+1) \leqslant (\alpha+\beta+1)^2 \leqslant e^{\alpha+\beta}$$

such terms. Since symmetrization does not increase our bound,

$$\|c_{\alpha\beta}^{L}\| \leqslant (4\, e\, M_{15})^{\alpha+\beta} e^{-m_0 D(L)} \|C\|,$$

which completes the proof of Lemma 4.3.

5. The vacuum self energy per unit volume is finite

The lower bound E_g on the operator $\hat{H}(g) = H_0 + H_{I,g}$ is called the vacuum energy, because

$$\hat{H}(g)\,\Omega_g = E_g \Omega_g, \tag{5.1}$$

and

$$E_g = (\Omega_g, \hat{H}(g)\,\Omega_g)$$

is the value of the energy operator $\hat{H}(g)$ in its vacuum state. This actually represents the *shift* in vacuum energy between the problem with and without interaction,

$$E_g = E_g - E_0,$$

where $E_0 = 0$ is the lower bound of H_0. The scale of $\hat{H}(g)$ is arbitrarily specified so that

$$(\Omega_0, \hat{H}(g)\,\Omega_0) = 0. \tag{5.2}$$

It is customary and convenient to add a constant to the Hamiltonian $\hat{H}(g)$ so that its vacuum energy is zero

$$H(g) = \hat{H}(g) - E_g, \tag{5.3}$$

and

$$H(g)\,\Omega_g = 0. \tag{5.4}$$

This shift is one of the standard renormalizations of quantum field theory, and we shall call $H(g)$ the renormalized Hamiltonian. We omit the renormalization to the mass m_0 and to the coupling constant λ since perturbation theory indicates that they are finite for our

$(\varphi^4)_2$ model, even in the limit $g(x)\to 1$. We note that it is the renormalized Hamiltonians $H(g)$, and not the unrenormalized Hamiltonians $\hat{H}(g)$ that have a limit as $g(x)\to 1$. In Section 2 we proved that as $g(x)\to 1$,

$$e^{-iH(g)t}\to e^{-iHt}$$

in a certain weak sense, and $0\leqslant H$.

In this section, we bound the rate at which E_g can diverge. According to perturbation theory, E_g is proportional to the volume of space in which the particles interact. However, the perturbation expansion for E_g can be shown by methods such as [18] not to converge and at best to be asymptotic. Thus we cannot get a rigorous bound on E_g from perturbation theory. Instead, we use a modification of Nelson's method [22]. It is this method which produced the estimates (cf. II, (2.1.16–18)) stating that the vacuum self energy for a fixed volume cutoff is finite [22, 8, 4].

THEOREM 5.1. *Let* $0\leqslant g(x)\leqslant 1$ *and for some constant* M_0,

$$\left|\frac{dg(x)}{dx}\right|\leqslant M_0.$$

Then

$$-M\leqslant E_g\leqslant 0, \tag{5.5}$$

where M *is a positive constant proportional to the volume of the set* $(\operatorname{supp} g)_1$, *that is the set of points within distance one of* $(\operatorname{supp} g)$.

Remarks 1. This theorem states that the average vacuum energy density, namely the vacuum self energy per unit volume, is finite for the theory with $g(x)=1$. Thus it shows that perturbation theory, though inapplicable, actually predicts the correct answer.

2. Furthermore, the theorem assures us that the expected number of (bare) particles per unit volume and the expected free energy per unit volume are both finite in the vacuum Ω. We define these expectation values as limits as $g(x)\to 1$ of the expectation value in Ω_g. Since

$$(\Omega_g, N\Omega_g)\leqslant\frac{1}{m_0}(\Omega_g, H_0\Omega_g), \tag{5.6}$$

it is sufficient to bound the free energy per unit volume. Since $H_0=2\hat{H}(g)-\hat{H}(2g)$,

$$(\Omega_g, H_0\Omega_g)\leqslant(\Omega_g,(2\hat{H}(g)-E_{2g})\,\Omega_g)=2E_g-E_{2g}. \tag{5.7}$$

In particular, if we choose g to be $g_n(x)$ defined in (2.3), then (5.5) says that for M independent of n,

$$-nM\leqslant E_{2g_n}. \tag{5.8}$$

Therefore, by (5.6–7)
$$(\Omega_{g_n}, N\Omega_{g_n}) \leqslant \text{const. } n, \tag{5.9}$$

$$(\Omega_{g_n}, H_0\Omega_{g_n}) \leqslant \text{const. } n, \tag{5.10}$$

$$\omega_n(N) \leqslant \text{const. } n, \tag{5.11}$$

and
$$\omega_n(H_0) \leqslant \text{const. } n. \tag{5.12}$$

3. A similar result holds for the Hamiltonians H_V defined in II, (2.1.7). Namely, there is a positive constant M independent of V, such that as $V \to \infty$

$$-MV \leqslant H_V. \tag{5.13}$$

Our proof yields (5.13) as well as (5.5), although a somewhat easier proof may be based on a modification of Federbush's proof for V fixed [4, 19].

4. The theorem also holds for any interaction Hamiltonian in two dimensions of the form

$$H_{I,g} = \int :P(\varphi(x)):g(x)\,dx,$$

where P is a polynomial of even degree whose leading coefficient is positive.

5. The same proof also shows that for $g(x)$ satisfying the hypotheses of the theorem, and for $\varepsilon > 0$, there exists a constant M proportional to $(\text{supp } g)_1$ such that

$$0 \leqslant \varepsilon N + H_{I,g} + M.$$

Proof of Theorem 5.1. We use a partition of unity $\zeta_i(x)$ constructed from a positive C^∞ function $\zeta(x)$ with support in the interval $|x| < 1$, such that

$$\zeta_i(x) = \zeta(x - i), \tag{5.14}$$

and
$$\sum_i \zeta_i(x) = 1. \tag{5.15}$$

Thus we decompose g and $H_{I,g}$ into a sum of local parts

$$g_i = \zeta_i g \tag{5.16}$$

$$H_{I,g} = \sum_i H_{I,g_i} \tag{5.17}$$

Since we use the Feynman Kac formula to bound $\hat{H}(g)$, we wish to approximate $\hat{H}(g)$ by a Hamiltonian with a finite number of modes. Thus we study the Hamiltonian

17 – 702903 *Acta mathematica*. 125. Imprimé le 26 Octobre 1970.

$$H(g)_{K,V} = H_{0,K,V} + H_{I,g,K,V} \tag{5.18}$$

$$H_{0,K,V} = \int_{-K}^{K} a^*(k)\, a(k)\, \mu(k_V)\, dk \tag{5.19}$$

$$H_{I,g,K,V} = \lambda \int :\varphi_{K,V}(x)^4 :\, g(x)\, dx \tag{5.20}$$

and

$$\varphi_{K,V}(x) = \left(\frac{1}{2V}\right)^{\frac{1}{2}} \sum_{k\in\Gamma_V} e^{-ikx}\,(a_V^*(k) + a_V(-k))\, \mu(k)^{-\frac{1}{2}}\, \xi(k/K). \tag{5.21}$$

See II, section 2.1 for the notation used above. Here $\xi(x) \geqslant 0$ is a fixed C_0^∞ function, equal to one on a neighborhood of zero, and vanishing for $|x| > 1$. As in II, section 2.2, we find that

$$\hat{H}(g)_{K,V} \,|\, \mathfrak{F}_{K,V}^{\perp} \geqslant E_{K,V}, \tag{5.22}$$

where $E_{K,V}$ is the lower bound of $\hat{H}(g)_{K,V}$ on the full Fock space $\mathfrak{F}$. Therefore, we need only find the lower bound of $\hat{H}(g)_{K,V} \,|\, \mathfrak{F}_{K,V}$, which we now study. For simplicity of notation, we sometimes suppress the V cutoffs and define

$$H_I(K, j) = H_{I,g_j,K,V}. \tag{5.23}$$

Thus it is sufficient to obtain a lower bound for

$$\{H_{0,K,V} + \sum_j H_I(K, j)\} \,|\, \mathfrak{F}_{K,V} \tag{5.24}$$

which is independent of K and V, and which is at most proportional to the number of nonzero terms $H_I(K, j)$. On $\mathfrak{F}_{K,V}$

$$H_{0,K,V} = \sum_{\substack{k\in\Gamma_V \\ |k|\leqslant K}} a_V^*(k)\, a_V(k)\, \mu(k) \tag{5.25}$$

We note that $H_I(K, j)$ is the sum of five monomials in creation and annihilation operators, each of which has the kernel proportional to

$$b_{K,j}(k_1, \ldots, k_4) = \lambda(2V)^{-2} (g\zeta_j)^{\sim} \left(\sum_{l=1}^{4} k_l\right) \prod_{l=1}^{4} [\mu(k_l)^{-\frac{1}{2}}\, \xi(k_l/K)], \tag{5.26}$$

corresponding to an expansion in creation and annihilation operators for volume V,

$$H_I(K, j) = \sum_{k_1,\ldots,k_4\in\Gamma_V} \sum_{\alpha=0}^{4} \binom{4}{\alpha} b_{K,j}(k_1, \ldots, k_4)\, a_V^*(k_1) \ldots a_V^*(k_\alpha)\, a_V(-k_{\alpha+1}) \ldots a_V(-k_4). \tag{5.27}$$

We decompose the single particle Fock space $\mathfrak{F}_{V,1}$ of $\mathfrak{F}_V$ into an orthogonal direct sum

$$\mathfrak{F}_{V,1} = \bigoplus_i \mathfrak{F}_{V,1,i}, \tag{5.28}$$

where $\mathfrak{F}_{V,1,i}$ consists of the Fourier series of functions vanishing off the interval

$$[i-\tfrac{1}{2}, i+\tfrac{1}{2}] \cap [-V/2, V/2]. \tag{5.29}$$

Furthermore, we have the Parseval equality

$$\int_{-V/2}^{V/2} f(x)\, h(x)\, dx = \sum_{k\in\Gamma_V} \tilde{f}(k)\, \tilde{h}(-k) = \sum_{k,k'\in\Gamma_V} \tilde{f}(k)\, \tilde{h}(k')\, \delta(k; -k'),$$

where

$$\delta(k;k') = \begin{cases} 1, & k=k' \\ 0, & k\neq k'. \end{cases}$$

From this, we see that

$$\begin{aligned} \sum_{k\in\Gamma_V} \tilde{f}(k)\, \tilde{h}(-k) &= \sum_{i,j} \int_{-V/2}^{V/2} f_i(x)\, h_j(x)\, dx = \sum_i \int_{-V/2}^{V/2} f_i(x)\, h_i(x)\, dx \\ &= \sum_i \sum_{k\in\Gamma_V} (f_i)^{\sim}(k)\, (h_i)^{\sim}(-k). \end{aligned} \tag{5.30}$$

Thus (5.30) provides a decomposition into localized parts,

$$f_i \in \mathfrak{F}_{V,1,i}.$$

Using this decomposition, we can decompose any r-fold tensor product

$$\mathfrak{F}_{V,1} \otimes_s \mathfrak{F}_{V,1} \otimes_s \cdots \otimes_s \mathfrak{F}_{V,1},$$

into an orthogonal direct sum, and corresponding to (5.30) we have

$$\sum_{k_j\in\Gamma_V} \tilde{f}(k_1,\ldots,k_r)\, \tilde{h}(-k_1,\ldots,-k_r) = \sum_i \sum_{k_j\in\Gamma_V} (f_i)^{\sim}(k_1,\ldots,k_r)\, (h_i)^{\sim}(-k_1,\ldots,-k_r), \tag{5.31}$$

where $i = \{i_1, \ldots, i_r\}$ is a localization index and

$$f_i \in \mathfrak{F}_{V,1,i_1} \otimes_s \cdots \otimes_s \mathfrak{F}_{V,1,i_r}.$$

We therefore can write

$$b_{\mathrm{K},j}(k_1,\ldots,k_4) = \sum_{i_1,\ldots,i_4} b_{\mathrm{K},j,i_1,\ldots,i_4}(k_1,\ldots,k_4) \tag{5.32}$$

corresponding to the direct sum (5.28), with

$$b_{K,j,i_1,\ldots,i_4} \in \mathfrak{F}_{V,1,i_1} \otimes_s \cdots \otimes_s \mathfrak{F}_{V,1,i_4}.$$

Let H_I (K, j, i_1, ..., i_4) be the part of H_I(K, j) with the kernels (5.32). We notice that the kernel (5.32) has five localization indices. If the indices are incompatible, that is, if any of the differences $|j-i_l|$ is large, then the kernel must be small. We need the same result for a somewhat more general kernel. For $k_l \in \Gamma_V$, let

$$b_{K,j,s}(k_1, \ldots, k_4) = b_{K,j}(k_1, \ldots, k_4) \exp\{-\sum_{l=1}^{4} s_l \mu(k_l)\}, \tag{5.33}$$

where $s = \{s_1, \ldots, s_4\}$.

Lemma 5.2. *For each integer r, each $\varepsilon > 0$, and each positive number T, there is a constant $c = c(r, T)$ such that for $0 \leqslant s_l \leqslant T$, and $1 \leqslant \Lambda \leqslant K$,*

$$\|b_{\Lambda,j,s,i} - b_{K,j,s,i}\|_{2,V} \leqslant c\Lambda^{-\frac{1}{2}+\varepsilon} \prod_{l=1}^{4} (1+|j-i_l|)^{-r}. \tag{5.34}$$

Note $i = \{i_1, ..., i_4\}$. Note that $b_{\Lambda,j,s,i}$ is the ith component of $b_{\Lambda,j,s}$ defined in (5.33). Here the norm $\|\cdot\|_{2,V}$ refers to the Fourier series L_2 norm, which equals $(V/2\pi)^2$ times the usual L_2 norm of (5.26) with each k_l replaced by k_{lV}.

Before proving this lemma, we make some general remarks. Let $f(x)$ be a function with a smooth Fourier transform. That is, assume that for all α,

$$\|D^\alpha \tilde{f}\|_2 < \infty. \tag{5.35}$$

This corresponds to a rapidly decreasing $f(x)$. Now let

$$f_V(x) \in L_2([-V/2, V/2]) \tag{5.36}$$

be a periodic function with Fourier coefficients

$$V^{-\frac{1}{2}} \tilde{f}(k), \quad k \in \Gamma_V.$$

Then

$$f_V(x) = \sum_{n=-\infty}^{\infty} f(x+nV), \tag{5.37}$$

as can be checked by computing the Fourier coefficient of the right side of (5.37). In fact (5.37) is valid for $f \in L_1$. Using the rapid decrease of $f(x)$, we have for any integer r, a bound on the norms,

$$\|(1+|x|)^r f_V\|_{2,V} \tag{5.38}$$

in $L_2([V/2, V/2])$, which is independent of V, as long as V is bounded away from zero. Furthermore, the norms (5.38) can be estimated in terms of the norms (5.35).

Proof of Lemma 5.2. We have $b_{K,j,s}$ given by (5.33). Multiplication of the $b_{K,j,s}$ by $\exp(ij\sum_{l=1}^4 k_l)$ maps the $(i_1, \ldots, i_4)$ component of $b_{K,j,s}$ into the $(i_1-j, \ldots, i_4-j)$ component of

$$\lambda(2V)^{-2}(g(\cdot+j)\zeta)^{\sim}(\sum_{l=1}^4 k_l)\prod_{l=1}^4 \mu(k_l)^{-\frac{1}{2}}\xi(k_l/K)\exp(-\sum_{l=1}^4 s_l\mu(k_l)).$$

These functions are of the form f_V, and we now see that the corresponding functions f satisfy estimates of the form (5.35), which are independent of K and j. Let $P = k_1 + \ldots + k_4$, and

$$F(K, \Lambda, s) = (1+P^2)^{-\frac{1}{2}}\prod_{l=1}^4[\mu(k_l)^{-\frac{1}{2}}e^{-s_l\mu(k_l)}]\{\prod_{j=1}^4\xi(k_j/K) - \prod_{j=1}^4\xi(k_j/\Lambda)\}.$$

Then $D^r F(K, \Lambda, s) \in L_2$ and

$$\|D^r\{(b_{K,j,s} - b_{\Lambda,j,s})\exp(ij\sum_{l=1}^4 k_l)\}\|_2$$

$$\leqslant \sum_\alpha \binom{r}{\alpha}\|D^\alpha[(1+P^2)^{\frac{1}{2}}(g(\cdot+j)\zeta)^{\sim}(P)]\|_\infty \times \|D^{r-\alpha}F(K,\Lambda,s)\|_2. \tag{5.39}$$

We observe that
$$\xi(k/K) = 1 = \xi(k/\Lambda)$$
unless
$$|k| > \text{const. }\Lambda.$$

Thus $D^{p-\alpha}F(K, \Lambda, s) = 0$ unless at least one of the k's has magnitude greater than const. Λ. By checking the order of convergence of the integrals, we see that for each r, the L_2 norm of $D^{r-\alpha}F(K, \Lambda, s)$ is $O(\Lambda^{-\frac{1}{2}+\varepsilon})$, independent of K and s as long as they lie in their allowed ranges. (See [26].) (1) In order to see that the L_∞ norms in (5.39) are bounded independently of j, we note that

$$|D^\beta(1+P^2)^{\frac{1}{2}}| \leqslant \text{const. }(1+|P|)$$

for a constant independent of β if $|\beta| \leqslant |r|$, and $|r|$ is fixed. Thus for $|\alpha| \leqslant r$.

$$\begin{aligned}\|D^\alpha\{(1+P^2)^{\frac{1}{2}}(g(\cdot+j)\zeta)(P)\}\|_\infty &\leqslant \sup_{\beta\leqslant|r|}\text{const. }\|(1+|P|)D^\beta(g(\cdot+j)\zeta)^{\sim}(P)\|_\infty \\ &\leqslant \sup_{\beta\leqslant|r|}\text{const. }(\|x^\beta g(\cdot+j)\zeta\|_1 + \|D(x^\beta g(\cdot+j)\zeta)\|_1) \\ &\leqslant \text{const. }\sup_{\beta\leqslant r}((1+M_0)\|x^\beta\zeta\|_1 + \|D(x^\beta\zeta)\|_1)\end{aligned} \tag{5.40}$$

by the assumed bounds on $g(x)$ and $|Dg(x)|$. Since $\zeta \in C^\infty$ and has compact support, (5.40) is dominated by $c = c(|r|, M_0, T)$. Thus (5.39) satisfies a bound

$$\|D^r\{(b_{K,j,s} - b_{\Lambda,j,s})\exp(i\sum_{l=1}^4 k_l j)\}\|_2 \leqslant c(r, T)\Lambda^{-\frac{1}{2}+\varepsilon},$$

(1) A direct calculation is also possible.

which is independent of j and K. This is an estimate of the form (5.35), so the corresponding norms (5.38) satisfy similar bounds. The rapid decrease of these functions could be written as the estimate (5.34), which completes the proof.

We now bound $|E_g|$ through the formula

$$|E_g| = t^{-1} \log \|e^{-tH(g)}\| = t^{-1} \log \{ \sup_{\|\theta\|=1=\|\chi\|} |(\theta, e^{-tH(g)}\chi)| \}. \tag{5.41}$$

Since

$$\lim_{\mathrm{K}, V\to\infty} E_{g,\mathrm{K},V} = E_g,$$

it is sufficient to bound $|E_{g,\mathrm{K},V}|$ uniformly in K and V. We use the Feynman Kac formula to represent the operator $\exp(-tH(g)_{\mathrm{K},V})$ on $\mathfrak{F}_{\mathrm{K},V}$. For $\theta, \chi \in \mathfrak{F}_{\mathrm{K},V}$

$$(\theta, e^{-tH(g)_{\mathrm{K},V}}\chi) = \int_{C_{\mathrm{K},V}} \exp\left[-\int_0^t H_{I,g,\mathrm{K},V}(q(s))\,ds\right] \bar{\theta}(q(0))\chi(q(t))\,dq(\cdot). \tag{5.42}$$

Here $C_{\mathrm{K},V}$ is the path space corresponding to $\mathfrak{F}_{\mathrm{K},V}$. It consists of continuous paths $q(\cdot)$ taking values $q(s)$ in a Euclidean space of high dimension—one dimension for each mode permitted by the cutoffs K and V. The integral is a Wiener type integral on $C_{K,V}$ coming from $H_{0,\mathrm{K},V}$. The transition probabilities of the Wiener process are the kernel of $\exp(-tH_{0,\mathrm{K},V})$. We bound the inner product (5.42) with Hölder's inequality, as in [22, 8]. For some sufficiently large numbers p and t chosen independently of g, K and V, we have

$$|E_{g,\mathrm{K},V}| \leq (pt)^{-1} \log \left\{ \int_{C_{\mathrm{K},V}} \exp - \left[p\int_0^t H_{I,g,\mathrm{K},V}(q(s))\,ds\right] dq(\cdot) \right\}. \tag{5.43}$$

Thus we must show that

$$\int_{C_{\mathrm{K},V}} \exp\left[-p\int_0^t H_{I,g,\mathrm{K},V}(q(s))\,ds\right] dq(\cdot) \tag{5.44}$$

grows at most exponentially in the number of j's, with the exponential constant independent of K and V. We have

$$:\varphi_{\Lambda,V}(x)^4: = (\varphi_{\Lambda,V}^2(x) - 3C_0)^2 - 6C_0^2,$$

where C_0 is the no particle expectation value

$$C_0 = (\Omega_0, \varphi_{\Lambda,V}(x)^2\Omega_0) = \frac{1}{2V}\sum_{k\in\Gamma_V} \frac{1}{\mu(k)} \xi^2(k/\Lambda) \leq C(\log \Lambda)$$

for $\Lambda \geq 2$. Thus

$$-6C^2\lambda\|\zeta\|_1(\log \Lambda)^2 \leq H_I(\Lambda, j),$$

and for a new constant c_1 depending on t

$$1 - c_1(\log \Lambda)^2 \leqslant \int_0^t H_I(\Lambda, j)\,(q(s))\,ds. \tag{5.45}$$

Let
$$\delta H(\Lambda, j) = H_I(\mathrm{K}, j) - H_I(\Lambda, j) \tag{5.46}$$

and
$$I(\Lambda, j) = \int_0^t \delta H(\Lambda, j)\,(q(s))\,ds. \tag{5.47}$$

Also let Pr denote the measure defined by the integral

$$\int_{C_{\mathrm{K},V}} dq(\,\cdot\,).$$

We define $\mathfrak{J}$ as the set of integers j for which $g\zeta_j \neq 0$. Let $\mathfrak{J}_0$ denote any subset of $\mathfrak{J}$. We consider a finite sequence $\{\Lambda_j : j \in \mathfrak{J}_0\}$ such that

$$2 \leqslant \Lambda_j \leqslant \mathrm{K}.$$

We define for such a sequence $\{\Lambda_j\}$

$$P(\{\Lambda_j\}) = Pr\{|I(\Lambda_j, j)| \geqslant 1, \text{ for all } j \in \mathfrak{J}_0\}. \tag{5.48}$$

The main step remaining is to prove

LEMMA 5.3. *For constants c, d independent of* K *and* V,

$$P(\{\Lambda_j\} \leqslant \prod_{j \in \mathfrak{J}_0} [c \exp(-d\Lambda_j^{1-\varepsilon})]. \tag{5.49}$$

Let us for the moment assume Lemma 5.3 and derive the bound on (5.37). We represent $C_{\mathrm{K},V}$ as a disjoint union of measurable subsets X_ν, and we estimate the integral by the maximum of its integrand on each X_ν.

$$\left| \int_{C_{\mathrm{K},V}} F(q(\,\cdot\,))\,dq(\,\cdot\,) \right| \leqslant \sum_\nu Pr\{X_\nu\} \sup_{q(\cdot) \in X_\nu} |F(q(\,\cdot\,))|. \tag{5.50}$$

Let us assign to each path $q(\,\cdot\,) \in C_{\mathrm{K},V}$ a sequence of integers ν_j with $j \in \mathfrak{J}$. We define

$$\nu_j(q(\,\cdot\,)) = \nu_j \tag{5.51}$$

as the smallest integer in the interval $[d_1, \mathrm{K}]$ such that

$$|I(\nu_j, j)\,(q(\,\cdot\,))| \leqslant 1. \tag{5.52}$$

Here $d_1 > 2$ is a constant which we choose depending only on the c and d in Lemma 5.3.

We take d_1 and K to be integers. Since $I(\mathrm{K}, j) = 0$, there is a smallest integer ν_j with the required property.

For each sequence of integers $\{\nu_j\}$, namely

$$\nu_j \in [d_1, \mathrm{K}], \quad j \in \mathfrak{J},$$

let $X(\{\nu_j\})$ be the set of all paths in $C_{\mathrm{K},V}$ which, by (5.51), correspond to the sequence $\{\nu_j\}$. Then $C_{\mathrm{K},V}$ is a disjoint union of measurable sets

$$C_{\mathrm{K},V} = \bigcup_{\{\nu_j\}} X(\{\nu_j\}), \tag{5.53}$$

where the various ν_j range over $[d_1, \mathrm{K}]$.

Furthermore, we note that

$$\int_0^t H_{I,g,\mathrm{K},V}(q(s))\, ds = \sum_{j \in \mathfrak{J}} \left\{ \int_0^t H_I(\nu_j, j)\,(q(s))\, ds + I(\nu_j, j) \right\}, \tag{5.54}$$

and by (5.45)

$$1 - c_1 (\log \nu_j)^2 \leqslant \int_0^t H_I(\nu_j, j)\,(q(s))\, ds.$$

Therefore for $q(\cdot) \in X(\{\nu_j\})$, (5.54) yields

$$-c_1 \sum_{j \in \mathfrak{J}} (\log \nu_j)^2 \leqslant \int_0^t H_{I,g,\mathrm{K},V}(q(s))\, ds. \tag{5.55}$$

To each path $q(\cdot) \in C_{\mathrm{K},V}$ we now assign a second sequence of integers $\{\Lambda_j\}$, where j ranges over a subset $\mathfrak{J}_0$ of $\mathfrak{J}$. Let $\mathfrak{J}_0$ consist of those k for which

$$\nu_k > d_1 + 1, \tag{5.56}$$

and for $k \in \mathfrak{J}_0$ define

$$\Lambda_k = \nu_k - 1.$$

To each subset $X(\{\nu_j\})$ of $C_{\mathrm{K},V}$, there corresponds one sequence $\{\Lambda_k\}$. For $k \in \mathfrak{J}_0 \subset \mathfrak{J}$,

$$|I(\Lambda_k, k)\,(q(\cdot))| \geqslant 1,$$

and so

$$Pr\{X(\{\nu_j\})\} \leqslant P(\{\Lambda_k\}). \tag{5.57}$$

Thus using (5.50), (5.55), (5.57) and Lemma 5.3, the desired integral (5.44) is bounded by

$$\sum_{\{\nu_j\}} Pr(X\{\nu_j\}) \exp\,[c_1 p \sum_{j \in \mathfrak{J}} \log^2 \nu_j] \leqslant \sum_{\{\nu_j\}} \prod_{j \in \mathfrak{J}_0} [c \exp\,(-d\Lambda_j^{\frac{1}{4}-\varepsilon})] \prod_{j \in \mathfrak{J}} [\exp\,(c_1 p \log^2 \nu_j)]. \tag{5.58}$$

Each ν_j in the above sum ranges over $d_1 \leqslant \nu_j \leqslant \mathrm{K}$, and we now treat the various cases in which some of the ν_j equal d_1. Let M be the number of elements of $\mathfrak{J}$ and M_0 be the number of elements of $\mathfrak{J}_0$, leaving $(M-M_0)$ of the ν_j equal to d_1. Since there are 2^M subsets of $\mathfrak{J}$, $\mathfrak{J}_0$ can be chosen in 2^M different ways. Restricting the sum (5.58) to $\{\nu_j\}$ which give one fixed subset $\mathfrak{J}_0$, this restricted sum is bounded by

$$\exp [(M-M_0)\, c_1 p \log^2 d_1] \sum_{\{\Lambda_j\}} \prod_{j\in\mathfrak{J}_0} [\exp (\log c + c_1 p \log^2 \nu_j - d\Lambda_j^{\frac{1}{4}-\varepsilon})]$$
$$\leqslant \exp [(M-M_0)\, c_1 p \log^2 d_1] \exp [M_0 d_2]. \tag{5.59}$$

Here we have chosen the arbitrary constant d_1 sufficiently large so that for $\nu > d_1$,

$$c_1 p \log^2 \nu + \log c < \tfrac{1}{2} d(\nu-1)^{\frac{1}{4}-\varepsilon} = \tfrac{1}{2} d\Lambda^{\frac{1}{4}-\varepsilon},$$

in which case

$$\sum_{\Lambda=d_1}^{K-1} \exp (-\tfrac{1}{2} d\Lambda^{\frac{1}{4}-\varepsilon}) \leqslant \sum_{\Lambda=0}^{\infty} \exp (-\tfrac{1}{2} d\Lambda^{\frac{1}{4}-\varepsilon}) = \exp [d_2].$$

We are considering the case of M_0 different Λ_j's so the bound (5.59) follows. Thus if

$$a = d_2 + c_1 p \log^2 d_1,$$

(5.59) is dominated by $\exp [aM]$ and (5.44) is less than

$$2^M \exp [aM],$$

which proves the theorem.

Proof of Lemma 5.3. For any set of positive even integers l_j,

$$P(\{\Lambda_j\}) \leqslant \int_{C_{K,V}} \prod_{j\in\mathfrak{J}_0} I(\Lambda_j, j)^{l_j} dq(\cdot). \tag{5.60}$$

We approximate the Riemann integral defining $I(\Lambda_j, j)$ by Riemann sums. In order to bound (5.60) it is sufficient to do so with $I(\Lambda_j, j)$ replaced by a Riemann sum. We replace all $L = \sum_j l_j$ factors by Riemann sums and then expand. The result will be a sum of M^L terms, if each Riemann sum has M terms. Each term is a product of L factors, and each factor occurs at a sharp time. To write down a typical term, we choose mesh times $s_1, \ldots, s_L$, and for the αth factor we choose an index $j(\alpha)$. The index j is chosen exactly l_j times. Then (5.60) is replaced by

$$P(\{\Lambda_j\}) \leqslant t^L \sup \int_{C_{K,V}} \delta H(\Lambda_{j(1)}, j(1))\,(q(s_1)) \ldots \delta H(\Lambda_{j(L)}, j(L))\,(q(s_L))\, dq(\cdot), \tag{5.61}$$

where the supremum is taken over all choices of mesh times, mesh lengths and all choices of $j(\alpha)$ consistent with the l_j. We can rearrange the factors in the integral, so that is no loss of generality to assume that

$$s_1 < s_2 < \ldots < s_L.$$

By the definition of the Wiener integral, (5.61) equals the no particle expectation value

$$t^L \sup (\Omega_0, \exp(-s_1 H_{0,\mathrm{K},V})\, \delta H(\Lambda_{j(1)}, j(1)) \exp(-(s_2 - s_1) H_{0,\mathrm{K},V}) \\ \ldots \exp(-(s_L - s_{L-1}) H_{0,\mathrm{K},V})\, \delta H(\Lambda_{j(L)}, j(L))\, \Omega_0). \qquad (5.62)$$

The next step is to evaluate (5.62) as a sum of elementary terms. Each elementary term will itself be a sum of explicit functions. This reduction of (5.62) is accomplished by repeated application of the commutation relation

$$[a_V(-k), a_V^*(k')] = \delta(-k; k') = \begin{cases} 1 & \text{if } -k = k' \\ 0 & \text{otherwise} \end{cases}$$

and

$$[a_V(-k), a_V(k')] = 0,$$

for $k, k' \in \Gamma_V$. We use this relation to bring all the annihilation operators $a_V(k)$ to the right and the creation operators $a_V(k)$ to the left. Furthermore, these commutation relations yield the relations for $k \in \Gamma_{\mathrm{K},V}$,

$$a_V(k) \exp(-sH_{0,\mathrm{K},V}) = \exp(-sH_{0,\mathrm{K},V})\, a_V(k) \exp[-s\mu(k)],$$

which are used to bring the annihilation operators to the right past the factors $\exp(-sH_{0,\mathrm{K},V})$. Since

$$a_V(k)\Omega_0 = 0,$$

moving the annihilation operators to the right does reduce (5.62) to the desired sum of elementary terms. Furthermore, each variable k must occur in a delta function $\delta(-k; k')$ in any non zero term.

In order to keep track of the distinct terms and the commutation rules used, we introduce a graph G for each term. The value of the term will be a number $N(G)$ assigned to G. Thus we shall write

$$(\Omega_0, \prod_{\alpha=1}^{L} \{\exp[-(s_\alpha - s_{\alpha-1}) H_{0,\mathrm{K},V}]\, \delta H(\Lambda_{j(\alpha)}, j(\alpha))\} \Omega_0) = \sum_G N(G), \qquad (5.63)$$

where $s_0 = 0$, and where the product of noncommuting operators is defined by

$$\prod_{\alpha=1}^{L} \{A_\alpha\} = A_1 A_2 \ldots A_L.$$

We construct G as follows. We start with L vertices, labeled by the index α, $1 \leqslant \alpha \leqslant L$. From each vertex we draw four lines (called legs), labeled by (α, ϱ), $1 \leqslant \varrho \leqslant 4$. Each factor

$\delta H(\Lambda_{j(\alpha)}, j(\alpha))$ in (5.63) is the difference between two interaction Hamiltonians, and as such is the sum of five monomials as in (5.27). For each factor we choose one such monomial with ν_α creators and $4-\nu_\alpha$ annihilators. This monomial will correspond to a vertex in G, and we draw ν_α legs from the αth vertex to the left, and $4-\nu_\alpha$ legs pointing to the right. Each leg (α, ϱ) corresponds to a momentum variable $k_{\alpha\varrho}$ in the kernel $B(\Lambda_{j(\alpha)}, j(\alpha))$ at vertex α. Since our factors are noncommuting, we order the vertices linearly from left to right, as in (5.63), with vertex 1 on the left and vertex L on the right. Each time that the use of the commutation relations introduces a delta function, we say that the variables $-k_{\alpha\varrho}$ and $k_{\alpha'\varrho'}$ occurring in it are *contracted*. Graphically we join together the annihilation leg (α, ϱ) and the creation leg (α', ϱ') which correspond to these variables. We have already noticed that for a nonzero term, each variable occurs in exactly one delta function, and so for the corresponding graph each creating leg (pointing left) is connected to a unique annihilating leg (pointing right). Thus only terms with $\nu_1=0$, $\nu_L=4$ will contribute. Since annihilating legs all moved to the right, the annihilating legs at vertex α are always contracted with creating legs on their right.

The possible graphs are all graphs which can be constructed in this manner. Each graph consists of L vertices and $2L$ lines (each line being two joined legs). Each vertex is the endpoint of four lines and each line joins two distinct vertices. This last fact results from Wick ordering the interaction. Evidently there are fewer than

$$(4L)!! = (4L-1)(4L-3)\dots 3\cdot 1$$

possible graphs.

We let $N(G)$ denote the contribution to (5.63) from a given graph G and this is obtained as follows: To the vertex α of G we assign the kernel

$$B(\Lambda_{j(\alpha)}, j(\alpha); k_{\alpha 1}, \dots, k_{\alpha 4}) = b_{K, j(\alpha)}(k_{\alpha 1}, \dots, k_{\alpha 4}) - b_{\Lambda_{j(\alpha)}, j(\alpha)}(k_{\alpha 1}, \dots, k_{\alpha 4}), \tag{5.64}$$

where $b_{K,j}$ is defined in (5.26). We write down the product of the L kernels (5.64) for G and multiply this by a product of $2L$ delta functions

$$\delta(-k_{\alpha\varrho}; k_{\alpha'\varrho'}),$$

one delta function for each line, obtained by pairing the annihilator (α, ϱ) with the creator (α', ϱ'), $\alpha' > \alpha$. We then multiply by a product of $2L$ energy factors

$$\exp[-(s_{\alpha'} - s_\alpha)\mu(k_{\alpha\varrho})],$$

one for each line, which arise from the commutations past $\exp(-sH_{0,K,V})$ factors. These factors could actually be included in the kernels $B(\Lambda_{j(\alpha)}, j(\alpha))$ by defining for each graph G

$$\hat{B}(\Lambda_{j(\alpha)}, j(\alpha)) = B(\Lambda_{j(\alpha)}, j(\alpha)) \exp\left[-\sum_{\varrho=1}^{4} (s_{\alpha'} - s_\alpha)\, \mu(k_{\alpha\varrho})/2\right]. \tag{5.65}$$

The numerical value of G is the sum of this expression over the $4L$ variables $k_{\alpha\varrho} \in \Gamma_{K,V}$ times some numerical factors.

$$N(G) = \sum_{\substack{k_{\beta\sigma} \in \Gamma_{K,V} \\ 1 \leq \beta \leq L \\ 1 \leq \sigma \leq 4}} \left[\sum_{\alpha=1}^{L} \binom{4}{\nu_\alpha} \hat{B}(\Lambda_{j(\alpha)}, j(\alpha); k_{\alpha\varrho})\right] \left[\prod_{\substack{1 \leq \alpha \leq L \\ 1 \leq \varrho \leq 4}} \delta(-k_{\alpha\varrho}; k_{\alpha'\varrho'})\right]. \tag{5.66}$$

Let us now motivate the type of bound we expect (in the simple case $s_1 = \ldots = s_L = 0$). We expect that different regions of localization will be independent, by the localization property of free fields (cluster property). Since each localization index j must be chosen exactly l_j times, exact localization would mean for $s_1 = \ldots = s_L = 0$ that (5.63) equals

$$\prod_{j \in \mathfrak{J}_0} (\Omega_0, \delta H(\Lambda_j, j)^{l_j} \Omega_0), \tag{5.67}$$

which can be shown to be dominated by

$$\prod_{j \in \mathfrak{J}_0} [c^{l_j} (2 l_j)!\, \Lambda_j^{-l_j/2}], \tag{5.68}$$

for some constant c independent of K, V, $\mathfrak{J}_0$, Λ_j and l_j. In fact, we will show that although the localization is not exact, it is approximate in the sense that (5.68) actually bounds (5.63), and this holds even in the case that the s_j are nonzero. In other words, the action of $\exp(-sH_0)$ does not destroy, in any essential way, the localization of the particles.

LEMMA 5.4. *The expectation value* (5.63) *is dominated by*

$$\prod_{j \in \mathfrak{J}_0} [c^{l_j} (2 l_j)!\, \Lambda_j^{-l_j(1-\varepsilon)/2}],$$

where the constant c depends only on T providing the bounds

$$0 \leq g(x) \leq 1,$$

$$|Dg(x)| \leq M_0,$$

and $$0 \leq s_\alpha \leq T$$

are satisfied.

Let us assume this lemma and complete the proof of Lemma 5.3. We apply Lemma 5.4 to (5.63), and get a bound on (5.61) independent of the variables in the supremum. Thus

$$Pr(\{\Lambda_j\}) \leqslant (ct)^L \prod_{j\in\mathfrak{J}_0} (2l_j)!\Lambda_j^{-l_j(1-\varepsilon)/2}.$$

We now choose l_j to be the positive even integer nearest to $(4ct/\Lambda_j^{\frac{1}{2}-\varepsilon})^{-\frac{1}{2}} = O(\Lambda_j^{(1-\varepsilon)/4})$. Then using Sirling's formula to estimate the factorials, with $d \leqslant (4ct)^{-\frac{1}{2}}$, with a new constant c_1, and with a new ε,

$$Pr(\{\Lambda_j\}) \leqslant \prod_{j\in\mathfrak{J}_0} [c_1 \exp(-d\Lambda_j^{\frac{1}{4}-\varepsilon})].$$

This completes the proof.

Proof of Lemma 5.4. By repeated application of the Schwarz inequality in the variables $k_{\alpha\varrho}$, $N(G)$ in (5.66) can be dominated by

$$|N(G)| \leqslant 2^{4L} \prod_{\alpha=1}^{L} \|\hat{B}(\Lambda_{j(\alpha)}, j(\alpha))\|_2 = 2^{4L} \prod_{j\in\mathfrak{J}_0} \|\hat{B}(\Lambda_j, j)\|_2^{l_j} \leqslant \prod_{j\in\mathfrak{J}_0} [c^{l_j}\Lambda_j^{-l_j(1-\varepsilon)/2}],$$

by Lemma 5.2. However, we need a sharper bound (5.82) which reflects the localization of particles. Thus we classify the graphs in a way which reflects this property.

For a particular graph G, each vertex α, $1 \leqslant \alpha \leqslant L$, has an associated localization number $j(\alpha) \in \mathfrak{J}_0$. For $\varrho = 1, 2, 3, 4$, let us define $j(\alpha, \varrho)$ by

$$j(\alpha, \varrho) = j(G; \alpha, \varrho) = j(\alpha'), \tag{5.69}$$

where the ϱth leg at vertex α is connected in G to the ϱ'th leg at vertex α'. We can also define

$$\delta(\alpha, \varrho) = j(\alpha) - j(\alpha, \varrho), \tag{5.70}$$

which measures the difference in configuration space localization between the two vertices connected by the pair of legs (α, ϱ), (α', ϱ') corresponding to a line in G.

The functions (5.69) uniquely specify the functions (5.70) and vice versa, for any graph G contributing to (5.63). However, (5.69) does not uniquely specify the graph. We now show that there are at most

$$\prod_{j\in\mathfrak{J}_0} [4^{l_j}(2l_j)!] \tag{5.71}$$

out of the (fewer than $(4L)!!$) different graphs G with the same function $j(G, \alpha, \varrho)$.

Let us suppose that in G there are

$$l_{ij} = l_{ij}(G)$$

lines connecting a j-localized vertex to an i-localized vertex, and to simplify the following formulae, we count each line with $i = j$ twice. Thus

$$\sum_{j\in\mathfrak{J}_0} l_{ij} = 4l_i, \tag{5.72}$$

and
$$\sum_{i,j\in\mathcal{J}_0} l_{ij}=4L. \tag{5.73}$$

Note that the G's which contribute to (5.65) are exactly those for which the off diagonal l_{ij}'s vanish, and $l_{ij}=4l_i\delta_{ij}$.

With the functions $j(\alpha,\varrho)$ given and with i and j held fixed, there are two sets of l_{ij} legs specified, and for the graphs we are considering each leg in the first set must be contracted to a leg in the second set. For a vertex α of a leg in the first set, $j(\alpha)=j$ and for a vertex α' of a leg in the second set, $j(\alpha')=i$. If $i\neq j$ then these two sets of legs are disjoint and the contractions can be made in at most $(l_{ij})!$ ways. If $i=j$ then the two sets of legs coincide and the contractions can be made in at most $(l_{ii})!!$ ways. Once the contractions have been made for all $i\leqslant j$, the graph G is uniquely determined, and so there are at most

$$\prod_{\substack{i<j\\ i,j\in\mathcal{J}_0}}[l_{ij}!]\prod_{i\in\mathcal{J}_0}[l_{ii}!!] \tag{5.74}$$

graphs with given functions $j(\alpha,\varrho)$.

If l is even
$$l!\leqslant 2^l[(\tfrac{1}{2}l)!]^2$$
$$l!!\leqslant 2^{l/2}(\tfrac{1}{2}l)!$$

and there are similar estimates for l_{ij} odd.

Using (5.72) – (5.73), we thus have that (5.74) is dominated by

$$\prod_{i,j\in\mathcal{J}_0}[2^{l_{ij}/2}(\tfrac{1}{2}l_{ij})!]=4^L\prod_{j\in\mathcal{J}_0}\{\prod_{i\in\mathcal{J}_0}[(\tfrac{1}{2}l_{ij})!]\}\leqslant 4^L\prod_{j\in\mathcal{J}_0}[(2l_j)!],$$

which is the desired bound (5.71).

Thus
$$\left|(\Omega_g,\prod_{\alpha=1}^{L}\delta H(\Lambda_{j(\alpha)},j(\alpha))\,\Omega_0)\right|\leqslant\{\prod_{j\in\mathcal{J}_0}[(4)^{l_j}(2l_j)!]\}\sum_{\{\delta(\alpha,\varrho)\}}\sup|N(G)|. \tag{5.75}$$

Here the supremum is over all graphs G with fixed $j(G,\alpha,\varrho)=j(\alpha,\varrho)$, or fixed $\delta(\alpha,\varrho)$, and the sum is over all possible integers $\delta(\alpha,\varrho)$ of graphs contributing to (5.63). We now bound $N(G)$ in (5.66).

Each of the $2L$ independent sums over momentum variables $k_{\alpha\varrho}\in\Gamma_V$ can be regarded as an inner product in the one particle space $\mathfrak{H}_{V,1}$, or the equivalent inner product in the space of Fourier series on $[-V/2,V/2]$. We now make an expansion by writing the inner product on $\mathfrak{H}_{V,1}$ as a sum of inner products on the $\mathfrak{H}_{V,1,i}$, using (5.28). Because of the support properties, many terms vanish giving as in (5.31)

$$N(G)=\sum_{\substack{i_{\beta\sigma},k_{\beta\sigma}\\ 1\leqslant\beta\leqslant L\\ 1\leqslant\sigma\leqslant 4}}\prod_{\alpha=1}^{L}\hat{B}(\Lambda_{j(\alpha)},j(\alpha),i_{\alpha\varrho};k_{\alpha\varrho})\binom{4}{\nu_\alpha}\prod_{\substack{1\leqslant\alpha\leqslant L\\ 1\leqslant\varrho\leqslant 4}}\delta(-k_{\alpha\varrho};k_{\alpha'\varrho'})\,\delta(i_{\alpha\varrho};i_{\alpha'\varrho'}). \tag{5.76}$$

The sum runs over all possible choices of the localization indices $i_{\beta\sigma}$ and the momenta $k_{\beta\sigma}$. We now apply the Schwarz inequality repeatedly in the $2L$ independent $k_{\alpha\varrho}$ variables. This gives the bound

$$|N(G)| \leqslant \sum_{\substack{i_{\beta\sigma} \\ 1\leqslant\beta\leqslant L \\ 1\leqslant\sigma\leqslant 4}} \prod_{\alpha=1}^{L} 16 \|\hat{B}(\Lambda_{j(\alpha)}, j(\alpha), i_{\alpha\varrho})\|_2 \prod_{\substack{1\leqslant\alpha\leqslant L \\ 1\leqslant\varrho\leqslant 4}} \delta(i_{\alpha\varrho}; i_{\alpha'\varrho'}). \tag{5.77}$$

We choose V sufficiently large to enclose the support of the spatial cutoff $g(x)$. Thus by Lemma 5.2 independently of V,

$$\|\hat{B}(\Lambda_{j(\alpha)}, j(\alpha), i_{\alpha 1}, \ldots, i_{\alpha 4})\|_2 \leqslant c\Lambda_{j(\alpha)}^{-(1-\varepsilon)/2} \prod_{1\leqslant\varrho\leqslant 4} (1+|j(\alpha)-i_{\alpha\varrho}|)^{-r}, \tag{5.78}$$

where c depends on λ, M_0, r and T. Using the fact that each index $j=j(\alpha)$ occurs l_j times in G, we have by (5.77)–(5.78),

$$|N(G)| \leqslant \prod_{j\in\mathcal{J}_0} [16 c^{l_j} \Lambda_j^{-l_j(1-\varepsilon)/2}] \sum_{\substack{i_{\beta\sigma} \\ 1\leqslant\beta\leqslant L \\ 1\leqslant\sigma\leqslant 4}} \prod_{\substack{1\leqslant\alpha\leqslant L \\ 1\leqslant\varrho\leqslant 4}} (1+|j(\alpha)-i_{\alpha\varrho}|)^{-r} \delta(i_{\alpha\varrho}; i_{\alpha'\varrho'}). \tag{5.79}$$

Since each leg (α, ϱ) is paired with another leg (α', ϱ'), the factors

$$(1+|j(\alpha)-i_{\alpha\varrho}|)^{-r}\delta(i_{\alpha\varrho}; i_{\alpha'\varrho'})$$

occur in pairs,

$$(1+|j(\alpha)-i_{\alpha\varrho}|)^{-r}(1+|j(\alpha')-i_{\alpha'\varrho'}|)^{-r}\delta(i_{\alpha\varrho}; i_{\alpha'\varrho'})^2$$
$$= (1+|j(\alpha)-i_{\alpha\varrho}|)^{-r}(1+|j(\alpha')-i_{\alpha\varrho}|)^{-r}\delta(i_{\alpha\varrho}; i_{\alpha'\varrho'}), \tag{5.80}$$

where
$$\delta(\alpha, \varrho) = j(\alpha) - j(\alpha') = j(\alpha) - j(\alpha, \varrho). \tag{5.81}$$

As
$$(1+|x|)^{-2}(1+|y|)^{-2} \leqslant (1+|x-y|)^{-1}(1+|x|)^{-1},$$

(5.80) is dominated by

$$(1+|\delta(\alpha,\varrho)|)^{-r/2}(1+|j(\alpha)-i_{\alpha\varrho}|)^{-r/2}.$$

Hence by (5.79)

$$|N(G)| \leqslant \prod_{j\in\mathcal{J}_0} (16\, cc_1)^{l_j} \Lambda_j^{-l_j(1-\varepsilon)/2} \prod_{\substack{1\leqslant\alpha\leqslant L \\ 1\leqslant\varrho\leqslant 4}} (1+|\delta(\alpha,\varrho)|)^{-r/2}, \tag{5.82}$$

where we have chosen $r>2$ and used the fact that

$$\sum_{\substack{i_{\beta\sigma} \\ 1\leqslant\beta\leqslant L \\ 1\leqslant\sigma\leqslant 4}} \prod_{\substack{1\leqslant\alpha\leqslant L \\ 1\leqslant\varrho\leqslant 4}} (1+|j(\alpha)-i_{\alpha\varrho}|)^{-r/2} = c_1^L,$$

with
$$c_1 = [\sum_{n=-\infty}^{\infty} (1+|n|)^{-r/2}]^4.$$

The bound (5.82) gives the dependence of G on the localization of its vertices. Thus by (5.75) and (5.82),

$$\left|(\Omega_0, \prod_{\alpha=1}^{L} \delta H(\Lambda_{j(\alpha)}, j(\alpha))\, \Omega_0)\right| \leq \prod_{j \in \mathcal{J}_0} [(64\, cc_1^2)^{l_j}(2l_j)!\, \Lambda_j^{-l_j(1-\varepsilon)/2}],$$

which establishes the lemma.

References

[1]. ARAKI, H., The type of von Neumann algebra associated with the free field. *Progr. of Theor. Phys.*, 32 (1964), 956–965.

[2]. DIXMIER, J., *Les algèbres d'opérateurs dans l'espace Hilbertien (algèbres de von Neumann)*. Gauthiers-Villars, Paris, 1957.

[3]. EPSTEIN, H., GLASER, V. & JAFFE, A., Nonpositivity of the energy density in quantized field theories. *Nuovo Cimento*, 36 (1965), 1016–1022.

[4]. FEDERBUSH, P., A partially alternate derivation of a result of Nelson. *J. Math. Phys.*, 10 (1969), 50–52.

[5]. FELL, J., The dual spaces of C^*-algebras. *Trans. Amer. Math. Soc.*, 94 (1960), 365–403.

[6]. GALINDO, A., On a class of perturbations in quantum field theory. *Proc. Nat. Acad. Sci. (U.S.A.)*, 48 (1962), 1128–1134.

[7]. GLIMM, J., A Stone–Weierstrass theorem for C^*-algebras. *Ann. of Math.*, 72 (1960), 216–244.

[8]. —— Boson fields with nonlinear selfinteraction in two dimensions. *Commun. Math. Phys.*, 8 (1968), 12–25.

[9]. —— *Varenna summer school-lectures, 1968*. Academic Press.

[10]. GLIMM, J. & JAFFE, A., A $\lambda(\phi^4)_2$ quantum field theory without cutoffs I. *Phys. Rev.*, 176 (1968), 1945–1951.

[11]. —— Singular perturbations of self adjoint operators. *Comm. Pure Appl. Math.*, 22 (1969), 401–414.

[12]. —— The $\lambda(\phi^4)_2$ quantum field theory without cutoffs II. The field operators and the approximate vacuum. *Ann. of Math.*, 91 (1970). 362–401.

[13]. —— An infinite renormalization of the Hamiltonian is necessary. *J. Math. Phys.*, 10 (1969), 2213–2214.

[14]. GRIFFIN, E., Some contributions to the theory of rings of operators II. *Trans. Amer. Math. Soc.*, 79 (1955), 389–400.

[15]. HAAG, R. & KASTLER, D., An algebraic approach to quantum field theory. *J. Math. Phys.*, 5 (1964), 848–861.

[16]. HALL, D. & WIGHTMAN, A., A theorem on invariant functions with applications to relativistic quantum field theory. *Mat.-Fys. Medd. Kongl. Danske Vidensk. Selskab*, 31 (1957) No. 5.

[17]. HEPP, K., Théorie de la renormalisation, Springer Verlag, New York, 1969,

[18]. JAFFE, A., Divergence of perturbation theory for bosons. *Commun. Math. Phys.*, 1 (1965), 127–149.

[19]. —— *Varenna summer school lectures, 1968*. Academic Press.

[20]. KRISTENSEN, P., MEJLBO, L. & POULSEN, E., Tempered distributions in infinitely many dimensions I. *Commun. Math. Phys.*, 1 (1965), 175–214.

[21]. NAIMARK, M., *Normed rings*. Noordhoff, Groningen, 1959.

[22]. NELSON, E., A quartic interaction in two dimensions. In *Mathematical theory of elementary particles*, edited by R. Goodman and I. Segal, M.I.T. Press 1966.

[23]. SEGAL, I., Nonlinear functions of weak processes I. *J. Functional Anal.*, 4 (1969), 404–456.

[24]. STREIT, L., A generalization of Haag's theorem. *Nuovo Cimento*, to appear.
[25]. WATSON, G., *A treatise on the theory of Bessel functions*. Cambridge 1944.
[26]. WEINBERG, S., High energy behavior in quantum field theory. *Phys. Rev.*, 118 (1960), 838–849.
[27]. CANNON, J. & JAFFE, A., Lorentz covariance of the $\lambda\phi_2^4$ quantum field theory. *Commun. Math. Phys.*, 17 (1970), 261-321.
[28]. ROSEN, L., A $\lambda\phi^{2n}$ field theory without cutoffs. *Commun. Math. Phys.*, 16 (1970), 157–183.

Received November 7, 1969

The Wightman axioms and particle structure in the $\mathcal{P}(\phi)_2$ quantum field model[1]

By JAMES GLIMM[2], ARTHUR JAFFE[3] and THOMAS SPENCER[4]

Abstract: The $\mathcal{P}(\phi)_2$ quantum field model with small coupling constant satisfies the Wightman axioms. The spectrum of the mass operator $M = (H^2 - P^2)^{1/2}$ contains two isolated eigenvalues: 0 and $m > 0$. Hence the model describes the scattering of particles of mass $m > 0$. The proof contains an n-particle cluster expansion and an analysis of the subspace of states with energy less than $(n + 1)m(1 - \varepsilon)$.

CONTENTS

1. The particle structure

Nontrivial quantum fields exist in two space time dimensions, [3]–[6], [8]–[9]. Physicists use similar fields in four dimensions to describe elementary particles. For this reason it is important to establish qualitative properties of these fields, in order to demonstrate that they are suitable for the description of physical processes, such as the scattering of particles.

We establish two such results. For sufficiently weak $\mathcal{P}(\phi)_2$ interactions we verify the Wightman axioms. These axioms guarantee the principles of

[1] This paper incorporates the results presented previously in two unpublished manuscripts, references [10], [11].

[2] Supported in part by NSF Grant GP-24003.

[3] Supported in part by NSF Grant GP-31239X and by the Air Force Office of Scientific Research, Contract AF 44620-70-C 0030.

[4] Supported in part by the Alfred P. Sloan Foundation.

quantum mechanics and special relativity. Secondly, we establish the existence of particles and the existence of a scattering matrix S for these models. Single particles appear as an isolated eigenvalue $m > 0$ of the mass operator M, as described below.

Our analysis of the spectrum of the energy and momentum operators involves an n particle cluster expansion. We establish estimates for this expansion that yield the existence of an infinite volume limit for the theory, the proof of the remaining Wightman axioms, and the analysis of the particle structure. These estimates are an outgrowth of a study of local Hamiltonians [6], [7]. However the results and estimates of this paper are independent of [3]-[7].

A general requirement of quantum mechanics, and one of the Wightman axioms, is that the states are given by unit vectors θ in a Hilbert space $\mathcal{H}$. We believe that the Hilbert space $\mathcal{H}$ for a weak $\mathcal{P}(\phi)_2$ interaction has a Schrödinger representation

$$\mathcal{H} = L_2(\mathcal{S}'(R^1), dq) . \tag{1.1}$$

Here the Schwartz distribution space $\mathcal{S}'(R^1)$ is the configuration space of a classical field, and dq is a countably additive non-Gaussian Borel measure on $\mathcal{S}'(R^1)$. In this representation the time zero field $\phi_0(f), f \in \mathcal{S}(R^1)$, is simply multiplication by the linear coordinate function $f \in \mathcal{S}(R^1)$. Thus for $F \in L_2(\mathcal{S}'(R^1), dq)$ and $q \in \mathcal{S}'(R^1)$

$$\phi_0(f)F(q) = \left(\int f(\vec{x})q(\vec{x})d\vec{x}\right)F(q) = \langle q, f\rangle F(q) . \tag{1.2}$$

Hence $\phi_0(f)$ measures the strength or position of the field in the coordinate direction f, just as the position operator q_j in ordinary quantum mechanics measures the position of the jth particle. The moments of dq, namely

$$W_0(\vec{x}_1, \cdots, \vec{x}_n) = \int \phi_0(\vec{x}_1) \cdots \phi_0(\vec{x}_n)dq \in \mathcal{S}'(R^n) , \tag{1.3}$$

are the time-zero vacuum expectation values.

Relativistic covariance is expressed in terms of a unitary representation U of the inhomogeneous Lorentz group on $\mathcal{H}$. The energy operator $P_0 \geqq 0$ and the momentum operator P_1 are defined to be self-adjoint infinitesimal generators of the time and space translation subgroups respectively. The space-time field operator $\phi(f)$, $f \in \mathcal{S}(R^2)$, is defined by a time evolution of ϕ_0 in the Heisenberg picture, namely,

$$\phi(f) = \int e^{itP_0}\phi_0(f(\cdot, t))e^{-itP_0}\, dt . \tag{1.4}$$

A particle may be thought of as a stable isolated disturbance of the quantum field. Let Ω denote the vacuum state vector in $\mathcal{H}$. Then $P_0\Omega = 0$

$= P_1\Omega$ and $e^{-itP_0}\Omega = \Omega$, so Ω is an equilibrium state under the time evolution subgroup. In the representation (1.1) above, Ω is the function

$$\Omega = 1 \in L_2(\mathcal{S}'(R^1), dq)\,, \tag{1.5}$$

so the expectation defined by Ω is the probability measure dq on the space $\mathcal{S}'(R^1)$ of classical field configurations. We think of Ω, or dq, as the undisturbed state of the field. Thus any state $\theta \neq \Omega$ is a disturbed state, and the distinguishing feature of a particle is its stability: In isolation from other particles, the structure of a particle is preserved under time evolution. According to this point of view, two or more particles can be identified in a state of a quantum field only if the particles are sufficiently separated so that they are not interacting. To find a particle in an arbitrary state θ, we consider the time dependent state

$$\theta(t) = U(t)\theta = e^{-itP_0}\theta$$

into which (or from which) θ evolves. As $|t| \to \infty$, we expect the particles to separate, and so they can be identified in $\theta(t)$. These are the particles before $(t \to -\infty)$ or after $(t \to +\infty)$ the interaction, and they are called the "in" particles and the "out" particles of the state θ. It follows that particles are properties of the $|t| \to \infty$ asymptotes of $\theta(t)$. In particular they provide a set of labels for these asymptotes. The Haag-Ruelle scattering theory [20] establishes the validity of this picture for certain states θ, under further hypotheses to be verified below.

From general spectral theory, however, the $|t| \to \infty$ asymptotes are determined by the (generalized) eigenfunctions of P_0. Since P_0 is the energy operator, particles provide a natural set of labels for the energy eigenfunctions. Since P_0 and P_1 commute we can choose the eigenfunctions to be joint eigenfunctions for P_0 and P_1. In this case the associated particles have definite momentum. The in and out particles are associated with two distinct sets of eigenfunctions and provide two distinct (in and out) representations of the spectral resolution of P_0. The in and out particles, eigenfunctions and spectral resolutions are interchanged by the scattering operator, i. e., S matrix.

Let $\sigma(P_0, P_1) \subset R^2$ denote the joint spectrum of the operators P_0, P_1. Positivity of the energy and Lorentz covariance ensure that $\sigma(P_0, P_1)$ is a Lorentz invariant subset of the forward cone,

$$0 \leq P_0\,, \qquad 0 \leq P_0^2 - P_1^2\,.$$

According to the Wightman axioms, the origin of R^2 is a simple eigenvalue of P_0 and P_1, and the corresponding eigenfunction is the vacuum state Ω.

By special relativity, single particles have a mass $(P_0^2 - P_1^2)^{1/2}$. Thus the mass operator, defined to be

$$M = (P_0^2 - P_1^2)^{1/2},$$

has an isolated eigenvalue $m > 0$ in the case the field describes a particle with positive mass. States with two or more asymptotic particles would have $M \geqq 2m$. If the in and out asymptotic particle states span $\mathcal{H}$, the theory is said to be asymptotically complete, and S is a unitary mapping of $\mathcal{H}$.

In this paper we establish this qualitative picture,* with the exception of asymptotic completeness, for weak $\mathcal{P}(\phi)_2$ interactions. We state our results in detail in the following section, and reduce the problem to estimates. In Chapters 2–4, we prove these estimates by an inductively defined expansion. We establish convergence of the expansion by introducing a family of norms on the terms in our expansion, and proving that each expansion step is norm decreasing. These techniques are related to expansions used in statistical mechanics.

The spatially cutoff $\mathcal{P}(\phi)_2$ model is defined by the Fock space Hamiltonian

$$H(g, \lambda, m_0) = H_0(m_0) + \lambda \int_{R_1} :\mathcal{P}(\phi_0(\vec{x}, m_0)): g(\vec{x}) d\vec{x} . \tag{1.6}$$

Fock space $\mathcal{F} = L_2(\mathcal{S}'(R^1), d\phi)$ is defined by the Gaussian measure $d\phi$ with mean zero and covariance $\mu^{-1} = (-d^2/dx_1^2 + m_0^2)^{-1/2}$, see [1]. Also : : denotes Wick ordering with respect to $d\phi$. We choose $\lambda\mathcal{P}(\cdot)$ bounded from below and $0 \leqq g(\cdot) \leqq 1$ with g measurable and of compact support. Then $H(g, \lambda, m_0)$ is essentially self-adjoint and inf spectrum H is a simple eigenvalue of H, with eigenstate Ω_g. We define

$$W(f_1, \cdots, f_n) = \langle \Omega_g, \phi(f_1) \cdots \phi(f_n)\Omega_g \rangle , \tag{1.7}$$

where $f_i \in \mathcal{S}(R^2)$ and where the time translation (1.4) is given by $H(g, \lambda, m_0)$.

We take advantage of the fact that positivity of the energy, $P_0 \geqq 0$, permits an analytic continuation to the region $\operatorname{Im} t \leqq 0$, and consequently a substitution $t \rightarrow -it$. This substitution converts the original hyperbolic field equation, e.g. $\Box\phi + m^2\phi + \phi^3 = 0$, into an elliptic equation, namely $-\Delta\phi + m^2\phi + \phi^3 = \delta/\delta\phi$, and it converts the Schrödinger equation $i\partial/\partial t = P_0$ into a heat equation $\partial/\partial t = -P_0$. The covariance group also undergoes an analytic continuation, and for the elliptic, or imaginary, time fields, the covariance group is the Euclidean group; hence these fields are called Euclidean fields [22], [15], [18], [13].

* Note added in proof: The presence or absence of bound states (depending on $\mathcal{P}$) is established in [24], [25], [26]. The Bethe Salpeter kernel, which relates to asymptotic completeness, will be studied in forthcoming papers.

The Feynman-Kac formula [22], [15], [2] provides an explicit construction of Euclidean field Φ in terms of the $t=0$ field ϕ_0. In the Feynman-Kac formula, we consider paths which are functions of the (imaginary) time parameter t, and which take values in the space of classical field configurations, namely $\mathcal{S}'(R^1)$. A convenient (and covariant) choice for the space of all such paths is $\mathcal{S}'(R^2)$. The Euclidean field, denoted $\Phi(f)$, $f \in \mathcal{S}(R^2)$, is simply the linear coordinate function

$$\Phi(f)(q) = \langle q, f\rangle \,, \qquad q \in \mathcal{S}'(R^2)$$

defined on the path space $\mathcal{S}'(R^2)$. Thus $\Phi(f)$ measures the position, or strength, of the field in the coordinate direction f. We will consider measures dq and $d\Phi$, defined on the path space $\mathcal{S}'(R^2)$. These measures have sufficient regularity to allow the definition of the sharp time field

$$\Phi(\tau, f) \equiv \Phi(\delta_\tau \otimes f) = \int \Phi(\tau, \vec{x}) f(\vec{x}) d\vec{x}$$

for $f \in \mathcal{S}(R^1)$. The connection between Φ and ϕ_0 is given by the formula

$$\Phi(0, \vec{x}) = \phi_0(\vec{x}) \,.$$

The Euclidean-Fock space is $L_2(\mathcal{S}'(R^2), d\Phi)$, where $d\Phi$ is the Gaussian measure with mean zero and covariance $(-\Delta + m_0^2)^{-1}$. Let h be a function with compact support in R^2, and let $0 \leq h \leq 1$. Define the action

$$V(h) = \lambda \int_{R^2} :\mathcal{P}(\Phi(x)): h(x) dx \tag{1.8}$$

as a function on $\mathcal{S}'(R^2)$, where $: \, :$ denotes Wick order with respect to the measure $d\Phi$, as in [9], [1]. Let dq_h be the cutoff measure defined by

$$dq_h = \frac{e^{-V(h)} d\Phi}{\int e^{-V(h)} d\Phi} \,.$$

If $h \to 1$ through a sequence of rectangular cutoffs, the limit

$$dq = \lim_{h \to 1} dq_h \,, \tag{1.9}$$

exists at least for λ/m_0^2 small, and defines dq [17].[5] The cutoff Schwinger function

$$S_h(x_1, \cdots, x_n) = \int \Phi(x_1) \cdots \Phi(x_n) dq_h \,, \tag{1.10}$$

is by definition the nth moment of the cutoff measure dq_h. The approximate Schwinger function S_h is well defined as a tempered distribution and, in

[5] The construction [17] of the measure dq assumes Theorem 1.1.7, and consequently the $n=0$ case of Theorem 1.1.11. We do not use dq in the proof of Theorem 1.1.7, but we assume its existence for our general exposition.

fact, integration over the space variable alone is sufficient to define S_h. Let $h_j = g(\bar{x})\chi_j(t)$, where χ_j is the characteristic function of the interval $[-j, j]$, and let $\hat{x} = (ix_0, \bar{x})$. Then the Feynman-Kac formula yields

$$W(\hat{x}_1, \cdots, \hat{x}_n) = \lim_{j\to\infty} S_{h_j}(x_1, \cdots, x_n) . \tag{1.11}$$

Finally, we note that a unitary scaling transformation on $L_2(\mathcal{S}'(R^2), d\Phi)$ carries $\Phi(x, m_0) \to \Phi(sx, s^{-1}m_0)$ and $V(\lambda, m_0) \to V(s^{-2}\lambda, s^{-1}m_0)$. Thus λ/m_0^2 is a dimensionless coupling constant, and a basic hypothesis of the entire paper is that λ/m_0^2 is chosen to be small.

1.1. The main results

The $\mathcal{P}(\phi)_2$ quantum field model is constructed as a limit of space cutoff theories, as the limit in $\mathcal{S}'(R^{2n})$ of the $W(x_1, \cdots, x_n)$ defined in (1.7) as $g \to 1$, or the limit $S_h(x_1, \cdots, x_n)$ defined in (1.10) as $h \to 1$.

THEOREM 1.1.1. *Let λ/m_0^2 be sufficiently small. The $\mathcal{P}(\phi)_2$ quantum field model exists as a limit of space cutoff theories and satisfies the Wightman axioms. Moreover the Wightman functions are continuous in λ ($\lambda \geqq 0$) in $\mathcal{S}'$ for λ sufficiently small.*

Remark. For the fourth order interaction $\mathcal{P}(\xi) = \xi^4 + m^2\xi^2 + \mu\xi$, with $\mu \neq 0$, Griffiths and Simon have eliminated the weak coupling hypothesis of this theorem [12], [21]. Our proof does not depend in an essential way on the boundary conditions while the Griffiths-Simon proof uses Dirichlet conditions.

Let

$$m = \inf \operatorname{spec} \cdot M \restriction \Omega^{\perp} .$$

We call m the physical mass.

THEOREM 1.1.2. *Let λ/m_0^2 be sufficiently small. The physical mass m is positive and m is an isolated discrete eigenvalue for M. Thus M has no spectrum in an interval (m, m_1), $m_1 > m$. As $\lambda \to 0$ with m_0 fixed, $m \to m_0$ and $m_1 \to 2m_0$.*

The eigenspace corresponding to the eigenvalue m of M is the space of one particle states. If $f \in \mathcal{S}(R^2)$ has fourier transform $f^\sim$ which satisfies

$$\operatorname{suppt} f^\sim(p_0, p_1) \subset \{p\colon 0 < p_0^2 - p_1^2 < m_1^2\} , \tag{1.1.1}$$

then $\phi(f)\Omega$ is a single particle state, i.e. an eigenstate of M with eigenvalue m.

THEOREM 1.1.3. *Let λ/m_0^2 be sufficiently small. The single particle states are spanned by the states $\phi(f)\Omega$, where f satisfies* (1.1.1).

According to conventional ideas, we expect only discrete spectrum for M in the interval $(m, 2m)$, and only continuous spectrum in the interval

$[2m, \infty]$.[6] The discrete spectrum (if any) in $(m, 2m)$ corresponds to bound states. Each discrete (nonzero) eigenvalue of M has infinite multiplicity, because its eigenspace carries a unitary representation of the Lorentz group. If the unitary representation is irreducible, then the eigenvalue refers to a single elementary particle or bound state.

THEOREM 1.1.4. *Let λ/m_0^2 be sufficiently small. The mass m representation of the Lorentz group on the single particle space is irreducible.*

Again let f satisfy (1.1.1) and define f_t by the equation

$$f_t^{\sim}(p) = \exp(it\mu(\bar{p}))f^{\sim}(p)$$

where $\mu(\bar{p}) = (\bar{p}^2 + m^2)^{1/2} = (p_1^2 + m^2)^{1/2}$. According to the Haag-Ruelle scattering theory [20] the multiparticle asymptotic in and out states are spanned by the vectors

$$\lim_{t\to\pm\infty} e^{-itP_0}\phi(f_t^{(1)}) \cdots \phi(f_t^{(n)})\Omega , \tag{1.1.2}$$

where the $f^{(j)}$ have nonoverlapping supports in momentum space. This restriction on the supports is necessary in two or three space-time dimensions for the proof of convergence of (1.1.2) as $t \to \pm\infty$, see [14].

Let $\mathcal{H}_{\text{in}}$ and $\mathcal{H}_{\text{out}}$ be the set of in and out states. By the Haag-Ruelle theory, Theorems 1.1.2 and 1.1.3 ensure the existence of an isometric scattering operator S from $\mathcal{H}_{\text{in}}$ and $\mathcal{H}_{\text{out}}$. A major open question is asymptotic completeness, that is, whether $\mathcal{H}_{\text{in}} = \mathcal{H} = \mathcal{H}_{\text{out}}$.

We derive Theorems 1.1.2–1.1.4 from an analysis of the bounded portions of the energy spectrum. We now consider test functions $h \in L_2(R^1)$ at the time zero, and let $\phi(h)$ denote the corresponding time-zero fields. Let

$$\phi_r(h_r) = \int :\phi(\bar{x}_1) \cdots \phi(\bar{x}_r): h_r(\bar{x}_1, \cdots, \bar{x}_r)d\bar{x} ,$$

$h_r \in L_2(R^r)$, be a time-zero Wick monomial of degree r. Let E_γ denote spectral projection for the energy region[7] $P_0 \leqq \gamma = (n+1)m_0(1-\varepsilon)$.

THEOREM 1.1.5. *Let n and $\varepsilon > 0$ be given, and let λ/m_0^2 be sufficiently small. Then $E_\gamma\mathcal{H}$ is spanned by vectors of the form*

$$\Omega,\ e^{tP_0}E_\gamma\phi(h_1)\Omega,\ \cdots,\ e^{tP_0}E_\gamma\phi(h_n)\Omega . \tag{1.1.3}$$

In the case $n = 1$, we prove a stronger result, by setting $t = 0$.

THEOREM 1.1.6. *Let $\delta > 0$ be given, and let λ/m_0^2 be sufficiently small.*

[6] For an even theory (e.g., ϕ^4), the symmetry $\phi \to -\phi$ generates a selection rule (the even and odd subspaces). On the odd subspace, discrete spectrum may be expected in the interval $(m, 3m)$.

[7] The energy bound and decay rate can be expressed intrinsically as $\gamma = (n+1)m(1-\varepsilon')$, since Theorem 1.1.2 ensures $m \to m_0$ as $\lambda/m_0^2 \to 0$.

Then $\mathcal{H}$ is spanned by vectors of the form

$$\Omega,\ \phi(h)\Omega \tag{1.1.4}$$

together with vectors of energy $P_0 \geqq 2m_0(1-\delta)$.

Remark. In Chapter 4 we establish a weaker result than Theorem 1.1.6, where we also include in the spanning states all vectors of total momentum $|P_1| \geqq \sqrt{3}\,m_0(1-\delta)$. This weaker version (Theorem 1.1.6′) is sufficient for the proof of Theorems 1.1.2, 1.1.3, and 1.1.4. Then using Theorem 1.1.2, Theorem 1.1.6 follows as stated.[8]

Proof of Theorems 1.1.2, 1.1.3, and 1.1.4, assuming Theorems 1.1.1, 1.1.5, and 1.1.6′. Let E be the projection onto states with energy $P_0 \leqq 2m_0(1-\varepsilon)$ and momentum $|P_1| \leqq \sqrt{3}\,m_0(1-\varepsilon)$. Let F be the projection onto states spanned by (1.1.4). By Theorem 1.1.6′

$$\mathcal{H} = \text{Range}\,(1-E) + \text{Range}\,F$$

and so Range E is spanned by the vectors

$$\Omega,\ E\phi(h)\Omega\,, \qquad h \in C_0^\infty(R^1)\,.$$

Let $h_a(x) = h(x-a)$. Then

$$e^{-iaP_1}\phi(h)\Omega = \phi(h_a)\Omega$$

and so

$$\phi(h_1 * h_2) = \int h_1(a)e^{-iaP_1}\phi(h_2)\Omega\,da$$

is in the span of the space translates of $\phi(h_2)\Omega$. Since E and P_1 commute, $E\phi(h_1 * h_2)\Omega$ is in the span of the space translates of $E\phi(h_2)\Omega$.

By Theorem 1.1.1 the two point function approaches the free field two point function as $\lambda \to 0$. If we express the truncated two point function in momentum space we see that for small λ the spectrum of P_0 is nonempty in the interval $(0, m_0+\varepsilon)$ and $E(\Omega^\perp) \neq 0$. We now choose h_2 so that $h_2^\sim(\bar{p}) > 0$. With h_2 fixed, the functions $h_1 * h_2$ are dense in $C_0^\infty(R^1)$ so for any $h \in \mathcal{S}(R^1)$, $E\phi(h)\Omega$ lies in the span of translates of $E\phi(h_2)\Omega$. Thus

$$E\phi(h_2)\Omega - \langle\Omega, \phi(h_2)\Omega\rangle\Omega$$

is a cyclic vector for the space translation group acting on $E(\Omega^\perp)$, and so the space translation group acting on this subspace has multiplicity one.

[8] The geometry of the one particle hyperboloid ensures that states with energy $P_0 \leqq 2m_0(1-\varepsilon)$ have momentum $|P_1| \leqq \sqrt{3}\,m_0(1-\varepsilon)$, as follows: By Theorem 1.1.2, $m \geqq m_0(1-\varepsilon)$. For states orthogonal to the vacuum Ω,

$$P_1^2 \leqq P_0^2 - m^2 \leqq 3m_0^2(1-\varepsilon)^2\,.$$

Since $P_1^2 = 0$ for the vacuum, $P_1^2 \leqq 3m_0^2(1-\varepsilon)^2$ in both cases. Thus the extra hypothesis in Theorem 1.1.6′ is not needed, and Theorem 1.1.6 follows.

Therefore the energy P_0 on this subspace is a function of the momentum. By Lorentz invariance, $P_0^2 = P_1^2 + (m')^2$ for some m', and we easily see that $m = m'$. The mass operator acting on $E(\Omega^\perp)$ therefore has a single eigenvalue m in $(0, 2m_0(1-\varepsilon))$. Since M is Lorentz invariant, M acting on $\mathcal{H}$ has a single eigenvalue in $(0, 2m_0(1-\varepsilon))$.

By Theorem 1.1.5 with $n = 0$, we see that $m \geqq m_0 - \varepsilon$ hence Theorem 1.1.2 is complete. Theorem 1.1.4 follows from the fact that reducibility produces multiplicity in the representation of the space translation subgroup. To prove Theorem 1.1.3, we observe that $E\phi(h)\Omega = \phi(f)\Omega$ for some $f \in L_2(R^2)$ satisfying (1.1.1). Thus the one particle states in the range of E are spanned by the states $\phi(f)\Omega$, where f satisfies (1.1.1). The theorem follows by Lorentz invariance.

The proof of Theorem 1.1.1 depends on a cluster property (uniform in h) for the measures dq_h or its moments S_h. The cluster property states that at widely separated points x and y, the strengths $\Phi(x)$ and $\Phi(y)$ of the field are independent up to some error $O(\exp[-m|x-y|])$. Two random variables $X(\omega)$ and $Y(\omega)$ are independent when the factorization property

$$\int F(X)G(Y)d\omega = \int F(X)d\omega \int G(X)d\omega$$

holds for functions F and G. The cluster property asserts that this identity holds up to an exponentially small error. The n-particle cluster expansion isolates explicitly n-particle corrections to the factorization; the remaining error $o(\exp[-nm|x-y|])$ is smaller than an n-particle decay. We use such cluster properties to establish Theorems 1.1.5–1.1.6′.

For functions of the Euclidean field in this identity, we consider monomials of the form

$$Q(x_0) = \int d\bar{x}\, w(x) \prod_{i=1}^{r} :\Phi(x_i)^{n_i}: \tag{1.1.5}$$

and

$$Q = \int dx_0 Q(x_0)\,. \tag{1.1.6}$$

Q is said to be localized if $w(x)$ is supported in a product $\Delta_{j_1} \times \cdots \times \Delta_{j_n} \subset R^{2r}$ where Δ_{j_i} is the unit lattice square in R^2 centered at the lattice side $j_i \in Z^2$. We assume a bound on the n_i in (1.1.5), $n_i < \bar{n}$.[9] Any choice $\bar{n} \geqq \deg P$ is sufficient. Since there is no restriction on r, Q still may have an arbitrarily large degree.

For a localized monomial Q or $Q(x_0)$, let $N(\Delta)$ be the local degree, i.e. the sum of the n_i for which $\Delta_{j_i} = \Delta$. The estimates in this paper are expres-

[9] We may eliminate this hypothesis, i.e., take $\bar{n}=\infty$, by replacing $N(\Delta)!$ by $(N(\Delta)!)^3$ in (1.1.7), (1.1.8), and (2.4.5).

sed in terms of norms. Define

$$\|Q\| = K_1[\textstyle\prod_\Delta K_2^{2N(\Delta)} N(\Delta)!]\|w\|_2 , \tag{1.1.7}$$

$$\|Q(x_0)\| = K_1[\textstyle\prod_\Delta K_2^{2N(\Delta)} N(\Delta)!]\|w(x_0, \cdot)\|_2 . \tag{1.1.8}$$

If Q is not localized, or is a general polynomial, it is written as a sum of localized monomials Q_i, and then we set

$$\|Q\| = \textstyle\sum_i \|Q_i\| . \tag{1.1.9}$$

For two localized monomials Q_1 and Q_2, with disjoint localization squares $\{\Delta_{j_i}^{(1)}\}$ and $\{\Delta_{j_l}^{(2)}\}$ respectively, let d be the width of the widest strip bounded by two parallel lines, and separating the $\{\Delta_{j_i}^{(1)}\}$ from the $\{\Delta_{j_l}^{(2)}\}$

THEOREM 1.1.7. *Let m_0, K_2 and K_1 be sufficiently large and let λ be sufficiently small. There are constants $0 < m_1$ and $O(1)$ independent of h, Q_1 and Q_2 such that*

$$\left|\int Q_1 Q_2 dq_h - \int Q_1 dq_h \int Q_2 dq_h\right| \leqq O(1) e^{-m_1 d} \|Q_1\| \|Q_2\| . \tag{1.1.10}$$

Proof of Theorem 1.1.1 assuming Theorem 1.1.7. In order to establish the Wightman axioms we verify the equivalent Osterwalder-Schrader axioms for the Schwinger functions [18], [23]. By the estimates of this paper, S_h is bounded in $\mathcal{S}'(R^{2n})$ uniformly as $h \to 1$. According to the discussion of [19, Section 6] it thus is sufficient to prove convergence of the cutoff Schwinger functions S_h. Convergence as $h \to 1$ of the real time vacuum expectation values (1.1.7) follows: The uniform bound on $W(x_1, \cdots, x_n)$ in $\mathcal{S}'(R^{2n})$ yields a uniform bound on their analytic continuations to Euclidean points. Thus convergence of the Schwinger functions and Vitali's theorem yield real-time convergence. The convergence (uniform in λ) of the Schwinger functions in λ ensures that the Schwinger functions are continuous in λ. Continuity of the Wightman functions follows from the uniform bound [6] on their analytic continuation and Vitali's theorem.

By the uniform boundedness of S_h in $\mathcal{S}'(R^{2n})$, we may prove convergence on a dense set of test functions. For test functions $f_i \in C_0^\infty(\mathcal{O})$ and for cutoff functions h_1 and h_2, we assert that

$$\left|\int \textstyle\prod_i \Phi(f_i) dq_{h_1} - \int \prod_i \Phi(f_i) dq_{h_2}\right| \leqq 0(1) e^{-m_1 d/2} \tag{1.1.11}$$

where $d = \text{dist}\,(\mathcal{O}, \text{suppt}\, h_1 - h_2)$, and then Theorem 1.1.1 follows. To prove (1.1.11) let $g = h_2 - h_1$, and define $g_i = g\chi_{\Delta_i}$, where χ_{Δ_i} is the characteristic function of Δ_i. Also we use the notation

$$\langle F \rangle_h = \int F dq_h ,$$

and set $Q = \prod_i \Phi(f_i)$. Then

$$\langle Q\rangle_{h_2} - \langle Q\rangle_{h_1} = \int_0^1 d\alpha \frac{d}{d\alpha}\langle Q\rangle_{h_1+\alpha g}$$

$$(1.1.12) \qquad = -\int_0^1 d\alpha[\langle QV(g)\rangle_{h_1+\alpha g} - \langle Q\rangle_{h_1+\alpha g}\langle V(g)\rangle_{h_1+\alpha g}]$$

$$= -\int_0^1 d\alpha \sum_{i\in Z^2}[\langle QV(g_i)\rangle_{h_1+\alpha g} - \langle Q\rangle_{h_1+\alpha g}\langle V(g_i)\rangle_{h_1+\alpha g}]\,.$$

We apply Theorem 1.1.7 to each term in the sum to obtain

$$|\langle Q\rangle_{h_2} - \langle Q\rangle_{h_1}| \leqq \sum_{i\in Z^2\cap \operatorname{supp} g} O(1)e^{-m_1 \operatorname{dist}(\mathcal{O},\Delta_i)}$$
$$\leqq O(1)e^{-m_1 d/2}\,.$$

As a special case, the cutoff Schwinger functions S_h converge as $h \to 1$.

Remark. The $h = 1$ fields can also be constructed from the Wightman reconstruction theorem starting from the Wightman functions as well as from the GNS construction using C^* algebra states. These latter constructions coincide with each other [6], [5] and with the Osterwalder-Schrader construction employed above.

We now reformulate Theorems 1.1.5 and 1.1.6 in terms of path space. In the monomial $Q(x_0)$ of (1.1.5), suppose $x_0 = (x_{1,0}, \cdots, x_{r,0})$ where $x_{1,0} \leqq \cdots \leqq x_{r,0}$. Let

$$\eta_l = x_{l+1,0} - x_{l,0} \geqq 0$$

and let

$$(1.1.13) \qquad \theta = \int d\vec{x}w(x){:}\,\phi_0(\vec{x}_{1,1})^{n_1}{:}\, e^{-\eta_1 P_0} \cdots e^{-\eta_{r-1}P_0}{:}\,\phi_0(\vec{x}_{r,1})^{n_r}{:}\,\Omega\,.$$

One version of the Feyman-Kac formula is the identity

$$\int Q(x_0)dq = \langle\Omega, \theta\rangle$$

where the inner product on the right side is in $\mathcal{H}$.

THEOREM 1.1.8. *For $Q(x_0)$ of the form* (1.1.5),

$$\left|\int Q(x_0)\,dq_h\right| \leqq O(1)\,||\,Q(x_0)\,||\,,$$

uniformly as $h \to 1$.

Remark. A special case of the theorem is the bound on the n-point Schwinger functions

$$|S_h(f_1 \otimes \cdots \otimes f_n)| \leqq n!\prod |f_i|_{\mathcal{S}}$$

where $|\cdot|_{\mathcal{S}}$ is a Schwartz space norm (in this case an L_2 norm) independent of h. This bound verifies the Osterwalder-Schrader axiom EO'' of [23] for weakly coupled $\mathcal{P}(\phi)_2$ models.

COROLLARY 1.1.9. *Let Q and θ be related as in* (1.1.13). *Then*

$$\|\theta\| \leqq O(1)\|Q(x_0)\| = O(1)[\prod_\Delta K_2^{2N(\Delta)} N(\Delta)!]\|w(x_0, \cdot)\| .$$

Proof. We write $\|\cdot\|_{K_2}$ to indicate the dependence of the norm on K_2. Apply the theorem to

$$Q_1 = Q(x_{i0} - x_{10})Q(x_{10} - x_{i0})^- ,$$

where Q has the form (1.1.5), to obtain

$$\|Q(x_0)\|^2 = \int Q_1 dq \leqq O(1)\|Q_1\|_{K_2} \leqq O(1)\|Q(x_0)\|^2_{2K_2} .$$

We let A and B denote monomials of type (1.1.5) of degree $|A|$ and $|B|$ and with kernels w_A and w_B respectively. We assume that A is localized at times $x_{i0} \leqq 0$, and that B is localized at times $x_{i0} \geqq 0$. Let B_T denote the translate of B at time $T > 0$, so that B_T is localized at times $\geqq T$.

In the next two results, we take all times $x_{1,0} = t = \cdots = x_{r,0}$ equal in $Q(x_0)$. To indicate this restriction, we write $Q(t)$ in place of $Q(x_0)$.

THEOREM 1.1.10. *Let $\varepsilon > 0$, $T > 0$ be given, and let B have support in momentum space in the region $|P_1| \leqq \sqrt{3}\, m_0(1 - \varepsilon)$. Then there is a*

$$Q(T) = \alpha(T) + \int \Phi(0, \vec{x})h_T(\vec{x})d\vec{x} ,$$

chosen independently of B, such that

$$\alpha(T) \in R^1 , \quad h_T \in L_2(R^1) , \quad \|Q(T)\| \leqq O(1)\|A\| ,$$

and

$$\left|\int (A - Q(T))B_T dq\right| \leqq O(1)\|A\|\,\|B\|e^{-2Tm_0(1-\varepsilon)} .$$

THEOREM 1.1.11. *Let n and $\varepsilon > 0$ be given and let $\gamma = (n + 1)m_0(1 - \varepsilon)$. For any A, we may choose n^{th} degree polynomials $Q_n(t_j)$ of the form* (1.1.5) *with $r = 1$, so that*

$$\sum_{t_j > 0} e^{\gamma t_j}\|Q_n(t_j)\| \leqq O(1)\|A\| \tag{1.1.14}$$

and so that

$$Q = \sum_{0 < t_j < T} Q_n(t_j) \tag{1.1.15}$$

satisfies

$$\left|\int (A - Q)B_T dq_h\right| \leqq O(1)\|A\|\,\|B\|e^{-\gamma T} , \tag{1.1.16}$$

uniformly in h. We shall choose $t_j = j$, $j \in Z$.

Proof of Theorem 1.1.7, assuming Theorems 1.1.8 and 1.1.11. Take

$n=0$ in Theorem 1.1.11 Then Q is a polynomial of degree zero, i.e., a constant. From the choice $B=I$, it follows that

$$Q=\int A dq_h+O(1)\|A\|e^{-m_0(1-\varepsilon)T}.$$

After rotating and translating the strip separating Q_1 and Q_2 in Theorem 1.1.7, and after choosing $A=Q_1$, $B_T=Q_2$ in Theorem 1.1.11, the proof follows from Theorem 1.1.8, (1.1.16) and the above determination of Q.

Proof of Theorem 1.1.5, assuming Theorem 1.1.11, Corollary 1.1.9. We choose

$$\beta\equiv(n+1)m_0(1-\delta)<\gamma\equiv(n+1)m_0(1-\varepsilon)\,.$$

Let $\theta(t_j)$ be the vector (1.1.13) corresponding to $Q_n(t_j)^-$ and let

$$\begin{aligned}\psi(t_j)&=e^{t_jP_0}E_\gamma\theta(t_j)\\ \psi&=\textstyle\sum_j\psi(t_j)\,.\end{aligned}$$

Since

$$\begin{aligned}\textstyle\sum_j\|\psi(t_j)\|&\leqq\textstyle\sum_j e^{t_j\gamma}\|\theta(t_j)\|\\ &\leqq\textstyle\sum_j e^{\gamma t_j}\|Q_n(t_j)\|\leqq O(1)\|A\|\,,\end{aligned}$$

the series defining ψ converges, and $\|\psi\|\leqq O(1)\|A\|$. To prove Theorem 1.1.5 it suffices to show that

$$E_\beta(\theta_A-\psi)=0$$

where θ_A is the vector (1.1.13) corresponding to A. Let

$$\psi_T=\textstyle\sum_{0<t_j<T}\psi(t_j)\,,$$

and then $\psi_T\to\psi$. Moreover, if $T_1\leqq T_2\leqq\infty$,

$$\begin{aligned}\|e^{-T_1P_0}(\psi_{T_1}-\psi_{T_2})\|&\leqq\textstyle\sum_{T_1\leqq t_j<T_2}\|e^{(t_j-T_1)P_0}E_\gamma\theta(t_j)\|\\ &\leqq\textstyle\sum_{T_1\leqq t_j<T_2}e^{(t_j-T_1)\gamma}\|\theta(t_j)\|\\ &\leqq O(1)e^{-\gamma T_1}\|A\|\,.\end{aligned}\tag{1.1.17}$$

Here we used Corollary 1.1.9 and (1.1.14). Consequently,

$$\begin{aligned}\|E_\beta(\theta_A-\psi)\|^2&\leqq e^{\beta T}\langle\theta_A-\psi,E_\beta e^{-TP_0}(\theta_A-\psi)\rangle\\ &\leqq e^{\beta T}\langle\theta_A-\psi,e^{-TP_0}(\theta_A-\psi)\rangle\\ &\leqq O(1)\|A\|^2e^{-(\gamma-\beta)T}+e^{\beta T}|\langle\theta_A-\psi_T,e^{-TP_0}\theta_A\rangle|\\ &\quad+e^{\beta T}|\langle\theta_A-\psi_T,e^{-TP_0}\psi_T\rangle|\,.\end{aligned}\tag{1.1.18}$$

We bound the second term above as follows:

$$\begin{aligned}\langle\theta_A-\psi_T,e^{-TP_0}\theta_A\rangle&=\int(A-Q)^-A_Tdq\\ &\quad+\textstyle\sum_{0<t_j<T}\langle\theta(t_j),(E_\gamma-I)e^{-(T-t_j)P_0}\theta_A\rangle\,.\end{aligned}\tag{1.1.19}$$

Hence by (1.1.14)-(1.1.16) and $\|(E_\gamma - I)e^{-(T-t_j)P_0}\| \leqq e^{-(T-t_j)\gamma}$,

$$|\langle \theta_A - \psi_T, e^{-TP_0}\theta_A\rangle| \leqq O(1)\|A\|^2 e^{-\gamma T} + \|A\| \textstyle\sum_{0<t_j<T} e^{-(T-t_j)\gamma}\|\theta(t_j)\| \leqq O(1)\|A\|^2 e^{-\gamma T} . \tag{1.1.20}$$

Similarly, we bound the third term in (1.1.18):

$$\begin{aligned}\langle \theta_A - \psi_T, e^{-TP_0}\psi_T\rangle = & \textstyle\sum_{0<t_j<T} \langle \theta_A - \psi_{T-t_j}, e^{-(T-t_j)P_0}\theta(t_j)\rangle \\ & - \textstyle\sum_{0<t_j<T} \langle \psi_T - \psi_{T-t_j}, e^{-(T-t_j)P_0}E_\gamma\theta(t_j)\rangle \\ & + \textstyle\sum_{0<t_j<T} \langle \theta_A - \psi_{T-t_j}, e^{-(T-t_j)P_0}(E_\gamma - I)\theta(t_j)\rangle .\end{aligned} \tag{1.1.21}$$

We bound the first term in (1.1.21) using (1.1.19) with $T - t_j$ in place of T and $\theta(t_j)$ in place of θ_A. In place of (1.1.20), we obtain

$$\begin{aligned}&|\textstyle\sum_{0<t_j<T} \langle \theta_A - \psi_{T-t_j}, e^{-(T-t_j)P_0}\theta(t_j)\rangle| \\ &\leqq \textstyle\sum_{0<t_j<T} O(1)\|A\|\,\|Q(t_j)\|e^{-(T-t_j)\gamma} \leqq O(1)\|A\|^2 e^{-\gamma T} .\end{aligned} \tag{1.1.22}$$

For the second term in (1.1.21), we use (1.1.17) with $T_1 = T - t_j$, $T_2 = T$, obtaining the bound $O(1)\|A\|^2 e^{-\gamma T}$. For the third term in (1.1.21), we use

$$\begin{aligned}&|\textstyle\sum_{0<t_j<T} \langle \theta_A - \psi_{T_j}, e^{-(T-t_j)P_0}(E_\gamma - I)\theta(t_j)\rangle| \\ &\leqq \textstyle\sum_{0<t_j<T} \{\|\theta_A\| + \|\psi_{T-t_j}\|\}e^{-(T-t_j)\gamma}\|\theta(t_j)\| \\ &\leqq O(1)\|A\|\textstyle\sum_{0<t_j<T} e^{-(T-t_j)\gamma}\|Q(t_j)\| \\ &\leqq O(1)\|A\|^2 e^{-T\gamma} .\end{aligned}$$

Combining these bounds,

$$|\langle \theta_A - \psi_T, e^{-TP_0}\psi_T\rangle| \leqq O(1)\|A\|^2 e^{-\gamma T} .$$

Thus using (1.1.20), we obtain for (1.1.18) the bound

$$\|E_\beta(\theta_A - \psi)\|^2 \leqq O(1)\|A\|^2 e^{-(\gamma-\beta)T} .$$

Thus $E_\beta(\theta_A - \psi) = 0$. Since the states θ_A are dense in $\mathcal{H}$, the proof is complete.

Proof of Theorem 1.1.6′, assuming theorem 1.1.10. We use Theorem 1.1.10 in place of Theorem 1.1.11. The only new complication is the restriction on the momentum support of B. Let A be supported in the region

$$|P_1| \leqq \sqrt{3}\, m_0(1 - \delta) . \tag{1.1.23}$$

The proof consists of approximating $E_\beta\theta_A$ by $E_\beta\theta_Q$ where Q is a linear function of the time-zero field. We choose a function $\zeta \in \mathcal{S}(R^1)$ with Fourier transform $\zeta^\sim(P_1)$ supported in the region

$$|P_1| \leqq \sqrt{3}\, m_0(1 - \varepsilon) , \tag{1.1.24}$$

and equal to one on (1.1.23). Let $Q(T)$, α_T, h_T be chosen by Theorem 1.1.10 and let

$$Q(T, \zeta) = \alpha_T + \int \Phi(0, \bar{x})(h_T * \zeta)(\bar{x}) d\bar{x} .$$

By Theorem 1.1.10

$$\left| \int (A - Q(T) Q(T, \zeta)_{\bar{T}}) dq \right| \leqq O(1) \| A \|^2 e^{-T\gamma} , \tag{1.1.25}$$

where $\gamma = 2m_0(1 - \varepsilon)$. We introduce the vectors $\theta_T = \theta_{Q(T)}$ and $\theta_{T,\zeta} = \theta_{Q(T,\zeta)}$. Then (1.1.25) can be written

$$|\langle \theta_A - \theta_T, e^{-TP_0} \theta_{T,\zeta} \rangle| \leqq O(1) \| A \|^2 e^{-T\gamma} . \tag{1.1.26}$$

Furthermore

$$Q(T) - Q(T, \zeta) = \Phi(0, h_T * (\delta - \zeta)) ,$$

is supported in momentum space on the complement of (1.1.23). For this reason, and because $P_1^2 \leqq P_0^2$, we see that $\theta_T - \theta_{T,\zeta}$ is a state whose energy spectrum satisfies the inequality

$$\nu \equiv \sqrt{3}\, m_0(1 - \delta) \leqq P_0 . \tag{1.1.27}$$

Then

$$\begin{aligned} &\langle \theta_A - \theta_T, e^{-TP_0}(\theta_A - \theta_T) \rangle \\ &\leqq O(1) \| A \|^2 e^{-\gamma T} + |\langle \theta_A - \theta_T, e^{-TP_0}(\theta_T - \theta_{T,\zeta}) \rangle| \\ &\quad + |\langle \theta_A - \theta_T, e^{-TP_0} \theta_{T,\zeta} \rangle| . \end{aligned} \tag{1.1.28}$$

Here the first term results from Theorem 1.1.10 with $B = A$. The third term is bounded by (1.1.26). For the second term, we use (1.1.27) and Theorem 1.1.8 to obtain the bound $O(1) \| A \|^2 e^{-\nu T}$. Thus for $\nu' < \nu$,

$$\begin{aligned} \| E_{\nu'}(\theta_A - \theta_T) \|^2 &\leqq e^{\nu' T} \| E_{\nu'} e^{-TP_0/2}(\theta_A - \theta_T) \|^2 \\ &\leqq e^{\nu' T} \| e^{-TP_0/2}(\theta_A - \theta_T) \|^2 \\ &\leqq O(1) \| A \|^2 e^{-(\nu - \nu')T} \to 0 \end{aligned} \tag{1.1.29}$$

as $T \to \infty$. This proves Theorem 1.1.6′ with $\beta = 2m_0(1 - \delta)$ replaced by ν, and thus it proves Theorem 1.1.2 with $m_1 \to 2m_0$ replaced by $m_1 \geqq \nu$. However, from Theorem 1.1.2 we still conclude that $m \to m_0$ as $\lambda \to 0$, and so we may suppose that $m > m_0(1 - \varepsilon)$. With this extra information, we can improve on (1.1.27) as follows:

$$\| e^{-TP_0}(\theta_T - \theta_{T,\zeta}) \| \leqq O(1) \| A \|^2 e^{-T\beta'} ,$$

where

$$\beta' = m_0((1 - \varepsilon)^2 + 3(1 - \delta)^2)^{1/2} > \beta .$$

Now we choose $\nu' = \beta$ in (1.1.29) and obtain

$$\begin{aligned} \| E_\beta(\theta_A - \theta_T) \|^2 &\leqq e^{\beta T} \langle \theta_A - \theta_T, E_\beta e^{-TP_0}(\theta_A - \theta_T) \rangle \\ &\leqq O(1) \| A \|^2 e^{-(\gamma - \beta)T} \to 0 . \end{aligned}$$

This completes the proof.

Proof of Theorem 1.1.8, assuming Theorem 1.1.10. Choose $T = 0$ and $B = I$ in Theorem 1.1.11. Then $Q = 0$, so Theorem 1.1.8 is a special case of Theorem 1.1.10.

We are now reduced to proving Theorems 1.1.10 and 1.1.11. These results are obtained from the cluster expansion, to which the remainder of the paper is devoted.

2. Inserting contours

2.0 Introduction

The proofs of Theorems 1.1.10 and 1.1.11 are based on an inductively defined expansion for $\int AB_T dq_k$, described in this and subsequent chapters. We establish the convergence of this expansion by introducing a norm on the set of terms in each expansion step and by showing that each expansion step is norm decreasing. In our expansion, we use a space-time cover by squares of unit area. Thus our expansions involve a characteristic length, and convergence of our expansions requires both sufficiently small λ and sufficiently large m_0. Our Theorems 1.1.1–1.1.6, the consequences of our expansions, involve no such length. Hence, by scaling, they apply whenever the dimensionless coupling λ/m_0^2 is sufficiently small.

In order to define the expansion we introduce measures on $\mathcal{S}'(R^2)$ which factor over disjoint regions of space-time. If C is a continuous positive definite bilinear form on $\mathcal{S}(R^2)$, let $d\Phi_C$ denote the Gaussian measure on $\mathcal{S}'(R^2)$ of mean zero and covariance C. The family of covariances $\mathcal{C}$ that we consider are constructed from the free covariance $C_0 = (-\Delta + m_0^2)^{-1}$ by the application of two elementary operations: Diagonalization and convex sums. Let $Y \subset R^2$ be a union of unit lattice squares Δ, centered at integer coordinates, and let Y also be the operator of multiplication by the characteristic function of the set Y. If C is a bounded operator on $L_2(R^2)$ we define its diagonalization with respect to Y to be

$$C_Y = YCY + (1 - Y)C(1 - Y) .$$

We say ∂Y is a complete contour in C_Y. The convex sum of two covariance operators C' and C'' is

$$C(s) = sC' + (1 - s)C'' , \qquad s \in [0, 1] . \tag{2.0.1}$$

If $s \in (0, 1)$ and C' or C'' has complete (or partial) contour ∂Y we say $C(s)$ has ∂Y as a partial contour.

If ∂Y is a complete contour in Y, the measure $d\Phi_C$ factors across ∂Y. This means that if polynomials A and B are localized in Y and $\sim Y$ respectively (i. e., their localizations are separated by ∂Y), then

$$\int ABe^{-V(h)}d\Phi_C = \int Ae^{-V(Yh)}d\Phi_C \cdot \int Be^{-V((1-Y)h)}d\Phi_C$$

$$= \frac{\int Ae^{-V(h)}d\Phi_C \cdot \int Be^{-V(h)}d\Phi_C}{\int e^{-V(h)}d\Phi_C} . \tag{2.0.2}$$

In other words, the Gaussian measure $d\Phi_C$ factors across each of its complete contours, for $C \in \mathcal{C}$. Also the measure

$$dq = dq_C = \frac{e^{-V}d\Phi_C}{\int e^{-V}d\Phi_C}$$

factors across each complete contour, and

$$\int ABdq_C = \int Adq_C \int Bdq_C .$$

We say that (2.0.2) "factors out" the polynomial (or vertices) A.

In this chapter we express $\int AB_T e^{-V}d\Phi$ as a sum of terms labeled by indices α or α',

$$\sum_\alpha b_\alpha \int \left(\prod_{i=1}^{n_\alpha} a_i\right) B_T e^{-V} d\Phi_{C_\alpha} + \sum_{\alpha'} b_{\alpha'} \int e^{-V} d\Phi_{C_{\alpha'}} . \tag{2.0.3}$$

Here C_α belongs to $\mathcal{C}$, b_α and $b_{\alpha'}$ are constants, the a_i are annihilation operators, to be defined later, and $n_\alpha \leqq n$. We call this part of our expansion *the increasing contour expansion*. We generate the expansion with the elementary identity

$$\int Fd\Phi_{C'} = \int Fd\Phi_{C''} + \int_0^1 ds \frac{d}{ds} \int Fd\Phi_{C(s)} , \tag{2.0.4}$$

where $C(s) = sC' + (1-s)C''$, as in (2.0.1). If C'' has ∂Y as a complete contour, we say that (2.0.4) inserts the complete contour ∂Y. If $C'' = C_0$, we say that (2.0.4) removes all contours. For suitably regular functions F on $\mathcal{S}'(R^2)$, integration by parts yields (see [1])

$$\frac{d}{ds}\int Fd\Phi_{C(s)} = \frac{1}{2}\int_0^1 ds \int (\delta C \cdot \Delta_\Phi) Fd\Phi_{C(s)} . \tag{2.0.5}$$

Here $\delta C = dC(s)/ds = C' - C''$,

$$(\delta C \cdot \Delta_\Phi)F = \int \delta C(x, y) \frac{\partial^2 F}{\partial\Phi(x)\partial\Phi(y)} dxdy , \tag{2.0.6}$$

and $\delta C(x, y)$ is the integral kernel of the operator δC.

Three general types of integrands F occur in our expansion. For the first class of terms,

$$F = R(\Phi)B_T e^{-V(h)}, \tag{2.0.7}$$

where $R(\Phi)$ is a polynomial in Φ localized at times earlier than T. Such polynomials $R(\Phi)$ arise, for example, from derivatives $(\delta C \cdot \Delta_\Phi)Ae^{-V}$. In these terms we factor out $R(\Phi)$ by inserting a contour that separates the localization of $R(\Phi)$ from the localization of B_T. The term in (2.0.4) with the complete contour contributes to the first sum in (2.0.3), while the second term in (2.0.4) is expanded further. The new contour is chosen to enclose previous contours, so the contours are "increasing".

The second type of integrand is

$$F = R(\Phi)e^{-V(h)},$$

where $R(\Phi)$ is a polynomial whose localizations include times greater than T. In this case, factoring out $R(\Phi)$, yields a contribution to the second sum in (2.0.3), plus a term to be expanded further. The third class of F has annihilators; we deal with these terms in a slightly more complicated way. We obtain (2.0.3) by an iteration of these steps, described in detail in Section 2.3.

Each application of (2.0.6) will yield a small factor in our estimates, either $O(m_0^{-1})$ or $O(\lambda)$, and this is the key mechanism yielding convergence of our expansion.

In Chapter 3 we give an expansion of the form (2.0.3) where now $C_\alpha = C_{\alpha'} = C_0$. We obtain this expansion by alternating the increasing contour expansion and the formula (2.0.4) for removing contours.

Finally, in Chapter 4, we define the full n-particle cluster expansion. For the full expansion, we apply the Chapter 3 expansion and we use an identity for the Euclidean field, $\Phi = a + a^*$. Integration by parts yields

$$a = \Phi + K\frac{\partial}{\partial\Phi} \tag{2.0.8}$$

where K is an integral operator defined in Chapter 4. Hence we obtain in (2.0.3),

$$\int\left(\prod_{i=1}^{n_\alpha} a_i\right) B_T e^{-V}\, d\Phi = \int\left[\prod_{i=1}^{n_\alpha}(\Phi + K\partial/\partial\Phi)\right] B_T e^{-V}\, d\Phi .$$

The pure Φ term has the desired form and contributes to the polynomial Q in Theorem 1.1.11. Terms involving $K\partial/\partial\Phi$ yield new vertices, and we apply the Chapter 3 expansion to these terms. Iterating this procedure, we obtain the desired cluster expansion. We summarize the structure of the expansion in the following block diagram:

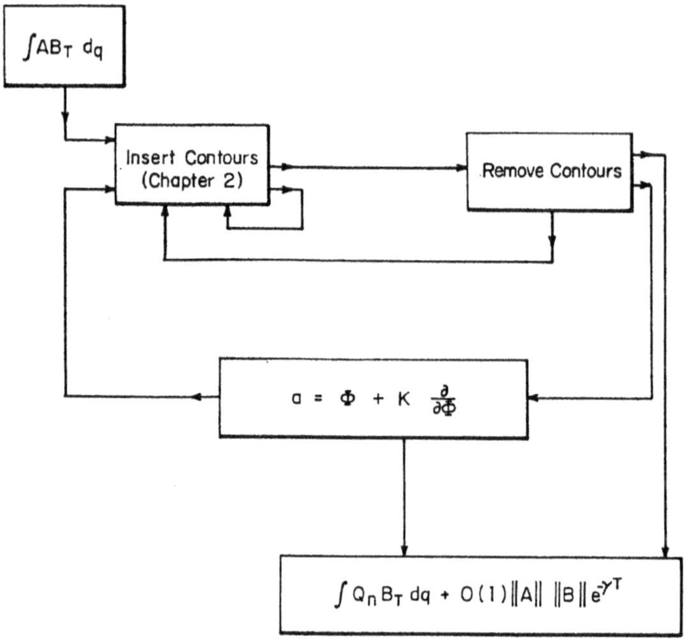

FIGURE 1. *The n-particle cluster expansion*, an example of an "inductive construction". The Chapter 3 expansion includes the two blocks for inserting contours and for removing contours. In Chapter 4, all blocks are combined.

2.1 The covariance operators

In this section we introduce notation for the covariance operators of class $\mathcal{C}$, and we prove estimates on the behavior of the kernels δC which appear in (2.0.5). We consider two subsets $\mathcal{C}_1$, $\mathcal{C}_2$ of $\mathcal{C}$. The first subset, $\mathcal{C}_1$, occurs in the increasing contour expansion of this chapter. Given an element $C_* \in \mathcal{C}_2$, and finite sequences

(2.1.1) $$\mathcal{Y}_j = \{Y_1, \cdots, Y_j\}, \qquad s = \{s_1, \cdots, s_j\},$$

of expanding contours ∂Y_i, $Y_i \subsetneq Y_{i+1}$, and parameters $s_i \in (0, 1)$, we define $C_*(\mathcal{Y}_j, s) = C_j$ by recursion:

$$C_1 = s_1 C_* + (1 - s_1)(C_*)_{Y_1},$$

and

(2.1.2) $$C_i = s_i C_{i-1} + (1 - s_i)(C_{i-1})_{Y_i}, \qquad i \geq 2.$$

Let $Y_0 = \emptyset$, and let $\delta Y_k = Y_k \sim Y_{k-1}$ be the increment added by Y_k to Y_{k-1}. Clearly $C_i(x, y) = C_{i-1}(x, y)$, unless x and y are separated by ∂Y_i. Let $y \notin Y_j$

and $x \in Y_j$. By induction we see for $x \in \delta Y_k$,

$$C_j(x, y) = S_{k,j} C_*(x, y) = \left(\prod_{i=k}^{j} s_i\right) C_*(x, y) . \tag{2.1.3}$$

Here

$$S_{k\ j} = \prod_{i=k}^{j} s_i . \tag{2.1.4}$$

A factor s_i enters the product $S_{k,j}$ for each contour ∂Y_i among $\partial Y_1, \cdots, \partial Y_j$ which separate x and y.

The increment $\delta C_j = (d/ds_j)C_j$ vanishes unless x and y are separated by δY_j. Let

$$\delta S_1 \equiv 1; \quad \delta S_{k,j} \equiv \frac{d}{ds_j} S_{k.j} = \prod_{i=k}^{j-1} s_i = S_{k.j-1} , \tag{2.1.5}$$

for $j > 1$. If $x \in Y_j \cap \delta Y_k$, $y \notin Y_j$, then

$$\delta C_j(x, y) = \delta S_{k.j} C_*(x, y) .$$

When possible, we suppress j in $S_{k,j} = S_k$. We say that C_* is a background covariance with contours Y_j superimposed.

PROPOSITION 2.1.1. *Let Δ and Δ' be squares of unit area or intervals of unit length in R^2. Let $d =$ dist.(Δ, Δ'), $\varepsilon > 0$, $p < \infty$ and let $C_j \in \mathcal{C}_1$ have the form above, with j contours ∂Y_i. Then*

$$\sum_{\Delta,\Delta'} \| \delta C_j \|_{L_p(\Delta\times\Delta')} e^{m_0(1-\varepsilon)d} \leqq O(1) \sum_{k=1}^{j} \delta S_{k,j} |\delta Y_k| , \tag{2.1.6}$$

and for $d \geqq 1$, $p = \infty$ is allowed in (2.1.6).

Proof. We note that $|C_*(x, y)| \leqq C_0(x, y)$ so we may bound C_* by C_0 in (2.1.2). Also note C_0 is locally L_p, and for $|x - y|$ bounded away from zero, $C_0(x - y) \leqq O(1) \exp(-m_0(1 - \varepsilon/2)|x - y|)$. Since δC_j vanishes unless δY_j separates x and y, we can choose $\Delta \in \delta Y_k$, $\Delta' \notin Y_j$. We fix Δ and sum over Δ'. By (2.1.6) and the exponential bound on C_0,

$$\sum_{\Delta'} \| \delta C_j \|_p e^{m_0(1-\varepsilon)d} \leqq O(1) \delta S_{k,j} .$$

Summing Δ over δY_k gives the area $|\delta Y_k|$, and summing k yields (2.1.6).

To complete our characterization of $\mathcal{C}_1$, we describe the subset $\mathcal{C}_2$ of $\mathcal{C}$ which contains C_*. The class $\mathcal{C}_2$ is generated by alternately diagonalizing the covariance C, and forming a convex linear combination of C with C_0. The first step is to diagonalize C_0. Thus given

$$\mathfrak{X}_l = \{X_1, \cdots, X_l\}; \qquad u = \{u_1, \cdots, u_l\}$$

with $u_i \in (0, 1)$, $u_l \in (0, 1]$, we define

$$C_j = (1 - u_j)C_0 + u_j(C_{j-1})_{X_j} , \tag{2.1.7}$$

and $C(\mathfrak{X}_l, u) = C_l$. If $u_l = 1$, we say that C_l has a complete contour X_l.

The family of sets $X_1, \cdots, X_l$ is not monotone in general, in contrast to the Y_i's in (2.1.6). Also $C_*(\mathcal{X}_j, s) \equiv C(\mathcal{X}_l, \mathcal{Y}_j, u, s) \equiv C(u, s)$.

The increment

$$\delta C_j = \delta C_j(x, y) = \frac{d}{du_j} C_j = (C_{j-1})_{X_j} - C_0$$

equals $- C_0$ if ∂X_j separates x and y. We define

$$U_{j,l} = U_j = \prod_{i=j}^{l} u_i; \quad \delta U_{j,l} = \prod_{i=j}^{l-1} u_i = U_{j,l-1}; \quad \delta U_{l,l} = 1 . \tag{2.1.8}$$

Then we determine the form for δC_l by induction: For $C_l = C(\mathcal{X}_l, u) \in \mathcal{C}_2$,

$$\delta C_l(x, y) = - \delta U_{j,l} C_0(x, y) \tag{2.1.9}$$

when x and y are separated by ∂X_j, but are not separated by ∂X_r for any $r > j$. Also $\delta C_l = 0$ if no contour separates x and y.

PROPOSITION 2.1.2. *Let $p < \infty$, $\varepsilon > 0$, $d = \mathrm{dist}(\Delta, \Delta')$ and let $C_l \in \mathcal{C}_2$ have the form above with l contours ∂X_i. Then*

$$\sum_{\Delta,\Delta'} \|\delta C_l\|_{L_p(\Delta\times\Delta')} e^{m_0(1-\varepsilon)d} \leqq O(1) \sum_{j=1}^{l} \delta U_{j,l} |X_j| . \tag{2.1.10}$$

Proof. We modify the proof of Proposition 2.1.1 as follows: Fix Δ inside X_j and Δ' outside X_j. We sum over all Δ, Δ' that are not separated by ∂X_r, for any $r > j$. Denoting this restricted sum $\sum^{(j)}$, we obtain as before,

$$\sum^{(j)}_{\Delta,\Delta'} \|\delta C_l\|_p e^{m_0(1-\varepsilon)d} \leqq O(1) \delta U_{j,l} |X_j| .$$

Since δC_l vanishes unless some contour separates Δ and Δ', the sums $\sum^{(l)}$, $\sum^{(l-1)}, \cdots, \sum^{(1)}$ exhaust all $R^2 \times R^2$ for which δC_l does not vanish.

2.2 Change of covariance

In this section we formulate the basic identity

$$\int F d\Phi_{C'} - \int F d\Phi_{C''} = \frac{1}{2} \int_0^1 ds \int (\delta C \cdot \Delta_\Phi) F d\Phi_{C(s)} , \tag{2.2.1}$$

where $\delta C = (d/ds) C(s) = C' - C''$, in terms of annihilation operators. We also specify the general form of the functions F that arise in the expansion.

For each $C \in \mathcal{C}_{11}$, let t_C be the largest time of any contour in C, i. e.,

$$t_C = \sup\{x_0 | x \in \partial Y_j \text{ or } x \in \partial X_j\} ,$$

where ∂Y_i and ∂X_j are contours of $C = C(\mathcal{Y}, \mathcal{X})$. Let

$$\mu = \mu_{\bar{x}} = \left(- \frac{d^2}{d\bar{x}^2} + m_0^2 \right)^{1/2} , \tag{2.2.2}$$

and let $\mu(\bar{p}) = (\bar{p}^2 + m_0^2)^{1/2}$. For an operator A on L_2, let $A(x, y)$ denote its integral kernel.

LEMMA 2.2.1. *For each z_0 such that $t_C \leqq z_0 \leqq y_0$ and $x_0 \leqq z_0$,*

$$\delta C(x, y) = \int \delta C(x, z)[e^{-(y_0-z_0)\mu}](\vec{z}, \vec{y})\, d\vec{z} . \tag{2.2.3}$$

Proof. By the definition of $\mathcal{C}_1$ in Section 2.1, if $t_C \leqq z_0 \leqq y_0$, then $\delta C(x, z)$ equals $C_0(x, z)$ times a function that is independent of z. Thus it suffices to establish (2.2.3) with δC replaced by C_0. For real p,

$$(2\mu(p_1))^{-1}\int_{-\infty}^{\infty} e^{ip_0t-\mu(p_1)|t|}\,dt = \mu(p_1)^{-1}\mathrm{Re}(ip_0 + \mu(p_1))^{-1} = (p_0^2 + p_1^2 + m_0^2)^{-1} ,$$

so

$$C_0(x, z) = \left[\frac{e^{-|z_0-x_0|\mu}}{2\mu}\right](\vec{x}, \vec{z})$$

and (2.2.3) follows.

It is convenient to introduce the time z_0 annihilation operator defined on $L_2(\mathcal{S}'(R^2), d\Phi)$ by

$$\begin{aligned} a(z) &= \int_{y_0 \geqq z_0} dy[(2\mu)^{-1}\exp(-\, y_0 - z_0)\mu](\vec{z}, \vec{y})\,\partial/\partial\Phi(y) \\ &= \int_{y_0 \geqq z_0} dy C_0(z, y)\,\partial/\partial\Phi(y) . \end{aligned} \tag{2.2.4}$$

The annihilation operator is simply a derivative, $\partial/\partial\Phi$, forward in time. We distinguish it from $\partial/\partial\Phi$ for the purpose of defining our expansion.

We note that $a(z)$ has a relativistic localization, so that $[a(z_0, \vec{z}), a(\vec{z}_0, \vec{y})^*]$ is proportional to $\mu^{-1}(\vec{y} - \vec{z})$. By Lemma 2.2.1 for each $z_0 \geqq t_{C(s)}$,

$$\begin{aligned} \frac{1}{2}\delta C\cdot\Delta_\Phi &= \int_{x_0 \leqq z_0 \leqq y_0} dxdy\,\delta C(x, y)\partial^2/\partial\Phi(x)\partial\Phi(y) \\ &\quad + \frac{1}{2}\int_{x_0, y_0 \leqq z_0} dxdy\,\delta C(x, y)\partial^2/\partial\Phi(x)\partial\Phi(y) \\ &= \int_{x_0 \leqq t_{C(s)}} dxd\vec{z}\,\delta\hat{C}(x, z)\partial/\partial\Phi(x)a(z) \\ &\quad + \frac{1}{2}\int_{x_0, y_0 \leqq z_0} dxdy\,\delta C(x, y)\partial^2/\partial\Phi(x)\partial\Phi(y) . \end{aligned} \tag{2.2.5}$$

Here $\delta\hat{C}(x, z) = (2\mu_{\vec{z}}\delta C)(x, z)$. We also note that $\delta C(x, z)$ and $\delta\hat{C}(x, z)$ vanish unless x_0 or z_0 is less than $t_{C(s)}$.

Suppose that δC_j (or δC_l) are as in Proposition 2.1.1 (or 2.1.2). Furthermore assume that if $x \in \Delta$, $y \in \Delta'$, then

$$x_0 \leqq t_{C(s)} \leqq z_0 - 1 \leqq y_0 - 1 .$$

PROPOSITION 2.1.1′ (2.1.2′). *Under the above hypotheses, $\delta\hat{C}_j$ (or $\delta\hat{C}_l$) satisfies* (2.1.6) (*or*(2.1.10)).

Proof. For Δ, Δ' as above, $\delta C(x, y)$ is a function of x, s, u times $C_0(x, y)$. Thus $\delta\hat{C}(x, y)$ is the same function of x, s, u times

$$[e^{-|y_0-x_0|\mu}](\vec{x}, \vec{y}) ,$$

which is bounded by $O(1)\exp(-m_0(1-\varepsilon/2)|x-y|)$. The propositions now follow as previously.

The integrands in (2.2.1) will all have the form

$$F_\alpha = \int dx d\vec{y}\left(\prod_{j=1}^{n_a} a(y_j)\right)\left(\prod_i :\Phi(x_i)^{n_i}:\right)e^{-V}w(\alpha, x, y) \tag{2.2.6}$$

and the kernels w may also depend on parameters s, u. In addition, certain of the vertices: $\Phi(x_i)^{n_i}$: may occur at sharp times (e. g., A or B_T vertices) and for such vertices the x_i integration extends only over the space variable $\vec{x}_i$. For convenience, we place all annihilation operators at equal times y_{0j} in $Z + 1/2$. (i. e., at the time boundary of some lattice square) with $y_{0_\alpha} = y_{01} = y_{02} = \cdots = y_{0n_\alpha}$. Moreover, the annihilators occur at a later time than any contour in C, that is, $y_0 > t_{C(u,s)}$.

In addition to the formulas (2.2.1), (2.2.5) for a change in the covariance, we have a second basic formula to advance the time of an annihilation operator. For $y_0 \leqq z_0$,

$$\begin{aligned} a(y) &= \int d\vec{z}\,[e^{-(z_0-y_0)\mu}](\vec{y}, \vec{z})a(z_0, \vec{z}) \\ &\quad + \int_{y_0}^{z_0} dx_0 \int d\vec{x}[(2\mu)^{-1}e^{-(x_0-y_0)\mu}](\vec{y}, \vec{x})\partial/\partial\Phi(x) . \end{aligned} \tag{2.2.7}$$

For a product of annihilation operators, as in (2.2.6), we apply (2.2.7) iteratively to each factor $a(y_i)$. Note that by advancing an annihilator a time $d = z_0 - y_0$, the first term in (2.2.7) contributes a factor $O(e^{-m_0 d})$ to the expansion, and the second term, after differentiating $e^{-V(h)}$, contributes a factor $O(\lambda)$. These factors yield convergence of the expansion.

It simplifies the expansion to allow (2.2.7) in the case of $n_a = 0$ annihilators. In this case, (2.2.7) yields the identity $1 = 1$, i. e., no change is produced by such an expansion step. We note that the inductive construction of Section 2.3 specifies a time y_0 for the annihilation operators even in the case $n_a = 0$. For the initial term, at the start of the expansion, $y_0 = 0$, by convention.

In order to take advantage of the fact that $C_0(x, y)$ vanishes exponentially at large values of $|x - y|$, we localize all variables x and $\vec{y}$. Let Δ_i denote the lattice square of unit length, centered at $i \in Z^2$, and let I_i denote the interval of unit length centered at $i \in Z$. We require that $w(x, y)$ in (2.2.6) be supported in a product of squares $x_\nu \in \Delta_{i_\nu}$ and intervals $\vec{y}_\nu \in I_{i_\nu}$.

Such an F is called localized, and a general F is a sum of localized integrands. The application of a derivative $\partial/\partial\Phi$ localized in $\Delta \in R^2$, to a localized integrand F, can be written as a sum of derivatives with respect to each of the variables in F localized in Δ. In the language of graphs, this is the sum over all possible contractions in Δ. In the use of formulas (2.2.1), (2.2.5), and (2.2.7), we imply a sum over all localizations and contractions for factors $\Phi(x)$, $\partial/\partial\Phi(x)$, and $a(x)$. Thus each term resulting from these formulas will have uniquely defined localizations and contractions.

2.3. The increasing contour expansion

The increasing contour expansion is based on an iterated application of (2.2.1), (2.2.5), and (2.2.7). The purpose of this expansion is to factor out vertices (polynomials in Φ) arising from A and from derivatives $\partial/\partial\Phi$ applied to e^{-V}. We achieve this factorization by inserting a contour enclosing the lattice squares in which the vertices are localized. We distinguish between the derivative $\partial/\partial\Phi$ and the annihilation operator a, in that we do not factor out vertices arising from

$$\left(\prod_{i=1}^{n_a} a(y_i)\right) B_T e^{-V} .$$

The expansion is defined so that a contour never encloses an annihilator. We achieve this by using (2.2.7) to advance the time of each annihilator beyond the localization times of all vertices that we wish to factor out. We now explain the expansion in detail.

Throughout the increasing contour construction, the terms have the form

$$m_\alpha = \int ds\, du \int F_\alpha(u, s)\, d\Phi_{C(u,s)} , \tag{2.3.1}$$

where F has the form (2.2.6) and where $C(u, s)$ is an element of $\mathcal{C}_1$, defined in Section 2.1. The integration over u, s ranges over the partial contours in $C(u, s)$. Each expansion step replaces a term of the form (2.3.1) by a sum of similar terms. This sum ranges over the localizations of F and the contractions arising from (2.2.5) and (2.2.7) described in the previous section.

We divide our terms into two categories: Leading terms (l-terms) and remainder terms (r-terms). We define an r-term by the condition $y_0 \geqq T$, where y_0 is the time of the annihilators in F (see (2.2.6) and Section 2.4). The expansion of an r-term yields only r-terms. A term is called a leading term if it is not an r-term. The expansion of an l-term will yield both l- and r-terms.

Since annihilators advanced beyond the time T do not contribute to convergence, we define

$$t = \min \{y_0, T\} ,$$

to aid the discussion of convergence in Section 2.4.

Our expansion is defined so that the kernel w of F factors in all l-terms, $w = w_1 \otimes w_B$, where w_1 is independent of B. At the start of the increasing contour expansion, $C(u, s)$ in (2.3.1) equals C_*, some element of $\mathcal{C}_2$, say $C_* = C(\mathfrak{X}_l, u)$. We let $t_* \geqq t_{C_*}$ be the time y_0 at the start of the construction. When the increasing contour expansion terminates for a particular term, the covariance for that term has the form $(C_*)_{Y_j} = C_*(Y_j)$ of an element of $\mathcal{C}_2$ with a complete contour. Thus $C_*(Y_j) = C(\mathfrak{X}_{l+1}, u)$ where $X_{l+1} = Y_j$ and $u_{l+1} = 1$.

The increasing contour construction then has the general form

$$\int du \int F(u) d\Phi_{C_*} = \sum_\alpha \int du \int F_\alpha(u) d\Phi_{C_*(Y_{j(\alpha)})} ,$$

which we now specify further.

Given a term m_α of the form (2.3.1), we give the next step of the expanding contour construction. The construction terminates for the term in question under one of the following two conditions:

T1. The construction terminates for an l-term m_α which contains no vertices at times earlier than T, contains at most n annihilators ($n_\alpha \leqq n$) and for which C_α contains a complete contour.

T2. The construction terminates for an r-term m_α which contains no vertices, contains no annihilators and for which C_α contains a complete contour.

If m_α does not satisfy T1–T2, we either:

I. Insert a contour using (2.2.1) and (2.2.5), or

II. Advance the time of annihilation operators using (2.2.7).

To apply (2.2.1) to a term m_α, let $C' = C_\alpha = C(\mathfrak{X}_l, \mathcal{Y}_j)$ and set $C'' = C'_{Y_{j+1}} = C(\mathfrak{X}_l, Y_{j+1})$. The new contour ∂Y_{j+1} is defined as follows: Let $N(\Delta)$ be the degree of the monomial in F_α localized in the lattice square Δ, as in Section 1.1. Let $S_j(\Delta)$ be the square centered at Δ of length $2[N_j(\Delta)^{1/4}]+1$, where $[x]$ denotes the integral part of x. Then

$$Y_{j+1} = Y_j \cup \bigcup_{\Delta \in \mathfrak{D}} S_j(\Delta) , \tag{2.3.2}$$

where

$$\mathfrak{D} = \{\Delta : N_j(\Delta) \neq 0\} ,$$

if m_α is an r-term, and

$$\mathfrak{D} = \{\Delta\colon N_j(\Delta) \neq 0, \text{ and time } \Delta \leqq T\},$$

if m_α is an l-term. Let t_{j+1} be the maximum time of Y_{j+1}.

We apply (2.2.1) and (2.2.7) in detail as follows:

I. Insert the contour ∂Y_{j+1} using (2.2.1) with $C'' = (C_\alpha)_{Y_{j+1}}$ as above if
a) $n_\alpha \leqq n$; and
b) $t_{j+1} < y_0$; and
c) $y_0 < T - 1$ or $y_0 = +\infty$.

For l-terms we express $\delta C \cdot \Delta_\Phi$ as in (2.2.5). For r-terms we use (2.0.5) so that no annihilators occur. Note (b) guarantees that ∂Y_{j+1} does not enclose any annihilators.

II. Use (2.2.7) to advance the time of $0 \leqq n_\alpha \leqq n+1$ annihilators from the time y_0 to the time z_0, if
a) $n_\alpha = n + 1$ and $y_0 < T - 1$; or
b) $y_0 < t_{j+1}$ and $y_0 < T - 1$; or
c) $T - 1 \leqq y_0 < +\infty$.

In cases (a) and (b), choose $z_0 = y_0 + 1$, and in case (c) choose $z_0 = +\infty$.

Remark. The contour ∂Y_{j+1} is defined not only to contain the localization of extraneous monomials, permitting them to be factored out, but also ∂Y_{j+1} is defined to avoid divergences resulting from monomials of high degree localized in a single square Δ. Although we can not avoid high values of $N(\Delta)$ in all terms, Y_{j+1} is constructed so that if $N(\Delta)$ is large, each additional contraction or derivative occurs over a long distance, giving rise to a compensating factor of $\exp(-m_0 d(\Delta, \partial Y_{j+1}))$.

Throughout the increasing contour expansion, note that $n_\alpha \leqq n + 1$. For a given elementary expansion step II note that new vertices are the only vertices localized in the time strip $R^1 \times [y_0, T]$. New vertices are defined to be those vertices arising from contraction to the exponent within a given expansion step. Hence there are at most n_α possible contractions to new vertices arising from the application of (2.2.7) in a single expansion step.

The expanding contour construction is completely defined by iterating I-II until termination. In the terms for which the expansion terminates (and thus have a complete contour) we factor out the vertices in the interior of the complete contour, using (2.0.2). In l-terms, this factorization leaves unchanged the vertices of B_T and the annihilators present. In r-terms, no vertices or annihilators remain after the factorization. Our expansion now takes the form

$$\int du \int F(u)\, d\Phi_{C_*}$$

$$(2.3.3) \qquad = \sum_\alpha \int du ds \int d\bar{x} d\bar{y}\, w_l'(\alpha, u, s, x, y) \int \left(\prod_{i=1}^{n_\alpha} a(y_i)\right) B_T(x) e^{-V} d\Phi_{C_\alpha}$$

$$+ \sum_\alpha \int du ds\, w_r'(\alpha, u, s) \int e^{-V} d\Phi_{C_\alpha} .$$

Here $C_\alpha = C_*(\mathcal{Y}_j, s)$ has a complete contour ∂Y_j, depending on α. Also $B_T(x)$ is the density for B_T.

$$B_T = \int d\bar{x}\, w_B(\bar{x}) B_T(x) .$$

The first sum in (2.3.3) is the sum of l-terms, while the second sum in (2.3.3) is the sum of r-terms. The summation over α indexes the various terms. For l-terms, it includes summation over n_α, $0 \leqq n_\alpha \leqq n$. For both l-terms and r-terms it labels summation over the various complete contours ∂Y_j and over the various contractions. The numerical kernels w_l', w_r' are functionals of the factored vertices, as given by (2.0.2). Namely,

$$(2.3.4) \qquad \begin{aligned} w_l'(\alpha, u, s, x, y) &= \frac{\int P_l(\alpha, u, s, x, y) e^{-V(Y_j)} d\Phi_{C_\alpha}}{\int e^{-V(Y_j)} d\Phi_{C_\alpha}} , \\ w_r'(\alpha, u, s) &= \frac{\int P_r(\alpha, u, s) e^{-V(Y_j)} d\Phi_{C_\alpha}}{\int e^{-V(Y_j)} d\Phi_{C_\alpha}} , \end{aligned}$$

and $P_l(\alpha, \cdot)$ and $P_r(\alpha, \cdot)$ are polynomials localized in $Y_j = Y_j(\alpha)$. We have used (2.0.1) to factor $\int \exp(-V(\sim Y_j)) d\Phi_{C_\alpha}$ from both numerator and denominator.

We obtain the desired expansion, by making the change

$$d\Phi_{C_\alpha} \longrightarrow d\Phi_{C_*(Y_j(\alpha))} ,$$

in (2.3.4), where $\partial Y_j(\alpha)$ is the complete contour in the term labeled by α. We write for the remainder terms

$$(2.3.5) \qquad \int e^{-V} d\Phi_{C_\alpha} = \int e^{-V} d\Phi_{C_*(Y_j(\alpha))} \left(\frac{\int e^{-V(Y_j)} d\Phi_{C_\alpha}}{\int e^{-V(Y_j)} d\Phi_{C_*(Y_j(\alpha))}} \right) ,$$

and for the l-terms,

$$\int \left(\prod_{i=1}^{n_\alpha} a(y_i)\right) B_T(x) e^{-V} d\Phi_{C_\alpha}$$

(2.3.6)
$$= \int \left(\prod_{i=1}^{n_\alpha} a(y_i)\right) B_T(x) e^{-V} d\Phi_{C_*(Y_j(\alpha))} \left(\frac{\int e^{-V(Y_j)} d\Phi_{C_\alpha}}{\int e^{-V(Y_j)} d\Phi_{C_*(Y_j(\alpha))}} \right).$$

We absorb the ratios in (2.3.5), (2.3.6) into the coefficients w' in (2.3.4), and we include this rearrangement as part of the elementary step producing the term with the complete contour ∂Y_j. Thus the increasing contour expansion is

$$\int du \int F(u) d\Phi_{C_*}$$

(2.3.7)
$$= \sum_\alpha \int du \int d\bar{x} d\bar{y}\, w_l(\alpha, u, x, y) \int \left(\prod_{i=1}^{n_\alpha} a(y_i)\right) B_T(x) e^{-V} d\Phi_{C_*(X_{l+1}(\alpha))}$$
$$+ \sum_\alpha \int du\, w_r(\alpha, u) \int e^{-V} d\Phi_{C_*(X_{l+1}(\alpha))},$$

where $\partial Y_j = \partial Y_j(\alpha) = \partial X_{l+1}(\alpha)$ is the inserted contour, and where

$$w_l(\alpha, u, x, y) = \frac{\int ds \int P_l(\alpha, u, s, x, y) e^{-V(Y_j)} d\Phi_{C_\alpha}}{\int e^{-V(Y_j)} d\Phi_{C_*(Y_j)}},$$

(2.3.8)
$$w_r(\alpha, u) = \frac{\int ds \int P_r(\alpha, u, s) e^{-V(Y_j)} d\Phi_{C_\alpha}}{\int e^{-V(Y_j)} d\Phi_{C_*(Y_j)}}.$$

2.4. Norms for convergence

In this section we establish the convergence of the increasing contour expansion. Each elementary step in the expansion expresses a term

(2.4.1)
$$m_\alpha = \int du\, ds \int dx\, dy \prod_{i=1}^{n_\alpha} a(y_i) \prod_{i=1}^{r} :\Phi(x_i)^{n_i}: e^{-V} w_\alpha d\Phi_{C_\alpha}$$

as a sum of terms of the same form. Here $w_\alpha = w(\alpha, u, s, x, y)$ is defined by (2.3.4), and $C_\alpha = C(\mathfrak{X}_l, \mathcal{Y}_j, u, s)$.

Let I_k be the set of indices α which label the terms m_α at the kth stage of the increasing contour expansion. Thus for each k we have

$$\int A B_T e^{-V} d\Phi = \sum_{\alpha \in I_k} m_\alpha .$$

Also let $M_k = \{m_\alpha : \alpha \in I_k\}$ be the set of terms in the sum. Each step of the expansion can be regarded as a mapping

$$M_k \longrightarrow M_{k+1}$$

of sets of expansion terms.

For each term m_α, we introduce a norm $\|m_\alpha\|$, and then we define a norm on M_k by

(2.4.2) $$\|M_k\| = \sum_{\alpha\in I_k} \|m_\alpha\| \,.$$

The proof of convergence consists in showing that for each k,

$$\|M_{k+1}\| \leqq \|M_k\| \,.$$

This decrease in norm is ensured by the small factors, $O(\lambda)$ or $O(m_0^{-1})$, introduced by each step in our expansion, as well as by an appropriate choice of norms $\|\cdot\|$. Unfortunately this simple structure in the convergence proof is made possible only by the choice of a complicated norm.

To define our norms, we introduce the free module W_k generated by the symbols $\alpha \in I_k$. In other words, W_k is the vector space of functions defined on I_k. For a function of the form $\alpha \to c_\alpha m_\alpha$ in W_k, we define the norm $\sum |c_\alpha| \, \|m_\alpha\|$, where $\|m_\alpha\|$ is given by the formula

(2.4.3) $$\begin{aligned} \|m_\alpha\| &= \mathcal{K} K_1^\nu e^{\gamma t} \mathfrak{N} \\ &\times \int du\, ds \exp\left[\sum_{k=1}^{l} U_{k,l} |X_k| + \sum_{k=1}^{j} S_{k,j} |\delta Y_k|\right] \| w(\alpha, u, s, y_0, \cdot)\|_2 \,. \end{aligned}$$

The constants $\mathcal{K}$, ν, etc., are explained below. Then

$$\|\alpha \longrightarrow c_\alpha m_\alpha\| = \sum_{\alpha\in I_k} |c_\alpha| \, \|m_\alpha\| \,.$$

For initial and final terms we shall see that there is a simple relation between this norm and the norm defined by (1.1.7), (1.1.8). The purpose of this section is to show that the increasing contour expansion is norm decreasing; that is, $\sum_{\alpha\in I_k} \|m_\alpha\|$ is decreasing in k. In Chapters 3 and 4 we show that the full n-particle expansion is norm decreasing so that the norm of the initial term

$$\int AB_T e^{-V} d\Phi$$

dominates the sum of the norms of the final terms. This bound implies Theorem 1.1.11, as we shall see in Chapter 4.

Now we explain (2.4.3) in detail. The indices α (or the corresponding terms m_α) are divided into five categories and for each category we specify values of $\mathcal{K}$, $\mathfrak{N}$, ν, t above. Two of these categories do not occur until later chapters. In the cases where w does not depend on u or s, the sum $\sum_{k=1}^{l}$ or $\sum_{k=1}^{j}$ runs over an empty set of indices, and equals zero, so the integral $\int du$ or $\int ds$ can be omitted from the norm.

The first category is the initial term $m_\alpha = \int AB_T dq$. There are three categories of partially expanded terms, denoted P_1, P_2, and P_3, and there are final terms. The P_1 terms arise in Chapter 2. They have a covariance $C = C_0$ or $C \neq C_0$, $C \in \mathcal{C}_1$, but no complete contour. The P_2 terms occur at the end of the Chapter 2 expansion. They have a covariance $C \neq C_0$, $C \in \mathcal{C}_2$, and C has one complete contour. The leading, or $l-P_2$ terms have no vertices before time T, and the remainder, or $r-P_2$ terms have no vertices at all. The P_3 terms occur at the end of the Chapter 3 expansion. They have covariance $C = C_0$, $n_a > 0$ annihilators, and vertices as in P_2 terms. The final terms occur at the end of the Chapter 4 expansion, and also at the end of the Chapter 3 expansion, in the case of $n_a = 0$ annihilators.

The time y_0 is chosen to increase as the expansion progresses. (The expansion step (4.5) is exceptional, and allows a decrease in y_0.) For initial terms, $y_0 = 0$, and for remainder terms, $y_0 = +\infty$. For leading terms, $y_0 \geqq 0$ is the largest time of any annihilator. Recall that y_0 is defined by step II of the inductive expansion which advances y_0 even when $n_a = 0$. Recall also $t = \min\{y_0, T\}$. We choose $\gamma = (n+1)m_0(1-\varepsilon)$.

We select ν to decrease as the expansion progresses. For each change in category, ν decreases by one. Thus for leading terms, $\nu = 6$ (initial), $\nu = 5\,(P_1)$, etc. For remainder terms, $\nu = 4\ (P_1)$, $\nu = 3\ (P_2)$, etc. There are no initial or P_3 remainder terms. The definition of the categories is summarized in Table 1 below.

TABLE 1. Categories of expansion terms.

	ν	Covariance	Vertices and Annihilators
Initial	6	$C=C_0$	AB_T not in form required for final terms
P_1	5 if l 4 if r	$C=C_0$; or $C\neq C_0$, $C\in\mathcal{C}_1$, no complete contour	
P_2	4 if l 3 if r	$C\neq C_0$, $C\in\mathcal{C}_2$ one complete contour	
P_3	3	$C=C_0$	$N_a>0$ annihilators
Final-l	2	$C=C_0$	$B_T\times\{$polynomial of degree $\leq n$ at sharp time $t<T\}$
Final-r	1	$C=C_0$	No vertices, no annihilators

We take $\mathcal{K} = 1$ except for P_1 terms. For a P_1 term, $\mathcal{K} = \mathcal{K}_1\mathcal{K}_2\mathcal{K}_3\mathcal{K}_4$ is a product of four factors, and each $\mathcal{K}_i$ has the role of transferring convergence from strongly convergent steps (i. e., norm decreasing steps) to steps which would otherwise have some limited increase in norm.

The factor $\mathcal{K}_1$ contains the convergence which arises from the contractions over long distances. Explicitly,

$$\mathcal{K}_1 = e^{\varepsilon m_0 D/2} ,$$

where D is the sum of the contraction distances arising since the last insertion of a complete contour. Thus if we consider the Chapter 2 expansion alone, D is the sum of all contraction distances in the expansion. Hence $\mathcal{K}_1$ increases throughout the increasing contour construction and returns to one at the end of it.

$\mathcal{K}_2$ and $\mathcal{K}_3$ decrease throughout the increasing contour construction, and return to one at the end of it. $\mathcal{K}_2$ compensates for the insufficient convergence which results from increasing the time t in the presence of $0 \leqq n_a < n+1$ annihilators in the expansion step II. Explicitly, if $\delta t = t' - t$ is the increase in t, then

$$\mathcal{K}_2 = e^{-(1-\varepsilon)m_0\Sigma\delta n} , \tag{2.4.4}$$

where $\delta n = (n + 1 - n_a)\delta t$, and the sum runs over all expansion steps since the last insertion of a complete contour. We also define

$$\mathcal{K}_3 = e^{-|Y_j|} ,$$

$$\mathcal{K}_4 = K_2^j ,$$

where j is the number of $\mathcal{Y}_j$ contours in

$$C = C(\mathcal{X}_l, \mathcal{Y}_j, u, s) ,$$

and we define $\mathfrak{N} = \prod_\Delta \mathfrak{N}(\Delta)$ and

$$\mathfrak{N}(\Delta) = \begin{cases} K_2^{2N(\Delta)}\, N(\Delta)! \\ K_2^{N(\Delta)}\, N(\Delta)! \, . \end{cases} \tag{2.4.5}$$

Here the first choice is made for all initial and final terms and also for leading terms with Δ beyond the time T. Otherwise the second choice is made.

Remark. Let $\|\cdot\|_0$ denote the norm defined in (1.1.7), (1.1.8). For initial and final terms m_α we have

$$\|m_\alpha\| = K_1^\nu e^{\gamma t} \|m_\alpha\|_0 ,$$

where ν, t depend on the term m_α.

THEOREM 2.4.1. *Let ε and n be given. Let m_0, K_2 and K_1 be sufficiently large and let λ be sufficiently small. Then the increasing contour expansion (2.3.7) is norm decreasing.*

Proof. It is sufficient to show that if $m_\alpha = \sum m'_{\alpha'}$ represents an elementary expansion of the term m_α as defined by I or II, we have

$$\sum \|m'_{\alpha'}\| \leqq \|m_\alpha\| .$$

We choose the constants of our theorem in the following order: m_0, K_2, K_1, and then λ.

LEMMA 2.4.2. *The elementary step* II *of advancing* n_α *annihilators from time* $y_0 < T - 1$ *to* $y'_0 = y_0 + 1$ *is norm decreasing.*

Proof. Since $y'_0 < T$, m_α, as well as each $m'_{\alpha'}$, is a leading term. A term $m'_{\alpha'}$ has $n'_{\alpha'} = k$ annihilators at the time $y'_0 = y_0 + 1$ and $n_\alpha - k$ contractions in the strip $[y_0, y_0 + 1]$. The sum extends over the various possible contractions and their localizations of the k annihilators and over the possible values of k, $k = 0, \cdots, n_\alpha$.

To obtain the kernel $w'_{\alpha'}$ of $m'_{\alpha'}$, we define the function

$$v = (\otimes_{i=1}^{k} e^{-\mu})(\prod_{i=1}^{n_\alpha - k} e^{-\sigma_i \mu})\, w_\alpha$$

with n_α factors $e^{-\sigma\mu}$ acting on the $\bar{y}$ variables of w_α. Of these factors, there is one factor $e^{-\mu}$ for each of the k annihilators advanced by one unit of time, and one factor $e^{\sigma_i \mu}$ for each of the $n_\alpha - k$ contractions, $0 \leqq \sigma_i \leqq 1$. The kernel $w'_{\alpha'}$ is obtained as follows: Multiply v by the characteristic function of a product L' of squares and intervals to provide the localization of each variable. If $k < n_\alpha$, at least one contraction to the exponent occurs in each term, so we multiply v by a numerical factor $\leqq O(\lambda)$ given by the coefficients of the interaction polynomial. If two or more contractions to the same vertex occur, identify the corresponding variables of v.

Let $\sum d'$ be the sum of the distances for this step, as determined by L' and the localization of w_α. Then

$$D' = D + \sum d' , \qquad \sum d' \geqq k. \tag{2.4.6}$$

Furthermore, $w'_{\alpha'}$ satisfies the estimate

$$\| w'_{\alpha'} \|_{L_2} \leqq O(1) e^{-m_0 (\Sigma d')(1-\varepsilon/4)} \| w_\alpha \|_{L_2} \tag{2.4.7}$$

where $O(1) \to 0$ as $m_0 \to \infty$, for $k = n_\alpha > 0$ and $O(1) \to 0$ as $\lambda \to 0$ for $k < n_\alpha$. This estimate follows from the exponential decay of $\exp(-\sigma\mu)$ (cf. [6, Lemma 3.2]).

From the definition (2.4.4),

$$\mathcal{K}'_2 e^{\gamma y'_0} = e^{n_\alpha m_0 (1-\varepsilon)} \mathcal{K}_2 e^{\gamma y_0} . \tag{2.4.8}$$

Moreover

$$\mathcal{K}'_1 = \mathcal{K}_1 e^{\varepsilon m_0 \Sigma d'/2} , \qquad \mathcal{K}'_3 = \mathcal{K}_3,\ \mathcal{K}'_4 = \mathcal{K}_4 .$$

It follows that

$$\mathcal{K}' e^{\gamma y'_0} \| w'_{\alpha'} \|_{L_2} \leqq O(1) e^{-m_0 \varepsilon \Sigma d'/4} \mathcal{K} e^{\gamma y_0} \| w_\alpha \|_{L_2} . \tag{2.4.9}$$

For $k = n_a$, we use $\sum d' \geq k$ in proving (2.4.9). For $k < n_a$, $O(1) \to 0$ as $\lambda \to 0$ and since λ is chosen after m_0, a factor $e^{n_a m_0 (1-\varepsilon)}$ can be absorbed into $O(1)$. For $k = n_a$, the factors $N(\Delta)$ remain unchanged so

$$||m'_{\alpha'}|| \leqq O(1) e^{-\varepsilon m_0 \Sigma d'/4} ||m_\alpha|| .$$

The sum over localizations is controlled using the exponential decay in $\sum d'$, and so for m_0 sufficiently large

$$\sum_{\alpha';\, k=n_a} ||m'_{\alpha'}|| \leqq \frac{1}{2} ||m_\alpha|| . \tag{2.4.10}$$

For $k < n_a$, all $n_a - k$ contractions occur in squares in which $N(\Delta) = 0$. Thus

$$\mathfrak{N}' \leqq K_2^{p(n_a-k)} (p(n_a - k))! \, \mathfrak{N} , \tag{2.4.11}$$

where $p = \deg$ $(\mathcal{P} = V(h))$ in the action (1.8). The new factor $K_2^{p(n_a-k)}(p(n_a - k))!$ can be absorbed by $O(1)$ in (2.4.9), since λ is chosen after K_2. Thus for λ sufficiently small,

$$\sum_{\alpha';\, k<n} ||m'_{\alpha'}|| \leqq \frac{1}{2} ||m_\alpha|| , \tag{2.4.12}$$

which combined with (2.4.10), completes the proof of the lemma, for $n_a > 0$. For $n_a = 0$ there is only one term in the expansion of m_α. The norm is unchanged because the decrease in $\mathcal{K}_2$ compensates for the increase in $e^{\gamma t}$.

LEMMA 2.4.3. *The elementary step* II *of advancing annihilators from a time* $y_0 \in [T - 1, T]$ *is norm decreasing.*

Proof. We describe the modifications of the proof above. The principal modification concerns contractions to B_T. Each term $m'_{\alpha'}$ is now a remainder term and so $k = 0$, i.e., the annihilators are advanced to time $+\infty$, and necessarily contract. The sum over contraction localizations includes all lattice squares with times greater than y_0. The kernel $w'_{\alpha'}$ of $m'_{\alpha'}$ satisfies the bound (2.4.7). For contractions to new vertices, the proof is as before. If an annihilator contracts to B_T, the kernel is obtained by identifying the contracted variables in

$$v = w_B \otimes \left(\bigotimes_{i=1}^{n_a} e^{-\sigma_i \mu} w_1 \right) .$$

Here we use the factorization property for the kernel of leading terms (e.g., m_α) to write w_α as

$$w_\alpha = w_B \otimes w_1 ,$$

where w_B is the kernel for the observable B_T. For contractions to B_T, $\sigma_i \geqq 1$ and for σ_i bounded away from zero, $e^{-\sigma_i \mu}$ is a bounded operator from

L_2 to L_∞, with norm $O(\exp(-\sigma_i m_0(1-\varepsilon))) = o(1)$ as $m_0 \to \infty$, and (2.4.7) follows from this fact.

For all remainder terms, $t' = T$, so $t' \leqq t + 1$ and (2.4.8) is replaced by

$$\mathcal{K}_2' e^{\gamma t'} \leqq e^{n_a m_0(1-\varepsilon)} \mathcal{K}_2 e^{\gamma t} , \tag{2.4.13}$$

while (2.4.6) and (2.4.9) are unchanged.

Let $j_1(\Delta)$ be the number of annihilators which contract in Δ, and let $j_0(\Delta)$ be the number which contract to B_T in Δ, so that

$$0 \leqq j_0(\Delta) \leqq j_1(\Delta)$$

and

$$\textstyle\sum_\Delta j_1(\Delta) = n_a \leqq n + 1 . \tag{2.4.14}$$

Thus

$$\begin{aligned} N'(\Delta) &\leqq N(\Delta) + (p-1)(j_1(\Delta) - j_0(\Delta)) - j_0(\Delta) \\ &\leqq N(\Delta) + p j_1(\Delta) - j_0(\Delta) , \end{aligned}$$

and using (2.4.14), and $(a + b)! \leqq 2^{a+b} a!\, b!$,

$$\textstyle\prod_\Delta N'(\Delta)! \leqq O(1) 2^{|B|} \prod_\Delta (N(\Delta) - j_0(\Delta))! . \tag{2.4.15}$$

Here $O(1)$ depends on p and n, but not on the degree of B,

$$|B| \equiv \textstyle\sum_{\Delta \subset [\nu_0, \infty] \times R} N(\Delta) .$$

Recall that definition (2.4.5) on $\mathfrak{N}$ changes from a leading term m_α to remainder term $m'_{\alpha'}$, and that the change is a factor K_2 per monomial degree, for squares with times greater that T. We take $K_2 \geqq 2$ so that the product of the K_2 factors dominates the factor $2^{|B|}$. Thus

$$\mathfrak{N}' \leqq O(1) \mathfrak{N} K_2^{p n_a} \prod_\Delta \frac{(N(\Delta) - j_0(\Delta))!}{N(\Delta)!} . \tag{2.4.16}$$

Here $O(1)$ is independent of $|B|$. This inequality is used in place of (2.4.11). Furthermore if $j_0(\Delta) = j_1(\Delta)$ for all Δ, i.e., if all contractions are to B_T, then the factor $K_2^{p n_a}$ can be omitted from (2.4.16).

For fixed values of $j_0(\Delta)$ and $j_1(\Delta)$, the $j_0(\Delta)$ annihilators contracting to B_T in Δ can be chosen in at most

$$\binom{n_a}{j_0} \leqq (n+1)^{n+1} = O(1)$$

ways, and the corresponding B_T vertices and contractions can be chosen in

$$\binom{N(\Delta)}{j_0} j_0! \leqq O(1) \frac{N(\Delta)!}{(N(\Delta) - j_0(\Delta))!}$$

ways. For fixed $j_1(\Delta)$, the sum over choices of $j_0(\Delta)$ yields at most

$$\prod_{\Delta}(j_1(\Delta)+1) \leqq \prod_{\Delta} e^{j_1(\Delta)} \leqq e^{n+1} = O(1)$$

terms. The value of $j_1(\Delta)$ is determined by the contraction localizations and the sum over localizations is bounded as before. The estimate now proceeds as before, with

$$O(1)\mathfrak{N}'\prod_{\Delta}\frac{N(\Delta)!}{(N(\Delta)-j_0(\Delta))!} \leqq O(1)\mathfrak{N} .$$

In case $j_0(\Delta) < j_1(\Delta)$ for some Δ, the constants $O(1)$ depend on K_2, but since λ is chosen after K_2, $O(1)$ may be absorbed by the $O(1)$ in (2.4.7). Thus for λ small,

$$\sum_{\alpha'} \|m'_{\alpha'}\| \leqq \|m_\alpha\| .$$

In case $j_0(\Delta) = j_1(\Delta)$ for all Δ, the constant $O(1)$ is independent of m_0, K_2, K_1 and λ, and so for m_0 large

$$\sum_{\alpha'} \|m'_{\alpha'}\| \leqq \|m_\alpha\| ,$$

which completes the proof of the lemma.

PROPOSITION 2.4.4. *The elementary step* I *of inserting a contour in* m_α *decreases norms by a factor of at least* 3/4.

Proof. The expansion takes the form

$$m_\alpha = m''_\alpha + \sum_{\alpha'} m'_{\alpha'}$$

where m''_α is the P_2 term (with one complete contour) and each m'_α is a P_1 term with only partial contours. The sum over α' extends over the two terms of (2.2.5), over the various contractions, and over their localizations.

Let the covariance m_α be $C = C(\mathfrak{X}_l, \mathcal{Y}_j, u, s)$. Then the covariance of each term $m'_{\alpha'}$ is

(2.4.17) $$C' = C(\mathfrak{X}_l, \mathcal{Y}_{j+1}, u, s') \in \mathcal{C}_1 ,$$

where

$$\mathcal{Y}_{j+1} = \{\mathcal{Y}_j, Y_{j+1}\} , \qquad s' = \{s, s_{j+1}\} .$$

The term m''_α has $\partial Y_{j+1} = \partial X_{l+1}$ as a complete contour, and it has the covariance

(2.4.18) $$C_2 = C(\mathfrak{X}_{l+1}, u) \in \mathcal{C}_2$$

where

$$\mathfrak{X}_{l+1} = \{\mathfrak{X}_l, X_{l+1} = Y_{j+1}\} , \qquad u_{l+1} = 1 .$$

In this term, (2.3.5) or (2.3.6) is used to replace $C(\mathfrak{X}_l, \mathcal{Y}_{j+1})$ by $C(\mathfrak{X}_{l+1})$. We now divide the proof of the proposition into two lemmas.

LEMMA 2.4.5. *Let $m'_{\alpha'}$ be as above. Then*

$$\sum_{\alpha'} \|m'_{\alpha'}\| \leqq \frac{1}{4}\|m_\alpha\| .$$

Consider first the P_1 terms $m'_{\alpha'}$, and divide these terms into four classes $m'_{\alpha',i}$, $i = 1, 2, 3, 4$, according to whether outside Y_{j+1} there is a new annihilator ($i = 1, 2$) or a new contraction ($i = 3, 4$) and whether the new contraction inside Y_{j+1} occurs to the exponent ($i = 1, 3$) or to an existing vertex ($i = 2, 4$). (Recall that $\delta C'(x, y)$ vanishes unless ∂Y_{j+1} separates x and y.) Consider first the case $i = 1$ (outside, annihilator; inside, contraction to the exponent (that is a new vertex).) The kernel of $m'_{\alpha',1}$ is

$$w'_{\alpha',1} = w(\alpha, u, s, x, y_0, \bar{y})\delta\hat{C}_{j+1}(x_{j+1}, y_{j+1})v(x_{j+1}, y_{j+1}) .$$

Here $v(x_{j+1}, y_{j+1})$ is a product of a numerical coefficient, $O(\lambda)$, from the new vertex, times the characteristic function of $\Delta_1 \times \Delta_2$ and Δ_1 and Δ_2 provide the localization for the variables x_{j+1} and y_{j+1}. Thus

$$\|w'_{\alpha',1}(u, s', y'_0, \cdot)\|_2 = \|v\|_\infty \|\delta\hat{C}\|_{L_2(\Delta_1\times\Delta_2)} \|w\|_2 .$$

Let $d = \text{dist.}(\Delta_1, \Delta_2)$ be the contraction distance. By construction, $d \geqq 1$. By Proposition 2.1.1′ and the bound $\|v\|_\infty \leqq O(\lambda)$,

$$\sum_{\Delta_1,\Delta_2} e^{m_0(1-\varepsilon)d}\|w'_{\alpha',1}\|_2 \leqq O(\lambda)\sum_{k=1}^{j+1}\delta S_{k,j+1}|\delta Y_k| \, \|w_\alpha\|_2 . \tag{2.4.19}$$

For all cases $i = 1, \cdots, 4$, either $y'_0 = y_0$ and $\delta n = 0$ or else $n_a = 0$ and $\delta n = (n+1)(t'-t)$. Thus $\mathcal{K}'_2 e^{rt'} = \mathcal{K}_2 e^{rt}$ and one can check that

$$\mathcal{K}' e^{rt'} = K_2 e^{\varepsilon m_0 d/2 - |\delta Y_{j+1}|}\mathcal{K} e^{rt} . \tag{2.4.20}$$

Returning to the case $i = 1$, we have

$$N'(\Delta) = N(\Delta) , \qquad \Delta \neq \Delta_1$$
$$N'(\Delta_1) \leqq N(\Delta_1) + p - 1 .$$

If $N(\Delta_1) = 0$ then $N'(\Delta_1) \leqq (p-1) = O(1)$ and if $N(\Delta_1) \neq 0$ then $N(\Delta_1) \leqq d^4$ and

$$\begin{aligned} N'(\Delta_1)! &\leqq (N(\Delta_1) + p)^p N(\Delta_1)! \\ &\leqq O(1)e^{\varepsilon m_0 d/2} N(\Delta_1)! . \end{aligned} \tag{2.4.21}$$

Combining this bound with (2.4.20) gives

$$\mathcal{K}' e^{rt'}\mathcal{N}' \leqq O(1)e^{-|\delta Y_{j+1}| + \varepsilon m_0 d}\mathcal{K} e^{rt}\mathcal{N} \tag{2.4.22}$$

where $O(1)$ depends on K_2 and m_0. Take $\varepsilon < 2/5$, and substitute (2.4.19) and (2.4.22) in the sum over $\|m'_{\alpha',1}\|$. With

$$A = \sum_{k=1}^{j} S_{k,j}|\delta Y_k| + |\delta Y_{j+1}| ,$$
$$B = \sum_{k=1}^{l} U_{k,l}|X_k| ,$$

$$\sum_{\alpha'} \| m'_{\alpha',1} \| \leq O(\lambda) e^{-|\delta Y_{j+1}|} \mathcal{K} e^{\gamma t} \mathfrak{N} \int du\, ds' A e^{A s_{j+1} + B} \| w_\alpha \|_2 .$$

Because w does not depend on s_{j+1}, and because

$$ds' = ds\, ds_{j+1} ,$$

the ds_{j+1} integration occurs in the form

$$\int_0^1 ds_{j+1} A e^{A s_{j+1}} \leq 2e^A ,$$

and it follows that

$$\sum_{\alpha'} \| m'_{\alpha',1} \| \leq O(\lambda) e^{-|\delta Y_{j+1}|} \mathcal{K} e^{\gamma t} \mathfrak{N} \int du\, ds\, e^{A+B} \| w_\alpha \|_2 = O(\lambda) \| m_\alpha \| . \tag{2.4.23}$$

Consider next the terms $m'_{\alpha',2}$ with an annihilator outside Y_{j+1} and a contraction to an existing vertex inside. The kernel $w'_{\alpha',2}$ is a localization of

$$w(\alpha, u, s, y_0, \bar{y}, x) \delta \hat{C}_{j+1}(x_i, y_{j+1})$$

where x_i is one of the variables from x. With the localizations $x_i \in \Delta_1$ and $y_{j+1} \in \Delta_2$, we use the bound

$$\| w'_{\alpha',2} \|_2 \leq \| \delta \hat{C}_{j+1} \|_{L_\infty(\Delta_1 \times \Delta_2)} \| w_\alpha \|_2 .$$

By construction, the existing vertices have a distance at least 1 to the contour ∂Y_{j+1}, and so the contraction distance d is at least 1.

By Proposition 2.1.1, with ε replaced by $3\varepsilon/8$ and $p = \infty$,

$$\sum_{\Delta_1, \Delta_2} e^{m_0(1-\varepsilon/2)d} \| w'_{\alpha',2} \|_2 \leq O(e^{-m_0 \varepsilon/8}) \Big(\sum_{k=1}^{j+1} \delta S_{k,j+1} | \delta Y_k | \Big) \| w_\alpha \|_2 . \tag{2.4.24}$$

For $i = 2$, (2.4.21) is replaced by

$$\begin{aligned} N'(\Delta) &= N(\Delta) , & \Delta \neq \Delta_1 \\ N'(\Delta_1) &= N(\Delta_1) - 1 , \end{aligned}$$

which yields, in place of (2.4.22),

$$\mathcal{K}' e^{\gamma t'} \mathfrak{N}' \leq O(1) N(\Delta_1)^{-1} e^{-|\delta Y_{j+1}| + \varepsilon m_0 d/2} \mathcal{K} e^{\gamma t} \mathfrak{N} . \tag{2.4.25}$$

The sum over α' includes a sum over the $N(\Delta_1)$ possible contractions in Δ_1, as well as the sum over Δ_1 and Δ_2. Using (2.4.24) and (2.4.25), we follow the derivation of (2.4.23) to obtain

$$\sum_{\alpha'} \| m'_{\alpha',2} \| \leq O(e^{-m_0 \varepsilon/8}) \| m_\alpha \| .$$

The remaining cases $i = 3, 4$ are bounded similarly, and we obtain

$$\sum_{\alpha'} \| m'_{\alpha'} \| \leq O(1) \| m_\alpha \| \tag{2.4.26}$$

where $O(1) \leq 1/4$ for m_0 large and $\lambda = \lambda(m_0, K_2)$ small.

We now consider m''_α in the expansion. The following lemma completes the proof of the proposition.

LEMMA 2.4.6.

$$\|m''_\alpha\| \leq \frac{1}{2}\|m_\alpha\| .$$

Proof. The term m''_α is a P_2 term, and so for this term $\mathcal{K}'' = 1$. Furthermore, $t'' = t$, and

$$\mathfrak{N}''(\Delta) = \begin{cases} 0, & \Delta \subset Y_{j+1} \\ \mathfrak{N}(\Delta), & \Delta \not\subset Y_{j+1} \end{cases}$$

since m''_α contains only B_T vertices. We write

$$\mathfrak{N}_1 = \prod\nolimits_{\Delta\subset Y_{j+1}} K_2^{N(\Delta)} , \qquad \mathfrak{N}_2 = \prod\nolimits_{\Delta\subset Y_{j+1}} N(\Delta)! .$$

Thus $\mathfrak{N} = \mathfrak{N}''\mathfrak{N}_1\mathfrak{N}_2$ and

$$\mathcal{K}''K_1^{\nu''}e^{rt}\mathfrak{N}'' \leq K_1^{\nu-1}e^{rt}\mathfrak{N}'' . \tag{2.4.27}$$

We claim (i) there is a constant $K_3 > 0$ such that if $m_0 \geq 1$, if $K_2 = K_2(m_0, K_3)$ is sufficiently large and if $\lambda \leq 1$, then

$$\|w''_\alpha(u, y_0, \cdot)\| \leq \mathfrak{N}_1^{1/2}\mathfrak{N}_2 e^{K_3|Y_{j+1}|}\int ds\, \|w_\alpha(u, s, y_0, \cdot)\|_2 , \tag{2.4.28}$$

(ii) for $K_2 = K_2(m_0, K_3)$ sufficiently large,

$$e^{(K_3+1)|Y_{j+1}|} \leq \frac{1}{2}\mathfrak{N}_1^{1/2}\mathcal{K} . \tag{2.4.29}$$

By (2.4.27)-(2.4.29), with $K_1 \geq 1$, $\|m''_\alpha\| \leq (1/2)\|m_\alpha\|$. Note that the factor $e^{|X_{l+1}|}$ which occurs in $\|m''_\alpha\|$ but not in $\|m_\alpha\|$ is controlled by (2.4.29), since $X_{l+1} = Y_{j+1}$.

We first establish (2.4.29). Recall that

$$\mathcal{K} = K_2^j \exp\left[\varepsilon m_0 D/2 - (1-\varepsilon)m_0 \sum \delta n - |Y_{j+1}|\right] . \tag{2.4.30}$$

Since at most $n + 1$ annihilators are advanced in time (by step II) to allow the insertion of some contour ∂Y_r with $|\delta Y_r| \geq 1$, one can check that

$$\sum \delta n \leq (n+1)\sum\nolimits_{r=1}^{j+1}|\delta Y_r| \leq (n+1)|Y_{j+1}| . \tag{2.4.31}$$

Let us consider a nonempty increment

$$\delta Y_r = Y_{r+1} \sim Y_r \neq \emptyset$$

in Y_r. Except when Y_r is enlarged to contain A or B_T vertices,

$$\begin{aligned} \delta Y_r &= \bigcup\nolimits_i (S_{r+1}(\Delta_i) \sim S_r(\Delta_i)) \\ &= \bigcup\nolimits_i \delta S_r(\Delta_i) . \end{aligned}$$

Hence such an increase occurs because in certain squares Δ_i, the length $2[N_r(\Delta_i)^{1/4}]+1$ of $S_r(\Delta_i)$ has increased; see Section 2.3. We define v_k to be the smallest integer such that $[v_k^{1/4}]=k$. Then the regions $\delta S_r(\Delta_i)$ contributing to δY_r are characterized by an increase in $N_r(\Delta_i)$, namely,

$$N_r(\Delta_i) < v_{k_i} \leqq v_{k_i'} \leqq N_{r+1}(\Delta_i) .$$

With this notation,

$$|S_r(\Delta_i)| = (2k_i+1)^2 \text{ and } S_{r+1}(\Delta_i) = (2k_i'+1)^2 ,$$

so

$$|\delta Y_r| \leqq O(1)\textstyle\sum_i((k_i')^2 - k_i^2) .$$

However, $N(\Delta_i)$ increases only for steps in which there is a contraction to the exponent in Δ, and for each contraction the increase in $N(\Delta_i)$ is at most $p-1$. Such contractions arise either from advancing the time of $n+1$ (or fewer) annihilators, or from the (at most two) contractions from $\delta C\cdot\Delta_\Phi$. Thus we see that there are at most $\max\{n+1, 2\} \leqq n+2$ squares Δ_i with $\delta S_r(\Delta_i) \neq \emptyset$. It follows that

$$\begin{aligned}\delta v_{k_i} &= v_{k_i'} - v_{k_i} \leqq \delta N(\Delta_i)\\ &= N_{r+1}(\Delta_i) - N_r(\Delta_i) \leqq (p-1)(n+2) ,\end{aligned}$$

and so for $k \geqq k_{\min}(p, n)$, we have $k_i' = k_i + 1$, and $(k_i')^2 - k_i^2 = 2k_i + 1$. Hence

$$|\delta Y_r| \leqq O(1)\textstyle\sum_i k_i \leqq O(1)\sum_i v_{k_i}^{1/4} .$$

For each Δ_i there must have been $(v_{k_i'} - v_{k_i})/p$ contractions to the exponent in Δ_i, whose contraction distances were at least k_i. Thus the corresponding contribution δD_r to D satisfies

$$\delta D_r \geqq \sum_i k_i \frac{|v_{k_i'} - v_{k_i}|}{p} \geqq \varepsilon_1 \sum_i v_{k_i} ,$$

where ε_1 depends only on p. Thus for any constant $\alpha > 0$,

$$|\delta Y_r| \leqq \alpha\delta D_r + O(\alpha^{-1}) . \tag{2.4.32}$$

The exceptions to the derivation of (2.4.32) occur in the steps in which Y_r is enlarged to contain either the A or the B_T vertices. We consider the two cases:

$$N_{j+1}(\Delta_i) \leqq \frac{1}{2} N_{r+1}(\Delta_i), \text{ and } N_{j+1}(\Delta_i) > \frac{1}{2} N_{r+1}(\Delta_i) .$$

In the first case, the reduction in the degree of the A or B_T monomial in Δ, namely

$$N_{r+1}(\Delta_i) - N_{j+1}(\Delta_i) \geqq \frac{1}{2} N_{r+1}(\Delta_i) ,$$

implies there were at least $(1/2)N_{r+1}(\Delta_i)$ contractions to vertices in Δ_i. Since each corresponding contraction distance is at least $\varepsilon_1 N_{r+1}(\Delta_i)^{1/4}$, we recover (2.4.32) in this case.

In the second case, $N_{j+1}(\Delta_i) \geqq (1/2)N_{r+1}(\Delta_i)$, recall that $N(\Delta_i) = N_{j+1}(\Delta_i)$ by definition. Thus

$$|S_{r+1}(\Delta_i) - S_r(\Delta_i)| \leqq O(N(\Delta_i)^{1/2})$$

and for $K_2 = K_2(K_3)$ sufficiently large,

$$\exp[(K_3 + 2 + (n+1)m_0)|\delta Y_r|] \leqq \mathfrak{N}_1^{1/2} .$$

Hence in all cases, summing (2.4.32) over r yields

(2.4.33) $\exp[(K_3 + 2 + (n+1)m_0)|Y_{j+1}|] \leqq \exp[\alpha D + O(j+1)\alpha^{-1}]\mathfrak{N}_1^{1/2}$.

Substituting (2.4.31) and (2.4.33) into (2.4.30) we obtain (2.4.29) if $K_2 = K_2(\varepsilon, m_0, K_3)$ is large enough.

We now establish (2.4.28). The kernel w''_α is given by (2.3.8), and for an l-term,

$$P''_l(u, s, x_B, y) = w_B(x_B)\int w_1(\alpha, u, s, y_0, \vec{y}, x)\prod_i : \Phi(x_i)^{n_i}: dx$$

is the path space monomial containing all vertices at times earlier than T, and factored out of the path space integral in (2.3.4) and (2.3.8). For $m_0 \geqq 1$, $\lambda \leqq 1$, there is a constant K_3, independent of C, m_0 and λ such that

(2.4.34) $$1 \leqq \|e^{-V(Y)}\|_{L_2(d\Phi_C)} \leqq e^{K_3|Y|}$$

and

(2.4.35) $\|P''_l(u, s, y, x_B)\|_{L_2(d\Phi_C)} \leqq [\prod_\Delta K_3^{N(\Delta)} N(\Delta)!]\,\|w(\alpha, u, s, y, x_B, \cdot)\|_{L_2}$.

These are standard estimates for such integrals (see [1]). Thus

$$|w''_\alpha(u, x_B, y)| = \left|\int ds \int P''_l e^{-V(Y_{j+1})} d\Phi_C\right|$$

$$\leqq e^{K_3|Y_{j+1}|}[\prod_\Delta K_3^{N(\Delta)} N(\Delta)!]\int ds\, \|w(\alpha, u, s, y, x_B, \cdot)\|_{L_2} ,$$

which yields (2.4.28), for $K_2 \geqq K_3^2$.

Since our expansion does not terminate after a finite number of steps, we must show that the sum (2.3.7) is convergent. We show that the sum of all terms not in final form tends to zero. This follows from two facts. First, for each term m_α,

$$|m_\alpha| \leqq \|m_\alpha\|_{2K_2} e^{O(\mathrm{Vol})} ,$$

where the volume of interaction, denoted Vol, is the area of the set of points within unit distance of suppt h. Here we use (2.4.34), (2.4.35) and a Schwarz inequality to factor out vertices. Also $O(\text{Vol}) \to 0$ as $\lambda \to 0$. Secondly, a fraction $\beta > 0$ of the elementary expansion steps is the step I of inserting a contour. By Propositions 2.4.4, after $O(L)$ expansion steps,

$$\sum' \|m_\alpha\| \leqq \left(\frac{3}{4}\right)^{O(L)} \|AB_T\| ,$$

where $\sum'$ denotes the sum over terms not in final form. The desired bound then follows.

3. Contours removed

In this chapter we remove all contours from the Chapter 2 expansion. In Chapter 2, we factored out the vertices at times earlier than T from the integrand F. Starting with the same integral, $\int F d\Phi_{C_*}$, the goal of the present expansion is still to factor out vertices at times earlier than T, but also to return to the Gaussian measure $d\Phi_{C_*}$. The new expansion is still defined inductively, and the terms still have the form (2.2.6). There are two elementary steps. The first is the Chapter 2 expansion, (2.3.7), the increasing contour construction. The second elementary step is the use of (2.2.1) and (2.2.5) to remove contours. These two steps are applied in alternation until the expansion terminates.

To describe the process of removing contours in more detail, we consider a P_2 term m_α, arising at the end of an increasing contour expansion, i.e., in (2.3.7). If $C_* = C(\mathfrak{X}_l, u)$, and if Y_j is the complete contour in m_α, then $X_{l+1} = Y_j$ and in (2.2.1) $s = u_{l+1}$, $C' = C_{*X_{l+1}}$ and $C'' = C_0$. The first term in (2.2.1) is a P_3 term if $n_a > 0$, and a final term if $n_a = 0$. The expansion of this chapter terminates for this term. The second term in (2.2.1) is written as a sum (over possible localizations and contractions), and for the purpose of defining norms the terms in this sum are regarded as P_1 terms. In these terms, we make the new choice $C_* = C(\mathfrak{X}_{l+1}, u)$, $\mathfrak{X}_{l+1} = \{\mathfrak{X}_l, X_{l+1}\}$, and we take $j = 0$ and $Y_0 = \varnothing$. Then $\mathcal{K}$, defined by m_α, equals one and

$$\mathcal{K}' = e^{\varepsilon m_0 d'/2} ,$$

where d' is the contraction distance in (2.2.1).

The expansion now has the form

$$\int F d\Phi_{C_0} = \sum_\alpha \int d\bar{x}\, d\bar{y}\, w_L(\alpha, x, y) \int \prod_{i=1}^{n_a} a(y_i) B_T(x) e^{-V} d\Phi_{C_0} \qquad (3.1)$$
$$+ \sum_\alpha w_R(\alpha) \int e^{-V} d\Phi_{C_0} .$$

The summation in (3.1) extends over the various terms generated by our expansion. This includes summation over $0 \leqq n_a \leqq n$ annihilators and over various space time localizations.

THEOREM 3.1. *Let* ε, n *be given. Let* m_0, K_2, K_1 *be sufficiently large and let* λ *be sufficiently small. Then the expansion* (3.1) *is norm decreasing.*

Remark. The expansion (3.1) is also norm decreasing if we start with a P_1 or P_2 term in place of an initial term.

Proof. It is sufficient to prove that each elementary step of the expansion decreases norms, and using Theorem 2.4.1 it is sufficient to show that removing the contours decreases norms. It is sufficient to consider this step applied to a single P_2 term m_α.

Consider the first (P_3 or final) term, m'_α, in (2.2.1). We have

$$\|m'_\alpha\| = K_1^{\nu'} e^{\gamma t'} \mathfrak{N}' \|w'\|_2 \leqq K_1^{-1} \|m_\alpha\| \leqq \frac{1}{2} \|m_\alpha\| , \tag{3.2}$$

since $\nu' \leqq \nu - 1$, $t' = t$, $\mathfrak{N}'(\Delta) = \mathfrak{N}(\Delta)$ and

$$\begin{aligned} |w'(\alpha, x, y)| &= \left| \int du\, w(\alpha, u, x, y) \right| \\ &\leqq \int du\, e^{A} |w(\alpha, u, x, y)| , \end{aligned} \tag{3.3}$$

with

$$A = \sum_{k=1}^{l} U_{k,l} |X_k| + |X_{l+1}| = \sum_{k=1}^{l=1} \delta U_{k,l+1} |X_k| . \tag{3.4}$$

Consider next the P_1 terms $m'_{\alpha'}$ which result from applying (2.2.1) and (2.2.5) to m_α as above. Consider the term which introduces an annihilator $a(y_{n_a+1})$ localized in a lattice interval Δ_2. The kernel $w'_{\alpha'}$ of $m'_{\alpha'}$ is

$$w'_{\alpha'}(u', y_0, \bar{y}', x) = w(\alpha, u, y_0, \bar{y}) \delta \hat{C}_{l+1}(x, y_{n_a+1}) v(x)$$

where $y' = \{y, y_{n_a+1}\}$, $\delta \hat{C}_{l+1}$ is given by the discussion following (2.1.7) and (2.2.5), and $v(x)$ is $O(\lambda)$ times the characteristic function of a lattice square Δ_1. As in (2.4.19), we obtain

$$\sum_{\Delta_1 \Delta_2} e^{m_0(1-\varepsilon/2)d'} \|w'_{\alpha'}\|_2 \leqq O(\lambda A) \|w_\alpha\|_2 . \tag{3.5}$$

Here Proposition 2.1.2 replaces 2.1.1 and A is defined in (3.4). Since

$$N(\Delta'_1) \leqq p - 1 = O(1) ,$$

$$\|m'_{\alpha'}\| = O(1) K_1^{\nu'} e^{\varepsilon m_0 d'/2} e^{\gamma t} \mathfrak{N} \int du' e^{A u_{l+1}} \|w'_{\alpha'}\|_2 ,$$

where $O(1)$ depends on K_2 and $du' = du\, du_{l+1}$. Taking $\varepsilon \leqq 1$ and substituting (3.5) in the above equation yields

(3.6)
$$\begin{aligned}\textstyle\sum_{\alpha'}\|m'_{\alpha'}\| &\leqq O(\lambda)K_1^{\nu'}e^{\bar{r}t}\mathfrak{N}\int du'Ae^{Au_{l+1}}\|w_\alpha\|_2\\ &\leqq O(\lambda)K_1^{\nu'}e^{\bar{r}t}\mathfrak{N}\int du\,e^{A}\|w_\alpha\|_2\\ &= O(\lambda)\|m_\alpha\|\,.\end{aligned}$$

Here $O(\lambda)$ depends on K_1 and K_2, but since λ is chosen after K_1 and K_2, $O(\lambda) \leqq 1/4$ for λ small.

The remaining terms $m'_{\alpha'}$ have an outside contraction $\partial/\partial\Phi(y_{n_\alpha+1})$ to a vertex, and are bounded similarly. Combined with (3.2) and (3.6), this completes the proof. We note that the. expansion (3.1) is convergent, as follows similarly to the proof of convergence of (2.3.7).

4. The n-particle cluster expansion

The n-particle cluster expansion leads to Theorems 1.1.8–1.1.11. In this expansion, the annihilators from the Chapter 3 expansion are replaced by Euclidean fields. In the case $n = 1$, the Euclidean fields occur at time zero only.

The n-particle expansion is defined inductively, by alternating two elementary expansion steps. The first expansion step is the Chapter 3 expansion. The second elementary step uses a modified form of (2.2.7), namely,

$$a(y) = \Phi(y) - \int_{z_0 \leqq y_0} dz[(2\mu)^{-1}e^{-(y_0-z_0)\mu}](\bar{y}, z)\frac{\partial}{\partial\Phi(z)}\,, \tag{4.1}$$

to convert an annihilator at the time y_0 into a field at the same time. The proof of (4.1) involves an integration by parts in the function space integral $\int F d\Phi_{C_0}$, and it depends on the explicit form of the covariance C_0. Hence (4.1) is valid when applied to an integrand in such an integral, but it is not valid as an operator on F. We write (4.1) as

$$a(y) = \Phi(y) + \Big(K\frac{\partial}{\partial\Phi}\Big)(y)$$

where K is the integral operator defined in (4.1).

In order to obtain a generalization of (4.1) valid for more than one annihilator, let

$$:\Phi(y_1, \cdots, y_r): = :\Phi(y_1)\cdots\Phi(y_r):\,,$$
$$(K\Phi)(y_1, \cdots, y_r) = \textstyle\prod_{i=1}^r\Big(K\frac{\partial}{\partial\Phi}\Big)(y_i)$$

for $r > 1$ and let $\Phi(\varnothing) = 1$,

$$\Big(K\frac{\partial}{\partial\Phi}\Big)(\varnothing) = 1\,.$$

For J a subset of $\{1, \cdots, n_a\}$ with complement $\sim J$, let y_J and $y_{\sim J}$ denote the corresponding subsets of $(y_1, \cdots, y_{n_a})$. Then

$$\sum_{i=1}^{n_a} a(y_i) = \sum_{J\subset\{1,\cdots,n_a\}} :\Phi(y_J):\left(K\frac{\partial}{\partial\Phi}\right)(y_{\sim J}) \,. \tag{4.2}$$

The derivatives in each monomial act on F, but not on the Wick ordered coefficient $:\Phi(y_J):$. Again integration by parts occurs in (4.2), and (4.2) is valid when applied to an integrand. The term with $J = \{1, 2, \cdots, n_a\}$ in (4.2) is a final term, and the other terms are P_1 terms. The full expansion, obtained by alternating these two steps, has the form

$$\begin{aligned}\int AB_T e^{-V} d\Phi_{C_0} &= \sum_{y_0 \leq T} \sum_{i=0}^{n} \int d\bar{y}\, w_i(y) \int :\Phi(y_1)\cdots\Phi(y_i):\, B_T e^{-V} d\Phi_{C_0} \\ &\quad + w_r \int e^{-V} d\Phi_{C_0}\,,\end{aligned} \tag{4.3}$$

once we group together terms with common values of i and y_0.

THEOREM 4.1. *Let ε and n be given. Let m_0, K_2 and K_1 be sufficiently large, and let λ be sufficiently small. Then the n-particle cluster expansion (4.3) is norm decreasing.*

Proof. We show that each expansion step, applied to a single term m_a, is norm decreasing. Since the Chapter 3 expansion is norm decreasing, it is sufficient to consider the step (4.2). This step is similar to, but easier than the analogous step in the proof of Theorem 4.2 below, so we do not give the details.

Proof of Theorems 1.1.8 *and* 1.1.11. Divide (4.3) by $\int e^{-V} d\Phi_{C_0}$ and set

$$Q_n(y_0) = \sum_{i=0}^{n} \int d\bar{y}\, w_i(y) \int :\Phi(y_1)\cdots\Phi(y_i): \,.$$

(By removal of the Wick ordering in $Q_n(y_0)$, we may apply the norms of (1.1.8).)
Then as $y_0 = t$ for $y_0 \leqq T$,

$$\int AB_T dq = \sum_{t\leq T} \int Q_n(t) B_T dq + w_r \,. \tag{4.4}$$

Let $\|\cdot\|_0$ denote the Chapter 1 norm (1.1.7), (1.1.8), and let $\|\cdot\|$ denote the norm (2.4.3). Then for $t \leqq T$,

$$\begin{aligned}\|Q_n(t) B_T\| &= e^{\gamma t} \|Q_n(t)\|_0 \|B\|_0\,, \\ \|w_r\| &= K_1 e^{\gamma T} |w_r|\,, \\ \|AB_T\| &= K_1^4 \|A\|_0 \|B\|_0\,.\end{aligned}$$

Hence by Theorem 4.1,

$$\sum_{t\leq T} e^{rt} ||Q_n(t)||_0 ||B||_0 + e^{rT}|w_r| \leq \sum_{t\leq T} ||Q_n(t)B_T|| + ||w_r||$$
$$\leq ||AB_T|| = K_1^4 ||A||_0 ||B||_0 .$$

Theorem 1.1.11 follows from this inequality, while Theorem 1.1.8 is the special case $B = I$, $n = 0$.

In the case $n = 1$, the identity

$$a(y) = \int d\bar{z}\, [e^{y_0 \mu}](\bar{y}, \bar{z})\Phi(\bar{z}, 0) - \int_{z_0 \leq y_0} dz [(2\mu)^{-1} e^{(y_0 - |z_0|)\mu}](\bar{y}, \bar{z}) \frac{\partial}{\partial \Phi}(z) , \tag{4.5}$$

for $y_0 \geq 0$, replaces (4.1). This identity also involves integration by parts, and is valid when applied to the integrand F in $\int F d\Phi_{C_0}$. The one-particle expansion is

$$\int AB_T dq = w_0 \int B_T dq + \int d\bar{y}\, w_1(\bar{y}) \int \Phi(\bar{y}, 0) B_T dq + w_r . \tag{4.6}$$

Theorem 1.1.10 is implied by the following result.

THEOREM 4.2. *Let B have support in the momentum range $|P_1| \leq \sqrt{3}\, m_0(1 - \varepsilon)$. Let m_0, K_2 and K_1 be sufficiently large and let λ be sufficiently small. Then the one-particle cluster expansion* (4.6) *is norm decreasing.*

The restriction on the momentum support of B applies to the Fourier transform

$$B^{\sim}(\bar{p}) = \int d\bar{x}\, e^{-i\bar{p}\bar{x}}\, U(\bar{x}) B U(\bar{x})^* .$$

It is important to work with a translation invariant Hamiltonian, in order that time translation preserve subspaces of bounded momentum. The estimates of Theorems 1.1.8, 1.1.11, 3.1 and 4.1 are uniform in the space cutoff h, and the path space integrals converge as $h \to 1$. Thus these theorems hold in the infinite volume limit, and we proceed to work in the limit $h = 1$. An alternative would be to use a periodic space cutoff; see [8] for example. In either case, the Hamiltonian is translation invariant and conserves momentum. By momentum conservation, a nonzero contribution to $\int AB_T dq$ from $A^{\sim}(\bar{p})$ arises only for $\bar{p}$ in the momentum range $|\bar{p}| \leq \sqrt{3}\, m_0(1 - \varepsilon)$. Thus, in (4.6), $\Phi^{\sim}(\bar{p}, 0)$ can be multiplied by a function $\zeta^{\sim}(\bar{p})$ which equals one on the set $|\bar{p}| \leq \sqrt{3}\, m_0(1 - \varepsilon)$. We take $\zeta^{\sim}$ to be smooth and supported on the set $|\bar{p}| \leq \sqrt{3}\, m_0(1 - \varepsilon/2)$.

The main difference between (4.5) and (4.1) is the occurrence of the unbounded operator kernel $\exp(y_0\mu)$, $y_0 \geq 0$, in (4.5). We use the fact that there is a gap between the energy bounding $\mu(\bar{p})$, for $|\bar{p}| \leq \sqrt{3}\, m_0(1 - \varepsilon/2)$, and the energy $2m_0(1 - \delta)$ given by the decay rate of Theorem 1.1.11 with

$n = 1$ and ε replaced by $\delta < \varepsilon$. In fact for such $\vec{p}$,

$$\mu(\vec{p}) \leqq m_0(1 + 3(1 - \varepsilon/2)^2)^{1/2} = 2m_0(1 - \varepsilon') \,, \tag{4.7}$$

defining $\varepsilon' = \varepsilon'(\varepsilon)$. Then

$$\varepsilon' = \frac{3}{8}\varepsilon - \frac{3}{32}\varepsilon^2 + \frac{1}{2}(\varepsilon')^2 > \frac{9}{32}\varepsilon \,,$$

so we choose $\delta = \varepsilon/4$, ensuring $\delta < \varepsilon'$.

Proof of Theorem 4.2. Each expansion step is norm decreasing. To show this, we again need consider only the step (4.5) applied to a single term m_α. This m_α is a P_3 term, of the form

$$m_\alpha = \int d\bar{y}\, w_\alpha(\bar{y}) \int a(y)\, B_T\, dq \,.$$

Let I_i denote the interval $|z - i| \leqq 1/2$, and also its characteristic function. The expansion step (4.5) is the identity

$$m_\alpha = \sum\nolimits_{i \in Z} m_i + \sum\nolimits_{\substack{j \in Z^2 \\ j_0 \leqq y_0}} m_j$$

where

$$m_i = \int_{I_i} w'_\alpha(\bar{z}, 0) \int \Phi(\bar{z}, 0)\, B_T\, dq$$

$$m_j = \int_{\Delta_j} dz\, w''_\alpha(z) \int \frac{\partial V}{\partial \Phi(z)} B_T\, dq$$

$$w'_\alpha(z) = [e^{(y_0 - |z_0|)\mu}](\zeta^* w_\alpha)(\bar{z})$$

$$w''_\alpha(z) = [(2\mu)^{-1} e^{(y_0 - |z_0|)\mu}](\zeta^* w_\alpha)(\bar{z})$$

and $\zeta^\sim$ is the C^∞ function chosen above. Here m_i is a final term and m_j is a P_1 term.

LEMMA 4.3. *Let ε'' satisfy $\delta < \varepsilon'' < \varepsilon'$ and let $w \in L_2(I_0)$. Let $\sigma = 0, 1$. Then there is a constant $K_4 = K_4(m_0)$ such that*

$$\| I_i \mu^{-\sigma} e^{t\mu}(\zeta^* w) \|_2 \leqq K_4(i^2 + 1)^{-1} e^{2t m_0(1-\varepsilon'')} \| w \|_2 \,.$$

Proof. By Fourier analysis, and (4.7),

$$\begin{aligned} \| I_i \mu^{-\sigma} e^{t\mu}(\zeta^* w) \| &\leqq O(1)(i^2 + 1)^{-1} \left\| \left(- \frac{\partial^2}{\partial \vec{p}^2} + 1 \right) \mu^{-\sigma} e^{t\mu(\vec{p})} \zeta^\sim w^\sim \right\| \\ &\leqq O(t^2)(i^2 + 1)^{-1} e^{2t m_0(1-\varepsilon')} \left\| \left(- \frac{\partial^2}{\partial \vec{p}^2} + 1 \right) w^\sim \right\|_2 \\ &= O(1)(i^2 + 1)^{-1} e^{2t m_0(1-\varepsilon'')} \| (\bar{z}^2 + 1) w \|_2 \\ &\leqq O(1)(i^2 + 1)^{-1} e^{2t m_0(1-\varepsilon'')} \| w \|_2 \,, \end{aligned}$$

because suppt $w \subset I_0$.

To compare the norms of m_i and m_j with m_α, we note that $\mathcal{K} = 1 = \mathcal{K}'$, and $\mathfrak{N}' = O(1)\mathfrak{N}$ where $O(1)$ is a constant depending on K_2. The value of t decreases in this expansion step, so that $t' = 0$ for the m_i term and $t' = j_0 + (1/2)$ for m_j. Thus using Lemma 4.3,

$$\begin{aligned} \|m_i\| &= O(1) K_1^2 (i^2+1)^{-1} e^{rt} \mathfrak{N} \|w_B\|_2 \|w_\alpha\|_2 \\ &= O(1) K_1^{-1} (i^2+1)^{-1} \|m_\alpha\| \end{aligned}$$

if w_α is supported in I_0, and a translate of this estimate in the general case. Here $O(1)$ depends on m_0 and K_2. Since K_1 is chosen after m_0 and K_2,

$$\sum_i \|m_i\| \leqq \frac{1}{2} \|m_\alpha\|$$

for K_1 sufficiently large.

The estimates on m_j are similar, with the observation that the new vertex $\partial V / \partial \Phi$ has a kernel containing a factor $O(\lambda)$. We use small λ, chosen after K_1, to ensure

$$\sum_j \|m_j\| \leqq \frac{1}{4} \|m_\alpha\| .$$

This completes the proof of Theorem 4.2.

Finally, we note that the expansions (4.3) and (4.6) converge, as follows from two facts. First, the $n = 0$ expansion converges, since the Chapter 3 expansion converges. Hence Theorem 1.1.8 applies and $|m_\alpha| \leqq O(1)\|m_\alpha\|$. Second, each application of the expansion (4.2) or (4.5) decreases by a factor 3/4, as does the Chapter 3 expansion. Thus the sum $\sum |m_\alpha|$ over terms not in final form is decreasing.

References

[1] J. Dimock and J. Glimm, Measures on Schwarz distribution space and applications to $\mathcal{P}(\phi)_2$ field theories, Adv. Math. **12** (1974), 58-83.

[2] J. Feldman, A relativistic Feynman-Kac formula, Nuclear Physics **B 52** (1973), 608-614.

[3] J. Glimm and A. Jaffe, The $\lambda\phi_2^4$ quantum field theory without cutoffs I, Phys. Rev. **176** (1968), 1945-1951.

[4] ———, The $\lambda\phi_2^4$ quantum field theory without cutoffs, II. The field operators and the approximate vacuum, Ann. of Math. **91** (1970), 362-401.

[5] ———, The $\lambda\phi_2^4$ quantum field theory without cutoffs, III. The physical vacuum, Acta Math. **125** (1970), 203-267.

[6] ———, The $\lambda\phi_2^4$ quantum field theory without cutoffs, IV. Perturbations of the Hamiltonian, J. Math. Phys. **13** (1972), 1558-1584.

[7] ———, Positivity of the ϕ_3^4 Hamiltonian, Fortschritte der Physik, **21** (1973), 327-375.

[8] ———, Quantum field models, in *Statistical Mechanics and Quantum Field Theory*, Ed. by C. de Witt and R. Stora, Gordon and Breach, New York, 1971.

[9] ———, Boson quantum field models, in *Mathematics of Contemporary Physics*, Ed. by R. Streater; Academic Press, New York, 1972.

[10] ———, The n-particle cluster expansion for the $\mathcal{P}(\phi)_2$ quantum field model, unpublished.

[11] J. GLIMM and T. SPENCER, The Wightman axioms and the mass gap for the $\mathcal{P}(\phi)_2$ quantum field theory, unpublished.

[12] R. GRIFFITHS and B. SIMON, The ϕ_2^4 field theory as acla ssical Ising model, Commun. Math. Phys. **33** (1973), 145-164.

[13] F. GUERRA, L. ROSEN and B. SIMON, The $\mathcal{P}(\phi)_2$ Euclidean quantum field theory as classical statistical mechanics, to appear in Ann. of Math.

[14] K. HEPP, On the connection between Wightman and LSZ quantum field theory, in *Axiomatic Field Theory*, Ed. by M. Chretien and S. Deser, Gordon and Breach, New York, 1966.

[15] E. NELSON, Quantum fields and Markoff fields, A.M.S. Symp. on Partial Differential Equations, Berkeley, 1971.

[16] ———, Construction of quantum fields from Markoff fields, J. Funct. Anal. **12** (1973), 97-112.

[17] C. NEWMAN, The construction of stationary two dimensional Markoff fields with an application to quantum field theory, J. Funct. Anal. **14** (1973), 44-61.

[18] K. OSTERWALDER and R. SCHRADER, Axioms for Euclidean Green's functions, Commun. Math. Phys. **31** (1973), 83-112.

[19] ———, Euclidean Fermi fields and a Feynman-Kac formula for boson-fermion models, Helv. Phys. Acta, **46** (1973), 277-302.

[20] D. RUELLE, On the axiomatic condition in quantum field theory, Helv. Phys. Acta **35** (1962), 147-163.

[21] B. SIMON, Correlation inequalities and the mass gap in $\mathcal{P}(\phi)_2$, II. Uniqueness of the vacuum for a class of strongly coupled theories, to appear in Ann. of Math.

[22] K. SYMANZIK, Euclidean quantum field theory, in *Local Quantum Theory*, 1968 Rendiconti della Scuola internazionale di Fisica, "E. Fermi", XLV Corso, R. Jost, Editor, Academic Press, New York, 1969.

[23] K. OSTERWALDER and R. SCHRADER, Axioms for Euclidean Green's function II, preprint.

[24] J. GLIMM, A. JAFFE and T. SPENCER, In: Lecture Notes in Physics Vol. 25, Ed. by G. Velo and A. Wightman, Springer Verlag, Berlin, 1973.

[25] J. FELDMAN, On the absence of bound states in the $\lambda(\varphi^4)_2$ quantum field model without symmetry breaking. To appear in Canadian Journal of Physics.

[26] T. SPENCER. The absence of even bound states for $\lambda(\varphi^4)_2$, to appear in Commun. Math. Phys.

(Received October 16, 1972)
(Revised June 28, 1973)

II Phase Cell Localization and φ_3^4 Stability

Introduction

The fundamental idea in all the work above was to understand in detail the consequences of certain Lagrangians. The methods always led to consideration of degrees of freedom in the interaction that could be localized approximately in phase space cells of unit size. These degrees of freedom could be analyzed in detail, and it was possible to prove that different regions of phase space were approximately independent. On the basis of this independence, renormalization could be studied independently in each local space-time volume, as well as the exponential decay rates in the cluster expansions of Section 1.

This philosophy emerged slowly as a systematic technique. One simple illustration of the idea was a new proof of the finite vacuum energy per unit volume in $P(\phi)_2$ Hamiltonians. The positivity bound involves localization in momentum space, and $O(1)$ localization in space-time. The independence of different space-time volumes yields a bound on the energy density. These methods were developed further to analyze perturbations of the $P(\phi)_2$ Hamiltonians by local field operators. In the fourth paper in the ϕ_2^4 series, the authors bound the vacuum enery of $H \pm \phi$. As a consequence, they prove self adjointness of the field operators in the representation defined by the physical vacuum. A general method to study perturbing operators of H, based on estimates of commutators between ϕ and H, is also developed in that paper.

These methods were practice to develop techniques sufficiently powerful to analyze the three-dimensional problem. The fundamental estimate on the positivity of the ϕ_3^4 Hamiltonian is proved in the last paper in this section, certainly the most complicated paper by the authors. In this construction the authors estimate those degrees of freedom which are localized simultaneously on specific length scales and on specific momentum scales. The estimates for $\exp(-tH)$ are defined inductively on successive length or momentum scales, and they apply for all values of the ϕ^4 coupling constant. This philosophy also appears at the basis of the renormalization group method. Using block spin transformations and scaling, the ideas in this paper have been implemented from such a point of view.

Bibliography

Positivity and self adjointness of the $P(\phi)_2$ Hamiltonian, Commun. Math. Phys. 22, 253–258 (1971).

The $\lambda\phi_2^4$ quantum field theory without cutoffs IV. Perturbations of the Hamiltonian, J. Math. Phys. 13, 1568–1584 (1972).

Positivity of the ϕ_3^4 Hamiltonian, Fort. Physik 21, 327–376 (1973).

Positivity and Self Adjointness of the $P(\phi)_2$ Hamiltonian

JAMES GLIMM*
Courant Institute of Mathematical Sciences, New York University, New York, New York

ARTHUR JAFFE**
Lyman Laboratory of Physics, Harvard University, Cambridge, Massachusetts

Received May 15, 1971

Abstract. We give a new proof that the locally correct Hamiltonian $H(g)$ is self adjoint, and that the vacuum energy $E(g) = \inf \text{spectrum} H(g)$ satisfies $-O(D) \leqq E(g)$, where $0 \leqq g \leqq 1$ and $D = \text{diam.supp.} g$.

An existence theorem has been proved for boson quantum fields with polynomial self interactions in two space time dimensions, and many basic properties of these quantum field models have been established. A self contained account of this theory is presented in [1]. A principal step in the construction of the field theory is to show that the Hamiltonian (energy operator) for a bounded space time region is bounded from below and self adjoint. The original proof of semiboundedness was given by Nelson [2] for the ϕ^4 theory. It was extended by Glimm [3] to a different type of space cutoff and to a positive polynomial $P(\phi)$ interaction. The authors [4] obtained a volume independent bound on the vacuum energy per unit volume. The original proof of self adjointness was given by the authors [5] for the ϕ^4 theory and by Rosen [6] for the $P(\phi)$ theory. Subsequent simplifications have been given [7–12]. In this note we present an easy proof of self adjointness and of the volume independent lower bound. The previous simplifications did not yield the volume independent lower bound. See [1] for notation.

Let

$$H(\kappa) = H_{o\,V} + H_I(\kappa) = H_{0,V} + \int :P(\phi_{\kappa,V})(x): g(x)\, dx\,,$$

where g is a function with compact support and $0 \leqq g \leqq 1$. We use the fact that the operators $H(\kappa)$, $H_{o,V}$, $H_I(\kappa)$ are each self adjoint and bounded from below.

---

* Supported in part by Air Force Office of Scientific Research, Contract AF49(638) -1719.

** On leave at Princeton University. Alfred P. Sloan Foundation Fellow. Supported in part by Air Force Office of Scientific Research, Contract F44620–70–C–0030.

Theorem 1. *$H(\kappa)$ is bounded from below, uniformly in κ, and V for $V \geqq 1$. Moreover* $\exp(-tH(\kappa))$ *converges in norm as $\kappa \to \infty$ and $V \to \infty$, uniformly in t, for t bounded away from zero and infinity.*

The proof of this result depends on the fact that certain perturbation expansions for $H(\kappa)$ converge in the limit of high energy. See also [11]. We use the Duhamel formula to generate such an expansion:

$$e^{-tH(\kappa)} = e^{-tH(\kappa')} - \int_0^t e^{-sH(\kappa')}(H(\kappa) - H(\kappa'))\, e^{-(t-s)H(\kappa)} ds\,. \qquad (1)$$

To simplify the formulas below, we define

$$H(\sigma, s) = \begin{cases} H(\kappa') & \text{for} \quad \sigma < s \\ H(\kappa) & \text{for} \quad \sigma > s \end{cases}$$

and write the integrand in (1) as

$$\left(e^{-\int_0^t H(\sigma,s)d\sigma} \delta H(s)\right)_+ \qquad (2)$$

where the subscript $+$ denotes a time ordering and $\delta H(s)$ inserts $\delta H = H(\kappa) - H(\kappa')$ at the time s. The formula (1) follows from differentiating $F(t, s) = \exp(-sH(\kappa'))\exp(-(t-s)(H(\kappa)))$ with respect to s and integrating the result from 0 to t. Since $F(t, s)$ is strongly continuous in s for $0 \leqq s \leqq t$, and it is norm differentiable with the bounded and norm continuous derivative (2) in $0 < s < t$, the equality (1) is valid whenever the integral in (1) exists. To prove Theorem 1, we use an iterated Duhamel formula, and we bound the resulting integrands. The bounds on the integrands are independent of the Duhamel formula and justify its use.

Let $p = \deg P$ and $\kappa_j = \exp(j^{2/p})$. By undoing the Wick ordering in H_I, we have the cutoff dependent bound

$$-O(j) = -O(\log \kappa_j)^{p/2} \leqq H_I(\kappa_j)\,. \qquad (3)$$

Let

$$h_j = \begin{cases} H(\kappa_j) & \text{if} \quad \kappa_j \leqq \kappa \\ H(\kappa) & \phantom{\text{if}} \quad \kappa_j \geqq \kappa \end{cases}$$

and let $\delta h_j = H(\kappa) - H(\kappa_j)$. Iterating (1) we obtain

$$\exp(-tH(\kappa)) = \sum_{n=0}^{\infty} (-1)^n \int ds \left(e^{-\int_0^t H(\sigma,s)d\sigma} \prod_{i=1}^{n} \delta h_i(s_i)\right)_+ \qquad (4)$$

where the time integration extends over the domain

$$0 \leqq s_1 \leqq s_2 \leqq \cdots \leqq s_n \leqq t \qquad (5)$$

and $H(\sigma, s) = h_j$ if $s_{j-1} < \sigma < s_j$. The series terminates for $\kappa_n \geqq \kappa$.

In the representation of Fock space for which the operators $H_I(\kappa)$ are all diagonal, $\exp(-tH_{o,V})$ is an integral operator with a positive kernel. We take absolute values of operator kernels in this representation. We use the Trotter product formula for each factor $\exp(-(s_{j+1}-s_j)h_j)$ in (4) together with the lower bound (3) and the upper bound

$$|\delta h_i| \leqq \kappa_i^{-1/4} + \kappa_i^{1/4}(\delta h_i)^2 = f_i$$

to obtain

$$\left\langle \theta, \left(e^{-\int_0^t H(\sigma,s)\,d\sigma} \prod_{i=1}^n \delta h_i(s_i) \right)_+ \theta \right\rangle \leqq \|\theta\|^2 e^{O(n)t} \left\| \left(e^{-\int_0^t H_{o,V}\,d\sigma} \prod_{i=1}^n f_i(s_i) \right)_+ \right\| . \tag{6}$$

The time ordered product above is easily seen to be a bounded operator, and it is norm continuous on the open domain $0 < s_1 < \cdots < s_n < t$. To obtain a bound on the closed domain (5), we use the estimate

$$\|f_j(N+I)^{-p}\| \leqq O(\kappa_j^{-1/4}) = 0(\exp(-\tfrac{1}{4}j^{2/p})) . \tag{7}$$

Here N is the number of particles operator, see [1]. The maximum length $s_j - s_{j-1}$ of the time intervals in the time ordered product is at least $t/(n+1)$. We commute np powers of N onto the maximum time interval, and use (7) to show

$$\left\| \left(e^{-\int_0^t H_{o,V}\,d\sigma} \prod_{i=1}^n f_i(s_i) \right)_+ \right\| \leqq a^n (n!)^p \|N^{pn} e^{-tH_o/(n+1)}\| \prod_{j=1}^n \kappa_j^{-1/4}$$
$$\leqq \left(\frac{c}{t}\right)^{pn} (n!)^{2p} \exp(-bn^{1+2/p})$$

for certain positive constants a, b, c. Hence (6) is bounded uniformly in s and so the integral in (4) exists. This justifies (4) and yields the bound

$$\exp(-tH(\kappa)) \leqq \sum_{n=1}^{\infty} \exp(an\log n - bn^{1+2/p}) < \infty , \tag{8}$$

with new constants a and b. These constants are independent of κ and are bounded uniformly if $V > 1$ and if t is bounded away from zero and ∞. This completes the proof that $H(\kappa)$ is bounded from below. The same estimates yield norm convergence of the semigroup $\exp(-tH(\kappa))$ as $\kappa \to \infty$, and the limit $V \to \infty$ is also easily controlled. It follows from Theorem 1 and the convergence of $H(\kappa)$ on the dense set $C^\infty(H_o)$, that the limit of $\exp(-tH(\kappa))$ is a semigroup $\exp(-tH)$ with self adjoint generator H, see [1].

We now establish a volume independent lower bound for $H(\kappa)$. Let $A(x)$ be the configuration space annihilation operator, so $A(x)$ annihilates a particle localized (in the Newton-Wigner sense) at x. We let

$$N_i = \int_{i-1/2}^{i+1/2} A(x)^* A(x) dx$$

and

$$N_{\text{loc}} = \sum_{i=-\infty}^{\infty} N_i e^{-m|i|/2}$$

be localized number operators. We assert that for g supported in a fixed interval B,

$$\| f_j(N_{\text{loc}} + I)^{-p} \| \leqq O(\kappa_j^{-1/4}) . \tag{9}$$

Our proof of Theorem 1 is valid with $\varepsilon N_{\text{loc}}$ replacing H_o and (9) replacing (7). Thus

$$0 \leqq \varepsilon N_{\text{loc}} + H_I(\kappa) + O(1)\, \mathrm{I} \tag{10}$$

Summing (10) over translates and using the bound $N \leqq \text{const} H_o$ yields

Theorem 2. *Let* $0 \leqq g \leqq 1$ *and let* $D = \text{diam.supp.}g$. *Then*

$$0 \leqq H(\kappa) + 0(D) ,$$

as $D \to \infty$, *and* $O(D)$ *is uniform in* κ *and* $V \geqq 1$.

Proof. We only need to establish (9), and without loss of generality we may assume that $\text{supp} g \subset (-\frac{1}{8}, \frac{1}{8})$. Let $w_j(x_1, \ldots, x_r)$ be the configuration space kernel of a Wick monomial W_j contributing to f_j, $0 < r \leqq 2p$. Let

$$w_{j,i_1,\ldots,i_r}(x) = w_j(x) \prod_{\nu=1}^{r} E_{i_\nu}(x_\nu)$$

where $E_i(x)$ is the characteristic function of the interval $[i-1/2, i+1/2]$. We assert that

$$\begin{aligned} \| W_{j,i_1,\ldots,i_r} \theta \| &\leqq \text{const} \| w_{j,i_1,\ldots,i_r} \|_2 \left\| \prod_{\nu=1}^{r} (N_{i_\nu} + I)^{1/2} \theta \right\| \\ &\leqq O(\kappa_j^{-1/4}) \exp(-m|i_1| - \cdots - m|i_r|) \left\| \prod_{\nu=1}^{r} (N_{i_\nu} + I)^{1/2} \theta \right\| . \end{aligned} \tag{11}$$

Since $\|(N_i + I)^{1/2} (N_{\text{loc}} + I)^{-1/2}\| \leqq \exp(m|i|/2)$, summing (11) over the i_ν proves (9). The first inequality in (11) is elementary, since N_i measures the number of particles in $[i - \frac{1}{2}, i + \frac{1}{2}]$. To bound the L_2 norm of

$w_{j,i_1,\ldots,i_r}$, we represent the kernel as

$$w_{j,i_1,\ldots,i_r} = \left(\prod_{v=1}^{r} K_{v,i_v}\right) w_j \tag{12}$$

where K_{v,i_v} is an integral operator on the variable x_v of w_j. Then

$$\begin{aligned}\|w_{j,i_1,\ldots,i_r}\|_2 &\leqq \|w_j\|_2 \prod_{v=1}^{r} \|K_{v,i_v}\| \\ &\leqq O(\kappa_j^{-1/4}) \prod_{v=1}^{r} \|K_{v,i_v}\| .\end{aligned}$$

Let $\mu_x^2 = \left(-\frac{d^2}{dx^2} + m^2\right)$ and let ζ be multiplication by a C^∞ function with support in $(-1/4, 1/4)$, such that for all x and y, $\zeta(x)\chi_\kappa(x-y)g(y) = \chi_\kappa(x-y)g(y)$. Here the ultraviolet cutoff in $H_I(\kappa)$ is introduced by the cutoff field $\phi_\kappa(x) = (\chi_\kappa * \phi)(x)$ and χ_κ is supported in $[-\kappa^{-1}, \kappa^{-1}]$. Then we take

$$K_{v,i_v} = E_{i_v}(x_v)\,\mu_{x_v}^{-1/2}\zeta(x_v)\,\mu_{x_v}^{1/2},$$

and (12) is valid. The operators $\mu_x^{\pm 1/2}$ are given by convolution with a distribution $k_\pm(x)$ that is C^∞ except at $x = 0$, with derivatives $0(1)\exp(-m|x|)$ as $|x| \to \infty$. An easy computation then shows that $\|K_{v,i}\| \leqq O(1)e^{-m|i|}$. This completes the proof of Theorem 2.

References

1. Glimm, J., Jaffe, A.: Field theory models. In: Dewitt, C., Stora, R. (Eds.) 1970 Les Houches Lectures. New York: Gordon and Breach Science Publ. 1971.
2. Nelson, E.: A quartic interaction in two dimensions. In: Goodman, R., Segal, I. (Eds.): Mathematical theory of elementary particles. Cambridge: MIT Press 1966.
3. Glimm, J.: Boson fields with nonlinear self interaction in two dimensions. Commun. math. Phys. **8**, 12—25 (1968).
4. — Jaffe, A.: The $\lambda(\phi^4)_2$ quantum field theory without cutoffs III: The physical vacuum. Acta Math. **125**, 203—267 (1970).
5. — — A $\lambda\phi^4$ quantum field theory without cutoffs. I. Phys. Rev. **176**, 1945—51 (1968).
6. Rosen, L.: A $\lambda\phi^{2n}$ field theory without cutoffs. Commun. math. Phys. **16**, 157—183 (1970).
7. Segal, I.: Construction of nonlinear local quantum processes: I. Ann. Math. **92**, 462—481 (1970).
8. Høegh-Krohn, R., Simon, B.: Hypercontractive semigroups and two dimensional self-coupled bose fields. J. Functional Analysis, To appear.
9. Masson, D., McClary, W.: On the self adjointness of the $(g(x)\phi^4)^2$ Hamiltonian. Commun. math. Phys. **21**, 71—74 (1971).

10. Konrady, J.: Almost positive perturbations of positive self adjoint operators. To appear.
11. Federbush, P.: A convergent expansion for the resolvent of $:\phi^4:_{1+1}$. To appear.
12. Masson, D.: Essential self adjointness of semi bounded operators: An extension of the Kato-Rellich theorem. To appear.

James Glimm
Courant Institute of Mathematical Sciences
New York University
251 Mercer Street
New York, N.Y. 10012/USA

Arthur Jaffe
Lyman Laboratory of Physics
Harvard University
Cambridge, Mass. 02138, USA

The $\lambda\varphi_2^4$ Quantum Field Theory without Cutoffs. IV. Perturbations of the Hamiltonian

James Glimm*
Courant Institute of Mathematical Sciences, New York University, New York, New York

and

Arthur Jaffe†
Lyman Laboratory of Physics, Harvard University, Cambridge, Massachusetts

(Received 3 March 1972)

We introduce an inductive method to estimate the shift δE in the vacuum energy, caused by a perturbation δH of the $\mathcal{P}(\varphi)_2$ Hamiltonian H. We prove that if δH equals the field bilinear form $\varphi(x, t)$, then δE is finite. We show that the vacuum expectation values of products of fields (Wightman functions) exist and are tempered distributions. They determine, via the reconstruction theorem, essentially self-adjoint field operators $\varphi(f)$, for real test functions $f \in \mathcal{S}(R^2)$. We also bound the perturbation of the $\mathcal{P}(\varphi)_2$ Hamiltonian by a polynomial $(\mathcal{P}_1(\varphi))(h) = \delta H$, so long as $\mathcal{P} + \mathcal{P}_1$ is formally positive. In that case, and with $\|h\|_\infty \leq 1$, δE is bounded by const$(1 + \text{diam supp } h)$.

1. THE MAIN RESULTS

We derive new estimates on the vacuum energy in the $\mathcal{P}(\varphi)_2$ quantum field model. These estimates permit us to establish one of the missing Wightman axioms for $\mathcal{P}(\varphi)_2$.

Theorem A: The Wightman functions or vacuum expectation values

$$\langle \Omega, \varphi(x_1, t_1)\varphi(x_2, t_2)\cdots\varphi(x_n, t_n)\Omega\rangle, \tag{1.1}$$

for the $\mathcal{P}(\varphi)_2$ model exist as tempered distributions in $\mathcal{S}'(R^{2n})$.

Our estimates bound the shift in the vacuum energy due to a local perturbation of the Hamiltonian H. The simplest such estimate is

Theorem B: The field $\varphi(x, t)$ is a bilinear form on the domain $\mathfrak{D}(H^{1/2}) \times \mathfrak{D}(H^{1/2})$, and $H + \varphi(x, t)$ is bounded from below.

Other new results for the $\mathcal{P}(\varphi)_2$ quantum field model are established in Theorem 1.1–Corollary 1.3. The methods and estimates of this paper appear to have other applications. For example, it should be possible to show that the Wightman functions (1.1) have exactly the singularities predicted by perturbation theory. Furthermore, the methods used here appear useful in the study of the more singular φ_3^4 model.

Solutions to the $\mathcal{P}(\varphi)_2$ model were constructed in earlier work of the authors and of Rosen. They are known to satisfy the Haag–Kastler axioms for quantum fields, and most of the Wightman axioms, as was shown by the authors, Cannon and Rosen. See Refs. 1 and 2 for references, notation, and proofs.

The Hamiltonian operator in the $\mathcal{P}(\varphi)_2$ theory is defined as the limit of approximate, or cutoff, Hamiltonians $H(g)$,

$$H(g) = H_0 + H_1(g) - E(g)I. \tag{1.2}$$

Here $H_0 = H(0)$ is the free Hamiltonian for bosons of mass $m > 0$, and

$$H_1(g) = \int :\mathcal{P}(\varphi(x)): g(x)dx. \tag{1.3}$$

The space cutoff g has compact support, and

$$0 \leq g(x) \leq 1. \tag{1.4}$$

The vacuum energy $E(g)$ is a finite constant (divergent as $g \to 1$) and chosen so that

$$0 = \inf \text{ spectrum } H(g). \tag{1.5}$$

Certain uniform estimates on $H(g)$ allow us to pass to a limit $g = 1$ and so obtain a Hamiltonian without cutoffs,

$$H = \lim H(g), \tag{1.6}$$

see Refs. 1 and 3.

We now describe our new uniform estimates for cutoff Hamiltonians. We use these estimates to establish properties of the theory without cutoffs ($g \equiv 1$), such as Theorems A and B.

Case 1. Perturbations of arbitrary degree: Let h be a real function $\|h\|_\infty \leq 1$, $D = 1 + \text{diam. supp. } h < \infty$. Let $p = \max\{\deg\mathcal{P}, \deg\mathcal{P}_1\}$, and let

$$0 \leq \mathcal{P}(\zeta)g(x), \quad 0 \leq \mathcal{P}(\zeta)g(x) + \mathcal{P}_1(\zeta)h(x) \tag{1.7}$$

for all real ζ, x. If $\deg\mathcal{P} = \deg\mathcal{P}_1$ and $\mathcal{P} \neq \mathcal{P}_1$, we assume that the coefficient of the degree p term in $\mathcal{P}g + \mathcal{P}_1 h$ is bounded away from zero on a neighborhood of supp h. Let

$$\delta H = \int :\mathcal{P}_1(\varphi(x)): h(x)dx = \delta H(h), \tag{1.8}$$

$$\delta E = \inf \text{ spectrum}\{H(g) + \delta H\} = \delta E(g, h). \tag{1.9}$$

Case 2. Linear perturbations: Let h be a real C^∞ function with

$$\|\mu^{-(1/2)-\epsilon}h\|_2 \le 1 \tag{1.10}$$

for $\epsilon = (8p)^{-1}$, $p = \deg \mathcal{P}$, and let $D = 1 + \text{diam supp } h < \infty$. Let

$$\delta H = \varphi(h) = \int \varphi(x)h(x)dx, \tag{1.11}$$

and let

$$\delta E = \delta E(g,h) = \inf \text{ spectrum } \{H + \varphi(h)\}. \tag{1.12}$$

Theorem 1.1: There exists a constant M, independent of g and h, such that in Cases 1 and 2

$$|\delta E| \le MD. \tag{1.13}$$

Remarks: In Case 2, if $1 \le \|\mu^{-1/2-\epsilon}h\|_2 < \infty$, we prove (1.13) with $M = M(\|\mu^{-1/2-\epsilon}h\|_2)$. Formal reasoning indicates that $\epsilon = \frac{1}{2}$ is possible; but a proof requires improvements in Sec. 3B.

Theorems A and B are corollaries to Theorem 1.1.

We consider space-time averaged fields

$$\varphi(f) = \int \varphi(x,t)f(x,t)dxdt$$

in the theory with the cutoff Hamiltonian $H(g)$ and vacuum Ω_g, $H(g)\Omega_g = 0$. The g-cutoff vacuum expectation values are tempered distributions.[4] The analog to Theorem A in the cutoff model is:

Theorem A_g: There exist translation invariant Schwartz space norms $|\cdot|_r$, independent of g, such that

$$|\langle \Omega_g, \varphi(f_1)\cdots\varphi(f_n)\Omega_g\rangle| \le |f_1|_1 \cdots |f_n|_n. \tag{1.14}$$

The new aspect of (1.14) is a bound that is independent of g. In order to ensure space translation invariance of a limiting vacuum, we average the vacuum expectation values $\langle \Omega_g, \varphi(f_1)\cdots\varphi(f_n)\Omega_g\rangle$ over space translations in an interval of length 0 (diam supp g). Since the norms $|\cdot|_r$ are translation invariant, the space averaged expectation values are also bounded by $|f_1|_1\cdots|f_n|_n$. By the methods of Refs. 3 and 5 we pass to a subsequence as $g \to 1$. As in Refs. 3 and 5, the limiting vacuum expectation values (1.1) satisfy the bound (1.14) with $g = 1$. Furthermore, the limiting vacuum expectation values (1.1) are obtained as limits as $g \to 1$ of a subsequence of the space averaged $\langle \Omega_g, \varphi(f_1)\cdots\varphi(f_n)\Omega_g\rangle$. Applying the Schwartz nuclear theorem to (1.14) proves that the vacuum expectation values are tempered distributions. Thus Theorem A follows from Theorem A_g, which we establish below.

We remark that the estimate of Theorem 1.1 is also valid with a space cutoff given by a periodic box of volume V. (See Refs. 1 and 5 for notation.) In Cases 1_V and 2_V we define δH and δE as in (1.8)–(1.12), with an additional dependence on V. We assume supp $h \subset [-\frac{1}{2}V, \frac{1}{2}V]$.

Theorem 1.1_V: There is a constant M independent of g, h, and V such that in Cases 1_V and 2_V

$$|\delta E| \le MD.$$

Our proof in this paper of Theorem 1.1, combined with the methods of Ref. 5, Sec. 2 that deal with estimates in a periodic box, prove Theorem 1.1_V. Likewise, we can establish a relevant Theorem A_V, and we pass to the $V = \infty$ limit in the vacuum expectation values. With this method, it is not necessary to average the vacuum expectation values

$$\langle \Omega_V, \varphi_V(f_1)\cdots\varphi_V(f_n)\Omega_V\rangle$$

over space translations in order to ensure translation invariance of the limiting vacuum Ω.

We now assume Theorem 1.1 and turn to the proof of Theorems A_g and B. We use

Lemma 1.1: Let $0 \le H = H^*$. Let A and $\dot{A} = [iH, A]$ be real bilinear forms with domain $\mathcal{C}^\infty(H) \times \mathcal{C}^\infty(H)$. Let $R = (H + I)^{-1}$, and let

$$\|R^{1/2}AR^{1/2}\| + \|R^{1/2}\dot{A}R^{1/2}\| \le M. \tag{1.15}$$

Then A uniquely determines a symmetric operator (also called A) with domain $\mathfrak{D}(H)$ and for a universal constant α,

$$\|AR\| \le \alpha M. \tag{1.16}$$

Proof: Let $R(y) = [H + (y+1)I]^{-1}$. Then

$$R^{1/2} = \frac{1}{\pi}\int_0^\infty \lambda^{-1/2}R(\lambda)d\lambda.$$

On $\mathcal{C}^\infty(H) \times \mathcal{C}^\infty(H)$,

$$\begin{aligned} AR^{1/2} &= R^{1/2}A + \frac{1}{\pi}\int_0^\infty \lambda^{-1/2}[A, R(\lambda)]d\lambda \\ &= R^{1/2}A - \frac{i}{\pi}\int_0^\infty \lambda^{-1/2}R(\lambda)\dot{A}R(\lambda)d\lambda. \end{aligned} \tag{1.17}$$

By (1.15), $R(\lambda)\dot{A}R(\lambda)$ is bounded, norm continuous in λ, with norm bounded by $(\lambda+1)^{-1}M$. Hence the integral over λ in (1.17) is norm convergent, and by (1.17)

$$\begin{aligned} |\langle\theta, AR\theta\rangle| &\le \|R^{1/2}AR^{1/2}\|\|\theta\|^2 + O(M)\|\theta\|^2 \\ &\le \alpha M\|\theta\|^2. \end{aligned}$$

The constant α depends only on the integral of $\lambda^{-1/2}(\lambda+1)^{-1}$. Lemma 1.1 then follows by the Riesz representation theorem.

Proof of Theorem A_g: Let g be fixed. We apply Theorem 1.1, Case 2, for an arbitrary real element $h/\|\mu^{-(1/2)-\epsilon}h\|_2$ of $\mathcal{C}_0^\infty(R^1)$. This yields

$$\begin{aligned} \pm\varphi(h) &\le \text{const}\,\|\mu^{-(1/2)-\epsilon}h\|_2[H(g) + I] \\ &\le \text{const}\,\|h\|_1[H(g) + I] \end{aligned} \tag{1.18}$$

with constants independent of g, but possibly depending on D. By taking a partition of unity, $1 = \sum \zeta_i$, where ζ_i is the translate of a fixed $\mathcal{C}_0^\infty$ function, we obtain

$$\pm\varphi(h\zeta_i) \le \text{const}\,\|h\zeta_i\|_1[H(g) + I],$$

and summing

$$\pm\varphi(h) \le \text{const}\,\|h\|_1[H(g) + I],$$

with a constant independent of g and h. For real $f \in \mathcal{S}(R^2)$, we integrate over time to obtain

$$\pm\varphi(f) \le \text{const}\,\|f\|_1[H(g) + I] \tag{1.18'}$$

with a constant independent of f, g. Likewise

$$\pm \dot{\varphi}(f) = \mp \varphi(D_t f) \leq \text{const} \|D_t f\|_1 [H(g) + I],$$

so by Lemma 1.1

$$\|\varphi(f) R_g\| \leq \alpha(\|f\|_1 + \|D_t f\|_1), \tag{1.19}$$

where $R_g = [H(g) + I]^{-1}$. On the domain $\mathcal{C}^\infty(H(g)) \times \mathcal{C}^\infty(H(g))$,

$$\langle \chi, [H(g) + I]^r \varphi(f)\theta\rangle$$
$$= \sum_{j=0}^{r} \binom{r}{j} i^j \langle \chi, \varphi(D_t^j f)[H(g) + I]^{r-j}\theta\rangle. \tag{1.20}$$

Thus by (1.19)

$$\|(H(g) + I)^r \varphi(f)\theta\| \leq |f|_{r+1} \|(H(g) + I)^{r+1}\theta\|, \tag{1.21}$$

where

$$|f|_{r+1} = \alpha 2^{r+1} \sum_{j=0}^{r+1} \|D_t^j f\|_1 \tag{1.22}$$

is a translation invariant Schwartz space norm. The bound (1.14), and hence Theorem A_g follows from (1.21) and (1.22).

The Hilbert space $\mathcal{H}_{\text{GNS}}$ associated with the infinite volume limit is defined by means of the GNS construction, see Refs. 3 and 5. The limiting vacuum expectation values (1.1) can also be used to construct an infinite volume Hilbert space $\mathcal{H}_W$, by the Wightman reconstruction theorem. The vacuum expectation values in these two theories coincide. Hence by the uniqueness portion of the reconstruction theorem, $\mathcal{H}_W$ can be identified as a subspace of $\mathcal{H}_{\text{GNS}}$.

Proposition 1.1: $\mathcal{H}_W = \mathcal{H}_{\text{GNS}}$: That is, the Wightman and GNS constructions yield the same infinite volume Hilbert space. In particular the cyclic subspace $\mathfrak{D}$ generated from Ω by polynomials in $\varphi(f)$, $f \in \mathcal{S}(R^2)$, is dense in $\mathcal{H}_{\text{GNS}}$.

Proof: Let $\mathcal{K}_{\alpha\beta}$ be the subset of the real elements $\mathcal{S}_R(R^2)$ of $\mathcal{S}(R^2)$ satisfying

$$\|D_t^n f\|_1 \leq \alpha\beta^n.$$

Clearly $\cup_{\alpha\beta}\mathcal{K}_{\alpha\beta}$ is dense in $\mathcal{S}_R(R^2)$. Let $\mathfrak{F}_{\alpha\beta}$ be the linear span of vectors

$$e^{i\varphi(f_1)} \cdots e^{i\varphi(f_n)}\Omega$$

for $f_j \in \mathcal{K}_{\alpha\beta}$, $n = 0, 1, \cdots$. By definition $\cup_{\alpha\beta}\mathfrak{F}_{\alpha\beta}$ is dense in $\mathcal{H}_{\text{GNS}}$. Using (1.19) and (1.20), we establish for $f_j \in \mathcal{K}_{\alpha\beta}$,

$$\|\varphi(f_1) \cdots \varphi(f_k)\Omega\| \leq B^k k! \tag{1.23}$$

with a constant $B = B(\alpha, \beta)$. (See Ref. 4, p. 389). We now prove that $\mathfrak{D}$ is dense. For any χ we define

$$f(\lambda) = \langle \chi, e^{i\lambda_1\varphi(f_1)} \cdots e^{i\lambda_n\varphi(f_n)}\Omega\rangle$$

as an iterated power series, summing first λ_1, then $\lambda_2, \cdots$. These series converge on account of (1.23). Thus if $\chi \perp \mathfrak{D}$, $f(\lambda) = 0$. In that case $\chi \perp \cup_{\alpha\beta}\mathfrak{F}_{\alpha\beta}$, so $\chi = 0$ and $\mathfrak{D}$ is dense.

We give three short corollaries to Theorem 1.1 which establish further properties of the space-time averaged fields $\varphi(f)$.

Corollary 1.1: Let $f, D_t f \in \mathcal{L}_1$. Then $\varphi(f)$ is an operator on $\mathfrak{D}(H)$ and

$$\|\varphi(f)(H + I)^{-1}\| \leq \alpha(\|f\|_1 + \|D_t f\|_1). \tag{1.24}$$

Corollary 1.2: Let f be real, and let $f, D_t f, D_{tt} f \in \mathcal{L}_1$. Then $\varphi(f)$ is essentially self-adjoint on any core for H.

Corollary 1.3: Let $f \in \mathcal{S}(R^2)$ be real, and let $\varphi(f)$ be the operator obtained from the vacuum expectation values (1.1) by the reconstruction theorem. [The domain of $\varphi(f)$ is $\mathfrak{D}$.] Then $\varphi(f)$ is essentially self-adjoint.

Proofs: By (1.18′), for real f,

$$\pm \langle \theta_g, \varphi(f)\theta_g\rangle \leq O(1)\|f\|_1 \langle \theta_g, [H(g) + I]\theta_g\rangle \tag{1.25}$$

for $\theta_g = \varphi(f_1) \cdots \varphi(f_n)\Omega_g \in \mathfrak{F}$. Let

$$\theta = \varphi(f_1) \cdots \varphi(f_n)\Omega \in \mathfrak{D} \subset \mathcal{H}_{\text{GNS}} = \mathcal{H}_W = \mathfrak{F}_{\text{ren}}.$$

We recall that the vacuum expectation value $\langle \theta, \varphi(f)\theta\rangle$ is obtained as a limit of a subsequence from space translation averages of $\langle \theta_g, \varphi(f)\theta_g\rangle$, and similarly $\langle \theta, H\theta\rangle$ is obtained as a limit from the cutoff expectation values $\langle \theta_g, H(g)\theta_g\rangle$. Thus we obtain in the limit $g = 1$,

$$\pm \langle \theta, \varphi(f)\theta\rangle \leq O(1)\|f\|_1 \langle \theta, (H + I)\theta\rangle.$$

Furthermore, we have $\dot{\varphi}(f) = -\varphi(D_t f)$ on $\mathfrak{D} \times \mathfrak{D}$, as a consequence of the corresponding identity in the cutoff theory. Hence Corollary 1.1 follows by Lemma 1.1.

By Proposition 1.1, the domain $\mathfrak{D}$ is dense. Since $\mathfrak{D}$ is invariant under $\exp(itH)$, $\mathfrak{D}$ is a core for H. Hence Corollary 1.3 is a consequence of Corollary 1.2. We infer Corollary 1.2 from Corollary 1.1 and a general result:

Theorem 1.2: Let $0 \leq H = H^*$ and $R = (H + I)^{-1}$. Let A be a symmetric operator with domain $\mathfrak{D}(H)$. Suppose that

$$\|AR\| + \|\dot{A}R\| < \infty, \tag{1.26}$$

where $\dot{A} = [iH, A]$ is defined in $\mathfrak{D}(H^2) \times \mathfrak{D}(H^2)$. Then A is essentially self-adjoint on any core for H.

Proof: Let $H = \int \lambda dE_\lambda$ be the spectral resolution of H. Let $C_n = E_n A E_n$. Then C_n is a bounded, self-adjoint operator with resolvent $K_n = (C_n - iy)^{-1}$. We first prove that K_n converges to the resolvent of a self-adjoint operator C. We then show that A^- extends C, so A is essentially self-adjoint on $\mathfrak{D}(H)$. The estimate $\|AR\| \leq O(1)$ then yields essential self-adjointness on any core for H.

(a) Graph convergence: We note that given (1.26), it follows by (1.17) that $R^{1/2}AR^{1/2}$ is bounded. Since $\|(E_n - I)R^{1/2}\| \leq O(n^{-1/2})$ as $n \to \infty$,

$$\|R(C_n - A)R\| \le O(n^{-1/2}). \tag{1.27}$$

Furthermore, using (1.27) and the identity $C_n R^2 = RC_n R - iR\dot{C}_n R$, we have

$$\|(C_n - A)R^2\| \le O(n^{-1/2}), \qquad n \to \infty. \tag{1.28}$$

Hence the operators C_n have a strong graph limit, a closed operator C. (See Ref. 1, Sec. 6.2). By (1.28) C agrees with A on $\mathfrak{D}(H^2)$. Since AR is bounded, and C is closed, C agrees with A on $\mathfrak{D}(H)$.

(*b*) *Smoothness preserving:* For real y, $|y|$ sufficiently large, we prove

$$\|(H + I)K_n R\| \le 1. \tag{1.29}$$

In fact, for $\theta \in \mathfrak{D}(H)$ and $|y| - \|\dot{A}R\| \ge 1$,

$$\begin{aligned}\|(H + I)(C_n - iy)\theta\| &\ge \|(C_n - iy)(H + I)\theta\| - \|\dot{C}_n\theta\|\\ &\ge (|y| - \|\dot{C}_n R\|)\|(H + I)\theta\|\\ &\ge \|(H + I)\theta\|,\end{aligned}$$

proving (1.29).

(*c*) *Resolvent convergence:* For $\theta \in \mathfrak{D}(H)$, $\|(K_n - K_m)\theta\|^2 = \langle K_n\theta, (K_n - K_m)\theta\rangle - \langle K_m\theta, (K_n - K_m)\theta\rangle$ By (1.29), for $|y|$ sufficiently large,

$$\begin{aligned}\|(K_n - K_m)\theta\|^2 &\le 2\|R(C_n - C_m)R\| \, \|(H + I)\theta\|^2\\ &\le o(1)\|(H + I)\theta\|^2\end{aligned}$$

as $n, m \to \infty$ by (1.27). Since $C_n = C_n^*$, we have $\|K_n\| \le |y|^{-1}$, for y real. Hence for $|y|$ sufficiently large,

$$\text{s-lim}\, K_n = K. \tag{1.30}$$

By (a) and (c), $C_n = C_n^*$ has a graph limit C and a strong resolvent limit. We infer from Ref. 1, Theorem 6.2.3 that $C = C^*$ and $\lim K_n = K$ is the resolvent of C. Furthermore, since H is self-adjoint, the uniform bound (1.29) ensures $K: \mathfrak{D}(H) \to \mathfrak{D}(H)$ and

$$\|(H + I)KR\| \le 1.$$

(*d*) *Essential self-adjointness:* We prove that $C \subset A^-$. Hence $C = A^-$. Let $\theta \in \mathfrak{D}(C)$ or $\theta = K\chi$. Choose $\chi_n \in \mathfrak{D}(H)$, $\chi_n \to \chi$. Since K is bounded, $\theta_n = K\chi_n \to \theta$. By (c), $\theta_n \in K\mathfrak{D}(H) \subset \mathfrak{D}(H) = \mathfrak{D}(A)$. By (a), C and A agree on $\mathfrak{D}(H)$. Thus

$$(A - iy)\theta_n = (C - iy)\theta_n = \chi_n \to \chi.$$

Hence $\theta \in \mathfrak{D}(A^-)$ and $A^-\theta = C\theta$.

Proof of Theorem B: In the case of the space-time averaged field operator $\varphi(f)$, we have identified $\varphi(f)$, acting on $\mathfrak{D} \subset \mathcal{H}_{\text{GNS}} = \mathcal{H}_W$, with a restriction of the self-adjoint field operator $\varphi(f)$ introduced in Ref. 3. In the case of the bilinear form $\varphi(x, t)$, we do not attempt such an identification. Bilinear forms, in contrast with self-adjoint operators, do not have natural maximal domains of definition. The domains for $\varphi(x, t)$ introduced in Ref. 3 could very possibly meet $\mathfrak{D}$ in the zero vector alone. Thus we proceed as follows.

By Theorem A and the space–time translation invariance of the vacuum, there exists a bilinear form $\varphi(x, t)$ on $\mathfrak{D} \times \mathfrak{D}$ such that

$$\langle\theta, \varphi(x, t)\chi\rangle \in \mathcal{C}^\infty(R^2), \tag{1.31}$$

and

$$\int \langle\theta, \varphi(x, t)\chi\rangle f(x, t)\,dx\,dt = \langle\theta, \varphi(f)\chi\rangle, \tag{1.32}$$

for all θ, $\chi \in \mathfrak{D}$ and all $f \in \mathcal{S}(R^2)$. It is with this definition of $\varphi(x, t)$ that we prove Theorem B.

Let χ be a real, positive $\mathcal{C}_0^\infty$ function of total integral one and $\chi_\kappa(x) = \kappa\chi(\kappa x)$. In one space dimension, $\|\mu^{-1/2-\epsilon}\delta_x\|_2 \le$ const, where δ_x is the Dirac measure concentrated at x. Then

$$\|\mu^{-1/2-\epsilon}(\chi_\kappa * \delta_x)\|_2 \le \|\mu^{-(1/2)-\epsilon}\delta_x\|_2 \le M, \qquad \|\chi_\kappa\|_1 \le 1, \tag{1.33}$$

$$\|\mu^{-(1/2)-\epsilon}(\chi_\kappa * \delta_x - \delta_x)\| \le O(\kappa^{-\epsilon/2}), \tag{1.34}$$

$$\|\mu^{-(1/2)-\epsilon}(\delta_x - \delta_{x+a})\| \le O(|a|^{\epsilon/2}). \tag{1.35}$$

We let $f_\kappa(x, t) = h_\kappa(x)\alpha_\kappa(t)$. By the remark below Theorem 1.1, we have for vectors θ_g as in (1.25)

$$\pm\langle\theta_g, \varphi(f_\kappa)\theta_g\rangle \le (2M)^{-1}\|\mu^{-(1/2)-\epsilon}h_\kappa\|_2\|\alpha_\kappa\|_1 \times \langle\theta_g, [H(g) + \text{const}]\theta_g\rangle. \tag{1.36}$$

As in the proof of Corollary 1.1, we space average and pass to a limit $g \to 1$ to obtain for $\theta \in \mathfrak{D}$

$$\pm\langle\theta, \varphi(f_\kappa)\theta\rangle \le (2M)^{-1}\|\mu^{-(1/2)-\epsilon}h_\kappa\|_2\|\alpha_\kappa\|_1 \times \langle\theta, (H + \text{const})\theta\rangle. \tag{1.37}$$

We now choose $h_\kappa = \chi_\kappa * \delta_x$ and $\alpha_\kappa = \chi_\kappa * \delta_t$. Passing to the limit $\kappa \to \infty$, we have (1.31)–(1.37)

$$\pm\langle\theta, \varphi(f_\kappa)\theta\rangle \to \pm\langle\theta, \varphi(x, t)\theta\rangle \le \tfrac{1}{2}\langle\theta, (H + \text{const})\theta\rangle. \tag{1.38}$$

By taking limits of θ, (1.38) extends to $\mathfrak{D}(H^{1/2}) \times \mathfrak{D}(H^{1/2})$, and defines $\varphi(x, t)$ on that domain. We note by (1.35) that $\varphi(x, t)$ so defined is continuous in (x, t), and hence satisfies (1.32). This completes the proof of Theorem B.

The remainder of this paper is devoted to proving Theorem 1.1. For convenience, we suppress the κ cutoff or g cutoff and write

$$H(h) = H(g) + \delta H(h), \quad \delta E = \text{inf spectrum } H(h).$$

Both H_0 and $H(h)$ are self-adjoint operators on Fock space, and each has a unique ground state vector:

$$H_0\Omega_0 = 0, \qquad H(h)\Omega(h) = \delta E\Omega(h).$$

The vectors Ω_0 and $\Omega(h)$ are not orthogonal. See Refs. 1 and 4 for proofs. It follows that $\delta E(h)$ is given by the exact formula

$$\delta E(h) = -\lim_{T\to\infty} (1/T)\log\langle\Omega_0, e^{-TH(h)}\Omega_0\rangle.$$

Our proof of Theorem 1.1. is based on estimating

$$\langle\Omega_0, e^{-TH(h)}\Omega_0\rangle \tag{1.39}$$

and bounding it by exp (MDT). The best bound previously established was $|\delta E| \le O(1 + \text{diam supp } g)$, which diverges as $g \to 1$. The stronger bound proved here estimates the finite correction δE to the divergent vacuum energy $E(g)$ of (1.2).

We base our estimate of (1.39) on function space integration and the Feynman-Kac formula. Our new estimates can also be proved on Fock space (Q space) without appealing to probability theory methods. With some simple modifications to the present proof, we can eliminate the path space methods and substitute methods of Refs. 6 and 2.

In our proof of a uniform bound on δE, a major idea is to take advantage of localization in configuration space, as well as in momentum space. We isolate the contributions to $\delta E(h)$ from parts of the interaction localized near the support of h. We bound these contributions by the inductive construction of Sec. 2. We bound (1.39) by a sum of terms; each term represented by a graph. Each step in the induction replaces one graph by a sum of graphs with new vertices and new lines. At the end of each inductive step we decide for each graph whether to terminate the induction or whether to continue the expansion for that graph. If the inductive construction continues, the new vertices and lines introduce convergence factors into the expansion. One possibility is that the new vertices have a large lower momentum cutoff κ in some new vertex and thus contribute the small tail $O(\kappa^{-\delta})$ of a convergent integral to our estimates. Such vertices are said to be localized in high momentum regions. The remaining new vertices are localized in space-time regions far from the vertices of the original graph, so that at least one line connects vertices with space-time localizations separated by a large Euclidean distance d. By construction, d grows rapidly for successive inductive steps. In Sec. 3, we show that each line so localized in space-time contributes a convergence factor exp $(-md/2)$ to our estimates of the corresponding graph. Here m is the mass of the particles in H_0 (bare mass). The small factors $\kappa^{-\delta}$ or $\exp(-md/2)$ lead to the proof of convergence of our inductive estimate. In particular they dominate the large number of graphs that occur and yield a convergent sum of small terms to bound δE.

2. THE INDUCTIVE CONSTRUCTION

A. Introduction

The proof of Theorem 1.1 is based upon an inductively constructed upper bound to (1.39). The inductive construction consists of repeating three basic substeps:

(a) the graph expansion,
(b) the squaring operation,
(c) the path space construction.

The estimates which prove convergence of the inductive construction are established in Secs. 3 and 4.

It is convenient to define

$$H_{\mathrm{II}} = H_{\mathrm{I}} + \delta H. \tag{2.1}$$

We also introduce a momentum cutoff indexed by κ into the interaction Hamiltonian H_{II}. Let

$$H_{\mathrm{I},\kappa} = \int :\mathcal{P}(\varphi_\kappa(x)): g(x)dx,$$
$$H_{\mathrm{II},\kappa} = H_{\mathrm{I},\kappa} + \int \mathcal{P}_1(\varphi_\kappa(x)): h(x)dx.$$

We now describe the localization in space and time of functions on path space. We index the space-time localization by $i = \{i_0, i_1\} \in \mathbb{Z} \times \mathbb{Z}$. We define the space-time square $\Delta_i = [i_0, i_0 + 1) \times [i_1, i_1 + 1)$. Here the first factor $[i_0, i_0 + 1)$ is a time interval, and the second factor is a space interval. We define the space-time localized interaction Hamiltonian $H_{I,i}$ as a time-dependent operator on Fock space

$$H_{\mathrm{I},i}(t) = H_{\mathrm{I}}[\zeta_i(\cdot, t)g(\cdot)]$$

or as a function on path space,

$$H_{\mathrm{I},i}(t, q(t)) = H_{\mathrm{I}}[\zeta_i(\cdot, t)g(\cdot)](q(t)).$$

Here ζ_i is the characteristic function of Δ_i. Thus

$$H_{\mathrm{I}} = \sum_i H_{\mathrm{I},i}. \tag{2.2}$$

Similarly we define δH_i, $H_{\mathrm{II},i}$. Also we may localize the interaction both in space-time and in momentum space. We define

$$H_{\mathrm{I},\kappa,i} = \int :\mathcal{P}(\varphi_\kappa(x)): \zeta_i g(x)dx,$$

and similarly we define $H_{\mathrm{II},\kappa,i}$.

Let X be a union of space intervals $[j, j+1)$, $j \in \mathbb{Z}$. We define

$$H_{\mathrm{I}}(X) = \sum_{i_1 \subset X} H_{\mathrm{I}}(\zeta_i g),\ H(X) = H_0 + H_I(X),$$

and

$$E(X) = \inf \text{spectrum}\{H_0 + H_{\mathrm{I}}(X)\}.$$

(Note that we now omit the vacuum energy from H.) Furthermore, we introduce sets $Y \subset R^2$ which are unions of squares $\Delta_i \subset R^2$. Let $Y(t)$ be the time slice of Y at time t. Let $|X|$ be the length of X and let $|Y|$ be the area of Y.

With the hypothesis of Theorem 1.1, we establish the inequality

$$-MD \le \delta E. \tag{2.3}$$

The constant M is independent of g and h. For Theorem 1.1, Case 2, the bound (2.3) ensures

$$0 \le H - E(g) \pm \varphi(h) + MD.$$

Thus if Ω is the vacuum for H,

$$\delta E \le \langle \Omega, \{H - E(g) + \varphi(h)\}\Omega\rangle$$
$$\le \langle \Omega, \varphi(h)\Omega\rangle \le \langle \Omega, \{H - E(g) + MD\}\Omega\rangle \le MD,$$

proving $|\delta E| \le MD$. For Case 1 of the theorem, the desired inequality $|\delta E| \le MD$ follows by an interchange in the roles of H and $H + \delta H$ in the proof of (2.3). The bound (2.3) is a consequence of two inequalities.

Proposition 2.1: There is a constant a_0 such that for all intervals X with $|X| \ge a_0$,

$$E(g) \le E(\sim X) + |X|. \tag{2.4}$$

There is a constant M such that for all intervals X_0 containing supp h,

$$\inf\{E(\sim X) + |X| : X_0 \subset X\} \le E(g) + \delta E(g, h) + M|X_0|. \quad (2.5)$$

The constants M and a_0 are independent of g and h, subject to the conditions of Theorem 1.1.

We choose $\max\{D, a_0\} \le |X_0| \le \max\{D, a_0\} + 1$. By Proposition 2.1,

$$E(g) \le E(g) + \delta E(g, h) + M|X_0|. \quad (2.6)$$

Thus with a new constant M, we have proved (2.3) and hence Theorem 1.1.

In Secs. 2 and 3 we prove (2.5). In Sec. 4 we modify this proof to establish (2.4). We now describe the inductive construction used to bound (1.39). At each step in the inductive construction, we dominate (1.39) by a sum of terms of the form

$$\int ds \int_Q R(s) \exp[-\int H_{\text{II}}(\sim Y(\sigma), \sigma)(q(\sigma)) d\sigma] dq(\cdot), \quad (2.7)$$

where R is a polynomial on path space and where $Y(\sigma)$ defines a localization of the exponent, as above. The connection between the Fock space inner product (1.39) and the path space integral (2.7) is given by the Feynman–Kac formula. Each inductive step replaces (2.7) by a sum of similar terms, and our final bound dominates (1.39) by the sum of a convergent series.

The first part of each inductive step, the graph expansion, is an expansion of (2.7) using the pull through formula. Each term in the expansion is represented by a graph. For some terms, the polynomial R is a constant. The associated graphs have no external legs and are called vacuum graphs. For these graphs, the inductive construction terminates. In terms for which the polynomial R is not a constant, new vertices have been introduced during the graph expansion, as will be explained below.

The second part of each inductive step is the squaring operation, in which we replace R by a positive polynomial R', $|R| \le R'$. The fact that R' is positive makes the next path space construction possible, and the fact that R' is a polynomial allows us to perform the next graph expansion step.

The third portion of each inductive step is the path space construction, which we use to bound (and remove) those parts of the exponent that are localized near the localization of R. We use the near positivity of $\int H_{\text{I},i}(q(s))ds$ to give an upper bound for localized portions of the exponent. This upper bound has the form of a convergent sum of polynomials on path space. Hence we remove part of the exponent and replace R by a sum of new polynomials R'.

In the first inductive step, we omit the graph expansion and the squaring operation and begin with the path space construction, see Sec. 2D. We then repeat the three substeps and arrive at a bound for (2.7) as a sum of vacuum graphs. Each vacuum graph is multiplied by a factor

$$\int_Q \exp(-\int_0^T H_{\text{I}}(\sim Y(\sigma))(q(\sigma)) d\sigma)$$

$$= \langle \Omega_0, e^{-H(\sim Y(0))} e^{-H(\sim Y(1))} \cdots e^{-H(\sim Y(T))} \Omega_0 \rangle$$

$$\le \exp(-\int_0^T E(\sim Y(\sigma)) d\sigma).$$

The vacuum graphs are small, for reasons given in Sec. 1.

B. The Graph Expansion

The graph expansion is generated by the pull through formula

$$a(k)e^{-tH} = e^{-t(H+\mu)}a(k) - \int_0^t ds\, e^{-s(H+\mu)}[a(k), H_{\text{I}}]\, e^{-(t-s)H}. \quad (2.8)$$

Here $H = H(X) = H_0 + H_{\text{I}}(X)$ and $\mu = (k^2 + m^2)^{1/2}$. To derive (2.8) we set

$$F(s) = \int e^{-s(H+\mu)} a(k) e^{-(t-s)H} f(k)\, dk,$$

where $f \in L_2$, and we evaluate

$$F(0) = F(t) - \int_0^t F'(s) ds.$$

By Rosen's higher-order estimates, $F(s)$ is continuously norm differentiable on $[0, t]$. From the more elementary first order estimates, one can show that $F(s)$ is strongly differentiable and norm bounded on $[0, t]$ and continuously norm differentiable on the open interval $(0, t)$.

We call our expansion the graph expansion because each term in it can be labeled by a graph; the graphs provide a convenient way to visualize the expansion. In order to accommodate our time-dependent Hamiltonians, we rewrite (2.8) as an anti-time-ordered integral

$$\overline{T}a(k, 0) \exp\left(-\int_0^t H(\sigma) d\sigma\right) = \overline{T}a(k, t) e^{-t\mu} \exp\left(-\int_0^t H(\sigma) d\sigma\right) - \overline{T}\int_0^t ds [a(k, s)e^{-s\mu}, H_{\text{I}}(X, s)] \exp\left(-\int_0^t H(\sigma) d\sigma\right). \quad (2.9)$$

Here $\overline{T}$ denotes the anti-time ordering. Later times, denoted by larger values of σ, are written to the right and precede in the operator product. The time parameter s in $a(k, s)$ and $H_{\text{I}}(X, s)$ indicate that the operators $a(k)$ or $H_{\text{I}}(X)$ occur at time s. We replace X by the space–time region Y, so that $H(\sigma) = H(Y(\sigma))$, and then (2.9) remains valid.

We say that (2.9) is the sum of a term with $a(k)$ "pulled through the exponent" and a term in which $a(k)$ "contracts to the exponent." When we use (2.9) to generate the graph expansion, (2.9) occurs multiplied by a polynomial in the fields R. New terms occur, arising from commutators between a and R, and we call these terms "contractions to existing vertices." For instance, if $R(s)$ equals a product of monomials $R_1(s_1) \cdots R_n(s_n)$, then for $t_1 \le t, s_1, \ldots, s_n \le t_2$,

$$\overline{T}a(k, t) R(s) \exp\left(-\int_{t_1}^{t_2} H(\sigma) d\sigma\right) = \overline{T}a(k, t_2) e^{-(t_2 - t)\mu} R(s) \exp\left(-\int_{t_1}^{t_2} H(\sigma) d\sigma\right) - \overline{T}R(s) \int_{t_1}^{t_2} ds' [a(k, s') e^{-(s'-t)\mu}, H_{\text{I}}(x, s')] \times \exp\left(-\int_{t_1}^{t_2} H(\sigma) d\sigma\right)$$

$$+ \overline{T} \sum_{j=1}^{n} [a(k, s_j)e^{-(s_j - t)\mu}, R_j(s_j)] \times \prod_{i \neq j} R_i(s_i) \exp\left(- \int_{t_1}^{t_2} H(\sigma)\, d\sigma\right). \quad (2.10)$$

We apply (2.10) to the basic term in the inductive construction, which we can write either as an integral on path space or as an anti-time ordered integral

$$\int ds \int_Q R(s) \exp\left[- \int H_I(Y(\sigma))\, d\sigma\right] dq(\cdot) = \int ds \left\langle \Omega_0, \overline{T}\, R(s) \exp\left(- \int_0^T H(Y(\sigma))\, d\sigma\right) \Omega_0 \right\rangle. \quad (2.11)$$

The integration over $s_1 \leq s_2 \leq \cdots \leq s_n$ in (2.11) arises from previous use of (2.10). In this section we consider the integrands of the ds integration in (2.11). We choose a linear factor φ in R and write it as the sum of an annihilation operator a and a creation operator a^*. Then we use (2.10) to move a to the right and the adjoint of (2.10) to move a^* to the left. The terms in which a or a^* are pulled through equal zero, since $a\Omega_0 = 0$. The remaining terms from (2.10) are functions of φ, but not π, and hence they can be written as integrals on path space.

We now explain the graph terminology. Vertices in our graphs arise from monomials φ^n which occur in the interaction Hamiltonian H_{II}. These vertices may be produced in R during the graph expansion, the squaring operation, or the path space construction. In applying our expansions, we always use the representation (2.2) to express H_I as a sum of parts $H_{I,i}$ localized in space-time cubes Δ_i. Hence each vertex in our graphs has a space-time localization index i. Each vertex also occurs at a definite time, corresponding to the time s' in the integrand of (2.10). Later vertices are placed to the right, in accordance with the anti-time ordering. A vertex arising from φ^n has n legs connected to it. Of these legs, j are contracted (connected) to legs of other vertices and $n-j$ are not contracted. The uncontracted legs have no specified directions and label linear fields φ in φ^n. The contracted legs necessarily point to the left or to the right, according to whether the vertex to which they contract lies at an earlier or at a later time. The legs pointing to the left are creation legs, and those pointing to the right are annihilation legs. These legs label creation operators $a(k)^*$ and annihilation operators $a(-k)$, respectively. Each use of the commutation relations $[a(-k), a(l)^*] = \delta(k+l)$, replaces R by a new monomial. The graph of the new monomial R' is formed by contracting the legs of $a(-k)$ and $a(l)^*$ in the graph of R. Two such contracted *legs* are called a *line*. Each use of the pull through formula (2.10) also replaces R by a sum of new monomials. The second term on the right of (2.10) has a new graph formed by introducing a new vertex at time s', and one leg of this vertex is contracted to the leg of $a(k, s')$. We also say that this new vertex results from the contraction of the leg of $a(k, s')$ to the exponent.

Associated with each vertex is a kernel. If the vertex comes from φ^n and if it has j contracted legs, then the kernel is

$$v(k_1, \ldots, k_j) = [a^\#(k_1), [a^\#(k_2), \cdots, [a^\#(k_j), \varphi^n] \cdots]],$$

where $a^\#(k) = a(-k)$ or $a^*(k)$. If $j = n$, then v is fully contracted.

We label the vertices of the graph of R by a variable ν, $\nu = 1, 2, \ldots, N$, and the momenta $k_{r,\nu}$ at vertex ν. For a fully contracted φ^n vertex we introduce the kernel

$$v_\nu(k_{1,\nu}, \ldots, k_{n,\nu}) = (\zeta_i g)^\sim \left(\sum_{r=1}^{n} k_{r,\nu}\right) \times \prod_{r=1}^{n} [\mu(k_{r,\nu})^{-1/2}\, \tilde{\chi}(k_{r,\nu}\kappa^{-1})], \quad (2.12)$$

where ζ_i gives the space-time localization of the vertex and $\tilde{\chi}(\cdot \kappa^{-1})$ is the upper momentum cutoff function for the $H_{I,\kappa,i}$ vertex. In the path space construction of Sec. 2D, we will encounter lower cutoff vertices of the form $H_{I,i} - H_{I,\kappa,i}$. For such a vertex, we replace

$$\prod_{r=1}^{n} \tilde{\chi}(k_{r,\nu}\kappa^{-1}) \quad \text{by} \quad 1 - \prod_{r=1}^{n} \tilde{\chi}(k_{r,\nu}\kappa^{-1}) \quad (2.13)$$

in (2.12). For such vertices we define the lower momentum cutoff $\kappa(v_\nu) = \max\{1, \kappa\}$.

A φ^n vertex which is not fully contracted has a kernel obtained as follows: Multiply (2.12) by a Wick ordered product of fields $\tilde{\varphi}(k)$, one field for each uncontracted momentum variable. Then integrate this product over the uncontracted momenta.

We also introduce energy factors $\mu(k_{r,\nu})^{\epsilon_{r,\nu}} = \mu_{r,\nu}^{\epsilon_{r,\nu}}$ and define the kernel

$$\bar{v}_\nu = v_\nu \prod_r \mu_{r,\nu}^{\epsilon_{r,\nu}}. \quad (2.14)$$

We use these factors μ^ϵ in order to transfer powers of the energy from one kernel to another. Hence they satisfy

$$\prod_{\nu, r} \mu_{r,\nu}^{\epsilon_{r,\nu}} = 1, \quad (2.15)$$

and then $(\prod_\nu v_\nu) = (\prod_\nu \bar{v}_\nu)$.

We choose $-\frac{1}{2} \leq \epsilon_{r,\nu}$ and

$$\epsilon_\nu = \sum_r \epsilon_{r,\nu}^+ < \tfrac{1}{2},$$

where $\epsilon_{r,\nu}^+ = \max\{0, \epsilon_{r,\nu}\}$. We note that for $\delta > 0$,

$$\|\bar{v}_\nu\|_2 \leq O[\kappa(v_\nu)^{-(1/2)+\epsilon_\nu+\delta}].$$

In addition to the vertices, each line (two contracted legs) has associated a function of the momenta and an integration over the momenta of the two contracted legs. Suppose the annihilation leg with momentum $k_{r,\nu}$ of vertex ν (localized at time $s_\nu = s_{r,\nu}$) contracts with the creation leg with momentum $k_{r',\nu'}$ of vertex ν' (localized at time $s_{r',\nu'}$). Associated with this line is the factor

$$\delta(k_{r,\nu} + k_{r',\nu'}) \exp[-(s_{r',\nu'} - s_{r,\nu})\mu_{r,\nu}] = \delta(k_{r,\nu} + k_{r',\nu'}) \exp(-\delta s_{r,\nu}\mu_{r,\nu}), \quad (2.16)$$

where $0 \leq \delta s_{r,\nu}$ is the time difference of the contraction. The factor $\exp(-\delta s \mu)$ introduces a time localization. We insert this factor in the kernel by defining

$$\hat{v}_\nu = \bar{v}_\nu \exp(-\tfrac{1}{2} \sum_r \delta s_{r,\nu}\mu_{r,\nu}). \quad (2.17)$$

Hence we assign half the time localization factor for a given line to each of the two vertices which the line connects.

A polynomial $\hat{R}(s)$ is then constructed from its graph by the formula

$$\hat{R}(s) = \mathcal{T}\int[\prod_\nu \hat{v}_\nu (k_{r,\nu}; q(\cdot))][\prod_{\text{lines}} \delta(k_{r,\nu} + k_{r',\nu'})]dk. \tag{2.18}$$

We define the polynomial

$$R(s) = BL\hat{R}(s), \tag{2.19}$$

where B and L are constants, depending on the inductive construction. B is a product of constants e^j, $e^{\mathcal{J}_0}$, and η from (2.32) or (2.34), while L is a product of distance localization factors l_ν, coming from (2.22) or (2.24).

The polynomial R and the kernels $\hat{v}_\nu$ contributing to it are unbounded operators in the anti-time ordered formula (2.11). When this formula is interpreted in path space, R and the kernels $\hat{v}_\nu$ become multiplication operators on path space. If $\hat{v}_\nu$ is fully contracted, then it is a constant as a function on path space. Similarly, if the graph of R has no uncontracted legs, then R is a constant as a function on path space. Such a graph is called a vacuum graph.

To complete the graph expansion, we must specify which legs of the polynomial are to be pulled through and which are not, since the infinite series expansion obtained by pulling through all legs appears to diverge. We treat on a different footing the vertices which were present at the start of the graph expansion part of the given inductive step. We pull through all these legs until they contract to yield lines. We do not pull through legs of vertices added in the course of the graph expansion of the given inductive step. Thus the graph expansion in a given inductive step yields a finite number of terms, and at the end of a given graph expansion step all vertices in R with uncontracted legs are new (pull through) vertices introduced during that graph expansion step.

C. The Squaring Operation

At the conclusion of the graph expansion for (2.11), we have replaced (2.11) by a sum of similar terms. We now replace R in each term of the form (2.11) by a positive polynomial on path space. We make this replacement for two reasons: During the PS construction we take absolute values, which could replace the polynomial R on path space by the nonpolynomial function $|R|$. In that case we could not perform the next inductive step, since the graph expansion is only defined for polynomial R. The second reason to modify R is to improve our estimates in Sec. 3 for the pull through vertices in R (vertices introduced by contractions to the exponent). For this reason we sometimes choose to replace R by a polynomial of high degree.

A simple estimate for R is obtained by repeated application of the Schwarz inequality to (2.18). We obtain

$$|R| \le \prod_\nu \|\hat{v}_\nu\|_2 \le \prod_\nu \|\bar{v}_\nu\|_2, \tag{2.20}$$

where the $\mathcal{L}_2$ norms $\|\hat{v}\|_2$ or $\|\bar{v}\|_2$ are $\mathcal{L}_2$ norms in the contracted momentum variables. Unless v_ν is fully contracted, $\|\bar{v}_\nu\|_2$ depends on the uncontracted legs as a function on path space. In Theorem 3.1 we improve (2.20) to take into account the space-time localization of the vertices. We prove

$$|R| \le \prod_\nu (\gamma e^{-md_\nu/3}\|\bar{v}_\nu\|_2), \tag{2.21}$$

where d_ν is the contraction distance for vertex v_ν, as defined below in Sec. 3B.

We use the bound (2.21) as the first step in our squaring operation. We give the graphical interpretation of $\|\bar{v}_\nu\|_2$. Suppose the set $\mathcal{C}$ of contracted legs $\bar{v}_\nu$ has $\bar{n}$ elements. We start with two identical $\bar{v}_\nu$ vertices and contract the $\bar{n}$ legs of the first set $\mathcal{C}$ to the $\bar{n}$ legs of the second set $\mathcal{C}$. We then take the square root, yielding the graph of $\|\bar{v}_\nu\|_2$. In case that $\bar{v}_\nu$ is fully contracted, $\|\bar{v}_\nu\|_2$ is the square root of a vacuum diagram. It is a constant on path space, so we do not need to square: Otherwise, $\|\bar{v}_\nu\|_2$ is the square root of a polynomial function on path space.

In the latter case, we use the bound $ab \le a(1 + b^n)$ with $a = \exp(-md_\nu/6) \le 1$ and $b = \gamma\exp(-md_\nu/6)$ $\|\bar{v}\|_2$ to obtain

$$\gamma e^{-md_\nu/3}\|\bar{v}_\nu\|_2 \le l_\nu + (l_\nu\|\bar{v}_\nu\|_2)^{n_\nu}. \tag{2.22}$$

Here $l_\nu = \gamma e^{-md_\nu/6}$, and the squaring exponent n_ν is a positive even integer, chosen below. Since n_ν is even, $\|\bar{v}_\nu\|_2^{n_\nu}$ is a polynomial on path space.

We define the graph of the constant term l_ν in (2.22) to be a single vertex with no legs. We define the kernel corresponding to this vertex to equal one. The constant l_ν becomes a factor in L of (2.19), reflecting the history of previous contractions to $\bar{v}_\nu$. It assigns a space-time localization factor for each such contraction.

The second term in (2.22) yields a graph in which n_ν pull through vertices have uncontracted legs. The contracted legs of this graph are all localized in a single square Δ, the localization of the original vertex $\hat{v}_\nu$. Therefore these lines contribute no small localization factors. Each vertex, however, is accompanied by l_ν, which becomes a factor in L for the polynomial R.

Let $\hat{v}_\nu$ be a pull through vertex with uncontracted legs. We define $d_c(v_\nu)$ as the (integer part of the) Euclidean distance of the contraction giving rise to the vertex $\hat{v}_\nu$. We choose n_ν as follows: Either

$$n_\nu = 2 \quad \text{or} \quad n_\nu = [4d_c(v_\nu)]^6. \tag{2.23}$$

To choose between the two possibilities, consider a fixed cube Δ and apply (2.22) successively, in order of decreasing value of $d_c(v_\nu)$, to the kernels $\hat{v}_\nu$ localized in Δ. Each application of (2.22) yields two terms. Thus the j-fold application to j vertices in Δ yields 2^j terms. In these terms n_ν enters only when the second term in (2.22) is selected at least once. For the $(j+1)$th application of (2.22) we choose $n_\nu = 2$ in the above case, that is when the second term in (2.22) has been selected at least once in the first j applications of (2.22). Otherwise, we choose $n_\nu = [4d_c(v_\nu)]^6$. With N vertices in Δ we obtain 2^N terms. In each of these terms we define $d_c(\Delta) = d_c(v_\nu)$, where $\hat{v}_\nu$ is the vertex for which $n_\nu = [4d_c(v_\nu)]^6$.

For uniformity of notation, we use the bound

$$\gamma e^{-md_\nu/3}\|\bar{v}_\nu\|_2 \le l_\nu\|\bar{v}_\nu\|_2, \tag{2.24}$$

for fully contracted vertices in (2.21).

D. The Path Space Construction

The path space integrals

$$\int R(s)\,\exp\Big[-\int H_{\rm II}(q(\sigma))\,d\sigma\Big]dq(\cdot) \qquad (2.25)$$

cannot be evaluated or estimated directly. On the other hand, integrals of polynomials on path space can be estimated easily. The purpose of the path space construction is to bound (2.25) by a sum of integrals of polynomials on path space. The term by term integration of the exponential series in (2.25) converges only if $p \le 2$. In the case of linear perturbations we make a power series expansion of $\exp[-\int \delta H(q(\sigma))\,d\sigma]$. Otherwise, we use an asymptotic expansion of $\exp[-\int H_{\rm II}(q(\sigma))d\sigma]$ to obtain our bound. We use the formal positivity of $H_{\rm II}$ and undo the Wick ordering in low momentum regions. This bounds the low momentum part of the exponent from above by a constant. The high momentum tail is bounded by a convergent sum of polynomials on path space.

More precisely, for Theorem 1.1, Case 1, we write

$$H_{{\rm II},i} = H_{{\rm II},i,\kappa} + \delta H_{{\rm II},i,\kappa}\,. \qquad (2.26)$$

By undoing the Wick ordering in the low momentum part of $H_{{\rm II},i,\kappa}$, we obtain

$$-M_1(\log\kappa)^{p/2} + 2 \le H_{{\rm II},i,\kappa}\,. \qquad (2.27)$$

The constant M_1 is independent of g, h, i and κ (see, e.g., Refs. 1 and 2.) It is convenient to choose a sequence of cutoffs $\kappa_j = \exp[(j/M_1)^{2/p}]$, so that by (2.27) and the localization of $H_{{\rm II},i}$ in a time interval of unit length,

$$-j + 2 \le \int H_{{\rm II},i,\kappa_j}(q(\sigma))\,d\sigma. \qquad (2.28)$$

We bound $\exp(-\int H_{{\rm II},i}d\sigma)$ by writing path space Q as a countable disjoint union $Q = \cup_{j\ge j_0} Q_j$, where for $j > j_0$,

$$Q_j = \{Q : -j \le \int H_{{\rm II},i}d\sigma \le -j+1\}. \qquad (2.29)$$

Then

$$Q_{j_0} = \{Q : -j_0 \le \int H_{{\rm II},i}d\sigma\}. \qquad (2.30)$$

By (2.28), paths in Q_j, for $j > j_0$, yield the bound

$$1 \le \Big|\int \delta H_{{\rm II},i,\kappa_j}\,d\sigma\Big| = R_{i,j}\,. \qquad (2.31)$$

Hence for $n \ge 1$, we have $1 \le (R_{i,j})^n$ on Q_j. For $j > j_0$, we choose $n = n(j)$ to be the largest even integer less than $\kappa_j^{1/8p}$. We define $n(j_0) = 0$. Then

$$\exp(-\int H_{{\rm II},i}d\sigma) \le \sum_{j\ge j_0} e^j R_{i,j}^{n(j)}. \qquad (2.32)$$

For Theorem 1.1, Case 2, we write

$$\int \delta H_i\,d\sigma = \int \varphi(h)_i\,d\sigma = \int \varphi(h)_i(q(\sigma))\,d\sigma$$

and expand

$$\exp(-\int \delta H_i\,d\sigma) \le e^{J_0} + \eta \sum_{k\ge J_0} \frac{1}{(2k)!}\Big[\int \varphi(h)_i\,d\sigma\Big]^{2k}. \qquad (2.33)$$

Here η is a universal constant, independent of J_0. We apply (2.32) to $H_{{\rm I},i}$ rather than $H_{{\rm II},i}$. Since $H_{{\rm II},i} = H_{{\rm I},i} + \delta H_i$, we have

$$\exp\Big(-\int H_{{\rm II},i}\,d\sigma\Big) \le \Big(e^{J_0} + \eta \sum_{k\ge J_0} \frac{1}{(2k)!}\big(\int \varphi(h)_i\,d\sigma\big)^{2k}\Big) \times \sum_{j\ge J_0} e^j R_{i,j}^{n(j)}. \qquad (2.34)$$

The "path space construction" is the use of (2.32) or (2.34) in squares Δ_i. We also say that we "remove portions of the exponent" from (2.25). Our choice of $j_0 = j_0(\Delta_i)$ will be made in this and later sections.

We begin the inductive construction (the first inductive step) by applying (2.32) or (2.34) in each square $\Delta_i \in \mathfrak{D}_0$, where $\mathfrak{D}_0$ is a given union of squares Δ. For our proof of (2.5), we choose

$$\mathfrak{D}_0 = X_0 \times [0, T],$$

so the perturbation δH is localized in $\mathfrak{D}_0$. In these squares we choose $j_0(\Delta_i) = J_0$. After this initial use of the PS construction, only $H_{\rm I}$ remains in the exponent, δH has been completely removed. Further application of the PS construction introduces new $H_{\rm I}$ vertices into graphs, but no new δH vertices.

For each term in the PS construction we decide when to terminate the PS portion of an inductive step. An individual term has the form (2.25) and is labeled by a single graph. For each term we make some definitions: We let $\mathfrak{D}_e$ be the set of squares Δ_i with a nonzero exponent. For a square Δ, we let

$$d_0(\Delta) = {\rm dist}(\Delta, \mathfrak{D}_0), \qquad d_e(\Delta) = {\rm dist}(\Delta, \mathfrak{D}_e),$$

where ${\rm dist}(X, Y)$ is the Euclidean distance between X and Y. We let $n(\Delta)$ equal the number of vertices with uncontracted legs that are localized in Δ. If $n(\Delta) \ne 0$, let $\Gamma(\Delta)$ be the square of side length $8n(\Delta)^{1/3}$, centered at Δ, and with sides parallel to the space and time axes. Note that $n(\Delta)$, $\mathfrak{D}_e$, and $d_e(\Delta)$ change during the inductive construction. We let $n(\Delta)_r$ denote the value of $n(\Delta)$ at the end of the rth inductive step, etc.

Our condition for terminating the PS construction of a given inductive step (and hence for terminating the inductive step) is the following:

$$\text{If } n(\Delta) \ne 0, \quad \text{then } \Gamma(\Delta)\cap\mathfrak{D}_e = \emptyset, \qquad (2.35)$$

for every square Δ in our space-time cover. If (2.35) is not met in some square Δ', we apply (2.32) to remove the exponent in each $\Delta \subset \Gamma(\Delta')\cap\mathfrak{D}_e$. We say that Δ' has forced the PS construction in Δ, and we choose $j_0(\Delta)$ in this case as

$$j_0(\Delta) = [J_1 n(\Delta')^{1/6}] + 1. \qquad (2.36)$$

Here $[\cdot]$ denotes "integral part." Also we choose J_1 sufficiently large so that for $x \ge 1$ we have

$$(15)^3 x^2 \le \tfrac{1}{2}\exp(x^{1/3p}M_2),$$

where $M_2 = (8p)^{-1}(J_1/M_1)^{2/p}$. By (2.36), if $n(\Delta) \ne 0$,

$$(15)^3 n(\Delta')^2 \le \tfrac{1}{2}\exp[n(\Delta')^{1/3p}M_2] \le \kappa_{j_0(\Delta)}^{1/8p} \le n_{ps}(\Delta) \le n(\Delta). \qquad (2.37)$$

Remark: Note $n(\Delta) = n_{pt}(\Delta) + n_{ps}(\Delta) + n_h(\Delta)$, where $n_{pt}(\Delta)$ is the number of PT vertices in Δ with uncontracted legs, $n_{ps}(\Delta)$ is the number of PS vertices in Δ

with uncontracted legs, and $n_h(\Delta)$ is the number of uncontracted $\varphi(h)$ vertices in Δ. Here we call the vertices from $R^n_{i,j}$ in (2.32) PS vertices in Δ_i, and we call the vertices from $\int \varphi(h)_i\, d\sigma$ in (2.34), the $\varphi(h)$ vertices in Δ_i.

Proposition 2.2: If $\mathfrak{D}_0 \cap \mathfrak{D}_e = \emptyset$ and $n(\Delta) \neq 0$ during the PS contruction, then

$$d_0(\Delta) \le n(\Delta)^{1/3}. \tag{2.38}$$

Proof: Step 1: Assume that the proposition holds at some point during the PS construction of the rth inductive step. We show that the proposition remains valid after the application of PS construction in any square Δ and, hence, at the end of the inductive step. If $n_{pt}(\Delta) \neq 0$, the PT vertices in Δ were present throughout the rth PS construction. Hence, by assumption, $d_0(\Delta) \le n_{pt}(\Delta)^{1/3} \le n(\Delta)^{1/3}$ and (2.38) is valid. On the other hand, suppose $n_{ps}(\Delta) \neq 0$ and Δ' forced the PS construction in Δ. Since $\Gamma(\Delta')$ has diagonal length $\sqrt{2}\,8n(\Delta')^{1/3} \ge \mathrm{dist}(\Delta', \Delta)$ and since $d_0(\Delta') \le n(\Delta')^{1/3}$, the triangle inequality yields

$$d_0(\Delta) \le n(\Delta') + \mathrm{dist}(\Delta, \Delta')$$
$$\le (1 + \sqrt{2}\,8)n(\Delta')^{1/3} \le 15n(\Delta')^{1/3}. \tag{2.39}$$

In this case (2.38) follows by (2.37). Since $\Delta \notin \mathfrak{D}_0$, $n_h(\Delta) = 0$ and step 1 is proved.

Step 2: The proposition holds for the first PS construction. In the configuration immediately after removing the exponent from $\mathfrak{D}_0$, if $n(\Delta) \neq 0$, then $\Delta \in \mathfrak{D}_0$ and $d_0(\Delta) = 0$. Hence the assertion follows by step 1.

Step 3: Assuming the proposition holds for the rth PS construction, we establish the proposition for the $(r + 1)$th PS construction. By the definition of $\Gamma(\Delta)$, whenever $n(\Delta)_r \neq 0$,

$$d_0(\Delta) \le n(\Delta)_r^{1/3} \le \tfrac{1}{2} d_e(\Delta)_r. \tag{2.40}$$

We consider the start of the $(r + 1)$th PS construction. Only PT vertices v formed in the graph expansion of the $(r + 1)$th inductive step can have uncontracted legs. These vertices have been squared during the squaring operation of the $(r + 1)$th step. Hence if such a vertex v is localized in Δ, we have $n(\Delta) \ge [4d_c(\Delta)]^6$ by (2.23). Suppose v was formed by the contraction to the exponent of v' localized in Δ'. By the definition of the graph expansion, the vertex v' must be present at the end of the rth PS construction. Thus $n(\Delta')_r \neq 0$, and by (2.40) we have $d_0(\Delta') \le d_e(\Delta')_r$. Note also $d_c(v) = \mathrm{dist}(\Delta, \Delta') \le d_c(\Delta)$ and $d_e(\Delta')_r \le d_c(\Delta)$. Hence by the triangle inequality,

$$d_0(\Delta) \le d_0(\Delta') + \mathrm{dist}(\Delta, \Delta') \le d_e(\Delta')_r + d_c(\Delta)$$
$$\le 2d_c(\Delta) \le n(\Delta)^{1/6} \le n(\Delta)^{1/3}, \tag{2.41}$$

establishing (2.38) for Δ. By step 1 the proof of the proposition is complete. We have also proved

Corollary 2.1: At the end of the PS construction of each inductive step, (2.40) holds in every Δ with $n(\Delta) \neq 0$.

Let $n(\Delta) \neq 0$ during the PS portion of some inductive step. We assign a square

$$\Delta\# \in \mathfrak{D}_0, \quad n(\Delta\#)_1 \neq 0, \tag{2.42}$$

and an integer $k(\Delta) \in \mathfrak{Z}^+$ to Δ in the following manner: If $\Delta \in \mathfrak{D}_0$ let $\Delta\# = \Delta$, $k(\Delta) = 0$. If $\Delta \notin \mathfrak{D}_0$, suppose we are considering the rth PS construction. In case $n_{ps}(\Delta) \neq 0$, let $\Delta^{(1)}$ be the square that forced the PS construction in Δ, and we consider $\Delta^{(1)}$ at the time (during the rth PS construction) that the exponent is removed in Δ. In case $n_{ps}(\Delta) = 0$, then $n_{pt}(\Delta) \neq 0$, and we choose any PT vertex v in Δ with the maximum contraction distance, i.e., $d_c(v) = d_c(\Delta)$. We define $\Delta^{(1)}$ to be the square with the vertex $v^{(1)}$ that contracted to the exponent forming v, and we consider $\Delta^{(1)}$ at the end of the $(r - 1)$th inductive step. Continuing in this fashion, we assign $\Delta^{(k)}$ to $\Delta^{(k-1)}$ until we arrive at a square $\Delta^{(k)} \in \mathfrak{D}_0$ (and the first inductive step). We let $\Delta\# = \Delta^{(k)}$, $k(\Delta) = k \ge r$.

Let

$$N_0 = \min\{n(\Delta)_1 : \Delta \in \mathfrak{D}_0, n(\Delta)_1 \neq 0\}. \tag{2.43}$$

We note that in the proof of (2.5), either $n_h(\Delta) \ge 2J_0$ or $n_{ps}(\Delta) \ge \frac{1}{2}\kappa J_0^{1/8p}$ for $\Delta \in \mathfrak{D}_0$, $n(\Delta)_1 \neq 0$. Thus $N_0 \to \infty$ as $J_0 \to \infty$.

Corollary 2.2: If $n(\Delta) \neq 0$ during the PS construction of the rth inductive step,

$$n(\Delta) \ge N_0^{2^r}. \tag{2.44}$$

Proof: If $n_{ps}(\Delta) \neq 0$, we have by (2.37) that $0 \neq n(\Delta^{(1)})^2 \le n(\Delta)$. If $n_{ps}(\Delta) = 0$, we have by Corollary 2.1 and Eq. (2.23),

$$0 \neq n(\Delta^{(1)})^2_{r-1} \le d_e(\Delta^{(1)})^6_{r-1} \le d_c(\Delta)^6 \le n(\Delta).$$

By induction we obtain

$$n(\Delta) \ge n(\Delta\#)_1^{2^k} \ge N_0^{2^r}, \tag{2.45}$$

since $n(\Delta\#)_1 \neq 0$.

Corollary 2.3: For any PT vertex v,

$$d_c(v) \ge N_0. \tag{2.46}$$

Proof: Let v be a PT vertex formed by the contraction of a vertex v' in Δ' during the rth inductive step. By Corollary 2.1, $d_c(v) \ge d_e(\Delta')_{r-1} \ge n(\Delta')^{1/3}_{r-1}$ and, by Corollary 2.2, $d_c(v) \ge N_0$.

E. Special Procedures for $\varphi(h)$ Vertices

We modify the inductive construction slightly in order to improve our estimates on $\varphi(h)$ vertices. These vertices are introduced with the first path space construction by applying (2.34). They are contracted during the first graph expansion. All the $\varphi(h)$ vertices are localized in $\mathfrak{D}_0$; but they may contract to vertices localized in any square Δ.

The Schwarz inequality of Sec. 2C introduces into our estimates the factors $\|\mu^{-1/2}h\|_2$. We modify Sec. 2C to arrive at constants $\|\mu^{-1/2-\epsilon}h\|_2$ required for Theorem 1.1. The extra factor $\mu^{-\epsilon}$ is obtained from the convergence of the $\mathfrak{L}_2$ norm of the kernels of H_I or from the integration over some times s. Formally the method yields $\epsilon = \frac{1}{2}$, while the bound should fail for $\epsilon > \frac{1}{2}$. We give the proof for $\epsilon = (8p)^{-1}$.

In the graph expansion, a $\varphi(h)$ vertex contracts either to an H_I vertex or to another $\varphi(h)$. If $\varphi(h)$ contracts to an H_I, we multiply the kernel $v_\nu(k)$ of the contracted $\varphi(h)$ vertex by $\mu(k)^{-\epsilon}$, and we multiply the kernel of H_I vertex by μ^ϵ. Hence our kernels, with the $\mu^{\pm\epsilon}$ factors, have the general form (2.17) that we estimate in Sec. 3. We choose $\epsilon = (8p)^{-1}$. [To deal with $\epsilon \geq (2p)^{-1}$, it would be necessary to integrate over times s in such graphs.]

The remaining $\varphi(h)$ vertices contract to each other. The polynomial of a corresponding connected component of the graph of $\hat{R}(s)$ is

$$r_{\nu\nu'} = \int v_\nu(k) v_{\nu'}(-k) e^{-\delta s \mu}\, dk ds_\nu ds_{\nu'}, \tag{2.47}$$

where the $s_\nu, s_{\nu'}$ integrals are restricted to the time localization of the vertices and also $0 \leqq \delta s$. We perform the s_ν integration. We obtain a denominator μ^{-1} and two terms coming from the two endpoints of integration. Thus we obtain two graphs, of the form (2.47), each with only one time integration and with the kernels $\mu^{-1/2} v_\nu$ replacing v_ν. The application of (2.21) to such graphs yields for the integrand of the $s_{\nu'}$ integration

$$|r_{\nu\nu'}(s_{\nu'})| \leq 2 l_\nu^2 \|\mu^{-1/2} v_\nu\|_2 \, \|\mu^{-1/2} v_{\nu'}\|_2. \tag{2.48}$$

3. ESTIMATES

A. Estimates on Kernels

In this section we derive estimates on kernels in the inductive construction. These estimates deal quantitatively with localization in configuration and in momentum space.

Lemma 3.1: Let $0 \leq \alpha, \beta$, $\alpha^2 + \beta^2 < m^2$, $0 < t$, $p \in R^s$, and

$$f(x,t) = D_t^j D_x^k \int \mu^\nu e^{-\mu t + ipx} dp.$$

Then for $\lambda = [\nu + j + |k| + s]$,

$$\left| \left(e^{\beta|x|} - \sum_{n=0}^{\lambda} \frac{1}{n!} (\beta|x|)^n \right) f(x,t) \right| \leq O(e^{-\alpha t}). \tag{3.1}$$

Remarks: The function $f(x,t)$ is smooth except possibly at $x = 0, t = 0$. Our estimates are uniform for $t > 0$. For $|x| > \epsilon > 0$ we have

$$|f(x,t)| \leq O(e^{-\alpha t - \beta|x|}) \tag{3.2}$$

by the lemma. For $t > \epsilon > 0$, our proof of (3.1) proves (3.2). We take the limit $t \to 0$ and define $f(x,0)$ which satisfies (3.1) and (3.2) with $\alpha = 0$, $\beta < m$. Furthermore, if $\lambda < 0$, then $f(x,t)$ is continuous and (3.2) holds for all (x,t).

Proof: We establish the case $s = 1$, $\nu = j = k = 0$. Then $\lambda = 1$. The proof in general is similar. We use the Cauchy formula to estimate the derivatives of $\exp[-t\mu(p)]$

$$D_p^n e^{-t\mu(p)} = \frac{n!}{2\pi i} \oint \frac{e^{-t\mu(\zeta)}}{(\zeta - p)^{n+1}} d\zeta, \tag{3.3}$$

and we integrate over a circle centered at p. For $|p| \leq m$, we choose a circle of radius $\beta + \epsilon$, where $\alpha^2 + (\beta + \epsilon)^2 < m^2$. We note that if $\mathrm{Re}\, z \geq 0$, then

$$(\mathrm{Re}\, z)^{1/2} \leq \mathrm{Re}(z^{1/2}). \tag{3.4}$$

Thus on our circle of integration $(\mathrm{Re}\,\mu^2)^{1/2} \leq \mathrm{Re}\,\mu$ and

$$\alpha \leq [p^2 + m^2 - (\beta + \epsilon)^2]^{1/2} \leq \mathrm{Re}\,\mu. \tag{3.5}$$

Thus (3.3) yields for $|p| \leq m$,

$$|D_p^n e^{-t\mu(p)}| \leq n!(\beta + \epsilon)^{-n} e^{-\alpha t}.$$

For $|p| > m$, we use a circle of radius $|p|$. On this circle $m \leq (\mathrm{Re}\,\mu^2)^{1/2} \leq \mathrm{Re}\,\mu$, and

$$|D_p^n e^{-t\mu(p)}| \leq n! |p|^{-n} e^{-mt}.$$

Expanding $e^{\beta|x|}$ in (3.1) in a power series and using $\|x^r f\|_\infty \leq \|D^r f^\sim\|_1$,

$$\begin{aligned}
|\{e^{\beta|x|} - (1 + \beta|x|)\} f(x,t)| &\leq \sum_{n=2}^{\infty} \frac{1}{n!} (\beta|x|)^n |f(x,t)| \\
&\leq \sum_{n=2}^{\infty} \frac{\beta^n}{n!} \|D_p^n e^{-t\mu(p)}\|_1 \\
&\leq \sum_{n=2}^{\infty} 2m \left(\frac{\beta}{\beta+\epsilon}\right)^n e^{-\alpha t} + 2\int_m^\infty dp \left(\frac{\beta}{p}\right)^2 \\
&\quad \times e^{-mt} \sum_{n=0}^{\infty} \left(\frac{\beta}{|p|}\right)^n \\
&\leq O(e^{-\alpha t}) + O(e^{-mt}) \int_m^\infty p^{-2} dp \leq O(e^{-\alpha t}).
\end{aligned}$$

Lemma 3.2: Let e_i be the characteristic function of $[i, i+1]$. Then for $0 \leq t$, $\alpha^2 + \beta^2 < m^2$, $\mu_x = \mathfrak{F}\mu(p)\mathfrak{F}^{-1}$, we have

$$\|e_i \exp(-t\mu_x) e_j\| \leq O(e^{-\alpha t - \beta|i-j|}).$$

Proof: If $|i - j| \leq 2$, the bound follows from $|e_i| \leq 1$, $\|e^{-t\mu_x}\| \leq e^{-tm}$. If $|i - j| > 2$, the kernel $k_t(x,y)$ of $e_i \exp(-t\mu_x) e_j$ is $k_t(x,y) = e_i(x) f_t(x - y) e_j(y)$ where f_t is the Fourier transform of $\exp[-t\mu(p)]$. Note that k_t has support in the region $|x - y| > 1$. We apply Lemma 3.1 with $\alpha^2 + (\beta + \epsilon)^2 < m^2$, $j = k = \nu = 0$,

$$\begin{aligned}
|k_t(x,y)| &\leq O(1) e^{-\alpha t - \beta|x-y| - \epsilon|x-y|} e_i(x) e_j(y) \\
&\leq O(e^{-\alpha t - \beta|i-j|}) e^{-\epsilon|x-y|}.
\end{aligned}$$

Since $\exp(-\epsilon|x - y|)$ is the kernel of a bounded operator, Lemma 3.2 follows.

Lemma 3.3: Let $\zeta \in \mathcal{C}_0^\infty$, $\beta < m$, $-1 \leq \tau \leq 1$. Then

$$\|e_i \mu_x^{-\tau} \zeta \mu^\tau\| \leq O(e^{-\beta|i|}).$$

Proof: We let $e = \sum_{|j| \leq N} e_j$, where $[-N, N]$ is sufficiently large to include the support of ζ. We write $w_i = e_i \mu^{-\tau} \zeta \mu^\tau$ as

$$w_i = w_i e + w_i(I - e)$$

and bound each term. Let $r(x,y)$ be the kernel of $w_i(I - e)$. By Lemma 3.1, for $|i|$ sufficiently large,

$$|r(x,y)| \leq O(e^{-\beta|i| - \epsilon|x-y|}).$$

Thus for $|i|$ large,

$$\|w_i(I-e)\| \le O(e^{-\beta|i|}).$$

We now assume $\tau \ge 0$, and note that $\|w_i\| \le$ const, since

$$w_i = e_i\zeta + e_i\mu^{-\tau}[\zeta, \mu^\tau],$$

and $[\zeta, \mu^\tau]$ is a bounded operator. In the momentum representation, the kernel $h(p,p')$ of $[\zeta, \mu^\tau]$ satisfies

$$|h(p,p')| = |\tilde{\zeta}(p-p')\{\mu(p)^\tau - \mu(p')^\tau\}|$$
$$\le O(1)\mu(p-p')^\tau|\tilde{\zeta}(p-p')|,$$

which is a rapidly decreasing function of $p - p'$ and, hence, the kernel of a bounded operator. We use this to bound $\|w_i\|$ and $\|w_i e\|$ for $|i| \le O(1)$. If $\tau < 0$, we use the representation

$$w_i = e_i\zeta + [\mu^{-\tau}, \zeta]\mu^\tau.$$

For $|i|$ large, we bound the kernel $r_i(x,y)$ of $w_i e$ as follows:

$$r_i(x,y) = e_i(x)\int f_-(x-z)\zeta(z)f_+(z-y)dz e(y),$$

where $f_\pm$ is the Fourier transform of $\mu(p)^{\pm\tau}$. For $|x|$ sufficiently large, $f_-(x-z)\zeta(z) = h_x(z)$ is a $\mathcal{C}_0^\infty$ function of z, and, by Lemma 3.1, $D_z^n h_x(z)$ is bounded by $O(e^{-\beta|x|})$, $|x| > R$. Furthermore,

$$\int f_-(x-z)\zeta(z)f_+(z-y)dz = \langle f_{+,y}, h_x\rangle$$

is the evaluation of a tempered distribution f_+, translated by y, with the $\mathcal{C}_0^\infty$ function $h_x(\cdot)$. Since translation is continuous in $\mathcal{S}'$, $f_{+,y}e(y)$ varies over a bounded set in $\mathcal{S}'$ as y varies over the support of e. Hence for some Schwartz space norm $\|\cdot\|_s$ independent of x, y,

$$|e(y)\langle f_{+,y}, h_x\rangle| \le \|h_x\|_s \le O(e^{-\beta|x|}),$$

for $|x|$ sufficiently large. Thus for $|i|$ sufficiently large

$$|r_i(x,y)| \le O(e^{-\beta|i|})e_i(x)e(y),$$

and, since $e_i(x)e(y)$ is the kernel of a bounded operator,

$$\|w_i e\| \le O(e^{-\beta|i|}),$$

to complete the proof.

We now apply these bounds to a φ^n vertex $\hat{v}_\nu$ of the form (2.17). Suppose $\hat{v}_\nu$ has $\bar{n}$ contracted legs and is localized in Δ_i. (Thus the time $s_{r,\nu}$ of the vertex ν lies in $[i_0, i_0+1)$.) Let E_j be the localization projection

$$E_j = \prod_{r=1}^{\bar{n}} e_{j_r}(x_r),$$

and $\sum E_j = I$.

Proposition 3.1: Let $\alpha_r^2 + \beta_r^2 < m^2$. Then

$$\|E_j\hat{v}_\nu\|_2 \le \|\bar{v}_\nu\|_2\gamma \prod_{r=1}^{\bar{n}} (e^{-(1/2)\alpha_r\delta s_{r,\nu} - (1/2)\beta_r|j_r - i_1|}). \quad (3.6)$$

Proof: Let $\delta S = \frac{1}{2}\sum_r \mu_{r,\nu}\delta s_{r,\nu}$. For $0 < \delta$, δ sufficiently small, we have by Lemma 3.2,

$$\|E_j\hat{v}_\nu\|_2 \le \sum_{j'} \|E_j e^{-\delta S}E_{j'}\bar{v}_\nu\|_2$$
$$\le O(1)\sum_{j'}\prod_{r=1}^{\bar{n}} (e^{-(1/2)\alpha_r\delta s_{r,\nu} - (1/2)(\beta_r+\delta)|j_r - j_r'|})\|E_{j'}\bar{v}_\nu\|_2. \quad (3.7)$$

In order to display the localization of $\bar{v}_\nu$, we write

$$\bar{v}_\nu = \left(\prod_{r=1}^{\bar{n}} K_{r,i_1}\right)\bar{v}_\nu = \bar{K}\bar{v}_\nu. \quad (3.8)$$

Here

$$K_{r,i_1} = \mu_{x_r}^{-(1/2)+\epsilon_{r,\nu}}\zeta_{i_1}(x_r)\mu_{x_r}^{(1/2)-\epsilon_{r,\nu}},$$

$\zeta_{i_1} \in \mathcal{C}_0^\infty$, and ζ_{i_1} equals one on a neighborhood of $[i_1, i_1+1]$, sufficiently large so (3.8) holds. In fact, we take ζ_{i_1} to be the translate of a fixed $\mathcal{C}_0^\infty$ function, so the $\mathcal{L}_p$ norms of its derivatives are independent of i_1. By Lemma 3.3,

$$\|E_j\bar{K}\| \le O(1)\prod_{r=1}^{\bar{n}} e^{-(1/2)(\beta_r+\delta)|j_r' - i_1|}.$$

Thus performing the sum in (3.7) yields (3.6).

We remark that for an optimal choice of (α, β) satisfying $\alpha^2 + \beta^2 = (m-\epsilon)^2$,

$$e^{-\alpha t - \beta|x|} \le e^{-(m-\epsilon)d},$$

where $d = (t^2 + x^2)^{1/2}$ is the Euclidean distance.

B. Estimates on Graphs

We use the estimates of Sec. 3A to establish estimates on a single polynomial $\hat{R}_\alpha$. We consider kernels $\hat{v}_\nu$ of the form (2.17), where $\nu = 1, 2, \ldots, N$ labels the vertices of the graph of R_α. Each kernel $\hat{v}_\nu(k_{1,\nu}, \ldots, k_{s,\nu}, q(\cdot))$ is a function of contracted lines with momenta $k_{r,\nu}$ and of the path space variables $q(\cdot)$ for uncontracted legs.

Each vertex has a space-time localization and a lower momentum cutoff $\kappa(v_\nu)$. The norm $\|\hat{v}_\nu\|_2 = \|\hat{v}(\cdot; q(\cdot))\|_2$ is the $\mathcal{L}_2$ norm of $\hat{v}$ in the contracted momentum variables, and it is a function on path space. (It is a constant on path space for fully contracted graphs.)

Let the line (i.e., the pair of contracted legs) with momentum $k_{r,\nu}$ connect vertices v_ν and $v_{\nu'}$ with space-time localizations Δ and Δ', respectively. We define the contraction distance $d_{r,\nu}$ of this line as the Euclidean distance between Δ and Δ'. We define the total contraction distance $d(v_\nu) = d_\nu$ of the vertex v_ν as

$$d_\nu = d(v_\nu) = \sum_r d_{r,\nu}, \quad (3.9)$$

where the sum extends over the contracted legs at vertex ν. Given $\epsilon > 0$, we introduce the localization factor l_ν of the vertex v_ν:

$$\gamma e^{-md_\nu/3}. \quad (3.10)$$

Here γ is a sufficiently large constant, chosen below.

Let $\hat{R}_\alpha(s)$ be a polynomial of the form (2.18).

Theorem 3.1: Given $0 < \epsilon$, there exists a constant γ such that

$$|\hat{R}_\alpha(s)| \le \left(\prod_{\text{vertices}} \gamma\|\bar{v}_\nu\|_2\right)\left(\prod_{\text{lines}} e^{-(m-\epsilon)d_{r,\nu}}\right)$$
$$\le \prod_\nu(\gamma e^{-md_\nu/3}\|\bar{v}_\nu\|_2). \quad (3.11)$$

Remarks: For a fully contracted graph, $\|\tilde{v}_\nu\|_2 \leqq O(\kappa(v_\nu)^{-(1/2)+\epsilon_\nu+\delta})$, $\delta > 0$. Thus for a new constant γ,

$$|\hat{R}_\alpha(s)| \leq \prod_\nu [\gamma\kappa(v_\nu)^{-(1/2)+\epsilon_\nu+\delta} e^{-md_\nu/3}]. \quad (3.12)$$

With our choice $\epsilon_\nu \leq \frac{1}{8}$ in Sec. 2E, fully contracted graphs arising in the inductive construction satisfy

$$|\hat{R}_\alpha(s)| \leq \prod_\nu [\gamma\kappa(v_\nu)^{-1/4} e^{-md_\nu/3}]. \quad (3.13)$$

The proof of Theorem 3.1 follows from Proposition 3.1 in an elementary fashion. We localize each leg in space, and apply the Schwarz inequality to obtain

$$|\hat{R}| \leq \sum_{i_\nu} \prod_\nu \|E_{i_\nu} \hat{v}_\nu\|_2.$$

Using the fact that each pair of contracted legs has the same space-time localization, we bound the summation over i_ν with Proposition 3.1. We choose $(\alpha_{r,\nu}, \beta_{r,\nu})$ to optimize our bound and so obtain (3.11).

C. Combinatoric Estimates

At the conclusion of the inductive construction, we obtain an upper bound for (1.39) as a sum of constants on path space, with each constant represented by a vacuum graph. We now estimate this sum, which involves among other things, counting the number of terms in the expansion. We assign a positive constant $c(v_\nu)$ to each vertex v_ν in the inductive construction; $c(v_\nu)$ is called a combinatoric factor.

One use of combinatoric factors is to estimate sums by supremums. For instance,

$$|\sum_\alpha a_\alpha| \leq \sup_\alpha |c_\alpha a_\alpha|,$$

provided $\sum c_\alpha^{-1} \leq 1$. We also use the combinatoric factors to absorb the constants e^j or η in (2.34), except those coming from squares $\Delta \in \mathfrak{D}_0$. Since each vertex is localized in some square Δ, the time integration in (2.11) extends over a region of volume at most one. Thus we replace the integral over s by the supremum norm of the integrand. We note that for each $\varphi(h)$-$\varphi(h)$ contraction, the integral over one $\varphi(h)$ time has already been estimated and is not a part of the ds integration in (2.11), see Sec. 2E. We have

$$\begin{aligned}
&\langle\Omega_0, e^{-T(H_0+H_I+\delta H)}\Omega_0\rangle \\
&= \int_Q \exp\left(-\int_0^T H_{II}(q(\sigma))d\sigma\right)dq(\cdot) \\
&\leq \sum_\alpha \int ds R_\alpha(s) \exp\left(-\int_0^T H_I(\sim Y_\alpha(\sigma))(q(\sigma))d\sigma\right) \\
&\leq \sum_\alpha \int ds |R_\alpha(s)| \exp\left(-\int_0^T E(\sim Y_\alpha(\sigma))d\sigma\right).
\end{aligned} \quad (3.14)$$

Each polynomial $R_\alpha(s)$ is defined in (2.19) and is estimated in Sec. B:

$$|R_\alpha(s)| \leq B_\alpha L_\alpha \prod_\nu (l_\nu \|\tilde{v}_\nu\|_2).$$

The constants in B_α, with the exception of factors e^{J_0} from applying (2.32) or (2.34) in squares $\Delta \in \mathfrak{D}_0$, are parts of combinatoric factors $c(v_\nu)$. With $c(v_\nu)$ as chosen below we will obtain

$$\langle\Omega_0, e^{-T(H_0+H_I+\delta H)}\Omega_0\rangle \leq e^{2J_0|X_0|T} \sup_\alpha \prod_\nu (c(v_\nu)l_\nu\|\tilde{v}_\nu\|_2) \times \exp\left(-\int_0^T E(\sim Y_\alpha(\sigma))d\sigma\right). \quad (3.15)$$

The factor $\exp(2J_0|X_0|T)$ bounds the product of the constants e^{J_0} from squares $\Delta \in \mathfrak{D}_0$. We also prove

Proposition 3.2: For N_0 of (2.43) sufficiently large, we have

$$\begin{aligned}
\prod_\nu (c(v_\nu)l_\nu\|\tilde{v}_\nu\|_2) &\leq e^{-N(\alpha)} \\
&\leq \exp\left(-\int_0^T |Y_\alpha(\sigma)|d\sigma - |X_0|T\right),
\end{aligned} \quad (3.16)$$

for every graph R_α in (3.14), where $N(\alpha)$ is the number of vertices in R_α.

The inequality (2.5) follows immediately. By (3.15) and (3.16), we have

$$\begin{aligned}
&\langle\Omega_0, e^{-T(H_0+H_I+\delta H)}\Omega_0\rangle \\
&\leq e^{(2J_0+1)|X_0|T} \sup_\alpha \exp\left(-\int_0^T \{E(\sim Y_\alpha(\sigma)) + |Y_\alpha(\sigma)|\}d\sigma\right) \\
&\leq e^{(2J_0+1)|X_0|T} \sup_{X\supset X_0} e^{-T\{E(\sim X)+|X|\}}.
\end{aligned} \quad (3.17)$$

Thus, by the formula above (1.39),

$$-E(g) - \delta E(g,h) \leq M|X_0| - \inf_{X\supset X_0} \{E(\sim X) + |X|\},$$

where $M = 2J_0 + 1$, which proves (2.5).

We now assign the combinatoric factors $c(v_\nu)$ to vertices v_ν. We denote these factors c_{pt}, c_{ps}, or c_h for PT, PS, or $\varphi(h)$ vertices, respectively.

Proposition 3.3: Let $\beta_0 > 0$ be given, and let N_0 of (2.43) be sufficiently large. There are constants β, N, independent of β_0 and N_0 such that

$$c_{pt}(v) \leq \beta d(v)^N, \quad (3.18)$$

$$c_{ps}(v) \leq \beta d(v)^N \kappa(v)^{1/8}, \quad (3.19)$$

$$c_h(v) \leq \beta_0 d(v)^N \quad (3.20)$$

and such that (3.5) is valid.

Proof: We assign a finite product of combinatoric factors to each vertex. Each factor in the product has the form $O(1)d(v)^n\kappa(v)^\epsilon$ and, hence, so does the product. Therefore, we only need to keep track of which powers $\kappa(v)^\epsilon$ occur, and we need to show that some factor for $\varphi(h)$ vertices can be made small.

1. *The Graph Expansion*

We start by assigning combinatoric factors to a given graph expansion step. This is our major assignment of combinatoric factors, since we count the number of graphs R_α that are formed during a graph expansion step. At the start of a graph expansion of a particular term, $n(\Delta)$ vertices in Δ have uncontracted legs. These "old" vertices contract to each other during the graph expansion, and they also contract to the exponent to form "new" PT vertices. The old vertices may contract further with the new vertices; but the new vertices do not contract to the exponent. A given PS or $\varphi(h)$ vertex can contract during at most two inductive steps. We assign combinatoric factors to vertices from the time they are introduced until the graph expansion when they fully contract.

If the rth leg of the νth vertex contracts to a vertex localized in Δ_j, we set $j = j(r, \nu)$. Consider the set of all graphs in which the contraction localizations $j = j(r, \nu)$ of the old vertices take on fixed values. We count the number of such graphs $N(\{j(r,\nu)\})$, and then we sum over the allowed $\{j(r,\nu)\}$. Suppose that vertices v and v', localized in Δ and Δ', respectively, are connected by a line with contraction distance dist $(\Delta, \Delta') \le \min\{d(v), d(v')\}$. Since

$$\sum_{\Delta'} \text{dist}(\Delta, \Delta')^{-3} = \beta_1 < \infty,$$

a combinatoric factor $\beta_1 d(v)^3$ assigned to each leg is sufficient to deal with each sum over contraction localizations.

Thus we may consider a fixed set of $\{j(r,\nu)\}$. Suppose that in a given graph expansion step l_{ij} old legs localized in Δ_i contract to legs localized in Δ_j, and that l'_{ij} new legs localized in Δ_i contract to legs localized in Δ_j. (Either l_{ij} or l'_{ij} is zero.) The contraction localizations $\{j(r,\nu)\}$ determine the $\{l_{ij}\}$ and the $\{l'_{ij}\}$. However the $\{j(r,\nu)\}$ do not determine the number of new vertices, their degrees, nor the number of contracted legs at each new vertex. Since new vertices need not be fully contracted, the number of new vertices in Δ_i with legs contracting to Δ_j lies between l'_{ij}/p and l'_{ij}. Each new vertex is a monomial φ^n, $1 \le n \le p$, and l'_{ij} values of n may be chosen in $p^{l'_{ij}}$ ways for a total of at most $l'_{ij}\, p^{l'_{ij}}$ possible choices of the number and degree of new vertices in Δ_i. We assign a combinatoric factor $\beta_2 = p^{2p}$ per new PT vertex, or at least $\beta_2^{l'_{ij}/p} = p^{2l'_{ij}}$ to count new vertices in Δ_i, and we assign a factor 2^p per vertex to count whether each of the (at most) p legs at a new vertex contracts. For new vertices, the localization factor $\beta_1 d(v)^3$ per leg deals with the sum over localizations, as above. Thus we now may fix the contraction localizations $\{j(r,\nu)'\}$ of the new vertices as well as $\{j(r,\nu)\}$. Let $l^{\#}_{ij}$ denote l_{ij} or l'_{ij}.

For new or old vertices, the $l^{\#}_{ij}$ legs from Δ_i to Δ_j can be connected in $l^{\#}_{ij}!$ ways if $i \ne j$ or in $l_{ii}!! \le (l_{ii}!)^{1/2}$ ways if $i = j$. We need to count the contractions from Δ_i to Δ_j or from Δ_j to Δ_i but not both, so

$$N\{j, j'\} \le \prod_{i,j} (l^{\#}_{ij}!)^{1/2}. \tag{3.21}$$

If $l'_{ij} \ne 0$, we divide $(l'_{ij}!)^{1/2} \le (l'_{ij})^{l'_{ij}/2}$ equally among the l'_{ij} legs and, thus, assign a factor $(l'_{ij})^{p/2}$ per new PT vertex. In the $(k+1)$th inductive step, by Corollary 2.1, any PT vertex v in Δ_i contributing to l'_{ij} satisfies

$$l'_{ij} \le pn(\Delta_j)_k \le pd_e(\Delta_j)_k^3 \le pd(v)^3.$$

Hence a factor $[pd(v)^3]^{p/2}$ per new PT vertex in Δ_i dominates the contribution of Δ_i to (3.21).

We now consider the case of old vertices $l_{ij} \ne 0$. Let m_{il}, $l = 1, 2, 3$ denote the number of uncontracted legs in Δ_i at the end of the kth inductive step and belonging to PS vertices ($l = 1$), $\varphi(h)$ vertices ($l = 2$), or PT vertices ($l = 3$). Then $m_{i1} \le pn_{ps}(\Delta_i)_k$, $m_{i2} \le n_h(\Delta_i)_k$, and $m_{i3} \le (p-1)n_{pt}(\Delta_i)_k$. Also

$$l_i = \sum_j l_{ij} = m_{i1} + m_{i2} + m_{i3},$$

$$\prod_j (l_{ij}!)^{1/2} \le (m_i!)^{1/2} \le \prod_{l=1}^{3} (9m_{il})^{m_{il}/2}. \tag{3.22}$$

We assign the factor $(9m_{il})^{m_{il}/2}$ among the vertices of type l localized in Δ_i. For $l = 1$, we have $m_{i1} \le pn_{ps}(\Delta_i) \le \kappa(v)^{1/8p}$. We divide the factor $m_{i1}^{1/2}$ equally among the PS vertices in Δ_i, yielding a factor

$$m_{i1}^{p/2} \le O(1)\kappa(v)^{1/16} \tag{3.23}$$

per vertex.

For $\varphi(h)$ vertices, we assign the factor

$$3m_{i2}^{1/2} \le 3n_h(\Delta_i)^{1/2} \tag{3.24}$$

per vertex, as well as a factor 2 per vertex for the two terms formed by the time integration for each φ–φ contraction (see Sec. 2E).

Finally, we consider PT vertices m_{i3}. We assert that there exists a constant β_3, depending only on p, such that

$$m_{i3} \le \beta_3 d_c(\Delta_i)^6/3. \tag{3.25}$$

Our combinatoric factor (3.22) for $l = 3$ is bounded by

$$[O(1)d_c(\Delta_i)]^{\beta_3 d_c(\Delta_i)^6}. \tag{3.26}$$

The squaring exponent (2.23) of Sec. 2C ensures that if $m_{i3} \ne 0$, then at least $d_c(\Delta_i)^6$ old PT vertices v, each with $d_c(\Delta_i) = d_c(v)$, are localized in Δ_i. We assign a combinatoric factor $O(1)d_c(v)^{\beta_3}$ to each old PT vertex, which dominates (3.26).

We now prove (3.25). Let N_k be the number of new PT vertices produced in the graph expansion of the kth inductive step, and which retain uncontracted legs after squaring. Hence by (2.23),

$$m_{i3} \le (p-1)n_{pt}(\Delta_i)_k \le (p-1)\{2N_k + [4d_c(\Delta_i)]^6\}.$$

But the vertices N_k are produced by contractions to the exponent, so

$$N_k \le p \sum_{\Delta_j} n(\Delta_j)_{k-1}, \tag{3.27}$$

where the sum extends over squares Δ_j satisfying $n(\Delta_j)_{k-1} \le d_e(\Delta_j)_{k-1}^3$, by Corollary 2.1. Since the vertices remain after squaring, $d_e(\Delta_j)_{k-1}^3 \le d_c(\Delta_i)^3$, and at most $4d_c(\Delta_i)^2$ squares Δ_j enter the sum (3.27). Hence $N_k \le 4pd_c(\Delta_i)^5$, proving (3.25).

2. *The Squaring Operation*

The squaring operation (2.22) introduces two terms; so a combinatoric factor 2 is sufficient for each application of (2.22). We assign this factor to the vertex v' which contracts to the exponent forming the new PT vertex v requiring the use of (2.22). Each vertex v' can form at most p new PT vertices during the graph expansion, and we assign a combinatoric factor 2^p per vertex to deal with the squaring operation.

3. *The Path Space Construction*

Let us first consider (2.32). If $j > j_0$, we assign a combinatoric factor $\exp(\beta_4)$ per PS vertex. This factor $\exp[\beta_4 n_{ps}(\Delta)]$ dominates both e^j and the factor $O(1)j^2$ for summation over j. In fact, $j = O(1)$

$(\log\kappa_j)^{p/2} \leq O(1)\kappa_j^{1/8p}$, and $n_{ps}(\Delta) \geq \kappa_j^{1/8p}$ if the lower cutoff in Δ is κ_j. Thus for β_4 sufficiently large,

$$O(1)j^2e^j \leq \exp(\beta_4\kappa_j^{1/8p}) \leq \exp[\beta_4 n_{ps}(\Delta)].$$

In case $j = j_0$, there are no PS vertices in Δ_i. For $\Delta_i \subset \mathfrak{D}_0$, we include the factor $\exp(J_0)$ explicitly in our bound (3.15). For $\Delta_i \not\subset \mathfrak{D}_0$, we assign $\exp[j_0(\Delta_i)]$ as a combinatoric factor for the vertices localized in Δ_j, where Δ_j forced the PS construction in Δ_i. By (2.35), the $n(\Delta_j)$ vertices in Δ_j with uncontracted legs force the PS construction in at most $64n(\Delta_j)^{2/3}$ squares Δ_i. [This is the area of $\Gamma(\Delta_j)$.] By (2.36), we assign at most $\exp(64n(\Delta_j)^{2/3}\{[J_1 n(\Delta_j)^{1/6}] + 1\})$ among the vertices in Δ_j. Hence a factor $O(1)$ per vertex is sufficient.

Finally we consider (2.33), the part of (2.34) not analyzed above. The factor $\exp(J_0)$ is kept explicitly in (3.15). For $2k$ $\varphi(h)$ vertices we have a factor $\eta/(2k)! \leq \eta k^{-2k}$ or a factor

$$O(1)n_h(\Delta_i)^{-1} \tag{3.28}$$

per vertex. A combinatoric factor $O(1)$ per vertex dominates the sum over k. We choose $N_0 \leq n_h(\Delta_i)$ sufficiently large so that (3.28) dominates (3.24), and their product is sufficiently small to ensure (3.20). This completes the proof of the proposition.

Proof of Proposition 3.2: We use Theorem 3.1 and Proposition 3.3. For a PT, PS, or $\varphi(h)$ vertex v_ν, we have

$$c(v_\nu)l_\nu\|\bar{v}_\nu\|_2 \leq \begin{cases} \beta\gamma d(v_\nu)^N e^{-md(v_\nu)/6}, & \text{PT} \\ \beta\gamma d(v_\nu)^N e^{-md(v_\nu)/6}\kappa(v_\nu)^{-1/8}, & \text{PS} \\ \beta_0\gamma d(v_\nu)^N e^{-md(v_\nu)/6}\|\mu^{-(1/2)-1/8p}h\|_2, & \varphi(h) \end{cases}$$

We use Theorem 3.1 to give the estimate $O[\kappa(v_\nu)^{-(1/2)+1/8+\delta}]$ for PS vertices, the $\kappa^{1/8}$ coming from Sec. 2E, and this dominates the $O[\kappa(v_\nu)^{1/8}]$ growth of (3.19). By Corollary 2.2, $\kappa(v_\nu)$ for PS vertices tends to infinity with N_0 of (2.43). Similarly, for PT vertices, $d(v_\nu) \to \infty$ as $N_0 \to \infty$ by Corollary 2.3. We choose β_0 sufficiently small and N_0 sufficiently large so

$$c(v_\nu)l_\nu\|\bar{v}_\nu\|_2 \leq e^{-1}.$$

To complete the proof of (3.16), we show that each polynomial R_α has a graph with $N(\alpha)$ vertices, where

$$N(\alpha) \geq \int_0^T |Y_\alpha(\sigma)|d\sigma - |X_0|T.$$

The right side is the area in $\sim\mathfrak{D}_0$ in which the exponent has been removed. Let $\Delta \not\subset \mathfrak{D}_0$ and let Δ' force the PS construction in Δ. We count the vertices in Δ'. The square $\Gamma(\Delta')$ has area $64n(\Delta')^{2/3}$ and if $N_0 > (64)^3$, then by Corollary 2.2, $64n(\Delta')^{2/3} < n(\Delta')$. Hence R_α has more vertices than squares in $\sim\mathfrak{D}_0$ with the exponent removed.

4. COMPLETION OF THE BOUND ON δE

Theorem 4.1: There exists a_0 such that for all intervals X with $|X| \geq a_0$,

$$E(g) \leq E(\sim X) + |X|. \tag{4.1}$$

We use the inductive construction to bound $E(\sim X)$ and prove (4.1). In this section we describe the modifications to Secs. 2 and 3 required to establish (4.1), i.e., (2.4) and, hence, to complete the proof of Proposition 2.1 and Theorem 1.1.

A. The Inductive Construction

Let $X_a = X_a(g)$ minimize $E(\sim X) + |X|$ under the restriction that X is an interval of length $|X| \geq a$. Then, for $|X| \geq a$,

$$-E(\sim X) \leq -E(\sim X_a) + |X| - |X_a|. \tag{4.2}$$

We assume that (4.1) is false, namely for $a \geq a_0$,

$$-E(g) \leq -E(\sim X_a) - |X_a|. \tag{4.3}$$

We use the inductive construction and (4.2) and (4.3) to bound $E(\sim X_a)$; we thereby obtain a contradiction, establishing (4.1). We bound

$$\int \exp\left(-\int_0^T H_I(\sim X_a)(q(\sigma))d\sigma\right)dq(\cdot). \tag{4.4}$$

The only change in the inductive construction occurs at the start of the first inductive step, i.e., the start of the first PS construction. Roughly, the idea is to begin the inductive construction by inserting and removing the factor $\exp(-H_I(X_a))$ in the integrand of (4.4). For this purpose we use the inequality

$$1 \leq e^{-r+s} + (r/s)^{2n}, \tag{4.5}$$

valid for real r, for $s > 0$ and for $n \in \mathfrak{Z}^+$. We choose $n = [|X_a|^{1/8p}]$, $s = \frac{1}{2}|X_a|$, and

$$r = \int_{i_0}^{i_0+1} H_I(X_a)(q(\sigma))d\sigma. \tag{4.6}$$

We begin the first PS construction by inserting (4.5) and (4.6) in the integrand of (4.4) at each time i_0 such that $0 \leq i_0 \leq T$, $i_0 \in \tau\mathfrak{Z}^+$, where τ is a positive integer to be determined later. Thus we apply (4.5) and (4.6) in the union of squares

$$\mathfrak{D}_1 = \{\Delta_i : i_1 \in X_a, i_0 \in \tau\mathfrak{Z}^+, i_0 \leq T\},$$

producing $2^{[T/\tau]}$ terms. For each such term, let $\mathfrak{D}_0$ be the subset of $\mathfrak{D}_1$ in which the exponent is removed, i.e., the squares in which the second term in (4.5) is selected. Let $\mathfrak{D}_0'$ be the complementary subset of $\mathfrak{D}_1$ in which the exponent remains. Then $\mathfrak{D}_1 = \mathfrak{D}_0 \cup \mathfrak{D}_0'$, and each square $\Delta_i \in \mathfrak{D}_0$ contains $2[|X_a|^{1/8p}]$ vertices, each of the form $2|X_a|^{-1}H_{I,i}$. We call these vertices PS vertices, we include the factor $2|X_a|^{-1}$ with the kernel of these PS vertices, and we assign the function $\kappa(v) = |X_a|$ to such PS vertices. Then $\kappa(v)$ plays the role of a lower cutoff $n = [\kappa(v)^{1/8p}]$ and $\|v\|_2 \leq O(\kappa(v)^{-1/2})$.

After starting the PS construction in this manner and obtaining PS vertices in $\mathfrak{D}_0$, we continue the inductive construction as in Sec. 2. With our definition of the PS vertices in $\mathfrak{D}_0$, the estimates of Sec. 2 and 3 concerning the inductive construction apply. In the case of (2.43), $N_0 = 2n \geq 2a_0^{1/8p} \to \infty$ as $a_0 \to \infty$.

At the end of the inductive construction, we will obtain $\exp[-E(\sim X)]$, which we bound by (4.2) or (4.3). Here X is the space region (at some time) in which the exponent is removed. The inequalities (4.2) or (4.3) apply if X is an interval and either $|X| \geq a$ or $X = \emptyset$, respectively. We now verify these conditions for any X appearing at the end of the inductive construction. If $\Delta_i \not\subset X_a \times [0, T]$, then $d_0(\Delta_i)$ exceeds the horizontal, or equal time distance from Δ_i to $X_a \times [0, T]$. By Corollary 2.1, if $n(\Delta_i)_r \neq 0$, then at the end of the rth PS construction $d_e(\Delta_i)_r > d_0(\Delta_i)$. Thus we ensure that X has the desired form if whenever the exponent is removed from $\Delta \in \mathfrak{D}_0'$, it is also removed from all other squares in $\mathfrak{D}_0'$ with the same time localization as Δ.

Suppose that Δ' forces the PS construction in $\Delta \in \mathfrak{D}_0'$. If $d_0(\Delta') \geq 8^{-1}\,\mathrm{dist}(\Delta, \Delta') + 8^{-1}|X_a|$, then by (2.38)

$$n(\Delta')^{1/3} \geq d_0(\Delta') \geq 8^{-1}\,\mathrm{dist}(\Delta, \Delta') + 8^{-1}|X_a|.$$

Thus Δ' forces the PS construction in a square $\Gamma(\Delta')$ with side length $8n(\Delta')^{1/3} \geq \mathrm{dist}(\Delta, \Delta') + |X_a|$ and centered at Δ'. This $\Gamma(\Delta')$ includes all squares in $\mathfrak{D}_0'$ that are localized at the same time as Δ.

On the other hand, if $d_0(\Delta') \leq 8^{-1}\,\mathrm{dist}(\Delta, \Delta') + 8^{-1}|X_a|$, let $d_0(\Delta') = \mathrm{dist}(\Delta', \Delta'')$, $\Delta'' \in \mathfrak{D}_0$. Then

$$\begin{aligned}\mathrm{dist}(\Delta, \Delta') &\geq \mathrm{dist}(\Delta, \Delta'') - \mathrm{dist}(\Delta'', \Delta')\\ &\geq \tau - d_0 \geq \tau - 8^{-1}\,\mathrm{dist}(\Delta, \Delta') - 8^{-1}|X_a|.\end{aligned}$$

Suppose $\tau = 10|X_a|$. Thus

$$|X_a| \leq 8^{-1}\,\mathrm{dist}(\Delta, \Delta'). \tag{4.7}$$

Let $d_1(\Delta')$ be the distance between the space localization of Δ' and the center of X_a. {If $\Delta' = \Delta_i$ and $X_a = [\alpha, \beta]$, then $d_1(\Delta') = |i_1 - \frac{1}{2}(\alpha + \beta)|$.} Half the side length of $\Gamma(\Delta')$ is greater than $2^{-1/2}\,\mathrm{dist}(\Delta, \Delta')$. We show that

$$d_1(\Delta') + \tfrac{1}{2}|X_a| \leq 2^{-1/2}\,\mathrm{dist}(\Delta, \Delta'), \tag{4.8}$$

so Δ' forces the PS construction in all squares in $\mathfrak{D}_0'$ localized at the same time as Δ. By the triangle inequality, $d_1(\Delta') \leq d_0(\Delta') + \frac{1}{2}|X_a|$; so by (4.7),

$$\begin{aligned}d_1(\Delta') + \tfrac{1}{2}|X_a| &\leq d_0(\Delta') + |X_a|\\ &\leq \tfrac{1}{8}\,\mathrm{dist}(\Delta, \Delta') + \tfrac{9}{8}|X_a|\\ &\leq \tfrac{3}{8}\,\mathrm{dist}(\Delta, \Delta') \leq 2^{-1/2}\,\mathrm{dist}(\Delta, \Delta').\end{aligned}$$

Thus we choose $\tau = \tau(a) = 10|X_a|$ to ensure the desired geometry.

B. The Combinatorial Estimates

The combinatoric factors for the graph expansion and the squaring operation are assigned as in chapter 3. In the path space construction, we now consider the combinatoric factors from the application of (4.5) and (4.6) in $\mathfrak{D}_1$. Let $\mathfrak{I}_1 = \{i_0 : 0 \leq i_0 \leq T, i_0 \in \tau\mathbb{Z}_+\}$. For a vacuum graph R_α at the end of the inductive construction, let $\mathfrak{I}_0 = \mathfrak{I}_0(\alpha) \subset \mathfrak{I}_1$ be the subset of times for which the second term of (4.5) was chosen. Let $\mathfrak{I}_0'(\alpha) \subset \mathfrak{I}_1$ be the complementary subset of times for which the first term of (4.5) was chosen. Let $\mathfrak{I}_0'' \subset \mathfrak{I}_0'$ be the subset of times in $\mathfrak{I}_0'$ when the PS construction was not used in X_a. We note that $Y_\alpha(\sigma)$ in (3.14) has the form

$$Y_\alpha(\sigma) = \begin{cases}\emptyset, & [\sigma] \in \mathfrak{I}_0''(\alpha)\\ X, X \supset X_a, & [\sigma] \notin \mathfrak{I}_0''(\alpha).\end{cases} \tag{4.9}$$

Thus $H_I(\sim Y_\alpha(\sigma)) = H_I(g), H_I(\sim X)$ in these respective cases.

Each application of (4.5) yields two terms, so we assign a combinatoric factor 2 to each time $i_0 \in \mathfrak{I}_1$. If $i_0 \in \mathfrak{I}_0$, there are vertices in X_a at time i_0 and we assign the factor 2 to any such vertex. If $i_0 \in \mathfrak{I}_0' \sim \mathfrak{I}_0''$, we assign the factor 2 to a vertex in the square forcing the PS construction, as with the factor e^s below. The remaining factor is $\exp(|\mathfrak{I}_0''|\log 2)$, where $|\mathfrak{I}_0''|$ is the number of elements of $\mathfrak{I}_0''$.

The other relevant factor is the constant $\exp(\frac{1}{2}|X_a|) = e^s$ from the first term of (4.5). If $i_0 \in \mathfrak{I}_0' \sim \mathfrak{I}_0''$, then $\exp(\frac{1}{2}|X_a|)$ is a combinatoric factor $\exp(\frac{1}{2})$ per square $\Delta = \Delta_{i_0, i_1}$ for $i_1 \in X_a$. We distribute this factor among the vertices in Δ' that forced the PS construction in Δ. Since $n(\Delta')$ vertices in Δ' force the PS construction in at most $64n(\Delta')^{2/3}$ squares, we assign at most a factor $O(1)$ per vertex in Δ'. For times $i_0 \in \mathfrak{I}_0''$, we obtain a factor $\exp(\frac{1}{2}|X_a||\mathfrak{I}_0''|)$ in our expansion.

Thus the inductive construction yields

$$\begin{aligned}\langle\Omega_0, e^{-T[H_0+H_I(\sim X_a)]}\Omega_0\rangle &= \int e^{-H_I[\sim(X_a)(q(\sigma))]d\sigma}\,dq(\cdot)\\ &\leq \sum_\alpha \int ds\,|R_\alpha(s)|\,\exp\Big(-\int_0^T H_I(\sim Y_\alpha(\sigma))(q(\sigma))d\sigma\Big)\\ &\leq \sup_\alpha e^{|\mathfrak{I}_0''(\alpha)|\{\log 2 + \frac{1}{2}|X_a|\}}\\ &\quad\times \prod_\nu (c(v_\nu)l_\nu\gamma\|\overline{v}_\nu\|_2)\,\exp\Big(-\int_0^T E(\sim Y_\alpha(\sigma))d\sigma\Big).\end{aligned} \tag{4.10}$$

For $|\mathfrak{I}_0''(\alpha)|$ time intervals $\sigma \in [i_0, i_0 + 1)$, we bound $-E(\sim Y_\alpha(\sigma)) = -E(g)$ by (4.3). For the remaining time intervals we use (4.2). Thus

$$\begin{aligned}&\exp\Big(-\int E(\sim Y_\alpha(\sigma))d\sigma\Big)\\ &\quad\leq e^{-|\mathfrak{I}_0''(\alpha)||X_a|}e^{-TE(\sim X_a)}\exp\Big(\int_0^T |Y_\alpha(\sigma) \sim X_a|d\sigma\Big).\end{aligned} \tag{4.11}$$

By Proposition 3.2, for N_0 sufficiently large,

$$\prod_\nu (c(v_\nu)l_\nu\gamma\|\overline{v}_\nu\|_2) \leq e^{-N(\alpha)}.$$

In our case

$$N(\alpha) \geq \int_0^T \{|Y_\alpha(\sigma) \sim X_a|\}d\sigma + \{T/\tau - |\mathfrak{I}_0''(\alpha)|\}. \tag{4.12}$$

The bound $N(\alpha) \geq \int_0^T \{|Y_\alpha(\sigma) \sim X_a|\}d\sigma$ follows as in the proof of Proposition 3.2, neglecting squares in $\mathfrak{D}_1$. In a strip $X_a \times [i_0, i_0 + 1]$, $i_0 \in \mathfrak{I}_0$, there is at least one vertex. For a square $\Delta \in X_a \times \mathfrak{I}_0' \sim \mathfrak{I}_0''$, we count one of the $n(\Delta')$ vertices in Δ', where Δ' forced the PS construction in Δ. Thus we count at least one vertex for each time interval in $\mathfrak{I}_1 \sim \mathfrak{I}_0''$. There are $T/\tau - |\mathfrak{I}_0''|$ such time intervals, establishing (4.12).

From (4.10)–(4.12), we find

$$\begin{aligned}&\langle\Omega_0, e^{-T[H_0+H_I(-X_a)]}\Omega_0\rangle\\ &\quad\leq \sup_\alpha e^{-TE(-X_a)}e^{-T/\tau}e^{-|\mathfrak{I}_0''(\alpha)|\{-\log 2 - 1 + \frac{1}{2}|X_a|\}}\\ &\quad\leq e^{-T[E(-X_a)+\tau^{-1}]},\end{aligned}$$

if a_0 (and hence $|X_a| \geq a \geq a_0$ and N_0) are sufficiently large.

Taking the logarithm, dividing by T and letting $T \to \infty$, the left side converges to $-E(\sim X_a)$. This gives the contradiction $0 \leq -\tau^{-1}$, and proves (4.1).

5. A CORRECTION

Our proof[5] of the spectrum condition $P^2 \leq H^2$ contains a gap, as was pointed out by Fröhlich and Faris. Namely, we required the Lorentz rotated Hamiltonian in a periodic box to have a simple ground state, which does not follow, as claimed, from standard methods. The remaining results in Ref. 5 are either independent of this gap or are proved (and improved) by the present paper, with the exception of the estimate

$$\pm \nabla\varphi(h) \leq \text{const}\|h\|_2(H + I).$$

This bound is Theorem 1.1, Case 2, $\epsilon = \frac{1}{2}$, and it should follow from the methods of the present paper.

---

* Supported in part by the National Science Foundation, Grant GP-24003.

† Supported in part by the Air Force Office of Scientific Research, Contract AF 44620-70-C0030 and the National Science Foundation, Grant GP-31239X.

1 J. Glimm and A. Jaffe, *Quantum Field Theory Models*, in the 1970 Les Houches lectures, edited by C. Dewitt and R. Stora (Gordon and Breach, New York, 1971).

2 J. Glimm and A. Jaffe, "Boson Quantum Field Models," in the 1971 London lectures, edited by R. Streater (to appear).

3 J. Glimm and A. Jaffe, Acta Math **125**, 203 (1970).

4 J. Glimm and A. Jaffe, Ann. Math. **91**, 362 (1970).

5 J. Glimm and A. Jaffe, J. Math. Phys. **11**, 3335 (1970); Commun. Math. Phys. **22**, (1971).

6 J. Glimm and A. Jaffe, Commun. Math. Phys. **22**, 253 (1971).

Positivity of the $\varphi_3{}^4$ Hamiltonian

JAMES GLIMM[1])

Courant Institute of Mathematical Science, New York University, New York, N. Y.

ARTHUR JAFFE[2])

Lyman Laboratory of Physics, Harvard University, Cambridge, Mass.

Abstract

The renormalized $\varphi_3{}^4$ Hamiltonian is bounded from below by a constant proportional to the volume.

Contents

---

[1]) Supported in part by the National Science Foundation, under Grant NSF GP 24002.
[2]) Supported in part by the National Science Foundation, under Grant NSF GP 31239 X and by the AFOSR under Contract AF 44620–70–C0030.

1. Introduction

1.1. Discussion

We prove that the renormalized φ_3^4 Hamiltonian is bounded from below by a constant proportional to the volume. The renormalization constants in our Hamiltonian $H(g)$, with a spatial cutoff g, are given by perturbation theory; they are the ultraviolet divergent contributions to the ground state energy and to the rest mass of a single particle. After subtraction of a finite vacuum energy constant $E(g)$, we obtain a positive, densely defined Hamiltonian. The analogous positivity result is well known for the two dimensional $\mathscr{P}(\varphi)_2$ and Yukawa$_2$ Hamiltonians. A subsequent development of the φ_3^4 model might proceed along the lines of the $\mathscr{P}(\varphi)_2$ and Yukawa$_2$ models. In this case the lower bound on $H(g)$ would be a major step toward constructing a self adjoint operator $H(g)$, which defines dynamics with finite propagation speed.

The main difficulty in the positivity proof lies with the divergences. Since the ultraviolet divergences become more severe with increasing numbers of space dimensions, our proof of positivity for the φ_3^4 Hamiltonian is considerably more difficult than for the two dimensional models. The only "superrenormalizable" boson self interactions are $\mathscr{P}(\varphi)_2$ and φ_3^4. The next most singular self interaction in three dimensions is φ_3^6 which is comparable in difficulty to φ_4^4. Our present methods seem to be marginal for such "renormalizable, but not superrenormalizable" interactions, and new methods are presumably needed for them.

The Hamiltonian $H(g)$ is defined as a limit (as $\varkappa \to \infty$) of approximate, momentum cutoff Hamiltonians $H(g, \varkappa)$. The cutoff operators are self adjoint and bounded from below on Fock space. The first result about the φ_3^4 Hamiltonian was the proof [4] that the sequence $\{H(g, \varkappa)\}$ yields a densely defined limiting operator $H(g)$. The domain of $H(g)$ lies in a Hilbert space $\mathscr{F}(g)$ disjoint from Fock space. The non Fock representation of the fields on $\mathscr{F}(g)$ was studied by Hepp [9], Fabrey [3], Eckmann and Osterwalder [2], and Eckmann [1].

In the present paper we show that $\{H(g, \varkappa)\}$ is bounded from below, uniformly in $\varkappa$, yielding a lower bound for $H(g)$. Let g be a $C_0^\infty(R^2)$ function, $0 \leqq g \leqq 1$, and let $A(g)$ denote the area of the set of points within distance 1 of the support of g. Let $E(g, \varkappa)$ be the vacuum energy of $H(g, \varkappa)$, namely

$$E(g, \varkappa) = \inf \text{spectrum } H(g, \varkappa). \tag{1.1.1}$$

Theorem 1.1.1. *There exists a positive constant, independent of g and $\varkappa$, such that*

$$-\text{const } A(g) \leqq E(g, \varkappa). \tag{1.1.2}$$

With g held fixed, it follows from [4] that $E(g, \varkappa)$ is bounded from above. We expect, but have not proved, that as $g \to 1$,

$$E(g, \varkappa) \leqq \text{const } A(g). \tag{1.1.3}$$

The bound (1.1.3) is used in the $\mathscr{P}(\varphi)_2$ and Yukawa$_2$ models, in passing to limit $g = 1$.

The cutoff operator $H(g, \varkappa)$ has a unique vacuum vector $\Omega(g, \varkappa)$, i.e. $E(g, \varkappa)$ is a simple eigenvalue of $H(g, \varkappa)$. The uniqueness proof [5] shows that $\Omega(g, \varkappa)$ is not orthogonal to the vacuum vector Ω_0 of H_0. Hence,

$$-E(g, \varkappa) = \lim_{t\to\infty} \frac{\log \langle \Omega_0, \exp(-t\, H(g, \varkappa))\, \Omega_0\rangle}{t}. \tag{1.1.4}$$

We prove Theorem 1.1.1, by showing that there exists a constant 0(1), independent of g, $\varkappa$ and t, such that for $t \geqq 1$,

$$\langle \Omega_0, \exp(-tH(g, \varkappa))\, \Omega_0\rangle \leqq \exp(0(1)\, tA(g)). \tag{1.1.5}$$

We use two basic expansions to estimate (1.1.5). Our high momentum expansion, the "perturbation-contraction" expansion, is given in Chapter 2. This expansion is a renormalized perturbation expansion for $\langle \exp(-tH(g, \varkappa))\rangle$, which exhibits the renormalization cancellations in an explicit fashion. The perturbation-contraction expansion is insufficient by itself, since the perturbation series for $\langle \exp(-t\, H(g, \varkappa))\rangle$ diverges. The second expansion, the Wick expansion, gives a bound on the low momentum part of the interaction. It is based on the Wick ordering bound

$$-0(\varkappa^2) \leqq\ :\varphi_x^4(\boldsymbol{x}):, \tag{1.1.6}$$

obtained from: $\varphi_\varkappa^4: = \varphi_\varkappa^4 - 6\varphi_\varkappa^2 \langle \varphi_\varkappa^2\rangle + 3\langle \varphi_x^2\rangle^2$, where $\langle \varphi_\varkappa^2\rangle = \langle \Omega_0, \varphi_x(\boldsymbol{x})^2\, \Omega_0\rangle = 0(\varkappa)$. Combining these two methods, we obtain (1.1.5).

The basic idea of our method is to perform the perturbation-contraction expansion for portions of the Hamiltonian that are localized in sufficiently small space regions and time intervals. We use (1.1.6) to obtain a convergent expansion for $\exp(-tH)$ in each region, and we show that different regions are essentially independent. We localize the fields $\varphi(\boldsymbol{x})$ in H_I, or in other words, we localize the particles that enter the interaction. For our present problem, we find it necessary to localize the particles simultaneously in configuration space and in momentum space, namely in phase space. Of course, such localization can not be exact, but it can be given by some function f so that both f and its Fourier transform $f^\sim$ are rapidly decreasing. We call such localization, polynomial localization in phase space. The exact form of our localization function $\zeta(x, k)$ is given in Section 1.2.

The diameter of our localization region is limited by the uncertainty principle

$$O(1) \leqq \delta x\, \delta p, \tag{1.1.7}$$

for each position and momentum coordinate. In particular, at high values of the momentum, $|p| > M$, we can impose a small spreading, $O(M^{-1})$, of the configuration space localization due to the momentum, space localization. Thus if $M = O(\varkappa)$, the minimum volume of space localization volume compatible with the uncertainty principle is $O(\varkappa^{-s})$, where s is the number of space dimensions. Since our actual estimates deal with the localized action, we allow for time intervals of length δt as well, where $0(1) \leqq \delta E\, \delta t$. The minimum space-time volume $|\Delta|$ for a space time localization cube Δ is

$$|\Delta| = O\big((\delta x)^s\, \delta t\big) = O(\varkappa^{-s-1}). \tag{1.1.8}$$

In (1.1.5), the exponent $S = O(1)\, tA(g)$ has the dimensions of action. In perturbation theory, for a superrenormalizable interaction, the vacuum energy satisfies

$$|E_\varkappa| \leqq A(g)\, O(\varkappa^s), \tag{1.1.9}$$

as $\varkappa \to \infty$. Since the vacuum energy is the most divergent renormalization constant, the bound on S, neglecting renormalization cancellations, is

$$|S_\varkappa| \leqq tA(g)\, O(\varkappa^s). \tag{1.1.10}$$

We show that $S_\varkappa$ is approximately additive,

$$S_\varkappa \approx \sum S_\varkappa(\Delta),$$
$$S_\varkappa(\Delta) \approx \big(|\Delta|/tA(g)\big)\, S_\varkappa \leqq 0\, (|\Delta|\, \varkappa^s), \tag{1.1.11}$$

by (1.1.10). We conclude from (1.1.8), that the cubes Δ of minimum size satisfy $S_\varkappa(\Delta) \leqq O(\varkappa^{-1})$, as $\varkappa \to \infty$. Associated with the approximate additivity of S_x is the approximate factorization of the physical vacuum state Ω with a single factor $\Omega(\Delta)$ for each region Δ. Corresponding to the small contribution, $S(\Delta)$, of Δ to S, is the fact that at most $O(1)$ bare particles, with momenta greater than $O(|\Delta|^{-1/(s+1)})$, contribute to the factor $\Omega(\Delta)$ and to Ω. We use the bound (1.1.11), along with convergence from the perturbation-contraction expansion in Δ. These ideas yield (1.1.5), and in principle they apply to superrenormalizable theories in general. For interactions that are renormalizable, but not superrenormalizable, these ideas become marginal, and hence their applicability is not clear. For instance, the vacuum energy in this case satisfies $|E| \leqq O\big(A(g)\, \varkappa^{s+1}\big)$ in perturbation theory. Hence $S_x(\Delta) \approx O(1)$ for cubes Δ of minimum size, and we lose control over the approximations in the above analysis. As in the present paper, it seems that the portions of H_I which couple high and low momentum modes will be the most difficult to control. (Various such problems arise in Chapters 2 and 3 here.) Indeed, this part of H_I gives rise to the mass, field strength and charge renormalizations. However, we note that this portion of H_I is superrenormalizable, although H_I itself is not.

Our bound for $\langle \exp(-tH(g, \varkappa))\rangle$ is defined by an inductive construction [6]. At each stage of the proof we dominate $\langle \exp(-tH)\rangle$ by a sum of terms of the form

$$\int ds \langle TR(s) \exp\big(-\int_0^t d\sigma\, H(\sigma, s)\big)\rangle, \tag{1.1.12}$$

where $R(s)$ is a polynomial in the fields, $H(\sigma, s)\big)$ is a time dependent Hamiltonian, and T is a time ordering operator for time parameters s and σ. Each inductive step or substep replaces (1.1.12) by a sum of similar terms bounding (1.1.12) from above. The induction continues for a particular term until $H(\sigma, s) \equiv 0$ for that term. If $H(\sigma, s) = 0$, then (1.1.12) has the form of a sum of vacuum Feynman graphs. Our expansion is defined in Chapters 2 and 3, and we estimate the sum of terms (1.1.12) in Chapters 4, 5 and 6.

1.2. Localization of Wick polynomials

For mathematical properties of operators and forms on Fock space, see [7—8]. We define a Euclidean field $\varphi(x)$, for $x \in R^3$, by

$$\varphi(x) = (2\pi)^{-3/2} \int e^{-ikx}\, \varphi(k)^\sim\, dk,$$
$$\varphi(k)^\sim = (k^2 + m^2)^{-1/2} \{A(k)^* + A(-k)\}.$$

Here the Euclidean annihilation and creation operators $A(k)$ satisfy $[A(k), A(k')] = 0$, $[A(k), A(k')^*] = \delta^3(k - k')$. This Euclidean field reduces at time zero to the usual time zero free field,

$$\varphi(\boldsymbol{x}) \equiv \varphi(\boldsymbol{x}, x_0 = 0) = (2\pi)^{-1} 2^{-1/2} \int e^{-i\boldsymbol{k}\boldsymbol{x}} \mu(\boldsymbol{k})^{-1/2} \{a(\boldsymbol{k})^* + a(-\boldsymbol{k})\} d\boldsymbol{k},$$

defining $a(\boldsymbol{k})$ in terms of $A(k)$. Then $[a(\boldsymbol{k}), a(\boldsymbol{k}')^*] = \delta^2(\boldsymbol{k} - \boldsymbol{k}')$.
For suitable functions ζ, defined on the relativistic phase space $R^3 \times R^3$, we define a localized Euclidean field $\varphi(\zeta)$ by

$$\varphi(\zeta) = (2\pi)^{-3/2} \int e^{-ikx} \zeta(x, k) \varphi(k)^{\sim} dk\, dx. \tag{1.2.1}$$

For a Wick polynomial $:\varphi^j:$ we permit j distinct mo mentum localizations, but only one space-time localization. We define

$$:\varphi^j(\zeta): = (2\pi)^{-3j/2} \int e^{-i\left(\sum_{l=1}^{j} k_l\right)x} \zeta(x, k) : \prod_{l=1}^{j} \varphi^{\sim}(k_l) : dk\, dx, \tag{1.2.2}$$

where $\zeta(x, k)$ is a function of $x \in R^3$ and $k = (k_1, \ldots, k_j) \in R^{3j}$. If (1.2.2) is hermitian, $\zeta(x, k) = \zeta(x, -k)$. In later sections of the pa per, we set $m = 1$ for convenience.
We choose our momentum function η to be a C^∞ function on R^1 satisfying

$$\eta(\varrho) = 1, \qquad |\varrho| \leq \frac{1}{2}$$

$$0 \leq \eta(\varrho) \leq 1, \qquad \frac{1}{2} \leq \varrho \leq 2$$

$$\eta(\varrho) = 0, \qquad 2 \leq \varrho.$$

For $k \in R^3$ and a vector cutoff $\alpha \in (R^+)^3$, we set

$$\eta_\alpha(k) = \prod_{l=0}^{2} \eta(k^{(l)}/\alpha^{(l)}), \tag{1.2.3}$$

where l indexes the components of the momentum. With $\alpha^{(l)} = 0$, we define $\eta(k^{(l)}/\alpha^{(l)}) = 0$. With $\alpha^{(l)} \leq \beta^{(l)}$, the momentum cutoff

$$\eta_{\alpha,\beta}(k) = \eta_\beta(k) - \eta_\alpha(k) \tag{1.2.4}$$

provides an upper and lower cutoff in momentum space. In space time, we localize in cubes $\Delta \subset R^3$, with characteristic function χ_Δ. We allow $\zeta(x, k)$ in (1.2.1) – (1.2.2) to be a linear combination of functions of the form

$$h(x)\, \chi_\Delta(x) \prod_{r=1}^{j} \eta_{\alpha_r,\beta_r}(k_r) \tag{1.2.5}$$

where $h(x)$ is a space time cutoff. We define for a polynomial $:\varphi^j:$,

$$\lambda = \max_{l,r} \{\alpha_r{}^{(l)}, 2\}$$

as the maximum lower momentum cutoff.

It is convenient to restrict the allowed classes of cubes Δ and intervals $[\alpha^{(l)}, \beta^{(l)}]$. The allowed cubes have the form $2^{-j}\Delta_0 + 2^{-j}z$, where $z \in Z^3$, and Δ_0 is the cube with volume $|\Delta_0| = 1$, centered at the origin. For a space time cutoff h, we define vol (h) as the number of translates $\Delta_0 + z$, $z \in Z^3$, that intersect support h.
The allowed momentum cutoffs $\alpha^{(l)}$, $\beta^{(l)}$ are chosen from the sequence $\{M_0, M_1, \ldots, \infty\}$ where $M_0 = 0$, M_1 is chosen sufficiently large, and

$$M_{j+1} = M_j^{l+\nu} = M_1^{(l+\nu)^j}. \tag{1.2.6}$$

In later chapters we choose $\nu > 0$, sufficiently small.
According to the uncertainty principle, particles created by (1.2.1) or (1.2.2) are localized in a cube with length

$$|\Delta|^{1/3} + M_j^{-1} \tag{1.2.7}$$

where M_j is the lower momentum cutoff for the particle in question and hence determines spreading. If

$$M_j^{-1} \leqq |\Delta|^{1/3}, \tag{1.2.8}$$

we call the localization proper, and otherwise we call it improper. We introduce the localization ratio

$$b = \max\{1, |\Delta|^{-1/3} M_j^{-1}\}, \tag{1.2.9}$$

and in both cases the particles are localized in cubes of length

$$0(1)\, b\, |\Delta|^{1/3}. \tag{1.2.10}$$

We are also interested in measuring the effective strength of the force between two particles localized in different cubes Δ, Δ', and in momentum ranges with lower cutoffs M_j, $M_{j'}$. This force is expressed in terms of the scaled distance d, which we define to be

$$d = \max\{1, \text{dist.}(\Delta, \Delta'), M_j \,\text{dist.}(\Delta, \Delta'), M_{j'} \,\text{dist.}(\Delta, \Delta')\}. \tag{1.2.11}$$

In case both lower momentum cutoffs are zero, the scaled distance d reduces to $\max\{1, \text{dist.}(\Delta, \Delta')\}$, where dist. (Δ, Δ') is the Euclidean distance. We have chosen a dimensionally inhomogeneous definition, to emphasize the ultraviolet aspects of d.

1.3. The renormalized Hamiltonian

We study Hamiltonians of the form

$$H(g, \varkappa) = H_0 + \int : \mathscr{P}\left(\varphi_\varkappa(\boldsymbol{x}), g(\boldsymbol{x})\right) : d\boldsymbol{x},$$

where $\mathscr{P}$ is a quartic interaction plus counterterms. In particular we bound $\langle \Omega_0, \exp(-t\, H(g, \varkappa))\, \Omega_0\rangle$. It is convenient to relate this expression to the Euclidean field $\varphi(x)$ of Section 1.2., for we then obtain a perturbation series directly in terms of Euclidean Feynman propagators, $(k^2 + m^2)^{-1}$. We use the "relativistic Feynman-Kac formula" investigated by SYMANZIK [11] and NELSON [10],

$$\begin{aligned} &\langle \Omega_0, \exp(-t\, H(g, \varkappa))\, \Omega_0\rangle \\ &\quad = \langle \Omega_0, \exp\left(-\int_0^t dx_0 \int d\boldsymbol{x} : \mathscr{P}(\varphi_x(x)\, g(\boldsymbol{x})) :\right) \Omega_0\rangle \\ &\quad = \langle \Omega_0, \exp(-V)\, \Omega_0\rangle . \end{aligned}$$

Thus we define the Hamiltonian by giving the renormalized, Euclidean action, $V = V_I + V_C$. Here V_I and V_C are Wick polynomials in the Euclidean field $\varphi(x)$. If the renormalization constants are time independent, if the space time cutoff of Section 1.2. has the form

$$h(x) = g(\boldsymbol{x})\, \chi_{[0,t]}(x_0), \tag{1.3.1}$$

and if the momentum cutoff η of Section 1.2. is independent of time and energy, then $\exp(-V)$ has an interpretation in terms of a time independent Hamiltonian. We call such a cutoff in V a "Hamiltonian cutoff", and we assume $g(\boldsymbol{x}) \in C_0^\infty(R^2)$, or else that $h(\boldsymbol{x})$ is the characteristic function of a union of translates of Δ_0. Our estimates for $\exp(-V)$ also apply for $h(x) \in C_0^\infty(R^3)$. We begin our proof of Theorem 1.1.1 with a Hamiltonian cutoff, but in the inductive construction we replace ζ by the more general cutoffs of Section 1.2.

It is convenient to define the counterterm V_C in a slightly different manner than suggested by the Hamiltonian. Rather, we use the vacuum energy renormalization usually associated with the Euclidean Green's functions

$$\langle \Omega_0, \varphi(x_1) \ldots \varphi(x_n) \exp(-V)\Omega_0 \rangle.$$

We show below that our Green's function renormalization differs from the Hamiltonian renormalization by a wave function renormalization constant, and a transient that vanishes as $t \to \infty$. Hence we obtain estimates on the Hamiltonian. The advantage of using a Green's function counterterm, rather than a local Hamiltonian counterterm, is that we obtain exact cancellation of the second and third order vacuum graphs in our expansion of $\exp(-V)$.

For a cutoff ζ, we define

$$V_I = :\varphi^4(\zeta):. \tag{1.3.2}$$

Then the vacuum energy counterterms are

$$E_2(\zeta) = -\frac{1}{2} \langle V_I^2 \rangle_{\Omega_0}, \tag{1.3.3}$$

$$E_3(\zeta) = (1/6) \langle V_I^3 \rangle_{\Omega_0},$$

where the expectation is in the (Euclidean) no particle state Ω_0. If

$$\zeta(x, k) = \zeta_0(\boldsymbol{x}, \boldsymbol{k})\, \chi_{[0,t]}(x_0) = h(x)\eta(\boldsymbol{k}), \tag{1.3.4}$$

then the energy integrals in (1.3.3) are elementary and can be evaluated by residues. This gives

$$E_2(\zeta) = t\bar{E}_2(\zeta_0) + \Lambda_2(\zeta_0) + T_2(\zeta_0, t) \tag{1.3.5}$$

where we now identify $\bar{E}_2$ as the Hamiltonian counterterm in second order, Λ_2 as the wave function renormalization in second order, and $T_2(\zeta_0, t)$ as a transient. In fact, for the cutoff (1.3.4),

$$V_I = \int_0^t dx_0 :\varphi^4(\zeta_0, x_0):, \tag{1.3.6}$$

and if $H_I = :\varphi^4(\zeta_0, x_0 = 0):$, we obtain

$$\begin{aligned} \bar{E}_2(\zeta_0) &= -\langle H_I H_0^{-1} H_I \rangle_{\Omega_0} \\ &= 4!2^{-4} (2\pi)^{-13/2} \int (\mu_1\mu_2\mu_3\mu_4)^{-1} (\mu_1 + \cdots \mu_4)^{-1} |\zeta_0(\textstyle\sum \boldsymbol{k}_i, \boldsymbol{k})|^2\, d\boldsymbol{k}, \end{aligned} \tag{1.3.7}$$

where $\mu(k) = (k^2 + m^2)^{1/2}$, and where $\tilde{\zeta}_0$ denotes the partial Fourier transform of $\zeta_0(\boldsymbol{x}, \boldsymbol{k})$ in $\boldsymbol{x}$. Similarly, we find that

$$A_2(\zeta_0) = \langle H_I H_0^{-2} H_I \rangle_{\Omega_0}, \tag{1.3.8}$$

$$T_2(\zeta_0, t) = -\langle H_I H_0^{-2} e^{-tH_0} H_I \rangle_{\Omega_0}. \tag{1.3.9}$$

Thus we see that $\bar{E}_2$ is the exact second order Hamiltonian counterterm for the vacuum energy. Since A_2 enters (1.3.5) independent of t, it does not provide a Hamiltonian counterterm. Rather, it is the second order contribution to the wave function renormalization. Let $\Omega(\zeta_0)$ be the ground state of

$$\begin{aligned} H(\zeta_0) &= H_0 + H_I(\zeta_0) + C(\zeta_0) \\ &= H_0 + H_I(\zeta_0) - \bar{E}_2(\zeta_0) - \bar{E}_3(\zeta_0) - \bar{M}_2(\zeta_0). \end{aligned} \tag{1.3.10}$$

Then we define the wave function renormalization as

$$Z(\zeta_0) = |\langle \Omega_0, \Omega(\zeta_0) \rangle|^2 = \exp\big(-A(\zeta_0)\big),$$

and in second order perturbation theory, $A(\zeta_0) = A_2(\zeta_0)$. Finally, we note that $T_2(\zeta_0, t)$ of (1.3.9) is bounded uniformly in $\varkappa$ for t bounded away from zero, and $T_2(\zeta_0, t) \to 0$ exponentially as $t \to \infty$.

Similarly, we find that

$$E_3(\zeta) = t\bar{E}_3(\zeta_0) + A_3(\zeta_0) + T_3(\zeta_0, t), \tag{1.3.11}$$

where $\bar{E}_3(\zeta_0)$ is the third order Hamiltonian counterterm for the vacuum energy $\langle H_I H_0^{-1} H_I H_0^{-1} H_I \rangle$, where $A_3(\zeta_0)$ is the third order wave function renormalization constant, and where $T_3(\zeta_0, t) \to 0$ exponentially as $t \to \infty$. The respective ultraviolet behavior is $0(\log \varkappa)$ for $\bar{E}_3$ and $0(1)$ for A_3 and T_3.

Next we consider the mass renormalization counterterm, which is logarithmically divergent in second order. In a cube Δ, the momentum cutoff $\eta = \eta(k)$ has the form

$$\eta(k) = \prod_{i=1}^{4} \eta_i(k_i),$$

cf. (1.2.5). Furthermore $\eta_1 = \eta_2 = \eta_3 = \eta_4$, as we will see in (2.7). We define

$$\delta m^2(\eta) = -4^2 \times 6 \times (2\pi)^{-9} \int \delta(k_2 + k_3 + k_4) \left(\prod_{i=2}^{4} \mu_i^{-2}\right) \eta_i(-k_i)\, \eta_i(k_i)\, dk_2\, dk_3\, dk_4 \tag{1.3.12}$$

and

$$M_2(\zeta) = \frac{1}{2} \Sigma_\Delta\, \delta m^2(\eta) :\varphi_{\eta_1}^2 (h^2 \chi_\Delta):. \tag{1.3.13}$$

In the case of a Hamiltonian cutoff, $M_2(\zeta)$ is an integral (from 0 to t), and the time zero integrand, denoted $\bar{M}_2(\zeta_0)$ is the Hamiltonian counterterm.

With this definition of $\bar{M}_2(\zeta_0)$, the renormalized Hamiltonian is given by (1.3.10). Similarly we define the renormalized action V by

$$\begin{aligned} V(\zeta) &= V_I(\zeta) + V_C(\zeta) \\ V_C(\zeta) &= -E_2(\zeta) - E_3(\zeta) - M_2(\zeta). \end{aligned} \tag{1.3.14}$$

Our bound for $\exp\big(-V(\zeta)\big)$ thus differs from the corresponding bound for

$\exp\left(-tH(\zeta_0)\right)$ by a scalar factor. With the definitions above,

$$\langle\Omega_0, e^{-tH(g,\varkappa)}\,\Omega_0\rangle \tag{1.3.15}$$
$$= \langle\Omega_0, e^{-V}\,\Omega_0\rangle \exp\left(-\Lambda_2(\zeta_0) - \Lambda_3(\zeta_0) - T_2(\zeta_0, t) - T_3(\zeta_0, t)\right) \leqq \langle\Omega_0, e^{-V}\,\Omega_0\rangle,$$

for $\varkappa$ sufficiently large.

Finally, we note that the Hamiltonian interpretation of $\exp(-V)$ requires that the momentum cutoff η in (1.3.4) be energy independent. For Theorem 1.1.1, we desire a uniform bound on $\exp\left(-V(\zeta)_\varkappa\right)$ where $\varkappa$ is the upper cutoff in the momentum only. In the remainder of the paper, however, it is convenient to have an overall upper cutoff $\varkappa$ in both the energy and in the momentum. The inductive proof of Chapters 2—6 is equally valid if we fix the cutoff in the momentum at $\varkappa$, and we let $\varkappa_0$ denote the energy cutoff. (Note our cutoffs (1.2.3) can be chosen separately in each component.) The resulting bound is uniform in $\varkappa_0$ and $\varkappa$, and

$$\exp\left(-V(\zeta)_{\varkappa_0,\varkappa}\right) \leqq \exp O\left(tA(g)\right). \tag{1.3.16}$$

Furthermore, with $\varkappa$ fixed, $V(\zeta_{\varkappa_0,\varkappa})$ is bounded from below, uniformly in $\varkappa_0$, by a standard Wick ordering bound. This yields $\exp\left(-V(\zeta)^{\varkappa_0,\varkappa} \to \exp\left(-V(\zeta_0)_\varkappa\right)\right.$ as $\varkappa_0 \to \infty$, and hence the bound (1.3.16) applies to the Hamiltonian as desired: $\exp\left(-V(\zeta_0)_\varkappa\right) \leqq \exp O\left(tA(g)\right)$.

2. A renormalized perturbation expansion

We use two simple identities, which we call the *perturbation formula* and the *contraction formula* to generate an asymptotic expansion for $\langle\exp(-tH)\rangle$. We call this the perturbation contraction $(P-C)$ expansion. The P formula generates a series in the renormalized interaction $V = V_I + V_C$. The C formula exhibits the cancellation of divergences between the unrenormalized interaction V_I and the counterterms V_C. The $P-C$ expansion is a partially renormalized expansion. In later stages of the expansion we allow certain uncancelled logarithmic divergences. They occur only in terms that have strongly convergent factors, introduced earlier in the expansion, which dominate these divergences.

We define our expansion for $\langle\exp(-V)\rangle$ inductively, and hence the name *inductive construction*. Each step consists of three main parts: A high momentum $(P-C)$ expansion given in this chapter, a low momentum (Wick) expansion in Chapter 3 and combinatoric estimates in Chapter 4 which bound the number of terms generated during these expansions. After these three steps, we decide for each term whether to terminate the expansion, or whether to return to the high momentum expansion and another inductive step. When the induction terminates for a particular term, we use the estimates of Chapters 5 and 6 to bound the term. Each such term is labeled by a Feynman graph G, and the term is an integral $I(G)$. These integrals are the familiar momentum space Feynman integrals that arise in perturbation calculations. These integrals occur, however, with a restricted momentum range and in particular we utilize the convergence from a lower cutoff. In this manner we convert an asymptotic series in the coupling constant into a convergent bound. Our final bound for $\langle\exp(-tH)\rangle$ is a convergent sum of upper bounds on Feynman graphs.

The purpose of the high momentum $(P-C)$ expansion is to lower the upper momentum cutoff in the exponent V. For this purpose, let ζ_s be a family of cutoffs ζ, where $0 \leqq s \leqq 1$, and ζ_0 has a lower value of the upper momentum cutoff than ζ_1.

The perturbation formula (P formula) is

$$e^{-V(\zeta_1)} = e^{-V(\zeta_0)} - \int_0^1 ds \frac{d}{ds} V(\zeta_s)\, e^{-V(\zeta_s)}, \tag{2.1}$$

valid since $V(\zeta_{s_0})$ and $V(\zeta_{s_1})$ are both functions of the Euclidean field, and hence commute. Thus the P formula expresses $\exp(-V)$ as the sum of a term with a lowered cutoff and a term involving dV/ds. We call dV/ds a perturbation vertex (P vertex). If we wish to assert that we applied (2.1) during the rth inductive step, we call dV/ds a P_r vertex, and we say that the vertex is introduced in the P_r—C_r expansion.

We exhibit the renormalization cancellations in $V = V_I + V_C$ only when V occurs as a P vertex. For this purpose we use the contraction formula (C formula),

$$A(k)\, e^{-V} = e^{-V} A(k) + [V, A(k)]\, e^{-V}, \tag{2.2}$$

or its adjoint. The C formula is valid because both V and $[V, A]$ are functions of the Euclidean field, and hence commute. We call $[V, A(k)]$ or $[A(k)^*, V]$ a contraction vertex (C vertex).

The perturbation formula (2.1) remains valid upon multiplication by a polynomial R in the Euclidean fields,

$$Re^{-V(\zeta_1)} = Re^{-V(\zeta_0)} - \int ds\, R \frac{dV(\zeta_s)}{ds} e^{-V(\zeta_s)}. \tag{2.3}$$

In the contraction formula, multiplication by R yields extra contracted terms, but no new vertices,

$$A(k)\, Re^{-V} = Re^{-V} A(k) + \{[A(k), R] + [V, A(k)]\, R\}\, e^{-V}. \tag{2.4}$$

We use (2.4) to move certain annihilation operators to the right, until they give $A(k)\,\Omega_0 = 0$ plus a sum of contracted terms. We use its adjoint to move certain $A(k)^*$'s similarly to the left in our expansion.

At each point in the inductive construction, we have an upper bound for $\langle \exp(-V)\rangle$ as a sum of terms. Each inductive step or substep replaces a single term by a sum of terms, each of which has the form

$$\sup_s |\langle R(s)\, e^{-V(\zeta_s)}\rangle|. \tag{2.5}$$

The P—C expansion is a finite sequence of substeps. In each substep we add a single P vertex, localized in a space time cube Δ, and we exhibit certain renormalization cancellations for this vertex. The expansion is specified once we give (a) the prescription for adding one P vertex and (b) the prescription for terminating the expansion.

Associated with each term (2.5) is a cover $\mathscr{D}$ of space time by nonoverlapping cubes Δ. For each cube $\Delta \in \mathscr{D}$ there is an upper cutoff $M_{u(\Delta)}$, a lower cutoff $M_{l(\Delta)}$ and an intensity $s(\Delta)$. The intensity lies in $[0, 1]$, and interpolates between the upper cutoffs $M_{u(\Delta)}$ and $M_{u(\Delta)-1}$. The s-dependent momentum cutoff for one leg is

$$\eta_{M_l,M_u}(s, k) = s\eta_{M_l,M_u}(k) + (1 - s)\, \eta_{M_l,M_{u-1}}(k), \tag{2.6}$$

while the cutoff ζ_s in $\exp\left(-V(\zeta_s)\right)$ is

$$\zeta(x, k) = \left(\prod_i \eta_{M_{l(\Delta)}, M_{u(\Delta)}}\left(s(\Delta), k_i\right)\right) (h\chi_\Delta)^\sim \left(\sum_i k_i\right) = \eta_s(k)\, (h\chi_\Delta)^\sim\, (\Sigma k_i) \tag{2.7}$$

in the cube Δ. At the beginning of the first inductive step we choose $s(\Delta) = 1$, $M_{l(\Delta)} = 0$, $M_{u(\Delta)-1} \leqq \varkappa \leqq M_{u(\Delta)}$.
The space time cover at the start of the first inductive step is

$$\{\Delta_0 + \alpha : \alpha \in Z^3\}. \tag{2.8}$$

Next we explain the substep of adding a single P vertex in Δ. With l, u, s and ζ given. we apply (2.1) with s replaced by

$$s'(\Delta) \in [0, s(\Delta)],$$

provided $s(\Delta) > 0$. If $s(\Delta) = 0$, the first term in (2.1) was selected in the previous application of (2.1) in Δ. We then replace $s(\Delta)$ by 1, $u(\Delta)$ by $u(\Delta) - 1$ and proceed as before.
Next we prescribe the renormalization cancellations. (A) First we use (2.4) to contract all legs of the new P vertex. (B) In most cases, the cancellation is completed by contracting all legs of all new C vertices formed by (A). We make three exceptions to (B):
(B 1) If the P vertex in (A) is a mass counterterm in V_C, then the legs of the new C vertices from A are not contracted, namely (B) is not performed.
(B 2) If the P vertex in (A) contracts three times to one new C vertex and once to a second new C vertex, a mass subdiagram is formed. In order to permit cancellation with the counterterm in (B 1), we contract the remaining leg of the mass subdiagram, i.e. the fourth leg of the first new C vertex, but do not contract the three remaining legs of the second new C vertex.
(B 3) If the P vertex contracts three times to old vertices and once to a new C vertex in (A), then we apply (2.4) "in reverse". We write this term as a term in which a single P leg is uncontracted, plus terms in which all four legs contract to old vertices. This completes the C expansion associated with a single P vertex.
Now we describe which terms are combined to cancel, yielding a convergent remainder. The second and third order vacuum energy dagrams formed from a P vertex and one or two C vertices cancel exactly with the δE_2 and δE_3 counterterm contributions to the P vertex, see (1.3.3). Graphically, these terms are

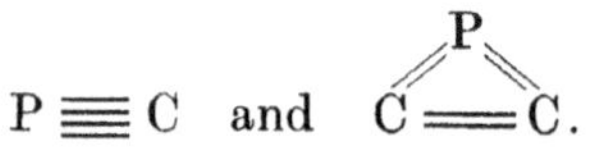

Turning to the mass renormalization, we combine the counterterm in the P vertex with the mass term containing a P vertex contracted three times to a new C vertex, graphically

$$—\mathrm{P}\equiv\mathrm{C}— \; + \; —\mathrm{M}—.$$

Let the P vertex be localized in Δ and the C vertex in Δ'. Since Δ and Δ' belong to the same cover, they do not overlap. If $\Delta \neq \Delta'$, the counterterm is zero and the combined mass subgraph has a kernel

$$w_{\Delta, \Delta'}(k_1, k_5) = \text{const.} \int \alpha\beta\gamma \, dk_2\, dk_3\, dk_4 .$$

If $\Delta = \Delta'$, it has a kernel

$$w_{\Delta,\Delta}(k_1, k_5) = \text{const.} \int (\alpha\beta\gamma - \bar{\alpha}\bar{\beta}\bar{\gamma})\, dk_2\, dk_3\, dk_4. \tag{2.10}$$

In both cases,

$$\alpha = \left(\frac{d}{ds}\eta_s\right)(k_1, k_2, k_3, k_4)\, \eta_s(k_2, k_3, k_4, k_5),$$

$$\beta = \bar{\beta} = (h\chi_\Delta)^\sim(-P - k_1)\,(h\chi_{\Delta'})^\sim(P + k_5)$$

$$\gamma = (\mu_1\mu_2^2\mu_3^2\mu_4^2\mu_5)^{-1}$$

$$\bar{\alpha} = \frac{1}{2}\frac{d}{ds}\eta_s(0, p_2, p_3, p_4)\,\eta_s(p_2, p_3, p_4, 0)\,\eta_1(k_1)\,\eta_1(k_5)$$

$$\bar{\gamma} = \left(\mu_1\,\mu(p_2)^2\,\mu(p_3)^2\,\mu(p_4)^2\,\mu_5\right)^{-1},$$

and $\mu_i = \mu(k_i)$, $\quad p_i = k_i - (1/3)P$, $\quad P = k_2 + k_3 + k_4$.

Finally, we specify when to terminate the $P-C$ expansion in a particular inductive step. Let $n_r(\Delta)$ be the number of P vertices that are introduced during the the rth inductive step and are localized in Δ. Let $\mathscr{D}_e$ be the set of cubes for which the exponent is nonzero. We permit a new P_r vertex in Δ provided Conditions (a)—(c) below are satisfied. If there are no cubes Δ satisfying Conditions (a)—(c), the $P-C$ expansion terminates for this inductive step, and the inductive construction continues according to Chapter 3.

(a) $\Delta \in \mathscr{D}_e$, $\quad u(\Delta) \geqq 2$

(b) $n_1(\Delta) \leqq M^\varepsilon_{u(\Delta)}$, if $r = 1$,

$\quad n_r(\Delta) \leqq |\Delta|^{-\varepsilon/2}$, if $r > 1$,

(c) $\sup_{\Delta' \in \mathscr{D}_e} M^\varepsilon_{u(\Delta')}\, d(\Delta, \Delta')^{-4} \leqq c M^\varepsilon_{u(\Delta)}$, if $r = 1$.

Here $d(\Delta, \Delta') = 1 + r(\Delta, \Delta')$, and $r(\Delta, \Delta')$ is the distance from the center of Δ to the center of Δ'. The constant c is

$$c = \sum_{\alpha \in Z^3} d(\Delta_0, \Delta_0 + \alpha)^{-4}.$$

The Condition (c) ensures that the upper cutoff does not vary too rapidly from cube to cube.

Let $\mathscr{D}_a$, $\mathscr{D}_c$ be the sets of cubes violating (a), (c). We claim that the sup over $\Delta' \in \mathscr{D}_e$ in Condition (c) equals the sup over $\Delta' \in \mathscr{D}_e \sim \mathscr{D}_c$. (We note that $\mathscr{D}_e \neq \mathscr{D}_c$, by nature of the expansion.) In fact, suppose $r = 1$, suppose a cube Δ' stops the expansion in Δ through Condition (c), and suppose Δ'' stops the expansion in Δ'. Then

$$d(\Delta, \Delta'') \leqq 1 + r(\Delta, \Delta') + r(\Delta', \Delta'') \leqq d(\Delta, \Delta')\, d(\Delta', \Delta''). \tag{2.11}$$

Thus

$$\begin{aligned} M^\varepsilon_{u(\Delta)} &\leqq \left(d(\Delta, \Delta')^4\, c\right)^{-1} M^\varepsilon_{u(\Delta')} \\ &\leqq \left(d(\Delta, \Delta')^4\, d(\Delta', \Delta'')^4\, c^2\right)^{-1} M^\varepsilon_{u(\Delta'')} \\ &\leqq \left(d(\Delta, \Delta'')^4\, c\right)^{-1} M^\varepsilon_{u(\Delta'')}, \end{aligned}$$

and so Δ'' also stops the expansion in Δ. This transitivity property proves our assertion.
At the end of the $P_1 - C_1$ expansion, each cube $\Delta' \in \mathscr{D}_e \sim \mathscr{D}_c$ has $M^{\varepsilon}_{u(\Delta')}$ vertices, or else $\Delta' \in \mathscr{D}_a$. If $\mathscr{D}_a \neq \mathscr{D}_e$, there exist $\Delta' \in \mathscr{D}_e \sim \mathscr{D}_c$ with $M^{\varepsilon}_{u(\Delta')}$ P_1 vertices. We "assign"

$$\frac{1}{2} M^{\varepsilon}_{u(\Delta')} \left(d(\Delta, \Delta')^4 c\right)^{-1} \geqq \frac{1}{2} M^{\varepsilon}_{u(\Delta)} \tag{2.12}$$

of the Δ' vertices to Δ, and we "assign" $1/2\ M^{\varepsilon}_{u(\Delta')}$ of the Δ' vertices to Δ'. In this way, each cube $\Delta \in \mathscr{D}_e$ has at least $1/2\ M^{\varepsilon}_{u(\Delta)}$ vertices assigned to it, each with a maximum lower cutoff of at least $M_{u(\Delta)-1}$. We count the vertices assigned to Δ in $n_1(\Delta)$ in Chapters 3 and 4, and we treat these vertices as if they were in Δ. The case $\mathscr{D}_a = \mathscr{D}_e$ occurs only in the first inductive step, and it is handled separately in Chapter 3.

3. The low momentum expansion

3.1. Introduction

The lower bound for H rests on two expansions. Each expansions bounds a term

$$\sup_s \langle R(s)\, e^{-V(s)} \rangle \tag{3.1.1}$$

by a sum of similar terms, where $R(s)$ and $V(s)$ are polynomials in the fields, and the expectation is taken in the no particle state.
The high momentum expansion ($P - C$ expansion) was given in Chapter 2. After cancellation of the renormalization counterterm, we obtain a convergence factor from each integration over a high momentum P vertex in the polynomial R.
The low momentum expansion of this chapter results in raising the lower momentum cutoff in the exponent V, and the expansion has two parts. The first, the Wick construction, is based on the fact that $:\varphi^4:$ is formally positive. By removing the Wick ordering in $:\varphi^4:$, we bound the low momentum part of $\exp(-:\varphi^4:)$ from above.
We obtain convergence of the Wick construction by introducing Wick vertices (polynomials) into R in order to estimate the subsets of path space on which $:\varphi^4:$ is very negative. The Wick construction gives an upper bound for (3.1.1) and it depends on positivity. The polynomial $R(s)$ in (3.1.1) may be assumed positive, for we can use the "squaring" bound

$$|R(s)| \leqq \frac{1}{2} \{\delta^{-1} + \delta R(s)^2\} \tag{3.1.2}$$

where the constant δ depends on $R(s)$. We postpone the choice of δ to Chapter 4. (See Case S.) It is sometimes convenient to regard the constant term $(2\delta)^{-1}$ in (3.1.2) as a graph: we define this term as a *squaring graph*. This graph has vertices, but no legs. It has one squaring vertex for each P vertexin $R(s)$ introduced since the previous squaring. These squaring vertices merely serve a bookkeeping purpose. By convention, if G is a squaring graph, we let $I(G) = 1$ and multiply it by the constant $(2\delta)^{-1}$.

The second part of the low momentum expansion, the low-momentum contraction, is performed after completion of the Wick expansion. This expansion reduces the degree of the polynomial $R(s)$. The purpose of this expansion is to avoid squaring the low momentum vertices too many times. The convergence associated with the low momentum regions in early inductive steps is limited, and each use of the squaring formula (3.1.2) increases the divergent combinatoric factors. These factors, introduced in Chapter 4, are used to estimate the number of terms in our expansion.

3.2. The Wick construction

For any space time cube Δ', we define $l(\Delta')$ to be the largest integer for which

$$|\Delta'|\, M^2_{l(\Delta')} \leqq 1. \tag{3.2.1}$$

Note that for $|\Delta'| = 1$, $l(\Delta') = 0$ and $M_{l(\Delta')} = 0$, as at the start of the induction. In the Wick construction, we are given a cube Δ', and a cutoff ζ in the exponent V in (3.1.1). Furthermore in Δ', ζ has a lower momentum cutoff at $M_{l(\Delta)}$. Here Δ is some cube containing Δ', and so $M_{l(\Delta)} \leqq M_{l(\Delta')}$. The purpose of the Wick construction is to increase the lower momentum cutoff in Δ' from $M_{l(\Delta)}$ to $M_{l(\Delta')}$. Let ζ' be the new cutoff function with the desired lower momentum cutoff in Δ', and define $\delta\zeta$ and δV by

$$\zeta = \zeta' + \delta\zeta$$

$$V(\zeta) = V(\zeta') + \delta V.$$

Our basic inequality is

$$e^{-V(\zeta)} \leqq K_1 e^{-V(\zeta')} + (1-\chi)\, e^{-V(\zeta)}. \tag{3.2.2}$$

Here K_1 is a constant, independent of V, ε and M_1 and chosen below (3.2.22), and χ is the characteristic function of the subset of path space on which

$$e^{-\delta V} \leqq K_1.$$

For cubes below the minimum size

$$|\Delta'|\, M^2_{u(\Delta')} \leqq 1, \tag{3.2.3}$$

$M_{l(\Delta')} = M_{u(\Delta')}$ and $\zeta' = O$ on Δ'. In this case

$$-O(1) = -O(|\Delta'|)\, M^2_{u(\Delta')} \leqq \delta V$$

and so $e^{-\delta V} \leqq K_1$ on all of path space. Thus for cubes satisfying (3.2.3), $\chi \equiv 1$ and the exponent is completely removed in Δ' by (3.2.2).

In the remaining case, with (3.2.3) violated, we bound $1-\chi$ from above by a polynomial. To do this, we analyse δV, using the decomposition

$$\varphi(x, t) = \varphi_1(x, t) + \varphi_2(x, t),$$

where φ_1 has lower and upper momentum cutoffs at $M_{l(\Delta)}$ and $M_{l(\Delta')}$ respectively, while φ_2 has lower and upper momentum cutoffs at $M_{l(\Delta')}$ and $M_{u(\Delta)}$ respectively. The estimate

$$-O(1) = -O(|\Delta'|)\, M^2_{l(\Delta')} \leqq \int_{\Delta'} :\varphi_1{}^4:\, dx\, dt, \tag{3.2.4}$$

which will govern the analysis of δV, provides the reason for choosing $M_{l(\Delta')} \sim |\Delta'|^{-1/2}$ as the crossover momentum between φ_1 and φ_2, and hence for the definition (3.2.1). On the other hand the requirement that φ_2 (i.e. the field in $V(\zeta')$) should be properly localized in Δ' is satisfied if $M_{l(\Delta')} \geqq |\Delta'|^{-1/3}$. The inequality

$$|\Delta'|^{-1/3} \leqq M_{l(\Delta')} \sim |\Delta'|^{-1/2}$$

is the formal reason that the Wick construction is useful. For the φ_4^4 interaction the corresponding inequality is marginal.

In the analysis of δV, we replace φ_1 by its average value $\bar{\varphi}_1$, plus an error. Explicitly,

$$\bar{\varphi}_1 = |\Delta'|^{-1} \int_{\Delta'} \varphi_1(x, t)\, dx\, dt \tag{3.2.5}$$

$$\delta\varphi_1(x, t) = \varphi_1(x, t) - \bar{\varphi}_1, \qquad (x, t) \in \Delta'. \tag{3.2.6}$$

(This method is similar to the treatment of some infrared problems in quantum field theory). We estimate the average field $\bar{\varphi}_1$ by classical methods, i.e. Hölder's inequality, and we find that the error term $\delta\varphi$ has an effective low momentum cutoff. In Chapter 6, we obtain a convergence factor $b^{-3(1+\nu)/2}$ associated with each $\delta\varphi_1$ leg in R.

Let

$$A = \Big(\int_{\Delta'} \varphi_1^4\, dx\, dt\Big)^{1/4} = \|\varphi_1\|_4. \tag{3.2.7}$$

By Hölder's inequality,

$$|\bar{\varphi}_1| = |\Delta'|^{-1} \Big|\int_{\Delta'} \varphi_1\, dx\, dt\Big| \leqq |\Delta'|^{-1/4} A. \tag{3.2.8}$$

We expand δV in terms of φ_1 and φ_2. For simplicity in our formulas below, we omit the space-time cutoff h and the measure $dx\, dt$ for integrations over the space-time domain Δ'.

$$\delta V = \delta V_I(\Delta') + \delta V_C(\Delta') \tag{3.2.9}$$

$$\begin{aligned}
\delta V_I &= \int :\varphi_1^4: + 4\int :\varphi_1^3\varphi_2: + 6\int :\varphi_1^2\varphi_2^2: + 4\int :\varphi_1\varphi_2^3: \\
\delta V_C &= O(\log M_{u(\Delta')}) \left(\int :\varphi_1^2: + 2\int :\varphi_1\varphi_2:\right) \\
&\quad + O(\log M_{l(\Delta')}) \left(\int :\varphi_1^2: + 2\int :\varphi_1\varphi_2: + \int :\varphi_2^2:\right) \\
&\quad + O(|\Delta'|\, M_{l(\Delta')} \log M_{u(\Delta')}).
\end{aligned}$$

The (positive) constant $O(\log M_{u(\Delta')})$ arises from the term $\delta m^2 \int :\varphi^2: - \delta m^2 \int :\varphi_2^2:$, while the (positive) constant $O(\log M_{l(\Delta')})$ arises from raising the lower cutoff in δm^2. The change in the second order vacuum energy counterterm is bounded by $O(|\Delta'|\, M_l \log M_l)$ see Proposition 6.3.1. This constant also dominates the change in the third order vacuum energy, $O(|\Delta'| \log M_l)$, as we will see in Chapter 6.

We require the relation

$$\log M_{u(\Delta')} \leqq \varepsilon^{-1}\, O(\log |\Delta'|^{-1}), \tag{3.2.10}$$

to be justified in Section 3.3. Also,

$$-O(M_{l(\Delta')}) \leqq\ :\varphi_1^2:,$$

so by (3.2.1)

$$-O(|\Delta'|^{1/2}) \leqq \int :\varphi_1^2:. \tag{3.2.11}$$

Using (3.2.10)–(3.2.11), we have

$$\begin{aligned}\delta V_C &\geqq \varepsilon^{-1} O(\log|\Delta'|^{-1})\{\int :\varphi_1\varphi_2: + O(1)\int :\varphi_2{}^2: + O(|\Delta'|^{1/2})\} \qquad (3.2.12)\\ &\geqq |\Delta'|^{-1/2}\{\int \varphi_1\varphi_2 + O(1)\int :\varphi_2{}^2: + O(|\Delta'|^{1/2})\}\end{aligned}$$

where we use

$$\int :\varphi_1\varphi_2: = \int \varphi_1\varphi_2 + O(|\Delta'|^{1/2}),$$

and we choose M_1 sufficiently large, depending on ε. Choosing M_1 large here forces $|\Delta'|$ to be small, see Section 3.3. Similarly, removing the Wick ordering in the φ_1 parts of δV_1 yields

$$\begin{aligned}\delta V_I \geqq \int\varphi_1{}^4 &+ 4\int \varphi_1{}^3\varphi_2 + 6\int \varphi_1 :\varphi_2{}^2: + 4\int \varphi_1 :\varphi_2{}^3:\\ &+ O(|\Delta'|^{-1/2})\int \varphi_1{}^2 + O(|\Delta'|^{-1/2})\int \varphi_1\varphi_2 \qquad (3.2.13)\\ &+ O(|\Delta'|^{-1/2})\int :\varphi_2{}^2: + O(1).\end{aligned}$$

We substitute the bound

$$\frac{1}{2}\int\varphi_1{}^4 + O(|\Delta'|^{-1/2})\int :\varphi_1{}^2 \geqq O(1) \qquad (3.2.14)$$

to remove the pure $\varphi_1{}^2$ term. Thus

$$\begin{aligned}\delta V \geqq \frac{1}{2}\int \varphi_1{}^4 &+ 4\int \varphi_1{}^3\varphi_2 + 6\int \varphi_1{}^2:\varphi_2{}^2: \qquad (3.2.15)\\ &+ 4\int \varphi_1:\varphi_2{}^3: + O(|\Delta'|^{-1/2})\int \varphi_1\varphi_2\\ &+ O(|\Delta'|^{-1/2})\int :\varphi_2{}^2: + O(1),\end{aligned}$$

so we bound δV from below by $(1/2)A^4 + O(1)$, plus cross terms between Wick ordered high momentum fields and unordered low momentum fields. In order to isolate the average low momentum field we insert $\varphi_1 = \bar{\varphi}_1 + \delta\varphi_1$ in the cross terms to obtain,

$$\delta V \geqq \frac{1}{2}\{A^4 + b_3(\bar{\varphi}_1)^3 + b_2(\bar{\varphi}_1)^2 + b_1\bar{\varphi}_1 + b_0 + O(1)\}. \qquad (3.2.16)$$

Here

$$\begin{aligned}b_3 &= 8\int \varphi_2 \qquad (3.2.17)\\ b_2 &= 24\int \delta\varphi_1\varphi_2 + 12\int :\varphi_2{}^2:\\ b_1 &= 24\int (\delta\varphi_1)^2\varphi_2 + 24\int \delta\varphi_1 :\varphi_2{}^2: + 8\int :\varphi_2{}^3: + O(|\Delta'|^{-1/2})\int \varphi_2\\ b_0 &= 8\int (\delta\varphi_1)^3\varphi_2 + 12\int (\delta\varphi_1)^2 :\varphi_2{}^2: + 8\int \delta\varphi_1 :\varphi_2{}^3:\\ &\quad + O(|\Delta'|^{-1/2})\int \delta\varphi_1\varphi_2 + O(|\Delta'|^{-1/2})\int :\varphi_2{}^2:.\end{aligned}$$

Each term in each b_i contains φ_2. Using (3.2.8), we bound (3.2.16) from below by

$$\delta V \geqq \frac{1}{2}\{A^4 - a_3A^3 - a_2A^2 - a_1A - a_0 - O(1)\}, \qquad (3.2.18)$$

where

$$a_i = a_i(\Delta') = |\Delta'|^{-i/4} |b_i|, \tag{3.2.19}$$

for $i = 0, 1, 2, 3$. We note that terms in a_i of total degree l in $\delta\varphi_1$ and φ_2 have a coefficient c satisfying

$$|c| \leqq O(|\Delta'|^{-1+l/4}). \tag{3.2.20}$$

Let $\sum_{i=0}^{3} a_i = \sum_{j=1}^{12} \alpha_j$, where each α_j is a monomial, and let

$$P(\Delta') = \sum_{i=1}^{12} \alpha_j(\Delta')^l,$$

where $l = 2(\log_2 |\Delta'|^{-1})^3$.

Lemma 3.2.1. *Let $\chi(\Delta')$ be as in* (3.2.1). *Then*

$$1 - \chi(\Delta') \leqq P(\Delta'). \tag{3.2.21}$$

Proof. By (3.2.18), δV is bounded from below on the set of path space on which

$$\sup_{1 \leqq j \leqq 12} |\alpha_j(\Delta')| \leqq 1, \tag{3.2.22}$$

with a lower bound K_2 independent of the cutoffs or the particular cube Δ'. Thus we define $K_1 = \exp(-K_2)$. Furthermore, on the complement of the set (3.2.22), some $|\alpha_j| > 1$, so

$$\{1 - \chi(\Delta')\} \leqq \{1 - \chi(\Delta')\}\, P(\Delta') \leqq P(\Delta').$$

3.3. The localization cubes

We apply (3.2.2) to (3.1.1) as follows: First we choose a refinement $\mathscr{D}_1$ of $\mathscr{D}$, and we apply (3.2.2) in the cubes Δ' in a subset $\mathscr{D}_1^0$ of $\mathscr{D}_1$. We obtain thereby a sum of terms, and we now consider a fixed term in this sum. If the first term is selected in Δ_1, the lower cutoff in the exponent in Δ_1 is $M_{l(\Delta_1)}$. For cubes Δ_1 not of this type, that is for which the second term in (3.2.2) is selected, the lower cutoff in the exponential remains at $M_{l(\Delta)}$, where

$$\Delta_1 \subset \Delta \in \mathscr{D}.$$

Let $\mathscr{D}_1'$ be the set of these cubes.

We subdivide the cubes in $\mathscr{D}_1'$, obtaining a new refinement $\mathscr{D}_2$ of $\mathscr{D}_1$, and we apply (3.2.2) to the cubes $\mathscr{D}_2 \sim \mathscr{D}_1$. We continue in this fashion with refinements $\mathscr{D}_3$, $\mathscr{D}_4, \ldots$. This process terminates for a given term with a cover $\mathscr{D}_j$ if for each cube $\Delta_j \in \mathscr{D}_j$ the lower momentum cutoff in Δ_j occurs at $M_{l(\Delta_j)}$. For cubes Δ' below the minimum size (3.2.3), the exponent is completely removed in Δ'. Thus the Wick expansion terminates after a finite number of steps.

In each cover, we require adjacent cubes to have nearly uniform size, namely for adjacent cubes Δ_α, Δ_β of the same cover,

$$|\Delta_\alpha| \,/\, |\Delta_\beta| = \frac{1}{8}, 1, \text{ or } 8, \tag{3.3.1}$$

provided the exponent is not zero in Δ_α and Δ_β.

Lemma 3.3.1. *Let Δ be a cube in a cover $\mathscr{D}$ satisfying* (3.3.1). *Let Δ be subdivided into 8 cubes, and let other cubes in $\mathscr{D}$ be divided in eighths as necessary in order to get a new cover $\mathscr{D}'$ satisfying* (3.3.1). *Then the number of cubes in $\mathscr{D}' \sim \mathscr{D}$ satisfies*

$$|\mathscr{D}' \sim \mathscr{D}| \leqq O(\log |\Delta|^{-1}).$$

Proof. Let $C(\Delta) = \mathscr{D} \sim \mathscr{D}'$ be the set of cubes that must be divided to obtain $\mathscr{D}'$. Then $|\mathscr{D}' \sim \mathscr{D}| = 8\ |C(\Delta)|$. Let $|\Delta| = 8^{-l}$, and for $0 \leqq i \leqq l$, let $C_i(\Delta)$ be the set of 27 cubes of volume 8^{-i} forming a $3 \times 3 \times 3$ cubic array on our lattice, with Δ in the center cube. (The position of Δ in the center cube Δ_c is determined by obtaining Δ_c by refinement of our original space time cover, and by further refining Δ_c to obtain Δ.) We see that

$$C(\Delta) \subset \bigcup_i C_i(\Delta),$$

as follows: A cube Δ' of volume $8^{-(l-1)}$ need be divided only if it borders Δ, and hence belongs to $C_{l-1}(\Delta)$. If dividing Δ' necessitates dividing Δ'' of volume $8^{-(l-2)}$, then Δ'' must border $C_{l-1}(\Delta)$, and hence belongs to $C_{l-2}(\Delta)$, etc. Hence

$$\begin{aligned} |\mathscr{D}' \sim \mathscr{D}| &\leqq \sum_{i=0}^{l} 8\ |C_i(\Delta)| \\ &\leqq 8 \times 27(l+1) \\ &= O(l) \leqq O(-\log |\Delta|). \end{aligned}$$

We construct $\mathscr{D}_1$ in steps, dividing Δ or its subcubes into eighths, and adding extra cubes to satisfy (3.3.1) at each step. In choosing which cube $\Delta' \subset \Delta \in \mathscr{D}$ to subdivide into eighths, we pick a Δ' of maximum size in Δ. We call this a *primary* subdivision, while the extra cubes added to satisfy (3.3.1) arise from a *secondary* subdivision. We continue this subdivision until each subcube $\Delta' \subset \Delta \in \mathscr{D}$ satisfies $|\Delta'| \leqq n(\Delta)^{-1/2}\ |\Delta|$, for $r > 1$. In the first inductive step we use the extra condition

$$|\Delta_1| \leqq \min \{M_3^{-2}, M_{u(\Delta)}^{-\varepsilon/2}\}, \tag{3.3.2}$$

so that $l(\Delta_1) \geqq 3$. This removes the exponent in cubes $\Delta \in \mathscr{D}_a$.

We define $\mathscr{D}_1{}^0$, the set of cubes in $\mathscr{D}_1$ in which (3.2.2) is applied, as the set of cubes for which $\delta V \neq 0$ and $V(\zeta) \neq V(\zeta')$, cf. (3.2.2).

We now turn to the construction of $\mathscr{D}_2, \mathscr{D}_3, \ldots$ By induction, suppose we are given a sequence of covers $\mathscr{D} = \mathscr{D}_0, \mathscr{D}_1, \ldots, \mathscr{D}_i$ with $\mathscr{D}_{\alpha+1}$ refining $\mathscr{D}_\alpha$, and suppose that we are given a term resulting from the application of (3.2.2) to the appropriate cubes in $\bigcup_{\alpha=1}^{i} \mathscr{D}_\alpha$. To carry out the next step in the Wick expansion, we define $\mathscr{D}_i{}^0$ to be the set of cubes, $\Delta_i \in \mathscr{D}_i$, to which (3.2.2) has been applied, and $\mathscr{D}_i' \subset \mathscr{D}_i{}^0$ the subset of cubes on which the second term was selected. Thus $\mathscr{D}_i'$ is exactly the set of cubes Δ_i in which the lower momentum cutoff lies below $M_{l(\Delta_i)}$. If $\mathscr{D}_i' = \varnothing$, the Wick expansion for the term we are considering is complete. If $\mathscr{D}_i' \neq \varnothing$, we divide each cube $\Delta_i \in \mathscr{D}_i'$ into eighths, and we subdivide other cubes of $\mathscr{D}_i$ as necessary, in order to achieve (3.3.1). The resulting new cover is $\mathscr{D}_{i+1}$. Next we apply (3.2.2) to each cube $\Delta_{i+1} \in \mathscr{D}_{i+1}$ for which the lower momentum cutoff in the exponent lies below $M_{l(\Delta_{i+1})}$. This condition defines $\mathscr{D}_{i+1}^0$. With

$\Delta_{i+1} \subset \Delta_i \in \mathscr{D}_i$, we have

$$M_{l(\Delta_{i+1})} = M_{l(\Delta_i)}$$

before the application of (3.2.2) to cubes in $\mathscr{D}_{i+1}$, and so (3.2.2) is certainly applied in the case $\Delta_{i+1} \subset \Delta \in \mathscr{D}_i'$. This completes the induction and defines a single Wick expansion step.

Next we bound the factor $1 - \chi$ in (3.2.2), in a term for which the Wick expansion is complete. Let $\mathscr{D}_1, \ldots, \mathscr{D}_j$ be the finite sequence of covers associated with this term. Note that $\mathscr{D}_j$ is the cover $\mathscr{D}$ for the P—C expansion of the following inductive step, and that $\mathscr{D}_j' = \varnothing$. We introduce the definitions

$$\mathscr{D}^0 = \bigcup_{i=1}^{j} \mathscr{D}_i^0$$

$$\mathscr{D}' = \bigcup_{i=1}^{j-1} \mathscr{D}_i'.$$

Then $\mathscr{D}^0$ is the set of cubes in which (3.2.2) has been applied and $\mathscr{D}'$ is subset consisting of cubes in which the second term of (3.2.2) was selected. Let $\mathscr{D}^*$ be the set of cubes in $\mathscr{D}'$ which do not contain smaller cubes in $\mathscr{D}'$. It follows that the cubes in $\mathscr{D}^*$ are disjoint. For $\Delta \in \mathscr{D}' \sim \mathscr{D}^*$, we substitute the upper bound

$$1 - \chi \leqq 1.$$

For $\Delta \in \mathscr{D}^*$, we use (3.2.21).

For combinatoric purposes, we must count the number of times we apply the expansion (3.2.2). For this reason, we construct a mapping of cubes in $\mathscr{D}^0$ (of the rth inductive step) into P_r and W_r vertices. In the case $r = 1$, we also map onto cubes in $\mathscr{D}_e$. For similar reasons, we map cubes in $\mathscr{D}_j \sim \mathscr{D}$ into P_r and W_r vertices.

Theorem 3.3.2. *The above assignment of cubes in $\mathscr{D}^0 \cup (\mathscr{D}_j \sim \mathscr{D})$ is one to one when mapping into vertices, and maps at most $O(1)$ cubes onto each cube of $\mathscr{D}_e$ for $r = 1$.*

Proof. We first bound the number of cubes in $\mathscr{D}_1 \sim \mathscr{D}$, for $r > 1$. Recall that the definition of $\mathscr{D}_1$ allows no primary subdivision of a cube Δ' with volume less than $|\Delta|\, n(\Delta)^{-1/2}$, where $\Delta \in \mathscr{D}$. Hence if Δ' has a primary subdivision and $|\Delta'| = 8^{-j}$,

$$j \leqq \log |\Delta|^{-1} + \frac{1}{2} \log n(\Delta),$$

and by Lemma 3.3.1

$$|\mathscr{D}_1 \sim \mathscr{D}| \leqq \sum_{\Delta \in \mathscr{D}_e} O(1)\, n(\Delta)^{1/2} \left(\log |\Delta|^{-1} + \frac{1}{2} \log n(\Delta)\right)^2.$$

If $\Delta \in \mathscr{D}_e$, then $\Delta \notin \mathscr{D}_a$ by (3.3.2). Condition (b) of Chapter 2 ensures $|\Delta|^{-\varepsilon} \leqq n(\Delta)$, and by (3.3.2), $|\Delta| \leqq M_1^{-2}$ for $\Delta \in \mathscr{D}_e$. Thus we obtain the bound

$$|\mathscr{D}_1 \sim \mathscr{D}| \leqq \sum_{\Delta \in \mathscr{D}_e} n(\Delta)^{2/3}$$

for $M_1 = M_1(\varepsilon)$ sufficiently large. For $r = 1$, $|\Delta| = 1$, and $\log |\Delta|^{-1} = 0$. Thus the analysis above yields

$$|\mathscr{D}_1 \sim \mathscr{D}| \leqq \sum_{\Delta \in \mathscr{D}_e} \left(\frac{1}{2} M^{\varepsilon}_{u(\Delta)} + O(1)\right),$$

where $O(1)$ depends on M_1 and arises from the sum over cubes in $\mathscr{D}_a$. Thus we can

29*

assign each cube in $\mathscr{D}_1 \sim \mathscr{D}$ to a distinct P_r vertex, with the exception for $r = 1$, noted above.

We now assign each cube Δ in $\mathscr{D}_i \sim \mathscr{D}_{i-1}$ to a cube in $\mathscr{D}^*$. We note that $\mathscr{D}^0$ is contained in $\bigcup_{i=1}^{j} (\mathscr{D}_i \sim \mathscr{D}_{i-1})$. We first assign Δ to Δ', where Δ' is the cube in $\mathscr{D}_i'$ whose primary subdivision led to the square Δ. By Lemma 3.3.1, at most $O(1)$ $\log |\Delta'|^{-1}$ cubes Δ are assigned to Δ' in this fashion. We assign Δ' to the smallest cube Δ^* in $\mathscr{D}^*$ contained in Δ'. Each cube Δ^* in $\mathscr{D}^*$ is contained in at most $O(1)$ $\log |\Delta^*|^{-1}$ larger cubes Δ', and any such cube Δ' satisfies $\log |\Delta'|^{-1} \leqq \log |\Delta^*|^{-1}$. Hence at most $O(1) (\log |\Delta^*|^{-1})^2$ cubes Δ are assigned to each $\Delta^* \in \mathscr{D}^*$, and every cube in $\mathscr{D}^0 \sim \mathscr{D}_1$ and in $\mathscr{D}_j \sim \mathscr{D}$ has been assigned to a cube in $\mathscr{D}^*$. For M_1 sufficiently large, $O(1) (\log |\Delta^*|^{-1})^2 \leqq (\log |\Delta^*|^{-1})^3$, and the Theorem follows.

Proposition 3.3.3. *There is an integer $n_1 = n_1(\varepsilon)$ with the following property: Let Δ_1 and Δ_2 be cubes associated with the term we are considering. If $|\Delta_1| < 1$ and if a vertex or exponent is localized in Δ_2, then*

$$|\Delta_1|^{n_1} \leqq |\Delta_2| \{2c + 2c \operatorname{dist} (\Delta_1, \Delta_2)\}^{n_1}$$

Proof. The cubes Δ_1 and Δ_2 may be introduced during different inductive steps. Let Δ_i' be the cube of volume 1 containing Δ_i. By the definition of the first refinement $\mathscr{D}_1$ of $\mathscr{D}$,

$$|\Delta_1| \leqq \frac{1}{2} M_{u(\Delta_1')}^{-\varepsilon},$$

and by condition (c) of Chapter 2,

$$M_{u(\Delta_1')}^{-\varepsilon} \leqq (M_{u(\Delta_2')}^{-\varepsilon})\, c\, d(\Delta_1', \Delta_2')^4 .$$

By (3.2.2), and the fact that $u(\Delta_2) \leqq u(\Delta_2')$,

$$M_{u(\Delta_2')} \geqq M_{u(\Delta_2)} \geqq |\Delta_2|^{-1/2}.$$

As

$$d(\Delta_1', \Delta_2') \leqq 1 + \text{dist.} (\Delta_1, \Delta_2),$$

$$|\Delta_1|^{2/\varepsilon} \leqq |\Delta_2| \{2c + 2c \text{ dist.} (\Delta_1, \Delta_2)\}^{8/\varepsilon},$$

and the proposition follows.

3.4. The low momentum contractions

Lemma 3.4.1. *Let ν be sufficiently small, depending on ε. Let $\mathscr{D}$ be a cover associated with a term at the end of the Wick construction of the rth inductive step. For $\Delta \in \mathscr{D}$, $l(\Delta) \geqq 3r$.*

Proof. In the first inductive step, $l(\Delta) \geqq 3$ by (3.2.1) and (3.3.2). We assume that the lemma is valid at the end of the rth inductive step. In the $(r + 1)^{\text{st}}$ P—C expansion, we have $n(\Delta) \geqq |\Delta|^{-\varepsilon} P_{r+1}$ vertices in Δ, by virtue of Condition (b) of Chapter 2 and (2.12). During the first refinement $\mathscr{D}_1$ of the $(r + 1)^{\text{st}}$ inductive step, Δ is divided into at least $n_{r+1}(\Delta)^{1/2}$ subcubes. If Δ_1 is any such subcube,

$$|\Delta_1| \leqq n_{r+1}(\Delta)^{-1/2} |\Delta| \leqq |\Delta|^{1+\varepsilon/2} \leqq M_{l(\Delta)}^{-2(1+\varepsilon/2)}.$$

Choosing ν so small that $(1 + \nu)^3 \leqq 1 + \varepsilon/2$, we have

$$M_{l(\Delta)}^{1+\varepsilon/2} \geqq M_{3r}^{1+\varepsilon/2} \geqq M_{3r}^{(1+\nu)^3} \geqq M_{3(r+1)} .$$

Thus $|\Delta_1| \leqq M_{r+1}^{-2}$ and the lemma follows by (3.2.1).
To specify the low momentum contractions, suppose that we are in the rth inductive step, and consider an uncontracted field in $R(s)$, created in the ith inductive step. In this field we specify a new cutoff

$$M_{r+2i-1} \tag{3.4.1}$$

for Wick vertices and at

$$M_{r+2i-2} \tag{3.4.2}$$

for P or C vertices. We then contract the part of the field below this cutoff, by moving the creation operator to the left and the annihilation operator to the right. The legs are considered in reverse inductive oder, i.e., in order of decreasing i, and W vertices before P or C vertices. The cutoffs (3.4.1)—(3.4.2) grow faster than 2^{r-i+1}, the number of vertices produced by squaring, but less rapidly than the lower cutoff in the exponent, by Lemma 3.4.1. For this reason, no contractions to the exponent occur.
If $V \neq 0$, the rth inductive step is now complete. If $V = 0$, we contract all legs in the polynomial R in (3.1.1), and then the entire expansion is complete for the term in question.

4. Combinatoric factors

The inductive construction of Chapters 2 and 3 is now complete, and each inductive step yields an upper bound for $\langle\exp(-tH)\rangle$ as a sum of terms of the form $\sup_s \langle R(s)\, e^{-V(s)}\rangle$. Those terms for which the inductive construction has terminated satisfy $V(s) = 0$. For these terms, $R(s)$ (as a function on path space) is a constant, expressed as a sum of elementary integrals $I(G)$ labeled by Feynman graphs G. The integrals $I(G)$ are estimated in Chapters 5 and 6. In this chapter we are concerned with the combinatoric problem of estimating the number of terms in the sum over graphs G, especially the sum for which $I(G)$ satisfies some given upper bound. These estimates, combined with the estimates of Chapters 5 and 6, complete the proof that the $\varphi_3{}^4$ Hamiltonian is bounded from below, Theorem 1.1.1.
We pose the combinatoric estimates of this chapter in terms of combinatoric factors $c(G)$, as explained in [6]. For example,

$$|\sum_G I(G)| \leqq \Big(\sum_G c(G)^{-1}\Big) \sup_G c(G)\, |I(G)|, \leqq \sup_G c(G)\, |I(G)|,$$

where the combinatoric factors $c(G)$ are positive numbers satisfying $\sum c(G)^{-1} \leqq 1$. We assign combinatoric factors $c(G)$ to certain vertices, lines and legs as the final part of each inductive step. The combinatoric factors are assigned to lines and vertices in each elementary substep of the inductive construction, and the total combinatoric factor for a line or vertex is the product of these elementary factors. In addition to using the combinatoric factors to count terms, we absorb into our upper bounds on the factors $c(G)$, the constant $\delta^{\pm 1}$ in formula (3.2) and the constant K_1 in (3.4).

Theorem 4.1. *The following are upper bounds on combinatoric factors associated with the inductive expansion of Chapters 2 and 3: Given ε, ν there exist n, M, L, β such that if $M_1 \geqq M$, we have the bounds*

d^n	per line,
$L\lambda^{\beta\varepsilon}$	per P_1 vertex,
$L\|\Delta\|^{-\beta\varepsilon}$	per C, W or P_r vertex, $r > 1$,
1	per squaring vertex,
$b^{3(1+\nu)/2}$	per W leg.

Here β is independent of ε and ν. In addition we obtain an overall factor exp $(O(\text{vol}\, h))$ for the expansion.

Proof. We assign the combinatoric factors separately for each of the various substeps that occur in the inductive construction. Here $\lambda \geqq M_{u(\Delta)-1}$ is the maximum lower cutoff for a P_1 vertex, Δ is the localization cube for a vertex, d is the scaled length for a line, and b is the localization ratio. See Section 1.2. for definitions. Since the bounds of the theorem are invariant in form when raised to a bounded power (except the factor per W leg) we need only establish the first four bounds for each of the substeps considered.

Case P. We first consider the substep of using the P expansion formula (2.1) to add a single P vertex localized in Δ. Formula (2.1) gives two terms, one with a decrease in the upper momentum cutoff in Δ, and one with a new P vertex in Δ. We assign a combinatoric factor $c_1 = O(1) \log \lambda$ to the new vertex, and $c_2 = (1 - c_1^{-1})^{-1}$ to the term with a decreased upper momentum cutoff. With this choice, $c_1^{-1} + c_2^{-1} = 1$. We now consider c_1 and c_2 separately. We bound c_1 as follows. For a P_n vertex, $n > 1$, let $\Delta' \supset \Delta$ be the cube in the first inductive step containing Δ. By Condition (b) of Chapter 2, and (3.3.2),

$$M^{\varepsilon}_{u(\Delta)} \leqq M^{\varepsilon}_{u(\Delta')} \leqq |\Delta|^{-2}. \tag{4.1}$$

Thus

$$\log M_{u(\Delta)} \leqq \frac{2}{\varepsilon} \log (|\Delta|^{-1}) + \frac{1}{\varepsilon} \log 2. \tag{4.2}$$

Given $\alpha > 0$, we have after the first Wick expansion,

$$(\log M_{u(\Delta)})^{\alpha} \leqq |\Delta|^{-\varepsilon}, \tag{4.3}$$

for M_1 sufficiently large (depending on ε and α). Our bound for c_1 becomes

$$c_1 \leqq O(1) \log \lambda, \qquad P_1 \text{ vertex}$$

$$c_1 \leqq O(1)\, |\Delta|^{-\varepsilon}, \qquad P_r \text{ vertex}, \quad r > 1.$$

We note that for a sequence λ_j of decreasing cutoffs equal to M_j,

$$\begin{aligned} \sum c_1^{-1} &= \sum_j O(1) (\log M_j)^{-1} \\ &= O(1) (\log M_1)^{-1} \sum_j (1 + \nu)^{-j} \leqq O(1), \end{aligned} \tag{4.5}$$

where the bound $O(1)$ is uniform for $2 \leqq M_1$, but diverges as $\nu \to 0$. (We assume $0 < \nu \leqq 1$.)

We now return to the factor c_2 that we associate with the term in (2.1) with a decreased upper cutoff. In a given inductive step, we may obtain a sequence of successively lower cutoffs from repeated application of (2.1) in a given cube Δ. The product of the c_2 factors for such a sequence is

$$\prod c_2 = \prod (1 - c_1^{-1})^{-1} \leqq \prod (1 + 2c_1^{-1})$$
$$\leqq \exp (2 \sum c_1^{-1}) \leqq O(1) \tag{4.6}$$

by virtue of (4.5). Here $O(1)$ diverges as $\nu \to 0$. The product (4.6) of c_2's is a combinatoric factor, and we assign it to the preceding P vertex introduced in Δ during the present inductive step. If there are no such P vertices in Δ, we assign the combinatoric factor as follows: Assume that we are in the rth inductive step. If $r = 1$, we assign the factor $O(1)$ to the cube Δ. Since $|\Delta| = 1$, such factors contribute at most the overall factor $\exp\big(0(\text{vol}\, h)\big)$.

If $r > 1$, then Δ belongs to $\mathscr{D}_j$ for the $(r-1)^{\text{st}}$ inductive step. Recall that $\mathscr{D}_j$ is the final cover for the Wick construction (3.4). By Theorem 3.3.2, we assign Δ to a P_{r-1} or W_{r-1} vertex.

Case C. Next we consider the case of a single contraction. This may be either the expansion of Chapter 2 that generates new C vertices and cancels the divergent terms, or else it may be the low momentum contraction of Section 3.4 which reduces the number of uncontracted legs.

In dealing with the $P-C$ expansion, it is convenient to associate each C vertex with generating vertices. A P vertex may contract to the exponent to form a C vertex, or it may contract to the exponent to form C', which contracts to the exponent to form the C vertex. We call the vertices P and C' (if it exists) *generating* vertices for C. By the $P-C$ expansion of Chapter 2, in a particular term, at most 16 C vertices have a given P vertex as a generating vertex, and at most 3 C vertices have a given C vertex as a generating vertex. We now divide the possible contraction steps into seven cases.

Case Cl. The low momentum contraction expansion of Chapter 3 defines a cutoff in each inductive step, below which legs of existing vertices contract at the end of the inductive step. The first subcase is the restriction of a leg to have momentum above or below this cutoff. Each such decomposition introduces a combinatoric factor 2 per leg or at most 2^4 per vertex.

By Lemma 3.4.1 and (3.4.1)—(3.4.2), every vertex with upper cutoff at M_j is fully contracted at the end of the jth inductive step. Thus the maximum combinatoric factor for a vertex with upper cutoff M_j is 2^{4j}.

For $0 < \nu < 1$,

$$j = \frac{\log\log M_j - \log\log M_1}{\log(1+\nu)} \leqq \frac{2}{\nu} \log\log M_j. \tag{4.7}$$

Thus

$$2^{4j} \leqq (\log M_j)^{8\nu^{-1}\log 2}. \tag{4.8}$$

If the vertex in question is a P_1 vertex,

$$2^{4j} \leqq M_{j-1}^{\varepsilon} = \lambda^{\varepsilon} \tag{4.9}$$

valid for M_1 sufficiently large (depending on ε and ν). We assign the combinatoric factor λ^{ε} to each P_1 vertex.

If the vertex in question is a C_1 vertex localized in Δ' and with upper cutoff $M_{j'}$, we obtain as above a bound $M^{\varepsilon}_{j'-1}$ on the combinatoric factor, which we assign to the generating P_1 vertex. Let the generating P_1 vertex be localized in Δ and have upper cutoff M_j. By Condition (c) of Chapter 2, applied when the cube Δ was selected for the P_1 vertex,

$$M^{\varepsilon}_{j'} \leqq O(1)\,\{1 + \text{dist.}\,(\Delta, \Delta')\}^4\, M^{\varepsilon}_{j}. \tag{4.10}$$

Hence

$$2^{4j'} \leqq O(1)\,\lambda^{\varepsilon}\,\{1 + \text{dist.}\,(\Delta, \Delta')\}^4, \tag{4.11}$$

where λ refers to the generating P_1 vertex. Thus (4.11) is dominated by a factor

$$O(1)\,\lambda^{\varepsilon} \tag{4.12}$$

assigned to the generating P_1 vertex and d^4 assigned to each line at the generating P_1 and C_1 vertices.

Finally, we consider the case that the vertex being restricted is a W, P_r or C_r vertex, $r > 1$. Let us assume the vertex has upper cutoff M_j, is localized in Δ, and $M_{u(\Delta)}$ is the upper cutoff in Δ when the vertex is introduced. As in (4.1)—(4.3), and using (4.7),

$$2^{4j} \leqq |\Delta|^{-\varepsilon}, \tag{4.13}$$

for M_1 sufficiently large (depending on ε and ν). In this case we assign the combinatoric factor $|\Delta|^{-\varepsilon}$ to the W, P_r or C_r vertex.

The following C subcases refer to the contraction of two legs to form a line. In every case, we assume that a leg of a vertex localized in Δ, and introduced in the rth inductive step, contracts to a leg of a vertex localized in $\Delta(\alpha)$ and introduced in the $r(\alpha)^{\text{th}}$ inductive step. We assume that the contraction occurs during the ith inductive step. We may refer to the leg in Δ as a "Δ leg", or a "$P_r(\Delta)$ leg", etc., to be more specific. In the following lemma, we consider the scaled distance d of (1.2.11) for the line produced by such a contraction.

Lemma 4.2 (a) *Let $n \geqq 4$. For contraction to a $P_{r(\alpha)}$ or $C_{r(\alpha)}$ leg*

$$\sup_{\substack{r(\alpha)\leqq r\\ \Delta, i}} \sum_{\Delta(\alpha)} d^{-n} \leqq O(1). \tag{4.14}$$

(b) *Let n be sufficiently large (depending on ε). For contraction to a $W_{r(\alpha)}$ leg, ($r > r(\alpha)$), (4.14) is valid.*

(c) *For contraction of a W_r leg to a W_r leg* i.e. $r(\alpha) = r$,

$$\sup_{\Delta, i, r} \sum_{\Delta(\alpha)} |\Delta(\alpha)|^{\varepsilon}\,(b_\alpha\, d)^{-3(1+\nu)} \leqq O(1). \tag{4.15}$$

Proof. (a) In the proof, let $\Delta(\alpha) = \Delta'$, $r(\alpha) = r'$, etc. Since $r' \leqq r$, we have by (3.3.1), with the exception of at most $O(1)$ cubes Δ' in each cover $\big(O(1) \leqq 57\big)$,

$$\text{dist.}\,(\Delta,\Delta') > \frac{1}{2}\,|\Delta'|^{1/3}. \tag{4.16}$$

We fix r'. According to (3.2.1.), the lower cutoff in a $P_{r'}$ or $C_{r'}$ leg is greater than $1/2|\Delta'|^{-1/2}$. By (1.2.11),

$$d \geqq \frac{1}{2}\,|\Delta'|^{-1/6}\,|\Delta'|^{-1/3}\,\text{dist.}\,(\Delta, \Delta'). \tag{4.17}$$

Let $\sum^{(j)}$ denote the sum over Δ' satisfying (4.16), and with size $|\Delta'| = 8^{-j}$. Then

$$\sum_{\Delta'} d^{-4} \leqq O(1) + \sum_{j=0}^{\infty} \sum^{(j)} d^{-4}$$
$$\leqq O(1) + \sum_{j=0}^{\infty} \sum^{(j)} (4^6 |\Delta'|)^{2/3} \left(|\Delta'|^{-1/3} \operatorname{dist.} (\Delta, \Delta')\right)^{-4}$$
$$\leqq O(1) + O(1) \sum_{j=0}^{\infty} 8^{-2j/3} \leqq O(1). \tag{4.18}$$

Here we use the fact that for cubes Δ' of a fixed size, and with $\sum'$ the sum over Δ' satisfying (4.16),

$$\sum' |\Delta'|^{4/3} \left(\operatorname{dist.} (\Delta, \Delta')\right)^{-4} \leqq O(1),$$

with $O(1)$ independent of Δ, $|\Delta'|$. Thus the constants in (4.18) are independent of Δ, i, r, $r(\alpha)$ and the proof of (a) is complete.
(b) For a $P_r(\Delta)$ or $C_r(\Delta)$ vertex, we have with $O(1)$ exceptions Δ',

$$\operatorname{dist.} (\Delta, \Delta') > \frac{1}{2} |\Delta|^{1/3}. \tag{4.19}$$

The lower cutoff in the $P_r(\Delta)$ or $C_r(\Delta)$ leg is greater than $1/2\ |\Delta|^{-1/2}$, and when (4.19) holds,

$$d > \frac{1}{2} |\Delta|^{-1/6} > \frac{1}{2} |\Delta'|^{-1/6n_1} (2cd)^{-1/6}$$

where the last inequality follows by Proposition 3.3.3. Here $n_1 = n_1(\varepsilon)$ is given by the lemma. Thus if $\sum'$ denotes the sum over cubes Δ' satisfying (4.19),

$$\sum_{\Delta'} d^{-n} \leqq O(1) + O(1) \sum' |\Delta'|\, d^{-n+7n_1}$$
$$\leqq O(1) \tag{4.20}$$

for $n \geqq 4 + 7n_1(\varepsilon)$. Here we use the fact that the sum over Δ' ranges over $\mathscr{D}^*$, and $\mathscr{D}^*$ is a set of disjoint cubes (not necessarily a cover).
For a $W_r(\Delta)$ leg, $r > r(\alpha)$, we consider Δ'', the cube from the $P_r - C_r$ expansion that contains Δ. We remark that Δ'' belongs to the same cover as Δ', or to an earlier cover than Δ'. Hence with the exception of $O(1)$ cubes Δ', we have dist. $(\Delta, \Delta') > 1/2\ |\Delta''|^{1/3}$. The lower cutoff for a $W_r(\Delta)$ vertex is greater than $1/2\ |\Delta''|^{-1/2}$. Hence with these $O(1)$ exceptions, by Proposition 3.3.3,

$$d \geqq \frac{1}{4} |\Delta''|^{-1/6} \geqq \frac{1}{4} |\Delta'|^{-1/6n_1} (2cd)^{-1/6}$$

and the above proof of (4.20) is valid.
(c) Let $M_{j'}$, b' be the lower momentum cutoff and localization ratio for the $W_r(\Delta')$ leg. Then

$$b'd \geqq b' \frac{1}{2} \left(1 + M_{j'} \operatorname{dist.} (\Delta, \Delta')\right)$$
$$\geqq \frac{1}{2} \left(1 + |\Delta'|^{-1/3} \operatorname{dist.} (\Delta, \Delta')\right).$$

Thus

$$\sum_{\Delta'} |\Delta'|^{\varepsilon} (b'd)^{-3(1+\nu)} \leqq O(1) \sum_{\Delta'} |\Delta'|^{\varepsilon} \left(1 + |\Delta'|^{-1/3} \operatorname{dist.}(\Delta, \Delta')\right)^{-3(1+\nu)}$$

$$\leqq O(1) \sum_{j=0}^{\infty} 8^{-\varepsilon j} \leqq O(1).$$

Here we use the fact that for cubes Δ' of a fixed size $|\Delta'| = 8^{-j}$,

$$\sum_{\Delta'} \left(1 + |\Delta'|^{-1/3} \operatorname{dist.}(\Delta, \Delta')\right)^{-3(1+\nu)} \leqq O(1),$$

with $O(1)$ independent of Δ, $|\Delta'|$. The sum over j accounts for cubes of various sizes.
Case C 2. The second subcase is the contraction to the exponent producing a C_i vertex. This has two terms (one from V_I and one from the mass counterterm) for each cube $\Delta(\alpha) \in \mathscr{D}_\rho$. By Lemma 4.2 (a), a combinatoric factor $O(1)\, d^4$, for the contracted line, is sufficient to fix the cube $\Delta(\alpha)$. We assign d^4 to the line and $O(1)$ to the new C vertex.
Case C 3. The third subcase is the contraction to a $C_{r(\alpha)}$ vertex with upper cutoff M_j. By Lemma 4.2 (a), a combinatoric factor d^4 assigned to the contracted line and $O(1)$ to the $C_{r(\alpha)}$ vertex is sufficient to fix $\Delta(\alpha)$.
We now bound the number of existing C legs in $\Delta(\alpha)$. Let γ be the generating P vertex for α, and β the generating C vertex (if one exists). We may fix the localization cubes $\Delta(\beta)$ and $\Delta(\gamma)$ of β and γ with a combinatoric factor $O(1)\, d_{\alpha\beta}^4 d_{\alpha\gamma}^4$ or $O(1)\, d_{\alpha\gamma}^4$. We next bound the number $n_{r(\alpha)}\left(\Delta(\gamma)\right)$ of P vertices introduced in $\Delta(\gamma)$ during the $r(\alpha)^{\text{th}}$ inductive step. In fact

$$n_1\left(\Delta(\gamma)\right) \leqq \lambda(\gamma)^{2\varepsilon} \tag{4.21}$$

$$n_{r(\alpha)}\left(\Delta(\gamma)\right) \leqq |\Delta(\gamma)|^{-\varepsilon},\ r(\alpha) > 1,$$

and each such P vertex generates at most $128 = O(1)$ C legs. The increase in C vertices due to squaring is bounded by a factor $2^{i-r(\alpha)+1}$, and a factor $\left(1 + r(\alpha)^2\right)$ covers the summation over $r(\alpha)$. As in Case C 1, $i \leqq j$. Thus

$$\left(1 + r(\alpha)^2\right) 2^{i-r(\alpha)+1} \leqq 2^{i+2} \leqq 2^{j+2}. \tag{4.22}$$

We assign the factor $O(1)\, n_{r(\alpha)}\left(\Delta(\gamma)\right)$ of (4.21) to the generating P vertex γ for α. We bound (4.22) as we bound the factor (4.11) in Case C 1, and as there, we assign it to the generating P vertex for α, and associated lines.
Case C 4. The fourth subcase is the contraction to a $P_{r(\alpha)}$ vertex. By Lemma 4.2. (a), a combinatoric factor d^4 for the line and $O(1)$ for the $P_{r(\alpha)}$ vertex fixes $\Delta(\alpha)$. As in Case C 3, we arrive at a combinatoric factor for the $P_{r(\alpha)}$ vertex of $\lambda(\alpha)^{3\varepsilon}$ if $r = 1$, or $|\Delta(\alpha)|^{-2\varepsilon}$ if $r > 1$.
Case C 5. The fifth subcase is contraction to a $W_{r(\alpha)}$ vertex, $r(\alpha) < r$. By Lemma 4.2 (b), a combinatoric factor $O(1)\, d^n$ assigned to the line is sufficient to fix $\Delta(\alpha)$. We bound the numbers of $W_{r(\alpha)}$ vertices in $\Delta(\alpha)$ by $O(1)\left(\log |\Delta(\alpha)|^{-1}\right)^3$ from applying (3.2.21), times a factor $2^{i-r(\alpha)+1}$ for the increase in vertices due to squaring. As

in Case C 1,

$$O(1)\left(\log|\Delta(\alpha)|^{-1}\right)^3 2^{i-r(\alpha)+1} \leq |\Delta(\alpha)|^{-\varepsilon}, \tag{4.23}$$

for M_1 sufficiently large, depending on ν and ε, and we assign (4.23) to the W vertex α.

Case C 6. The sixth subcase is contraction to a $W_r(\Delta(\alpha))$ leg. By the low momentum contraction expansion, the Δ leg must be a $W_r(\Delta)$ leg, i.e. if $r = r(\alpha)$ only a W leg contracts to a W. We wish to reorder the sum over cubes so that legs contract from smaller to larger cubes. For each W_r leg, we introduce a factor 2^{i-r+1} which fixes the choice of whether that leg contracts to another W_r leg in the ith inductive step. For the legs that do contract, we have another factor 2 to determine whether they contract to a smaller cube or to a larger cube. The sum over all contractions can now be written as a sum over contractions from small cubes to large cubes, $|\Delta| \leq |\Delta(\alpha)|$. It is an iterated sum, with a single sum for the possible contractions at each smaller cube. We bound the combinatoric factor 2^{r-i+2} by $|\Delta|^{-\varepsilon}$ per $W_r(\Delta)$ vertex, as in Case C 5 above.
By Lemma 4.2 (c), a combinatoric factor $O(1)\,|\Delta(\alpha)|^{-\varepsilon}\,(db_\alpha)^{3(1+\nu)}$ is sufficient to fix the sum over the cubes $\Delta(\alpha)$. We assign $O(1)\,|\Delta(\alpha)|^{-\varepsilon}$ to the $W_r(\Delta(\alpha))$ vertex, and $d^{3(1+\nu)}$ to the line. Since $|\Delta| \leq |\Delta(\alpha)|$, we have $b_\alpha \leq b$. We assign $b_\alpha{}^{3(1+\nu)/2}$ to the $W_r(\Delta(\alpha))$ leg and $b_\alpha{}^{3(1+\nu)/2} \leq b^{3(1+\nu)/2}$ to the $W_r(\Delta)$ leg.
As in Case C 5, a combinatoric factor $|\Delta(\alpha)|^{-\varepsilon}$ bounds the number of $W_r(\Delta(\alpha))$ legs during the ith inductive step.

Case S. The squaring operation (3.1.2) introduces two terms for each application. We assign a combinatoric factor 2 to each term, which cancels the 1/2 in (3.1.2). If we are in the ith inductive step, we let

$$\delta = O(1)^N \prod_{R(s)} c, \tag{4.24}$$

where N is the number of P_i vertices in $R(s)$, and where the product ranges over the various combinatoric factors c assigned to the vertices, lines and legs of $R(s)$, prior, to the application of (3.1.2). With this choice, the coefficient of each squaring vertex as it occurs in the inductive expansion is $O(1)^{-1}$. We choose this factor $O(1)^{-1}$ sufficiently small, by choosing $O(1)$ in (4.24) sufficiently large, to dominate the combinatoric factors $O(1)$ that are assigned in Cases P and W 1 to squaring vertices. Since a bounded number of such assigments occurs, we may make the total coefficient of a squaring vertex, namely $O(1)^{-1}$ times these combinatoric factors, less than 1. With this choice of δ, the second term in (3.1.2), multiplied by its combinatoric coefficient, is

$$\delta\Big(\prod_{R(s)} c\Big)\, R(s)^2 = O(1)^N \Big(\prod_{R(s)} c\Big)^2 R(s)^2. \tag{4.25}$$

Thus for the purpose of our estimates, we regard the combinatoric factors for vertices, lines and legs of $R(s)$ unchanged in this term, except for an increase by a factor $O(1)$ for each P_i vertex.
Case W 1. This subcase is the application of the Wick bound in (3.2.2) in Δ. Each application of (3.2.2) produces two terms. In order to absorb the constant K_1 in (3.2.2), we bound the combinatoric factor for each such term by $2K_1 = O(1)$. By Theorem 3.2.2, we assign this factor $O(1)$ for Δ to a P vertex or W vertex localized in Δ' in $\mathscr{D} \cup \mathscr{D}^*$.

Case W 2. We finally consider the subcase of applying the bound (3.2.21) in cubes $\Delta^* \in \mathscr{D}^*$. To facilitate the estimates of Chapters 5—6, we express each W vertex introduced by (3.2.21) as a sum of vertices with momenta of each leg restricted to an interval $[M_j, M_{j+1}]$. To control the sum over j, we assign a combinatoric factor $O(1)\,(1+j^2)^4$ to the W vertex. This factor is bounded by $|\Delta|^{-\varepsilon}$, by (4.13). A factor $O(1) = 12$ controls the sum over the twelve terms $\alpha_j{}^l$ in P in (3.2.21).

5. Estimating large graphs

5.1. Basic estimates and proof of Theorem 1.1.1.

Our bound for $\langle \exp(-tH)\rangle_0$ now has the form

$$\langle \Omega_0, e^{-tH}\,\Omega_0\rangle \leqq \sup_G c(G)\,|I(G)| \tag{5.1.1}$$

where $I(G)$ is the integral corresponding to a graph G, where G is a vacuum graph occurring in the inductive construction of Chapters 2—3, and where $c(G)$ is the combinatorial coefficient given in Theorem 4.1. In this chapter we establish estimates on $I(G)$, by reducing such bounds to estimates on certain small subgraphs with one to four vertices. We establish the required bounds on the small subgraphs in Chapter 6.

Our main result is an estimate which, combined with (5.1.1) and Theorem 4.1 proves the postivity of the $\varphi_3{}^4$ Hamiltonian, namely Theorem 1.1.1. Let λ, b, d be as in Section 1.2. For each line, we let $s = \max\{1, M\}$, where $M = \min\{\alpha^{(l)}\}$ is the lower momentum cutoff in the line, c.f. (1.2.4). Note that if $d > 1$, then $d < |r|\, s$ where r is the vector displacement between the centers of the localization cubes Δ, Δ' connected by the line. Let ε, $\nu = \nu(\varepsilon)$ and $M_1 = M_1(\nu, \varepsilon)$ be as in previous chapters.

Theorem 5.1., *Given $\varepsilon_1 > 0$, sufficiently small, given n_1 and given ε, ν positive but sufficiently small depending on ε_1, there exists constants $\alpha_1 = \alpha_1(\varepsilon_1, n_1, \varepsilon, \nu)$ and $M_1(\varepsilon_1, \varepsilon)$ such that*

$$|I(G)| \leqq \Big(\prod_{\text{lines}} d^{-n_1}\Big)\Big(\prod_{P_1\text{vertices}} \lambda^{-\varepsilon_1}\Big)\Big(\prod_{\text{vertices}} \alpha_1\,|\Delta|^{\varepsilon_1}\Big)\Big(\prod_{w\text{legs}} b^{-3(1+\nu)/2}\Big), \tag{5.1.2}$$

whenever M_1 of (1.2.6) *is larger than $M_1(\varepsilon_1, \varepsilon)$.*

Proof of Theorem 1.1.1. The inductive construction yields a bound

$$\langle \Omega_0, \exp\big(-tH(g,\varkappa)\big)\,\Omega_0\rangle \leqq \sup_G c(G)\,|I(G)|$$

and by Theorem 4.1 and Theorem 5.1,

$$\langle \Omega_0, \exp\,(-tH(g,\varkappa))\,\Omega_0\rangle$$
$$\leqq \exp\big(O(\mathrm{vol}(h))\big)\sup_G \Big(\prod_{\text{lines}} d^{n-n_1}\Big)\Big(\prod_{P_1\text{vertices}} \lambda^{\beta\varepsilon-\varepsilon_1}\Big)\Big(\prod_{\text{vertices}} L\alpha_1|\Delta|^{-\beta\varepsilon+\varepsilon_1}\Big).$$

(1) Choose ε_1 sufficiently small, then choose $\varepsilon = \varepsilon(\varepsilon_1)$, $\nu = \nu(\varepsilon)$ sufficiently small and $M_1(\varepsilon_1, \varepsilon)$ sufficiently large so that Theorem 5.1 holds.

(2) Choose ε sufficiently small so that $\beta\varepsilon < \varepsilon_1$ in Theorem 4.1. This fixes n, L, M of Theorem 4.1, and $M_1(\varepsilon_1, \varepsilon)$ of Theorem 5.1.

(3) Choose n_1 sufficiently large, so $n < n_1$. This fixes α_1 of Theorem 5.1.
(4) Pick M_1 sufficiently large so that $M_1 \geqq M_1(\varepsilon_1, \varepsilon)$ and $M_1 \geqq M$ as required above. Furthermore large M_1 forces λ to be large and (for $r > 1$) $|\Delta|$ to be small. Thus we can also require that

$$(L\,\alpha_1)^{17}\,\lambda^{-(\varepsilon_1-\beta\varepsilon)} \leqq 1\,, \qquad P_1 \text{ vertices}$$

$$L\,\alpha_1\,|\Delta|^{\varepsilon_1-\beta\varepsilon} \leqq 1\,, \qquad W, P_r, C_r, r > 1\,, \text{ vertices.}$$

For C_1 vertices, $|\Delta| = 1$ and we assign $L\alpha_1$ for each C_1 vertex to the generating P_1 vertex. In Chapter 4, we remarked that each P vertex is the generating P vertex for at most 16 C vertices. Thus our choice of M_1 above is sufficient.
Making these choices yields the bound $\langle\Omega_0, \exp(-tH(g,\varkappa))\,\Omega_0\rangle \leqq \exp\big(O(t)\,A(g)\big)$, where for a Hamiltonian cutoff vol $(h) = O(t)\,A(g)$, and hence by (1.1.4),

$$-O(1)\,A(g) \leqq E(g,\varkappa)\,,$$

as desired.

5.2. Localization factors

We devote the rest of Chapters 5 and 6 to the proof of Theorem 5.1. First we deal with the localization factor d^{n_1} per line, providing polynomial space time localization in the scaled distance. We note that our momentum cutoffs have compact support, so we cannot obtain exponential space-time localization. For convergent graphs with no momentum cutoff, our proof yields the localization factor $\exp\big((1-\varepsilon)\,|r|\big)$.
We now obtain a majorizing function $\bar{I}(G)$, and constants a_1', n_1' such that

$$|I(G)| \leqq \Big(\prod_{\text{lines}} \alpha_1'\,d^{-n_1}\Big)\,\bar{I}(G)\,, \tag{5.2.1}$$

and

$$\bar{I}(G) \leqq \Big(\prod_{\text{lines}} d^{n_1'}\Big)\Big(\prod_{P_1\text{vertices}} \lambda^{-\varepsilon_1}\Big)\Big(\prod_{\text{vertices}} \alpha_1'\,|\Delta|^{\varepsilon_1}\Big)\Big(\prod_{W\text{legs}} b^{-3(1-r)\,2}\Big). \tag{5.2.2}$$

The theorem follows from (5.2.1)—(5.2.2), since n_1' is independent of the choice of n_1.
Let us assume that the graph G contains only P or C vertices. For each vertex in G we define $d_{\max}$ as the maximum of the scaled distances d_j for lines contracting to the vertex. If $d_{\max} > 1$, then $d_{\max} \leqq |r_{\max}|\,s_{\max}$ where $r_{\max}$ and $s_{\max}$ are defined as r and s for the line yielding $d_{\max}$. We let $\mathscr{L}$ be the set of vertices of G with $d_{\max} > 1$.
We now prove that,

$$|I(G)| \leqq \Big(\prod_{\mathscr{L}} \alpha_2 d_{\max}^{-4n}\Big)\,\bar{I}(G)\,, \tag{5.2.3}$$

from which (5.2.1) follows. (We here set $n_1 = n$ for simplicity.)
Suppose a vertex in $\mathscr{L}$ is localized in Δ. If $s_{\max} > 2\,|\Delta|^{-1/3}$, we refine Δ by cubes $\{\Delta'\}$, where

$$|\Delta'|^{-1/3} \leqq s_{\max} \leqq 2\,|\Delta'|^{-1/3}\,. \tag{5.2.4}$$

We write the vertex localized in Δ as a sum of vertices localized in each Δ'. The localization cube enters the kernel for the vertex through the linear dependence of the kernel on the momentum conservation factor $(h\chi_\Delta)^\sim$. Thus the sum over

$O(|\Delta|\,|\Delta'|^{-1})$ translates of $h\chi_{\Delta'}$ yields the required refinement. In this manner we obtain $\prod_{\mathscr{L}} O(|\Delta|\,|\Delta'|^{-1})$ topologically identical graphs G_α whose integrals $I(G_\alpha)$ have the same cutoffs, the same s's as $I(G)$, but where each G_α has a different set of r's and d's. Then

$$I(G) = \sum_\alpha I(G_\alpha). \tag{5.2.5}$$

We label the lines of G by j, and the corresponding line of G_α by (j, α). Likewise we let $r_{j,\alpha}$ be the vector displacement of the jth line in G_α, etc. Since each Δ' is contained in its Δ, for each line $d_j \leqq d_{j,\alpha}$, and for each vertex, $d_{\max} \leqq d_{\max,\alpha}$. Thus for each vertex, if $d_{\max} > 1$, then $d_{\max} < |r_{\max,\alpha}|\, s_{\max}$, for all α.
The kernel of the various $I(G_\alpha)$, for a given $I(G)$, has the form

$$m_\alpha(k) = w(k) \prod_{\text{lines}} \exp\,(ik_j r_{j\alpha})\,, \tag{5.2.6}$$

where $w(k)$ is independent of α and of the $r_{j\alpha}$. Then

$$I(G_\alpha) = \int m_\alpha(k)\, dk\,.$$

Multiplication of $I(G_\alpha)$ by a component $r_{j\alpha}^{(l)}$ of $r_{j\alpha}$ is equivalent to replacing w by $i\,\partial w/\partial k_{j\alpha}^{(l)}$. Thus

$$\left|\left(\prod_{\mathscr{L}} d_{\max}\, r_{\max,\alpha}\, s_{\max}\right)^{4n} I(G_\alpha)\right| \leqq \left|\int \left(\prod_{\text{lines}} e^{ik_j r_{j\alpha}}\right) \left(\prod_{\mathscr{L}} s_{\max}^2\, \Delta_{\max}^2\right)^{4n} w(k)\, dk\right| \tag{5.2.7}$$

where $\Delta_{\max}^2$ is the Laplacian $\sum \partial^2/\partial k_{\max}^{(l)2}$ for the momentum $k_{\max}$ of the line yielding $d_{\max}$.
We now estimate the derivatives of the factor μ^{-2}, η and $(h\chi_{\Delta'})_0^{\sim}$ in $w(k)$ that appear in (5.2.7). The propagator $\mu(k)^{-2} = (k^2+1)^{-1}$ satisfies for each n_1,

$$|D_j^{n_1} \mu(k_j)^{-2}| \leqq O(1)\, \mu(k_j)^{-2-|n_1|}\,,$$

where

$$D_j^{n_1} = \prod_{l=0}^{2} \left(\frac{\partial}{\partial k_j^{\,l}}\right)^{n_{1,l}}, \sum_l n_{1,l} = |n_1|\,.$$

This estimate follows from the Cauchy integral representation of μ^{-2}, and the constant is bounded by $O(n_1!)$. On the support of the kernel w,

$$|D_j^{n_1} \mu(k_j)^{-2}| \leqq O(s_j^{-|n_1|})\, \mu(k_j)^{-2} \tag{5.2.8}$$

Likewise, on the support of w, the $\mathscr{S}$ valued momentum cutoff in the jth line $\eta_{\alpha_j,\beta_j}(k_j)$ satisfies

$$|D_j^{n_2}\, \eta_{\alpha_j,\beta_j}(k_j)| \leqq O(1)\, s_j^{-|n_2|} \tag{5.2.9}$$

by (1.2.4) and the fact that $M_1 \geqq 1$. Here $O(1)$ in (5.2.9) depends on $|n_2|$, but is uniform in (α_j, β_j).
Finally, we note that the conservation of momentum factor $(h\chi_{\Delta'})_0^{\sim}$ that enters w

is $(h\chi_{\Delta'})^{\sim}$ translated to the origin. With the comparison function

$$F_\Delta(k) = \prod_{l=0}^{2} (1 + |\Delta|^{1/3} |k^{(l)}|)^{-1}, \tag{5.2.10}$$

we have

$$|D^{n_3}(h\chi_{\Delta'})^{\sim}(k) \leq O(1)|\Delta'|^{1+|n_3|/3} F_{\Delta'}(k) \leq O(1)\,(s_{\max}{}^{-|n_3|})\,|\Delta|\,F_\Delta(k), \tag{5.2.11}$$

since $|\Delta'| \leq |\Delta|$, in which case $|\Delta'|\,F_\Delta \leq O(1)\,|\Delta|\,F_\Delta$.
Combining (5.2.7)—(5.2.11), we obtain an estimate for $I(G_\alpha)$, namely

$$\left|\left(\prod_{\mathscr{L}} d_{\max}^{4n}\right) I(G_\alpha)\right| \leq \left[\prod_{\mathscr{L}} (r_{\max,\alpha}\, s_{\max})^{-4n}\right] \bar{I}(G), \tag{5.2.12}$$

$$\bar{I}(G) = \int \bar{w}(k)\, dk,$$

where $\bar{w}$ majorizes w and where $\bar{I}(G)$ will be shown to satisfy (5.2.2). Here $\bar{w}(k)$ is a product of the following factors: (i) a factor $\mu_i^{-2} = \mu(k_i)^{-2}$ for each factor μ_i^{-2} in $w(k)$; (ii) a factor $O(1)\,|\Delta|\,F_\Delta(q)$ bounding each factor $(h\chi_\Delta)^{\sim}(q)$ in w; (iii) the characteristic function of the support of w; (iv) any constants, such as uncancelled renormalization constants, not absorbed in the combinatoric factors of Chapter 4. In order to bound $I(G)$, we use (5.2.5) and sum the bounds (5.2.12) over α. Each line with $d = d_{\max,\alpha} > 1$, contracting to a vertex in the set $\mathscr{L}$ and localized in Δ', satisfies

$$|r_{j,\alpha}| > \frac{1}{2}\,|\Delta'|^{1/3} = O(s_{\max}^{-1}), \tag{5.2.13}$$

where $s_{\max}$ is determined by Δ'. The sum over α is a sum over translates of individual vertices in $\mathscr{L}$. For each such vertex, by (5.2.13),

$$\sum_\alpha r_{\max,\alpha}^{-4n} \leq O(s_{\max}^{4n}),$$

and for all vertices,

$$\sum_\alpha \prod_{\mathscr{L}} (r_{\max,\alpha}^{-4n}) \leq \prod_{\mathscr{L}} O(s_{\max}^{4n}). \tag{5.2.14}$$

From (5.2.12) and (5.2.14) we infer (5.2.3), and thus (5.2.1), for a diagram with P and C vertices only.
For a diagram containing W vertices, we follow a similar procedure. The only additional complication comes from the more complicated majorizing function $\bar{w}$ for w, due to the more complicated kernels for W vertices. A single W vertex with $\mathscr{L}$ legs has a kernel $(\mu_1 \cdots \mu_l)^{-1}\, v(k)$, where $v(k)$ is given by (6.2.6—7) and (6.2.12). With $s_{\max}$ defined as above, we define Δ' as in (5.2.4), when $s_{\max} > 2\,|\Delta|^{-1/3}$. Note that if $s_{\max}$ at the W vertex is associated with *a* $\delta\varphi_1$ leg, $s_{\max} \leq |\Delta|^{-1/3}$, and we define $\Delta' = \Delta$. We expand the "overall" conservation of momentum factor $(h\chi_\Delta)^{\sim}(\sum_{I_1}\lambda_j k_j + \sum_{I_2} k_j)$ as a sum over $(h\chi_{\Delta'})^{\sim}$ of the same momentum. We then proceed as above in the proof of (5.2.3) and (5.2.1), using the bound

$$|D^{n_3}(h\chi_\Delta)^{\sim}| \leq O(1)\, s_{\max}^{-|n_3|}\, |\Delta| F_\Delta,$$

if $s_{\max} \leqq |\Delta|^{-1/3}$, in place of (5.2.11). Furthermore, in the kernel $\overline{w}$, we use the bound (6.2.16) for $|v(k)|$ from the kernel of a W vertex. Thus (5.2.1) follows in all cases. In applying our estimates, we include any counterterm used to cancel a divergence (in Chapter 2) as part of the graph being estimated. Thus the exactly cancelled vacuum energy terms

 and 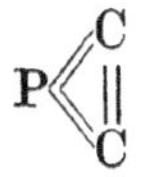

do not appear in the expansion. The mass term

may occur by itself in uncancelled form, or in the cancelled form

where the second graph denotes the mass counterterm in V_C, a M_2 type P vertex. The cancelled form of the mass term is convergent, and occurs only if the localization cubes of the P and C vertices coincide, see Proposition 5.3.4. If the cubes of the P and C vertices do not coincide, the uncancelled mass diagram is convergent.

5.3. Decomposing large graphs

The major part of our estimates for large graphs involves decomposing such a graph G into small subgraphs $\{g_j\}$, $j = 1, 2, \ldots$, and estimating $I(G)$ or $\bar{I}(G)$ in terms of the $I(g_j)$ or the $\bar{I}(g_j)$. Each subgraph g_j consists of a set of vertices, a set of "internal lines" (lines that connect these vertices) and a set of "external legs" (legs that either connect g_j to other subgraphs or which are external to G). Each line or leg has a monentum variable. We say $\{g_j\}$ is a decomposition of G into disjoint subgraphs, if $g_j \subset G$, and if each vertex of G lies in exactly one g_j. The integral $I(g_j)$ is a function of the external momenta. We divide these external momenta into two sets, initial (i) and final (f) momenta, according to whether they contract to g_{j+n} with $n < 0$ (initial), or with $n > 0$ (final). Legs external to G may be arbitrarily assigned. (The sets (i) and (f) should not be confused with a and a^* legs of Friedrichs diagrams.) We let $\|I(g)\|_{i,f}$ denote the operator norm of $I(g)$ from L_2 of the initial variables to L_2 of the final variables (Lebesque measure). Diagramatically we denote $I(g_j)$ by a vertex with initial legs i_j and final legs f_j leaving opposite sides,

$I(g_j) = i_j$ 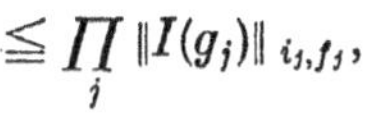 f_j

Our basic estimate for decomposing $I(G)$ is

$$\|I(G)\|_{i,f} \leqq \prod_j \|I(g_j)\|_{i_j,f_j}, \tag{5.3.1}$$

valid since $\|A \otimes I\| = \|A\|$ and $\|AB\| \leqq \|A\| \, \|B\|$. Furthermore, each $\|\cdot\|_{i,f}$ can be dominated by the Hilbert Schmidt norm $\|\cdot\|_{H.S.}$ and equals the Hilbert Schmidt norm if $i = \varnothing$ or $f = \varnothing$. Sometimes it is convenient to label the operator norms

$\|\cdot\|_{r,s}$ where r and s are the number of momenta in i and f respectively. Diagramatically, for example,

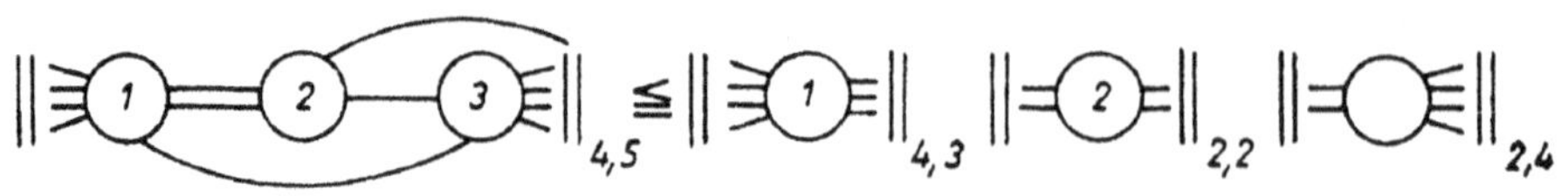

(5.3.2)

We now partition a vacuum diagram G into a union of disjoint subgraphs $\{g_j'\}$. We number these subgraphs in the order that they were formed in the inductive construction, i.e. from earlier to later subgraphs. Each subgraph g_j' consists of either
1) a single W vertex,
or
2) a P vertex, and the new C vertices it generates. If a g_j' consists of a single P vertex and it contracts four times to earlier subgraphs, we call this g_j' *exceptional.* We define a new partition $\{g_j\}$ by combining each exceptional subgraph g_j' with the latest subgraph g_j' to which it contracts.
We call g_j a P_r subgraph if its nonexceptional g_l' contains a P_r vertex (Case 2 above), and this P is called the *leading* P vertex. Otherwise, we call g_j a W subgraph. We further classify the P_l subgraphs as convergent P_C, or logarithmically divergent P_D, according to the number l_j of initial legs on the leading P vertex. Diagrams are P_C for $l_j \leqq 2$ and P_D for $l_j = 3$. According to the $P-C$ expansion of Chapter 2, the only two possible P_D subgraphs are

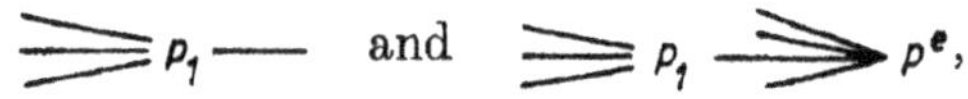

(5.3.3 a, b)

where P^e denotes an exceptional P vertex.
Lemma 5.3.1. Let ε_2 be sufficiently small, and let M_1 of (1.2.6) be sufficiently large, depending on ε, ε_2. There exist constants α_2 and n_2 such that for a P_C subgraph

$$\|\bar{I}(g_j)\|_{i_j,f_j} \leqq \lambda^{-\varepsilon_2} \Big(\prod_{\text{lines}} d^{n_2}\Big) \Big(\prod_{\text{vertices}} \alpha_2 \,|\Delta|^{\varepsilon_2}\Big), \tag{5.3.4}$$

for a P_D subgraph

$$\|\bar{I}(g_j)\|_{i_j,f_j} \leqq (\log \lambda)^4 \Big(\prod_{\text{lines}} d^{n_2}\Big) \Big(\prod_{\text{vertices}} \alpha_2 \,|\Delta|^{\varepsilon_2}\Big), \tag{5.3.5}$$

and for a W or P_r, $r > 1$, subgraph

$$\|\bar{I}(g_j)\|_{i_j,f_j} \leqq \Big(\prod_{\text{vertices}} \alpha_2 \,|\Delta|^{\varepsilon_2}\Big) \Big(\prod_{w\text{legs}} b^{-3(1+\nu)/2}\Big). \tag{5.3.6}$$

The factor λ refers to the leading vertex of the subgraph.

Proof of Theorem 5.1.1. We establish (5.2.2) to complete the proof of the theorem. We use (5.3.1) to bound $\bar{I}(G)$ by a product of factors of the form (5.3.4)—(5.3.6). The bound (5.3.6) on P_r, $r > 1$, or on W subgraphs already has the correct form required by (5.2.2), provided ε_1 is chosen sufficiently small. We note that P_C subgraphs contain at most 53 vertices: 1 leading P_1 vertex, 16 C vertices and 36 exceptional P vertices. Let λ_1 be the maximum lower cutoff for an exceptional $P_1(\Delta_1)$ vertex in a P_C subgraph. By conditon (c) of Chapter 2, applied at the time the leading P_1 vertex is introduced (and thus before the exceptional $P_1(\Delta_1)$. vertex is introduced),

$$\lambda^{-\delta} \lesssim (cd^4)^{\delta/(1+\nu)\varepsilon} M_{(u\Delta_1)}^{-\delta/(1+\nu)} \leqq (cd^4)^{\delta/(1+\nu)\varepsilon} \lambda_1^{-\delta}. \tag{5.3.7}$$

Thus we may bound the convergence factor $\lambda^{-\varepsilon_2}$ on the leading vertex in (5.3.4) by a product of factors $\lambda_1^{-\varepsilon_2/37}$, with one factor for each P_1 vertex, times a factor d^{n_2} per line and $O(1)$ per vertex.

For P_D subgraphs, we transfer factors $\lambda^{-\delta}$ from earlier P_C subgraphs in order to dominate the $(\log \lambda)^4$ in (5.3.5) and to obtain convergence for its P_1 vertices. Basically, this method works because a P_D subgraph has at least 3 initial legs and at most 1 final leg, and consequently the number of P_D subgraphs is bounded in terms of the number of P_C subgraphs. We note that initial legs of P_D subgraph must contract to P_C or P_D subgraph.

We define a *chain* to be a finite, connected sequence of lines and vertices in G, chosen so that each line joins two consecutive vertices in the sequence, and so that the order of the vertices in the sequence coincides with the order in which these vertices were created in the inductive construction. We consider chains that start with the final leg of a P_C subgraph, that terminate with an initial leg of a P_D subgraph, and otherwise pass through P_D subgraphs (necessarily of type (5.3.3a). Let l be the number of lines in the chain, and let $\varrho = 3^{-l+1}$ be the *weight* of a chain of length l. We let $\sum^i$ denote the sum over chains in G with a fixed initial leg, and $\sum^f$ denote a sum over chains in G with a fixed final leg.

Lemma 5.3.2. *For any fixed initial leg, or any fixed final leg,*

$$\textstyle\sum^i \varrho \leqq 2, \qquad \sum^f \varrho = 1. \tag{5.3.8i, f}$$

Proof. The first inequality is clear, since a chain with a fixed initial leg cannot branch. Thus there is exactly one such chain for each l, and

$$\textstyle\sum^i \varrho \leqq \sum_{l=1}^{\infty} 3^{-l+1} = \frac{3}{2}.$$

Let T be the set of P_D vertices that occur in the set of chains with a fixed final leg. We prove (5.3.8f) by induction on the length l of the chain. If $l = 1$, there is exactly one chain of length one, and (5.3.8f) holds. Let $n(l)$ be the number of chains of length l with a fixed final leg. Assume (5.3.8f) holds for chains of length at most k and add a P_D vertex to chains of length k. Then

$$\sum_{l=1}^{k} \varrho(l)\, n(l) = 1,$$

and after adding a vertex to j chains of length k,

$$\begin{aligned}\sum_{l=1}^{k+1} \varrho(l)\, n(l) &= \sum_{l=1}^{k-1} \varrho(l)\, n(l) + \varrho(k)\big(n(k) - j\big) + 3j\varrho(k+1) \\ &= \sum_{l=1}^{k} \varrho(l)\, n(l) + j\big(3\varrho(k+1) - \varrho(k)\big).\end{aligned}$$

Since $3\varrho(k+1) = \varrho(k)$, the lemma follows.

We now return to the proof of Theorem 5.1. above. For each chain, we take a factor $\lambda^{-\varrho\delta/3}$ from the leading P vertex of the initial P_D subgraph in the product

(5.3.1). By the bound (5.3.8i), each initial P_C subgraph has at most a factor $\lambda^{2\delta/3}$ assigned in this manner among the various P_D subgraphs on chains initiating at the P_C subgraph. We choose δ sufficiently small so that $2\delta/3 < \varepsilon_2/74$; thus the total factor from the initial P_C subgraph that must be assigned is $\lambda^{-\varepsilon_2/74}$, leaving a similar factor for the leading vertex in the P_C subgraph. As in the case of P_C above, we use (5.3.7) to convert $\lambda^{-\varrho\delta/3}$ into a factor $d^{n_2\varrho}$ per line in the chain, $O(1)^\varrho$ per vertex in the chain, and $\lambda_D^{-\varrho\delta/3}$ for the final P_D subdiagram. Again, by (5.3.8i), this results in at most a total factor d^{n_2} per line and $O(1)$ per vertex. The identity (5.3.8f) assures that a total factor $\lambda_D^{-\delta}$ is assigned to (5.3.3a) and $\lambda_D^{-2\delta}$ to (5.3.3b).
In the latter case, $P_D = P_1 - P_1$ and we divide $\lambda_D^{-2\delta}$ equally between the two P_D vertices, as above. This convergence dominates the logarithmic divergence in (5.3.5), and completes the proof of (5.2.2).
Next we state the basic lemmas on small subdiagrams to be used in proving lemma 5.3.1. These estimates are proved in Chapter 6. We let $M(1), \ldots, M(4)$ denote the upper cutoffs in the legs of a P or C vertex, numbered so that $M(1) \leq M(2) \leq \cdots$.

Proposition 5.3.3. *Let w be the kernel of a φ^4P or C vertex in Δ.*

(a) 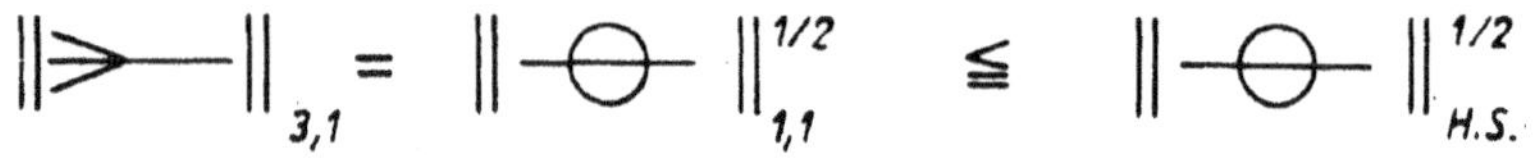

$$\leq (|\Delta|^{\varepsilon_3}) \left(\log M(2)\right)^{1/2},$$

for $\varepsilon_3 \leq 1/4$.

(b) 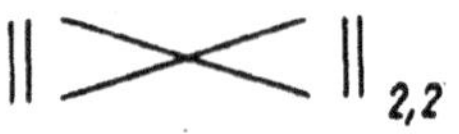 $\leq O(|\Delta|^{\varepsilon_3})\, \lambda^{-\varepsilon_3}$,

for $\varepsilon_3 < 1/16$.

(c) 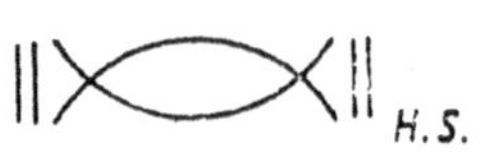 $\leq O(|\Delta|^{\varepsilon_3})\, \lambda^{-\varepsilon_3}$,

for $\varepsilon_3 < 1/16$.

Proposition 5.3.4. *The uncancelled mass renormalization diagram*

$$M = M(\Delta, \Delta') = -\,\mathrm{P}\equiv\mathrm{C}\,-$$

and the mass counterterm

$$\delta M = \delta M(\Delta, \Delta')\, \delta_{\Delta,\Delta'}$$

($\delta_{\Delta,\Delta'} = 0$ for $\Delta \neq \Delta'$ and 1 otherwise) have Hilbert Schmidt norms that are $O(|\Delta|^{1/4} |\Delta'|^{1/4} \log M(2)^{1/2})$. Their sum with kernel w, satisfies

$$\|w\|_{H.S.} \leq O(|\Delta'|^{\varepsilon_3} |\Delta|^{\varepsilon_3} \lambda^{-\varepsilon_3} d^{-n})$$

for some $\varepsilon_3 > 0$. Here d is the scaled distance from Δ to Δ'.

Proposition 5.3.5. *Let w be the kernel of a W vertex, as given by (3.2.17), (3.2.19), with each momentum leg restricted to an interval $[M_j, M_{j+1}]$. Let $\varepsilon_3 < 1/4$ and tet ν of (1.2.6) be sufficiently small. Then*

$$\|w\| \leq O(|\Delta|^{\varepsilon_3}) \prod_{\text{legs}} b^{-3(1+\nu)/2}.$$

Proposition 5.3.6. *The two vertex diagram*

$$P\equiv W$$

for contraction between a P vertex in Δ and a W vertex ix Δ' is bounded by

$$O\Big(|\Delta|^{\varepsilon_3}\,|\Delta'|^{\varepsilon_3}\,\lambda^{-\varepsilon_3}\prod_{\text{legs}} b^{-3(1+\nu)/2}\Big)$$

for some $\varepsilon_3 > 0$, provided ε and ν are sufficiently small.

Proof of Lemma 5.3.1. We list the various possible P_C, P_D, P_r or W subgraphs, and estimate each of them. We first consider a W subgraph. This consists of a single W vertex, and possibly some exceptional P vertices contracted to it. In the case of a single exceptional P vertex contracted four times to the W vertex, we use Proposition 5.3.6. In the remaining cases, we bound the operator norm of the W subdiagram by a Hilbert Schmidt norm.

If a graph G is the union of two graphs G_1 and G_2, then a special case of (5.3.1) is

$$\|I(G)\|_{\text{H.S.}} \leqq \|I(G_1)\|_{\text{H.S.}}\,\|I(G_2)\| \tag{5.3.9}$$

where the norm on $I(G_2)$ is an operator norm, acting from variables internal in G but external in G_2 to variables external in both G and G_2.

As a special case of (5.3.9), we let G_1 be the W vertex in the W subgraph, and G_2 the complementary subgraph of the exceptional P vertices. By construction each P vertex in G_2 contracts to the W subgraph G_1 and has legs external to G. Thus the operator norm of $I(G_2)$ is bounded by (a) and (b) of Proposition 5.3.3. We bound $\|I(G_1)\|_{H.S.}$ by Proposition 5.3.5, yielding

$$\|\bar{I}(g_j)\|_{\text{H.S.}} \leqq \Big(\prod_{W\text{legs}} \log\lambda\Big)\Big(\prod_{\text{vertices}} \alpha_3\,|\Delta|^{\varepsilon_3}\Big)\Big(\prod_{W\text{legs}} b^{-3(1+\nu)/2}\Big). \tag{5.3.10}$$

We recall that after the first Wick construction (and hence for P^e vertices in a W subgraph) we have by (4.2),

$$\begin{aligned}\log\lambda &\leqq \log M_{u(\Delta)} \leqq O(\varepsilon^{-1})\log(2|\Delta|^{-1}) \\ &\leqq |\Delta|^{\varepsilon_3/2}.\end{aligned} \tag{5.3.11}$$

Here $u(\Delta)$ refers to the point in the inductive construction when the P^e vertex is introduced, and the last inequality follows for M_1 sufficiently large, depending on ε, ε_3. We thus obtain (5.3.6) with $\varepsilon_2 < \varepsilon_3/2$, and our proof for W subgraphs is complete.

We next consider the two P_D subgraphs. The desired bound on (5.3.3a) follows by Proposition 5.3.3 (a). In the case of (5.3.3b), the operator norm of the P_D subgraph is the Hilbert Schmidt norm, and this norm squared is represented by the graph.

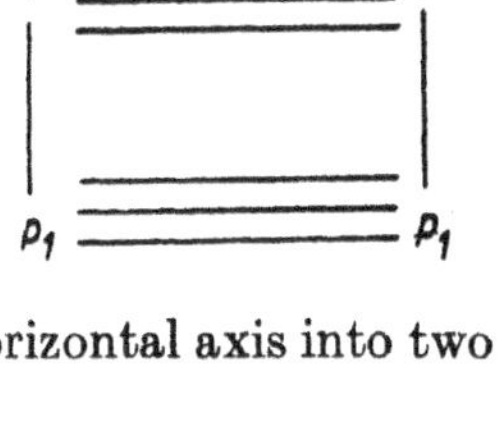

(5.3.12)

Splitting (5.3.12) across the horizontal axis into two subgraphs, we bound (5.3.12) by

$$\| - P_1 \equiv P_2 - \|_{\text{H.S.}} \| - P^e \equiv P^e - \|_{\text{H.S.}} \leqq \| - P_1 \equiv \|_{\text{H.S.}} \| \equiv P^e - \|_{\text{H.S.}} \tag{5.3.13}$$

using (5.3.1) two times. We now use Proposition 5.3.3 (a). If P^e is a P_r vertex with $r > 1$, we bound $\log \lambda_e$ by (5.3.11). If P^e is a P_1 vertex, we use Condition (c) of Chapter 2, as in the proof of (5.3.7), to establish $\log \lambda_e \leqq O(1)\, d \log \lambda$, where d is the distance for the line in the P_D subgraph connecting P_1 and P^e. In either case, we obtain (5.3.5).

We next consider P_C subgraphs; these have fewer than 53 vertices. In our estimates below, we obtain a convergence factor $\lambda^{-\varepsilon_3}$ from the leading P vertex (denoted by p) and $O(1)\,|\Delta|^{\varepsilon_3}$ from every vertex localized in Δ. In addition, certain nonleading vertices yield an ultraviolet divergent bound $O(\log \bar{\lambda}_j)$, where $\bar{\lambda}_j$ is the upper cutoff in the P or C vertex labelled by j. For a P_r or C_r vertex, $r > 1$, we have as in (5.3.11),

$$\log \bar{\lambda}_j \leqq \log M_{u(\Delta)} \leqq |\Delta|^{\varepsilon_3/2} \tag{5.3.14}$$

for M_1 sufficiently large. For a $P_1(\Delta)$ or $C_1(\Delta)$ vertex, by Condition (c) of Chapter 2,

$$\log \bar{\lambda}_j \leqq O(1) \log \big(1 + \text{dist.}\ (\Delta, \Delta_l)\big) \log \lambda,$$

where Δ_l is the localization cube for p. Thus by (2.11), and the definition (1.2.11),

$$\log \bar{\lambda}_j \leqq \big(\textstyle\prod O(1)\, d\big) \log \lambda, \tag{5.3.15}$$

where the product ranges over the scaled distances d for a sequence of lines connecting p to the vertex labelled by j. Since there are at most 52 such vertices j in the P_C subgraph, our logarithmically divergent bounds can be absorbed into a factor d^{n_2} per line, $O(1)$ per vertex, and $\lambda^{\varepsilon_3/2}$ for the vertex p, leaving a convergence factor $\lambda^{-\varepsilon_3/2}$ for p.

Aside from the leading vertex p, our P_C subgraph consists of a set of C vertices to which p contracts (inner C's), a set of C vertices to which the inner C's contract (outer C's) and exceptional P vertices contracting to the inner and outer C vertices. We choose a subgraph $C(g_j)$, called a *core*, contained in g_j. If p contracts two or three times to a single C, the core consists of p and this C vertex; otherwise it consists of p and two inner C vertices. (Since g_j is a P_C subgraph, p has at most two external legs and at least two contractions.) Each C vertex is either a φ^4 vertex or a M vertex (mass renormalization vertex from V_C). We choose C vertices in the core to be φ^4 vertices, as far as possible. Among possible cores with an equal number of vertices and φ^4 vertices, we choose a core with the maximum number of internal lines. The purpose of these restrictions is to reduce the number of cases to be considered when bounding $I(g_j)$. We note that an M type p vertex does not occur alone, but is part of a cancelled mass diagram.

We dominate $\|I(g_j)\|_{\text{H.S.}}$ for a P_C subgraph. We first use (3.5.9) to remove "outer layers" of g_j; while leaving its core. We take $C(g_j) \subset G_1$, and we remove G_2, allowing logarithmically divergent estimates on the norm $\|I(G_2)\|$. We first remove, one by one, P^e or noncore C vertices with external legs, bounding these by Proposition 5.3.3 (a), (b), or Proposition 5.3.4. We then remove an P^e—C or C—C pair, contracting to each other, but not part of the core. If a single φ^4 vertex contracts four times to the pair, we include that vertex with the pair in G_2. We bound the

Hilbert Schmidt norm of $I(G_2)$ using Proposition 5.3.3 (a), (c) or Proposition 5.3.4, and possibly (5.3.1). This yields a convergent or a logarithmically divergent bound. We continue removing noncore singlet P^e or C vertices, and noncore P^e-C or $C-C$ pairs, as far as possible. By the construction of Chapter 2 and the definition of the core, no P^e vertex contracts exclusively to the core. Thus all P^e vertices are removed at this stage, and noncore C vertices contract completely to the core. We finally remove any noncore, M type C vertex with a Hilbert Schmidt norm, using Proposition 5.3.4.

For a two vertex core, there are only two possible cases

$$= \mathrm{p} = \mathrm{C}, \qquad - \mathrm{p} \equiv \mathrm{C} -, \tag{5.3.16}$$

because the third order vacuum diagram

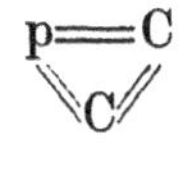

is cancelled exactly and does not occur. The Hilbert Schmidt norm in these cases is bounded by Proposition 5.3.3 (c) or Proposition 5.3.4.

For a three vertex core, $C(g_j)$ must be

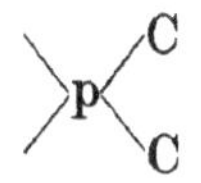

(5.3.17)

with possibly additional contractions between the two C vertices. For two φ^4 type C vertices, an additional φ^4 type C vertex can contract four times to the core, yielding

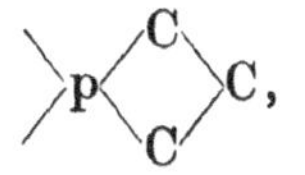

suppressing additional internal lines. We remove the p vertex with a convergent operator norm, using Proposition 5.3.3 (b), with the exception of the case

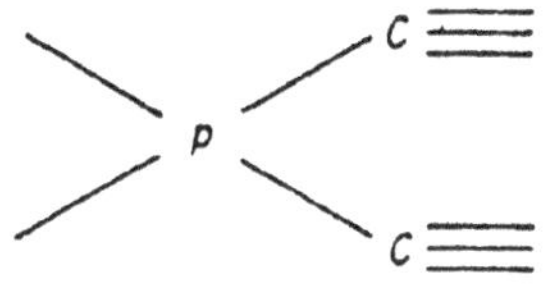

(5.3.18)

considered below. The remaining graph consists of 2 or 3 C vertices only and is connected. By removing one singlet C with an external leg from the 3 C graph, we are reduced to a $C-C$ pair and we obtain a logarithmically divergent bound by Proposition 5.3.3 (a), (b).

The Hilbert Schmidt norm of (5.3.18) squared is

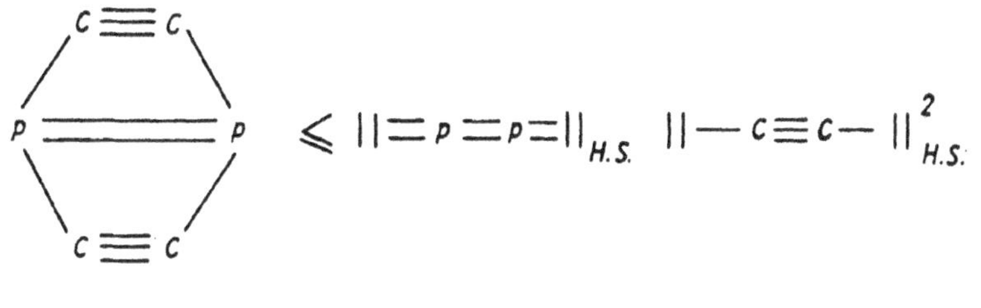

The desired bound follows by Proposition 5.3.3 (a), (c) and (5.3.1). This concludes our analysis of P_C subgraphs and proves (5.3.4).
Remaining are P_r subgraphs, $r > 1$. We treat these as with P_C or P_D subgraphs, but never use (5.3.15) to bound logarithmic divergences. Since P_r subgraphs contain $P_{r'}$ or $C_{r'}$ vertices, $r' \geq r$, we use (5.3.14) in every case. This yields the bound (5.3.6), and concludes the proof of Lemma 5.3.1.

6. Estimates of small graphs

6.1. Convolution estimates

We prove the basic estimates on small graphs in this chapter. We start with elementary estimates on convolutions, useful in more general circumstances. In the second section, we bound the Hilbert Schmidt norm of a single W vertex, while in the final section we deal with our remaining estimates.
Proposition 6.1.1. Let $p \in R^n$, $\delta + \lambda > n$, $\nu = \min\{\delta, \lambda, \delta + \lambda - n\}$, $\mu(p) = (p^2 + 1)^{1/2}$. Then

$$\mu^{-\delta} * \mu^{-\lambda} \leqq O(1) \begin{cases} \mu^{-\nu}, \text{ if } \max\{\delta, \lambda\} \neq n \\ \mu^{-\nu} \log \mu, \text{ if } \max\{\delta, \lambda\} = n. \end{cases}$$

Proposition 6.1.2. *Let* $p \in R^1$, $0 \leqq \delta \neq 1$, $0 \leqq \lambda \neq 1$, $\delta + \lambda > 1$, $\nu = \min\{\delta, \lambda, \delta + \lambda - 1\}$, $0 < a \leqq 1$ *and* $\mu_a(p) = \mu(ap)$. *Then*

$$\mu_a^{-\delta} * \mu^{-\lambda} \leqq \begin{cases} O(a^{-\delta})\, \mu^{-\nu}, & \text{if } \delta < 1 \\ O(a^{-1})\, \mu^{-\nu}, & \text{if } \lambda < 1 < \delta \\ O(a^{-\nu})\, \mu^{-\nu}, & \text{if } 1 < \lambda,\ 1 < \delta. \end{cases}$$

By scaling, we obtain from Proposition 6.1.2 the
Corollary 6.1.3: *Let* $p \in R^1$,

$$0 < a \leqq b,\ 0 \leqq \delta \neq 1, \qquad 0 \leqq \lambda \neq 1, \delta + \lambda > 1, \nu = \min\{\delta, \lambda, \delta + \lambda - 1\}.$$

(a) *If* $\delta < 1$, *then*

$$a^{\delta} b^{1-\delta} \mu_a^{-\delta} * \mu_b^{-\lambda} \leqq O(1)\, \mu_b^{-\nu}.$$

(b) *If* $\lambda < 1 < \delta$, *then*

$$a \mu_a^{-\delta} * \mu_b^{-\lambda} \leqq O(1)\, \mu_b^{-\lambda}.$$

(c) *If* $1 < \lambda$, $1 < \delta$, $\nu = \min\{\delta, \lambda\}$, *then*

$$\left(\frac{a}{b}\right)^{\nu} b \mu_a^{-\delta} * \mu_b^{-\lambda} \leqq O(1)\, \mu_b^{-\nu}.$$

Proofs. (6.1.1) It is sufficient to consider $0 < \delta \leqq \lambda$, for if $\delta < 0$ the proposition follows from the fact that $\nu = \delta$ and

$$\mu(p - q)^{-\delta} \leqq \mu(p)^{-\delta} \mu(q)^{-\delta}.$$

We consider three regions of integration,

$$\text{I:} \quad |q| \leqq \frac{1}{2}|p|; \qquad \text{II:} \quad \frac{1}{2}|p| \leqq |q| \leqq 2|p|;$$

$$\text{III:} \quad 2|p| \leqq |q|.$$

In region I, $\mu(p-q) \geqq O(1)\,\mu(p)$, so

$$\int_I \mu(p-q)^{-\delta}\,\mu(q)^{-\lambda}\,dq \leqq O\big(\mu(p)^{-\delta}\big) \int_I \mu(q)^{-\lambda}\,dq$$

$$\leqq \begin{cases} O(\mu^{-\delta}), & \lambda > n \\ O(\mu^{-\delta}\log\mu), & \lambda = n \\ O(\mu^{-\delta+n-\lambda}), & \lambda < n, \end{cases}$$

the desired bounds for region I.
In region II, $|q-p| \leqq |q| + |p| \leqq 3|p|$ and $|p| \leqq 2|q|$, so

$$\int_{II} \mu(p-q)^{-\delta}\,\mu(q)^{-\lambda}\,dq \leqq O\big(\mu(p)^{-\lambda}\big) \int_{II} \mu(p-q)^{-\delta}\,dq$$

$$\leqq \begin{cases} O\big(\mu(p)^{-\lambda}\big) & \delta > n \\ O(\mu^{-\lambda}\log\mu) & \delta = n \\ O(\mu^{-\lambda+n-\delta}) & \delta < n. \end{cases}$$

If $\delta = n$ and $\lambda > n$, $\mu^{-\nu}\log\mu \leqq O(\mu^{-\nu})$. If $\delta = n = \lambda = \nu$, the bound $O(\mu^{-\lambda}\log\mu)$ also has the desired form.
In region III, $|p-q| \geqq 1/2\,|q| \geqq |p|$. Thus

$$\int_{III} \mu(p-q)^{-\delta}\,\mu(q)^{-\lambda}\,dq \leqq \int_{III} \mu(q)^{-\lambda-\delta}\,dq \leqq O(1)\,\mu(p)^{-\lambda-\delta+n},$$

to complete the proof of (6.1.1).
We note that

$$\mu_a(p)^{-1} \leqq O(a^{-1})\,\mu(p)^{-1} \tag{6.1.2}$$

and so the case $\delta < 1$ follows from Prop. 6.1.1. For $1 < \lambda$, δ we have

$$\mu_a^{-\delta} * \mu^{-\lambda} \leqq \mu_a^{-\nu} * \mu^{-\lambda} \leqq O(a^{-\nu})\,\mu_a^{-\nu} * \mu^{-\lambda},$$

since $\mu^{-\delta} \leqq \mu^{-\lambda}$. Thus this case also follows from Prop. 6.1.1. We consider the remaining case, $\lambda < 1 < \delta$. In the region of integration I above,

$$\int_I \mu_a(p-q)^{-\delta}\,\mu(q)^{-\lambda}\,dq \leqq O(1)\,\mu_a(p)^{-\delta} \int_I \mu(q)^{-\lambda}\,dq$$

$$\leqq O\big(\mu_a(p)^{-\delta}\,\mu(p)^{1-\lambda}\big)$$

$$\leqq O(a^{-\delta})\,\mu(p)^{-(\delta+\lambda-1)}$$

which is the desired bound in region I.
In region II, the integral is bounded by

$$O\big(\mu(p)^{-\lambda}\big) \int_{II} \mu_a(p-q)^{-\delta}\,dq \leqq O(a^{-1})\,\mu(p)^{\lambda}$$

if $\delta > 1$, which is the desired bound.

In region III, the integral is bounded by

$$\int_{III} \mu_a(q)^{-\delta}\, \mu(q)^{-\lambda}\, dq \leq O(a^{-1})\, \mu(p)^{-\lambda} \int \mu(q)^{-\delta}\, dq \leq O(a^{-1})\, \mu(p)^{-\lambda},$$

to complete the proof of (6.1.2)

Proposition 6.1.4. *Let* $0 < \delta$, $0 < \lambda$, $\delta + \lambda = n$. *Then for* $p \in R^n$,

$$\int_{|q| \leq \varkappa} \mu(p - q)^{-\delta}\, \mu(q)^{-\lambda}\, dq \leq O(\log \varkappa).$$

Proof. In regions I, II of the proof (a) above, the above estimates yield a bound $O(1)$, while in region III the bound is $O(\log \varkappa)$.

We introduce comparison functions

$$F(k) = \prod_{l=0}^{2} \mu(k^{(l)})^{-1},$$

$$F_\Delta(k) = \prod_{l=0}^{2} \mu(|\Delta|^{1/3}\, k^{(l)})^{-1},$$

where the product extends over the three components of the energy and momentum in three space time dimensions. We note that

$$|\chi_\Delta^\sim| \leq O(|\Delta|)\, F_\Delta.$$

In addition, F_Δ is practically L_1, and Proposition 6.1.1 and Corollary 6.1.3 yield

Proposition 6.1.5 (a) *For* $\delta + \lambda > 1$, $\nu = \min\{\delta, \lambda, \delta + \lambda - 1\}$, $\bar{\nu} = \max\{\delta, \lambda\}$, $\varepsilon > 0$,

$$|\Delta|\, F_\Delta^\delta * F_\Delta^\lambda \leq \begin{cases} O(1)\, F_\Delta^\nu, \text{ if } \bar{\nu} \neq 1 \\ O(1)\, F_\Delta^{\nu-\varepsilon}, \text{ if } \bar{\nu} = 1 \end{cases}$$

(b) *For* $0 \leq \delta \neq 1$, $0 \leq \lambda \neq 1$, $\delta + \lambda > 1$, $\nu = \min\{\delta, \lambda, \delta + \lambda - 1\}$, $|\Delta| \leq |\Delta'|$,

$$|\Delta|^\delta\, |\Delta'|^{1-\delta}\, F_\Delta^\delta * F_{\Delta'}^\lambda \leq O(1)\, F_{\Delta'}^\nu, \text{ if } \delta < 1,$$

$$|\Delta|\, F_\Delta^\delta * F_{\Delta'}^\lambda \leq O(1)\, F_{\Delta'}^\lambda, \text{ if } \lambda < 1 < \delta,$$

$$|\Delta|^\nu\, |\Delta'|^{1-\nu}\, F_\Delta^\delta * F_{\Delta'}^\lambda \leq O(1)\, F_{\Delta'}^\nu, \text{ if } 1 < \lambda,\ 1 < \delta.$$

The last inequality also yields

$$|\Delta|^\nu\, |\Delta'| F_\Delta^\delta * F_{\Delta'}^\lambda \leq O(1)\, F^\nu,$$

if $1 < \lambda$, $1 < \delta$. We next give bounds for the integral of a product of three F's. Other such inequalities follow using Proposition 6.1.5 and Hölder inequalitites.

Corollary 6.1.6. *Let* $0 < \varepsilon < 1/2$, $0 \leq \alpha_i < 1/4$,

$$\alpha = \alpha_1 + \alpha_2, \qquad |\Delta| \leq |\Delta'| \leq |\Delta''| \leq 1.$$

Then

$$\int F_\Delta(k + p)\, F_{\Delta''}(p)^\alpha\, F_{\Delta'}(k' - p)\, dp$$

is bounded by $O(1)$ times any of

$$|\Delta|^{-3/4-\varepsilon/2}\,|\Delta'|^{-1/4+\varepsilon/2}\,F_{\Delta'}(k+k')^{(1+\varepsilon)/2}\,F_{\Delta''}(k)^{\alpha_1}\,F_{\Delta''}(k')^{\alpha_2},$$

$$|\Delta|^{-3/4-\varepsilon/2}\,|\Delta'|^{-3/4}\,F(k+k')^{(1+\varepsilon)/2}\,F_{\Delta''}(k)^{\alpha_1}\,F_{\Delta''}(k')^{\alpha_2},$$

or

$$|\Delta|^{-3/4-\varepsilon/2}\,|\Delta'|^{-3/4}\,|\Delta''|^{-\alpha}\,F(k+k')^{(1+\varepsilon)/2}\,F(k)^{\alpha_1}\,F(k')^{\alpha_2}.$$

Proof. By the Schwarz inequality and the second and third inequalities of Proposition 6.1.5 (b)

$$\int F_\Delta(k+p)\,F_{\Delta''}(p)^\alpha\,F_{\Delta'}(k'-p)\,dp$$

$$\leqq \left(\int F_\Delta(k+p)^{1+\varepsilon}\,F_{\Delta'}(k'-p)^{1+\varepsilon}\,dp\right)^{1/2}\left(\int F_\Delta(k+p)^{1-\varepsilon}\,F_{\Delta''}(p)^{2\alpha}F_{\Delta'}(k'-p)^{1-\varepsilon}\,dp\right)^{1/2}$$

$$\leqq O(|\Delta|^{-(1+\varepsilon)/2}\,|\Delta'|^{\varepsilon/2})\,F_{\Delta'}(k+k')^{(1+\varepsilon)/2}\times$$

$$\times\left(\int F_\Delta(k+p)^{2(1-\varepsilon)}\,F_{\Delta''}(p)^{4\alpha_1}\,dp^1\right)^{1/4}\left(\int F_{\Delta'}(k'-p)^{2(1-\varepsilon)}\,F_{\Delta''}(p)^{4\alpha_2}\,dp\right)^{1/4}$$

$$\leqq O(\Delta|^{-3/4-\varepsilon/2}\,|\Delta'|^{-1/4+\varepsilon/2})\,F_{\Delta'}(k+k')^{(1+\varepsilon)/2}\,F_{\Delta''}(k)^{\alpha_1}\,F_{\Delta''}(k')^{\alpha_2}.$$

The second and third inequalities follow from the first, and from the bound $|\Delta|\,F_\Delta \leqq F$, valid for $|\Delta| \leqq 1$.
In case $|\Delta'| = |\Delta''|$, we obtain
Corollary 6.1.7. *As in Corollary* 6.1.6, *if we assume $|\Delta'| = |\Delta''|$ but allow $-1/4 + \varepsilon/2 < \alpha_2 < 1/4$, the bounds of Corollary* 6.1.6 *follow.*

6.2. Bounds on a W vertex

The kernel of an individual vertex has the form

$$w(k_1, \ldots, k_l) = (\mu_1 \ldots \mu_l)^{-1}\,v(k_1, \ldots, k_l) \tag{6.2.1}$$

where v denotes the various cutoff and localization factors. For contraction vertices arising from φ^4, $l = 4$ and

$$v = (h\chi_\Delta)^\sim\left(\sum_i k_i\right)\eta(k_1, \ldots, k_l), \tag{6.2.2}$$

where χ_Δ is the characteristic function of the space time localization region Δ, and η is the momentum cutoff, a product of factors (1.2.4). In the case of a mass renormalization C vertex, $l = 2$ and v includes the mass shift,

$$v = -\frac{1}{2}\,\delta m^2\,(h\chi_\Delta)^\sim\,(k_1 + k_2)\,\eta(k_1, k_2). \tag{6.2.3}$$

Perturbation vertices from V_I are characterized by $l = 4$ and $v = (h\chi_\Delta)^\sim\,\delta\eta$, where $\delta\eta$ denotes the change in ultraviolet cutoffs. Mass renormalization P vertices may also occur, and for them the increment $\delta(\delta m^2\eta)$ replaces the factor $\delta m^2\eta$ in the C vertex kernel.

The Wick vertices have several possible kernels, namely those from (3.2.17) and (3.2.19); these may have $l = 1, 2, 3$ or 4. For a particular l, let I_1 denote the set of $\delta\varphi_1$ legs and I_2 the set of φ_2 legs, $|I_I| + |I_2| = l$. Expanding each $\delta\varphi_1$ leg as $\delta\varphi_1 = \varphi_1 - \bar{\varphi}_1$ contributes two terms to the kernel for v. Thus we obtain $2^{|I_1|}$ terms in all, with each term labeled by a subset J_1 of I_1 denoting the legs for which $\bar{\varphi}_1$ is chosen. The momenta of the $\bar{\varphi}_1$ legs do not occur in the momentum conservation factor $(h\chi_\Delta)^\sim$ for the vertex. Rather, the average of the low momentum field φ_1 over the space time cube Δ yields a factor $\chi_\Delta{}^\sim(k)/\chi_\Delta{}^\sim(O)$ which occurs in the kernel for each φ_1 leg. Thus the function $v(k)$ for W vertices equals

$$v(k) = c \sum_{J_1 \subset I_1} \left\{ \prod_{i \in J_1} \frac{\chi_\Delta{}^\sim(k_i)}{\chi_\Delta{}^\sim(0)} \right\} (h\chi_\Delta)^\sim \left(\sum_{i \notin J_1} k_i \right) \eta(k), \tag{6.2.4}$$

where $c = c(l, |I_1|)$ are the constants that arise in (3.2.19), and by (3.2.20)

$$|c| \leqq O(|\Delta|^{-1+l/4}). \tag{6.2.5}$$

We define for $\lambda_i \in [0, 1]$,

$$v_\lambda(k) = c(h\chi_\Delta)^\sim \left(\sum_{i \in I_1} \lambda_i k_i + \sum_{i \in I_2} k_i \right) \eta(k) \prod_{i \in I_1} \frac{\chi_\Delta{}^\sim\big((1 - \lambda_i)\, k_i\big)}{\chi_\Delta{}^\sim(0)}. \tag{6.2.6}$$

Then (6.4) can be written as a sum of $2^{|I_1|}$ evaluations at $\lambda_i = 0$ or $\lambda_i = 1$,

$$v(k) = v_\lambda(k) \,\big|_0^1\big|_0^1 \cdots \big|_0^1 . \tag{6.2.7}$$

Proof of Proposition 5.3.5. We estimate the Hilbert Schmidt norm of the kernel $w = w(l, |I_1|)$ for a W vertex, and obtain for $\varepsilon > 0$,

$$\|w\|^2_{\text{H.S.}} \leqq 0(|\Delta|^{1/2-\varepsilon}) \prod_j b_j{}^{-3(1+\nu)}. \tag{6.2.8}$$

We note that localization ratio b_i equals 1 for φ_2 legs.

Each momentum variable is restricted to an interval $|k^{(l)}| \in [M_i, M_{i+1}]$, where $M_i = M_{i+1}^{1/1+\nu}$, or possibly $|k^{(l)}| \in [0, M_1]$. If $b > 1$, then $b = M_i{}^{-1}\, |\Delta|^{-1/3}$ and

$$M_{i+1} \leqq (b^{-(1+\nu)} |\Delta|^{-1/3})\, |\Delta|^{-\nu/3}. \tag{6.2.9}$$

We use the estimate (6.2.9) a bounded number of times in establishing (6.2.8). Thus it is convenient to suppress the factor $|\Delta|^{-\nu/3}$, which we absorb into $|\Delta|^{-\varepsilon}$ of (6.2.8), for each use of (6.2.9). We must then choose ν sufficiently small. We let $M(j)$ denote the upper endpoint of the momentum interval for the *jth* momentum variable (leg) k_j.

We divide the I_1 legs into two disjoint sets

$$I_1 = I_1' \cup I_1''. \tag{6.2.10}$$

Here I_1' consists of those legs for which $b_j > 1$ (low momentum legs satisfying $M(j)^{1/1+\nu} < |\Delta|^{-1/3}$), while I_1'' consists of those $\delta\varphi_1$ legs for which $b_j = 1$ (high momentum legs satisfying $|\Delta|^{-1/3} \leqq \mu_j \leqq |\Delta|^{-1/2}$). Here and later we suppress factors $|\Delta|^{-\nu/3}$, so we replace (6.2.9) by

$$M(j) \leqq b_j{}^{-(1+\nu)}\, |\Delta|^{-1/3},\, j \in I_1';\, b_j = 1,\, j \in I_1''. \tag{6.2.11}$$

We exhibit the low momentum cancellation implicit in (6.2.7) by writing $v(k)$ as an integral over derivatives of $v_\lambda(k)$.

$$v(k) = \int_0^1 d\lambda_1 \cdots \int_0^1 d\lambda_\alpha \frac{\partial}{\partial \lambda} \cdots \frac{\partial}{\partial \lambda_\alpha} v_\lambda(k) \Big|_0^1 \cdots \Big|_0^1, \tag{6.2.12}$$

where the integral extends over $\{\lambda_j : j \in I_1'\}$, and the evaluation extends over $\{\lambda_j : j \in I_1''\}$. We note that

$$|D^n(h\chi_\Delta)^\sim| \leq O(|\Delta|^{1+n/3})\, F_\Delta(k), \tag{6.2.13}$$

and

$$|D^n(h\chi_\Delta)^\sim(k)/\chi_\Delta{}^\sim(0)| \leq O(|\Delta|^{n/3})\, F_\Delta(k). \tag{6.2.14}$$

We define

$$F_\lambda(k) = \Big[\prod_{i \in I_1'' \sim J_1''} F_\Delta(k_i)\Big]\, F_\Delta(\xi + P) \tag{6.2.15}$$

where $J_1'' \subset I_1''$ is the subset of indices from (6.2.12) for which λ_i, $i \in I_1''$ is evaluated at $\lambda_i = 1$, and where

$$\xi = \sum_{i \in I_1'} \lambda_i k_i, \qquad P = \sum_{i \in J_1'' \cup I_2} k_i.$$

On the support of w, $|\xi| \leq O(|\Delta|^{-1/3})$. The momentum P is approximately conserved. By (6.2.5)—(6.2.6) and (6.2.12)—(6.2.14) we obtain,

$$|v(k)| \leq \sum_{J_1''} O\left(|\Delta|^{(l/4)+(|I_1'|/3)}\right) \Big(\prod_{i \in I_1'} \mu_i\Big) \int_0^1 d\lambda_1 \cdots \int_0^1 d\lambda_\alpha\, F_\lambda(k). \tag{6.2.16}$$

We let $w(l, |I|; I_1'', J_1'')$ denote the kernel for a particular choice of J_1'', Suppressing l, $|I_1|$, and I_1'',

$$w = \sum_{J_1''} w(J_1''),\ \|w\|_{\text{H.S.}} \leq \sum_{I_1'', J_1''} \|w(J_1'')\|_{\text{H.S.}}$$

Suppressing the cutoffs $\eta_{\alpha,\beta}(k)$, we have

$$\|w(J_1'')\|^2_{\text{H.S.}} \leq O(|\Delta|^{(l/2)+(2|I_1'|/3)}) \sup_\lambda \int dk \frac{F_\lambda(k)^2}{D},$$

where $dk = dk_1 \cdots dk_l$ and $D = \prod_{i \in I_1'' \cup I_2} \mu_i^2$. Integrating the I_1' momenta yields a factor per I_1' leg of

$$\int_{|k_i| \leq M(i)} d^3k_i \leq O\big(M(i)^3\big) \leq O(b_i^{-3(1+\nu)}\, |\Delta|^{-1}),$$

where the last inequality follows by (6.2.11). Integrating the momenta in $I_1'' \sim J_1''$ yields a factor per leg

$$\int_{|\Delta|^{-1/3} \leq |k_i|} F_\Delta(k_i)^2\, \mu_i^{-2}\, dk_i \leq O(|\Delta|^{2/3}) \int F_\Delta(k_i)^2\, dk_i \leq O(|\Delta|^{-1/3}\, b_i^{-3/(1+\nu)})$$

using Proposition 6.1.5 (b), the lower cutoff $|\Delta|^{-1/3}$ for I_1'' legs, and the fact that $b_i = 1$ for $i \in I_1''$. We show below that for $\epsilon > 0$, integrating the remaining momenta in D gives,

$$\int_{J_1'' \cup I_2} dk \sup_{\xi} \left\{ \frac{F_\Delta(\xi + P)^2}{\prod_{i \in J_1'' \cup I_2} \mu_i^2} \right\} \leqq O(|\Delta|^{(1/2)(1-|I_2|-|J_1''|)-\epsilon}), \tag{6.2.17}$$

so that using $l = |I_1'| + |I_1''| + |I_2|$, we obtain

$$\|w(J_1'')\|^2_{\text{H.S.}} \leqq O(|\Delta|^{(1/2)+(1/6)(|I_1''|-|J_1''|)-\epsilon} \Big(\prod_{i \in I_1} b_i^{-3(1+\nu)}\Big) \leqq O(|\Delta|^{(1/2)-\epsilon}) \Big(\prod_i b_i^{-3(1+\nu)}\Big). \tag{6.2.18}$$

To establish (6.2.17), we note that $|I_2| = 1$, 2 or 3 for all W vertices, see (3.2.17). We check (6.2.17) for the various values of $|J_1''| + |I_2|$. If $|J_1''| + |I_2| = 1$, we use the lower cutoff $|\Delta|^{-1/2}$ on the I_2 leg to give

$$\int_{|P| > |\Delta|^{-1/2}} dP \sup_{\xi} \frac{F_\Delta(\xi + P)^2}{\mu(P)^2} \leqq O(|\Delta|) \int dP \sup F_\Delta(\xi + P)^2 \leqq O(1),$$

since $|\xi| \leqq O(|\Delta|^{-1/3}) \leqq O(|P|^{2/3})$ on the region of integration. If $|J_1''| + |I_2| = 2$, Proposition 6.1.5 (b) similarly yields $O(|\Delta|^{-1/2})$. If $|J_1''| + |I_2| = 3$, Proposition 6.1.4 yields

$$\int dP \sup_{\xi} F_\Delta(\xi + P)^2 \int \frac{dk_2\, dk_3}{\mu(P - k_2 - k_3)^2 \mu(k_2)^2 \mu(k_3)^2} \leqq O\big(|\Delta|^{-1} \log M_u(\Delta)\big).$$

As in Chapter 5 estimates, we use $\log M_{u(\Delta)} \leqq O(\log |\Delta|^{-1})$, since W vertices occur in Chapter 3 in small cubes. This yields the desired estimate for $|J_1''| + |I_2| = 3$. If $|J_1''| + |I_2| = 4$, we proceed similarly with variables P, k_1, k_2, k_3. Here $M(1)$, the upper cutoff in k_1 is the smallest upper cutoff, and since $|I_2| = 4$, this momentum is a J_1'' leg with an upper cutoff at $|\Delta|^{-1/2}$. Integrating k_2 and k_3 yields $O(\log M_{u(\Delta)})$ as above. The k_1 integral is

$$\int_{|k_1| < |\Delta|^{-1/2}} \mu(k_1)^{-2}\, dk_1 \leqq O(|\Delta|^{-1/2})$$

so the case $|J_1''| + |I_2| = 4$ is dominated by $O(|\Delta|^{-3/2} \log |\Delta|^{-1})$. This completes the proof of (6.2.19) and Proposition 5.3.5.

6.3. Estimates on other small subgraphs

The following proofs of Proposition 5.3.3 and Proposition 5.3.4 establish stronger bounds than claimed in Chapter 5.

Proof of Proposition 5.3.3. (a) We set

$A =$  $\mu^{1/4-\epsilon}$, and bound

$$\|A\|_{3,1} = \|A^*A\|_{1,1}^{1/2} \leqq \|A^*A\|_{\text{H.S.}}^{1/2},$$

where

$A^*A =$ 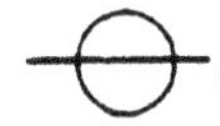, $\|A^*A\|_{\text{H.S.}}^2 =$ 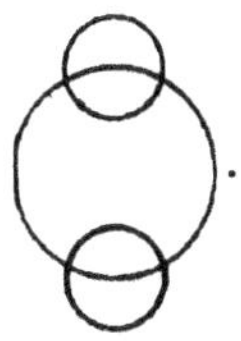.

The kernel of A^*A is bounded by

$$v(k_1, k_5) = |\Delta|^2 \int dk_2\, dk_3\, dk_4 \, \frac{F_\Delta(k_1 + k_2 + k_3 + k_4)\, F_\Delta(-k_2 - k_3 - k_4 + k_5)}{(\mu_1 \mu_5)^{(3/4)+\varepsilon}\, \mu_2{}^2\, \mu_3{}^2\, \mu_4{}^2}.$$

We let $P = k_2 + k_3 + k_4$, k_2 and k_3 be the variables of integration, where the upper momentum cutoff $M(2)$ for k_2 is the smallest upper cutoff among k_2, k_3, k_4. Thus by Propositions 6.1.1 and 6.1.4,

$$\int dk_2\, dk_3 \, \frac{1}{\mu_2{}^2\, \mu_3{}^2\, \mu(P - k_2 - k_5)^2} \leqq O\big(\log M(2)\big) \tag{6.3.1}$$

and by Proposition 6.1.5, for $\delta > 0$,

$$v(k_1, k_5) \leqq O\big(\log M(2)\big)\, |\Delta|\, F_\Delta(k_1 + k_2)^{1-\delta}\, (\mu_1 \mu_5)^{-(3/4+\varepsilon)}.$$

Taking L_2 norms gives

$$\|v\|_2{}^2 \leqq O\big(\log^2 M(2)\big)\, |\Delta|^2 \int \frac{F_\Delta(k_1 + k_5)^{2-2\delta}}{(\mu_1 \mu_5)^{(3+\varepsilon)/2}}\, dk_1\, dk_5 \leqq O\big(\log^2 M(2)\, |\Delta|\big),$$

proving

$$\|w\mu^{1/4-\varepsilon}\|_{3,1} \leqq O\big(|\Delta|^{1/4} \log^{1/2} M(2)\big).$$

(b—c) We first estimate the kernel $v(k_1, k_2, k_3, k_4)$ of the diagram

with a factor μ^ν multiplying each external line, and a factor $\mu^{2\lambda}$ multiplying each internal line. We let $K_1 = k_1 + k_2$, $K_2 = k_3 + k_4$, $P = k_5 + k_6$, and assume $\nu < 1$.

$$|v(k_1, k_2, k_3, k_4)| \leqq O(|\Delta|^2)\, (\mu_1 \mu_2 \mu_3 \mu_4)^{-1+\nu} \int dk_5\, dk_6 \, \frac{F_\Delta(K_1 + P)\, F_\Delta(-P + K_2)}{(\mu_5 \mu_6)^{2-2\lambda}}$$

$$\leqq O(|\Delta|^2)\, (\mu_1 \mu_2 \mu_3 \mu_4)^{-1+\nu} \int dP \, \frac{F_\Delta(K_1 + P)\, F_\Delta(-P + K_2)}{\mu(P)^{1-4\lambda}}$$

and by Corollary 6.6, for $\lambda < 1/4$,

$$\leqq O(|\Delta|)\, (\mu_1 \mu_2 \mu_3 \mu_4)^{-1+\nu}\, F_\Delta(K_1 + K_2)^{(1+\delta)/2}\, F(K_1)^{(1-4\lambda)/6}\, F(K_2)^{(1-4\lambda)/6}.$$

The L_2 norm of v is dominated by

$$\|v\|_2{}^2 \leqq O(|\Delta|^2) \int dK_1\, dK_2\, dk_1\, dk_4 \frac{F_\Delta(K_1 + K_2)^{1+\delta}\, F(K_1)^{(1-4\lambda)/3}\, F(K_2)^{(1-4\lambda)/3}}{[\mu_1 \mu_4 \mu(K_1 - k_1)\mu(K_2 - k_4)]^{2(1-\nu)}}$$

$$\leqq O(|\Delta|^2) \int dK_1\, dK_2 \frac{F_\Delta(K_1 + K_2)^{1-\delta}\, F(K_1)^{(1-4\lambda)/3}\, F(K_2)^{(1-4\lambda)/3}}{[\mu(K_1)\, \mu(K_2)]^{1-4\nu}}$$

$$\leqq O(|\Delta|^2) \int dK_1\, dK_2\, F_\Delta(K_1 + K_2)^{1-\delta}\, [F(K_1)\, F(K_2)]^{(2/3)-(4/3)(\lambda+\nu)}$$

$$\leqq O(|\Delta|^2) \int dK_1\, dK_2\, F_\Delta(K_1 + K_2)^{1+\delta}\, F(K_1)^{(4/3)-(8/3)(\nu+\lambda)} \leqq O(|\Delta|),$$

provided that $(\lambda + \nu)$ is sufficiently small, namely $4/3 - (8/3)\,(\lambda + \nu) > 1$, or

$$(\lambda + \nu) < \frac{1}{8}.$$

In particular, if $\lambda = \nu$, and if w is the kernel of a P or C vertex,

$$\|w \prod_{i=1}^{4} \mu_i{}^{1/16-\varepsilon}\|_{2,2} \leqq O(|\Delta|^{1/4}),$$

using

$$\|A\|_{2,2} = \|A^* A\|_{2,2}^{1/2} \leqq \|A^* A\|_{\text{H.S.}}^{1/2}.$$

If $\lambda = \nu$, and w is the kernel of the diagram

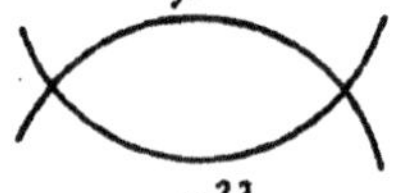

with either P or C vertices, then

$$\|w\left(\prod_{i=1}^{4} \mu_i{}^{1/16-\varepsilon}\right)\|_{\text{H.S.}} \leqq O(|\Delta|^{1/2}).$$

Proof of Proposition 5.3.4. We note that in the notation of Chapter 2, the kernel of the mass renormalization diagram is

$$w_{\Delta,\Delta'}(k_1, k_5) = \text{const.} \int (\alpha\beta\gamma - \bar{\alpha}\bar{\beta}\bar{\gamma})\, dk_2\, dk_3\, dk_4. \tag{6.3.2}$$

For $\Delta = \Delta'$, this is just (2.10), while for $\Delta \neq \Delta'$ we note that the contribution from $\int \bar{\alpha}\bar{\beta}\bar{\gamma}\, dk$ is zero, reducing (6.3.2) to (2.9). This is a consequence of $\chi_\Delta \chi_{\Delta'} = 0$, yielding

$$\int \bar{\beta}_{\Delta,\Delta'}\, dP = \text{const.}\, (h^2 \chi_\Delta \chi_{\Delta'})^\sim\, (k_5 - k_1) = 0.$$

We note that $P = k_2 + k_3 + k_4$, $k_i = p_i + P$,

$$|\mu(k_i)^{-2} - \mu(p_i)^{-2}| \leqq O(1)\, \mu(P)^{3\varepsilon}\, \{\mu(k_i)^{-2-\varepsilon} + \mu(p_i)^{-2-\varepsilon}\}$$

yielding for $0 \leqq \varepsilon \leqq 1/3$,

$$|\gamma-\bar{\gamma}| \leqq O(1)\,(\mu_1\mu_5)^{-1}\,\mu(P)^{3\varepsilon}\,\{\mu(k_2)^{-2-\varepsilon}+\mu(p_2)^{-2-\varepsilon}\big)\,\mu(k_3)^{-2}\,\mu(k_4)^{-2} + \\ +\mu(p_2)^{-2}\big(\mu(k_3)^{-2-\varepsilon}+\mu(p_3)^{-2-\varepsilon}\big)\,\mu(k_4)^{-2} + \\ +\mu(p_2)^{-2}\,\mu(p_3)^{-2}\big(\mu(k_3)^{-2-\varepsilon}+\mu(p_3)^{-2-\varepsilon}\big)\}.$$

In addition we need a bound on $(\alpha-\bar{\alpha})$. The P vertex in a mass diagram arises from lowering the upper cutoff in Δ. Let λ be the corresponding maximum lower cutoff in the P vertex, due to this change of cutoffs $\delta\eta = d/ds\ \eta_s$. For $\delta \geqq 0$, on the support of the cutoffs,

$$|\delta\eta - \delta\bar{\eta}| \leqq O(|P|^\delta\,\lambda^{-\delta}).$$

Since

$$(\alpha-\bar{\alpha})_s = \frac{d}{ds}\prod_{\text{legs}}(\eta_s\eta_{s'} - \bar{\eta}_s\bar{\eta}_{s'})|_{s=s'},$$

we have

$$|\alpha-\bar{\alpha}| \leqq O(P^\delta\lambda^{-\delta}).$$

Since $\lambda \geqq M(2)^{1/1+\nu} \geqq M(2)^{1/2}$,

$$|\alpha-\bar{\alpha}| \leqq O\big(|P|^\delta\,\lambda^{-\delta/2}\,M(2)^{-\delta/4}\big) \\ \leqq O\,|\Delta'|^{-\delta/3}\,\mu(|\Delta|^{1/3}\,P)^\delta\,\lambda^{-\delta/2}\,M(2)^{-\delta/4}. \tag{6.3.4}$$

We now estimate the Hilbert Schmidt norm of (6.3.2). We assume $|\Delta| \leqq |\Delta'|$. Though our kernel w is asymmetric in Δ and Δ', our estimates are symmetric. Thus

$$|w_{\Delta,\Delta'}(k_1,k_5)| \leqq O(1)\int|\alpha-\bar{\alpha}|\,\bar{\beta}\bar{\gamma}\,dk_2\,dk_3\,dk_4 + \\ + O(1)\int\bar{\beta}\,|\gamma-\bar{\gamma}|\,dk_2\,dk_3\,dk_4, \tag{6.3.5}$$

using $|\delta\eta| \leqq O(1)$.
We estimate the first term, using (6.3.4), by

$$O(|\Delta|\,|\Delta'|)\,\lambda^{-\delta/2}\,M(2)^{-\delta/4}\int dP\,dp_2\,dp_3\,\frac{F_\Delta(k_1+P)\,F_{\Delta'}(P)^{-\delta}\,|\Delta'|^{-\delta/3}\,F_{\Delta'}(-P+k_5)}{\mu_1\mu_5\mu(p_2)^2\,\mu(p_3)^2\,\mu(p_4)^2}.$$

Using $p_4 = -(p_2+p_3+2P)$ and (6.3.1), we bound this by

$$O(|\Delta|\,|\Delta'|^{1-\delta/3})\,\lambda^{-\delta/2}\,M(2)^{-\delta/4}\log M(2)\,(\mu_1\mu_5)^{-1}\times \\ \times\int dP\,F_\Delta(k_1+P)\,F_{\Delta'}(P)^{-\delta}\,F_{\Delta'}(-P+k_5) \\ \leqq O\,(|\Delta|^{1/4-\varepsilon/2}\,|\Delta'|^{1/4-\delta/3})\,\lambda^{-\delta/2}\,F(k_1+k_5)^{(1+\varepsilon)/2}\,F(k_5)^{-\delta}\,(\mu_1\mu_5)^{-1}. \tag{6.3.6}$$

The last inequality follows from Corollary 6.1.7 with $\alpha_1 = 0$, $\alpha_2 = -\delta$. In (6.3.6) we dominate the logarithmic divergence in M_i by $M_i^{-\delta/4}$, leaving the convergence factor $\lambda^{-\delta/2}$.

The L_2 norm of (6.3.5), multiplied by $(\mu_1\mu_5)^{1/4-\nu}$, is dominated by

$$O(|\Delta|^{1/4-\varepsilon/2}|\Delta'|^{1/4-\delta/3})\,\lambda^{-\delta/2}\left[\int F(k_1+k_5)^{1+\varepsilon}(\mu_1\mu_5)^{3/2+6\delta-2\nu}\,dk_1\,dk_5\right]^{1/2}$$

$$\leq O(\Delta^{1/4-\varepsilon/2}\,|\Delta'|^{1/4-\delta/3})\,\lambda^{-\delta/2}$$

for $\delta(\nu)$ sufficiently small.
Using (6.3.2)—(6.3.3), the second term in (6.3.5) yields a similar bound. Thus given $\varepsilon > 0$, there exists $\delta = \delta(\varepsilon) > 0$ such that

$$\|(\mu_1\mu_5)^{1/4-\varepsilon}\,w_{\Delta,\Delta'}\|_2 \leq O(|\Delta|^{1/8}\,|\Delta'|^{1/8}\,\lambda^{-\delta})\,. \tag{6.3.7}$$

Proof of Proposition 5.3.6. First assume that $|\Delta| \leq |\Delta'|$. By the Schwartz inequality

$$|P \equiv\equiv\equiv W|^2 \leq P \equiv\equiv\equiv P \quad W \equiv\equiv\equiv W$$

$$\leq \left(O|\Delta|\,M(1)\,M(2)^{3\varepsilon_1}\,\lambda^{-2\varepsilon_1}\,|\Delta'|^{1/2-\varepsilon}\right)\prod_i b_i^{-3(1+\nu)},$$

where we use Proposition 5.3.3 (d) and 5.3.5. Since the W vertex has at least one $\delta\varphi_1$ leg, and that leg has an upper cutoff $O(|\Delta'|^{-1/2})$, we have $M(1) \leq O(|\Delta'|^{-1/2})$. As in the proof of Proposition 5.3.5, there exists ε_2 such that

$$M(2)^{3\varepsilon_1} \leq O(|\Delta'|^{-\varepsilon_2})\,.$$

Hence for ε_1 sufficiently small, and using $|\Delta| \leq |\Delta'|$, there exists $\varepsilon_3 > 0$ such that

$$|P \equiv\equiv\equiv W|^2 \leq O(|\Delta|^{2\varepsilon_3}\,|\Delta'|^{2\varepsilon_3})\,\lambda^{-2\varepsilon_1}\prod_i b_i^{-3(1+\nu)}\,.$$

In case $|\Delta'| \leq |\Delta|$, we modify the above proof as follows: The kernel of the P vertex contains a factor $(h\chi_\Delta)^\sim(P)$, where P is the total momentum. In the estimate on the Hilbert Schmidt norm of the P vertex, we use only $[(h\chi_\Delta)^\sim]^{(1+\varepsilon_3)/2}$. Thus we can, without change of that estimate, include a factor $[(h\chi_\Delta)^\sim(P)]^{(1-\varepsilon_3)/2}$ in the kernel of the W vertex. Following our proof of Proposition 5.3.5, with our modified kernel, this yields an estimate

$$\|(h\chi_\Delta)^\sim(P)^{(1-2\varepsilon_3)}\,w_W\|^2_{\text{H.S.}} \leq O(|\Delta|^{-1+2\varepsilon_3}\,|\Delta'|^{3/2-\varepsilon-2\varepsilon_3})\prod_i b_i^{-3(1+\nu)}.$$

Thus

$$|P \equiv\equiv\equiv W|^2 \leq O\left(|\Delta|^{2\varepsilon_3}\,|\Delta'|^{3/2-\varepsilon-2\varepsilon_3}\,M(1)\,M(2)^{3\varepsilon_1}\,\lambda^{-2\varepsilon_1}\right)\prod_i b_i^{-3(1+\nu)}$$

and for ε, ε_1, and δ sufficiently small,

$$|P \equiv\equiv\equiv W|^2 \leq O(|\Delta|^{2\varepsilon_3}\,|\Delta'|^{2\varepsilon_3}\,\lambda^{-2\varepsilon_1})\prod_i b_i^{-3(1+\nu)}$$

as above.
Proposition 6.3.1. *The changes $\delta E_2(\Delta)$, and $\delta E_3(\Delta)$ in the second and third order counterterms in δV of* (3.2.9) *are bounded by* $O(|\Delta|^{1-\delta}\,M(1)\log M(2) \leq O(1)$.
Proof. The second order vacuum energy $E_2(\Delta)$ is a summation of $E_2(\Delta, \Delta')$ over Δ', where $E_2(\Delta, \Delta') = -1/2\,\langle V_I(\Delta)\,V_I(\Delta')\rangle$, and where Δ' ranges over a cover of $\mathscr{Q}_\varrho$.

As in (5.2.1), we obtain

$$|E_2(\Delta, \Delta')| \leq O(1) \left(\prod_{i=1}^{4} d_i^{-n} \right) I,$$

where

$$\bar{I} = |\Delta| \, |\Delta'| \int \frac{F_\Delta(P) \, F_{\Delta'}(P) \, dP \, dk_1 \, dk_2 \, dk_3}{\mu(P - k_1 - k_2 - k_3)^2 \, \mu_1^2 \, \mu_2^2 \, \mu_3^2}.$$

By (6.3.1) and Proposition 6.1.5 (b),

$$\bar{I} \leq |\Delta| \, |\Delta'| \, O(\log M(2)) \int F_\Delta(P) \, F_{\Delta'}(P) \, \mu_1^{-2} \, dk_1 \, dP$$
$$\leq O(M(1) \log M(2)) \, |\Delta|^{1-\delta}.$$

During the Wick expansion (3.2.9) of δV, $M(1) \leq O(|\Delta|^{-1/2})$, see (3.2.1), and log $M(2) \leq |\Delta|^{-\varepsilon}$, see (5.3.14). Thus $\bar{I} \leq O(1)$, and summing over d^{-4} yields the desired bound on $\delta E_2(\Delta)$, see Lemma 4.2 (a).
The bound on $\delta E_3(\Delta)$ is proved similarly.

References

[1] J.-P. ECKMANN, Commun. Math. Phys., **25**, 1 (1972).
[2] J.-P. ECKMANN and K. OSTERWALDER, Helv. Phys. Acta, **44**, 864 (1971).
[3] J. FABREY, Commun. Math. Phys., **19**, 1 (1970).
[4] J. GLIMM, Commun. Math. Phys., **10**, 1 (1968).
[5] J. GLIMM and A. JAFFE, Ann. Math., **91**, 362 (1970).
[6] J. GLIMM and A. JAFFE, J. Math. Phys., **13**, 1568 (1972).
[7] J. GLIMM and A. JAFFE, Field Theory Models, in *Statistical Mechanics and Quantum Field Theory*, C. Dewitt and R. Stora, editors, 1970 Les Houches Lectures, Gordon and Breach, New York 1971.
[8] J. GLIMM and A. JAFFE, Boson Field Models, in *Mathematics of Contemporary Physics*, R. Streater, editor, Academic Press, New York, 1972.
[9] K. HEPP, Théorie de la Renormalization, Springer Verlag, Berlin 1969.
[10] E. NELSON, Quantum Fields and Markoff Fields, American Mathematical Society Symposium on Partial Differential Equations, 1971, to appear.
[11] K. SYMANZIK, Euclidean Quantum Field Theory, in *Local Quantum Theory*, R. Jost, editor 1968 Varenna Lectures, Academic Press, 1969.

Herausgeber: Prof. Dr. Frank Kaschluhn, Prof. Dr. Artur Lösche, Prof. Dr. Rudolf Ritschl und Prof. Dr. Robert Rompe. Manuskripte sind zu richten an die Schriftleitung: Dr. Lutz Rothkirch, Sektion Physik der Humboldt-Universität zu Berlin, 104 Berlin, Hessische Str. 2. Verlag: Akademie-Verlag GmbH, 108 Berlin, Leipziger Str. 3/4, Fernruf: 220441, Telex-Nr. 0112020, Postscheckkonto: Berlin 35021. Die Zeitschrift „Fortschritte der Physik" erscheint monatlich; Bezugspreis dieses Heftes 10,–. Bestellnummer dieses Heftes: 1027/21/7. – Satz und Druck: VEB Druckhaus „Maxim Gorki", 74 Altenburg, Bez. Leipzig, Carl-von-Ossietzky-Str. 30/31. – Veröffentlicht unter der Lizenznummer 1324 des Presseamtes beim Vorsitzenden des Ministerrates der Deutschen Demokratischen Republik.

III Phase Transitions Exist

Introduction

While phase transitions were known to occur Ising-type lattice spin systems since Peierls' famous 1936 paper, they were believed to exist but had not been established in the corresponding statistical mechanics of classical fields or in continuum quantum field theory. The proof that phase transitions in such systems do exist in these systems is presented in the paper Phase transitions for ϕ_2^4 quantum fields with T. Spencer. The specific example worked out is the $\lambda\phi^4 - \phi^2$ model, with $0 < \lambda \ll 1$. The idea is to treat the low momentum degrees of freedom in each unit space-time volume as an Ising spin system, while the high momentum degrees of freedom are shown to have small effect for $\lambda \ll 1$. The technical tool which enabled the implementation of this phase cell localization idea is the "multiple reflection bounds" developed in that article. (These bounds are also known as "chessboard" estimates.)

The end result of the study is the nonuniqueness of the model for small λ. This nonuniqueness has an intuitive interpretation in terms of the two minima of the W-shaped potential function $\lambda\phi^4 - \phi^2$. Each minimum corresponds to an approximate quantum mechanics ground state. Hence quantum field theory admits two solutions, in distinction to quantum mechanics of a finite number of degrees of freedom which has a nodeless eigenfunction for a unique ground state. The choice of one solution of a mixture is dictated by some boundary condition or other parameter.

The other articles in this group show that the cluster expansion method can be extended to apply to the multiple phase ϕ^4 model. Hence it is possible to obtain information on the mass gap and other spectral issues in the two phase region. This requires an expansion in local phases, as well as the cluster expansion which decouples different volumes. Both these methods have been extended to more general classes of models. Eventually generalizations of these points of view have been used to settle important questions in the theory of phase transitions in statistical physics.

Bibliography

Phase transitions for ϕ_2^4 quantum fields, with T. Spencer, Commun. Math. Phys. 45, 203–216 (1975).

A convergent expansion about mean field theory I. The expansion, with T. Spencer, Ann. Phys. 101, 610–630 (1976).

A convergent expansion about mean field theory II. Convergence of the expansion, with T. Spencer, Ann. Phys. 101, 631–669 (1976).

Phase Transitions for φ_2^4 Quantum Fields*

James Glimm**
Rockefeller University, New York, N.Y. 10021, USA

Arthur Jaffe*** and Thomas Spencer***, †
Harvard University, Cambridge, Mass. 02138, USA

Abstract. Phase transitions for the quantum field interaction $\lambda\varphi^4 + m_0^2\varphi^2$, $m_0^2/\lambda \ll 1$ are established in two dimensional space time.

1. Introduction

We present a direct proof of the existence of phase transitions in quantum field theory. We consider here the simplest interaction for which a phase transition is expected, namely the

$$\lambda\varphi^4 + \tfrac{1}{2} m_0^2\varphi^2, \quad m_0^2/\lambda \ll 1, \tag{1.1}$$

perturbation of the free field of mass m_0. We give a complete proof in space time dimension $d=2$. Our same methods apply in principle to arbitrary even $P(\varphi)_2$ models without cutoff.

To define the interaction (1.1) for $d=2$ we require Wick ordering. We denote Wick ordering of P with respect to the covariance $(-\Delta + m_0^2)^{-1}$ by $:P:_{m_0}$. Then scaling and re Wick ordering leads to an equivalent theory with the bare mass $O(\sigma)^{-1}$ and the interaction which we study, see [13], is

$$:P(\varphi):_{\sigma^{-1}} = :(\varphi^2 - \sigma^2)^2/\sigma^2:_{\sigma^{-1}}, \quad \sigma \gg 1. \tag{1.2}$$

It is the occurrence of two distinct minima, separated by a large barrier, which suggests the occurrence of phase transitions for the interaction (1.2). The two pure phases are ground states localized (in φ space) near the two minima $\varphi = \pm\sigma$.

In the case we consider, the polynominal $P(\varphi)$ is invariant under the symmetry transformation $\varphi \to -\varphi$, while the pure phases are interchanged by the symmetry. We note, however, that symmetry breaking is a distinct issue from the existence of phase transitions. Just as in statistical mechanics, where phase transitions may occur without symmetry breaking [11], we expect phase transitions in field theory for certain $P(\varphi)$ models which do not possess a symmetry group, such as

---

* Presented at the International Colloquium on Mathematical Methods of Quantum Field Theory, Marseille, June 23–27, 1975.

** Supported in part by the National Science Foundation under Grant MPS 74-13252.

*** Supported in part by the National Science Foundation under Grant MPS 73-05037.

† On leave from Rockefeller University, New York, NY 10021, USA.

for the interaction

$$\sigma^2 P(\varphi) = (\varphi^2 - \sigma^2)^4 + \varepsilon\varphi^3 - \mu\varphi ,$$

for $\sigma \gg 1$, $\varepsilon \ll 1$, and $\mu = \mu(\varepsilon, \sigma)$.

The method we use here to study the interaction (1.2) is to generalize the φ^j bounds [3, 8, 2, 6, 5] in order to estimate perturbations of the measures for the interacting ground state. These estimates follow from the Osterwalder-Schrader positivity and the Euclidean invariance of the theory, combined with bounds we prove in Section 2 on the vacuum energy per unit volume. The existence of long range order (phase transitions and symmetry breaking) follow by a variant of a statistical mechanics argument due to Peierls. As a consequence, we find that the Dirichlet φ^4 theory (1.2) has a non-unique vacuum for $\sigma \gg 1$. The two phases are separated by an external field $-\mu\varphi$ added to $P(\varphi)$ of (1.2).

In a separate article [7], we generalize our convergent weak coupling expansion to yield a cluster expansion for the interaction (1.2). This expansion allows us to introduce boundary conditions (other than zero Dirichlet) which yield two distinct pure phases. In each pure phase we obtain from the cluster expansion a model with a unique vacuum, which satisfies the Wightman axioms with a mass gap. Of course, it would be of interest to pursue the particle structure of the model (1.2) via the cluster expansion.

In contrast to our detailed study based on the cluster expansion, our present paper gives a simple, direct proof that phase transitions occur. An alternative approach to the problem of phase transitions has been announced in [1], but the proof has not appeared.

We now introduce some notation and explain the Peierls argument. Let $\Lambda \subset R^2$ be a square, centered at the origin, and let Δ_Λ be the Laplace operator with zero Dirichlet boundary conditions on the boundary $\partial\Lambda$ of Λ. Let $d\varphi_{\varepsilon,\Lambda}$ denote the Gaussian measure over $\mathscr{S}'(R^2)$ wit mean zero and covariance $(-\Delta_\Lambda + \varepsilon^2)^{-1}$. Let

$$V(\Lambda) = \int_\Lambda :P(\varphi(x)):dx , \tag{1.3}$$

where $:P:$ denotes Wick ordering with respect to the covariance $(-\Delta_\Lambda + \varepsilon^2)^{-1}$. We let

$$\langle A \rangle = \lim_{\Lambda \nearrow R^2} \frac{\int A(\varphi) e^{-V(\Lambda)} d\varphi_{\varepsilon,\Lambda}}{\int e^{-V(\Lambda)} d\varphi_{\varepsilon,\Lambda}} , \tag{1.4}$$

and in order to ensure that P defined by (1.2) is equivalent to (1.1), we also choose $\varepsilon = \sigma^{-1}$.

For P given by (1.2), plus a linear term, the limit $\Lambda \nearrow R^2$ in (1.4) exists for

$$A = \int \varphi(x_1) \ldots \varphi(x_r) v(x_1, \ldots, x_r) dx , \qquad v \in \mathscr{S} ,$$

and also for $A = \exp Q$ with Q defined by (1.9), as well as for

$$A = \prod \chi_\pm(\Delta_j)$$

defined below.

For each unit lattice square $\Delta = \Delta_j$, centered at $j \in Z^2$, we define the averaged field

$$\varphi(\Delta) = \int_\Delta \varphi(x) dx .$$

Essentially $\varphi(\Delta)$ is the low momentum field in Δ, so we expect that $\varphi(\Delta)$ characterizes phase transitions in Δ. We define a localizing function of the averaged field

$$\chi_{(a,b)}(\Delta) = \begin{cases} 1 & \text{if } \varphi(\Delta) \in (a, b) \\ 0 & \text{otherwise.} \end{cases}$$

We use the particular functions

$$\chi_+ = \chi_{(0,\infty)} \quad \text{and} \quad \chi_- = \chi_{(-\infty,0)}$$

to separate phases and to establish the existence of long range order.

Lattice squares Δ_i, Δ_j are nearest neighbors if $|i-j| = 1$. Let $\mathcal{N}$ be a collection of nearest neighbor pairs and let $|\mathcal{N}|$ denote the cardinality of $\mathcal{N}$. Assume there exists a constant $K > 0$ such that

$$\langle \prod_{(\Delta, \Delta') \in \mathcal{N}} \chi_-(\Delta) \chi_+(\Delta') \rangle \leqq e^{-K|\mathcal{N}|} . \tag{1.5}$$

Theorem 1.1. *Consider a $P(\varphi)_2$ quantum field theory defined by* (1.3)–(1.4). *If* (1.5) *holds for K sufficiently large and for all $\mathcal{N}$, then for any two lattice squares Δ_i and Δ_j*

$$\langle \chi_+(\Delta_i) \chi_-(\Delta_j) \rangle \leqq 1/8 .$$

Corollary 1. *If the field φ has a $\varphi \to -\varphi$ symmetry* (*for example if P and the boundary conditions are both even*) *then $\langle \chi_\pm(\Delta) \rangle = \frac{1}{2}$. In this case,*

$$|\langle \chi_+(\Delta_i) \chi_-(\Delta_j) \rangle - \langle \chi_+(\Delta_i) \rangle \langle \chi_-(\Delta_j) \rangle| \geqq 1/8 ,$$

which establishes the existence of long range order.

In order to establish symmetry breaking, we consider the model

$$\begin{aligned} :P(\varphi, \mu): &= :(\varphi^2 - \sigma^2)^2/\sigma^2: - \mu\varphi \\ &= :P(\varphi): - \mu\varphi , \end{aligned} \tag{1.6}$$

with $:P:$ given by (1.2). The estimates in this paper are uniform for $0 \leqq \mu < \sigma^{-2}$. Then we find

Corollary 2. *For the interaction defined by* (1.3), (1.4), *and* (1.6),

$$\lim_{\mu \searrow 0} \langle \chi_+(\Delta) \rangle_\mu > 0.8 .$$

Proof. For $\mu \neq 0$, the $:P(\varphi, \mu):$ model has a unique vacuum, as a corollary of the Lee-Yang theorem [12]. Thus as $\text{dist}(\Delta, \Delta') \to \infty$,

$$\begin{aligned} \langle \chi_+(\Delta) \chi_-(\Delta') \rangle_\mu &\to \langle \chi_+(\Delta) \rangle_\mu \langle \chi_-(\Delta') \rangle_\mu \\ &= \langle \chi_+(\Delta) \rangle_\mu (1 - \langle \chi_+(\Delta) \rangle_\mu) , \end{aligned}$$

where we use clustering and translation invariance. Thus by the theorem,

$$\langle \chi_+(\Delta)\rangle_\mu(1-\langle \chi_+(\Delta)\rangle_\mu)\leqq 1/8\,,$$

which has the solution $\langle \chi_+(\Delta)\rangle_\mu > 0.8$. By FKG inequalities, $\langle \chi_+(\Delta)\rangle_\mu$ is monotone in μ, as $\mu\searrow 0$. Thus $\lim_{\mu\searrow 0}\langle \chi_+(\Delta)\rangle_\mu$ exists and is greater than $0.8>\frac{1}{2}$.

Theorem 1.2. *Let K be given, and let a φ^4 quantum field be defined by* (1.3), (1.4), *and* (1.6). *For σ sufficiently large,* (1.5) *holds for all collections $\mathcal{N}$ of nearest neighbor pairs.*

Proof of Theorem 1.1. Let Λ_0 be a square, centered at the origin, containing Δ_i, Δ_j and large enough (depending on i and j) so that

$$2+\operatorname{dist}(\Delta_i,\Delta_j)\leqq \frac{1}{9}|\partial\Lambda_0|\leqq \operatorname{dist}(\partial\Lambda_0,\{\Delta_i,\Delta_j\})\,. \tag{1.7}$$

A configuration is defined as a map from lattice squares in Λ_0 to ± 1. Given a configuration c, a subset $Y\subset c^{-1}(+1)$ is said to be + connected if any two lattice squares of Y can be connected by a path of nearest neighbor squares belonging to Y. The definition of − connected is analogous. Each configuration introduces a decomposition of Λ_0 into + and — connected components $\{X_k(c)\}$. Let X_i and X_j be components containing Δ_i and Δ_j respectively. Let γ_i be the outermost contour in ∂X_i. Technically, γ_i is defined to be the boundary of the component of $R^2\backslash X_i$ which contains the point at ∞. At least one of the contours γ_i or γ_j must separate Δ_i from Δ_j. Supposing this contour is γ_i, we define

$$\gamma=\gamma_i\backslash\partial\Lambda_0\,,$$

i.e. the portion of γ_i not contained in $\partial\Lambda_0$.

Letting $|\gamma|$ denote the length of γ, we assert that

$$|\gamma_i|+\operatorname{dist}(\gamma,\{\Delta_i,\Delta_j\})\leqq 11|\gamma|\,. \tag{1.8}$$

If $\gamma_i\cap\partial\Lambda_0=\emptyset$, then there is nothing to prove, since the shortest contour with a given radius ($\geqq\operatorname{dist}(\gamma,\{\Delta_i,\Delta_j\})$) is a square, and for this case, the circumference is eight times the radius. Now suppose $\gamma_i\cap\partial\Lambda_0\neq\emptyset$. Then

$$\begin{aligned}\operatorname{dist}(\gamma,\{\Delta_i,\Delta_j\})&\leqq \tfrac{1}{2}\operatorname{dist}(\Delta_i,\Delta_j)\leqq\frac{1}{18}|\partial\Lambda_0|\\&\leqq\tfrac{1}{2}\operatorname{dist}(\partial\Lambda_0,\{\Delta_i,\Delta_j\})\\&\leqq\tfrac{1}{2}|\gamma|\end{aligned}$$

by (1.7). Thus

$$\begin{aligned}|\gamma_i|+\operatorname{dist}(\gamma,\{\Delta_i,\Delta_j\})&\leqq|\gamma|+|\partial\Lambda_0|+\tfrac{1}{2}|\gamma|\\&\leqq 11|\gamma|\,.\end{aligned}$$

Since

$$\textstyle\prod_{\Delta\subset\Lambda_0}(\chi_+(\Delta)+\chi_-(\Delta))=1\,,$$

we have

$$\langle \chi_+(\Delta_i)\chi_-(\Delta_j)\rangle = \sum_c' \langle \chi_+(\Delta_i)\chi_-(\Delta_j)\prod_\Delta \chi_{c(\Delta)}(\Delta)\rangle ,$$

where $\sum_c'$ is the sum over configurations c with $c(\Delta_i)=1$, $c(\Delta_j)=-1$ and $\prod_\Delta$ omits Δ_i and Δ_j. Let $\gamma=\gamma(c)$ be defined as above, and let $\mathcal{N}=\mathcal{N}(\gamma)$ be the set of nearest neighbor pairs bordering γ. Then

$$\begin{aligned}\langle \chi_+(\Delta_i)\chi_-(\Delta_j)\rangle &= \sum_\gamma \sum_{\{c:\gamma(c)=\gamma\}} \langle \chi_+(\Delta_i)\chi_-(\Delta_j)\prod_\Delta \chi_{c(\Delta)}(\Delta)\rangle \\ &\leqq \sum_\gamma \langle \chi_+(\Delta_i)\chi_-(\Delta_j)\prod_{\Delta,\Delta'\in\mathcal{N}(\gamma)} \chi_+(\Delta)\chi_-(\Delta')\rangle \\ &\leqq \sum_\gamma e^{-K|\gamma|} .\end{aligned}$$

The sum ranges over γ's resulting from contours γ_i enclosing Δ_i and separating Δ_i from Δ_j. There are at most $3^{|\gamma_i|}$ such contours of length $|\gamma_i|$, with a fixed starting point. We choose as starting point the point of γ_i lying on the line (Δ_i, Δ_j) and closest to Δ_i. With $|\gamma_i|$ fixed, this can be chosen in at most $11|\gamma|$ ways, by (1.8). Thus $\langle \chi_+(\Delta_i)\chi_-(\Delta_j)\rangle < 1/8$ for K sufficiently large.

In order to prove Theorem 1.2, we study perturbations of the interacting Euclidean measure defined by polynomials of the form

$$Q(\xi, X) \equiv \sum_{\nu=1}^4 Q_\nu(\xi^{(\nu)}, X) , \tag{1.9}$$

where

$$Q_1(\xi^{(1)}, X) = \sigma^{-1}\sum_{\Delta_j\subset X} \xi_j^{(1)} \int_{\Delta_j} (:\varphi^2(x): - \sigma^2)dx , \tag{1.10}$$

$$Q_2(\xi^{(2)}, X) = (\ln\sigma)^{-1}\sum_{\Delta_j\subset X} \xi_j^{(2)} \int_{\Delta_j} :\varphi^2(x) - \varphi(\Delta_j)^2: dx , \tag{1.11}$$

$$Q_3(\xi^{(3)}, X) = \sum_{\substack{\Delta_i,\Delta_j\subset X \\ |i-j|=1}} \xi_{ij}^{(3)}(\varphi(\Delta_i) - \varphi(\Delta_j)) , \tag{1.12}$$

$$Q_4(\xi_j^{(4)}, \kappa, \kappa', X) = \sigma^{-4}(\kappa+1)^\delta \sum_{\Delta_j\subset X} \xi_j^{(4)} \sum_{j=1}^4 \int_{\Delta_i} :\varphi_\kappa^j(x) - \varphi_{\kappa'}^j(x): dx . \tag{1.13}$$

In (1.13), κ and κ' denote ultraviolet cutoffs, and

$$0 \leqq \kappa \leqq \kappa' \leqq \infty , \qquad \kappa < \infty .$$

For $\kappa=0$, $:\varphi_\kappa^j: \equiv 0$. Let $|X|$ denote the area of the subset $X\subset R^2$. Our main technical estimate is the following theorem, which will be proved in Section 3.

Theorem 1.3. *There is a constant K_1 and a $\delta>0$ such that for all Q defined by* (1.9)–(1.13), *with $|\xi_\alpha^{(\beta)}|\leqq 1$ and for $\langle\cdot\rangle$ defined by* (1.2)–(1.4), *and* (1.6) *with σ sufficiently large*

$$|\langle e^{Q(\xi, X)}\rangle| \leqq e^{K_1(\log\sigma)^2|X|} .$$

Proof of Theorem 1.2, assuming Theorem 1.3. For each pair $(\Delta_i, \Delta_j)\in\mathcal{N}$ we write

$$\chi_+(\Delta_i)\chi_-(\Delta_j) = (\chi_{(0,\sigma/2)} + \chi_{(\sigma/2,\infty)})(\chi_{(-\infty,-\sigma/2)} + \chi_{(-\sigma/2,0)})$$

as a sum of four terms. We shall bound each term for any even M by $[\sigma^{-1}Q_\nu(\varphi)]^M$, $\nu=1, 2, 3$. By using the Cauchy formula for derivatives in the variable ξ as in [2] we have

$$\sigma^{-M}\langle Q_\nu^M\rangle = \sigma^{-M}\left[\frac{d}{d\xi^{(\nu)}}\right]^M \langle e^Q\rangle \leqq \frac{M!\sigma^{-M}}{2\pi}\left|\oint \frac{\langle e^Q\rangle}{[\xi^\nu]^{M+1}} d\xi\right| . \tag{1.15}$$

In each case we choose $M=2[\sigma/8]$ to optimize the above bound. Here $[x]$ denotes the integer part of x.

For instance, we bound the term $\chi_{(\sigma/2,\infty)}(\Delta_i)\chi_{(-\infty,-\sigma/2)}(\Delta_j)$ by

$$[\sigma^{-1}(\varphi(\Delta_i)-\varphi(\Delta_j))]^M=\sigma^{-M}\left[\frac{d}{d\xi_{ij}^{(3)}}\right]^M e^{Q_3}\Bigg|_{\xi=0}.$$

Each of the remaining three terms is bounded by $\chi_{(0,\sigma/2)}(\Delta_i)$ or $\chi_{(-\sigma/2,0)}(\Delta_j)$. Note that on the support of $\chi_{(0,\sigma/2)}(\Delta)$,

$$\begin{aligned}\chi_{(0,\sigma/2)}(\Delta)&\leqq 1\leqq 4/3\left[1-\left(\frac{\varphi(\Delta)}{\sigma}\right)^2\right]\\ &\leqq\frac{4}{3}\sigma^{-2}[\sigma^2-\textstyle\int_\Delta:\varphi(x)^2:dx]+\frac{4}{3}\sigma^{-2}[\textstyle\int_\Delta:\varphi(x)^2:dx-:\varphi(\Delta)^2:+O(\ln\sigma)].\end{aligned}$$

Here $O(\ln\sigma)$ is a Wick ordering constant. Since $\sigma^{-2}O(\ln\sigma)\to 0$ one has for sufficiently large σ either

$$\chi_{(0,\sigma/2)}(\Delta_i)\leqq\sigma^{-M}\left[\frac{4}{\sigma}\left(\textstyle\int_{\Delta_i}:\varphi(x)^2:dx-\sigma^2\right)\right]^M$$

or

$$\chi_{(0,\sigma/2)}(\Delta_i)\leqq\sigma^{-2M}[4\textstyle\int_{\Delta_i}:\varphi(x)^2:dx-:\varphi(\Delta)^2:]^M.$$

The factor $\chi_{(-\sigma/2,0)}(\Delta_j)$ is bounded by the same expressions with Δ_j replacing Δ_i.

Now we can apply (1.15) to each pair of squares. Since a single square Δ_i occurs in Q_1 or Q_2 in at most four pairs $(\Delta_i,\Delta_j)\in\mathcal{N}$, the factorials are bounded by

$$(4M!)^{|\mathcal{N}|/4}\leqq([\sigma]!)^{|\mathcal{N}|/4}.$$

The theorem now follows from

$$\begin{aligned}\langle\textstyle\prod\chi_+(\Delta_i)\chi_-(\Delta_j)\rangle&\leqq e^{O(\log\sigma)^2|\mathcal{N}|}[[\sigma]!\sigma^{-\sigma}]^{|\mathcal{N}|/4}\\ &\leqq e^{O(\log\sigma)^2|\mathcal{N}|}e^{-\sigma|\mathcal{N}|/4}.\end{aligned}$$

In the last line we have used Stirling's formula (with $n=[\sigma]$),

$$n!\leqq(2\pi n)^{\frac{1}{2}}(n/e)^n.$$

Note that the weakly divergent bound $\exp(O(\ln^2\sigma))$ per square from Theorem 1.3 is bounded by the strongly convergent factor $\exp(-O(\sigma))$.

2. Bounds on Vacuum Energy per Unit Volume

In this section we give upper and lower bounds on the vacuum energy per unit volume. The upper bound is more elementary, and we consider it first. Let $|\partial\Lambda|$ denote the length of the boundary of Λ.

Proposition 2.1. *There is a constant $O(1)$ independent of σ and Λ such that*

$$e^{-O(1)|\Lambda|}\leqq\int e^{-V(\Lambda)}d\varphi_{\sigma^{-1},\Lambda}\qquad(2.1)$$

for all Λ satisfying $\sigma|\partial\Lambda|\leqq|\Lambda|$.

Proof. We translate φ by the function $\sigma f(x)$, where

$$0 \leqq f(x) = \sigma^{-2} \int_A (-\Delta_A + \sigma^{-2})^{-1}(x, y)dy \leqq 1 .$$

(Note that the Dirichlet Green's function is less than the free Green's function.) By convexity of the exponential,

$$\int e^W d\varphi \geqq \exp(\int W d\varphi) .$$

Hence for $W(\varphi)$ a Wick ordered polynomial without a constant term,

$$\int W d\varphi = 0 \quad \text{and} \quad \int e^W d\varphi \geqq 1 .$$

Thus we need only compute the constant in the exponent after translation. The purpose of the translation is to eliminate the leading term $O(\sigma^2|\Lambda|)$. There is a constant term associated with $V(\Lambda)$ and a constant associated with the change of measure $d\varphi \to d(\varphi + \sigma f)$. The constant associated with $V(\Lambda)$ of (1.3) is

$$\begin{aligned}\int_A P(\sigma f(x))dx &= \sigma^2 \int_A (f(x)^2 - 1)^2 dx \\ &\leqq 2\sigma^2 \int_A (1 - f(x))dx \\ &= 2 \int_{A\times A} [(-\Delta + \sigma^2)^{-1} - (-\Delta_A + \sigma^{-2})^{-1}](x, y)dxdy \\ &\quad + 2 \int_{R^2\setminus A} \int_A (-\Delta + \sigma^{-2})^{-1}(x, y)dxdy .\end{aligned}$$

In the first term, Δ and Δ_A agree except on $\partial\Lambda$, so that the difference kernel contributes only along the boundary $\partial\Lambda$. Scaling the length scale σ back to one shows that this term is $O(\sigma|\partial\Lambda|)$. The second term is integrated with one variable in and one variable outside of Λ. Again scaling the length back to one we obtain a contribution $O(\sigma|\partial\Lambda|)$. Thus

$$\int_A P(\sigma f(x))dx \leqq O(\sigma|\partial\Lambda|) \leqq O(|\Lambda|) ,$$

where the last inequality follows by our restriction on the volumes.

The constant associated with the translation of the Gaussian measure $d\varphi_{\sigma^{-1},\Lambda}$ is

$$\tfrac{1}{2}\sigma^2\langle f, (-\Delta_A + \sigma^{-2})f\rangle = \tfrac{1}{2}\int_A f(x)dx = O(|\Lambda|) ,$$

and the proof for $\mu = 0$ is complete. For $\mu > 0$, we use the Griffiths inequality

$$\frac{d}{d\mu}\int e^{-[V(\Lambda) - \mu\varphi(\Lambda)]}d\varphi \geqq 0$$

to complete the proof.

The lower bound on the vacuum energy concerns the perturbed Euclidean measure

$$\int e^{-[V(\Lambda) + Q(\xi, \Lambda)]}d\varphi_{\sigma^{-1},\Lambda} . \tag{2.2}$$

This bound, which is slowly divergent as $\sigma \to \infty$ (see Theorem 1.3), is sufficient for the present paper. Bounds uniform in σ may be found in [7]. By Schwarz' inequality, it is sufficient to establish separately the bounds

$$\int e^{4Q_v}d\varphi_{\sigma^{-1},\Lambda} \leqq e^{O(1)|\Lambda|} , \qquad v = 2, 3 , \tag{2.3}$$

$$\int e^{-2V(\Lambda)}e^{4Q_v}d\varphi_{\sigma^{-1},\Lambda} \leqq e^{O(1)\ln^2\sigma|\Lambda|} , \qquad v = 1, 4 . \tag{2.4}$$

Proposition 2.2. *There is a constant independent of σ and Λ such that* (2.3) *and* (2.4) *hold.*

Proof. The main point is that the low momentum modes of Q_2 and Q_3 are suppressed, thereby introducing an effective mass into the covariance. In technical terms, this idea emerges as the following fact: The Laplace operator of a bounded region, with Neumann boundary data, has $\lambda_1 = 0$ as the lowest eigenvalue, and the next eigenvalue λ_2 is strictly positive. The eigenspace corresponding to $\lambda_1 = 0$ is the space of functions which are constant on each connected component of the bounded region.

We write Q_3 (and $\xi^{(3)}$) as a sum of four terms so that in a single term, each lattice square Δ occurs only once in the sum over nearest neighbor pairs. By Hölders inequality, it is sufficient to consider a single such term. Let $\mathcal{N} = \{(\Delta, \Delta')\}$ be the corresponding set of nearest neighbor pairs of lattice squares, and let $\Delta_{\mathcal{N}}$ be the Laplace operator with Neumann data on

$$\bigcup_{(\Delta, \Delta') \in \mathcal{N}} \partial(\Delta \cup \Delta').$$

Let $d\varphi_{\mathcal{N}}$ be the Gaussian measure with covariance $(-\Delta_{\mathcal{N}} + \sigma^{-2})^{-1}$. By conditioning [10] and factorization of $d\varphi_{\mathcal{N}}$ across boundary lines, it is sufficient to consider a single pair of lattice squares. The required bound reduces to

$$\int e^{16 \operatorname{Re} \xi(\varphi(\Delta) - \varphi(\Delta'))} d\varphi_{\mathcal{N}} \leqq \exp[\tfrac{1}{2}(16 \operatorname{Re} \xi)^2 \langle \chi, (-\Delta_{\mathcal{N}} + \sigma^{-2})^{-1} \chi \rangle] = O(1),$$

where $\chi = \chi_\Delta - \chi_{\Delta'}$ is the difference of the characteristic functions of Δ and Δ'. To establish the bound $O(1)$, independent of σ, we use the fact that χ is perpendicular to the functions which are constant on $\Delta \cup \Delta'$, and so the lowest eigenvalue $\lambda_1 = 0$ of $\Delta_{\mathcal{N}}$ does not contribute.

For the case $\nu = 2$, it suffices to estimate

$$\int e^{4\xi :\varphi^2(\Delta) - \varphi(\Delta)^2: / \log \sigma} d\varphi_{\mathcal{N}},$$

where the Gaussian measure $d\varphi_{\mathcal{N}}$ has a covariance corresponding to Neumann boundary conditions on the boundary $\partial\Delta$ of the lattice square Δ. Let P_Δ be the orthogonal projection onto the subspace of L_2 spanned by χ_Δ and let

$$A = (\log \sigma)^{-1} (-\Delta_{\mathcal{N}} + \sigma^{-2})^{-1} (\chi_\Delta - P_\Delta).$$

Then by a standard computation (see also [7])

$$\int e^{4\xi :(\varphi^2(\Delta) - \varphi(\Delta)^2): / \log \sigma)} d\varphi_{\mathcal{N}} = |e^{-\frac{1}{2} \operatorname{tr} \ln(I - A)} e^{-\frac{1}{2} \operatorname{tr} A}| \leqq e^{\frac{1}{2} \|A\|_2^2},$$

where $\|\cdot\|_2^2$ is the Hilbert-Schmidt norm and we use $0 \leqq A < I$, true for $\sigma \gg 1$. In fact, since $\chi_\Delta - P_\Delta$ annihilates the eigenspace of $\Delta_{\mathcal{N}}$ corresponding to the eigenvalue $\lambda_1 = 0$, we see that for large σ

$$\|A\|_2^2 \leqq \left(\sum_{i=2}^{\infty} \lambda_i^{-2}\right) / \ln^2 \sigma < \tfrac{1}{2}.$$

This completes the proof.

For the case $\nu = 1$, we note that when $|\xi| \leqq 1$,

$$-O(1)\sigma^2(1 + \ln^2 \kappa) \leqq :(\varphi_\kappa^2(x) - \sigma^2)^2: + \xi\sigma :\varphi_\kappa^2(x) - \sigma^2:, \tag{2.5}$$

with Wick ordering with respect to a unit mass. By standard methods the vacuum energy per unit volume corresponding to the polynomial on the right is $O(\sigma^2 \ln^2 \sigma)$, see [9]. The proof of (2.4) for $\mu=0$ then reduces to scaling by σ^{-2}. We introduce the external field $\mu\varphi$ directly into (2.5). Again, for μ sufficiently small, we obtain the lower bound (2.5) by removing Wick ordering, and this completes the proof of (2.4). We omit the discussion of $v=4$.

3. Local Perturbations of the Interaction

In this section we extend the φ^j bounds to localized perturbations of the type $Q(\xi, X)$, and thereby prove Theorem 1.3. We make two modifications to the basic φ^j bound argument [3, 8, 2, 5]. The first is that we consider perturbations of the measure such as

$$\varphi(\Delta)^2 - \int_\Delta :\varphi(x)^2: dx\,,$$

which, while localized in a bounded region Δ, cannot be expressed as a local perturbation of the Hamiltonian. Secondly, by using Euclidean invariance of the physical measure ($\Lambda = R^2$), we show that it suffices to consider the case where X is a large rectangle. In the case X = rectangle, we work in a finite volume and obtain bounds uniform in Λ. This allows the passage to the limit $\Lambda \nearrow R^2$. The φ bounds have the form

$$\pm\varphi(f) \leqq H + O(1)|supp\, f|$$

rather than the cruder estimate

$$\pm\varphi(f) \leqq H + O(1)|\mathrm{diam}\; supp\, f|\,.$$

We do not require estimates on β_∞ [cf. 8, 2] which measures surface effects on the vacuum energy due to removing or adding parts of the interaction associated with a bounded region. (We remark that estimates on β_∞ are poor because of the small mass σ^{-1} in the covariance.) Dirichlet data provide a further complication to such estimates.

Let

$$H_l(\sigma) = H_{0,l}(\sigma) + \sigma^{-2} \int_{-l}^{l} :(\varphi(x)^2 - \sigma^2)^2: dx \tag{3.1}$$

be the Hamiltonian with Dirichlet boundary conditions at $x = \pm l$. Here $H_{0,l}(\sigma)$ is the free Hamiltonian with the same boundary conditions and mass σ^{-1}. Then $H_{0,l}(\sigma)$ and $H_l(\sigma)$ act on the time zero Fock space

$$\mathscr{F} = L^2(\mathscr{S}'(R^1), d\mu_{0,l})\,. \tag{3.2}$$

Here $d\mu_{0,l}$ is the Gaussian measure defined by the condition that the vacuum $\Omega_{0,l}$ for $H_{0,l}$ is represented by the function 1. Let Ω_l be the vacuum for $H_l(\sigma)$ and let $E_l(\sigma)$ be the vacuum energy, so that

$$(H_l(\sigma) - E_l(\sigma))\Omega_l = 0\,.$$

Recall that $H_l(\sigma)$ is a direct sum of a free Hamiltonian associated with the region $|x| \geqq l$ and an operator with compact resolvent associated with the region $|x| \leqq l$, so that the existence of Ω_l follows.

Lemma 3.1. *There is a constant independent of σ and a sequence $l_i \to \infty$ (possibility dependent on σ) such that*

$$-E_{l_i-a}(\sigma) \leqq -E_{l_i}(\sigma) + \text{const}\, a\,, \tag{3.3}$$

for any $a \geqq 0$.

Proof. By Proposition 2.1, there is a constant independent of σ and l such that

$$l \leqq e(l) \equiv -E_l + \text{const}\, l \tag{3.4}$$

for l sufficiently large. Choose l_i to be a one-sided maximum of e, so that for positive $l \leqq l_i$,

$$e(l) \leqq e(l_i)\,.$$

Since $e(l) \to \infty$ as $l \to \infty$, there is a sequence of such l_i, $l_i \to \infty$. This completes the proof.

The Euclidean Fock space is

$$\mathscr{K} = \mathscr{K}(-l, l) = L^2(\mathscr{S}'(R^2), d\varphi_{\sigma^{-1}, \Lambda_l})\,,$$

where $\Lambda_l = R \times [-l, l]$. For each time t there is a natural imbedding $J_t : \mathscr{F} \to \mathscr{K}$ from $\mathscr{F}$ to the time t subspace of $\mathscr{K}$. We also introduce spacetime rectangles

$$Y_{T,l} = [-T, 0] \times [-l, l]$$
$$X_{T,l} = [-T, T] \times [-l, l]\,.$$

For each $a \in (-l, l)$, we also introduce the Fock spaces

$$\mathscr{F}(a, l) = L_2(\mathscr{S}'(R^1), d\mu_{a,l})$$

and

$$\mathscr{K}(-l+2a, l),\ a>0$$
$$\mathscr{K}(-l, l-2a),\ a>0\,.$$

The space $\mathscr{K}(-l+2a, l)$ is determined by placing zero Dirichlet data on the line $x=l$ and also on the symmetric line $x=-l+2a$ obtained by reflecting the line $x=l$ about the line $x=a$, for $a>0$. The definition of $\mathscr{K}(-l, l-2a)$, $a>0$, is similar. The measure $d\mu_{a,l}$ is Gaussian, and its covariance is given explicitly in [5]. It is characterized by the following property: Let I_a be the natural injection from polynominals on $\mathscr{S}'(R^1)$ to generalized polynomials on $\mathscr{S}'(R^2)$ defined by the formula

$$I_a \phi(f) = \phi(f \otimes \delta_a)\,.$$

Here $f \in \mathscr{S}(R^1)$ and ϕ on the left above is a field (linear function) defined on $\mathscr{S}'(R^1)$, while ϕ on the right is a field on $\mathscr{S}'(R^2)$. We extend I_a to polynomials,

$$I_a(Q(\phi(f_1), \ldots, \phi(f_n)) = Q(I_a(\phi(f_1)), \ldots I_a(\phi(f_n)))\,,$$

so that I_a can be regarded as an identification map. Then $d\mu_{a,l}$ is characterized by the fact that I_a is isometric from $\mathscr{F}(a, l)$ into $\mathscr{K}(-l+2a, l)$ $(a>0)$ or $\mathscr{K}(-l, l+2a)$ $(a<0)$. We use the fact that conditional expectation onto the line $x=a$, $a>0$, of a function supported in the region $a \leqq x \leqq l$, is independent of the location of Dirichlet boundary data in the region $x<a$. In particular these conditional expectations, defined in $\mathscr{K}(-l+2a, l)$ and in $\mathscr{K}(-l, l)$ coincide, see [5]. In order to give this statement a meaning, we identify the $x=a$ functions of the fields in these two spaces. However, this identification is not a unitary map between the Hilbert spaces.

Proposition 3.2. *Let $l=l_i$ be as in Lemma 3.1. There is a constant independent of ξ, a, σ, and i such that for $a \geqq O(1)\sigma$ and for l sufficiently large,*

$$\|J_{-2}^* e^{-V(Y_{2,l})} e^{+Q(\xi, Y_{2,a})} J_0\|_{\mathscr{F}}^2 e^{E_l} \leqq e^{\text{const } a(\ln\sigma)^2} .$$

Proof. Let

$$A = J_{-2}^* e^{-V(Y_{2,l})} e^{-Q(\xi, Y_{2,a})} J_0 ,$$

and let $Q^{\sim}(\xi, X_{2,a})$ be the sum of $Q(\xi, Y_{2,a})$ and its time reflexion about the x-axis. Then

$$AA^* = J_{-2}^* e^{-V(X_{2,l})} e^{-Q^{\sim}(\xi, X_{2,a})} J_2 .$$

Since AA^* is self adjoint and positivity preserving, there is a unit vector $v \in L_\infty(\mathscr{S}'(R^1), d\mu_{0,l})$ such that

$$\begin{aligned} \|A\|^2 = \|AA^*\| &= \lim_{N\to\infty} \langle v, (AA^*)^N v\rangle_{\mathscr{F}}^{1/N} \\ &\leqq \lim_{N\to\infty} \|v\|_{L_\infty}^{2/N} \langle \Omega_{0,l}, (AA^*)^N \Omega_{0,l}\rangle_{\mathscr{F}}^{1/N} \\ &= \lim_{N\to\infty} \langle \Omega_{0,l}, (AA^*)^N \Omega_{0,l}\rangle_{\mathscr{F}}^{1/N} . \end{aligned}$$

The inner product can be written in Euclidean form, yielding

$$\begin{aligned} \|A\|^2 &\leqq \lim_{N\to\infty} \left(\int e^{-V(X_{2N,l})} e^{Q^{\sim}(\xi, X_{2N,a})} d\varphi_{\sigma^{-1}, \Lambda_l}\right)^{1/N} \\ &\leqq \lim_{N\to\infty} \left(\int e^{-V(X_{2N,l-a})} d\varphi_{\sigma^{-1}, \Lambda_{l-a}}\right)^{1/N} \\ &\quad \cdot \|I_{-a}^* e^{-V(X_{2N,a})} e^{Q^{\sim}(\xi, X_{2N,a})} I_a\|^{1/N} \\ &\leqq e^{-E_{l-a}} \lim_{N\to\infty} \|\ldots\|^{1/N} \\ &\leqq e^{O(a)-E_l} \lim_{N\to\infty} \|\ldots\|^{1/N} , \end{aligned} \tag{3.5}$$

where we use Lemma 3.1. Here the norm $\|\ldots\|$ is the operator norm from $\mathscr{F}(a, l)$ to $\mathscr{F}(-a, l)$. The norm is estimated by an integral. For vectors $v_\pm \in \mathscr{F}(\pm a, l)$, we have by the Schwarz inequality

$$\begin{aligned} |\langle I_{-a} v_-, e^{-V(X_{2N,a})} e^{Q^{\sim}(\xi, X_{2N,a})} I_a v_+\rangle| &\leqq \|(I_{-a}v_-)(I_a v_+)\|_{L_2(d\varphi_{\sigma^{-1}, \Lambda_l})} \\ &\quad \cdot \|e^{-V(X_{2N,a})} e^{Q^{\sim}(\xi, X_{2N,a})}\|_{L_2(d\varphi_{\sigma^{-1}, \Lambda_l})} . \end{aligned}$$

By Proposition 2.2, the second factor is bounded by $\exp[\text{const}(\ln\sigma)^2 Na]$. We bound the first factor as in [5]. For a $\geqq O(1)\sigma$ and l large, we have the hypercontractive bound

$$\int |(I_{-a}v_-)(I_a v_+)|^2 d\varphi_{\sigma^{-1}, \Lambda_l} \leqq \prod_{+,-} \|I_{\pm a} v_\pm\|_{L_2(d\varphi_{\sigma^{-1}, \Lambda_l})}^2$$

Thus the norm $\|\ldots\|$ in (3.5) satisfies

$$\|\ldots\|^{1/N} \leqq \exp[O(a)(\ln\sigma)^2]$$

and using (3.5), the proof is complete.

Theorem 3.3. *The limit $\Lambda \nearrow R^2$ in* (1.4) *exists for*

$$A = e^{zQ(\xi, X)} \quad \text{and} \quad A = \prod_{\Delta_j} \chi_\pm(\Delta_j).$$

Proof. We prove convergence for $\exp(zQ)$ in two steps. First in the ultraviolet cutoff case $Q = Q_\kappa$, the existence of the limit $\Lambda \nearrow R^2$ for $A = Q_\kappa^n$ exists by Griffiths' inequality. (Because of the use of full Dirichlet data, this argument is special to φ^4.) We expand $\exp(zQ_\kappa)$ in a power series in z. Using term by term convergence, and a φ^j bound to give a uniform bound on the sum, we obtain convergence for $\exp(zQ_\kappa)$. We then remove the momentum cutoff by considering the difference $\exp(zQ_\kappa) - \exp(zQ_{\kappa'})$ as a power series. By Proposition 2.2 (and 3.2) we have a uniform bound which is $o(1)$ as $\kappa, \kappa' \to \infty$. Here we use the bound on Q_4, cf. [6, Chapters 2 and 5].

To prove the convergence for $A = \Pi\chi_\pm(\Delta)$ note that since $\langle \prod_j \exp(i\xi_j\varphi(\Delta))\rangle_\Lambda$ converges so does $\langle \psi(\varphi(\Delta))\rangle$ for $\tilde{\psi} \in L_1$. Using the fact that $\chi_\pm \leqq 1$ we need only establish

$$\langle |(\chi_+ - \psi_r)(\varphi(\Delta))| \rangle_\Lambda \leqq \langle (\chi_+ - \psi_r)^2(\varphi(\Delta))\rangle_\Lambda^{\frac{1}{2}} \to 0$$

uniformly in Λ for a suitable sequence $\tilde{\psi}_r \in L_1$. To deal with the singularities of $\tilde{\chi}_\pm$ it suffices to show

(a) $\dfrac{d^n}{d\xi^n} \langle e^{i\xi\varphi(\Delta)}\rangle_\Lambda$

(b) $\xi^n \langle e^{i\xi\varphi(\Delta)}\rangle_\Lambda$

are bounded uniformly in ξ and Λ. (a) is an immediate consequence of the φ bound. (b) follows from integration by parts, e.g.

$$\begin{aligned}\langle \xi e^{i\varphi(\Delta)}\rangle_\Lambda &= -i \int_\Delta dx \left\langle \frac{\delta}{\delta\varphi(x)} e^{i\xi\varphi(\Delta)} \right\rangle_\Lambda \\ &= -i \int_\Delta dx \langle e^{i\xi\varphi(\Delta)} \int dy (-\Delta + \sigma^{-2})^{-1}(x, y) : P'(\varphi(y)) : \rangle_\Lambda\end{aligned}$$

and Proposition 3.2.

For the remainder of this paper, we take $\Lambda = R^2$. The bounds of Proposition 3.2 extend to this case, and prove Theorem 1.3 for (sufficiently long) rectangles (cf. [4]). The problem is to allow X to be an arbitrary union of lattice squares. Let T_a denote time translation, $(t, x) \to (t + a, x)$ and let $\mathcal{R}$ denote time inversion, $(t, x) \to (-t, x)$, and let T_a and $\mathcal{R}$ act on functions of the Euclidean field $\varphi(t, x)$.

Proposition 3.4. *Let A, B, and C be functions of φ localized in the time intervals* $(-\infty, -1)$, $(-1, 0)$, *and* $(0, \infty)$ *respectively, and suppose $B \geqq 0$. Then*

$$|\langle ABC\rangle| \leqq \langle C\mathcal{R}C^-\rangle^{\frac{1}{2}} \langle (T_1 A)\mathcal{R}(T_1 A)^-\rangle^{\frac{1}{2}} \lim_{N\to\infty} \langle \prod_{i=1}^N T_{2i}(B\mathcal{R}B^-)\rangle^{1/2N}$$

Proof. We use the positivity condition of Osterwalder-Schrader (positivity of the inner product in the physical Hilbert space). The result then follows from the Schwarz inequality and the density of L_∞ in L_2.

Proof of Theorem 1.3. Let $i=(i_0, i_1)\in Z^2$. Given ξ, let $\xi(i_0, N)$ and $\xi(i, N)$ be multiindices defined by

$$\xi(i_0, N)_j = \begin{cases} \xi_i & \text{if } i_1 = j_1 \text{ and } |i_0 - j_0| \leqq N \\ 0 & \text{otherwise} \end{cases}$$

$$\xi(i, N)_j = \begin{cases} \xi_i & \text{if } |i_0 - j_0| \leqq N \text{ and } |i_1 - j_1| \leqq N \\ 0 & \text{otherwise}. \end{cases}$$

Consider the case $Q_3 = 0$. By repeated use of Proposition 3.4 in the x_0 direction and again after Euclidean rotation in the x_1-direction we conclude that when X_N is an $N \times N$ rectangle

$$\begin{aligned}\langle e^{Q(\xi, X_N)}\rangle &\leqq \textstyle\prod_{i_0} \lim_{N\to\infty} \langle e^{Q(\xi(i_0, N))}\rangle^{1/N} \\ &\leqq \textstyle\prod_{i\in X} \lim_{N\to\infty} \langle e^{Q(\xi(i,N))}\rangle^{1/N^2}.\end{aligned}$$

By Theorem 3.3 with $\Lambda = \Lambda_l = [-l, l] \times R$ we have

$$\langle e^{Q(\xi(i,N))}\rangle = \lim_{l\to\infty} \langle e^{Q(\xi(i,N))}\rangle_{\Lambda_l}.$$

To express the last term in Fock space let

$$A = J_{-2}^{*} e^{-V(Y_2, l)} e^{-Q(\xi(i,N), Y_2, N)} J_0 .$$

Then

$$\begin{aligned}\langle e^{Q(\xi(i,N))}\rangle_{\Lambda_l} &= \langle \Omega_l A^{N/2} \Omega_l\rangle_{\mathscr{F}} \\ &\leqq \|A\|^{N/2} \\ &\leqq e^{\text{const}\, N^2 (\log\sigma)^2}.\end{aligned}$$

Here we have applied Proposition 3.2 with $a = N$. Hence combining these estimates we have

$$\langle e^{Q(\xi, X)}\rangle \leqq \textstyle\prod_{i\in Z^2\cap X} e^{K_1(\log\sigma)^2} = e^{K_1(\log\sigma)^2 |X|}.$$

In the case of Q_3 we apply the same argument with appropriate reflections, as in Proposition 3.2.

References

1. Dobrushyn, R., Minlos, R.: Construction of a one-dimensional quantum field via a continuous Markov field. Funct. Anal. Appl. **7**, 324—325 (1973)
2. Fröhlich, J.: Schwinger functions and their generating functionals. II. Adv. Math. (to appear)
3. Glimm, J., Jaffe, A.: The $\lambda(\varphi^4)_2$ quantum field theory without cutoffs. IV. J. Math. Phys. **13**, 1558—1584 (1972)
4. Glimm, J., Jaffe, A.: On the approach to the critical point. Ann. Inst. Henri Poincaré **22**, 13—26 (1975)
5. Glimm, J., Jaffe, A.: φ^j bounds in $P(\varphi)_2$ quantum field models. Proc. of the Colloq. on Math. Methods of Quantum Field Theory, Marseille, June 1975

6. Glimm, J., Jaffe, A.: Two and three body equations in quantum field models. Commun. math. Phys. **44**, 293—320 (1975)
7. Glimm, J., Jaffe, A., Spencer, T.: A cluster expansion for the φ_2^4 quantum field theory in the two phase region (in preparation)
8. Guerra, F., Rosen, L., Simon, B.: Nelson's symmetry and the infinite volume behavior of the vacuum in $P(\varphi)_2$. Commun. math. Phys. **27**, 10—22 (1972)
9. Guerra, F., Rosen, L., Simon, B.: The vacuum energy for $P(\varphi)_2$ infinite volume limit and coupling constant dependence. Commun. math. Phys. **29**, 233—247 (1973)
10. Guerra, F., Rosen, L., Simon, B.: The $P(\varphi)_2$ Euclidean quantum field theory as classical statistical mechanics. Ann. Math. **101**, 111—259 (1975)
11. Pirogov, S. A., Sinai, Ya. G.: Phase transitions of the first kind for small perturbations of the Ising model. Funct. Anal. Appl. **8**, 21—25 (1974) (Engl. trans.)
12. Simon, B., Griffiths, R.: The φ_2^4 field theory as a classical Ising model. Commun. math. Phys. **33**, 145—164 (1973)
13. Glimm, J., Jaffe, A., Spencer, T.: Existence of phase transitions for φ_2^4 quantum fields. Proc. of the Colloq. on Math. Methods of Quantum Field Theory, Marseille, June 1975

Communicated by K. Hepp

(Received June 29, 1975)

A Convergent Expansion about Mean Field Theory
I. The Expansion

JAMES GLIMM*

Rockefeller University, New York, New York 10021

ARTHUR JAFFE†

Harvard University, Cambridge, Massachusetts 02138

AND

THOMAS SPENCER*

Rockefeller University, New York, New York 10021

Received June 2, 1976

We give a convergent expansion for nearly Gaussian quantum field theory in the multiphase region. The expansion combines (1) an expansion in phase boundaries, (2) a cluster expansion, and (3) a perturbation expansion to isolate dominant behavior. We study in detail the ground state of the $\mathscr{P}(\phi)_2 = (\lambda\phi^4 - \phi^2 - \mu\phi)_2$ model, with $|\mu| < \lambda^2 \ll 1$. The ground state is close to the classical free field, obtained by replacing $\mathscr{P}(\phi)$ by the quadratic mean field polynomial $\mathscr{P}_c(\phi)$, tangent to $\mathscr{P}$ at a global minimum. Selecting one minimum gives a pure phase (ergodic ground state) satisfying the Wightman–Osterwalder–Schrader axioms with a positive mass. We also establish analyticity in λ for $\mu = 0$ in the sector $|\operatorname{Im}\lambda| < \epsilon \operatorname{Re}\lambda \ll 1$, for $\epsilon \ll 1$.

1. FORMAL IDEAS AND THE MEAN FIELD PICTURE

In [18], the authors established a nonuniqueness theorem for quantum fields: Consider an even polynomial $:\lambda\mathscr{P}(\phi) - \phi^2:$ with bare mass $m = 1$, Wick ordering mass 1, and space-time dimension $d = 2$. For λ sufficiently small, there is a phase transition. In other words, there are at least two vacuum states corresponding to distinct phases [18]. In this paper we develop a convergent expansion in two distinct pure phases. We establish the Wightman and the Osterwalder–Schrader

* Supported in part by the National Science Foundation under Grant MPS74-13252.
† Supported in part by the National Science Foundation under Grant MPS75-21212.

axioms with a mass gap in each of two pure phases. We show that the pure phase theory is approximately Gaussian (free), and we give bounds on the deviation of the vacuum state from a Gaussian.

The leading term in our expansion is given by the mean field approximation, in this case a direct sum of two Gaussian (free) theories with mean $\pm\phi_c = \xi_\pm$ and mass m_c. Let us start with the interaction polynomial

$$:\mathscr{P}_1(\phi): = :\lambda\phi^4 - \tfrac{1}{4}\phi^2 + (64\lambda)^{-1}:, \qquad \lambda \ll 1. \tag{1.1}$$

We note that by a scale transformation, a re-Wick ordering, and the addition of a constant, $:\mathscr{P}_1(\phi):$ is reparameterized as the polynomial

$$:\mathscr{P}_2(\phi): = :\lambda\phi^4 + \tfrac{1}{2}\phi^2:, \qquad \lambda \gg 1, \tag{1.2}$$

again with Wick mass 1 (see, e.g., [17]). Thus the strong bare coupling ϕ^4 model (1.2) is equivalent to weak coupling ϕ^4 with a negative mass term. The advantage of the parameterization (1.1) over (1.2) is the small coefficient of ϕ^4, relative to the mass term. Thus we can expect higher-order effects will be small for (1.1), but large if we study (1.2). The λ's in (1.1) and (1.2) do not coincide, and $\lambda_{(1.2)} \to \infty$ as $\lambda_{(1.1)} \to 0$. We study $\mathscr{P}_1$ of (1.1).

The classical values $\phi_c = \pm(8\lambda)^{-1/2}$ minimize $\mathscr{P}_1(\xi)$, and the classical mass m_c is defined by

$$m_c^2 = \mathscr{P}_1''(\phi_c) = 1. \tag{1.3}$$

Expanding $:\mathscr{P}_1(\phi):$ about $\phi_c = \pm(8\lambda)^{-1/2}$ gives

$$:\mathscr{P}_1(\phi): = \lambda:(\phi - \phi_c)^4: \pm(2\lambda)^{1/2} :\phi - \phi_c)^3: + \tfrac{1}{2}:(\phi - \phi_c)^2:.$$

Between the minima $:\mathscr{P}_1(\phi): = 0$ at $\phi = \phi_c = \pm(8\lambda)^{-1/2}$, the polynomial $:\mathscr{P}_1(\phi):$ forms a high potential barrier, since

$$:\mathscr{P}_1(0): = (64\lambda)^{-1} \gg 0.$$

As $\lambda \to \infty$, the width of the barrier increases because $|\phi_c(\lambda)| \to \infty$.

We show here that each classical minimum $\phi_c = \pm(8\lambda)^{-1/2}$ gives rise to a vacuum state. We select one minimum, $\phi_c = (8\lambda)^{-1/2}$, by imposing a positive external field outside a box Λ, and then we take the infinite volume limit. The transition probability between the two minima is small because of the large barrier at $\mathscr{P}_1(0)$ and because the kinetic energy makes large local fluctuations unlikely. For λ sufficiently small, our external field boundary condition chooses the vacuum state localized with $\langle\phi\rangle \sim (8\lambda)^{-1/2}$. The details of our proof are patterned in part after Peierls' proof of phase transition in the Ising model. By symmetry, we also have a solution localized with $\langle\phi\rangle \sim -(8\lambda)^{-1/2}$, and the two solutions are exchanged by the automorphism $\phi \to -\phi$.

Our expansion combines a Peierls-type method with the cluster expansion, developed previously to analyze regions of weak bare coupling, i.e., $\lambda \ll 1$ in (1.2), or large external field [15, 16, 27]. In both previous cases, the cluster expansion was used far from the region of phase transitions. In this single phase region, the cluster expansion has been an effective tool for the detailed investigation of superrenormalizable quantum fields [1, 2, 9, 10, 22, 23] including a description of the mass spectrum and scattering states (up to some threshold) [15, 28, 29], analyticity in the coupling constants [16], the Borel summability of the perturbation series [7], and nontriviality of the S-matrix [4, 6, 25]. Using the expansion presented here, similar questions can be studied in the multiphase weak coupling region ($\lambda \ll 1$ in (1.1)) or the (equivalent) strong bare coupling region ($\lambda \gg 1$ in (1.2)).

In the multiphase region we expect numerous bound states and resonances. As a special feature of two dimensions, soliton states accompany phase transitions [3, 8, 11, 19]. Correspondingly the soliton–antisoliton state should introduce a continuum threshold into the vacuum sector and possible bound state spectrum. Introduction of a small external field breaks the phase transition and should make the soliton sector unstable. It should transform these continuum (two soliton) states into bound states and/or resonances.

In particular, we note that in the two-dimensional Ising model with $T < T_c$, the truncated two-point function has a decay rate $r^{-2}e^{-mr}$ [24, 299], indicative of a continuum threshold. An isolated particle would yield an Ornstein-Zernike $r^{-1/2}\exp(-mr)$ decay in two dimensions. The continuum behavior arises from the fact that the solitons do not form a bound state. (The associated Fermi field is a free field.) Thus the Ising model particle spectrum is missing its "elementary particle," i.e., the isolated hyperboloid-like curve at the bottom of the nonzero energy-momentum spectrum for $T < T_c$.

The single soliton appears in a disjoint state space. Its presence can be seen directly from the fermion solution of I_2 [21] by replacing the periodic boundary conditions of [21] with antiperiodic boundary conditions. Since it is generally believed that the ϕ^4-soliton–antisoliton pairs do form bound states, one could still expect an elementary (boson) particle and an upper mass gap for the two-phase ϕ_2^4 theory, as well as a conventional decay rate, for its two-point function. Returning to the Ising model: As the external field $\mu \searrow 0+$, one could expect only bound state spectrum between m and $2m$ (for $T < T_c$), filling in densely to become a continuum at $\mu = 0$.

Let us now consider the mean field picture for a general interaction

$$\sum_{i=3}^{2n} \lambda_i(\phi - \phi_c)^i + \tfrac{1}{2}m_c^2(\phi - \phi_c)^2 - \mu(\phi - \phi_c),$$

with Wick ordering mass m_c. In the region of parameter space $(|\mu| + \Sigma|\lambda_i|)m_c^{-2} \ll 1$, we expect that the mean field approximation is valid both qualitatively and quantitatively. Qualitatively, the mean field picture should determine the topology of the phase transition hypersurface. Branches (multiple points) of the hypersurface should occur on surfaces of lower dimension, where more than two phases coexist in equilibrium. In general, branches with coexistence of up to n phases should occur for $\mathscr{P}(\phi)_2$ theories with $\deg \mathscr{P} = 2n$. These singularities of the phase transition hypersurface should occur for $\mathscr{P}(\phi)_2$ inside (as well as outside) the domain of convergence of the cluster expansion. Aside from these branches, given topologically by the mean field approximation, the phase transition should be regular (real analytic) within the domain of convergence of the cluster expansion. The boundaries (edges) of the phase transition hypersurface have a special significance; they are critical points. They should lie outside the domain of convergence of the cluster expansion. Quantitatively, the validity of the mean field picture means that a perturbation theory, in powers of λ_i/m_c^2, should be asymptotic as $\lambda_i/m_c^2 \to 0$.

As a special case of the above ideas, we mention two problems: (1) To demonstrate the existence of a triple point for the interaction $\phi^6 + \lambda\phi^4 - \sigma\phi^2$ with $\sigma \gg 1$ and $\lambda = \lambda(\sigma)$. (2) To prove the existence of phase transitions without symmetry breaking, for the interaction $\phi^6 + \epsilon\phi^3 - \sigma\phi^2 - \mu\phi$ for $\sigma \gg 1$, $\epsilon \ll 1$ and $\mu = \mu(\epsilon, \sigma) \neq 0$.

2. The Main Results

Our expansion applies to $\lambda\mathscr{P}_{\text{even}}(\phi) - \phi^2 - \mu\phi$ models with $\lambda > 0$ sufficiently small. For simplicity of notation, we restrict attention to the ϕ^4 model in a weak external field. Define the real interaction polynomial

$$\mathscr{P}(\xi) = \lambda\xi^4 - \tfrac{1}{4}\xi^2 - \mu\xi - E_c, \tag{2.1a}$$

where

$$0 < \lambda \ll 1, \qquad |\mu| \leqslant \lambda^2, \qquad \text{and} \qquad \inf_\xi \mathscr{P}(\xi) = 0. \tag{2.1b}$$

We define the interaction in a region Λ by

$$:\mathscr{P}_\Lambda: = \int_\Lambda :\mathscr{P}(\phi(x)): dx. \tag{2.2}$$

Here : : denotes Wick ordering mass equal to one. Let $d(\Phi - \mu_b)$ be the Gaussian measure with mean μ_b and covariance

$$C = (-\Delta + 1)^{-1}. \tag{2.3}$$

We define the finite volume measure $d\phi_\Lambda$ by

$$d\phi_\Lambda \equiv \exp(-:\mathscr{P}_\Lambda: + \tfrac{1}{2}:(\phi - \mu_b)^2_\Lambda:)\, d(\Phi - \mu_b). \tag{2.4}$$

Note that the second term in the exponent of (2.4) cancels the mean and the mass of the measure $d(\Phi - \mu_b)$ in the region Λ. However, a positive external field remains in the region exterior to Λ, which defines our + boundary conditions (for $\mu_b > 0$) as $\Lambda \nearrow R^2$.

For $\mu = 0$, the polynomial (2.1) has two absolute minima $\xi_\pm$ satisfying

$$\phi_c = \xi_+ = -\xi_- = (8\lambda)^{-1/2}. \tag{2.5}$$

For $\mu > 0$, $\mathscr{P}$ has a unique absolute minimum at

$$\phi_c = \xi_+ = \xi_+(\lambda, \mu) = (8\lambda)^{-1/2} + \mu + O(\mu^2). \tag{2.6a}$$

and a relative minimum at

$$\xi_- = \xi_-(\lambda, \mu) = -(8\lambda)^{-1/2} + \mu + O(\mu^2), \tag{2.6b}$$

uniformly in λ. At the absolute minimum ξ_+, the classical mass m_c is given by

$$m_c = 1 + (72\lambda)^{1/2}\mu + O(\mu^2). \tag{2.7}$$

Since $m_c \sim 1$, we have used Wick ordering and bare mass 1 in (2.2)–(2.4). We remark that the classical vacuum energy per unit volume is

$$E_c = -\{(64\lambda)^{-1} + \mu(8\lambda)^{-1/2}\} + O(\mu^2). \tag{2.8}$$

We now take Λ to be a square centered near the origin, such that $\mathrm{dist}(0, \partial\Lambda) \geqslant \frac{1}{4} | \Lambda |^{1/2}$. We state the main results proved by our new expansion.

THEOREM 2.1. *Let $d\phi_\Lambda$ be as above with $0 \leqslant \mu$ and $\mu_b = \xi_+$. For λ sufficiently small, the infinite volume limit of Schwinger functions*

$$S_n(x_1, \dots, x_n) = \left\langle \prod_{i=1}^{n} \phi(x_i) \right\rangle = \lim_{\Lambda \nearrow R^2} \frac{\int (\prod_{i=1}^{n} \phi(x_i))\, d\phi_\Lambda}{\int d\phi_\Lambda} \tag{2.9}$$

exist in $\mathscr{S}'(R^{2n})$ and satisfy the Osterwalder–Schrader axioms with exponental clustering. The corresponding quantum field model obeys all the Wightman axioms. It has a unique vacuum, a positive field expectation $\langle\phi\rangle$ and a positive mass $m = m(\lambda, \mu) > 0$.

For $\mu \leqslant 0$ and $\mu_b = \xi_-$, the same results hold except that the field expectation $\langle\phi\rangle$ is negative. For $\mu = 0$, the two choices $\mu_b = \xi_\pm$ give two solutions.

THEOREM 2.2. *Let $0 \leqslant \mu$, $\mu_b = \xi_+$. Then for λ sufficiently small,*

$$\langle\phi(x)\rangle = \xi_+ + O(\lambda^{3/2}), \tag{2.10a}$$

$$\langle\phi(x)\phi(y)\rangle - \langle\phi(x)\rangle\langle\phi(y)\rangle = C(x-y) + \lambda M_2(x, y), \tag{2.10b}$$

$$\langle\phi(x_1),..., \phi(x_n)\rangle_T = \lambda^{1/2} M_n(x_1, ..., x_n), \qquad n \geqslant 3, \tag{2.10c}$$

$$\left\langle\left(\int (\phi(x) - \xi_+) f(x)\, dx\right)^n\right\rangle$$

$$= (n-1)!!\, C(f, f)^{n/2} + O(\lambda^{1/2})(\|f\|_{L_1} + C(f, f)^{1/2})^n. \tag{2.10d}$$

Here $C = (-\Delta + 1)^{-1}$, and the functions $M_n(x_1, ..., x_n)$ are bounded and continuous in λ, x_1, $x_2, ..., x_n$, including $\lambda = 0$. Also, $\langle\cdot\rangle_T$ denotes the truncated (connected part of the) expectation value. By Theorem 2.1, the M_n are exponentially decreasing functions of the difference variables $x_j - x_{j+1}$. The $O(\lambda^{1/2})$ in (2.10d) depends on n.

For $\mu \leqslant 0$ and $\mu_b = \xi_-$, the above results hold, if we replace ξ_+ by ξ_- in (2.10).

The generalized Schwinger functions satisfy corresponding local regularity estimates:

THEOREM 2.3. *Let $0 \leqslant \mu$, $\mu_b = \xi_+$, and λ be sufficiently small. Then*

$$\left\langle\prod_{i=1}^n :(\phi(x_i) - \xi_+)^{m_i}:\right\rangle_T = \left\langle\prod_{i=1}^n :\psi(x_i)^{m_i}:\right\rangle_{0,T} + \lambda^{1/2} N_n(x_1, ..., x_n), \tag{2.11}$$

where $\langle\cdot\rangle_0$ denotes a Gaussian expectation of ψ with mean zero and covariance $C = (-\Delta + 1)^{-1}$. Also $N_n(x_1, ..., x_n)$ is bounded and continuous in λ, $x_1, ..., x_n$, including $\lambda = 0$, and the N_n are exponentially decreasing functions in the difference variables $x_j - x_{j+1}$.

THEOREM 2.4. *Under the hypothesis of Theorem 2.1, there is a measure $d\phi$ on $\mathscr{S}'(R^2)$ such that*

$$\langle\phi(x_1) \cdots \phi(x_n)\rangle = \int \phi(x_1) \cdots \phi(x_n)\, d\phi, \tag{2.12}$$

and $d\phi$ is ergodic under space or time translations.

Remark 1. The estimates (2.10)–(2.11) give a precise statement bounding the deviation (as $\lambda \to 0$) of $d\phi$ from a Gaussian measure with mean ξ_+ of ξ_- and covariance $C = (-\Delta + 1)^{-1}$.

Remark 2. The bound (2.10b) shows that $d\langle\phi\rangle/d\mu = 1 + O(\lambda)$. Integration yields

$$\langle\phi\rangle_\mu = \langle\phi\rangle_{\mu=0} + \mu + O(\lambda\mu), \tag{2.13}$$

where $\langle\phi\rangle_{\mu=0}$ is positive or negative according to whether μ, μ_b are positive or negative.

THEOREM 2.5. *The Schwinger functions* (2.9) *for* $\mu \equiv 0$ *are analytic functions of* λ *in the interior of a sector:*

$$|\operatorname{Im}\lambda| \leqslant \epsilon \operatorname{Re}\lambda \ll 1, \tag{2.14}$$

for ϵ *sufficiently small.*

3. THE EXPANSION IN PHASE BOUNDARIES

For the remainder of Part I we give the expansion which yields our results. In Part II we prove the convergence of our expansion and the theorems stated here. The expansion has three main ingredients: the expansion in phase boundaries given here, the cluster expansion of Section 5, and the asymptotic expansion of Section 6. The expansion in phase boundaries is motivated by low-temperature techniques in classical statistical mechanics, first introduced by Peierls in 1936 [26]. To obtain a mapping of the field variable $\phi(x)$ onto an Ising spin variable σ_i, we cover space-time R^2 by a lattice of unit squares Δ_i, centered at the lattice sites $i \in Z^2$. The boundaries of the Δ_i are (unit) lattice lines. Let $\sigma(\cdot)$ be a function on R^2 which is constant on lattice squares and assumes only the values $\pm$. In other words,

$$\sigma_i \equiv \sigma(\Delta_i) \equiv \sigma(x), \qquad i \in Z^2,\ x \in \Delta_i. \tag{3.1}$$

We impose "plus" boundary conditions on σ_i, namely,

$$\sigma_i = + \qquad \text{for } \Delta_i \not\subset \Lambda. \tag{3.2}$$

A configuration

$$\Sigma = \{\sigma_i\} \tag{3.3}$$

of the spin variable belongs to the $2^{|\Lambda|}$-dimensional state space. Each configuration Σ is in 1–1 correspondence with the set of lattice lines on which $\sigma(x)$ is discontinuous. We call this set of lattice lines the Peierls contour, or phase boundary, and also denote it by Σ. The $+$ boundary conditions ensure that the phase boundary Σ is a union of closed curves, and we let $|\Sigma|$ denote the total length of these curves.

We now consider the averaged field in Δ (the low momentum field in Δ) which we define by

$$\phi(\Delta) \equiv \int_\Delta \phi(x)\,dx \equiv \bar{\phi}(x), \qquad x \in \Delta. \tag{3.4a}$$

We define the fluctuation field by

$$\delta\phi(x) = \phi(x) - \bar{\phi}(x), \tag{3.4b}$$

which we use in later sections to bound the deviations of ϕ from its mean value $\bar{\phi}$.

The basic idea behind the expansion in phase boundaries is to choose $\sigma_i = \operatorname{sgn} \phi(\Delta_i)$ and to write $d\phi_\Lambda$ as a sum of $2^{|\Lambda|}$ measures $d\phi_{\Lambda,\Sigma}$. Here $d\phi_{\Lambda,\Sigma}$ is the restriction of $d\phi_\Lambda$ to the subset of paths with a fixed $\Sigma = \{\sigma_i\}$.

For technical reasons, we smooth out this definition of $d\phi_{\Lambda,\Sigma}$ as follows. Let $\chi_\pm$ be a smooth partition of unity on R^1 defined by

$$\chi_+(\xi) = \pi^{-1/2} \int_0^\infty e^{-(\xi-z)^2}\, dz = \chi_-(-\xi),$$
$$\chi_+(\xi) + \chi_-(\xi) = 1. \tag{3.5}$$

We set

$$\chi_\Sigma = \prod_{\Delta_i \subset \Lambda} \chi_{\sigma_i}(\phi(\Delta_i)) \tag{3.6}$$

and note that

$$\sum_{\{\sigma_i\}} \chi_\Sigma = 1.$$

The expansion in phase boundaries is merely the identity

$$d\phi_\Lambda = \sum_{\{\sigma_i\}} d\phi_{\Lambda,\Sigma} \tag{3.7}$$

where $d\phi_{\Lambda,\Sigma} = \chi_\Sigma\, d\phi_\Lambda$.

We remark that for a more general $\mathscr{P}(\phi)$ model, $\mathscr{P}$ may have up to $\frac{1}{2}\deg \mathscr{P}$ minima, and there may be up to $\frac{1}{2}\deg \mathscr{P}$ equilibrium phases. We require a spin with $\frac{1}{2}\deg \mathscr{P} = 2s+1$ components, i.e., at each lattice site we have the possibility

$$\sigma = -s, -s+1, \ldots, (s-1), s. \tag{3.8}$$

Thus, we have an analogy with a "spin s" Ising model, the case $s = \frac{1}{2}$ considered above.

4. Translation of ϕ

For each spin configuration Σ, we translate $\phi(x)$ by a function $g(x) = g(x, \Sigma)$. We choose g so that the translated field

$$\psi(x) = \phi(x) - g(x) \tag{4.1}$$

has a small mean in the measure $d\phi_{\Lambda,\Sigma}$. Ignoring the shift in kinetic energy, we would translate in each square Δ_i to one of the relative mimima, $\xi_\pm$ of $\mathscr{P}(\xi)$. Thus we define an external field

$$h(x) = \sigma(x)\,\xi_+ .$$

The classical magnetism (the response to the external field $h(x)$) is given by

$$g_c(x) = (\eta(-\Delta + \eta)^{-1}h)(x) = \xi_+ + \eta(-\Delta - \eta)^{-1}(h - \xi_+). \tag{4.3}$$

For convenience, we choose $\eta \neq 1$ to be a small constant defined later, independent of λ. Translation by the function $g_c(x)$ causes some technical inconvenience, because $g_c(x)$ has an exponential tail. We modify (4.3) by choosing a C_0^∞ function $\zeta(x)$ satisfying $0 \leqslant \zeta(x) \leqslant 1$ with

$$\begin{aligned} \zeta(x) &= 0, \qquad |x| > \tfrac{1}{2} \\ &= 1, \qquad |x| \leqslant \tfrac{1}{4}. \end{aligned}$$

For a given length scale L, define an approximate classical magnetism by

$$g(x) = \eta_\zeta \int (-\Delta + \eta)^{-1}(x - y)\,\zeta((x - y)/L)\,h(y)\,dy, \tag{4.4}$$

where

$$\eta_\zeta^{-1} = \int (-\Delta + \eta)^{-1}(y)\,\zeta(y/L)\,dy.$$

LEMMA 4.1. *Let* $\operatorname{dist}(x, y) \geqslant \frac{1}{2}L$. *Then* $g(x)$ *and* $(-\Delta + \eta)(g - g_c)(x)$ *are independent of* $h(y)$. *In particular, if* $\operatorname{dist}(x, \Sigma) \geqslant \frac{1}{2}L$,

$$g(x) = h(x) \qquad \text{and} \qquad (-\Delta + \eta)(g - g_c)(x) = 0. \tag{4.5}$$

Proof. The locality of $g(x)$ and $(-\Delta + \eta)g(x)$ follow from the localization of the kernel $(-\Delta + \eta)^{-1}\zeta$ in (4.4). Furthermore, $(-\Delta + \eta)\,g_c(x) = \eta h(x)$ is local, and (4.5) then follows.

Since $g(x) = h(x)$ for $\operatorname{dist}(x, \Sigma) \geqslant \frac{1}{2}L$, the length scale L provides an upper bound on the width of the phase boundaries. We use L as a characteristic length for the effects of phase boundaries.

Let $d\psi_a$ denote the Gaussian measure with mean zero and covariance $(-\Delta + a)^{-1}$.

LEMMA 4.2. *With* ψ *defined by* (4.1), (4.4),

$$d\phi_{\Lambda,\Sigma} = \chi_\Sigma e^{-F} \exp\left(-{:}\mathscr{P}_\Lambda(\phi){:} + \frac{\eta}{2}\int_\Lambda {:}(\phi - h)^2{:}\,dx + \left(\frac{1 - \eta}{2}\right)\int_\Lambda {:}\psi^2{:}\,dx\right) d\psi_1, \tag{4.6}$$

where

$$F = F(\Sigma) = F_1 + F_2 + F_3 + F_4 ,$$

and

$$F_1 = \tfrac{1}{2}\eta \int (h - g)^2 \, dx + \tfrac{1}{2} \int (\nabla g)^2 \, dx \tag{4.7a}$$

$$F_2 = \left(\frac{1 - \eta}{2}\right) \int_{\sim\Lambda} (g - \xi_+)^2 \, dx \tag{4.7b}$$

$$F_3 = \int \psi(x)(-\Delta + \eta)(g - g_c) \, dx \tag{4.7c}$$

$$F_4 = (1 - \eta) \int_{\sim\Lambda} \psi(x)(g(x) - \xi_+) \, dx \tag{4.7d}$$

To simplify the notation we define

$$V(W, \Sigma) = :\mathscr{P}_{\Lambda\cap W}(\phi): - \frac{1}{2}\eta \int_{\Lambda\cap W} :(\phi - h)^2: dx - \left(\frac{1 - \eta}{2}\right) \int_{\Lambda\cap W} :\psi^2: dx \tag{4.8a}$$

and

$$Q(W,\Sigma) = V(W,\Sigma) + F(W,\Sigma), \tag{4.8b}$$

where $F(W,\Sigma)$ *is obtained by restricting the range of integration in* (4.7a), (4.7c) *to* W *and in* (4.7b), (4.7d) *to* $W \sim \Lambda$. *Thus in* (4.8), *and throughout the rest of the paper, the* Λ *dependence is not explicit in the notation. We also define* $Q(\Sigma) = Q(\mathbb{R}^2,\Sigma)$. *By Lemma* 4.1, *if*

$$\operatorname{dist}(\partial W,\Sigma) \geqslant \tfrac{1}{2}L, \tag{4.8c}$$

then

$$Q(W,\Sigma) = Q(W,\Sigma \cap W). \tag{4.8d}$$

Since we are only concerned with $Q(W,\Sigma)$ *in case* (4.8c), *we always have* (4.8d), *which expresses the locality of* Q.

In this notation, (4.6) *becomes*

$$d\phi_{\Lambda,\Sigma} = \chi_\Sigma e^{-Q(\Sigma)} d\psi_1 . \tag{4.9}$$

Proof. The basic identity for translation of the Gaussian measure $d\psi_1$ is

$$d(\psi + f)_1 = \exp(-\tfrac{1}{2}\langle f, (-\Delta + 1)f\rangle - \langle \psi, (-\Delta + 1)f\rangle) \, d\psi_1 . \tag{4.10}$$

Hence

$$\begin{aligned} d\psi_1 &= d(\phi - g)_1 = d(\phi - \xi_+ + \xi_+ - g)_1 \\ &= \exp(-\tfrac{1}{2}\langle(\xi_+ - g), (-\Delta + 1)(\xi_+ - g)\rangle - \langle(\phi - \xi_+), (-\Delta + 1)(\xi_+ - g)\rangle) d(\phi - \xi_+)_1 . \end{aligned}$$

Comparison with (2.4) and (4.6) leads to the identity

$$F = \frac{\eta}{2}\int_{\Lambda} :(\phi - h)^2: dx + \left(\frac{1-\eta}{2}\right)\int_{\Lambda} :(\phi - g)^2: dx - \frac{1}{2}\int_{\Lambda} :(\phi - \xi_+)^2: dx$$
$$- \frac{1}{2}\langle(\xi_+ - g), (-\Delta + 1)(\xi_+ - g)\rangle - \langle \phi - \xi_+, (-\Delta + 1)(\xi_+ - g)\rangle.$$

The term in F linear in ψ is

$$\int_{\Lambda} \psi[-\eta h - (1 - \eta)\, g + \xi_+]\, dx + \int \psi[-\xi_+ + (-\Delta + 1)\, g]\, dx,$$

which by (4.3) and the identity $h \equiv \xi_+$ on $\sim\Lambda$ equals

$$(1 - \eta)\int_{\sim\Lambda} \psi(g - \xi_+)\, dx + \int \psi(-\Delta + \eta)(g - g_c)\, dx.$$

The term in F independent of ψ is

$$\int_{\Lambda}\left\langle\left(\frac{1}{2}\eta h^2 + \frac{1-\eta}{2}\right) g^2 - \frac{1}{2}\xi_+^2\right\rangle dx$$
$$- \frac{1}{2}\langle(\xi_+ - g), (-\Delta + 1)(\xi_+ - g)\rangle + \langle \xi_+, (-\Delta + 1)(\xi_+ - g)\rangle$$
$$+ (1 - \eta)\int_{\sim\Lambda} g(g - \xi_+)\, dx + \int g(-\Delta + \eta)(g - g_0)\, dx$$
$$= \left(\frac{1-\eta}{2}\right)\int_{\sim\Lambda} (\xi_+ - g)^2\, dx + \frac{1}{2}\eta\langle h - g, h - g\rangle + \frac{1}{2}\langle \nabla g, \nabla g\rangle.$$

5. The Cluster Expansion

The cluster expansion allows us to take $\Lambda \nearrow R^2$. In order to establish convergence of the cluster expansion, we introduce a localization length scale $l \gg m_0^{-1} \approx m^{-1} \approx 1$. Since the length scale L of the previous section bounds the width of phase boundaries, we also choose $l \ll L$ to display the structure of the pure phase. Thus we choose

$$1 \ll l \ll L. \tag{5.1}$$

We will apply the expansion on l-lattice segments b which bound l-lattice squares joining nearest neighbor points of lZ^2. Let $\mathscr{B}$ denote the set of all l lattice segments (bonds) b, and let Γ denote a subset of $\mathscr{B}$, with $|\Gamma|$ elements. Let

$$C_\Gamma = (-\Delta_\Gamma + 1)^{-1}, \tag{5.2}$$

where Δ_Γ is the Laplace operator with zero Dirichlet boundary conditions on Γ.

For each spin configuration Σ, we define a subset $\mathscr{B}(\Sigma)$ of $\mathscr{B}$ on which we perform the cluster expansion [16]. We interpolate between free and Dirichlet boundary conditions by introducing a parameter $s_b \in [0, 1]$ for each $b \in \mathscr{B}(\Sigma)$. Let $s = \{s_b\}$ and

$$C(s) = \sum_{\Gamma \subset \mathscr{B}(\Sigma)} \left(\prod_{b\in\Gamma^c} s_b\right)\left(\prod_{b\in\Gamma} (1 - s_b)\right) C_\Gamma \tag{5.3}$$

where

$$\Gamma^c = \mathscr{B}(\Sigma)\backslash\Gamma.$$

Also let $d\psi(s)$ denote the Gaussian measure with mean zero and covariance $C(s)$. Then $s_b = 0$ corresponds to zero Dirichlet data on the l-lattice segment b, while $s = 1$ denotes the absence of Dirichlet data on b. If $s_b = 0$ on Γ, we say that Γ is a Dirichlet contour. Let

$$s_\Gamma = \{s_b: b \in \Gamma\},$$

$$ds_{\Gamma^c} = \prod_{b\in\Gamma^c} ds_b \Big|_{s_\Gamma=0},$$

$$\partial_s^{\Gamma} = \prod_{b\in\Gamma} (\partial/\partial s_b).$$

The idea of the cluster expansion is to apply the fundamental theorem of calculus to each s_b,

$$F(s_b = 1) = F(s_b = 0) + \int_0^1 ds_b\, \partial_{s_b} F(s),$$

for $b \in \mathscr{B}(\Sigma)$. Thus we obtain the fundamental expansion formula

$$d\phi_{\Lambda,\Sigma} = \sum_{\Gamma \subset \mathscr{B}(\Sigma)} d\phi_{\Lambda,\Sigma,\Gamma} \tag{5.5}$$

where

$$d\phi_{\Lambda,\Sigma,\Gamma} = \int ds_{\Gamma^c}\, \partial_s^{\Gamma^c} \int \chi_\Sigma e^{-Q(R^2,\Sigma)}\, d\psi(s). \tag{5.6}$$

Here χ and Q are defined by (3.6) and (4.8). Note that $s_b = 0$ in (5.6), for $b \in \Gamma$, by the definition of ds_{Γ_c}. Recall that the Λ dependence of $Q(R^2, \Sigma)$ is suppressed. See (4.8).

It remains to define $\mathscr{B}(\Sigma)$, the set of l-lattice lines for the cluster expansion. We choose $\mathscr{B}(\Sigma)$ to be those islands of $+$ or $-$ spins, separated from a phase boundary by at least the width L. Thus on $\mathscr{B}(\Sigma)$ we expect the measure $d\phi_{\Lambda,\Sigma}$ to be nearly Gaussian, with mean $\sim\xi_\sigma$. More precisely, let

$$X = X(\Sigma) = \{\Delta_i : \operatorname{dist}(i,\Sigma) \geqslant L\}, \tag{5.7}$$

and

$$\mathscr{B}(\Sigma) = \mathscr{B} \cap X. \tag{5.8}$$

We apply the expansion (5.5)–(5.6) to a partition function

$$Z(\Lambda) = \int d\phi_\Lambda = \sum_{\Sigma} \sum_{\Gamma \subset \mathscr{B}(\Sigma)} \int d\phi_{\Lambda,\Sigma,\Gamma} \tag{5.9}$$

or to an expectation

$$\begin{aligned} \langle A \rangle &= Z(\Lambda)^{-1} \int A(\phi)\, d\phi_\Lambda, \\ &= Z(\Lambda)^{-1} \sum_{\Sigma} \sum_{\Gamma \subset \mathscr{B}(\Sigma)} \int A\, d\phi_{\Lambda,\Sigma,\Gamma}. \end{aligned} \tag{5.10}$$

In the remainder of the paper we study (5.9)–(5.10).

It is convenient to generalize (5.7)–(5.8) and to define

$$X_r^{\pm} = \{\Delta_i : \sigma_i = \pm, \operatorname{dist}(i,\Sigma) \geqslant rL\}, \tag{5.11}$$

and

$$X_r = X_r^{+} \cup X_r^{-}, \qquad X_r^{0} = \mathscr{R}^2 \backslash X_r. \tag{5.12}$$

In general we omit the subscript for $r = 1$. Also let

$$\mathscr{B}(\Sigma, Y) = \mathscr{B}(\Sigma) \cap Y = \mathscr{B} \cap X \cap Y. \tag{5.13}$$

6. The Asymptotic Expansion

We give the asymptotic expansion for $\langle A \rangle_\Lambda = \int A d\phi_\Lambda$ as a function of λ. We use the asymptotic expansion to isolate the dominant contribution as $\lambda \to 0$. The asymptotic expansion is performed before the expansion in phase boundaries and before the cluster expansion.

Let

$$A(\phi - \xi_+) \equiv \prod_{i=1}^{n} \int :(\phi(x) - \xi_+)^{m_i}: f_i(x)\, dx, \tag{6.1}$$

where : : denotes $(-\Delta + 1)^{-1}$ Wick ordering.

In order to state precisely the f dependence of the estimates, we introduce a norm $|||\, A\, |||$ on the functions of the form (6.1). In case suppt. $f_i \subset \Delta_i$, where Δ_i is a unit lattice square, let

$$N(\Delta, A) = \left\{\sum m_i \colon \Delta_i = \Delta\right\},$$

$$N(A) = \sum_{\Delta} N(\Delta, A) = \sum_{i} m_i .$$

For some constant K we define

$$|||\, A\, ||| = e^{KlN(A)} \left(\prod_{\Delta} N(\Delta, A)!\right)\left(\prod_{i} \| f_i \|_{L_2(\Delta_i)}\right). \tag{6.2}$$

For general f_i with no restrictions on the support, $|||\, A\, |||$ is defined by linearity, as a sum over $\Delta_1, \ldots, \Delta_n$ of localized A's.

THEOREM 6.1. *Let $0 \leqslant \mu$, $\mu_b = \xi_+$, $r \in Z_+$, and A as above. Then there are constants $\alpha_j(A, \Lambda, \lambda, \mu)$ and a finite sum of functions A^k of the form* (6.1) *such that*

$$\int A\, d\phi_\Lambda = \sum_{j=0}^{r} \alpha_j \lambda^{j/2} \int d\phi_\Lambda + \lambda^{(r+1)/2} \int \sum_{k} A_k\, d\phi_\Lambda .$$

Here α_j and A_k are continuous in $\lambda \in [0, \lambda_0]$, uniformly in Λ, and convergent as $\Lambda \nearrow R^2$, for λ_0 sufficiently small. In (6.1) we may also allow $f_i = \delta(x - a_i)$ in case the a_i are all unequal.

We show later that for functions A of the form (6.1), $\lim \int A d\phi_\Lambda / \int d\phi_\Lambda$ exists as $\Lambda \nearrow R^2$. Thus the theorem shows that

$$\int A\, d\phi = \lim_{\Lambda \nearrow R^2} \frac{\int A\, d\phi_\Lambda}{\int d\phi_\Lambda} = \sum \alpha_j \lambda^{j/2} + O(\lambda^{(r+1)/2}). \tag{6.3}$$

We generate the asymptotic expansion by integration by parts on path space. Before integration by parts, we choose a new variable $\psi = \phi - \xi_+$. We thus translate by the constant function $g = g_0 = h = \xi_+$, and (6.1) can be written

$$A = A(\psi) = \prod_{i=1}^{n} \int {:}\psi(x)^{m_i}{:}\, f_i(x)\, dx. \tag{6.4}$$

The function $F = 0$ in (4.6), so

$$\int A\, d\phi_\Lambda = \int A(\psi)\, e^{-V}\, d\psi_1 \tag{6.5}$$

where

$$V = \int_{\Lambda} V(x)\,dx \tag{6.6}$$

and where

$$V(x) = \lambda{:}\psi(x)^4{:} + \alpha(2\lambda)^{1/2}{:}\,\psi(x)^3{:} + \beta\lambda^{5/2}{:}\psi(x)^2{:}. \tag{6.7}$$

Computation shows that

$$\begin{aligned} \alpha &= (8\lambda)^{1/2}\,\xi_+ = 1 + O(\lambda^{1/2}\mu) \\ \beta &= (18)^{1/2}\,\mu\lambda^{-2}\alpha^{-1} = (18)^{1/2}\,\mu\lambda^{-2} + O(\mu^2\lambda^{-3/2}) = O(1), \end{aligned} \tag{6.8}$$

so the coefficients of V are small for small λ. Note if $\mu = 0$, then $\alpha = 1$, $\beta = 0$.

The formula for integration by parts is

$$\int {:}\psi(x)^m{:}\; R(\psi)\,d\phi_\Lambda = \iint C(x-y)\left[{:}\psi(x)^{m-1}{:}\,\frac{\delta R}{\delta\psi(y)} - R\,\frac{\delta V}{\delta\psi(y)}\right] dy\,d\phi_\Lambda\,. \tag{6.9}$$

Here

$$\delta V/\delta\psi(y) = 4\lambda{:}\psi(y)^3{:} + \alpha(18\lambda)^{1/2}{:}\psi(y)^2{:} + 2\beta\lambda^{5/2}\psi(y). \tag{6.10}$$

To generate the asymptotic series (6.3), we apply (6.9) in order to integrate by parts each factor ψ in the integrand of $\int A d\phi_\Lambda$. We also integrate by parts each factor of ψ produced by such integration, and we continue inductively until either (i) the integrand is constant on path space, or (ii) the integrand has a coefficient $O(\lambda^{(r+1)/2})$ and each Wick power has degree $\leqslant \deg \mathscr{P}$. Finally, we invert the transformation from ψ to ϕ. Thus we replace A by $\sum A_k$ plus constant terms.

The terms from the expansion which are constant on path space (i.e., the α_j) have several interesting properties. First we remark that

$$\alpha_0 = \int A(\psi)\,d\psi \tag{6.11}$$

where $d\psi$ is the Gaussian measure with mean zero and covariance $(-\Delta + 1)^{-1}$. Thus the terms independent of λ are the free field expectations of $A(\psi)$.

The terms α_j for $j > 0$ have an interpretation in terms of Feynman diagrams. In particular, the asymptotic series $\sum \alpha_j\lambda^{j/2}$ is exactly equal to the usual perturbation series for $\langle A(\psi)\rangle$, where the interaction density equals (6.7). (For $-$ boundary conditions, we reverse the sign of the cubic term in (6.7) and take $\psi = \phi - \xi_-$). We note in the special case

$$A = \phi(x) - \xi_+ = \psi(x), \tag{6.12a}$$

that

$$\alpha_0 = \alpha_1 = \alpha_2 = 0. \tag{6.12b}$$

An explicit calculation of α_3 (in the case $\Lambda = R^2$) shows that

$$\alpha_3 = 144\sqrt{2} \int C(x)^3\, dx - 864\sqrt{2} \int C(x)\, C(y)\, C(x-y)^2\, dx\, dy. \tag{6.13}$$

Here we have used the fact that $\int C(x)\, dx = 1$. In the conventional notation, α_3 is represented by the perturbation diagrams

$\alpha_3 =$ 

The numerical value of α_3 can be calculated from (6.13). As a corollary of Theorem 6.1, we find

$$\begin{aligned} \langle \phi(x) \rangle &= \xi_+ + \alpha_3 \lambda^{3/2} + O(\lambda^2) \\ &= \phi_c + \alpha_3 \lambda^{3/2} + O(\lambda^2) \end{aligned} \tag{6.14}$$

Thus $\alpha_3 \lambda^{3/2}$ is the first "quantum correction" to the classical field expectation ϕ_c.

We apply our estimates to the functions A obtained from the asymptotic expansion above. Thus it is no loss of generality to assume $m_i \leqslant 4$.

THEOREM 6.2. *Consider* A *of the form* (6.1) *with* $m_i \leqslant 4$. *Then for* λ *sufficiently small,*

$$\left| \int A\, d\phi_\Lambda \right| \leqslant |||\, A \,|||\, (1 + \lambda^{-N(A)/2} e^{-O(\lambda^{-1 \cdot 2})}) \int d\phi_\Lambda . \tag{6.15}$$

Theorems 2.2, 2.3, and 6.1 follow from Theorem 6.2 and the above discussion. In addition the bounds on (2.9) and (2.12) uniform in Λ follow from Theorem 6.2. To obtain the λ independent bounds required for Theorems 2.2 and 2.3 from the l (and hence λ) dependent norm $|||\cdot|||$ in (6.15), it is necessary first to use the asymptotic expansion of this section, and extract enough factors $\lambda^{1/2}$. The resulting bounds are uniform in λ for fixed $N(A)$ and uniform in $N(A)$ for fixed (small) λ.

7. FACTORIZATION AND PARTIAL RESUMMATION

We resume parts of the expansion (5.10) to factor out and cancel vacuum energy contributions. Let $\mathscr{L}(A)$ denote the localization of A, i.e., the set of lattice squares Δ_i contributing to (6.1) and let $\mathscr{L}(A) \subset W \cap \Lambda$. Define

$$R''(A, W, \Sigma, \Gamma) = \int_0^1 ds_{\Gamma^c}\, \partial^{\Gamma^c \cap \mathscr{B}(\Sigma, W)} \int \chi_\Sigma A e^{-Q(W,\Sigma)}\, d\psi(s) \tag{7.1}$$

and

$$Z''(W, \Sigma, \Gamma) = R''(A = I, W, \Sigma, \Gamma). \tag{7.2}$$

Then (5.10) can be written as

$$\langle A \rangle = Z(\Lambda)^{-1} \sum_{\Sigma} \sum_{\Gamma \subset \mathscr{B}(\Sigma)} R''(A, R^2, \Sigma, \Gamma). \tag{7.3}$$

We now exhibit the factorization in (7.3). Let $\{Y_1, ..., Y_\nu\}$ denote the set of connected components of $R^2 \sim \Gamma$. For $s_\Gamma = 0$, the covariance $C(s)$ defined in (5.3) is a direct sum over the Y_i, i.e., it is reduced by the direct sum decomposition.

$$L_2(R^2) = \sum_{i=1}^{\nu} L_2(Y_i).$$

Consequently, the measure $d\psi(s)$ of (7.1) factors. We now see that the integrand factors also.

For convenience, we consider A in (6.1) to be localized, so that $\operatorname{suppt} f_i \subset \Delta_i$. The general case is then handled by summation over localizations. With this restriction A factors.

Because the Dirichlet contour Γ is by construction bounded away from the phase boundary Σ, we apply Lemma 4.1 and the discussion following it to express $Q(R^2, \Sigma)$ as a local sum

$$Q(R^2, \Sigma) = \sum_{i=1}^{\nu} Q(Y_i, \Sigma \cap Y_i). \tag{7.4}$$

Finally, defining

$$\chi_{\Sigma \cap Y_i} = \prod_{\Delta_j \subset Y_i} \chi_{\sigma_j}(\phi(\Delta_j)),$$

we have

$$\chi_\Sigma = \prod_{i=1}^{\nu} \chi_{\Sigma \cap Y_i}. \tag{7.5}$$

Thus the integrand in (7.1) factors as asserted, and

$$R''(A, R^2, \Sigma, \Gamma) = \prod_{i=1}^{\nu} R''(A_i, Y_i, \Sigma \cap Y_i, \Gamma \cap Y_i). \tag{7.6}$$

Now we turn to the partial resummation. Fix Σ and Γ in (5.6), (7.3), and define $Y = Y(\Gamma)$ as the closure of the union of all components γ_i which meet the localization of A from (6.1):

$$Y = U_i\{Y_i : \operatorname{suppt} f_j \subset Y_i, \text{ some } j\}.$$

We also define

$$\partial Y^{\pm} = \partial Y \cap X^{\pm}$$

and then ∂Y, ∂Y^{+}, ∂Y^{-} each consist of a finite number of nonoverlapping closed contours.

By applying (7.3)–(7.6), we isolate A from the partition function for $\sim Y$. We perform the first partial resummation by holding $Y = Y(\Gamma)$ fixed and summing over all Dirichlet contours Γ compatible with $Y = Y(\Gamma)$. We call this condition a *constraint* on the sum over Γ. Thus we consider

$$\sum_{\{\Gamma\subset\mathscr{B}(\Sigma):Y(\Gamma)=Y\}} R''(A, R^2, \Sigma, \Gamma)$$

$$= \sum_{\{\Gamma\subset\mathscr{B}(\Sigma):Y(\Gamma)=Y\}} R''(A, Y, \Sigma\cap Y, \Gamma\cap Y)\, Z''(\sim Y, \Sigma\sim Y, \Gamma\sim Y). \tag{7.7}$$

The constrained sum (7.7) factors into an inside sum over $\mathscr{B}(\Sigma, Y)$ and an outside sum over $\mathscr{B}(\Sigma, \sim Y)$. The inside sum is

$$R'(A, Y, \Sigma\cap Y) = \sum_{\{\Gamma\subset\mathscr{B}(\Sigma,Y):Y(\Gamma)=Y\}} R''(A, Y, \Sigma\cap Y, \Gamma\cap Y) \tag{7.8}$$

and the outside sum is

$$Z'(\sim Y, \Sigma\sim Y) = \sum_{\{\Gamma\subset\mathscr{B}(\Sigma,\sim Y):\Gamma\supset\partial Y\}} Z''(\sim Y, \Sigma\sim Y, \Gamma\sim Y) \tag{7.9}$$

The outside sum can be evaluated exactly, using (5.5). In fact (7.9) is the partition function

$$Z'(\sim Y, \Sigma\sim Y, \partial Y) \equiv \int e^{-Q(\sim Y,\Sigma\sim Y)}\chi_{\Sigma\sim Y}\, d\psi_{\partial Y}, \tag{7.10}$$

where the subscript ∂Y in the measure $d\psi$ denotes Dirichlet data. Thus we write (7.5) as

$$\langle A\rangle = \sum_{\Sigma}\sum_{\partial Y} R'(A, Y, \Sigma\cap Y)\frac{Z'(\sim Y, \Sigma\sim Y, \partial Y)}{Z(R^2, \varnothing, \varnothing)}. \tag{7.11}$$

Let $N(W)$ denote the set of lattice squares within distance L of a set W. The second partial resummation is over Σ. We hold ∂Y^{+} and ∂Y^{-} fixed ($\partial Y^{\pm} = \partial Y\cap X^{\pm}$), and consequently

$$\Sigma \restriction N(\partial Y^{\pm}) = \pm \tag{7.12}$$

is the constraint on the sum over Σ. With this constraint, the Σ sum also factors across $N(\partial Y)$ into an inside sum and an outside sum. We define

$$R(A, Y, \partial Y^{\pm}) = \sum_{\{\Sigma\cap Y:\sigma_i\restriction N(\partial Y^{\pm})=\pm\}} R'(A, Y, \Sigma\cap Y) \tag{7.13}$$

and

$$Z(Y, N^+, N^-, \Gamma) = \sum_{\{\Sigma \cap Y : \sigma_i \upharpoonright N^\pm = \pm\}} Z'(Y, \Sigma \cap Y, \Gamma \cap Y). \tag{7.14}$$

Here we assume $N^+ \cap N^- = \varnothing$. In particular,

$$Z(Y, N(\partial Y^+), N(\partial Y^-), \partial Y) = \sum_{\{\Sigma \cap Y : N(\delta Y^\pm) = \pm\}} \int e^{-Q(Y, \Sigma \cap Y)} \chi_{\Sigma \cap Y}\, d\psi_{\delta Y}, \tag{7.15}$$

and (7.11) becomes

$$\langle A \rangle = \sum_{\partial Y^\pm} R(A, Y, \partial Y^\pm) \frac{Z(\sim Y, N(\partial Y^+), N(\partial Y^-), \partial Y)}{Z(\Lambda, \varnothing, \varnothing, \varnothing)}. \tag{7.16}$$

Again we can evaluate the outside sum, (7.15). We say that Σ is compatible with a pair $N^\pm$ if $\sigma \upharpoonright N^\pm = \pm$.

LEMMA 7.1. *Let A be a function of ϕ. For any two configurations Σ_1, Σ_2 compatible with $N(\Gamma^\pm)$ we have*

$$\int A e^{-Q(Y, \Sigma_1)}\, d\psi_\Gamma = \int A e^{-Q(Y, \Sigma_2)}\, d\psi_\Gamma. \tag{7.17}$$

Proof. We express Q and $d\psi_\Gamma$ in terms of ϕ. If we approximate Dirichlet boundary conditions by a large local mass $\delta m^2(x)$ concentrated in a small neighborhood of Γ, we see that the right and left sides of the above equations become

$$\frac{\int A \exp(-\int \delta m^2(x) :(\phi(x) - g_i(x))^2: dx)\, d\phi_{\Lambda \cap Y}}{\int \exp(-\int \delta m^2(x) :\psi(x)^2: dx)\, d\psi_1}, \tag{7.18}$$

$i = 1, 2$. Near Γ, the compatibility condition and Lemma 4.1 yield $g_1 = g_2$. Since $\delta m^2(x)$ is supported near Γ, the two expressions in (7.18) are equal.

Let $d\phi_{Y \cap \Lambda, \Gamma}$ denote the measure $\exp[-Q(Y, \Sigma)]\, d\psi_\Gamma$ in (7.17). This measure can be represented as in (2.4) as a Gaussian measure with nonzero Dirichlet data $\phi = \pm\xi$ on $\Gamma^\pm$ times a $\mathscr{P}(\phi)$ action factor. Then (7.14) becomes

$$Z(Y, N^+, N^-, \Gamma) = \int \Big[\prod_{\Delta \subset N^+} \chi_+(\phi(\Delta))\Big]\Big[\prod_{\Delta \subset N^-} \chi_-(\phi(\Delta))\Big]\, d\phi_{Y \cap \Lambda, \Gamma}. \tag{7.19}$$

REFERENCES

1. D. BRYDGES, Cluster expansions for fermion fields by the time-dependent Hamiltonian approach, *J. Math. Phys.*, to appear.

2. A. Cooper and L. Rosen, "The Weakly Coupled Yukawa$_2$ Field Theory: Cluster Expansion and Wightman Axioms," preprint.
3. R. Dashen, B. Hasslacher, and A. Neveu, Nonperturbative methods and extended Hadron models in field theory, I, II, III. *Phys. Rev. D.* **10** (1974), 4114–4141.
4. J. Dimock, "The $\mathscr{P}(\phi)_2$ Green's Functions: Asymptotic Perturbation Expansion," preprint.
5. J. Dimock and J. Glimm, Measures on the Schwartz distribution space and applications to quantum field theory, *Adv. Math.* **12** (1974), 58–83.
6. J.-P. Eckmann, H. Epstein, and J. Fröhlich, Asymptotic perturbation expansion for the S-matrix and the definition of time ordered functions in relativistic quantum field models.
7. J.-P. Eckmann, J. Magnen, and R. Sénéor, Decay properties and Borel summability for the Schwinger functions in $\mathscr{P}(\phi)_2$-theories, *Commun. Math. Phys.* **39** (1975), 251–271.
8. L. Faddeev, "Quantization of Solitons," IAS preprint.
9. J. Feldman and K. Osterwalder, The Wightman axioms and the mass gap for weakly coupled ϕ_3^4. *Ann. Phys.*, to appear.
10. J. Feldman and K. Osterwalder, The construction of $\lambda\phi_3^4$ quantum field models, *in* "Proceedings of Marseille Conference," June 1975.
11a. J. Fröhlich, New super selection sectors (Soliton-states) in two dimensional Bose quantum field models, *Commun. Math. Phys.* **47** (1976), 269–310.
11b. J. Fröhlich and B. Simon, "Pure States for General $P(\phi)_2$ Theories: Construction, Regularity, and Variational Equality," preprint, 1976.
12. J. Glimm and A. Jaffe, A $\lambda(\phi^4)_2$ Quantum field theory without cutoffs. I. *Phys. Rev.* **176** (1968), 1945–1951.
13. J. Glimm and A. Jaffe, The $\lambda(\phi^4)_2$ quantum field theory without cutoffs. III. The physical vacuum, *Act. Math.* **125** (1970), 203–267.
14. J. Glimm and A. Jaffe, The Yukawa$_2$ quantum field theory without cutoffs, *J. Functional Anal.* **7** (1971), 323–357.
15. J. Glimm, A. Jaffe, and T. Spencer, The Wightman axioms and particle structure in the $\mathscr{P}(\phi)_2$ quantum field model. *Ann. Math.* **100** (1974), 585–632.
16. J. Glimm, A. Jaffe, and T. Spencer, The particle structure of the weakly coupled $\mathscr{P}(\phi)_2$ model and other applications of high temperature expansions. Part II, The cluster expansion, *in* "Constructive quantum field theory" (G. Velo and A. Wightman, Eds.), Lecture notes in Physics, Vol. 25, Springer-Verlag, Berlin, 1973.
17. J. Glimm, A. Jaffe, and T. Spencer, Existence of phase transitions for quantum fields, 1975 Marseille conference.
18. J. Glimm, A. Jaffe, and T. Spencer, Phase transition for ϕ_2^4 quantum fields, *Commun. Math. Phys.* **45** (1975), 203–216.
19. J. Goldstone and R. Jackiw, Quantization of nonlinear waves, *Phys. Rev. D* **11** (1975), 1486–1498.
20. F. Guerra, L. Rosen, and B. Simon, The $\mathscr{P}(\phi)_2$ Euclidean quantum field theory as classical statistical mechanics, *Ann. Math.* **101** (1975), 111–259.
21. E. Lieb, D. Mattis, and T. Schultz, Two-dimensional Ising model as a soluble problem of many fermions. *Mod. Phys.* **36** (1964), 856–871.
22. J. Magnen and R. Sénéor, The infinite volume limit of the ϕ_3^4 model, *Ann. Inst. Henri Poincaré*, to appear.
23. J. Magnen and R. Sénéor, The Wightman axioms for the weakly coupled yukawa model in two dimensions, *Commun. Math. Phys.*, to appear.
24. B. McCoy and T. T. Wu, "The Two-Dimensional Ising Model," Harvard Univ. Press, Cambridge, Mass., 1973.

25. K. Osterwalder and R. Sénéor, The scattering matrix is nontrivial for weakly coupled $\mathscr{P}(\phi)_2$ models, *Helv. Phys. Acta*, to appear.
26. R. Peierls, Ising's model of ferromagnetism, *Proc. Cambridge Philos. Soc.* **32** (1936), 477–481.
27. T. Spencer, A mass gap for the $\mathscr{P}(\phi)_2$ quantum field model with a strong external field, *Commun. Math. Phys.* **39** (1974), 63–76.
28. T. Spencer, The decay of the Bethe-Salpeter kernel in $\mathscr{P}(\phi)_2$ quantum field models, *Commun. Math. Phys.* **44** (1975), 143–164.
29. T. Spencer and F. Zirilli, Scattering states and bound states in $\lambda\mathscr{P}(\phi)_2$," *Commun. Math. Phys.* **49** (1976), 1–16.

A Convergent Expansion about Mean Field Theory
II. Convergence of the Expansion

JAMES GLIMM*

Rockefeller University, New York, New York 10021

ARTHUR JAFFE†

Harvard University, Cambridge, Massachusetts 02138

AND

THOMAS SPENCER*

Rockefeller University, New York, New York 10021

Received June 2, 1976

We prove convergence of the mean field cluster expansion defined in the preceding paper and establish exponential clustering in a pure phase selected by suitable boundary conditions.

1. INTRODUCTION

1.1. *Outline of the Estimates*

Convergence of the mean field cluster expansion is established in this paper. The expansion defined in Part I (referred to here as I) has the form

$$\langle A \rangle = \lim_{\Lambda \nearrow R^2} Z(\Lambda)^{-1} \sum_{\Sigma,\Gamma} \int A \, d\phi_{\Lambda,\Sigma,\Gamma}, \qquad (1.1.1)$$

cf. (I.5.10). The summation ranges over phase boundaries Σ and Dirichlet contours Γ, and A is a function of the field.

For each term in the expansion, there are two sources of convergence. The

* Supported in part by the National Science Foundation under Grant MPS74-13252.
† Supported in part by the National Science Foundation under Grant MPS 75-21212.

Gaussian approximation provides the first source of convergence. Far from phase boundaries, i.e., in the region X of (I.5.7.), the interaction V has the form

$$V(x) = :\lambda\psi(x)^4: \pm (2\lambda)^{1/2}\alpha :\psi(x): + \lambda^{5/2}\beta :\psi(x)^2: \qquad (1.1.2)$$

with coefficients $O(\lambda^{1/2})$. Thus $e^{-V}d\psi_1$ is a small perturbation of the Gaussian ground-state measure $d\psi_1$ with mean zero and mass 1. The convergence of the approximation $e^{-V}d\psi_1$ by $d\psi_1$ is introduced in (1.1.1) by factors λ, $\lambda^{1/2}$, or $\lambda^{5/2}$ in (1.1.2).

The second source of convergence arises from the presence of a phase boundary Σ. For λ small, the existence of a phase boundary means the existence of a large fluctuation in the value of the field ϕ. In fact, only with small probability can $|\phi| \leqslant (32\lambda)^{-1/2}$, due to the potential barrier $\mathscr{P}(0) = O(\lambda^{-1})$ in the energy density $\mathscr{P}(\xi)$, near $\xi = 0$. The resulting small probability is $\exp(-O(\lambda^{-1}))$. This means that when a phase boundary Σ occurs, the field (with large probability) changes from order $\lambda^{-1/2}$ on one side of Σ to order $-\lambda^{-1/2}$ on the other side. This fluctuation yields a large derivative $\nabla\phi$, which results in a small factor $\exp(-O(\nabla\phi)^2) = \exp(-O(\lambda^{-1}))$ arising from the gradient term in the Euclidean action, cf. Section 2.4.

Associated with each convergence factor there is a length scale. The unit squares are used to analyze the fluctuations of the field, whereas the squares of length l are introduced to exhibit the decoupling of the field in the Gaussian approximation. Since $m \sim m_c \sim 1$, we exhibit the exponential clustering of the ground state in X by choosing

$$l = |\log\lambda|^{1/4} \gg 1. \qquad (1.1.3)$$

This choice of l gives coupling at most

$$e^{-m_c l} \ll 1$$

between nonadjacent l-lattice squares. (Compare with [15, 16][1] where $m_c = m_0 \gg 1$ and $l = 1$, but again $m_c l \gg 1$.)

The parameter L, chosen as

$$L = |\log\lambda|^2 \gg l, \qquad (1.1.4)$$

is defined so that the functional derivatives $\delta/\delta\psi$ which contract to the exponent in $X^0 = R^2 \backslash X$ occur over a long distance $L/2$. We choose L large to compensate for the fact that the coefficients of V in X^0 are not small. Thus L provides an upper bound for the distance scale on which phase boundaries influence the Gaussian approximation.

[1] See Part I for reference list.

The key estimate required to establish convergence is

$$\left| \int A \, d\phi_{\Lambda,\Sigma,\Gamma} \right| \leqslant \exp(a\lambda^{1/2} \mid \Lambda \mid - \delta l \mid \Gamma^c \mid - \delta\lambda^{-1/2} \mid \Sigma \mid), \tag{1.1.5}$$

where a, δ are constants independent of λ, Λ, Σ. We establish this bound in Section 3. The main idea is as follows. Evaluate the s derivatives in (I.5.6) (using integration by parts) to obtain a sum of terms of the form

$$\int A R e^{-Q} \chi_\Sigma \, d\psi(s).$$

Here R is a polynomial in ψ obtained from derivatives $\delta/\delta\psi$ applied to e^{-Q} and χ_Σ' denotes either χ_Σ or a possible derivative of χ_Σ. The polynomial R has coefficients of order $\mid \lambda^{1/2} \mid$ since the coefficients of V are of this size within X, where the functional derivatives are localized. By Hölder's inequality such a term is bounded by

$$\left| \int (AR)^q \, d\psi(s) \right|^{1/q} \left[\int e^{-pQ} (\chi_\Sigma')^p \, d\psi(s) \right]^{1/p}. \tag{1.1.6}$$

For illustrative purposes, let us consider the case where there are no phase boundaries, i.e., $\Sigma = \{\sigma_i = 1\}$, and where no derivatives occur, i.e., $\chi_\Sigma' = \chi_\Sigma$. We choose p near 1, independent of λ, and so that q is a large even integer. The first factor in (1.1.6) is a straightforward Gaussian integral which is bounded by $O(e^{-\delta l \mid \Gamma^c \mid})$ as in [16]. For the second factor the key estimate is a bound on the vacuum energy per unit volume, uniform as $\lambda \searrow 0$. By shifting a portion of the (unit) mass of $d\psi_1$ to the exponent the desired vacuum energy bound takes the form

$$\int \exp\left(-p\left[V(\Lambda) + \left(\frac{1-\eta}{2}\right) \int_\Lambda :\psi:^2(x)\, dx\right]\right) d\psi_{(\eta p+1-p)} \leqslant \exp(\text{const} \mid \Lambda \mid).$$

For fixed momentum cutoff κ and fixed $\eta > 0$, the above exponent (as a function of $\phi_\kappa(x)$),

$$:P(\phi_\kappa(x)): - \eta :(\phi_\kappa(x) - \xi_+)^2: \tag{1.1.7}$$

is not uniformly bounded from below as $\lambda \searrow 0$. In fact, when $\phi_\kappa(x) \approx -\xi_+$, (1.1.7) approximately equals $-\eta(2\xi_+)^2 = -(\eta/2)\lambda^{-1}$. In order to obtain a uniform bound on (1.1.7) we must consider the contribution of the factor χ_Σ which constrains the averaged field to be positive. In addition, it is necessary to include two error terms: The first error term occurs because χ_+ is the smoothed out $\bar{\phi}$ localization of (I.2.5). The second error term is $\zeta :\delta\phi^2:$, with ζ small, and measures the deviation of $\phi(x)$ from the averaged field $\bar{\phi}$ constrained by χ_+. In Theorem 3.1.1 we show that

$$:P(\phi_\kappa(x)): - \eta :(\phi_\kappa(x) - \xi_+)^2: - \log \chi_+(\bar{\phi}) \geqslant b \log^2 \kappa - \zeta :(\delta\phi_\kappa):^2 (x). \quad (1.1.8)$$

Here $\zeta > 0$ is given, $\eta \leqslant \eta(\zeta)$ is sufficiently small and $b = b(\zeta)$ is large. The constants η, b, ζ do not, however, depend on λ, $0 \leqslant \lambda \leqslant 1$ and ζ is chosen so that

$$\int \exp(4\zeta:\delta\phi_\kappa{}^2(\Delta):)\, d\phi_\epsilon \leqslant e^{O(1)|\Delta|} \quad (1.1.9)$$

uniformly in the momentum cutoff κ and the mass $\epsilon > 0$. This bound is obtained in the following section and relies on the fact that since $\delta\phi(x)$ is orthogonal to the zero momentum mode,

$$\int_\Delta \delta\phi^2(x)\, dx \leqslant \pi^{-2} \int_\Delta (\nabla\phi)^2 (x)\, dx.$$

1.2. *An Estimate on Schwinger Functions*

In this section we state two estimates, which yield the proof of Theorem I.6.2.

PROPOSITION 1.2.1. *Let A have the form* (I.6.1). *There is a $\delta > 0$ such that for λ sufficiently small,*

$$| R(A, Y, \partial Y^\pm)| \leqslant \lambda^{-N(A)/2} ||| A ||| e^{-\delta|Y \cap A|/l}. \quad (1.2.1)$$

If $\partial Y^- = \varnothing$, the factor $\lambda^{-N/2}$ can be replaced by

$$(1 + \lambda^{-N/2} e^{-O(\lambda^{-1/2})}),$$

provided $N(A) > 0$.

Remark. The divergent factor $\lambda^{-N/2}$ arises from the possibility of localizing $\phi(x)$ near ξ_-, i.e., in the minus well of the potential.

PROPOSITION 1.2.2. *Let λ be sufficiently small. Let $N^\pm = N(\partial Y^\pm)$. There is a constant $c < \infty$ such that*

$$\left| \frac{Z(\sim Y, N^+, N^-, \partial Y)}{Z(R^2, \varnothing, \varnothing, \varnothing)} \right| \leqslant \exp(c\lambda^{1/2}(| Y \cap \Lambda | + | N | + l^2 | \partial Y |)). \quad (1.2.2)$$

If $N^- \neq \varnothing$, we can multiply the right side by $\exp(-O(\lambda^{-1/2}))$.

Remark. The convergence factor $\exp(-O(\lambda^{-1/2}))$ arises from the small probability for occurrence of a phase boundary.

Proof of Theorem I.6.2. By a simple combinatoric argument (see [16]) there are at most $\exp(O(r/l^2))$ allowed choices of Y such that $|Y| = r$. With ∂Y fixed, there are at most $2^{|Y|/l^2}$ choices of ∂Y^+, ∂Y^-. Thus the series (I.7.15) is majorized by

$$||| A ||| (1 + \lambda^{-N/2} e^{-O(\lambda^{-1/2})}) \sum_{r=1}^{\infty} e^{-\delta r/l} \leqslant ||| A ||| \, O(l^2)(1 + \lambda^{-N/2} e^{-O(\lambda^{-1/2})}).$$

The factor $O(l^2)$ is absorbed into $||| A |||$ by an increase in the constant K in (I.6.2), and the proof is complete.

2. Estimates on Fluctuations

2.1. *The Momentum Cutoff*

We introduce a momentum cutoff (smoothing) operator $\rho_\kappa : f \to f_\kappa$. In order to preserve the notion of localization in space-time squares Δ, before and after smoothing, we require several properties of the cutoff:

(i) $\rho_\kappa \to I$ as $\kappa \to \infty$.

(ii) ρ_κ is self-adjoint.

(iii) If $x \in \Delta$, then $(\rho_\kappa f)(x)$ depends only on $f(y)$ for $y \in \Delta$.

(iv) $\rho_\kappa e_\Delta = e_\Delta$, where e_Δ is the characteristic function for Δ.

We achieve these properties with ρ_κ defined by the symmetrization of a non-negative integral kernel $\rho_\kappa'(x, y)$, i.e.,

$$f_\kappa(x) = \int \rho_\kappa'(x, y) f(y) \, dy, \tag{2.1.1}$$

such that for all Δ,

$$\int_\Delta \rho_\kappa'(x, y) \, dx = e_\Delta(y) = \int_\Delta \rho_\kappa'(y, x) \, dx. \tag{2.1.2}$$

Then (ii)–(iv) follow. To obtain ρ_κ', we first choose an even, positive, C^∞ function $w(t)$, supported in the interval $[-\frac{1}{2}, \frac{1}{2}]$, and with total integral 1. Define $\rho_\kappa(x) = \kappa^2 w(\kappa x_0) \, w(\kappa x_1)$ as an approximate Dirac measure. For $x \in \Delta$, let $x_\Delta^{(i)}$, $i = 1, 2, 3, 4$ be the four points (including $x \equiv x_\Delta^{(1)}$) obtained by reflecting about the two sides of Δ nearest to x. Then define

$$\rho_\kappa'(x, y) = \sum_\Delta \sum_{i=1}^{4} \rho_\kappa(x_\Delta^{(i)} - y) \, e_\Delta(x) \, e_\Delta(y). \tag{2.1.3}$$

We check that

$$\int_\Delta \rho_\kappa'(x, y)\, dx = \sum_{i=1}^{4} \int \rho_\kappa(x_\Delta^{(i)} - y)\, e_\Delta(y)\, dx$$
$$= \int \rho_\kappa(x - y)\, e_\Delta(y)\, dx = e_\Delta(y).$$

Similarly

$$\int_\Delta \rho_\kappa'(y, x)\, dx = e_\Delta(y),$$

so (2.1.2) holds. The inequality

$$\|(f_\kappa - f)\, e_\Delta\|_{L_p(\Delta)} \leqslant O(\kappa^{-\epsilon})\, \|f\|_S \tag{2.1.4}$$

expresses the precise form in which (i) holds. Here $2 \leqslant p < \infty$ and $\|f\|_S$ denotes a Sobolev norm with fractional derivatives in $L_{p'}(\Delta)$. If $\text{dist}(\text{suppt} f, \partial\Delta) \geqslant \kappa^{-1}$ then the operator ρ_κ is a convolution operator and (2.1.4) follows, e.g., using a Fourier transform. Let $\partial = \{x \mid \text{dist}(x, \partial\Delta) \leqslant \kappa^{-1}\}$. By Hölder's inequality

$$\|f\|_{L_p(\partial)} \leqslant \kappa^{-\epsilon}\, \|f\|_{L_{p'}(\partial)}$$

for $p' > p$, which establishes (2.1.4).

The Wick ordering constant for the cutoff field ϕ_κ is

$$c_\kappa(x) = \int \phi_\kappa(x)^2\, d\Phi$$
$$= \int \rho_\kappa(x, y)\, \rho_\kappa(x, z)\, C(y - z)\, dy\, dz = (\rho_\kappa C \rho_\kappa)(x, x). \tag{2.1.5}$$

Furthermore,

$$c_\kappa(x) \leqslant O(\ln \kappa), \tag{2.1.6}$$

since

$$c_\kappa(x) \leqslant \sup_y \int \rho_\kappa(x, z)\, C(y - z)\, dz$$
$$\leqslant 4 \sup_y \int \rho_\kappa(y - z)\, C(z)\, dz = O(\ln \kappa).$$

Finally, we remark that as an operator on $L_2(\Delta)$,

$$\|e_\Delta \rho_\kappa e_\Delta\| \leqslant 4. \tag{2.1.7}$$

Here we use (2.1.3) and the fact that convolution by $\rho_\kappa(x)$ has operator norm 1. Thus

$$\int_\Delta |f_\kappa|^2 \, dx \leqslant 16 \int_\Delta |f|^2 \, dx. \tag{2.1.8}$$

2.2. *Space Dependent Mass*

In this section we consider covariance operators for a Gaussian measure with a space-dependent mass. Let $\omega(x)$ satisfy

$$0 < \omega(x) \leqslant 1, \qquad \inf_x \omega(x) = \mathrm{\omega} > 0, \tag{2.2.1}$$

and let $1 - \omega(x)$ have compact support. In (2.2.2) below, $\omega(x)^{1/2}$ plays the role of a mass. Consider

$$d\psi_\omega(s) = Z^{-1} \exp\left(\tfrac{1}{2} \int (1 - \omega(x)) :\psi(x)^2: dx\right) d\psi_1(s) \tag{2.2.2}$$

where

$$Z = \int \exp\left(\tfrac{1}{2} \int (1 - \omega(x)) :\psi(x)^2: dx\right) d\psi_1(s). \tag{2.2.3}$$

By Jensen's inequality, $Z \neq 0$. In fact,

$$\int e^V \, d\psi \geqslant e^{\int V d\psi}$$

and so

$$Z^{-1} \leqslant \exp\left[O(1) \int (1 - \omega) \, dx\right]. \tag{2.2.4}$$

Thus $d\psi_\omega(s)$ is a Gaussian measure with mean zero and covariance

$$\begin{aligned} C_\omega(s) &= \sum_{j=0}^{\infty} C(s)^{1/2} \left[C(s)^{1/2} (1 - \omega) C(s)^{1/2}\right]^j C(s)^{1/2} \\ &= [C(s)^{-1} - (1 - \omega)]^{-1}. \end{aligned} \tag{2.2.5}$$

The convergence follows from the fact that $0 \leqslant 1 - \omega(x) \leqslant 1 - \mathrm{\omega} < 1$ and $\| C(s) \| \leqslant 1$. For $s = 1$,

$$(-\Delta + 1 - \omega(x))^{-1}$$

$$= \sum_{j=0}^{\infty} (-\Delta + 1)^{-1/2} [(-\Delta + 1)^{-1/2} (1 - \omega)(-\Delta + 1)^{-1/2}]^j (-\Delta + 1)^{-1/2} \tag{2.2.6}$$

Let Δ_N be the Laplace operator with zero Neumann data on some set of (lattice) lines in R^2, and let

$$C_\omega(s, t) = (1 - t)\, C_\omega(s) + t(-\Delta_N + \omega)^{-1}. \tag{2.2.7}$$

We note that

$$-\Delta_N \leqslant -\Delta_\Gamma \tag{2.2.8}$$

so

$$C(s) \leqslant (-\Delta_N + 1)^{-1} \tag{2.2.9}$$

and by (2.2.5)

$$C_\omega(s) \leqslant C_\infty(s) \leqslant (-\Delta_N + \omega)^{-1} \tag{2.2.10}$$

where

$$C_\infty(s) = [C(s)^{-1} - (1 - \omega)]^{-1}.$$

It follows that

$$\frac{d}{dt}\, C_\omega(s, t) = (-\Delta^N + \omega)^{-1} - C_\omega(s) \geqslant 0. \tag{2.2.11}$$

2.3. *The Fluctuation Field*

In this section we bound the fluctuation field $\delta\phi(x)$ and its translate $\delta\psi(x) = \delta\phi(x) - \delta g(x)$. Here $g(x)$ is the approximate classical magnetization (1.4.4). We use the method of conditioning [20], and we start with a direct proof of the conditioning inequality.

PROPOSITION 2.3.1. *Let V be a fixed (s, t independent) polynomial in ψ, and let $:V:$ denote $C_\omega(s, t)$ Wick ordering. Then*

$$\int e^{-:V:}\, d\psi_\omega(s, t) \tag{2.3.1}$$

is increasing in s and t.

Proof. We give the proof for t only. Let $Z(s, t)$ denote (2.3.1). Then

$$\begin{aligned}\frac{dZ(s, t)}{dt} &= -\int e^{-:V:}\, \frac{d{:}V{:}}{dt}\, d\psi_\omega(s, t) + \int e^{-:V:}\, \frac{d}{dt}\, d\psi_\omega(s, t)\\ &= \int \left(-\frac{d{:}V{:}}{dt} + \frac{1}{2}\,\Delta_\psi \cdot \frac{dC_\omega(s, t)}{dt}\right) e^{-:V:}\, d\psi_\omega(s, t)\end{aligned}$$

where

$$\Delta_\psi \cdot K \equiv \int K(x, y)\, \frac{\delta^2}{\delta\psi(x)\, \delta\psi(y)}\, dx\, dy$$

as in [16]. The term $-d{:}V{:}/dt$ cancels with $2^{-1}(\Delta_\psi \cdot dC_\omega/dt){:}V{:}$. Thus

$$\frac{dZ(s,t)}{dt} = \frac{1}{2}\int \left[\int \frac{\delta{:}V{:}}{\delta\psi(x)} \frac{d}{dt} c(x,y) \frac{\delta{:}V{:}}{\delta\psi(y)}\, dx\, dy\right] e^{-{:}V{:}}\, d\psi_\omega(s,t).$$

Here $c(x,y)$ is the kernel of $C_\omega(s,t)$. By (2.2.11), $dC_\omega/dt \geqslant 0$, so dc/dt is a positive bilinear form and $[\cdots] \geqslant 0$. Thus Z is increasing in t.

PROPOSITION 2.3.2. *Let* $\zeta \leqslant \pi^2/32$. There is a constant K (independent of ω, $\boldsymbol{\omega}$, s, t, ζ, κ) such that

$$\int \exp\left(\int_Y \zeta{:}(\delta\psi_\kappa)^2{:}\, dx\right) d\psi_\omega(s,t) \leqslant \exp(K\zeta \mid Y \mid).$$

Proof. We may suppose that the exponent is Wick ordered in the covariance $C_\omega(s,t)$. In fact, the modification $s = 1 \to s \leqslant 1$ decreases Wick subtractions and increases the integral. The further changes $t = 0 \to t \geqslant 0$ and $\omega = 1 \to \omega \leqslant 1$ are monotonic in the other direction, and so it is sufficient to examine the effect on the Wick constant when $\omega = 0$, $t = 1$. In this case we use the fact that $\delta\psi_\kappa$ is orthogonal to the constant function e_Δ to conclude that the change in the Wick constant is still $O(\zeta \mid Y \mid)$. Thus Proposition 2.3.1 is applicable. Choose Δ_N to have Neumann data on all unit lattice lines. The value $t = 1$ then provides an upper bound. However, for $t = 1$, the integral factors into a product over unit lattice squares, and so it is sufficient to consider a single lattice square.

Let P_Δ denote the projection in L_2 onto e_Δ. The integral for a single lattice square is given by

$$\exp(-\tfrac{1}{2}\operatorname{tr}\ln(1-A) - \tfrac{1}{2}\operatorname{tr} A) \leqslant \exp(\text{const} \| A \|_{H.S.}^2)$$

provided $\| A \| \leqslant \frac{1}{2}$ where A is defined by

$$A = \zeta[-\Delta_N + \boldsymbol{\omega}]^{-1}[\rho_\kappa{}^2 - P_\Delta].$$

Since $\rho_\kappa{}^2 - P_\Delta$ annihilates the ground state e_Δ of Δ_N and $-\Delta_N \upharpoonright \{e_\Delta\}^\perp \geqslant \pi^2$, the inverse is uniformly bounded for $0 \leqslant \boldsymbol{\omega} \leqslant 1$. Moreover, because $\| \rho_\kappa{}^2 \| \leqslant 16$ the condition $\| A \| \leqslant \frac{1}{2}$ is satisfied for $\zeta \leqslant \pi^2/32$. The bound on the Hilbert–Schmidt norm of A uniform in $\boldsymbol{\omega}$ follows similarly. (The restriction $\| A \| \leqslant 1 - \epsilon$ was unfortunately omitted in [18].)

Recall that $\psi = \phi - g$ and let $\delta g = g(x) - \int_\Delta g(x)\, dx$, $x \in \Delta$.

PROPOSITION 2.3.3. *Let* $\zeta \leqslant \pi^2/64$. *There is a constant* K (*independent of* ω, $\boldsymbol{\omega}$, s, t, ζ, κ) *such that*

$$\int \exp\left(\int_Y \zeta{:}(\delta\phi_\kappa)^2{:}\, dx\right) d\psi_\omega(s,t) \leqslant \exp\left(\zeta K \left(\int_Y (1 + \mid \nabla g \mid^2)\, dx\right)\right).$$

Proof. We expand

$$:\delta\phi_\kappa(x)^2: = :\delta\psi_\kappa(x)^2: + 2\delta\psi_\kappa(x)\,\delta g_\kappa(x) + \delta g_\kappa(x)^2. \tag{2.3.2}$$

We claim

$$(16)^{-1}\int_Y (\delta g_\kappa)^2\,dx \leqslant \int_Y (\delta g)^2\,dx \leqslant \pi^{-2}\int_Y (\nabla g)^2\,dx. \tag{2.3.3}$$

The first inequality follows by (2.1.8) while the second inequality follows from the bound

$$\pi \leqslant -\Delta_N \restriction (\{e_\Delta\}^\perp)$$

as in the proof of Proposition 2.3.2.

A Schwarz inequality allows us to consider separately the $:\delta\psi^2:$ and $\delta\psi$ terms from expansion (2.3.2) if we replace ζ by 2ζ. The $:\delta\psi^2:$ integral is bounded by Proposition 2.3.2. The $\delta\psi$ integral can be explicitly evaluated as

$$\exp\left[8\zeta^2\int_{Y\times Y}\delta g_\kappa(x)(\rho_\kappa{*}C\rho_\kappa)(x,y)\,\delta g_\kappa(y)\,dx\,dy\right] \leqslant \exp\left[O(1)\,\zeta^2\int_Y(\nabla g)^2\,dx\right]. \tag{2.3.4}$$

Here ρ_κ is the momentum cutoff operator and $C = C_\omega(s,t)$. We have used in (2. 3. 4), the bound

$$0 \leqslant E\rho_\kappa C\rho_\kappa E \leqslant E\rho_\kappa(-\Delta_N)^{-1}\rho_\kappa E \leqslant (4/\pi)^2 E$$

valid on $\{e_\Delta\}^\perp$. Here E is the operator of multiplication by the characteristic function of Y

2.4. *Phase Boundaries: The Leading Term*

The occurrence of phase boundaries, i.e., $\Sigma \neq \varnothing$, is reflected in the factor $\exp(-F)$ in the ground state measure $d\phi_{\Lambda,\Sigma}$ of (4.6). This factor is the effect of the change of variables $\phi \to \psi = \phi - g$ on the kinetic energy term $|\nabla g|^2$ in the action density.

The functions F_i of (I.4.7) are local, when $\mathrm{dist}(\partial Y, \Sigma) \geqslant L/2$. Locality means that in the notation of (I.4.8d),

$$F_i(Y, \Sigma) = F_i(Y, \Sigma \cap Y)$$

is independent of $\Sigma \sim Y$. See also Lemma I.4.1. The dominant contribution to F is F_1, which we now relate to the length $|\Sigma(Y)|$ of the phase boundaries in Y.

PROPOSITION 2.4.1. *Let $\partial Y \subset X$. Then*

$$\lambda^{-1}|4\ln\lambda|^{-4}|X^0 \cap Y| \leqslant (\xi_+ - \xi_-)^2|\Sigma \cap Y| \leqslant 12\eta^{-1}F_1(Y,\Sigma). \tag{2.4.1}$$

Proof. The first inequality follows from the definition (I.5.12) of X^0. For each segment $\gamma \subset \Sigma \cap Y$, let $X^0(\gamma)$ be the set of points within distance L of γ. Then

$$X^0 \cap Y \subseteq \bigcup_{\gamma \in \Sigma \cap Y} X^0(\gamma).$$

In fact, for $\Delta \subset X^0$, we have $\Delta \subset X^0(\gamma)$ for some $\gamma \in \Sigma$ by definition of X^0. Moreover if $\Delta \subset Y$ also, then $\gamma \subset Y$, because $\partial Y \subset X$ implies $\partial Y \cap X^0 = \phi$. Thus

$$(2L)^{-2} \mid X^0 \cap Y \mid \leqslant \mid \Sigma \cap Y \mid. \tag{2.4.2}$$

Since $L = (\ln \lambda)^2$, and since each jump has size $\xi_+ - \xi_- = (2\lambda)^{-1/2} + O(\mu)$, the first inequality holds.

To prove the second inequality, we note

$$g(x) - g(x') = \int_{x'}^{x} \nabla g \, ds$$

where $x \in \Delta$, $x' \in \Delta'$, and Δ, Δ' are neighboring squares with a common side γ. We integrate along a path parallel to the sides of Δ, Δ'. By the triangle and Schwarz inequalities,

$$\begin{aligned} \mid h(x) - h(x')\mid^2 &\leqslant (\mid h(x) - g(x)\mid + \mid g(x) - g(x')\mid + \mid g(x') - h(x')\mid)^2 \\ &\leqslant 3 \left(\mid h(x) - g(x)\mid^2 + \mid h(x') - g(x')\mid^2 + \int_{x}^{x} \mid \nabla g \mid^2 ds\right). \end{aligned}$$

We now integrate over Δ, Δ' and sum over $\mathcal{N}(Y)$, the nearest neighbor squares in Y. Since

$$\sum_{\mathcal{N}(Y)} (h(\Delta) - h(\Delta'))^2 = (\xi_+ - \xi_-)^2 \mid \Sigma \cap Y \mid, \tag{2.4.3}$$

the second inequality follows by (1.4.7a).

2.5. *Phase Boundaries: Error Estimates*

In the above section we found

$$e^{-F_1(Y)} \leqslant e^{-a\eta\lambda^{-1}|\Sigma \cap Y|} \tag{2.5.1}$$

for a constant a. Now we consider the error terms F_2, F_3, and F_4. The main results are contained in two estimates.

PROPOSITION 2.5.1. *Let $\partial Y \subset X$, let ∂Y be a Dirichlet contour, let $d\psi_\omega(s)$ be given by* (2.2.2), *let $0 < \eta_0 \leqslant 1$, and let $\lambda \leqslant \lambda(\eta_0, \omega)$ be sufficiently small. Then*

$$\int e^{-O(\eta_0^{-1}) F_3(Y)} \, d\psi_\omega(s) \leqslant e^{\lambda F_1(Y)}. \tag{2.5.2}$$

PROPOSITION 2.5.2. *Assume the hypotheses of Proposition* 2.5.1 *and in addition let* $0 \leqslant \zeta < \pi^2/2^7$, *let* $\omega \restriction \sim\Lambda = 1$, $1 \leqslant p$, *and let* $Y_t \subset Y \cap \Lambda$. *Then there exists a constant* K *(independent of* ω, $\mathbf{\omega}$, s, t, p, ζ, κ, η, η_0*) such that*

$$\int \exp\left[\zeta \int_{Y_t} :\delta\phi_\kappa^2(x): dx - pF_4(Y)\right] d\psi_\omega(s)$$

$$\leqslant \exp(K(|Y_t| + \zeta F_1(Y)) \exp(p^2(1-\eta) F_2(Y)). \tag{2.5.3}$$

LEMMA 2.5.3. *Let* $0 < \eta_0 \leqslant \eta \leqslant 1$, *and let* $\lambda \leqslant \lambda(\eta_0)$ *be sufficiently small. Then*

$$0 \leqslant \eta_\zeta - \eta \leqslant \lambda^2 \tag{2.5.4}$$

Proof. Recall that $\zeta = \zeta(x)$ is introduced below (I.4.3). Since $0 \leqslant \zeta \leqslant 1$,

$$\eta_\zeta - \eta = \eta\eta_\zeta(\eta^{-1} - \eta_\zeta^{-1}) = \eta\eta_\zeta \int (-\Delta + \eta)^{-1}(x)(1 - \zeta(x/L))\, dx \geqslant 0.$$

The integration above is supported in $|x| \geqslant L/4$, so

$$\eta_\zeta - \eta \leqslant \eta_\zeta\eta \int_{|x|\geqslant L/4} (-\Delta + \eta)^{-1}(x)\, dx$$

$$\leqslant O(e^{-\eta L/4}) \leqslant O(\lambda^2)$$

for small λ, since $L = |\ln \lambda|^2$.

LEMMA 2.5.4. *Let* $0 < \eta_0 \leqslant \eta \leqslant 1$ *and let* $\lambda \leqslant \lambda(\eta_0)$ *be sufficiently small. Then*

$$\| g - g_c \|_\infty + \| \nabla(g - g_c)\|_\infty \leqslant \lambda. \tag{2.5.5}$$

Proof. We estimate $g - g_c$.

$$(g - g_c)(x) = (\eta_\zeta - \eta)((-\Delta + \eta)^{-1} h)(x)$$

$$+ \eta_\zeta \int (-\Delta + \eta)^{-1}(x - y)\left[\zeta\left(\frac{x-y}{L}\right) - 1\right] h(y)\, dy.$$

Thus

$$\| g - g_c \|_\infty \leqslant (\eta_\zeta - \eta) \|(-\Delta + \eta)^{-1} h \|_\infty$$

$$+ \text{const} \| h \|_\infty \eta_\zeta\eta^{-1}(1 + \eta L^2)\, e^{-(\eta^{1/2}L/4)}. \tag{2.5.6}$$

By Lemma 2.5.3 and

$$\|(-\Delta + \eta)^{-1} h \|_\infty \leqslant \eta^{-1} \| h \|_\infty \leqslant \text{const}\, \eta_0^{-1}\lambda^{-1/2},$$

the first term in (2.5.6) is bounded by $\lambda/4$ for λ small. For the second term in (2.5.6), we follow the proof of Lemma 2.5.3 to bound it by

$$\text{const } \lambda^{-1/2}\eta_0^{-1}\lambda^2 < \lambda/4$$

for small λ, so $\| g - g_c \| \leqslant \lambda/2$. The bound on $\nabla(g - g_c)$ is similar, using the fact $\nabla(-\Delta + \eta)^{-1}$ is a bounded operator from L_∞ to L_∞.

Proof of Proposition 2.5.1. Write $Y = \bigcup_{i=1}^{\nu} Y_i$ as a union of its connected components. Both sides of (2.5.2) factor, so we can assume without loss of generality that Y is connected. Then ∂Y consists of a finite number of closed curves, and each component of $R^2 \sim Y$ has exactly one of these curves for its boundary. In this case, $\Sigma \cap Y$ is a set of phase boundaries in R^2, which determines a new choice of spins, and this new choice of spins agrees with $\{\sigma_i\} = \Sigma$ in Y. Define h_Y, $g_{Y,c}$, and g_Y as in Section I.4, using $\Sigma \cap Y$ in place of Σ. By Lemma I.4.1,

$$\begin{aligned}(-\Delta + \eta)(g_Y - g_{Y,c}) &= (-\Delta + \eta)(g - g_c) \qquad \text{on } Y \\ &= 0 \qquad \text{on } {\sim}Y.\end{aligned} \tag{2.5.7}$$

Thus

$$F_3(Y) = \langle \psi, (-\Delta + \eta)(g_Y - g_{Y,c})\rangle$$

and

$$\int e^{aF_3(Y)}\, d\psi_\omega(s) = \exp[2^{-1}a^2 \| C_\omega(s)^{1/2}(-\Delta + \eta)(g_Y - g_{Y,c})\|^2] \tag{2.5.8}$$

Since

$$C_\omega(s) \leqslant (-\Delta + \omega)^{-1},$$

(2.5.8) is majorized by

$$\exp[2^{-1}a^2\omega^{-1}\langle g_Y - g_{Y,c}, (-\Delta + \eta)(g_Y - g_{Y,c})\rangle].$$

By (2.5.1) and Lemma 2.5.4, the inner product above is bounded by $2\lambda^2 \mid X^0 \cap Y \mid$, which by Proposition 2.4.1, completes the proof.

Proof of Proposition 2.5.2. By conditioning we can take $t = 1$, so that the integral factors. The factors associated with squares $\Delta \subset Y_t$ are bounded by Proposition 2.3.3 because F_4 does not contribute. On squares $\Delta \not\subset \Lambda$, $\delta\phi$ does not contribute since $Y_t \subset \Lambda$, and ω does not contribute since $\omega = 1$ on ${\sim}\Lambda$. Thus

$$\begin{aligned}&\int e^{-pF_4(Y)}\, d\psi_\omega(s, t = 1) \\ &\quad = \exp\left(\tfrac{1}{2}p^2(1 - \eta)^2 \int_{\sim\Lambda} (g - \xi_+)(-\Delta^N + 1)^{-1}(g - \xi_+)\, dx\right) \\ &\quad \leqslant \exp(p^2(1 - \eta)\, F_2(Y)).\end{aligned}$$

2.6. *Spin Localization*

We give some simple estimates on the spin localization functions $\chi_\pm$ defined in (1.3.5). In fact, we bound the functions,

$$\chi_{a,b}(\xi) = \pi^{-1/2} \int_a^b e^{-(\xi-z)^2}\, dz,$$

which can be used for the spin localization of a more general interaction polynomial.

LEMMA 2.6.1. *Let $\theta < 1$. There is a constant $K = K(\theta)$ such that*

$$\chi_{a,b}(\xi) \leqslant K e^{-\theta(\xi-a)^2}, \qquad \xi \leqslant a \tag{2.6.1}$$

$$\chi_{a,b}(\xi) \geqslant 1 - K[e^{-\theta(\xi-a)^2} + e^{-\theta(\xi-b)^2}], \qquad a \leqslant \xi \leqslant b \tag{2.6.2}$$

$$\chi_{a,b}(\xi) \leqslant K e^{-\theta(\xi-b)^2}, \qquad b \leqslant \xi. \tag{2.6.3}$$

Proof. For $\xi \leqslant a$ (or similarly for $\xi > b$),

$$\begin{aligned}\chi_{a,b}(\xi) &\leqslant e^{-\theta(\xi-a)^2} \pi^{-1/2} \int_a^b e^{-(1-\theta)(\xi-z)^2}\, dz \\ &\leqslant (1-\theta)^{-1/2}\, e^{-\theta(\xi-a)^2}.\end{aligned}$$

For $a \leqslant \xi \leqslant b$, we use $\chi_{a,b} = 1 - \chi_{-\infty,a} - \chi_{b,\infty}$ and the above.

LEMMA 2.6.2. *For $\theta < 1$, there is a constant $K = K_\theta < \infty$ such that for all integers $n \geqslant 1$,*

$$\left| \frac{d^n}{d\xi^n} \chi_{a,b}(\xi) \right| \leqslant Kn!\, (e^{-\theta(\xi-a)^2} + e^{-\theta(\xi-b)^2}).$$

Proof. For $\theta < \eta < 1$, we obtain from Cauchy's integral formula the bound

$$\left| \frac{d^n}{d\xi^n} e^{-(\xi-z)^2} \right| \leqslant Kn!\, e^{-\eta(\xi-z)^2}$$

for real ξ, z. For $\xi \leqslant a$, or $\xi \geqslant b$, the bound of the lemma follows, as in Lemma 2.6.1. For $a \leqslant \xi \leqslant b$ we use the identity

$$\frac{d^n}{d\xi^n} \chi_{a,b} = -\frac{d^n}{d\xi^n} (\chi_{-\infty,a} + \chi_{b,\infty})$$

and proceed as before.

3. The Vacuum Energy Bound

3.1. *The Wick Ordering Lower Bound*

In this section we consider the interaction polynomial

$$:\mathscr{P}(\phi_\kappa(x)): = \lambda :\phi_\kappa(x)^4: - \tfrac{1}{4} :\phi_\kappa(x)^2: - \mu\phi_\kappa(x) - E_c \tag{3.1.1}$$

under the restrictions $0 < \lambda \leqslant \frac{1}{2}$, $0 \leqslant \mu \leqslant \lambda^2$, $E_c = -(64\lambda)^{-1} + O(\lambda^2)$, and with Wick ordering given by the covariance $(-\Delta + 1)^{-1}$. See I.2 for further details.

THEOREM 3.1.1. *Let $\eta \leqslant 10^{-2}$ and $81\eta \leqslant \zeta$. Then there is a constant $b = b(\zeta)$ such that for any $:\mathscr{P}(\phi_\kappa(x)):$ as above, any $x \in \Delta$ and any Σ,*

$$:\mathscr{P}(\phi_\kappa(x)): - \eta :(\phi_\kappa(x) - h(x))^2: - \ln \chi_{\sigma(\Delta)}(\bar{\phi}(\Delta)) \geqslant -b(\zeta) \ln^2\kappa - \zeta :\delta\phi_\kappa(x)^2:. \tag{3.1.2}$$

Proof. We define τ by

$$\begin{aligned}\tau(x) &= :\mathscr{P}(\phi_\kappa(x)): - \eta :(\phi_\kappa(x) - h)^2: + \zeta :\delta\phi_\kappa(x)^2: \\ &= \mathscr{P}(\phi_\kappa(x)) - \eta(\phi_\kappa(x) - h)^2 + \zeta\delta\phi_\kappa(x)^2 - 6\lambda c_\kappa \phi_\kappa^2 \\ &\quad + 3\lambda c_\kappa^2 + \eta c_\kappa - \zeta\delta c_\kappa,\end{aligned}$$

where c_κ, $\delta c_\kappa = O(\ln \kappa)$ are the Wick constants for ϕ_κ and $\delta\phi_\kappa$. We establish

$$\tau(x) - \ln \chi_{\sigma(\Delta)}(\Delta) \geqslant -O(1) \ln^2\kappa \tag{3.1.3}$$

and the theorem then follows.

Note for any $\epsilon > 0$ (and using $\mu\lambda \leqslant \lambda^3 \leqslant \lambda^2$),

$$\begin{aligned}\lambda(6c_\kappa + \epsilon/4)\, \phi_\kappa^2 + \epsilon\mu\lambda\phi_\kappa &\leqslant \epsilon\lambda^2\phi_\kappa^4 + (2\epsilon)^{-1}(6c_\kappa + \epsilon/4)^2 + 2\epsilon\lambda^3 \\ &\leqslant \epsilon\lambda^2\phi_\kappa^4 + O(\epsilon^{-1}) \ln^2\kappa.\end{aligned}$$

Also $| \epsilon\lambda E_c | \leqslant O(1)$, so

$$\tau(x) \geqslant (1 - \epsilon\lambda)\, \mathscr{P}(\phi_\kappa(x)) - \eta(\phi_\kappa - h)^2 + \zeta(\delta\phi_\kappa)^2 - O(\epsilon^{-1}) \ln^2\kappa. \tag{3.1.4}$$

We now fix $x \in \Delta$ and specify $h(\Delta)$ and $\phi(\Delta)$. This results in a number of cases to establish (3.1.3). We consider in detail $h(\Delta) = -\xi_+$. The estimates for $h(\Delta) = \xi_+$ are identical if we exchange ξ_+ and $-\xi_+$ throughout.

Case 1. $h(\Delta) = -\xi_+$ and $| \phi_\kappa(x) - \xi_+ | > \xi_+/3$. In this case $h(\Delta)$ is localized in one well of the W-shaped potential, while $\phi_\kappa(x)$ is localized outside the other well. Write $\mathscr{P}(\xi)$ in the form

$$\mathscr{P}(\xi) = \lambda(\xi - (8\lambda)^{-1/2})^2\, (\xi + (8\lambda)^{-1/2})^2 - \mu\xi - \delta E_c,$$

where $\delta E_c = E_c(\mu) - E_c(\mu = 0)$ is the shift in classical energy density for $\mu > 0$. By (I.2.8),

$$| \delta E_c | \leqslant O(\lambda^{3/2}).$$

Using (I.2.6), we see that if $| \xi - \xi_+ | > \xi_+/3$, then for some $\delta > 0$

$$\begin{aligned} \mathscr{P}(\xi) &\geqslant \delta(\xi + \xi_+)^2 - O(\lambda)(\xi - \xi_-) - O(\lambda), \\ &\geqslant \delta(1 - \epsilon\lambda)(\xi + \xi_+)^2 - O(\epsilon^{-1}). \end{aligned}$$

Substituting this bound into (3.1.4) yields

$$\begin{aligned} \tau &\geqslant [\delta(1 - \epsilon\lambda)^2 - \eta](\phi_\kappa(x) + \xi_+)^2 + \zeta(\delta\phi_\kappa)^2 - O(\epsilon^{-1}) \ln^2 \kappa \\ &\geqslant -O(\epsilon^{-1}) \ln^2 \kappa. \end{aligned} \tag{3.1.5}$$

In the last bound we choose $\epsilon = 10^{-3}$ and $\eta < 10^{-2}$ so the coefficient of $(\phi + \xi_+)^2$ is positive. Since $\chi_\sigma \leqslant 1$ and $-\ln \chi_\sigma \geqslant 0$, (3.1.3) follows.

Case 2. $h(\Delta) = -\xi_+$, $| \phi_\kappa(x) - \xi_+ | \leqslant \xi_+/3$. In this case h and $\phi_\kappa(x)$ are localized in different wells. Recall $\phi_\kappa(x) = \bar{\phi}(x) + \delta\phi_\kappa(x)$; thus in the region with $| \phi_\kappa - \xi_+ | \leqslant \xi_+/3$, either

$$\text{(a) } \delta\phi_\kappa \geqslant \xi_+/3 \qquad \text{or} \qquad \text{(b) } \bar{\phi} \geqslant \xi_+/3.$$

Subcase (a). We bound (3.1.4) from below, omitting the positive term $\mathscr{P}(\phi_\kappa)$. Then

$$\tau \geqslant -\eta(\phi_\kappa + \xi_+)^2 + \zeta(\delta\phi_\kappa)^2 - O(\epsilon^{-1}) \ln^2 \kappa.$$

Using $(\phi_\kappa - \xi_+)^2 \leqslant \xi_+^2/9$,

$$(\phi_\kappa + \xi_+)^2 \leqslant 2(\phi_\kappa - \xi_+)^2 + 8\xi_+^2 \leqslant 9\xi_+^2.$$

Thus for $\eta \leqslant \zeta/81$,

$$\begin{aligned} \tau &\geqslant -9\eta\xi_+^2 + \zeta(\delta\phi_\kappa)^2 - O(\epsilon^{-1}) \ln^2 \kappa \\ &\geqslant (9^{-1}\zeta - 9\eta)\, \xi_+^2 - O(\epsilon^{-1}) \ln^2 \kappa \\ &\geqslant -O(\epsilon^{-1}) \ln^2 \kappa. \end{aligned} \tag{3.1.6}$$

Subcase (b). We bound (3.1.4) from below (using $\mathscr{P}(\phi_\kappa) \geqslant 0$) by

$$\begin{aligned} \tau &\geqslant -\eta(\phi_\kappa + \xi_+)^2 + \zeta(\delta\phi_\kappa)^2 - O(\epsilon^{-1}) \ln^2 \kappa \\ &\geqslant -2\eta(\bar{\phi} + \xi_+)^2 + (\zeta - 2\eta)(\delta\phi_\kappa)^2 - O(\epsilon^{-1}) \ln^2 \kappa. \end{aligned} \tag{3.1.7}$$

Note by hypothesis (b) that

$$(\bar{\phi} + \xi_+)^2 \leqslant 20\bar{\phi}^2.$$

Thus for $\eta \leqslant \zeta/2$

$$\tau \geqslant -40\eta\bar{\phi}^2 - O(\epsilon^{-1}) \ln^2\kappa. \tag{3.1.8}$$

We nów use Lemma 2.6.1, and the fact that χ_σ and ϕ_κ are localized in opposite wells. Thus for $\theta < 1$,

$$-\ln \chi_{\sigma(\Delta)} \geqslant \theta\bar{\phi}^2 - O(1).$$

Choose $\eta < 1/80$ and $\theta \geqslant \frac{1}{2}$. Then

$$\tau - \ln \chi_{\sigma(\Delta)} \geqslant -O(\epsilon^{-1}) \ln^2\kappa. \tag{3.1.9}$$

The bounds (3.1.5), (3.1.6), and (3.1.9) establish (3.1.3) and the theorem.

We obtain a stronger Wick bound in case χ_σ is replaced by a derivative χ_σ'. In this case $\chi_\sigma'(\bar{\phi})$ has its maximum near the boundary of the localization region, i.e., near near $\bar{\phi} = 0$. However, this region is near the relative maximum $\mathscr{P}(0) = O(\lambda^{-1})$ of the W-shaped potential $\mathscr{P}$. Let $n(\Delta)$ denote the order of the derivative of χ_σ in Δ.

THEOREM 3.1.2. *Let $\zeta > 0$. There exist constants $a = a(\zeta) > 0$ and $b = b(\zeta)$ such that for $\mathscr{P}$, η, σ, and χ as in Theorem 3.1.1 and $n(\Delta) > 0$,*

$$\begin{aligned} &:\mathscr{P}(\phi_\kappa(x)): - \eta:(\phi_\kappa(x) - h(x))^2: - \ln |\chi'_{\sigma(\Delta)}(\bar{\phi}(\Delta))| \\ &\qquad \geqslant a\lambda^{-1} - b \ln^2 \kappa - \zeta:\delta\phi_\kappa(x)^2: - \ln(n(\Delta)!) \end{aligned} \tag{3.1.10}$$

Proof. We modify the proof of Theorem 3.1.1. As before, it is sufficient to consider in detail $h(\Delta) = -\xi_+$.

Case 1. $h(\Delta) = -\xi_+$ and $|\phi_\kappa(x) - \xi_+| \geqslant \xi_+/3$. As in (3.1.5) there is a constant $a_1 > 0$, independent of λ, μ, η, κ such that

$$\tau \geqslant a_1(\phi_\kappa + \xi_+)^2 + \zeta(\delta\phi_\kappa)^2 - O(1) \ln^2\kappa. \tag{3.1.11}$$

We consider two subcases

$$\text{(i) } |\phi_\kappa(x) + \xi_+| \geqslant \xi_+/3, \qquad \text{(ii) } |\phi_\kappa(x) + \xi_+| \leqslant \xi_+/3.$$

In subcase (i),

$$\tau \geqslant a_1\xi_+^2/9 - O(1) \ln^2\kappa \geqslant a\lambda^{-1} - O(1) \ln^2\kappa. \tag{3.1.12}$$

In subcase (ii), either

$$\text{(a) } |\delta\phi_\kappa| \geqslant \xi_+/3 \qquad \text{or} \qquad \text{(b) } |\bar{\phi}| \geqslant \xi_+/3.$$

If (a) holds, then

$$\tau \geqslant \zeta \xi_+^2/9 - O(1) \ln^2 \kappa \geqslant a\lambda^{-1} - O(1) \ln^2 \kappa. \tag{3.1.13}$$

If (b) holds, then by Lemma 2.6.2, for $\theta < 1$,

$$\begin{aligned} -\ln | \chi'_{\sigma(\Delta)} | &\geqslant \theta \bar{\phi}^2 - O(1) - \ln(n!) \\ &\geqslant \theta \xi_+^2/9 - O(1) - \ln(n!) \end{aligned}$$

where $n = n(\Delta)$ is the order of the derivative of χ in Δ. Thus

$$\tau - \ln | \chi'_{\sigma(\Delta)} | \geqslant a\lambda^{-1} - O(1) - \ln(n!) \tag{3.1.14}$$

Case 2. $h(\Delta) = -\xi_+$, $| \phi_\kappa(x) - \xi_+ | \leqslant \xi_+/3$. Either

$$\text{(a) } \delta\phi_\kappa \geqslant \xi_+/3 \qquad \text{or} \qquad \text{(b) } \bar{\phi} \geqslant \xi_+/3.$$

Subcase (a) follows as in Theorem 3.1.1 and subcase (b) follows as in subcase (b) above. This completes the proof in all cases.

3.2. *The Vacuum Energy*

The main technical estimate of this chapter is the vacuum energy bound. With Σ and $n(\Delta)$ given let X' be the union of the lattice squares Δ in which χ_Δ is differentiated, and thus for which $n(\Delta) \neq 0$. For any set Y let

$$Y_c = Y \cap (X^0 \cup X') \cap \Lambda, \qquad Y_d = (Y \sim Y_c) \cap \Lambda$$

denote convergent and weakly divergent regions in the vacuum energy estimates.

THEOREM 3.2.1. *Let $\eta > 0$ be sufficiently small. There are strictly positive constants $a = a(\eta)$, $b = b(\eta)$, and $c = c(\eta)$ with the following property. Let $Q(Y)$ be defined by* (I.4.8). *Let $1 \leqslant p \leqslant 1 + \eta/30$. Then*

$$\begin{aligned} &\int e^{-pQ(Y)} | \chi'_{\Sigma \cap Y} |^p \, d\psi_1(s) \\ &\qquad \leqslant \prod_{\Delta \subset Y} (n(\Delta)!) \exp[-aF_1 - b\lambda^{-1/2} | Y_c | + c\lambda^{1/2} | Y_d |]. \end{aligned}$$

The idea of the proof is to express the high momentum tail $\delta V_\kappa \equiv V - V_\kappa$ in terms of ψ, and then to apply to $V_\kappa \equiv V(\phi_\kappa)$ the Wick bounds of Theorems 3.1.1–3.1.2. The resulting bound is useful, but too crude, and is refined by including effects of first-order perturbation theory in Y_d. In this part of the proof, the perturbation V is expressed in terms of ψ. First we obtain detailed properties of the coefficients of ψ in V, δV_κ. Writing

$$V(\Delta) = \sum_{j=0}^{4} \int_{\Delta} k_j(x) :\psi^j(x):_1 \, dx, \tag{3.2.1}$$

where: $:_1$ denotes a Wick ordering mass squared of 1, we compute

$$\begin{aligned} &k_4 = \lambda, \qquad k_3 = 4\lambda g, \qquad k_2 = \tfrac{3}{4}(8\lambda g^2 - 1), \\ &k_1 = \tfrac{1}{2} g(8\lambda g^2 - 1) - \mu - \eta(g - h), \\ &k_0 = g^2(\lambda g^2 - \tfrac{1}{4}) - E_c - \mu g - \tfrac{1}{2}\eta(g - h)^2. \end{aligned}$$

LEMMA 3.2.2. *The following bounds are satisfied:*

$$\begin{aligned} &| k_j(x)| \leqslant O(\lambda^{1/2}) + O(1) \, | g(x) - h(x)|, \qquad j > 0 \\ &| k_0(x)| \leqslant O(\lambda) + O(1) \, | g(x) - h(x)|^2 \\ &| k_j(x)| \leqslant O(\lambda^{1/2}) \qquad \text{on } X. \end{aligned}$$

Proof. Note by (I.2.6) and (I.4.2),

$$\| g \|_\infty \leqslant \| h \|_\infty \leqslant O(\lambda^{-1/2}). \tag{3.2.2}$$

Thus $| k_4(x)|$ and $| k_3(x)|$ are $O(\lambda^{1/2})$. We next show

$$| 8\lambda g(x)^2 - 1 | \leqslant O(\lambda) + O(\lambda^{1/2}) \, | g(x) - h(x)|. \tag{3.2.3}$$

In fact, by (I.2.6) and $0 \leqslant \mu \leqslant \lambda^2$,

$$8\lambda h(x)^2 = 1 + O(\lambda).$$

So by (3.2.2)

$$8\lambda g^2 - 1 = 8\lambda(g - h)(g + h) + O(\lambda) \leqslant O(\lambda^{1/2}) \, | g - h | + O(\lambda),$$

proving (3.2.3). Using (3.2.2)–(3.2.3) we bound $| k_2(x)|$, $| k_1(x)|$ as desired. By (I.2.6),

$$E_c + (64\lambda)^{-1} = O(\mu\lambda^{-1/2}) \leqslant O(\lambda).$$

Thus

$$g^2(\lambda g^2 - \tfrac{1}{4}) - E_c = (64\lambda)^{-1}(8\lambda g^2 - 1)^2 + O(\lambda),$$

and using (3.2.3) we bound $| k_0(x)|$, to complete the proof.

For a lattice square $\Delta \subset \Lambda$, the basic polynomial for the Wick bound estimates is,

$$\begin{aligned} W(\Delta) &= V(\Delta) + \tfrac{1}{2}(1-\eta)\int_\Delta :\psi^2:_1 \, dx \\ &= \int_\Delta :\mathscr{P}(\phi):_1 \, dx - \tfrac{1}{2}\eta \int_\Delta :(\phi - h)^2:_1 \, dx \end{aligned} \tag{3.2.4}$$

and we define $W_\kappa(\Delta)$ by the substitution $\phi \to \phi_\kappa$. Let $\delta W_\kappa(\Delta) = W(\Delta) - W_\kappa(\Delta)$. Then

$$\delta W_\kappa(\Delta) = \sum_{j=1}^{4} \int_\Delta \hat{k}_j(x)(:\psi^j(x): - :\psi_\kappa{}^j(x):)\, dx + \sum_{j=1}^{3} \int_\Delta \delta\hat{k}_j(x, \kappa) :\psi_\kappa{}^j(x): d, \tag{3.2.5}$$

where

$$\begin{aligned} \delta\hat{k}_3 &= \lambda(g - g_\kappa) \qquad \delta\hat{k}_2 = 6\lambda(g^2 - g_\kappa{}^2) \\ \delta\hat{k}_1 &= \tfrac{1}{2}g(8\lambda g^2 - 1) - \tfrac{1}{2}g_\kappa(8\lambda g_\kappa{}^2 - 1) - \eta(g - g_\kappa) \\ \delta\hat{k}_0 &= g^2(\lambda g^2 - \tfrac{1}{4}) - g_\kappa{}^2(\lambda g_\kappa{}^2 - \tfrac{1}{4}) - \mu(g - g_\kappa) \\ &\quad - \tfrac{1}{2}\eta[(g-h)^2 - (g_\kappa - h)^2], \\ \hat{k}_j &= k_j, \qquad j \neq 2, \qquad \hat{k}_2 = 6\lambda g^2 - \tfrac{1}{4} - (\eta/2). \end{aligned}$$

LEMMA 3.2.3. *For each $p < \infty$ there is a constant $\epsilon = \epsilon(p) > 0$ such that the following bounds are satisfied*

$$\begin{gathered} |\hat{k}_j(x)| \leqslant O(1)(1 + |g - h|) \leqslant O(\lambda^{-1/2}) \\ \|\delta\hat{k}_j\|_{L_p(\Delta)} \leqslant O(\lambda^{-1}\kappa^\epsilon) \\ \delta\hat{k}_j = 0 \text{ on } X; \qquad |\hat{k}_j(x)| \leqslant O(1) \text{ on } X. \end{gathered}$$

Proof. The bounds on $\hat{k}_j$ are a consequence of the preceding lemma while the L_p bound on $\delta\hat{k}_j$ follows from (2.1.4). Since $g = h$ on X by (I.4.5), $k_j = \text{const}$ on lattice squares $\Delta \subset X$ and so by (iv) of Section 2.1, $\delta\hat{k}_j = 0$ on X.

Recall that ω, $d\psi_\omega(s)$, and $C_\omega(s)$ are defined in (2.2.1)–(2.2.5). We require an L_q bound on some fractional derivative of $C_\omega(s)$ as follows from a convergent Neumann series expansion combined with [16, Propositions 7.3–7.4]. In fact,

$$C_\omega(s) = C + C(1-\omega)\, C_\omega(s), \tag{3.2.6}$$

where C belongs to the class $\mathscr{C}$ of [16] of covariance operators with fractional derivatives in L_q (see also [2]). After differentiating the left variable of $C_\omega(s)$ (i.e., C in (3.2.6)), we use the majorization $0 \leqslant 1 - \omega \leqslant 1$ and

$$0 \leqslant C_\omega(s) \leqslant (-\Delta + \omega^2)^{-1}$$

(pointwise as an inequality between kernels of operators) and the fact that $(-\Delta + \omega^2)^{-1}$, as an L_1 convolution operator, preserves L_q norms.

LEMMA 3.2.4. *Let ω, $d\psi_\omega$, and δW_κ be as above. There are positive constants $K(\omega)$ and δ with the following property. Let $\{m(\Delta):\Delta \subset Y \cap \Lambda\}$ be a set of nonnegative integers and let $\{\kappa(\Delta):\Delta \subset Y \cap \Lambda\}$ be a set of positive numbers with $\kappa(\Delta)^\delta \geqslant \lambda^{-1}$ for $\Delta \subset X^0$. Then*

$$\left| \int \left[\prod_{\Delta \subset Y \cap \Lambda} \delta W_{\kappa(\Delta)}(\Delta)^{m(\Delta)} \right] d\psi_\omega(s) \right| \leqslant \prod_{\Delta \subset Y \cap \Lambda} [m(\Delta)!^2 \, (K(\omega)\, \kappa(\Delta)^{-\delta})^{m(\Delta)}].$$

Proof. As in [13, 5, 16].

LEMMA 3.2.5. *There is a constant K such that for q a positive even integer and for Y_t a subset of X,*

$$\left| \int \prod_{\Delta \subset Y_t} V(\Delta)^q \, d\psi_1(s) \right| \leqslant ((q!)^2 \cdot (K\lambda)^{q/2})^{|Y_t|}. \tag{3.2.7}$$

Proof. For $x \in X$, $g = h$ by Lemma I.4.1. Thus by Lemma 3.2.2, the kernels $k_j(x)$ satisfy $\| k_j \|_\infty \leqslant O(\lambda^{1/2})$. Hence, the lemma follows from the standard bound

$$\prod_{\Delta \subset Y_t} \left(K_1{}^q (q!)^2 \sum_{j=1}^{4} \| k_j \|_\infty \right)$$

on (3.2.7).

Next we establish Theorem 3.2.1 with the factor $\lambda^{1/2} \mid Y_d \mid$ replaced by the weaker bound $\mid Y_d \mid$. To take advantage of the fact that the coefficients of V are $O(\lambda^{1/2})$ in X and, ultimately, to recover the improved bound $O(\lambda^{1/2}) \mid Y_d \mid$ of Theorem 3.2.1, we perturb V in each lattice square $\Delta \subset Y_d \subset X$. For this purpose we define

$$V(t, Y) = \sum_{\Delta \subset Y_d} t(\Delta)\, V(\Delta) + V(Y_c).$$

Let $Y_t = \{\Delta : t(\Delta) \neq 0\}$. Since $T_t \subset Y_d \subset X$, the previous lemma applies to this choice of Y_t.

LEMMA 3.2.6. *Under the hypothesis of Theorem* 3.2.1,

$$\int e^{-p(F(Y)+V(t,Y))} \mid \chi'_{\Sigma \cap Y} \mid^p d\psi_1(s)$$

$$\leqslant \left(\prod_{\Delta \subset Y \cap \Lambda} n(\Delta)! \right) \exp[-aF_1(Y) - b\lambda^{-1/2} \mid Y_c \mid + O(1) \mid Y_t \mid).$$

Proof. Let

$$W(t, Y) = V(t, Y) + \tfrac{1}{2}(1 - \eta) \int_{Y_t} :\psi^2:_1 \, dx.$$

Then with $\omega = 1 - p(1 - \eta)\chi_{Y_t}$,

$$\begin{aligned} &\int e^{-p(F(Y)+V(t,Y))} \mid \chi'_{\Sigma\cap Y} \mid^p d\psi_1(s) \\ &\quad = Z^{-1} \int e^{-p(F(Y)+W(t,Y))} \mid \chi'_{\Sigma\cap Y} \mid^p d\psi_\omega(s) \qquad (3.2.8) \\ &\quad \leqslant e^{-p(F_1+F_2)} Z^{-1} \left(\int e^{-q' p F_3} \, d\psi_\omega(s)\right)^{1/q'} \left(\int e^{-pq(F_4+W)} \mid \chi'_{\Sigma\cap Y} \mid^{pq} d\psi_\omega(s)\right)^{1/q}. \end{aligned}$$

Here q and q' are dual Hölder exponents. We choose

$$q = 1 + \frac{\eta_0}{10}, \qquad p < 1 + \frac{\eta}{10}$$

so that qp, $q^2p^2 < 1 + \eta/2$, and $pq' = 0(\eta_0^{-1})$. The F_3 integral is bounded by Proposition 2.5.1 and Z^{-1} is bounded by (2.2.4), giving

$$\begin{aligned} &\left| \int e^{-p(F+V(t,Y))} \mid \chi'_{\Sigma\cap Y} \mid^p d\psi_1(s) \right| \\ &\quad \leqslant e^{-(1-\lambda)F_1 - pF_2} e^{K|Y_t|} \left(\int e^{-pq(F_4+W-\ln|\chi'_{\Sigma\cup Y}|)} \, d\psi_\omega(s)\right)^{1/q} \end{aligned}$$

We bound the W-integral using the standard proof of the linear lower bound on the vacuum energy, see [13, 5]. The basic idea is: Consider the subset of paths for which

$$\begin{aligned} &W(t, \Delta) + t(\Delta)\, \zeta{:}\delta\phi_\kappa\,(\Delta){:} - \ln \mid \chi'_{\sigma(\Delta)})| \\ &\quad \leqslant \inf_\phi \, [W_\kappa(t, \Delta) + t(\Delta)\, \zeta{:}\delta\phi_\kappa^2(\Delta){:} - \ln \mid \chi'_{\sigma(\Delta)} \mid - 1]. \end{aligned}$$

On this set it is also true that

$$1 \leqslant | \, \delta W_\kappa(t, \Delta)| \leqslant \delta W_\kappa(t, \Delta)^m$$

for any positive even integer m. Thus

$$\begin{aligned} I(\kappa, \Delta) &\equiv \inf_\phi \, [W_\kappa(t, \Delta) + t(\Delta)\, \zeta{:}\delta\phi_\kappa^2(\Delta){:} - \ln \mid \chi'_{\sigma(\Delta)} \mid] \\ &\geqslant \inf_\phi \Big[t(\Delta) \Big({:}\mathscr{P}_\Delta(\phi_\kappa){:}_1 - \tfrac{1}{2}\eta \int_\Delta {:}(\phi_\kappa - h)^2{:} \, dx + \zeta \int_\Delta {:}\delta\phi_\kappa^2{:} \, dx \Big) \\ &\qquad - \ln \mid \chi'_{\sigma(\Delta)} \mid \Big] \\ &\geqslant \begin{cases} -O(\ln^2 \kappa), & n(\Delta) = 0 \\ a\lambda^{-1} - O(\ln^2 \kappa) - \ln(n(\Delta)!), & n(\Delta) > 0 \end{cases} \end{aligned}$$

by Theorems 3.1.1 and 3.1.2.

We choose $\kappa_i = 2^i$, $i = i(\Delta)$ a positive integer depending on Δ. Let $\sum_i'$ denote the sum over all such $i(\Delta)$, subject to the restriction $\kappa_\Delta^{i(\Delta)} \geqslant \lambda^{-1}$ if $\Delta \subset X^0$. This restriction is made to offset the $O(\lambda^{-1})$ divergence of the coefficients in X^0. Then

$$\int e^{-pq(F_4+W-\ln|x_\sigma'|)}\, d\psi_\omega(s)$$

$$\leqslant \sum_i{}' \int d\psi_\omega(s)\Big(\prod_{\Delta\subset Y_t} (\delta W_{\kappa_i(\Delta)}(t,\Delta)^{m_i(\Delta)}$$

$$\times \exp[-pq(I(2\kappa_{i(\Delta)},\Delta) - \zeta{:}\delta\phi_\kappa^2(\Delta){:})]\Big)\, e^{-pqF_4}$$

$$\leqslant \Big(\prod_\Delta n(\Delta)!\Big)\, e^{-a\lambda^{-1}|X'\cap Y|} \sum_i{}' \int d\psi_\omega(s)$$

$$\times \prod_{\Delta\subset Y_t} (\delta W_{\kappa_{i(\Delta)}}^{m_i(\Delta)} \exp[(pq(O \ln^2 \kappa_{i(\Delta)} + \zeta{:}\delta\phi^2_{\kappa_{i(\Delta)}}(\Delta)])\, e^{-pqF_4}.$$

We separate $e^{pq(\zeta:\delta\phi^2:-F_4)}$ from the rest of the integrand with a Hölder inequality and bound it by Proposition 2.5.2. This gives as an upper bound for (3.2.8),

$$e^{(-(1-\lambda)+K\zeta)F_1(Y)} e^{K|Y_t|} \Big(\prod_\Delta n(\Delta)!\Big)\, e^{-b\lambda^{-1}|X'\cap Y|}$$

$$\times \sum_i{}' \Big(\int d\psi_\omega(s) \prod_\Delta \delta W_{\kappa_{i(\Delta)}}^{q' m_i(\Delta)}\Big)^{1/q'} e^{O\ln^2\kappa_{i(\Delta)}}.$$

Here if $i(\Delta)$ assumes its minimum value in Δ, then $m_{i(\Delta)}$ must be zero. Otherwise we choose

$$m_{i(\Delta)} = \kappa_{i(\Delta)}^{\delta/2}/q'.$$

We choose $\zeta < \pi^2/4$ and small enough so that the coefficient of F_1 above is bounded by $-\frac{3}{4}$. The δW integral is estimated by Lemma 3.2.4, and the factorials are bounded by Stirling's formula, leading to the upper bound for (3.2.8),

$$\exp(-\tfrac{3}{4}F_1(Y))\exp(K\,|\,Y_t\,| - b\lambda^{-1}\,|\,X'\cap Y\,|)$$

$$\times \Big(\prod_\Delta n(\Delta)!\Big) \sum_i{}' \prod_{\Delta\subset Y_t} \exp(-\kappa_{i(\Delta)}^{\delta/2}/q' + O\ln^2 \kappa_{i(\Delta)}),$$

where it is understood that $\kappa_{i(\Delta)}^{\delta/2}/q'$ is omitted from the exponent when $i(\Delta)$ assumes its minimum value. Recall that for $\Delta\subset X^0$, $\kappa_{i(\Delta)}^{\delta} \geqslant \lambda^{1/2}$, so that

$$\sum_i{}' \cdots \leqslant O(1)^{|Y_t|}\exp(O\ln^2(\lambda^{-1})\,|\,Y\cap X^0\,|)$$

$$\leqslant O(1)^{|Y_t|}\exp(\lambda^{1/2}F_1(Y))$$

using Proposition 2.4.1. Substitution into the above upper bound, another use of Proposition 2.4.1, and the definitions of Y_c and Y_d complete the proof.

Proof of Theorem 3.2.1. We use the identity

$$e^{-pV(\Delta)} = 1 - pV(\Delta) \int_0^1 e^{-pt(\Delta)V(\Delta)}\, dt.$$

Thus with $t \equiv \{t(\Delta): \Delta \subset Y_t\}$ we have

$$\int e^{-pQ(Y)} \,|\, \chi'_{\Sigma\cap Y} \,|^p\, d\psi_1(s)$$

$$\leqslant \sum_{Y_t \subset Y_d} \int_0^1 \int \Big| \prod_{\Delta\subset Y_d} pV(\Delta) \Big| \, e^{-p[F(Y)+V(t,Y)]} \,|\, \chi'_{\Sigma\cap Y} \,|^p\, d\psi_1(s)\, ds\, dt$$

$$\leqslant \sum_{Y_t\subset Y_d} \int \Big| \prod_{\Delta\subset Y_t} pV(\Delta) \Big|^{q'} d\psi_1(q)^{1/q'} \sup_t \Big| \int e^{-pq[F(Y)+V(t,Y)]} \,|\, \chi'_{\Sigma\cap Y} \,|^{pq}\, d\psi_1(s) \Big|^{1/q}. \tag{3.2.9}$$

Here $q \simeq 1 + \eta_0/30$ so that $pq \leqslant 1 + \eta/10$ and $q' = O(\eta_0^{-1})$ is an even integer. Since $Y_d \subset X$, using Lemmas 3.2.5 and 3.2.6, we bound (3.2.9) by

$$\sum_{Y_t\subset Y_d} (K(\eta_0)\, \lambda^{1/2})^{|Y_t|} \Big(\prod_\Delta n(\Delta)!\Big) \exp(-aF_1 - b\lambda^{-1/2} \,|\, Y_c \,|)$$

$$= \Big(\prod_\Delta n(\Delta)!\Big)\Big(\prod_{\Delta\subset Y_d} (1 + K(\eta_0)\, \lambda^{1/2})\Big) \exp(-aF_1 - b\lambda^{-1/2} \,|\, Y_c \,|)$$

$$\leqslant \Big(\prod_\Delta n(\Delta)!\Big) \exp(-aF_1 - b\lambda^{-1/2} \,|\, Y_c \,| + c\lambda^{1/2} \,|\, Y_d \,|).$$

This completes the proof.

3.3. *Deviation from the Gaussian*

To prove convergence of the cluster expansion we need a bound of the form (1.1.5). Consider a monomial A of the form (I.6.1), localized in unit lattice squares Δ_i:

$$A = A(\psi) = \int \prod_i :\psi(x)^{m_i}: f_i(x)\, dx. \tag{3.3.1}$$

Recall that $|||\, A \,|||$ is defined by (I.6.2).

THEOREM 3.3.1. *There are positive constants b, c, d, and a constant K in* (I.6.2) *such that*

$$\left|\partial_s^{\Gamma}\int e^{-Q(Y)}\chi_\Sigma A\, d\psi_1(s)\right| \leqslant |||\, A\, |||\, e^{-dl|\Gamma|}e^{-b\lambda^{-1/2}|X^0\cap Y|}e^{c\lambda^{1/2}|Y\cap\Lambda|} \tag{3.3.2}$$

for all λ *sufficiently small and for any consistent choice of* Γ, Σ, *and* Y.

Proof. Let $\mathscr{P}(\Gamma)$ be the set of partitions of the set Γ of l-lattice lines. As in [16], the left side of (3.3.2) equals

$$\sum_{\pi\in\mathscr{P}(\Gamma)}\int\left[\prod_{\gamma\in\pi}\tfrac{1}{2}(\partial_s{}^{\gamma}C(s))\cdot\Delta_\psi\right]e^{-Q(Y)}\chi_\Sigma A\, d\psi_1(s). \tag{3.3.3}$$

We write $\partial_s{}^{\gamma}C(s)$ as a sum of its l-lattice localizations $j=(j_1, j_2)\in lZ^4$,

$$\partial_s{}^{\gamma}C(s)=\sum_{j_\gamma\in lZ^4}\partial_s{}^{\gamma}C(s, j_\gamma)\equiv\sum_{j_\gamma\in lZ^4}\chi_{\Delta_{j_{\gamma,1}}}\partial_s{}^{\gamma}C(s)\,\chi_{\Delta_{j_{\gamma,2}}}\,.$$

Here $\Delta_{j_{\gamma,1}}$ and $\Delta_{j_{\gamma,2}}$ are l-lattice squares. Consider a term, indexed by π and $J\equiv\{j_\gamma\}_{\gamma\in\pi}$ in the resulting sum. Each such term has the form

$$\sum{}'\int A'W(\pi, J)\, e^{-Q(Y)}\chi_\Sigma'\, d\psi_1(s) \tag{3.3.4}$$

where $\sum'$ ranges over possible derivatives of the various linear factors of A, of the factors $\chi_{o(\Delta)}$ of χ_Σ and of $e^{-Q(Y)}$. Hölder's inequality yields as a bound on (3.3.4)

$$\sum{}'\,\|\,A'W(\pi, J)\|_{p'}\,\|\,\chi_\Sigma' e^{-Q(Y)}\,\|_p \tag{3.3.5}$$

where p and p' are dual indices. Using Theorem 3.2.1 for $p\simeq 1+\eta/30$ and for p' an even integer, the second factor is bounded by

$$\left(\prod_\Delta n(\Delta)!\right)\exp[-aF_1-b\lambda^{-1/2}\,|\,Y_c\,|+c\lambda^{1/2}\,|\,Y_d\,|]. \tag{3.3.6}$$

To bound the first factor in (3.3.5), we need to estimate the derivatives of $C(s)$ as in Proposition 8.1 of [16]. Let

$$d(j,\gamma)=\max\{\text{dist}(j_1, b)+\text{dist}(j_2, b):b\in\gamma\}$$

and let

$$\|f(x, y)\|_{L^{x,y}_{q_1,q_2}}=\left(\int\left|\int|f(x, y)|^{q_2}\,dy\right|^{q_1/q_2}dx\right)^{1/q_1}.$$

Let γ^* be a shortest path passing through $b\in\gamma$ and let $|\,\gamma^*\,|$ be its length.

LEMMA 3.3.2. *Let* $1 \leqslant q_1,\ q_2 \leqslant q_0 < \infty$. There are constants $K_6(q_0, \gamma)$, $K_7(q_0)$, and $\delta > 0$ such that for l sufficiently large (λ sufficiently small)

$$\| \partial_s{}^\gamma C(s) \|_{L_{q_1 q_2}(\Delta_{j_1} \times \Delta_{j_2})} \leqslant l^2 K_6(q_0, \gamma) \exp(-\delta(d(j, \gamma) + | \gamma^* |)) \tag{3.3.7}$$

and

$$\sum_{\pi \in \mathscr{P}(\Gamma)} \prod_{\gamma \in \pi} K_6(q_0, \gamma) \leqslant e^{K_7(q_0)|\Gamma|}. \tag{3.3.8}$$

Proof. Follow [16, 27], except for modifications arising from the use of a large-scale distance and a fixed bare mass equal to one. We can extract a factor $e^{-\delta|\gamma^*|}$ from [16, (8.8)] since $| \gamma^* | \leqslant | l |$ for all $l \in \mathscr{L}(\gamma)$, in the notation of [16].

LEMMA 3.3.3. *Let p, p′ be given. There is a* $K = K(p')$ in (I.6.2), *a* q_0 *in Lemma* 3.3.2, *and a* $\delta' > 0$ *such that for all* λ *sufficiently small,*

$$\| A' W(\pi, J) \|_{p'} \leqslant ||| A ||| \, e^{-\delta' l |\Gamma|} e^{l | Y \cap X' |} \prod_{\gamma \in \pi} K_6(q_0, \gamma) \, e^{-\delta' d(j_\gamma, \gamma)}. \tag{3.3.9}$$

Proof of Theorem 3.3.1. Since $Y \cap X' \subset Y_c$, the factor $\exp(l | Y \cap X' |)$ is controlled by the factor $\exp(-b\lambda^{-1/2} | Y_c |)$ in (3.3.6). We substitute (3.3.6) and (3.3.9) in (3.3.5). The sum $\sum'$ over derivatives in (3.3.4) and the sum over localizations J is bounded by [16, Lemmas 10.1–10.2]. The sum over $\pi \in \mathscr{P}(\Gamma)$ is dominated by (3.3.8).

Proof of Lemma 3.3.3. We modify [16, p. 239]. The L_{q_1, q_2} norms of (3.3.7), $1 \leqslant q_1, q_2 \leqslant q_0$ allow the use of [16, Lemma 9.2] even for l-lattice squares Δ_{j_ν} of area $l^2 \neq 1$. The other new issues are (a) the factor $l^2 = | \log \lambda |^{1/2}$ in (3.3.7), arising from the use of large lattice squares, (b) the large coefficients of $Q(Y)$ in case $\Delta_{j_\nu} \subset X^0$, and (c) another factor l^4, which is equal to the number of unit lattice square localizations for $\partial^\gamma C$, with the l-lattice localization j_γ held fixed.

The minimum nonzero value taken on by $| \gamma^* |$ and $d(j, \gamma)$ is l. Since $l^6 e^{-\delta l/2} \to 0$ as $\lambda \to 0$, the l^6 factor may be replaced by a small factor $e^{-\delta' l |\gamma|}$ if $| \gamma^* |$ or $d(j, \gamma)$ are nonzero, by a decrease in δ. Now suppose $| \gamma^* | = 0 = d(j, \gamma)$. Then $| \gamma | \leqslant 4$. Also $\Delta_j \subset X$ and so a derivative of e^{-Q} localized in Δ_j has only contributions from V(F_3 and F_4 do not contribute). For $\Delta_j \subset X$, the coefficient of V is $O(\lambda^{1/2})$ by Lemma 3.2.2 and since

$$l^6 \lambda^{1/2} \leqslant e^{-4\delta' l} \leqslant e^{-\delta' l |\gamma|}$$

for small λ, we obtain the desired factor $e^{-\delta' l |\gamma|}$ in this case. If the contraction is to $\chi_{\sigma(\Delta)}$, for Δ a unit lattice square in Δ_j, then $\Delta \subset Y \cap X'$ and we use the inequality

$$l^6 \leqslant e^{-\delta l |\gamma|} e^{l |\Delta|},$$

i.e., we use the contribution of $\Delta \subset Y \cap X'$ to the factor $\exp(l | Y \cap X' |)$ in (3.3.9).

For differentiation of a linear factor of A, we use the factor e^l from $e^{KlN(A)}$ in (I.6.2) so that

$$l^6 \leqslant e^{-\delta' l |\gamma|} e^l.$$

This shows that l^6 is offset a factor by $e^{-\delta l|\gamma|}$ in all cases.

Next we consider the $O(\lambda^{-1})$ coefficient resulting from differentiating e^{-Q} in a square $\Delta \subset X^0$. In this case $d(j, \gamma) \geqslant L = |\log \lambda|^2$ and

$$O(\lambda^{-1}) e^{-\delta d(j,\gamma)} \leqslant {}^{-\delta d(j,\gamma)/2}.$$

This controls the modifications (a)–(c) mentioned above.

Let $M(\Delta)$ be the number of contractions in a unit lattice square Δ. As in [16],

$$\| A' W(\pi, J) \|_{p'} \leqslant \prod_i \| f_i \|_{L_2} \prod_\Delta [p'(N(\Delta) + 4M(\Delta))]!^{1/p'} e^{O(1)(N(\Delta)+M(\Delta))}$$

$$\times \prod_{\gamma \in \pi} (K_6 e^{-\delta'(d(j,\gamma)+|\gamma|)}) e^{l|Y \cap X'|} e^{N(A)}. \tag{3.3.10}$$

To complete the proof, we must bound the factorials. We use the inequality,

$$(p'(N + 4M))!^{1/p'} \leqslant e^{\ln p'(N+4M)} (N + 4M)! \leqslant e^{\ln p'(N+4M)} N! (4M)^{4M} N^{4M}.$$

The factor $e^{\ln p' N}$ is bounded by $e^{K(p')lN}$ of (I.6.2). By the proof of [16, Lemma 10.2],

$$\exp(M^{3/2}) \leqslant e^{O|\Gamma|} e^{O\Sigma_\gamma d(j,\gamma)}$$

Thus the factor $(4M)^{4M}$ is harmless. The $N!$ reappears in $|||\, A\, |||$, and so we need only analyze N^{4M}. First suppose $N \leqslant M^{4/3}$. Then

$$N^{4M} \leqslant \exp(4M \ln(M^{4/3})) = \exp((16/3)\, M \ln M) \leqslant O(1) \exp(\delta' M^{3/2}).$$

Next suppose $N \geqslant M^{4/3}$. Then $M \leqslant N^{3/4}$ and

$$N^{4M} \leqslant \exp(4N^{3/4} \ln N) \leqslant O(1)\, e^N.$$

Thus, the factorials in (3.3.10) can be bounded in all cases and the proof of Lemma 3.3.3 is complete.

4. Convergence of the Expansion

4.1. *The "Inside" Sum*

In this section we prove Proposition 1.2.1. Write

$$\phi - \xi_+ = (\phi - g) + (g - \xi_+) = \psi + (g - \xi_+). \tag{4.1.1}$$

Since

$$|g(x) - \xi_+| \leqslant O(\lambda^{-1/2}),$$

we can express $A(\phi - \xi_+) = A(\psi + (g - \xi_+))$ as a finite sum of terms C_j, each of the form (3.3.1), with

$$\sum |||C_j||| \leqslant \lambda^{-N/2} |||A|||.$$

Apply Theorem 3.3.1 in each connected component Y_i of Y to give

$$|R''(A, Y, \Sigma, \Gamma)| \leqslant |||A||| \lambda^{-N/2} e^{-dl|\Gamma^c|} e^{-b\lambda^{-1/2}|X^0 \cap Y|} e^{c\lambda^{1/2}|Y \cap \Lambda|}. \tag{4.1.2}$$

Each l-lattice square of $X \cap Y$ which does not meet $\mathscr{L}(A)$ must contain at least one derivative line from Γ^c, since otherwise the square is a component which contributes to the ratio of partition functions in (I.7.11) and not to Y. Thus we have a "density of derivative lines" in $X \cap Y$, as reflected in the inequality

$$2l^2 |\Gamma^c| \geqslant |X \cap Y| - Nl^2. \tag{4.1.3}$$

Hence

$$|Y \cap \Lambda| = |Y \cap X^0| + |Y \cap X| \leqslant |Y \cap X^0| + 2l^2 |\Gamma^c| + Nl^2. \tag{4.1.4}$$

We combine (4.1.2) and (4.1.4) to yield

$$|R(A, Y, \partial Y^\pm)| \leqslant \lambda^{-N/2} |||A||| \sum_{\{\Sigma \cap Y, \Gamma\}} \exp[-2\delta l^{-1} |Y \cap \Lambda| - b(2\lambda)^{-1/2} |X^0 \cap Y|], \tag{4.1.5}$$

where $\delta > 0$ is independent of λ. In (4.1.5) we remove the constraint on the Γ sum, and we increase K in (I.6.2) in order to bound the Nl^2 term from (4.1.4).

Since there are at most $4|Y \cap \Lambda|/l^2$ l-lattice lines in $\mathscr{B}(\Sigma, Y)$ and hence at most $2^{4|Y \cap \Lambda|/l^2}$ terms in the sum over Γ,

$$|R(A, Y, \partial Y^\pm)| \leqslant \lambda^{-N/2} |||A||| \sum_{\{\Sigma \cap Y\}} \exp[-\delta l^{-1} |Y \cap \Lambda| - b(2\lambda)^{-1/2} |X^0 \cap Y|]. \tag{4.1.6}$$

The sum over $\Sigma \cap Y$ is bounded by the observation that $\Sigma \upharpoonright X^\pm = \pm$, and since the same is true for any l-lattice square bordering $X^\pm$, thus $\Sigma \cap Y$ is uniquely specified by $X^0 \cap Y$ and $\Sigma \cap (X^0 \cap Y)$. With $|X^0 \cap Y| = r$ fixed, there are 2^r choices of $\Sigma \cap X^0 \cap Y$ and at most

$$\binom{|Y \cap \Lambda|}{r}$$

choices of $X_0 \cap Y$. Thus

$$\sum_{\Sigma \cap Y} \exp[-b(2\lambda)^{-1/2} \mid X^0 \cap Y \mid] \leqslant \sum_{r=0}^{|Y \cap \Lambda|} \binom{|Y \cap \Lambda|}{r} (2e^{-b(2\lambda)^{-1/2}})^r$$

$$= (1 + 2e^{-b(2\lambda)^{-1/2}})^{|Y \cap \Lambda|} \leqslant \exp(\lambda \mid Y \cap \Lambda \mid).$$

Substitution in (4.1.6) completes the proof in the case $\partial Y^- \neq \varnothing$.

Suppose $\partial Y^- = \varnothing$. The unique $\Sigma \cap Y$ for which $X^0 \cap Y = \varnothing$ is $\sigma \equiv 1$ (and hence $g \equiv \xi_+$). Thus $g - \xi_+$ does not appear in (4.1.1) and $\lambda^{-N/2}$ does not appear in (4.1.2). For the remaining $\Sigma \cap Y$'s, $| X^0 \cap Y | \geqslant 1$ and an extra convergence factor $\exp[-b(2\lambda)^{-1/2}]$ results.

4.2. *Ratios of Partition Functions*

In this section we prove a generalization of Proposition 1.2.2 and extend this bound to complex λ in a sector $| \operatorname{Im} \lambda | \leqslant \epsilon \operatorname{Re} \lambda$. For sets A and B, let $A \Delta B = (A \cup B) \sim (A \cap B)$ be the symmetric difference. For pairs $N_i = \{N_i^+, N_i^-\}$ let

$$N_{\text{plus}} \equiv \{N^+ \cup N^-, \varnothing\}. \tag{4.2.1}$$

It is always assumed that $N(\Gamma^\pm) \subset N^\pm$ and $\partial Y \subset \Gamma$ in the expression $Z(Y, N, \Gamma)$.

THEOREM 4.2.1. *Suppose $N_2^- = \varnothing$. There are constants c and ϵ such that for*

$$| \operatorname{Im} \lambda | \leqslant \epsilon \operatorname{Re} \lambda \leqslant \epsilon^2, \qquad \mu = 0, \tag{4.2.2}$$

as well as for

$$0 \leqslant \mu \leqslant \lambda^2 \leqslant \epsilon^2,$$

we have

$$| Z(Y, N_1, \Gamma_1) Z(Y, N_2, \Gamma_2)^{-1} | \leqslant \exp[c \mid \lambda \mid^{1/2}(| N_{1\text{plus}} \Delta N_2 | + l^2 \mid \Gamma_1 \Delta \Gamma_2 \mid]. \tag{4.2.3}$$

If $N_1^- \neq \varnothing \neq N_1^+$, the right side of (4.2.3) *can be multiplied by* $\exp(-b \mid \lambda \mid^{-1/2})$, *for some $b > 0$.*

Remark. For $N_2 = \Gamma_2 = \Gamma_1 = \varnothing$ and $N_1^- = \Delta$, the theorem gives $\langle \chi_-(\Delta) \rangle \leqslant \exp(O(- \mid \lambda \mid^{-1/2}))$, which implies symmetry breaking.

LEMMA 4.2.2. *Let Δ^l be an l-lattice square. Let $N^l = \{N(\partial \Delta^l), \varnothing\}$ and $\Gamma^l = \partial \Delta^l$. There is a constant c such that for λ as in Theorem* 4.2.1

$$| Z(\Delta^l, N^l, \Gamma^l)^{\pm 1} | \leqslant \exp[c \lambda^{1/2} l^2].$$

Proof. By definition

$$Z(\Delta^l, N^l, \Gamma^l) = \int e^{-V(\Delta^l)} \prod_{\Delta\subset\Delta^l} \chi_+(\Delta)\, d\psi_\Gamma$$
$$= 1 + \int (e^{-V(\Delta^l)} - 1)\Big(\prod_{\Delta\subset\Delta^l} \chi_+(\Delta)\Big)\, d\psi_\Gamma + \int\Big(\prod_{\Delta\subset\Delta^l}\chi_+(\Delta) - 1\Big)\, d\psi_\Gamma$$

where V is given by (1.1.2). Here we have used the fact that $g \equiv h \equiv \xi_+$ and $F = 0$. By the Hölder inequality, used as in (3.2.9), the second term is bounded by $O(|\lambda|^{1/2} l^2)$. To bound the third term, we note that

$$0 \leqslant 1 - \prod_{\Delta\subset\Delta^l} \chi_+(\Delta) \leqslant \sum_{\Delta\subset\Delta^l} \chi_-(\Delta).$$

Since

$$\chi_-(\Delta) = -\pi^{-1/2} \int_{-\infty}^{-\xi_+} e^{-(\psi(\Delta)-a)^2}\, da,$$

a Schwarz inequality gives

$$\int \chi_-(\Delta)\, d\psi_\Gamma \leqslant \Big(\int e^{-2\psi(\Delta)^2}\, d\psi_\Gamma\Big)^{1/2} e^{-O(|\lambda|^{-1})} = e^{-O(|\lambda|^{-1})},$$

to complete the proof.

Proof of Proposition 1.2.2. We assume Theorem 4.2.1 and use it together with Lemma 4.2.2 to bound each factor on the right side of the following inequalities:

$$\frac{Z(\sim Y, N^+, N^-, \partial Y)}{Z(R^2, \varnothing, \varnothing, \varnothing)} = \frac{Z(\sim Y, N^+, N^-, \partial Y)}{Z(\sim Y, N_{\text{plus}}, \partial Y)} \frac{Z(\sim Y, N_{\text{plus}}, \partial Y)}{Z(R^2, \varnothing, \varnothing, \varnothing)},$$
$$\leqslant \exp(2c\,|\lambda|^{1/2}\,|N^-| + |\Gamma|)\, \frac{Z(\sim Y, N_{\text{plus}}, \partial Y)}{Z(R^2, \varnothing, \varnothing, \varnothing)}.$$

Since

$$Z(\sim Y, N_{\text{plus}}, \partial Y) = \frac{Z(R^2, N_{\text{plus}} \cup (Y \cap \Lambda), (\Gamma \cup \mathscr{B}(\Lambda \cap Y)))}{\prod_{\Delta^l\subset Y\cap\Lambda} Z(\Delta^l, N^l, \Gamma^l)},$$
$$\leqslant Z(R^2, \varnothing, \varnothing, \varnothing) \exp 2c\,|\lambda|^{1/2} (|\Lambda \cap Y| + |N| + l^2\,|\Gamma|),$$

the desired inequality (1.2.2) follows (with a new choice of c). If $N^- \neq \varnothing$, then by Theorem 4.2.1, the right side can be multiplied by $\exp(-b\,|\lambda|^{-1/2})$. Because $\sim Y$ is unbounded, and thus meets the $+$ region $\sim\Lambda$, the restriction $N_1^+ \neq \varnothing$ is not required, as we will see below.

Next we establish a special case of Theorem 4.2.1.

LEMMA 4.2.3. *For $0 \leqslant \mu \leqslant \lambda^2$ and λ sufficiently small positive,*

$$Z(Y, N, \Gamma) Z(Y, N_{\text{plus}}, \Gamma)^{-1} \leqslant 1.$$

Proof. Since Z factors across ∂Y, we assume without loss of generality that Y is connected. As in Section I.7, we introduce the approximate Dirichlet measures

$$d\phi_{Y\cap\Lambda,g} = \frac{\exp(-\int \delta m^2(x) :(\phi(x) - g(x))^2 dx)\, d\phi_{Y\cap\Lambda}}{\int \exp(-\int \delta m^2(x) :\psi(x)^2: dx)\, d\psi_1}, \tag{4.2.4}$$

where $g = g_\Sigma$ with Σ compatible with N and where $\delta m^2(\cdot)$ has support near Γ. Let

$$Z(Y, g, \mu) = \int d\phi_{Y\cap\Lambda,g}$$

and

$$E_{g,\mu}(A(\phi)) = Z(Y, g, \mu)^{-1} \int A\, d\phi_{Y\cap\Lambda,g}.$$

Let $g' = -g$, $g'' \equiv \xi_+$. Note that

$$|g(x)| = |g'(x)| \leqslant g''(x) \tag{4.2.5}$$

for all x, with equality for $x \in \operatorname{suppt} \delta m^2$, by Lemma I.4.1.

We assert that

$$Z(Y, g, \mu) \leqslant Z(Y, g'', \mu). \tag{4.2.6}$$

In fact the change $g \to g''$ in the exponent in (4.2.4) affects only the linear term in g, $\int \delta m^2(x)\phi(x)g(x)\, dx$, by (4.2.5). The exponent of this linear term can be expanded as a power series. Each term in the corresponding expansion for $Z(Y, g, \mu)$ is a Schwinger function for the measure $d\phi_{Y\cap\Lambda}$. By Griffiths' first inequality, these Schwinger functions increase if the test functions are increased pointwise. By (4.2.5), $g \to g''$ increases the test functions, hence (4.2.6).

Because $\prod \chi_+$ is monotone increasing and $\prod \chi_-$ is monotone decreasing, the FKG inequalities imply

$$\begin{aligned} &E_{g,\mu}\left(\left[\prod_{\Delta\subset N^+} \chi_+(\phi(\Delta))\right]\left[\prod_{\Delta\subset N^-} \chi_-(\phi(\Delta))\right]\right) \\ &\quad \leqslant E_{g,\mu}\left(\prod_{\Delta\subset N^+} \chi_+(\phi(\Delta))\right) E_{g,\mu}\left(\prod_{\Delta\subset N^-} \chi_-(\phi(\Delta))\right). \end{aligned} \tag{4.2.7}$$

Differentiation with respect to μ and another application of the FKG inequalities shows that $E_{g,\mu}(\prod \chi_+)$ is monotone increasing in μ. This statement holds for

$\mu < 0$ as well as $\mu > 0$, and as above, the same reasoning shows that $E_{g,\mu}(\prod \chi_+)$ is increasing in g, as long as $|g|$ is held fixed on suppt δm^2. Combining these inequalities with the $\phi \to -\phi$, $\mu \to -\mu$ symmetry allows us to bound (4.2.7) as follows:

$$
\begin{aligned}
(4.2.7) &\leqslant E_{g^*,\mu}\Big(\prod_{\Delta \subset N^+} \chi_+(\phi(\Delta))\Big) E_{g,\mu}\Big(\prod_{\Delta \subset N^-} \chi_-(\phi(\Delta))\Big) \\
&= E_{g^*,\mu}\Big(\prod_{\Delta \subset N^+} \chi_+(\phi(\Delta))\Big) E_{g',-\mu}\Big(\prod_{\Delta \subset N^-} \chi_+(\phi(\Delta))\Big) \\
&\leqslant E_{g^*,\mu}\Big(\prod_{\Delta \subset N^+} \chi_+(\phi(\Delta))\Big) E_{g^*,\mu}\Big(\prod_{\Delta \subset N^-} \chi_+(\phi(\Delta))\Big).
\end{aligned}
$$

Again by FKG inequalities, the above quantity is bounded by

$$E_{g^*,\mu}\Big(\prod_{\Delta \subset N_{\text{plus}}} \chi_+(\phi(\Delta))\Big).$$

We multiply the resulting inequality by (4.2.6), and take a limit as $\delta m^2 \to \infty$, suppt $\delta m^2 \to \Gamma$ and the proof is complete.

Proof of Theorem 4.2.1. The proof could be based on a Neumann series to invert a contraction operator defined by Kirkwood–Salsburg equations. Instead, we take advantage of the triangular nature of the equations to give an inductive proof.

As a simple reduction, we begin by imposing Dirichlet conditions on the border of a large square S of diameter

$$r \geqslant 8(|\Lambda|^2 + |\Gamma_1| + |\Gamma_2|)$$

centered at the origin and containing Γ. Let us assume the theorem in case $\partial S \subset \Gamma$, and establish the general case. Using (I.5.6) to insert Dirichlet data on ∂S and Theorem 3.3.1 to estimate error terms, one can show that

$$|Z(Y, N, \Gamma_i \cup \partial S) - Z(Y, N, \Gamma_i)| \leqslant e^{-2(|\Lambda^2| + |\Gamma_1| + |\Gamma_2|)} \tag{4.2.8}$$

Here we use convergence from a long contraction distance $O(r)$ to dominate the volume divergence $\exp(O|\lambda|^{1/2}|\Lambda|)$. Furthermore, Lemma 4.2.2 combined with the special case of the theorem we are assuming implies

$$|Z(Y, N_2, \Gamma_2 \cup \partial S)^{-1}| \leqslant e^{O(|\lambda|^{1/2})|\Lambda|}.$$

Combined with (4.2.8), this yields

$$|Z(Y, N_2, \Gamma_2 \cup \partial S)^{-1} - Z(Y, N_2, \Gamma_2)^{-1}| \leqslant e^{-(|\Lambda|^2 + |\Gamma_1| + |\Gamma_2|)}$$

for λ small. The general case follows from these bounds and the identity

$$a^{-1}b = a^{-1}(b - B) + (a^{-1} - A^{-1})B + A^{-1}B,$$

where $a = Z(Y, N_2, \Gamma_2)$, $A = (Y, N_2, \Gamma_2 \cup \partial S)$, etc.

We now assume $\partial S \subset \Gamma_1 \cap \Gamma_2$, $\Gamma_1 \cup \Gamma_2 \subset S$, $Y \subset S$. We use two induction variables. The first induction is an increasing induction on $|Y|$; if $|Y| = \Delta^l$ there is nothing to prove. Thus we assume the theorem for all $Y_0 \subset Y$. The second induction is a decreasing induction on the variable

$$J = \min\{|\Gamma_1|, |\Gamma_2|\} + \min\{|N_1|, |N_2|\}. \tag{4.2.9}$$

If J assumes its maximum value then the ratio Z/Z in (4.2.3) equals or is bounded by one, proving the theorem in this case.

Before proceeding, we reduce to three separate cases:

Case 1. $\Gamma_1 = \Gamma_2$, $N_1^- = \varnothing = N_2^-$

Case 2. $N_1^+ = N_2^+$, $N_1^- = \varnothing = N_2^-$

Case 3. $\Gamma_1 = \Gamma_2$, $N_2 = N_{1\text{plus}}$.

The general case follows from these three cases by the factorization

$$\frac{Z(Y, N_1, \Gamma_1)}{Z(Y, N_2, \Gamma_2)} = \frac{Z(Y, N, \Gamma_1)}{Z(Y, N_{1\text{plus}}, \Gamma_1)} \frac{Z(Y, N_{1\text{plus}}, \Gamma_1)}{Z(Y, N_{1\text{plus}} \cup N_2, \Gamma_1)}$$
$$\times \frac{Z(Y, N_{1\text{plus}} \cup N_2, \Gamma_1)}{Z(Y, N_{1\text{plus}} \cup N_2, \Gamma_2)} \frac{Z(Y, N_{1\text{plus}} \cup N_2, \Gamma_2)}{Z(Y, N_2, \Gamma_2)}.$$

Consider first Case 1, and assume the theorem has been proved in Cases 1–3 for smaller Y and for larger J. Suppose $|N_1| \leqslant |N_2|$. (The case $|N_1| \geqslant |N_2|$ is similar.) Also assume $N_1 \neq N_2$, for otherwise (4.2.3) equals one. Choose $\Delta \in N_2 \sim N_1$ and define

$$N_3 = \{N_3^+, N_3^-\} = \{N_1, \Delta\}.$$

Then

$$\left|1 - \frac{Z(Y, N_1, \Gamma_1)}{Z(Y, N_{3\text{plus}}, \Gamma_1)}\right| = \left|\frac{Z(Y, N_{3\text{plus}}, \Gamma_1) - Z(Y, N_1, \Gamma_1)}{Z(Y, N_{3\text{plus}}, \Gamma_1)}\right|$$
$$= \left|\frac{Z(Y, N_3, \Gamma_1)}{Z(Y, N_{3\text{plus}}, \Gamma_1)}\right|. \tag{4.2.10}$$

This ratio is in Case 3, and since

$$|N_1| = \min\{|N_1|, |N_2|\} < \min\{|N_3|, |N_{3\text{plus}}|\} = |N_1| + 1,$$

the inductive hypothesis applies, and (4.2.10) is bounded by $\exp(-b\,|\lambda|^{-1/2}/2)$. Here we assume $N_1 \neq \varnothing$. However, if $N_1 = \varnothing$, then $\Gamma_1 \cap \Lambda = \varnothing$ also and so $Y \cap (R^2 \backslash \Lambda) \neq 0$. In this case, the restriction $N_1^+ \neq \varnothing$ is not required for the insertion of the factor $\exp(-b\,|\lambda|^{-1/2})$. Thus without restriction on N_1,

$$[Z(Y, N_1, \Gamma_1)Z(Y, N_1 \cup \Delta, \Gamma_1)^{-1}]^{\pm 1} \leqslant \exp(c\,|\lambda|^{1/2})$$

for some constant c. Using this bound and the induction hypothesis again, we have

$$\left|\frac{Z(Y, N_1, \Gamma_1)}{Z(Y, N_2, \Gamma_2)}\right| \leqslant \left|\frac{Z(Y, N_1, \Gamma_1)}{Z(Y, N_1 \cup \Delta, \Gamma_1)}\right| \left|\frac{Z(Y, N_1 \cup \Delta, \Gamma_1)}{Z(Y, N_2, \Gamma_2)}\right|$$
$$\leqslant \exp(c\,|\lambda|^{1/2}\,|N_1 \Delta N_2|),$$

completing the proof in this case.

Next we consider Case 2. This case involves the insertion of Dirichlet data, and the methods are related to those of [16]. For $b \subset \Gamma_2 \sim \Gamma_1$, we show that

$$\left|1 - \frac{Z(\Gamma_1)}{Z(\Gamma_1 \cup b)}\right| = \left|\frac{Z(\Gamma_1 \cup b) - Z(\Gamma_1)}{Z(\Gamma_1 \cup b)}\right| \leqslant O(|\lambda|^{1/2}\, l^2) \qquad (4.2.11)$$

so that

$$|Z(\Gamma_1)Z(\Gamma_2)^{-1}|^{\pm 1} \leqslant (1 + O\,|\lambda|^{1/2} l^2)\,|Z(\Gamma_1 \cup b)Z(\Gamma_2)^{-1}|^{\pm 1}$$
$$\leqslant \exp(c\,|\lambda|^{1/2} l^2\,|\Gamma_1 \Delta \Gamma_2|),$$

where we have used the induction hypothesis for the last inequality and the notation $Z(\Gamma) = Z(Y, N_1, \Gamma)$. We expand $Z(\Gamma_1 \cup b) - Z(\Gamma_1)$ as in (I.5.5). For each term $Z''(Y, \Sigma, \Gamma)$ of the expansion, let $W(\Gamma)$ be the connected component of $Y \sim \Gamma$ which contains b. Each term in this construction factors,

$$Z'' = Z''(Y \sim W, \Sigma \sim W, \Gamma \sim W)\, Z''(W, \Sigma \cap W, \Gamma \cap W).$$

As in Section I.7, the Γ and Σ sums also factor (with ∂W fixed). The partial sums are subject to constraints and for the outside region $Y \sim W$, the partial sum can be resummed to yield

$$Z(\Gamma_1 \cup b) - Z(\Gamma_1) = \sum_{\partial W^{\pm}} Z(Y \sim W, N_3, \Gamma_3) \sum{}' Z''(W, \Sigma \cap W, \Gamma \cap W) \qquad (4.2.12)$$

where

$$N_3^{\pm} = [N_1^{\pm} \cup N(\partial W^{\pm})] \sim W \qquad \Gamma_3 = (\Gamma_1 \sim W) \cup \partial W.$$

Also the sum $\sum'$ over $\Sigma \cap W$ and $\Gamma \cap W$ is subject to the constraint $(\Sigma \cap W)^{\pm} \supset N_1^{\pm} \cap W$, $\Gamma \supset (\Gamma_1 \cap W) \cup \partial W$, and $W = W(\Gamma)$.

As in the proof of Proposition 1.2.1 given in Section 4.1, we have

$$\sum{}' \mid Z''(W, \Sigma \cap W, \Gamma \cap W)\mid \leqslant c \mid \lambda \mid^{1/2} l^2 e^{-2\delta \mid W \mid / l}.$$

In fact, there is a density of derivative lines in $X \cap W$, as in (4.1.4) (with $N = N(A) = 0$), and further convergence is produced by the existence of at least one derivative $\partial/\partial s_b$ which produces either χ_Δ' and the convergence $\exp(-O \mid \lambda \mid^{-1/2})$ or a vertex and the convergence $O(\mid \lambda \mid^{1/2})$. By Lemma 4.2.2, the same bound holds if we multiply the right side by

$$\mid Z^*(W)\mid = \prod_{\Delta^l \subset W} \mid Z(\Delta^l, N^l, \partial \Delta^l)\mid.$$

Substitution in (4.2.12) gives

$$\left| 1 - \frac{Z(\Gamma_1)}{Z(\Gamma_1 \cup b)} \right|$$
$$\leqslant c \mid \lambda \mid^{1/2} l^2 \sum_{\partial W^\pm} e^{-2\delta \mid W \mid / l} \mid Z(Y \sim W, N_3, \Gamma_3)\, Z^*(W)\, Z(\Gamma_1 \cup b)^{-1} \mid.$$

We express $Z(Y \sim W, N_3, \Gamma_3)\, Z^*(W)$ as a single partition function $Z(Y, N_3 \cup W, \Gamma_3 \cup \mathscr{B}(W^*))$. Because $\mid \Gamma_1 \cup b \mid > \mid \Gamma_1 \mid$ and $\mid N_3 \cup W \mid \geqslant \mid N_1 \mid$, the induction hypothesis applies and bounds the ratio of Z's above by $\exp(O(\mid \lambda \mid^{1/2} \mid W \mid))$.

To control the combinatorics in the above sum over $\partial W^\pm$, we observe that since W is a connected union of l-lattice squares containing b, the choice of W with $\mid W \mid$ fixed can be made in at most $\exp(O(\mid W \mid / l^2))$ ways, while the choice of $\pm$ for the contours of ∂W can also be made in at most $\exp(O(\mid W \mid / l^2))$ ways. Substitution in the above inequality yields (4.2.11) and completes the proof of Case 2.

Finally we consider Case 3. If $N_1^- = \varnothing$ there is nothing to prove. If $N_1^+ = \varnothing$, the theorem follows from Lemma 4.2.3 for real couplings and from the $\phi \to -\phi$ symmetry for complex coupling (with $\mu = 0$). Thus we suppose $N_1^- \neq \varnothing \neq N_1^+$. (To establish the mass gap, this remaining portion of Case 3 is not required.) In addition we suppose without loss of generality that Y is connected, and since $Y \subset S$, that Y is bounded. It follows that ∂Y has unique outer contour consisting of a simple closed curve. This curve is either $+$ or $-$.

As in I.5.5, we expand $Z(Y, N_1, \Gamma_1)$ as a sum of terms $Z''(Y, \Sigma, \Gamma)$. Choose a distinguished lattice square $\Delta \subset Y$, and with Δ fixed we define

$$Z^{(\pm)} = \Sigma^{(\pm)} Z'',$$

where $\Sigma^{(\pm)}$ denotes the sum over all terms Z'' which have a closed $+$ (resp. $-$) contour enclosing Δ. Observe that if the outer contour of Y is $+$ (resp. $-$) then $Z = Z^{(+)}$ (resp. $Z^{(-)}$). For each term Z'' of $Z^{(\pm)}$, we define $\gamma^\pm = \gamma^\pm(\Sigma, \Gamma, \Delta) = \gamma$ to be the closed $+$ (resp. $-$) contour of smallest area, enclosing Δ. Let W be the

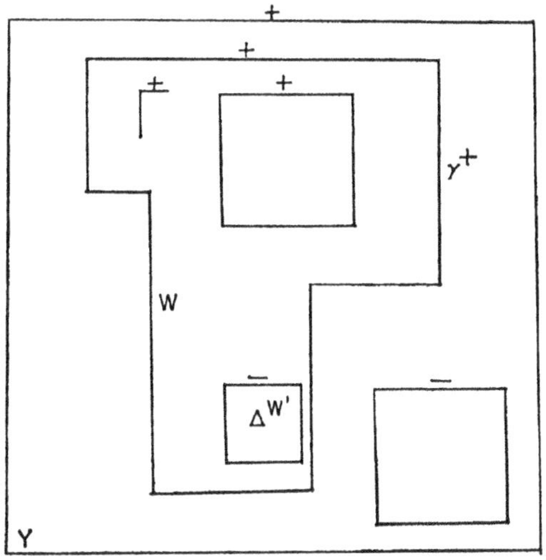

FIG. 1. The regions W and W' are indicated. The lines denote $\pm$Dirichlet lines which determine $\Gamma^{\pm}$.

connected component of $S \sim \Gamma$ which borders on γ and lies inside γ. There is only one such component, as we see from the minimality of γ, and $W \subset Y$. Let W' be the component of $Y \sim W$ which contains Δ_x if $\Delta \subset W$, let $W' = \varnothing$. Also set $U = W \cup W'$. See Fig. 1.

As before,

$$Z'' = Z''(Y \sim U, \Sigma \sim U, \Gamma \sim U) Z''(W, \Sigma \cap W, \Gamma \cap W) Z''(W', \Sigma \cap W', \Gamma \cap W'),$$

and as before, a partial resummation yields

$$Z^{(\pm)}(Y, N_1, \Gamma_1) = \sum_{\partial W^{\pm}} Z(Y \sim U, N_3, \Gamma_3) R(W)(Z - Z^{(\pm)})(W', N_4, \Gamma_4) \quad (4.2.13)$$

where

$$Z - Z^{\pm} = 1, \quad \text{when} \quad W' = \varnothing,$$

$$N_3^{\pm} = [N_1^{\pm} \cup N(\partial U^{\pm})] \sim U, \qquad \Gamma_3 = (\Gamma_1 \sim U) \cup \partial U,$$

$$N_4^{\pm} = [N_1^{\pm} \cup N(\partial W'^{\pm})] \cap W', \qquad \Gamma_4 = (\Gamma_1 \cap W') \cup \partial W'$$

and $R(W)$ denotes a constrained sum over the region W. In the region W', the absence of $+$ (resp. $-$) contours enclosing Δ is a constraint, leading to the difference $Z - Z^{(\pm)}$ above. (Note that all contours, by definition are connected.)

The occurrence of $Z^{(\pm)}$ on the right-hand side of (4.2.13) requires an augmentation of our inductive hypothesis. We assume inductively that

$$| Z^{(\pm)}(Y, N_1, \Gamma_1) Z(Y, N_{1\text{plus}}, \Gamma_1)^{-1} | \leqslant e^{-b|\lambda|^{-1/2}} \quad (4.2.14)$$

where $\Delta = \Delta^{\mp} \subset N_1^{\mp}$ determines $Z^{(\pm)}$. Since $Z^{(+)} = Z$ or $Z^{(-)} = Z$ and since $N_1^+ \neq \varnothing \neq N_1^-$, (4.2.14), proved inductively, also completes the proof of Case 3.

The region W must meet both N_1^- and N_1^+. In fact, one sign is achieved on γ and the other on $\partial W'$ if $W' \neq \varnothing$ and on $\Delta^{\mp}$ if $W' = \varnothing$. Here we use minimality of γ to infer that γ and $\partial W'$ have opposite signs. Thus for each term Z'' in the $R(W)$ sum, the convergence region $X^0 \cap W$ is not empty. Further convergence results from the density of derivative lines in $X \cap W$, leading to the estimate

$$|R(W)| \leqslant e^{-b|\lambda|^{-1/2}} e^{-2\delta|W|/l},$$

and as in Case 2, an estimate of the same form holds for $|R(W)\, Z^*(W)^{-1}|$. Substitution in (4.2.13) gives

$$|Z^{(\pm)}(Y, N_1, \Gamma_1)| \leqslant e^{-b|\lambda|^{-1/2}} \sum_{\partial W^\pm} e^{-2\delta|W|/l} \,|\, Z(Y \sim U, N_{3\text{plus}}, \Gamma_3)$$

$$\times Z^*(W)\, Z(W', N_{4\text{plus}}, \Gamma_4)|. \tag{4.2.15}$$

Here we have applied the inductive hypothesis in the regions $Y \sim U$ and W'. Because $W \neq \varnothing$ by construction, these regions are necessarily smaller than Y. However, the inductive hypothesis in Y also applies to the product of Z's in (4.2.15). In fact, since $X^0 \cap W \neq \varnothing$, there are more Dirichlet lines in $Z^*(W)$ than in $\Gamma_1 \cap W$, resulting in an equal J-value for previously considered cases 1 and 2, and the inequality

$$|Z^{\pm}(Y, N_1, \Gamma_1)\, Z(Y, N_{1\text{plus}}, \Gamma_1)^{-1}| \leqslant e^{-b|\lambda|^{-1/2}} \sum_{\partial W^\pm} e^{[O(|\lambda|^{1/2}) - 2\delta/l]|W|}.$$

Substitution into (4.2.15) combined with combinatorics as in Case 2 control the sum over $\partial W^\pm$. In this argument there is one new issue which did not arise in Case 2, namely, the choice of a distinguished point in W. We choose this point to be the first intersection of γ with some ray starting at $\Delta^{\mp}$. That choice can be made in at most $|\gamma| \leqslant |W|$ ways. This completes the proof of (4.2.14), Case 3, and Theorem 4.2.1.

4.3. *Exponential Clustering*

Let A and B be Wick polynomials of the form (I.6.1) and localized in regions $\mathscr{L}(A)$ and $\mathscr{L}(B)$, respectively. Let

$$D = \text{dist}(\mathscr{L}(A), \mathscr{L}(B)).$$

THEOREM 4.3.1. *There is $\delta > 0$ and a constant c_{AB} such that for λ sufficiently small,*

$$|\langle AB\rangle - \langle A\rangle\langle B\rangle| \leqslant c_{AB} e^{-\delta D} \tag{4.3.1}$$

uniformly in Λ.

Remark. An immediate corollary to this theorem is the unique vacuum and positive mass of Theorem I.2.1.

Proof. As in [16], we introduce duplicate variables ϕ, $\phi^{\sim}$ so that

$$\langle AB \rangle - \langle A \rangle \langle B \rangle = \tfrac{1}{2}\langle (A - A^{\sim})(B - B^{\sim}) \rangle. \tag{4.3.2}$$

With duplicate variables, there are independent Σ, $\Sigma^{\sim}$ expansions, and the terms in the double expansion are labeled by a pair $\{\Sigma, \Sigma^{\sim}\}$. Given $\{\Sigma, \Sigma^{\sim}\}$ we define the regions X, $X^{\pm}$ and $X^{\sim}$, $X^{\sim\pm}$ as before (see Section I.5) and set $\mathscr{B}(\Sigma, \Sigma^{\sim})$ to be the set of l-lattice lines in $(X \cap X^{\sim}) \sim (\mathscr{L}(A) \cup \mathscr{L}(B))$. As in [16], the cluster expansion treats ϕ and $\phi^{\sim}$ symmetrically, so that there is a single parameter s_b for both ϕ and $\phi^{\sim}$ variables, and a single label $\Gamma \subset \mathscr{B}(\Sigma, \Sigma^{\sim})$ for the terms of the expansion. Because the Σ and $\Sigma^{\sim}$ expansions are independent, the individual terms in this expansion are not invariant under the $\phi \leftrightarrow \phi^{\sim}$ symmetry. In particular, we consider four types of contours: $++$, $+-$, $-+$ and $--$, and in general

$$\Gamma = \Gamma^{++} \cup \Gamma^{+-} \cup \Gamma^{-+} \cup \Gamma^{--}.$$

With this notation the Γ^{++}, Γ^{--} contours are preserved by the $\phi \leftrightarrow \phi^{\sim}$ symmetry, while the Γ^{+-} and Γ^{-+} contours are interchanged.

For fixed Σ, $\Sigma^{\sim}$ and Γ, we define $\gamma = \gamma(\Sigma, \Sigma^{\sim}, \Gamma)$ to be the smallest $++$ or $--$ connected closed contour enclosing $\mathscr{L}(A)$. Because of $+$ boundary conditions, γ exists for nonzero terms in the expansion. Let Int γ denote the region enclosed by γ, and let W be the connected component of (Int γ) $\sim \Gamma$ which borders on γ. Also let W' be the component of $R^2 \sim W$ which contains $\mathscr{L}(A)$, if $\mathscr{L}(A) \not\subset W$. Otherwise let $W' = \varnothing$.

As a first step, we hold W fixed and resum the rest of the expansion. The result is a term (i) which depends only on ∂W, (ii) factors across γ, (iii) is symmetric under $\phi \leftrightarrow \phi^{\sim}$, and (iv) is symmetric under $\phi \leftarrow \phi^{\sim}$ in Int γ while ϕ and $\phi^{\sim}$ are fixed in $R^2 \sim$ Int γ. Because of (iv) and the subtractions in (4.3.2), this term is zero unless $\mathscr{L}(B) \subset$ Int γ, which we now assume. Consequently $|W|/l \geqslant D$.

We now proceed as in [16] and Section 4.2. The resummation in $R^2 \sim (W \cup W')$ is free from constraints and yields a partition function, while the resummation in W' involves contributions from A and a difference of partition functions, as in (4.2.13). Thus we obtain

$$|\langle AB \rangle - \langle A \rangle \langle B \rangle| \leqslant \sum_W |R(W)| \frac{Z(\sim W, N(\partial W), \partial W)}{Z(\Lambda, \varnothing, \varnothing)}$$

where $R(W)$ is the constrained sum associated with the region W. The theorem now follows as in Section 1.2 and [16], using a bound on $R(W)$, given below:

LEMMA 4.2.2. *There is a* $\delta > 0$ *such that for* λ *sufficiently small*

$$|R(W)| \leqslant |||A|||\,|||B|||\,e^{-\delta|W|/l}.$$

The summation over W is as in the proof of Theorem 4.2.1.

4.4. *Analyticity and the Infinite Volume Limit*

By the Lee–Yang theorem and results of Fröhlich and Simon [11b], the infinite volume limit of the Schwinger functions exists and is Euclidean invariant provided $\mu > 0$. Moreover, if A is a product of field operators, then in each finite volume Λ we have

$$|\langle A\rangle_{\mu_1} - \langle A\rangle_{\mu=0}| = \left|\int_0^{\mu_1} \frac{d}{d\mu}\langle A\rangle(\mu)\, d\mu\right| \leqslant \text{const}\ \mu_1$$

with the constant independent of Λ provided μ and λ are real. The bound on the above derivative is an immediate consequence of the mass gap as formulated in Theorem 4.3.1.

Thus the $\mu \searrow 0$ limit converges uniformly in Λ and so the Schwinger functions converge for $\mu = 0$ and λ real as $\Lambda \nearrow R^2$. To obtain analyticity in λ (for $\mu = 0$) we note that the finite volume Schwinger functions are analytic and uniformly bounded in Λ in the sector $0 < \text{Re}\,\lambda \leqslant \epsilon|\lambda|$.

An application of Vitali's theorem yields analyticity for infinite volume. This completes the proof of Theorems I.2.1 and I.2.5.

Convergence of the $\Lambda \nearrow R^2$ limit also follows from the expansion methods of this paper and since the expansion method established the stronger result that the series expansion (I.7.15) converges term by term and is valid in the infinite volume limit, we sketch the method. Let Δ be a lattice square not in Λ, and let Δ_t be a strip of Δ of width t, $0 \leqslant t \leqslant 1$. Let Z_9 denote the Λ dependence of Z, and let $Z_t = Z_{\Lambda\cup\Delta_t}$. Then we must show that

$$\left|\frac{Z_0(R^2, N(\Gamma), \Gamma)}{Z_0(R^2, \varnothing, \varnothing)} - \frac{Z_1(R^2, N(\Gamma), \Gamma)}{Z_1(R^2, \varnothing, \varnothing)}\right|$$

is small exponentially in $D = \text{dist}(\Gamma, \Delta)$. However,

$$\frac{d}{dt}\frac{Z_t(R_2, N(\Gamma), \Gamma)}{Z_t(R_2, \varnothing, \varnothing)}$$

is a truncated vacuum expectation value (Γ and $N(\Gamma)$ being regarded as observables) and the methods of Section 4.3 apply.

IV Phase Transitions and Critical Behavior

Introduction

The model with potential $\phi^4 + \sigma\phi^2$ can be studied as a function of the real parameter σ. For σ large in magnitude, expansion methods can be used to study the spectrum of the Hamiltonian H in detail. However the difficult region to study corresponds to intermediate values of σ, close to the critical value σ_c dividing the single phase region from the two phase region. The gap in the spectrum of H vanishes at the critical point, and the spectrum of H becomes a half-line. One of the most difficult problems of physics is to obtain detailed information about models in the neighborhood of the critical point. Some mathematically complete results and some conjectures in this direction are contained in these papers.

Bibliography

Critical point dominance in quantum field models, Ann. l'Inst. H. Poincaré A21, 27–41 (1974).

ϕ_2^4 quantum field model in the single phase region: differentiability of the mass and bounds on critical exponents, Phys. Rev. 10, 536–539 (1974).

A remark on the existence of ϕ_4^4, Phys. Rev. Lett. 33, 440–442 (1974).

On the approach to the critical point, Ann. l'Inst. H. Poincaré A22, 13–26 (1975).

Critical exponents and elementary particles, Commun. Math. Phys. 52, 203–209 (1977).

Critical point dominance in quantum field models

by

James GLIMM (*)
Courant Institute. New York University.
New York 10012

and

Arthur JAFFE (**)
Lyman Laboratory. Harvard University
Cambridge, Massachusetts

ABSTRACT. — We study the generating functional $\Gamma\{A\}$ for vertex functions in $\mathscr{P}(\varphi)_2$ models and prove the Callan-Symanzik equations for the weakly coupled $\lambda_0\varphi_2^4$ model. Under reasonable (but unproved) assumptions, we show that the dimensionless physical coupling constant g in $\varphi_{2,3}^4$ models achieves its maximum value at the critical point. We discuss a more general picture of « critical point dominance ».

1. THE CALLAN-SYMANZIK EQUATIONS

The Euclidean $\mathscr{P}(\varphi)_2$ quantum field models are parameterized by (m_0, g_0). Here m_0 is the bare mass and $g_0 = \lambda_0/m_0^2$ is the dimensionless bare charge. The bare charge λ_0 is the vector of coefficients of the interaction polynomial

$$\mathscr{P}(\varphi) = \sum_{j=1}^{2N} \lambda_0^{(j)} \frac{1}{j!} : \varphi^j : . \tag{1.1}$$

(*) Supported in part by the National Science Foundation, Grant NSF-GP-24003.
(**) Supported in part by the National Science Foundation, Grant NSF-GP-40354X.

We discuss the variation of the solution with respect to these bare parameters, and the transformation from bare parameters to a physical parameter space (m, g). Here m is the physical mass, inf (spectrum (M) $\sim \{0\}$) where the mass operator $M = (H^2 - P^2)^{\frac{1}{2}}$. A physical coupling constant g may be defined by the momentum space vertex function (for $l \geqq 3$) as

$$g^{(l)} = -\Gamma^{(l)}(p)/m^2 \,|_{p=0} \tag{1.2}$$

where

$$\Gamma^{(l)}(p)\delta\left(\sum_i p_i\right) = \int e^{i\Sigma p_j x_j}\tilde{\Gamma}^{(l)}(x)dx .$$

For $j = 1$, we take $g^{(1)} = \langle \varphi \rangle$. The field strength renormalization constant Z is

$$Z = (p^2 + m^2)\langle \Phi^{\sim}(p)\Phi(0)\rangle \,|_{p^2=-m^2} \tag{1.3}$$

where $\langle \quad \rangle$ denotes expectation in the Euclidean ground state measure. For increased flexibility we consider theories without the field strength renormalization, i. e. with $Z \neq 1$. Then the Green's functions and vertex functions can be considered as functions of Z as well as (m_0^2, g_0). We note that

$$Z^{l/2}\Gamma^{(l)}(p, m_0, g_0, Z) = \Gamma^{(l)}(p, m_0, g_0, 1), \tag{1.3a}$$

or in terms of the generating functional

$$\Gamma\{Z^{\frac{1}{2}}A, m_0, g_0, Z\} = \Gamma\{A, m_0, g_0, 1\}, \tag{1.3b}$$

and the field strength renormalization amounts to choosing the theory with $Z = 1$.

Under scaling, one can check that Z is invariant (dimensionless) and that

$$\Gamma^{(l)}(\alpha p, \alpha m_0, g_0, Z)(\alpha m)^{-2} = \Gamma^{(l)}(p, m_0, g_0, Z)m^{-2}, \tag{1.4}$$

see the appendix. Thus the definition of $g^{(j)}$ in (1.2) is scale invariant.

The Callan-Symanzik equations [*11*] describe a one parameter family of perturbations of the form $\mathscr{P}(\varphi)_2 + \sigma \int :\varphi^2: d\bar{x}$. After a new choice of bare mass $m_0(\sigma)$ (see the appendix), this family defines a smooth curve $(m_0(\sigma), g_0(\sigma), Z(\sigma))$. The Callan-Symanzik equations express the variation of the theory with respect to σ. One side of the equation is an expression for $(d/d\sigma)\Gamma\{A, \sigma\}$, which can be recognized as a prescription for φ^2 vertex insertions into the lines of the vertex functions. The other side of the equation is the chain rule,

$$\frac{d}{d\sigma}\Gamma\{A, \sigma, Z\} = \frac{dm(\sigma)^2}{d\sigma}\frac{\partial\Gamma}{\partial m(\sigma)^2} + \frac{dg(\sigma)}{d\sigma}\frac{\partial\Gamma}{\partial g(\sigma)} + \frac{dZ(\sigma)}{d\sigma}\frac{\partial\Gamma}{\partial Z(\sigma)} \tag{1.5}$$

Here Γ is the generating function for the (amputated, one particle irreducible) vertex parts.

We now study (1.5) for weak coupling. In [5] we prove the existence of Γ, for weak coupling, and its analyticity in A and σ. Let $J_A(\sigma)$ be the solution of the equation

$$G_x\{J, \sigma\} - G_x\{0, \sigma\} = A(x)$$

where $G\{J, \sigma\}$ is the generating function for the Euclidean Green's functions and $G_x = \delta G/\delta J(x)$. From the Legendre transformation

$$\Gamma\{A, \sigma\} = G\{J_A, \sigma\} - A(J_A(\sigma)) - G^{(1)}(J_A(\sigma)),$$

we find

$$\frac{d}{d\sigma}\Gamma\{A, \sigma\} = \frac{d}{d\sigma}G\{J, \sigma\}|_{J=J_A(\sigma)},$$

for $A \in H_1$, $J \in H_{-1}$ and $\|A\|_1$ and $\|J\|_{-1}$ small. We use the notation

$$\mathscr{Z}\{J, \sigma, Z_0\} = \langle e^{Z_0^{-\frac{1}{2}}\Phi(J)}\rangle_\sigma$$
$$\mathscr{Z}_{:xx:}\{J, \sigma, Z_0\} = Z_0^{-1}\langle :\Phi(x)^2: e^{Z_0^{-\frac{1}{2}}\Phi(J)}\rangle_\sigma$$

(with $m_0(\sigma)$-Wick ordering). The field strength renormalization then amounts to choosing Z_0 to be the Z of (1.3), but for the time being we regard Z_0 as an independent variable.

We have

$$G\{J, \sigma, Z/Z_0\} = \ln \mathscr{Z}\{J, \sigma, Z_0\}$$

and

$$\frac{d}{d\sigma}\Gamma = \frac{d}{d\sigma}G = \frac{d}{d\sigma}\ln \mathscr{Z} = \mathscr{Z}^{-1}\frac{d}{d\sigma}\mathscr{Z}.$$

In order to simplify subsequent formulas, we assume that $\mathscr{P}$ is even, so $G^{(1)}(J) = 0$, and then

$$(1.6)\quad \frac{d}{d\sigma}\Gamma\{A, \sigma\}$$
$$= \mathscr{Z}^{-1}Z_0\int[-\mathscr{Z}_{:xx:}\{J_A, \sigma\} + \mathscr{Z}\{J_A, \sigma\}\mathscr{Z}_{:xx:}\{0, \sigma\}]dx.$$

For J regular, we establish (1.6) by Dimock's asymptotic calculations [3]. First we keep m_0 fixed, and consider the interaction $\mathscr{P}(\varphi) + \sigma :\varphi^2:$. Then the mass renormalization transformation (see appendix) is used to obtain (1.6) with $m_0(\sigma)$ Wick order. For the techniques to prove convergence of the integral in (1.6) with $J \in H_{-1}$, see [5, Chapter 2]. The key point is that since H_{-1} is an L_2 norm, we must be careful to eliminate terms linear in J, or in localized portions J_α of J. Since $\mathscr{P}$ is even, these terms do not contribute to the cluster expansion of (1.6).

We note that the integrand

$$F(x) = -\langle :\Phi(x)^2: e^{Z_0^{-\frac{1}{2}}\Phi(J)}\rangle_\sigma + \langle :\Phi(x)^2:\rangle_\sigma\langle e^{Z_0^{-\frac{1}{2}}\Phi(J)}\rangle_\sigma,$$

which clusters can be written

$$F(x) = \lim_{a \to 0} F(x, x + a)$$

where

$$(1.7)\quad F(x, y) = - \langle \Phi(x)\Phi(y)e^{Z_0^{-\frac{1}{2}}\Phi(J)} \rangle_\sigma + \langle \Phi(x)\Phi(y) \rangle_\sigma \langle e^{Z_0^{-\frac{1}{2}}\Phi(J)} \rangle_\sigma ,$$

so

$$\frac{1}{Z_0} \frac{d\mathscr{Z}\{ J, \sigma \}}{d\sigma} = \lim_{a \to 0} \int F(x, x + a)dx .$$

Recall that $G\{ J, \sigma \} = \ln \mathscr{Z}\{ J, \sigma \}$, the generating function for connected parts. Then regarding Z_0, σ and λ_0 as independent variables,

$$(1.8)\quad \frac{d}{d\sigma}\Gamma\{ A, \sigma \} = \frac{d}{d\sigma}G\{ J, \sigma \} = \mathscr{Z}\{ J, \sigma \}^{-1} \frac{d}{d\sigma}\mathscr{Z}\{ J, \sigma \}$$

$$= - \lim_{a \to 0} Z_0 \int [G_{x,x+a}\{ J, \sigma \} + G_x\{ J, \sigma \}^2 - G_{x,x+a}\{ 0, \sigma \}]dx$$

We eliminate the J dependence in (1.8) by using $G_x\{ J, \sigma \} = A(x)$ and the fact that $-\Gamma_{xy}\{ A, \sigma \}$ is the kernel of the operator inverse to $G_{xy}\{ J, \sigma \}$. Thus by (1.8),

$$(1.9)\quad \frac{d}{d\sigma}\Gamma\{ A, \sigma \} = \lim_{a \to 0} Z_0 \int [\Gamma^{-1}_{x,x+a}\{ A, \sigma \} - \Gamma^{-1}_{x,x+a}\{ 0, \sigma \} - A(x)^2]dx$$

In order to equate (1.9) to the right side of (1.5), the only problem is to show that the indicated derivatives exist. We specialize further to the case of a pure $\lambda_0\varphi^4$ interaction. For small coupling, there are no multiple phases, and so m and g are functions of m_0 and g_0. The existence of derivatives $\partial\Gamma/\partial m_0^2$ and $\partial\Gamma/\partial g_0$ follows from the analyticity of $\Gamma\{ A, \sigma \}$ in the bare parameters [5], while $\partial\Gamma/\partial Z_0$ exists by (1.3*b*). The fact that $m_0(\sigma)^2$ is differentiable with a bounded inverse is the second transformation of the appendix. We regard Z as a function of m_0 and g_0, defined by (1.3), and we take $Z_0 = Z$. By scaling $\partial g/\partial m_0 = \partial Z/\partial m_0 = 0$ and $\partial m/\partial m_0 = 1$, while $\partial Z/\partial\lambda_0$ and $\partial m/\partial\lambda_0$ exist by [9]. Thus the derivatives $dZ(\sigma)/d\sigma$ and $dm(\sigma)/d\sigma$ taken at constant λ_0 exist also. We show below for weak coupling that $\partial g/\partial\lambda_0$ (m_0 constant) and $\partial g/\partial m_0$ (λ_0 constant) (i. e. $dg/d\sigma$) exist and are not zero. Furthermore, it follows that the Jacobian $\partial(m, g)/\partial(m_0, g_0)$ is not zero. The proof of the Callan-Symanzik equations for weak $\lambda_0\varphi_2^4$ models, namely

$$(1.10)\quad \frac{dm(\sigma)^2}{d\sigma}\frac{\partial\Gamma}{\partial m(\sigma)^2}\{ A, \sigma \} + \frac{dg(\sigma)}{d\sigma}\frac{\partial\Gamma}{\partial g(\sigma)}\{ A, \sigma \} + \frac{dZ(\sigma)}{d\sigma}\frac{\partial\Gamma}{\partial Z(\sigma)}\{ A, \sigma \}$$

$$= \lim_{a \to 0} Z_0 \int [\Gamma^{-1}_{x,x+a}\{ A, \sigma \} - \Gamma^{-1}_{x,x+a}\{ 0, \sigma \} - A(x)^2]dx$$

is reduced to the following result:

THEOREM 1.1. — For a pure φ^4 interaction with weak coupling, the coordinate transformation $(m, g) \leftrightarrow (m_0, g_0)$ is differentiable with a nonvanishing Jacobian, for real bare parameters and for g defined by (1.4).

Proof. — We have already seen that it is sufficient to show $\partial g/\partial\lambda_0$ exists and is nonzero. With m_0 fixed, $\partial g/\partial\lambda_0$ is proportional to

$$(1.11) \qquad -\left.\frac{\partial\Gamma^{(4)}(p)}{\partial\lambda_0}\right|_{p=0} = -\frac{\partial}{\partial\lambda_0}\int \left. G_a^{(4)}(x)\right|_{x_1=0} dx_2 dx_3 dx_4 ,$$

where the amputated Green's function G_a equals

$$G_a(x) = \int dx' G(x') \prod_{i=1}^{4} \tilde{\Gamma}^{(2)}(x_i - x_i') .$$

Using Dimock's proofs [3] of the asymptotic formulas for the Euclidean Green's functions, (1.11) can be evaluated as

$$-\frac{\partial\Gamma^{(4)}(0)}{\partial\lambda_0} = 1 + 0(\lambda_0)$$

which is $\neq 0$ for small coupling.

2. CRITICAL POINT DOMINANCE

In this section we discuss the global nature of the transformation $(m, g) \leftrightarrow (m_0, g_0)$ for the φ^4 model. Unlike the mathematical proofs of Section 1, our discussion here is a heuristic picture, based on reasonable assumptions about the physical parameters (m, g). In a multiple phase region we consider a pure phase. We conclude that the charge g achieves its maximum value g_c for the φ_2^4 or φ_3^4 model at the critical point ($m = 0$), which we presume exists. Here g_c may be infinite, but corresponds to finite values in the bare parameter space (m_0, g_0). We relate this behavior to $\beta(g)$, one of the coefficients in the Callan-Symanzik equation, and are led in the superrenormalizable cases ($d = 2, 3$) to:

THE PICTURE OF CRITICAL POINT DOMINANCE. — The curve $m(\sigma)^2$ has a unique zero at $\sigma = \sigma_c$, and increases monotonically to $+\infty$ on each side of this zero. The curve $g(\sigma)$ decreases monotonically to zero as $|\sigma - \sigma_c| \to \infty$, with a (possibly infinite) maximum g_c at $\sigma = \sigma_c$, see Figure 1.

We now discuss our assumptions and conclusions. For $d = 2$, the cluster expansion deals with the $\sigma \to +\infty$ limit [6]. After scaling (to ensure that the mass remains bounded) the model converges as $\sigma \to \infty$ to a free theory. In this limit $\lambda \approx \lambda_0$, $m \approx m_0$, and $g \approx g_0$; the monotonicity of $g(\sigma)$ or $m(\sigma)$ follows by perturbation theory, see [3]. The more general result that $m(\sigma)$ is monotonic for $\sigma \geq \sigma_c$ was established in [7]. We remark that in the Ising model, a stronger result holds for the inverse correlation length ξ^{-1}. For Ising_2, ξ^{-1} is monotonic in both the single phase region and the two phase region, see, e. g. [8], and ξ^{-1} vanishes at the critical point, near which there is a linear power law behavior $\xi^{-1} = a\,|\sigma - \sigma_c|$.

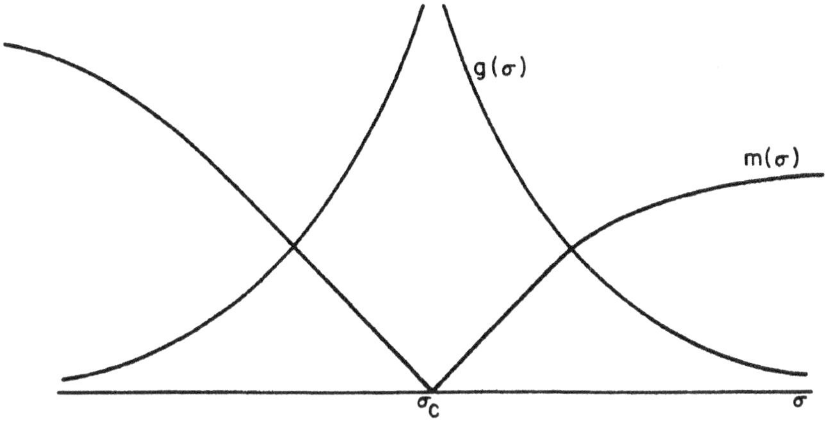

FIG. 1. — Critical Point Dominance.

For $d = 3$, the σ-perturbation produces no new ultraviolet divergences in a finite volume [4], and we expect the same behavior in an infinite volume. Thus for $d = 3$ the same picture should hold for $m(\sigma)$ as in $d = 2$. The dimensionless coupling constant in three dimensions is $g = -\Gamma^{(4)}(0)/m$.

This picture of monotonicity of $m(\sigma)$ is related to a choice of variables made by Symanzik in analyzing the characteristics of the Callan-Symanzik equations [11]. Symanzik chooses a dimensionless parameter s to reparameterize the $m(\sigma)$ curve, $\sigma = \sigma(s)$, starting from a point $m(0)$, and such that

$$\frac{d}{ds} m(s)^2 = m(s)^2 . \tag{2.1}$$

Since (2.1) can be integrated to yield $m(s)^2 = m(0)^2 e^s$, it is a valid assumption only on a branch of the σ curve for which $m(\sigma)^2$ varies monotonically between 0 and ∞. Since $m(\sigma)^2$ is finite for σ finite, and (assuming continuity) is bounded for σ bounded, $m(\sigma)^2 \to \infty$ can occur at most for $\sigma \to \pm\infty$. In terms of the new parameter, an s-curve corresponds to a monotone branch $\sigma > \sigma_c$ or $\sigma < \sigma_c$ of a σ-curve.

Let us assume that on each s-curve the use of physical coordinates $m(s)$ and $g(s)$ is justified. We now argue that g is monotone in s on each side of the critical point. Let $\beta = dg/ds$. By scaling, β is a function $\beta(g)$ of g alone. (Here we use the fact that the σ-curves and s-curves are mapped into one another by scaling. Then the tangent lines

$$dg(s)/dm(s)^2 = (dg(s)/ds) \times (dm(s)^2/ds)^{-1}$$

transform as m^{-2} and β is scale invariant.) Thus the equation

$$\frac{dg}{ds} = \beta(g) \tag{2.2}$$

is a first order ordinary differential equation.

PROPOSITION. — Assume g and $d\beta(g)/dg$ are bounded for $\sigma - \sigma_c \geqq \varepsilon > 0$, for all $\varepsilon > 0$. Then $g(s)$ is monotone in s and $\beta(g) < 0$.

Proof. — By the assumed bound, $\beta(g)$ satisfies

$$|\beta(g_1) - \beta(g_2)| \leq M|g_1 - g_2|,$$

giving existence and uniqueness for the solution of the equation (2.2). Let $(\bar{s}, \infty)$ be the maximum interval on which $g(s)$ is strictly monotonic in s. Then $\beta(g(\bar{s})) = 0$ and by the uniqueness of the solution to (2.2), $g(s) = g(\bar{s}) = \text{const.}$ for $s \in [\bar{s}, \infty)$. However, the $d = 2$ cluster expansion [6] concerns the $\sigma \to \infty$ $(s \to \infty)$ limit. In this limit $g \approx g_0$, $m \approx m_0$ and $g_0(s) \to 0$ as $s \to \infty$. By (2.1) and perturbation theory (which is asymptotic [3])

$$\beta(g) = -g + O(g^2) \tag{2.3}$$

as $g \to 0$. We conclude for small g that $\beta(g) < 0$, a contradiction to $\beta(g) = \beta(g(s)) = 0$, so $s = -\infty$, completing the proof for $d = 2$. For $d=3$, we expect a similar result with $\beta = -\text{const}\ g + O(g^2)$, but the required estimates have not yet been proved.

We now argue that for $\sigma < \sigma_c$, $g(s) \to 0$ as $s \to \infty$ $(\sigma \to -\infty)$. In that case, the proposition also applies for $\sigma < \sigma_c$, i. e. in the two phase region. We appeal to the Goldstone picture of the ground state, in which the ground state is determined by the minimum of the Euclidean action function V, and where the curvature at the minimum approximates the mass. For the φ^4 model with $\sigma \ll 0$, the minimum of $V(\cdot)$ occurs at $\pm(6|\sigma|/\lambda_0)^{\frac{1}{2}} \equiv \pm\bar{x}$. At the minimum, the curvature if $V^{(2)}(\pm\bar{x}) = 2|\sigma|$, while $V^{(3)}(\pm\bar{x}) = \pm(6\lambda_0|\sigma|)^{\frac{1}{2}}$ and $V^{(4)}(\pm\bar{x}) = \lambda_0$. Hence the dimensionless couplings at the minimum $\pm\bar{x}$ are for $d = 2$,

$$g^{(j)} = V^{(j)}(\pm\bar{x})/V^{(2)}(\pm\bar{x}) = \begin{cases} \pm(3\lambda_0/2|\sigma|)^{\frac{1}{2}}, & j = 3 \\ \lambda_0/2|\sigma|, & j = 4 \end{cases} \tag{2.4}$$

We conclude that as $\sigma \to -\infty$, and after scaling so that the mass $(2|\sigma|)^{\frac{1}{2}}$ remains bounded, the model converges to a free field (or a sum of two free fields, depending on the boundary conditions). By (2.4), we also see that the

Goldstone picture suggests $g(s) \approx \lambda_0/m(s)^2$ in this limit, and by (2.1), $\beta(g) \approx -g$ in this limit. In other words (2.3) holds for $\sigma \to -\infty$ as well as for $\sigma \to +\infty$. A similar picture holds for the quartic coupling $g^{(4)}$ (but not $g^{(3)}$) in three dimensions.

With this picture of critical point dominance, the function $\beta(g)$ has a zero at $g = 0$, and a second zero at $g = g_c$ (possibly infinity). The coupling constant is bounded by g_c, and β is represented in Figure 2. The zero at $g = 0$ governs the canonical (short distance, ultraviolet) behavior of the theory, while the presumed zero at $g = g_c$ governs the infrared (long distance) behavior.

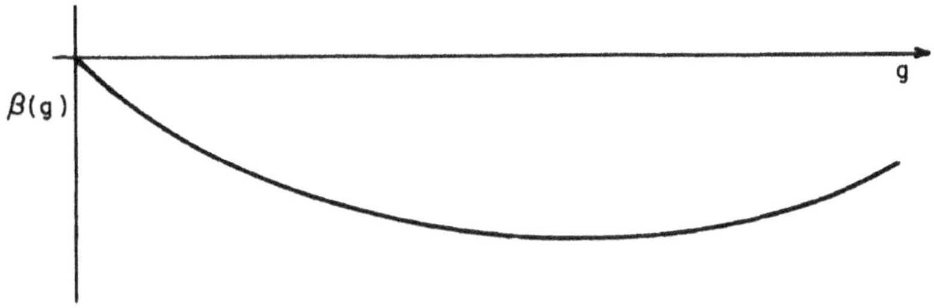

FIG. 2. — The curve $\beta(g)$ for $d = 2$.

This question remains open whether $g_c = \infty$, in which case the function $\beta(g) < 0$ for $g > 0$. In order for g_c to be finite, it is necessary that $\lambda_c = 0$ and that λ vanish as $\sigma \to \sigma_c$ in such a way that $\lambda/m^2 \to g_c$. If $g_c < \infty$, then our picture of critical point dominance says that g_c is the maximum value of g, and in the two phase region $\beta(g)$ returns to zero along another curve $\beta < 0$, see Figure 3.

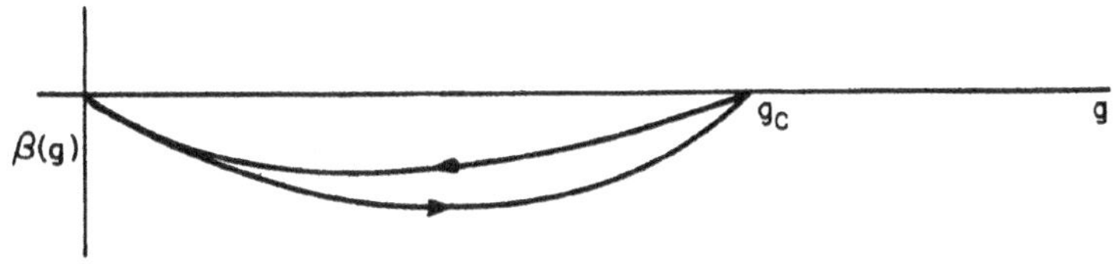

FIG. 3. — The trajectory of $\beta(g(\sigma))$, assuming $g_c < \infty$.

Critical theories ($g = g_c$) need not be scale invariant, but lie on the critical line of bare parameters $g_{0,c}$ (see the appendix). This line is transformed into itself by scale transformations. If a scale invariant φ_2^4 model exists, it is obtained by an infinite scaling of a critical (but not scale invariant) theory.

In four dimensions, as opposed to $d < 4$, the coupling constant λ is dimensionless, i. e. $g = \lambda$. Hence the first order contribution to $g(g_0)$ does not contribute to $\beta(g)$ and explicit calculation shows that the second order contribution is positive. Hence for weak coupling,

$$\beta(g) = \alpha g^2 + O(g^3), \qquad \alpha > 0 .$$

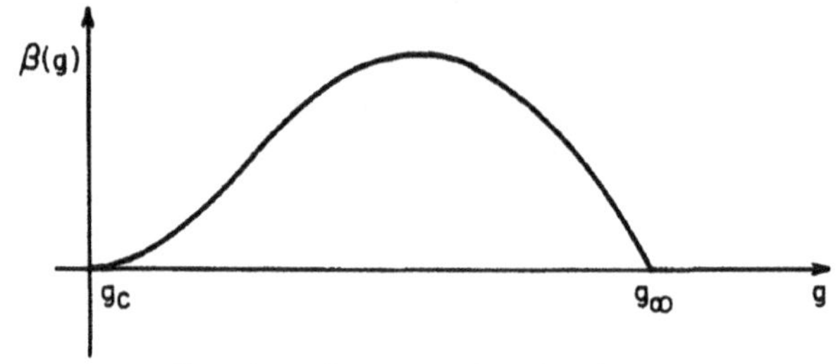

FIG. 4. — The curve $\beta(g)$ for $d = 4$.

The change in the sign of $\beta(g)$ from $d = 2$ or $d = 3$ reverses the roles of the zeros of $\beta(g)$. Thus the zero at $g = 0$ is the critical zero, $g_c = 0$, and governs the infrared (long distance) behavior of the theory. On the other hand, a zero at $g = g_\infty > 0$ would govern ultraviolet behavior (and is expected to be noncanonical). The arguments from the two dimensional case, applied to $\beta > 0$, yield characteristic curves drawn in Figure 5.

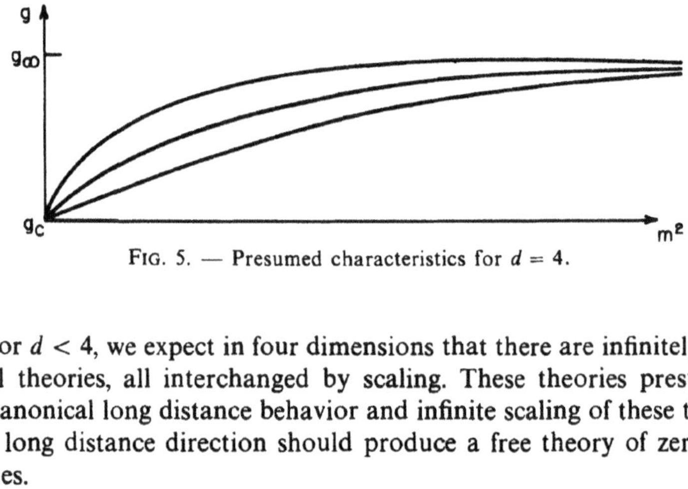

FIG. 5. — Presumed characteristics for $d = 4$.

As for $d < 4$, we expect in four dimensions that there are infinitely many critical theories, all interchanged by scaling. These theories presumably have canonical long distance behavior and infinite scaling of these theories in the long distance direction should produce a free theory of zero mass particles.

Finally we reconsider the behavior of

$$g = - Z^2 G^{(2)\sim}(0)^{-4} G^{(4)\sim}(0) m^{-4+d}$$

as we approach the critical point from the single phase region. By the definition of Z, $Z \leqq m^2 G^{(2)\sim}(0)$, so

$$g \leqq - G^{(2)\sim}(0)^{-2} G^{(4)\sim}(0) m^d .$$

Each of $G^{(2)\sim}$, $G^{(4)\sim}$ and m^2 is assumed to have behavior characterized by a critical exponent. We use the terminology of statistical mechanics, see [10]: the mass m is the inverse correlation length ξ^{-1}, and $G^{(2)\sim}(0)$ is proportional to susceptibility $(\partial M/\partial H)_T = \chi_T$. If $\varepsilon = (T/T_c - 1)$, the ε-dependence in the neighborhood of the critical point is given by

$$m = \xi^{-1} \sim \varepsilon^{\nu},$$
$$G^{(2)\sim}(0) \sim \chi_T \sim \varepsilon^{-\gamma}.$$

In field theory, we use $\varepsilon = \left(\frac{m_0^2}{m_{0,c}^2} - 1\right)$. The four point correlation function $G^{(4)}$ has the gap exponent Δ_4, which relates its critical behavior to the behavior of $G^{(2)}$,

$$-G^{(4)} \sim \varepsilon^{-2\Delta_4} G^{(2)} \sim \varepsilon^{-2\Delta_4 - \gamma}.$$

Thus

$$g \leqq O(\varepsilon^{\gamma - 2\Delta_4 + d\nu}).$$

For the Ising model in the single phase region ($\varepsilon > 0$) the exponents are

$$d = 2: \quad \gamma = \frac{7}{4}, \quad \nu = 1, \quad \Delta_4 = \frac{15}{8},$$

$$d = 3: \quad \gamma = \frac{5}{4}, \quad \nu = \frac{5}{8}, \quad \Delta_4 = \frac{25}{16},$$

so

$$\gamma - 2\Delta_4 + d\nu = \begin{cases} 0, & d = 2 \\ 0, & d = 3 \end{cases}$$

Here we use exponents as in [10], and the $d = 3$ exponents are approximate. While there is no known relation between the Ising model exponents and the field theory exponents, Wilson and others have argued that they may coincide. These exponents do not distinguish for $d = 2, 3$ whether g has a finite limit at the critical point. Furthermore, these exponents indicate that the dimensional coupling constant $\lambda_c = 0$.

The critical exponent η describes the rate of decay of the pair correlation function at the critical point,

$$G(r) \sim r^{-(d-2+\eta)},$$

and $\eta \neq 0$ is the case of anomalous (noncanonical) dimensions. In the I_2 model, $\eta = \frac{1}{4}$. This anomalous dimension indicates that the propagator decays faster than it would in the presence of zero mass particles (i. e. $\eta = 0$). Thus we do not expect zero mass particles in the critical theory and we expect $Z_c = 0$. If $Z \sim \varepsilon^\zeta$ near the critical point, the bound $Z \leqq m^2 G^{(2)\sim}(0)$ shows

$$\zeta \geqq 2\nu - \gamma .$$

By the known inequality $(2 - \eta)\nu \geqq \gamma$, we have $\zeta \geqq \eta\nu$. Furthermore, $\zeta = \frac{1}{4}$ for I_2, using a recent calculation of Tracy and McCoy [*12*], and thus for I_2 the inequality is an identity. Thus anomalous dimensions, $\eta > 0$, indicate $Z_c = 0$.

Rigorous bounds on the two point function [*6*] exclude zero mass particles in the two point function, and assuming $m \to 0$ as the critical point is approached, these same bounds then prove $Z \to 0$. For I_3, $\eta \approx .04 \neq 0$, so again we expect $Z_c = 0$ and no zero mass particles in the φ_3^4 field theory. It is expected (up to logarithms) that the exponents are given by their canonical values in I_4.

APPENDIX

REVIEW OF SCALE AND MASS SHIFT TRANSFORMATIONS

We assume $d = 2$ in the appendix. The corresponding results for $d > 2$ are easily derived.

SCALE TRANSFORMATIONS. — On Fock space, the scale transformation is a unitary operator U defined on an n-particle state f by

$$(U^*f)(\vec{k}_1, \ldots, \vec{k}_n) = \alpha^{n/2} f(\alpha\vec{k}_1, \ldots, \alpha\vec{k}_n),$$

for $\alpha > 0$. The free field φ_{m_0} of mass m_0, and its Hamiltonian H_{0,m_0}, transform as

$$U\varphi_{m_0}(\vec{x})U^* = \varphi_{\alpha m_0}(\vec{x}/\alpha),$$
$$UH_{0,m_0}U^* = \alpha^{-1}H_{0,\alpha m_0}.$$

Also the no particle state Ω_0 is invariant.

$$U\Omega_0 = \Omega_0,$$

and $Ua(\vec{k})U^* = \alpha^{\frac{1}{2}}a(k\alpha)$.

Furthermore, on Wick monomials, scale transformations map

$$U : \varphi_{m_0}(x)^j : U^* = : \varphi_{\alpha m_0}(\vec{x}/\alpha)^j : .$$

With an ultraviolet cutoff κ and box of length L,

$$UH_1(m_0, \lambda, \kappa, L)U^* = H_1(\alpha m_0, \alpha\lambda, \alpha\kappa, \alpha^{-1}L),$$

since the Wick ordering constant is invariant under scaling in two dimensions:

$$c(m_0, \kappa) = c(\alpha m_0, \alpha\kappa).$$

The vacuum energy for

$$H_{0,m_0} + H_1(m_0, \lambda_0, \kappa, L)$$

is the same as the vacuum energy for

$$\begin{aligned} U[H_{0,m_0} + H_1(m_0, \lambda_0, \kappa, L)]U^* &= \alpha^{-1}H_{0,\alpha m_0} + H_1(\alpha m_0, \alpha\lambda_0, \alpha\kappa, \alpha^{-1}L) \\ &= \alpha^{-1}[H_{0,\alpha m_0} + H_1(\alpha m_0, \alpha^2\lambda_0, \alpha\kappa, \alpha^{-1}L)]. \end{aligned}$$

Thus

$$E(m_0, \lambda_0, \kappa, L) = \alpha^{-1}E(\alpha m_0, \alpha^2\lambda_0, \alpha\kappa, \alpha^{-1}L)$$

and writing $H = H_0 + H_1 - E$,

$$UH(m_0, \lambda_0, \kappa, L)U^* = \alpha^{-1}H(\alpha m_0, \alpha^2\lambda_0, \alpha\kappa, \alpha^{-1}L).$$
$$U\Omega(m_0, \lambda_0, \kappa, L) = \Omega(\alpha m_0, \alpha^2\lambda_0, \alpha\kappa, \alpha^{-1}L).$$

Thus for a cutoff $\mathscr{P}(\varphi)_2$ field theory, scale transformations on Fock space provide a unitary map between Hamiltonians with different cutoffs, but the same g_0. Since the vacuum expectation values converge as κ, L $\rightarrow \infty$, we obtain in the limit an isomorphism of theories with constant g_0. We may define a unitary transformation

$$U : \mathscr{H}_1 \rightarrow \mathscr{H}_2$$

between the Hilbert spaces $\mathscr{H}_i$ of two such theories, where $U\Omega_1 = \Omega_2$,

$$UH_1U^* = \alpha^{-1}H_2,$$

and

$$U\varphi_1(x)U^* = \varphi_2(x/\alpha).$$

Theories described by different values of the parameter α are physically equivalent, and in the (m_0^2, λ_0) plane correspond to various points on lines of constant g_0, see Figure 6.

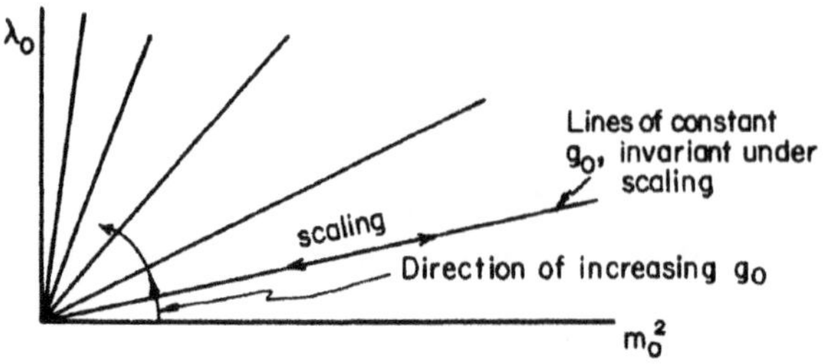

FIG. 6. — Scaling in the bare parameter space.

MASS SHIFT TRANSFORMATION. — The second transformation arises from a unitary operator on a Fock space over a finite volume (with e. g. periodic boundary conditions). In the infinite volume limit, we obtain an isomorphism of free fields of different mass, and of interacting fields as follows: we define in the Q-representation the scale (Bogoliubov) transformation,

$$U\varphi^{\sim}(\vec{k})U^{*} = \varphi^{\sim}(\vec{k})\alpha(\vec{k}),$$

where $\alpha(\vec{k}) = (\vec{k}^2 + m_0^2)^{\frac14}(\vec{k}^2 + m_1^2)^{-\frac14} = \mu_0^{\frac12}\mu_1^{-\frac12}$. Then by a standard calculation, see for example [1],

$$U\varphi_{m_0}(\vec{x})U^{*} = \varphi_{m_1}(\vec{x}),$$

$$U : \varphi_{m_0}(\vec{x})^j : U^{*} = \sum_{r=0}^{[j/2]} \frac{j!}{(j-2r)!\,r!\,2^r}(-\delta c)^r : \varphi_{m_1}(\vec{x})^{j-2r} : ,$$

with

(A1) $$\delta c = \frac{1}{4\pi}\int\left[\frac{1}{k^2 + m_0^2} - \frac{1}{k + m_1^2}\right]dk = \frac{1}{2}\ln\left(\frac{m_1}{m_0}\right).$$

We note that δc varies over $(-\infty, \infty)$ as m_1/m_0 varies over $(0, \infty)$. Also

$$UH_{0,m_0}U^{*} = H_{0,m_1} + \frac{1}{2}(m_0^2 - m_1^2)\int : \varphi_{m_1}^2 : d\vec{x} - EL$$

where E is the (finite) vacuum energy density

$$E = -\int\left[\frac{1}{2}(\mu_0 - \mu_1) + \frac{1}{4}\frac{m_0^2 - m_1^2}{\mu_1}\right]d\vec{k}.$$

Let $\mathscr{P}$ be an interaction polynomial for mass m_0,

$$\mathscr{P}(\varphi_{m_0}) = \sum_{j=1}^{2N} \lambda_0^{(j)} : \varphi_{m_0}^j : .$$

Define $\hat{\mathscr{P}}$ as the transformed polynomial,

$$\hat{\mathscr{P}}(\varphi_{m_1}) = U\mathscr{P}(\varphi_{m_0})U^{*} = \sum_{j=0}^{2N} \hat{\lambda}_0^{(j)} : \varphi_{m_1}^j : ,$$

and note $\lambda_0^{(j)} = \hat{\lambda}_0^{(j)}$ for $j = 2N, 2N - 1$. Let H_1 and $\hat{H}_1$ be the respective interaction Hamiltonians. Then for $H(m_0, \lambda) = H_{0,m_0} + H_1(m_0, \lambda) - E(m_0, \lambda)$, where E is the vacuum energy, we have

(A2) $$UH(m_0, \lambda_0)U^{*} = H(m_1, \tilde{\lambda}_0).$$

Here

(A3)
$$\tilde{\lambda}_0^{(j)} = \hat{\lambda}_0^{(j)}, \qquad j \neq 2,$$
$$\tilde{\lambda}_0^{(2)} = \hat{\lambda}_0^{(2)} + \frac{1}{2}(m_0^2 - m_1^2).$$

We note in particular that $\lambda_0^{(2N)} > 0$ and

(A4)
$$\tilde{\lambda}_0^{(2)} = \lambda_0^{(2)} + \frac{1}{2}(m_0^2 - m_1^2) + N(2N-1)!!\lambda_0^{(2N)}(-\delta c)^{N-1} + O(|\delta c|^{N-2}).$$

By (A1), $-\delta c \to +\infty$ as $m_1 \to 0$. Also
$$|\delta c| \leqq O\left(\left|\ln \frac{m_1}{m_0}\right|\right),$$
so $\tilde{\lambda}_0^{(2)} \to -\infty$ as $m_1 \to \infty$. Thus for m_0 fixed, $\tilde{\lambda}_0^{(2)}$ ranges over $(\infty, -\infty)$ as m_1 varies over $(0, \infty)$.

Remark 1. — Given $H(m_0, \lambda_0)$, there exists a bare mass m_1 for which
$$UH(m_0, \lambda_0)U^* = H(m_1, \tilde{\lambda}_0)$$
with $\tilde{\lambda}_0^{(2)} = 0$.

For the $\lambda_0 : \varphi_{m_0}^4 : + \sigma : \varphi_{m_0}^2 :$ interaction, $\tilde{\lambda}_0^{(2)}$ is monotonic in m_1, so the zero is unique. (More generally, $\mathscr{P}(\varphi_{m_0}) + \sigma\varphi_{m_0}^2$ has a unique zero for $|\sigma|$ large.) Hence with λ_0 fixed, we obtain a one parameter family of theories, parameterized by σ, with the interaction $\lambda_0 : \varphi_{m_1(\sigma)}^4 :$. In our diagram of couplings, this σ line is horizontal, with its height determined by λ_0, see Figure 7.

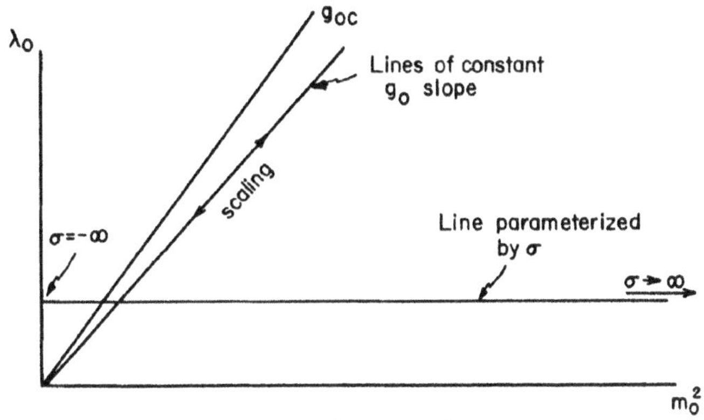

FIG. 7. — The $\lambda_0 : \varphi^4 : + \sigma : \varphi^2 :$ interaction.

For a pure φ^4 interaction, (A4) becomes

(A5)
$$\tilde{\lambda}_0^2 = \lambda_0^{(2)} + \frac{1}{2}(m_0^2 - m_1^2) - 3\lambda \ln\left(\frac{m_1}{m_0}\right).$$

With the choice $\tilde{\lambda}_0^{(2)} = 0$, $\lambda_0^{(2)} = \sigma$, m_0^2 fixed, $m_1^2 = m_1^2(\sigma)$, we then have

(A6)
$$\frac{dm_1^2(\sigma)}{d\sigma} = \left(\frac{1}{2} + \frac{3}{2}\frac{\lambda}{m_1^2(\sigma)}\right)^{-1}$$

Remark 2. — For the $\lambda\varphi_2^4$ interaction, $m_1^2(\sigma)$ is a differentiable function of σ, bounded away from zero for σ bounded away from $-\infty$.

Remark 3. — The finite volume may be introduced by periodic boundary conditions. Within the region of convergence of the cluster expansion, the infinite volume theory is independent of the choice of boundary conditions.

REFERENCES

[*1*] R. BAUMEL, Private communication.

[*2*] S. COLEMAN, Dilatations. *1971 Erice Lectures.*

[*3*] J. DIMOCK, Perturbation series asymptotic to Schwinger functions in $\mathcal{P}(\Phi)_2$. *Commun. Math. Phys.*, t. **35**, 1974, p. 347-356.

[*4*] J. FELDMAN, The Φ_3^4 field theory in a finite volume. *Commun. Math. Phys.*

[*5*] J. GLIMM and A. JAFFE, The entropy principle for vertex functions in quantum field models. *Ann. Institut Henri Poincaré*, t. **21**, 1974, p. 1-25.

[*6*] J. GLIMM, A. JAFFE and T. SPENCER, The particle structure of the weakly coupled $\mathcal{P}(\Phi)_2$ model and other applications of high temperature expansions. Part I. Physics of quantum field models. Part II. The cluster expansion. In: *Constructive quantum field theory*, Ed. by G. Velo and A. Wightman, Lecture notes in physics, Vol. 25, Springer Verlag, Berlin, 1973.

[*7*] F. GUERRA, L. ROSEN and B. SIMON, The $\mathcal{P}(\Phi)_2$ Euclidean quantum field theory as classical statistical mechanics. *Ann. Math.*

[*8*] B. MCCOY and T. T. WU, *The two dimensional ising model.* Harvard University Press, 1973.

[*9*] T. SPENCER, The mass gap for the $\mathcal{P}(\Phi)_2$ quantum field model with a strong external field. *Commun. Math. Phys.*

[*10*] H. E. STANLEY, *Introduction to phase transitions and critical phenomena.* Oxford University Press, 1971.

[*11*] K. SYMANZIK, Small distance behavior in field theory and power counting. *Commun. Math. Phys.*, t. **18**, 1970, p. 227-246.

[*12*] C. TRACY and B. MCCOY, Neutron scattering and the correlation functions of the Ising model near T_c. *Phys. Rev. Lett.*, t. **31**, 1973, p. 1500-1504.

(Manuscrit reçu le 11 février 1974)

ϕ_2^4 quantum field model in the single-phase region: Differentiability of the mass and bounds on critical exponents

James Glimm*† and Arthur Jaffe‡§

Institut des Hautes Etudes Scientifiques, Bures-sur-Yvette, Essonne, France

(Received 25 March 1974)

We prove that the physical mass m is a differentiable function of the bare mass m_0 down to the critical point (defined by the occurrence of $m = 0$ or of the onset of symmetry breaking). We derive an upper bound on $m'(m_0)$. We show that the critical exponents η, ν, γ, ζ are bounded from below by their classical values.

The physical mass m in the ϕ_2^4 quantum field model depends on the bare mass $m_0 > 0$, as well as on the other parameters of the theory. The cluster expansion converges for large m_0, and shows that $m > 0$ (Refs. 1 and 2) and that m is analytic in m_0.[3] As a consequence of correlation inequalities, m decreases as m_0 decreases for m_0 in an interval $(m_{0,c}, \infty)$.[4] According to the standard picture of symmetry breaking in the ground state (see Ref. 5), we expect that the lower end point $m_{0,c}$ of this interval is characterized by the condition $m = 0$. Two other alternatives, however, have not been excluded: $m_{0,c} = 0$ or $\langle \phi \rangle \neq 0$. In the Goldstone picture $m_0 > m_{0,c}$ corresponds to the single-phase region, and $m_{0,c} = 0$ would mean the absence of phase transitions. We expect that in a pure phase $\langle \phi \rangle$ is a continuous function of m_0, so we expect $\langle \phi \rangle = 0$ for $m_0 = m_{0,c}$. The two-dimensional Ising model is known to behave in this manner, while the open questions for the three-dimensional Ising model are similar to those discussed above for ϕ_2^4.

One purpose of this note is to prove that m is differentiable in m_0, for $m_0 > m_{0,c}$. We define $m_{0,c}$ to be the supremum of the m_0 for which $m = 0$ or $\langle \phi \rangle \neq 0$, and we define the ϕ_2^4 quantum field theory using Dirichlet boundary conditions. The convergence as $V \to \infty$ is based on monotonicity together with an upper bound on the Schwinger functions. For free boundary conditions, the upper bound follows from Ref. 6. The generalization of this upper bound to Dirichlet boundary conditions follows from the GKS inequality.[4,7] The monotonicity is established in Refs. 4 and 8. Our basic estimate is the correlation inequality that the four-point connected Euclidean Green's function is negative for $m_0 > m_{0,c}$.[5,9] We call this inequality the Lebowitz inequality.

To define m, choose $D > 0$ and consider the Euclidean expectation

$$F(D) = \int_{|\vec{x}|, |\vec{y}| < D} \langle \phi(x)\phi(y)\rangle_T \, d\vec{x}\, d\vec{y}\,\Big|_{x_0=0,\, y_0=D}, \qquad (1)$$

where $\langle\ \rangle_T$ denotes the truncated vacuum expectation value. For $m_0 > m_{0,c}$, $\langle\ \rangle_T = \langle\ \rangle$ in (1). Then

$$m = \lim_{D\to\infty} -\ln F(D)/D. \qquad (2)$$

The validity of (2) follows from the spectral representation of the two-point function and from standard properties of the two-point function.[6,10] We fix the Wick-ordering constants and vary $m_0 = m_0(\sigma)$ by the addition of a quadratic (mass) perturbation, $\frac{1}{2}\sigma \int :\phi(x)^2: dx$.[11] We define

$$m' \equiv \frac{dm}{d\sigma}. \qquad (3)$$

To bound the derivative m', we derive a bound on $-d[\ln F(D)/D]/d\sigma$ which is uniform in D and thus obtain the following:

Theorem 1. For each $\epsilon > 0$, $m(\sigma)$ is Lipschitz-continuous on the interval $\sigma \geq \sigma_c + \epsilon$. If $m(\sigma) \to 0$ as $\sigma \to \sigma_c$, then $m \leq (\sigma - \sigma_c)^{1/2}$.

Proof. Assuming $\sigma > \sigma_c$, we apply the GKS[4] inequality and the Lebowitz inequality[5] to obtain

$$\begin{aligned} 0 &\leq \frac{d}{d\sigma}[-\ln F(D)/D] \\ &= \frac{1}{2DF(D)} \int [\langle \phi(x) : \phi(z)^2 : \phi(y)\rangle \\ &\qquad - \langle \phi(x)\phi(y)\rangle\langle :\phi(z)^2:\rangle]\, d\vec{x}\, d\vec{y}\, dz \\ &\leq D^{-1}F(D)^{-1} \int \langle \phi(x)\phi(z)\rangle\langle\phi(z)\phi(y)\rangle\, d\vec{x}\, d\vec{y}\, dz. \end{aligned} \qquad (4)$$

To apply these inequalities we use a momentum cutoff as an intermediate step, in order to eliminate the Wick ordering.

The contribution of the region $z_0 \notin [0, D]$ or $|\vec{z}| \geq D$ to the upper bound in (4) is negligible, as $D \to \infty$ with σ fixed. As a consequence,

$$m' \leq \limsup_{D\to\infty} \sup_{0 \leq z_0 \leq D} \frac{\int \langle \phi(x)\phi(z)\rangle\langle\phi(z)\phi(y)\rangle\, d\vec{x}\, d\vec{y}\, d\vec{z}}{\int \langle \phi(x)\phi(y)\rangle\, d\vec{x}\, d\vec{y}}, \qquad (5)$$

where the $\vec{z}$ integration is restricted to $|\vec{z}| \leq D$.

We now justify the exclusion of these regions. We first consider the region I defined by $0 \leq z_0 \leq D$, $|\vec{z}| \geq D + D^{3/4}$. The spectral representation for $\langle \phi(x)\phi(z)\rangle\langle \phi(z)\phi(y)\rangle$ is

$$\int da\, db\, \rho(a)\rho(b)C_a(x-z)C_b(z-y),$$

where $C_a(x-y)$ is the kernel of $(-\Delta + a)^{-1}$. We choose a point w on the line xy with $w_0 = z_0$. Then

$$|x-z| \geq |x-w| + O(1)\min\{|z-w|^2/D, |z-w|\}$$
$$\geq |x-w| + O(D^{1/2}) + O(|\vec{z}|^{1/2}),$$

as $|\vec{z}| \to \infty$. For a given mass $a \geq m^2$,

$$C_a(x) \sim |x|^{-1/2} e^{-\sqrt{a}|x|} \quad \text{as } |x| \to \infty.$$

Thus,

$$C_a(x-z)C_b(z-y) \leq C_a(x-w)C_b(w-y)e^{-O(\delta)},$$

where $\delta = D^{1/2} + |\vec{z}|^{1/2}$. Thus we see that the integral over region I is bounded by $e^{-O(\delta)}$ times the integral over a strip of unit width about the line xy. Hence the integral over region I is negligible.

Furthermore, we see from the translation invariance of the integrands that the numerator and denominator both have magnitude $O(D^2 e^{-mD})$. Thus the region II defined by $0 \leq z_0 \leq D$, $D \leq |\vec{z}| \leq D + D^{3/4}$ contributes to the numerator at most the magnitude $O(D^{7/4}e^{-mD})$, and is negligible in the limit $D \to \infty$. Similarly, the region III defined by $z_0 \geq D + D^{3/4}$ and the region IV defined by $D + D^{3/4} \geq z_0 \geq D$ are negligible in (4) in the limit $D \to \infty$.

Let χ_D be the characteristic function of the interval $|\vec{z}| \leq D$ and let

$$f_D(z) = \int \langle \phi(x)\phi(z)\rangle \chi_D(\vec{z}) d\vec{x},$$

$$g_D(z) = \int \langle \phi(z)\phi(y)\rangle \chi_D(\vec{z}) d\vec{y},$$

so that the numerator in (5) is the L_2 inner product $\langle f, g\rangle$. Again, as $D \to \infty$, the angle between f_D and χ_D or between g_D and χ_D tends to zero. Thus as $D \to \infty$,

$$|\langle f, g\rangle| \leq \frac{\langle f, \chi\rangle\langle \chi, g\rangle}{\|\chi\|^2} + o(1),$$

or

$$m' \leq \limsup_{D\to\infty} \sup_{z_0 = w_0} \frac{\int\langle \phi(x)\phi(z)\rangle d\vec{x}\, d\vec{z} \int \langle \phi(w)\phi(z)\rangle d\vec{w}\, d\vec{z}}{2D\int\langle \phi(x)\phi(y)\rangle d\vec{x}\, d\vec{y}}, \tag{6}$$

where the integrals extend over $|\vec{x}|$, $|\vec{y}|$, or $|\vec{z}| \leq D$. We now show that we may replace the sup in (6) by the value at $z_0 = 0$ or D. Let

$$|t\rangle = e^{-tH/2}\int_{|\vec{x}| \leq D} \phi(\vec{x})\Omega, \quad F(t) = \langle t|t\rangle,$$

so the numerator in (6) is $G(t) = F(t)F(D-t)$, with $0 \leq t = z_0 = w_0 \leq D$. Note

$$(-1)^n F^{(n)}(t) = \langle t|H^n|t\rangle > 0,$$

and

$$G''(t) = F''(t)F(D-t) + F(t)F''(D-t)$$
$$-2F'(t)F'(D-t). \tag{7}$$

By the Schwarz inequality, $-F'(t) \leq F(t)^{1/2}F''(t)^{1/2}$, so from (7) we have $G''(t) \geq 0$. Thus $G(t)$ has its maximum value at an end point $T = 0$ or D. It follows that (6) factors to give

$$m' \leq \int \langle \phi(\vec{x})\phi(\vec{0})\rangle d\vec{x} = \tfrac{1}{2}\int \rho(a)a^{-1/2}da \leq (2m)^{-1}. \tag{8}$$

Here $\rho(a)$ is the spectral measure for the two-point function. The existence of a Hilbert space ensures that $\rho(a)da$ is a positive measure, and the canonical commutation relations ensure $\int \rho da = 1$. Thus integration of (8) down to $\sigma = \sigma_c$ completes the proof of the theorem.

Furthermore, if the mass is defined for a limit of Dirichlet data in a strip of finite width L, the mass $m(\sigma)$ is upper semicontinuous. This follows since $m(\sigma, L)$ is continuous and decreasing to $m(\sigma)$ as $L \to \infty$. Thus

$$m(\sigma) = \lim_{\epsilon \to 0_+} m(\sigma_c + \epsilon).$$

We remark that our bound (8) also yields an inequality for the mass m as a function of the bare mass m_0.

Theorem 2. Assume a pure $\phi_2{}^4$ interaction which is Wick-ordered in the bare vacuum for a field of mass m_0, the bare mass.

(a) Let $\epsilon > 0$. The mass $m(m_0)$ is Lipshitz-continuous on the interval $m_0 > m_{0,c} + \epsilon$.

(b) Assume $m_{0,c} \neq 0$, and $m(m_0) \to 0$ as $m_0 \to m_{0,c}$. Then the mass m is bounded by its classical value, with a logarithmic correction,

$$m \leq \left(m_0{}^2 - m_{0,c}{}^2 + 3\lambda \ln \frac{m_0{}^2}{m_{0,c}{}^2}\right)^{1/2}. \tag{9}$$

This theorem concerns the change of variables $\sigma \to m_0{}^2$, since we now Wick-order in the bare mass m_0. In Ref. 11 we saw that the change in Wick ordering constants due to the change in bare mass yields

$$\frac{d\sigma}{dm_0^2}=1+3\lambda m_0^{-2}. \qquad (10)$$

From this combined with theorem 1, we infer theorem 2a. From (8), we then obtain

$$\frac{dm^2}{dm_0^2}=2m\frac{dm}{d\sigma}\frac{d\sigma}{dm_0^2}\leq 1+3\lambda m_0^{-2},$$

which on integration yields (9).

The inequality (9) yields an inequality on the critical exponent ν governing the mass. Assuming that $m(m_{0,c}+\epsilon)\sim\epsilon^{\nu}$ as $\epsilon\to 0_+$, we see that $\nu\geq\frac{1}{2}$. (We give another elementary proof of this bound below.) The value $\nu=\frac{1}{2}$ is the classical value predicted by mean field theory or the Goldstone picture, while the value $\nu=1$ occurs for the two-dimensional Ising model (see Ref. 12). We believe that $\nu=1$ is more likely for ϕ_2^4 than $\nu=\nu_{cl}=\frac{1}{2}$, but we believe that $\nu\geq\frac{1}{2}$ is the best which can be obtained by methods which do not depend strongly on dimension, such as those used here. We note that Symanzik[13] has also used the Lebowitz inequality to bound an exponent.

We note that other exponents can also be bounded by their canonical values. The long-distance anomalous dimension η is greater than its classical value zero,

$$\eta\geq 0=\eta_{cl},$$

as follows in two dimensions from the observation that the two-point function is a monotonic decreasing function of $|x-y|$ (see Ref. 10). In greater than two dimensions, the free field has an $m\to 0$ limit, so the proof that $\eta\geq 0$ is a straightforward consequence of the Källén-Lehmann representation.

If we assume the existence of an isolated single-particle state of mass m (this is proved for λ small[1]) then the field-strength renormalization constant Z is defined by

$$G^{(2)}(p)=\int\langle\Phi(x)\Phi(0)\rangle e^{ipx}dx$$

$$=\frac{Z}{p^2+m^2}+\int_{m^2+\delta}^{\infty}\frac{\rho(a)da}{p^2+a}.$$

Let ζ be the critical exponent for Z, $Z\sim\epsilon^{\zeta}$. Since $0\leq Z\leq 1$, we have $\zeta\geq 0$. As discussed in Ref. 1, $\zeta=0$ is the classical value and is compatible with the existence of zero-mass particles. Since ρ is positive with total integral 1,

$$Z/m^2\leq G^{(2)}(0)\leq 1/m^2. \qquad (11)$$

Let γ be the critical exponent for the susceptibility $\chi=G^{(2)}(0)$,

$$G^{(2)}(0)=\int\langle\Phi(x)\Phi(0)\rangle dx\sim\epsilon^{-\gamma}.$$

Thus from (11) we obtain

$$2\nu-\zeta\leq\gamma\leq 2\nu. \qquad (12)$$

It is a reasonable hypothesis (apparently not accessible at present to proof) that $m'\leq O(1)Zm^{-1}$ and $\gamma=2\nu-\zeta$, improving the bounds (8) and (12), respectively.

From the Lebowitz inequality we obtain, as in (4), (9),

$$0\leq-\frac{dG^{(2)}(0)}{dm_0^2}=-(1+3\lambda m_0^{-2})\frac{dG^{(2)}(0)}{d\sigma}$$

$$\leq(1+3\lambda m_0^{-2})$$

$$\times\int\langle\Phi(x)\Phi(z)\rangle\langle\Phi(0)\Phi(z)\rangle dx\,dz$$

$$=(1+3\lambda m_0^{-2})G^{(2)}(0)^2. \qquad (13)$$

Integrating (13) down to the critical point yields

$$G^{(2)}(0)^{-1}\leq m_0^2-m_{0,c}^2+3\lambda\ln\frac{m_0^2}{m_{0,c}^2},$$

from which we obtain $\gamma\geq 1=\gamma_{cl}$, and by the upper bound in (12), $\nu\geq\frac{1}{2}=\nu_{cl}$. We summarize these bounds as follows:

Theorem 3. Suppose $m_{0,c}>0$ and $m\to 0$ as $m_0\to m_{0,c}$. Then the exponents η, ν, γ, and ζ are bounded from below by their classical values,

$$\eta\geq 0,\quad \nu\geq\tfrac{1}{2},\quad \gamma\geq 1,\quad \zeta\geq 0.$$

We raise the question whether other field-theory exponents are also bounded by their classical values.

Finally, we remark a simple inequality on α, the exponent for the specific heat C:

$$\epsilon^{-\alpha}\sim C\equiv\tfrac{1}{4}\int[\langle:\Phi(x)^2::\Phi(y)^2:\rangle$$

$$-\langle:\Phi(x)^2:\rangle\langle:\Phi(y)^2:\rangle]dy$$

$$\leq\tfrac{1}{2}\int\langle\Phi(x)\Phi(y)\rangle^2dy=\tfrac{1}{2}\int G^{(2)}(p)^2dp,$$

where again we use the Lebowitz inequality. From

$$G^{(2)}(p)=\int\frac{\rho(a)da}{p^2+a},$$

we find that

$$\int G^{(2)}(p)^2dp=\pi\int\frac{\rho(a)\rho(b)\ln(b/a)}{b-a}da\,db$$

$$\leq O(m^{-2}).$$

Here we employ an elementary estimate on $(b-a)^{-1}\ln(b/a)$, and hence we infer

$$\alpha\leq 2\nu. \qquad (14)$$

*Work supported in part by the National Science Foundation under Grant No. NSF-GP-24003.

†Permanent address: Rockefeller University, New York, New York 10021.

‡Work supported in part by the National Science Foundation under Grant No. NSF-GP-40354X.

§Permanent address: Lyman Laboratory of Physics, Harvard University, Cambridge, Massachusetts 02138.

[1]J. Glimm, A. Jaffe, and T. Spencer, Ann. Math. (to be published).

[2]J. Glimm, A. Jaffe, and T. Spencer, in *Constructive Quantum Field Theory*, edited by G.Velo and A. Wightman (Springer, Berlin, 1973).

[3]T. Spencer, Commun. Math. Phys. (to be published).

[4]F. Guerra, L. Rosen, and B. Simon, Ann. Math. (to be published).

[5]J. Glimm, A. Jaffe, and T. Spencer, in *Constructive Quantum Field Theory*, edited by G. Velo and A. Wightman (Ref. 2).

[6]J. Glimm and A. Jaffe, J. Math. Phys. 13, 1568 (1972).

[7]J. Fröhlich, Ann. Inst. Henri Poincaré (to be published).

[8]E. Nelson, in *Constructive Quantum Field Theory*, edited by G. Velo and A. Wightman (Ref. 2).

[9]J. Lebowitz, Commun. Math. Phys. 35, 87 (1974); B. Simon and R. Griffiths, *ibid.* 33, 145 (1973).

[10]B. Simon, Commun. Math. Phys. 31, 127 (1973); Ann. Math. (to be published).

[11]J. Glimm and A. Jaffe, Ann. Inst. Henri Poincaré 21, (1974).

[12]H. E. Stanley, *Introduction to Phase Transitions and Critical Phenomena* (Oxford Univ. Press, Oxford, England, 1971).

[13]K. Symanzik, Cargèse lectures, 1973 (unpublished).

Remark on the Existence of φ_4^4

James Glimm
Rockefeller University, New York, New York 10021

and

Arthur Jaffe*
Harvard University, Cambridge, Massachusetts 02138
(Received 10 June 1973)

The bound $|S_\epsilon^{(2)}(f\otimes g)| \leq \text{const}|f|_{\mathcal{S}}|g|_{\mathcal{S}}$, uniform as the lattice spacing $\epsilon \to 0$, is sufficient to yield construction of a φ^4 field theory. In perturbation theory, this bound holds in space-time dimensions $d \leq 4$. The bounds on the n-point Schwinger functions follow from the given bound by the use of Lebowitz's inequalities. They yield the growth $|S^{(n)}| \leq O(n!)$, consistent with the Osterwalder-Schrader axioms, which allow reconstruction of a real-time theory.

We remark that a uniform bound on the two-point Schwinger function (with a lattice approximation) yields the existence of a $\varphi_4^4 - \mu\varphi$ model without cutoffs, including the case $\mu = 0$. We assume that the two-point functions are labeled by a cutoff parameter ϵ, the lattice spacing, and that for some $\mathcal{S}$ norm, independent of ϵ,

$$|S_\epsilon^{(2)}(f\otimes g)| \leq |f|_{\mathcal{S}}|g|_{\mathcal{S}}. \tag{1}$$

Theorem 1.—With the assumption of (1), there exists a set of Schwinger functions $S^{(n)}(x_1, \ldots, x_n) = \lim_{\epsilon_j} S_{\epsilon_j}^{(n)}(x_1, \ldots, x_n)$ which satisfy the Osterwalder-Schrader axioms,[1] with the possible exception of Euclidean covariance and clustering. Here ϵ_j denotes a subsequence of lattice cutoffs, and $\epsilon_j \to 0$.

Remark.—As Wilson observed,[2] the nontriviality of the φ_4^4 model may be more difficult to establish than its existence. In particular, the hypothesis of Theorem 1 does not require the correct charge or mass renormalization, even in perturbation theory. Thus the limiting theory could describe a free field, and, still worse, the possibility $S^{(n)} \equiv 0$ is not excluded. Correct mass (and field-strength) renormalization prevents the vanishing of $S^{(2)}$.

We obtain uniform bounds on the n-point function in terms of the two-point function, and convergence of the $S_{\epsilon_j}^{(n)}$ by compactness. The uniform bounds depend on correlation inequalities which were proved in statistical mechanics by Lebowitz,[3] and which follow in φ^4 field theory by the spin-$\frac{1}{2}$ approximation of Simon and Griffiths.[4] Our use of these inequalities in field theory was suggested by the work of Feldman[5] and especially the work of Spencer[6] on the elimination of even bound states in the φ^4 model. See also Glimm and Jaffe,[7] where the inequality on the connected four-point function, $S_T^{(4)} \leq 0$, yields bounds on critical exponents by their mean-field values.

Proposition 1.—Under the above hypothesis (1), with $f_i \geq 0$ and $\mu \geq 0$,

$$0 \leq S_\epsilon^{(n)}(f_1 \otimes \cdots \otimes f_n) \leq 2^{n-1}(n-1)!\prod_{i=1}^{n}|f_i|_{\mathcal{S}}, \tag{2}$$

where $|f_i|_{\mathcal{S}}$ is the same norm as in (1).

Proof.—The proof involves an induction hypothesis. We assume $\mu \geq 0$ and use the Lebowitz inequalities[3]

$$\langle qt\rangle - \langle q\rangle\langle t\rangle \leq 0. \tag{3}$$

Here

$$t(x) = \tfrac{1}{2}(\varphi + \tilde{\varphi}), \quad q(x) = \tfrac{1}{2}(\varphi - \tilde{\varphi}),$$

$$t = \Pi_i t(x_i), \quad q = \Pi_i q(x_i),$$

and $\tilde{\varphi}$ denotes a duplicate Euclidean field variable. The restriction $\mu \geq 0$ is no loss of generality, by the $\varphi \to -\varphi$ symmetry. We take

$$t = \prod_{i=1}^{n} t(x_i), \quad q = q(x_{n+1})q(x_{n+2}),$$

and we write (3) as

$$\textstyle\sum_X S(\sim X)[S(X, x_{n+1}, x_{n+2}) - S(X)S(x_{n+1}, x_{n+2}) + S(X)S(x_{n+1})S(x_{n+2})] \leq \sum_X S(X, x_{n+1})S(\sim X, x_{n+2}), \tag{4}$$

where X denotes a subset of $x_1, \ldots, x_n$, and $\sim X$ the complement of X. By the Griffiths-Kelly-Sherman inequalities,[8] each term on the left-hand side of (4) is positive. Thus, in particular,

$$0 \leq S^{(n+2)}(x_1, \ldots, x_{n+2}) \leq S^{(n)}(x_1, \ldots, x_n)S^{(2)}(x_{n+1}, x_{n+2}) + \textstyle\sum_X S(X, x_{n+1})S(\sim X, x_{n+2}), \tag{5}$$

and we have bounded $S^{(n+2)}$ by a sum of products of lower-order Schwinger functions.

The inductive hypothesis we use is (2) for $f_i \geq 0$ and $n \geq 1$. For $n=2$, the hypothesis follows from (1), while for $n=1$, it follows from the Schwarz inequality and from (1):

$$S^{(1)}(f) \leq S^{(2)}(f \otimes f)^{1/2} \leq |f|_{\mathcal{S}}.$$

Assuming the inductive hypothesis for $S^{(1)}, \ldots, S^{(n+1)}$, we prove it for $S^{(n+2)}$. By (5),

$$S^{(n+2)}(f_1 \otimes \cdots \otimes f_n) \leq 2^n[(n-1)! + \sum_X |X|!(n-|X|)!] \prod_{i=1}^{n+2} |f_i|_{\mathcal{S}} \leq 2^{n+1}(n+1)! \prod_{i=1}^{n+2} |f_i|_{\mathcal{S}}.$$

Here we use the fact that with $|X|=j$ there are $\binom{n}{j}$ choices for the subset. This completes the proof of the induction hypothesis.

We consider the norms on $\mathcal{S}(R^d)$,

$$|f|_r = \sup_x |(1+x^2)^r f(x)|.$$

Proposition 2.—With the assumption of (1), there exists $r < \infty$ such that

$$|S_\epsilon^{(2)}(f \otimes g)| \leq \text{const} |f|_r |g|_r,$$

with the constant uniform in ϵ.

Proof.—Because $S^{(2)}$ is positive, and defined on $\mathcal{S} \otimes \mathcal{S}$, it is a positive measure. Let $0 \leq \chi_j(x) = \chi(x-j)$ be a C_0^∞ partition of unity, with $j \in Z^d$. Then

$$|S^{(2)}(f \otimes g)| \leq \sum_{i,j} (\sup_{\text{supp}\chi_i \otimes \chi_j} |f \otimes g|) S^{(2)}(\chi_i \otimes \chi_j).$$

Since the χ_i are translates of a particular function, for some $m < \infty$,

$$S^{(2)}(\chi_i \otimes \chi_j) < |\chi_i|_{\mathcal{S}} |\chi_j|_{\mathcal{S}} \leq \text{const}(1+i^2)^m(1+j^2)^m,$$

with the constant uniform in ϵ. By choosing $r = m+d$, the proposition follows.

By linearity, we then obtain from Propositions 1 and 2 the following:

Corollary.—With the assumption of (1), there exists $r < \infty$, and $c < \infty$, such that

$$|S_\epsilon^{(n)}(f_1 \otimes \cdots \otimes f_n)| \leq c^n n! \prod_{i=1}^{n} |f_i|_r. \tag{6}$$

Here $f_i \in \mathcal{S}(R^d)$, and the constants are uniform in ϵ.

Proof of Theorem 1.—The existence of a convergent subsequence, as $\epsilon \to 0$, follows from the Corollary to Proposition 2 by compactness, and the same bounds hold in the limit. The symmetry and Osterwalder-Schrader positivity of the lattice Schwinger functions are known for a finite-volume theory[9] and easily extended to the infinite-volume and $\epsilon = 0$ limits. This completes the proof.

Corollary.—With the assumption of (1), the generating function

$$Z\{f\} = \sum_{n=0}^{\infty} S^{(n)}(f \otimes \cdots \otimes f)/n!$$

is an analytic function for f in a neighborhood of the origin in $\mathcal{S}(R^d)$. It is also the characteristic function for a unique measure dq on $\mathcal{S}'(R^d)$, i.e.,

$$Z\{f\} = \int e^{\varphi(f)} dq, \quad S^{(n)}(f_1, \ldots, f_n) = \int \varphi(f_1) \cdots \varphi(f_n) dq. \tag{7}$$

Proof.—The existence and analyticity of Z follow immediately from (6), while the existence and uniqueness of dq follow by Minlos's theorem.

We remark that a number of other known properties of $(\varphi^4 - \mu\varphi)_2$ are also a consequence of hypothesis (1). In two and three dimensions, Theorem 1 may provide an alternative method for proof of existence of the φ^4 theory. In three dimensions, the finite-space-time-volume φ^4 theory exists[10-11] and is a limit as $\epsilon \to 0$ of a lattice theory.[12] It is sufficient to verify hypothesis (1) in the finite-volume, $\epsilon = 0$ theory, since the Lebowitz inequality (3) and the proof of (2) extend to the $\epsilon = 0$ limit. Then the norm $|f|_{\mathcal{S}}$ must be uniform in the volume.

*Work supported in part by the National Science Foundation under Grant No. NSF GP 40354X.

[1]K. Osterwalder, in *Constructive Quantum Field Theory*, edited by G. Velo and A. Wightman (Springer, Berlin, 1973). See also J. Fröhlich, to be published; K. Osterwalder and R. Schrader, to be published, Axiom (E0‴).

[2]K. Wilson and J. Kogut, to be published.

[3]J. Lebowitz, Commun. Math. Phys. 35, 87 (1974).

[4]B. Simon and R. Griffiths, Commun. Math. Phys. 33, 145 (1973).

[5]J. Feldman, "On the Absence of Bound States in the φ^4 Quantum Field Model without Symmetry Breaking" (to be published).

[6]T. Spencer, "The Absence of Even Bound States in φ_2^4" (to be published).

[7]J. Glimm and A. Jaffe, Phys. Rev. D (to be published).

[8]J. Ginibre, Commun. Math. Phys. 16, 310 (1970).

[9]F. Guerra, L. Rosen, and B. Simon, to be published; J. Fröhlich, to be published.

[10]J. Glimm and A. Jaffe, Fortschr. Phys. 21, 327 (1973).

[11]J. Feldman, to be published.

[12]Y. Park, "Lattice Approximation of the $(\varphi^4 - \mu\varphi)_3$ Field Theory in a Finite Volume" (to be published).

[13]We have been informed that J. Feldman has independently proved Proposition 1.

On the approach to the critical point [1]

by

James GLIMM [2]

Rockefeller University New York, New York 10021

and

Arthur JAFFE [3]

Harvard University Cambridge, Massachusetts 02138

ABSTRACT. — The mass perturbation yields a new construction for $\mathscr{P}(\varphi)_2$ = Even + Linear quantum field models. For φ^4 models, the Schwinger functions are differentiable in the parameter $\sigma = \delta m^2$, in the single phase region, and are continuous as $\sigma \searrow \sigma_c$. For the Ising$_d$ model, $d \geq 2$ classical bounds on the critical exponents, *e. g.* $1 \leq \gamma \leq 2\nu$, are a consequence of an Ornstein-Zernike type upper bound on the two point function.

1. INTRODUCTION

For certain critical values of the parameters, the Schwinger functions of quantum field theory or correlation functions of statistical mechanics are expected to exhibit nonexponential decay,

$$S^{(2)}(x) \equiv \int \varphi(x)\varphi(0)dq \sim |x|^{-d+2-\eta}$$

(1) We thank the Aspen Center for Physics for hospitality.

(2) Supported in part by the National Science Foundation under Grant MPS 74-13252.

(3) Supported in part by the National Science Foundation under Grant MPS 73-05037.

where $0 \leq \eta \leq 2$ and with the result that the corresponding integrated quantities, for example the susceptibility,

$$\chi \equiv \int S^{(2)}(x)dx$$

become infinite. Let $S^{(n)}(\sigma; x)$ denote the dependence of the n-point Schwinger function in an even $\mathscr{P}(\varphi)_2$ model on the mass perturbation

$$\frac{1}{2}\sigma \int :\varphi(x)^2: dx.$$

An elementary application of Griffiths' inequalities gives three ranges of parameter values as follows: $\sigma < \sigma_c^-$ defines a multiple phase theory, $\sigma_c^- < \sigma < \sigma_c^+$ defines a zero mass (critical) theory, and $\sigma_c^+ < \sigma$ defines a positive mass, single phase theory. It is expected that $\sigma_c^- = \sigma_c = \sigma_c^+$; however in some cases, σ_c may not define a zero mass theory, see chapter 2.

As $\sigma \to +\infty$, the limit $S^{(n)}(\sigma, x) \to 0$, $n \geq 1$, follows from the cluster expansion [15] [16]. Callan and Symanzik, see [2] [29], propose the mass perturbation, as in the formula

$$S^{(n)}(\sigma_c, x) = -\int_{\sigma_c}^{\infty} \partial_\sigma S^{(n)}(\sigma, x)d\sigma, \tag{1.1}$$

for the study of critical phenomena. In fact with g denoting the physical charge, hypotheses *e. g.* on

$$\beta(g) \equiv \frac{m}{2}\left(\frac{dg}{d\sigma}\frac{d\sigma}{dm}\right)_{\mathscr{P}=\text{const.}}$$

near $\sigma = \sigma_c$ lead to the scaling law and calculation of critical exponents in terms of *e. g.* $\beta'(g_c)$, see [3] [24].

In this paper, we show that the derivatives in (1.1) exist for φ^4 models in the single phase region, as a consequence of the Lebowitz inequalities [12]. The Lebowitz inequalities dominate $\partial S^{(n)}/\partial\sigma$ by a sum of products of two point functions,

$$0 \leq -\partial S^{(n)}/\partial\sigma \leq \sum \prod_{i,j\neq k,l} S^{(2)}(x_i - x_j) \times \int S^{(2)}(x_k - y)S^{(2)}(x_l - y)dy, \tag{1.2}$$

see chapter 3.

2. CONSTRUCTION OF THE $\mathscr{P}$ = EVEN − $\mu\varphi$ MODELS

We give a new construction of the $\mathscr{P}$ = Even + Linear model in two dimensions. We start from a polynomial $\mathscr{P}$ with a sufficiently large bare mass, so that it yields a Wightman theory as a consequence of a convergent cluster expansion [15] [16]. We study the associated Schwinger functions under a two parameter family of perturbations, of the form $\frac{1}{2}\sigma\varphi^2 - \mu\varphi$. Using a Griffiths inequality, we show that for $\sigma \leq 0$ and $\mu \geq 0$, the Schwinger functions converge monotonically to Schwinger functions for an Osterwalder-Schrader theory [23], with the possible exception of clustering. In the case of mixed phases, the decomposition into pure phases satisfying all the Osterwalder-Schrader axioms is given in [1] [5]. Each pure phase then gives rise to a $\mathcal{P}(\varphi)_2$ Wightman theory.

We consider the perturbation for the action,

$$\delta V = \frac{1}{2}\sigma \int_{\Lambda_1} :\varphi^2: dx - \mu \int_{\Lambda_2} \varphi dx, \tag{2.1}$$

where $\mu, -\sigma \geq 0$, and where Λ is a bounded space-time region with volume $|\Lambda|$. We suppose that the linear term in $\mathcal{P}$ has a nonpositive coefficient.

PROPOSITION 2.1. — Under the perturbation δV, the $\mathcal{P}(\varphi)_2$ Schwinger functions increase monotonically in $-\sigma$, μ, $|\Lambda_i|$.

Proof [18]. — By the first Griffiths inequality, $S^{(n)}(X) \geq 0$, and by the second Griffiths inequality

$$-\frac{d}{d\sigma} S^{(n)}(X) = \int_{\Lambda_1} [S^{(n+2)}(Xzz) - S^{(n)}(X)S^{(2)}(zz)]dz \geq 0. \tag{2.2}$$

For the lattice approximation to field theory, the Griffiths inequalities follow from [6], for $\mathcal{P} = \text{even} - \mu\varphi$, $\mu \geq 0$. These inequalities follow in field theory from convergence of the lattice approximation ([18] for $\mathcal{P}(\varphi)_2$, [25] for φ_3^4). Here we note the Wick ordering constant cancels between the two terms on the right of (2.2). This proves that $S^{(n)}$ increases monotonically in $-\sigma$, $|\Lambda_1|$. The proof of monotonicity in μ, $|\Lambda_2|$ is similar.

We remark that the Schwinger functions are uniformly bounded in Λ_i, σ, and μ, so long as σ and μ remain bounded (This is a consequence of Hamiltonian φ-bounds [8]. See Fröhlich [5] for the translation of these bounds to the Schwinger functions). Hence the Schwinger functions converge as $|\Lambda_i| \to \infty$, as $\sigma \searrow \bar{\sigma} > -\infty$, and as $\mu \nearrow \bar{\mu} < \infty$. Clearly the limiting Schwinger functions are Euclidean covariant, and the remaining Osterwalder-Schrader axioms (except clustering) follow from the properties of the approximate Schwinger functions.

THEOREM 2.2. — For each polynomial $\mathcal{P}(\varphi)$ with a convergent cluster expansion, we obtain a two parameter family of (possibly mixed phase) Osterwalder-Schrader theories, with Schwinger functions $S^{(n)}(\sigma, \mu; X)$.

Each theory $\{ S^{(n)}(\sigma, \mu; X) \}$ can be decomposed into a direct integral of pure phases, with each component an Osterwalder-Schrader theory.

REMARK 2.3. — We note that the theories $S^{(n)}(\sigma, \mu; X)$ are uniquely defined by the bare parameters, together with the bare mass. Thus with $\sigma_i \leq 0$, $\mu_i \geq 0$, we can add successively the perturbations

$$\delta V_i = \frac{1}{2}\sigma_i \int_{\Lambda_i} :\varphi^2: dx - \mu_i \int_{\Lambda_i} \varphi dx,$$

taking the limit $\Lambda_i \nearrow R^2$ in any order, and obtain $S^{(n)}(\sigma, \mu; X)$, where $\sigma = \Sigma\sigma_i$, and $\mu = \Sigma_i\mu_i$. The fact that the limit $S^{(n)}(\sigma, \mu; X)$ is independent of the order in which the perturbations are added has the following interpretation. If a phase transition occurs at the parameter values σ, μ, we can regard $S^{(n)}(\sigma, \mu; X)$ as defined by a δV_i perturbation, starting from the boundary conditions $S^{(n)}(\sigma^{(i)}, \mu^{(i)}; X)$, where $\sigma^{(i)} = \Sigma_{j \neq i}\sigma_j$ and $\mu^{(i)} = \Sigma_{j \neq i}\mu_j$. Since $S^{(n)}(\sigma, \mu; X)$ is independent of the order in which the δV_i perturbations are added, the various boundary conditions $\sigma^{(i)}$, $\mu^{(i)}$ define the same theory. Allowing for an infinite series σ_j, μ_j, we see that

$$S^{(n)}(\sigma, \mu ; X) = \lim_{\sigma \searrow \sigma', \mu' \nearrow \mu} S^{(n)}(\sigma', \mu' ; X)$$

Because of the possibility of phase transitions, the iterated limits with some or all $\mu_i \leq 0$ may not coincide with the $S^{(n)}(\sigma, \mu; X)$. In fact the iterated limits with each $\sigma_i \mu_i \leq 0$ are always defined (in any order) by monotonicity, and should provide a general set of boundary conditions. In particular, for φ^4, both pure phases as well as the even mixture could be constructed by this method. From this point of view, the construction of Theorem 2.2 with a single δV perturbation could be called *weak coupling boundary conditions.*

REMARK 2.4. — For φ^6 and higher order $\mathscr{P}(\varphi)$ theories, it is possible that curves of phase transitions can occur in the σ, μ plane for $\mu \neq 0$. Because the above iterated limits coincide, any such curve in the $\mu > 0$ half plane must have a positive slope in the σ, μ plane, at least in the case of a first order phase transition with respect to order parameter $\langle \varphi \rangle$. Standard mean field approximations suggest that there should be a phase transition only at $\mu = 0$ if the φ^4, φ^6, ... coefficients in an even $\mathscr{P}$ are fixed and $\sigma \ll 0$. This picture also suggests that the curves for $\mu > 0$ or $\mu < 0$ move in toward and meet the $\mu = 0$ line as σ decreases. It is consistent with this picture that $\sigma_c^+ = \sigma_c^-$ (as defined in chapter 3) could give the value of σ for which two such curves meet the $\mu = 0$ line. In this case it is reasonable to suppose that $m(\sigma_c) > 0$. See figure 1. The variation of the φ^4 and higher coefficients is expected to lead to tricritical points, see [26]. There heuristic calculations indicate classical tricritical exponents for $d = 3$, in contrast to an ordinary critical point.

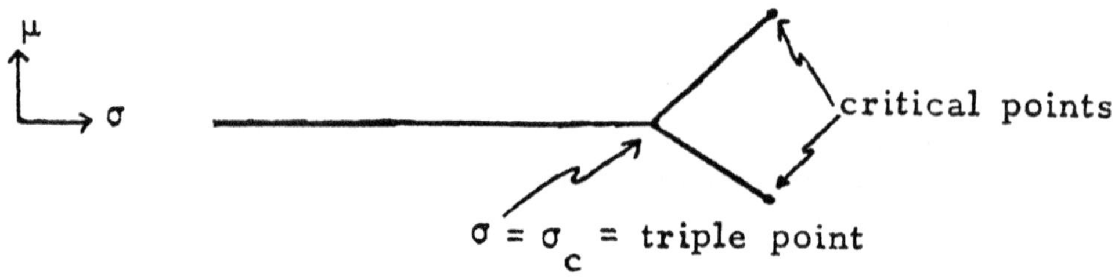

FIG. 1. — Postulated phase transition curves for $\varphi^6 + \lambda\varphi^4$ models, $\lambda \ll 0$.

REMARK 2.5. — The above construction (weak coupling boundary conditions) coincides with the theory defined by Dirichlet boundary data. In fact if the weakly coupled $\mathscr{P}(\varphi)_2$ theory defined above by a cluster expansion has Dirichlet data on the boundary of a region Λ_0, then $S(\sigma)$ is monotone increasing in Λ_0. The limits $\Lambda_0 \nearrow R^2$, $\Lambda_1 \nearrow R^2$, $\Lambda_2 \nearrow R^2$ may be taken in any order. With $\Lambda_0 \nearrow R^2$ first, we have weak $\mathscr{P}(\varphi)_2$ boundary conditions, since the cluster expansion defines a theory which is independent of boundary conditions. With $\Lambda_0 = \Lambda_1 = \Lambda_2 \nearrow R^2$, we have the conventional Dirichlet theory.

The weak coupling boundary conditions also coincide with free boundary data, provided only that $m_0 \gg 1$ and that the space cutoff in the mass term $\frac{1}{2}\sigma\int :\varphi^2: dx$ is removed after the space cutoff in the other terms of $H_I = \int \mathscr{P}(\varphi)dx$. It follows that the estimates of [*8*], established for the C*-algebra construction of $\mathscr{P}(\varphi)_2$ field theory [*7*], also have consequences for the theory constructed here; see section 5. An extensive analysis of even boundary conditions is given in [*19*]. It is an open question whether the periodic mass converges in the infinite volume limit to the mass of the infinite volume theory. A positive answer to this question would show that for φ^4, $m(\sigma) \searrow = 0$ as $\sigma \searrow \sigma_c$.

3. THE CRITICAL POINT IN EVEN $\mathscr{P}(\varphi)_2$ MODELS

The variation of the mass parameter σ in an even $\mathscr{P}(\varphi)_2$ model (see chapters 1, 2) yields three characteristic intervals:

a) a single phase interval (σ_c^+, ∞),

b) a critical interval (σ_c^-, σ_c^+),

c) a multiple phase interval $(-\infty, \sigma_c^-)$ with symmetry breaking.

The endpoints $\sigma_c^\pm$ may also be included in the intervals *a*) *c*).

We let $\mu = 0$, and let

$$M(\sigma)^2 = \lim_{|x|\to\infty} S^{(2)}(\sigma\,;\,x).$$

We define $m(\sigma)$ as the exponential decay rate of $S^{(2)}(\sigma; x) - M(\sigma)^2$. Consider the two conditions,

$$m(\sigma) > 0 \tag{3.1}$$

$$M(\sigma) = 0. \tag{3.2}$$

DEFINITION 3.1. — Let σ_c^+ be the infimum of the values of σ satisfying both (3.1) and (3.2). Let σ_c^- be the supremum of the values of σ which violate (3.2).

REMARK 3.2. — With the above definitions:

(*i*) $\sigma_c^- \leq \sigma_c^+ < \infty$.
(*ii*) On $(-\infty, \sigma_c^-)$, (3.2) is violated.
(*iii*) On (σ_c^-, σ_c^+), (3.2) holds, but $m = 0$.
(*iv*) On (σ_c^+, ∞), both (3.1) and (3.2) hold.

By scaling, $\sigma \to +\infty$ is the weak coupling limit. For weak coupling, the cluster expansion [*15*] [*16*] yields (3.1) and (3.2), so $\sigma_c^+ < \infty$. The remaining statements follow by the monotone decrease of $S^{(2)}(x; \sigma)$ in σ. The bound $|\sigma_c^-| < \infty$ is announced in [*4*]; see also [*5*].

THEOREM 3.3. — Let $\mu = 0$. Then $m(\sigma)$ is continuous on (σ_c^+, ∞) at all points of continuity of $S^{(2)}(\sigma; x)$.

THEOREM 3.4. — Let $\mu = 0$, and suppose (3.2) holds for $\sigma = \sigma_c^+$. Then $m(\sigma) \searrow m(\sigma_c^+)$ as $\sigma \searrow \sigma_c^+$.

THEOREM 3.5. — Let $\mu = 0$ and suppose $m(\sigma) \not\to 0$ as $\sigma \searrow \sigma_c^+$. Then (3.2) holds for $\sigma = \sigma_c^+$, and so $m(\sigma) \searrow m(\sigma_c^+)$ as $\sigma \searrow \sigma_c^+$.

REMARK 3.6. — Assuming φ_3^4 to be a Wightman theory, $M(\sigma_c) = 0$. This follows from the fact that $S^2(\sigma; x) \leq |x|^{-1}$. Thus $m(\sigma) \searrow m(\sigma_c^+)$.

REMARK 3.7. — See Figure 1. In that case, the limit $\sigma \searrow \sigma_c$ should yield the pure phase associated with $M(\sigma_c) = 0$, but with $m(\sigma_c) > 0$.

REMARK 3.8. — $\mathscr{P}(\varphi)_2$ critical theories, *i. e.* $m = 0 = M$, are not scale invariant. The continuum of such theories is parameterized by a scale length, see ([*16*], p. 155).

Proof of Theorems 3.3-3.5. — Since $S^{(2)}$ is monotone decreasing for $\sigma \in (-\infty, \infty)$, and continuous from above, it follows that $m(\sigma)$ is monotone increasing on (σ_c^-, ∞). If σ_0 satisfies (3.2), then $m(\sigma) \searrow m(\sigma_0)$ as $\sigma \searrow \sigma_0$, for the exponential decay rate of $S^{(2)}$ defines $m(\sigma_0)$. This proves Theorem 3.4. The same argument applied to convergence from below completes the proof of Theorem 3.3. Theorem 3.5 follows from the observation that if $m \not\to 0$, then $S^{(2)}$ has a uniform exponential decay as $\sigma \searrow \sigma_c$.

THEOREM 3.9. — For a φ_2^4 theory with $\mu = 0$, the derivatives $\partial S^{(n)}(\sigma; x)/\partial\sigma$ exist on (σ_c^+, ∞), and are bounded by (1.2). Furthermore $\partial Z(\sigma, f)/\partial\sigma$ exists, where $Z(\sigma, f) = \langle e^{\Phi(f)} \rangle_\sigma$, and $| f |_{L_p} \ll 1$.

Proof. — We use the Lebowitz inequalities

$$S^{(n+2)}(12U) - S^{(n)}(U)S^{(2)}(12) \leq \sum_{X\,\mathrm{odd}} S(1X)S(2 \sim X), \tag{3.3}$$

where X is a subset of U, see [*12*, eq. 5]. By induction, the right side of (3.3) can be bounded by a sum of products of two point functions of the stated form. The sum in (1.2) ranges over the $\leq (n+1)!$ terms which arise from this bound. The inductive hypothesis we use is the following: $S^{(2n)}$ is bounded by a sum of at most $(2n - 1)!$ products of n two point functions.

We first differentiate with respect to a change in σ in a finite volume Λ. By Remark 2.3, with $\delta\sigma \leq 0$,

$$S^{(n)}(\sigma) - S^{(n)}(\sigma + \delta\sigma)$$
$$= \lim_{\Lambda\to\infty} \int_{\sigma+\delta\sigma}^{0} \int_{\Lambda} [S^{(n+2)}(s, \Lambda, z, z) - S^{(n)}(s, \Lambda, U)S^{(2)}(s, \Lambda, z, z)]dz ds$$

where $S(s, \Lambda)$ has a mass perturbation σ in all of R^2, and a mass perturbation $s - \sigma$, $\delta\sigma \leq s - \sigma \leq 0$, in Λ. We substitute (3.3) and (1.2) as an upper bound for the integrand. By the Lebesgue monotone convergence theorem, we may take the limit $\Lambda \to \infty$ under the integral sign. Convergence of the z integration follows from the positive mass and exponential decay of $S^{(2)}$.

The derivative of $Z(\sigma, f)$ exists by summation of the exponential series.

REMARK 3.10. — The existence of the derivatives of the connected Euclidean Green's functions $S_T^{(n)}(x) \equiv G^{(n)}(x)$ follows from this result. The derivative $dG^{(n)}/d\sigma$ is bounded in the norms of [*9*].

THEOREM 3.11. — For a φ_2^4 theory with $\mu = 0$, the Euclidean vertex functions $\Gamma^{(n)}(x)$ have bounded derivatives $d\Gamma^{(n)}(x)/d\sigma$, on the interval (σ_c^+, ∞).

Proof. — By the definition of the $\Gamma^{(n)}(x)$ in terms of the $G^{(n)}(x)$ and $\Gamma^{(2)} = -G^{(2)-1}$, and Remark 3.10, it is sufficient to prove the boundedness of $d\Gamma^{(2)}/d\sigma$ as a map from H_1 to H_{-1}. By definition,

$$\frac{d\Gamma^{(2)}}{d\sigma} = \Gamma \frac{dG^{(2)}}{d\sigma} \Gamma.$$

Since Γ is a bounded transformation from H_1 to H_{-1}, we need only show that $dG^{(2)}/d\sigma$ is bounded from H_{-1} to H_1. However, as in (1.2) or [11],

$$0 \leq -\frac{dG^{(2)}(x)}{d\sigma} \leq \int G^{(2)}(x - z)G^{(2)}(z)dz. \tag{3.4}$$

For $\sigma \in (\sigma_c^+, \infty)$, we have $m(\sigma) > 0$, so the Fourier transform $G^{(2)\sim}(p)$ is bounded by $m(\sigma)^{-2}$. Since $G^{(2)}(x)$ is bounded fiom H_{-1} to H_1, it follows that $G^{(2)} {}_* G^{(2)}$ is also. Hence so is $dG^{(2)}/d\sigma$, by (3.4).

4. THE ISING MODEL CRITICAL POINT

In this section we obtain bounds on critical exponents in Ising models. The analysis of Section 3 may also be applied to the Ising model, with $\beta \nearrow \beta_c$ replacing $\sigma \searrow \sigma_c$. The bound on $dS^{(n)}/d\beta$ by a product of two point functions is known [20]. In our 'analysis of the Ising model we assume an Ornstein-Zernike upper bound on the two point function,

$$S^{(2)}(\beta \,;\, x) \leq \text{const.} \int \frac{e^{-ipx}}{p^2 + m^2} dp = S_0^{(2)}(m^2 \,;\, x) \tag{4.1}$$

where the constant is independent of β and where $m = \xi^{-1}$ is the inverse correlation length. In field theory, (4.1) follows from the Lehmann spectral formula. From this hypothesis, we have the bound $\eta \geq 0$ on the anomalous dimension. We now use (4.1) to derive other bounds on critical exponents.

THEOREM 4.1. — Assume the bound (4.1). Then:

(*a*) For the ferromagnetic Ising model with $\beta < \beta_c$, $d \geq 2$,

$$0 \leq \frac{d\chi}{d\beta} \leq 0(1)\chi^2. \tag{4.2}$$

(*b*) Assuming the existence of critical exponents, they satisfy

$$1 \leq \gamma \leq 2\nu \tag{4.3}$$

$$\Delta_4 \leq \frac{1}{2}(d\nu + \gamma) \tag{4.4}$$

$$\Delta_4 + \Delta_6 \leq d\nu + \gamma \tag{4.5}$$

REMARK 4.2. — The inequality

$$0 \leq -\frac{dm}{d\beta} \leq 0(m^{-1}) \tag{4.6}$$

also follows by slight modification of [11], using a transfer matrix formalism [22]. The bounds (4.1-6) hold for a φ^4 field theory.

Proof. — The inequality (4.2) follows as in [*11*], and yields $1 \leq \gamma$. Combined with (4.1) this yields (4.3). The inequalities (4.4) and (4.5) are a consequence of (4.1) and the bounds of [*13*].

5. PRESERVATION OF ESTIMATES

In this section, we sketch how to transfer uniform estimates, *e. g.* estimates of the form

$$e^{-(H\pm A)} \leq \text{const.}$$

established for $V < \infty$, to the infinite volume limit. This justifies the statement in Remark 2.5 that the Hamiltonian estimates [*8*] can be used in the infinite volume limit with weak coupling boundary conditions.

We are concerned with perturbations of the form

$$A_j = \int : \varphi^j : (x) h_j x dx, \tag{5.1}$$

with $j \leq \deg \mathscr{P} \equiv d$, with suppt. $h_j \subset [-a, a]$, and with $h_j \in L_{d/d-j}$, see [*5*]. In case $j = d$, we also require $\| h_j \|_\infty$ small. We also introduce the mass perturbation

$$B_2 = \frac{1}{2}\sigma \int_{-b}^{b} : \varphi^2 : (x) dx. \tag{5.2}$$

The estimates will be formulated in semigroup language, because the properties [*5*] of

$$z \to e^{-t(H + z_2 B_2 + z_j A_j)} \tag{5.3}$$

permit control of the semigroups without considering questions of operator domains. We use the path space notation implicit in the fact that the theories constructed here are Osterwalder-Schrader reconstructions [*23*].

THEOREM 5.1. — The Feynman-Kac density

$$e^{-\int_0^T [B_2(q(t)) + A_j(q(t))]dt}$$

defines a semigroup on the physical Hilbert space $\mathscr{H}$.

Proof. — This statement follows by limits from the approximate theories, given the convergence of the weak coupling Schwinger functions with φ^j vertices, and the uniform bounds on the analytic functions (5.3).

In particular, with b and a fixed, the bounds of [*8*] are transfered into the weak coupling infinite volume limit by the argument of this section. See Remark 2.9.

The semigroup of Theorem 5.1 will be called $\exp[-tH(\sigma, j)]$. We identify

$$H(\sigma, j) = H + B_2 + A_j \tag{5.4}$$

on the domain

$$\mathscr{D} = \{ e^{-sH}\mathscr{R}\Omega : s > 0 \}$$

where $\mathscr{R}$ is a polynomial in fields and Ω is the ground state of H. By the reconstruction theorem [23]. $\mathscr{D} \subset D(H)$ and by the cluster expansion $\mathscr{D} \subset D(B_2) \cap D(A_j)$. The limits of the difference quotients from finite volume then establish (5.4) on $\mathscr{D} \times \mathscr{D}$, and hence by the above domain inclusions (5.4) holds on $\mathscr{D}$. In the case of the mass perturbation, $H + B_2$, (5.4) extends to $D(H) \times D(H)$ by Spencer's local N_τ estimate [28]. The fact that $N_{\tau,\mathrm{loc}} \geq 0$ allows the transfer of the first order estimate on $N_{\tau,\mathrm{loc}}$ to the infinite volume limit [1]. Let

$$\delta E(\sigma, b, h_j) = - \ln \| \exp [- H(\sigma, j)] \|, \tag{5.5}$$

$$\delta E(\sigma, b) = \delta E(\sigma, b, \phi) = \inf \text{spect.} (H + B_2). \tag{5.6}$$

We normalize the Wick ordering so that

$$\langle : \varphi^2 : \rangle = 0, \tag{5.7}$$

with expectations taken in the weak coupling physical Hilbert space. Then $\langle : \varphi^j : \rangle = 0$, with the standard definition of Wick order. The absence of domain difficulties follows from the cluster expansion, see [27].

Let $\delta E^*(\sigma, b, h_j)$ be the lim sup of the finite volume (weak coupling) approximates to $\delta E(\sigma, b, h_j)$. Then

$$- \delta E \leq - \delta E^*, \tag{5.8}$$

as in the proof of Theorem 5.1. Our main result is

THEOREM 5.2. — With constants uniform in b,

$$- \delta E(\sigma, b, h_j) \leq - \delta E(\sigma, b) + 0\left(a + \| h_j \|_{d/d-j}^{d/d-j}\right). \tag{5.9}$$

LEMMA 5.3.

$$- \delta E(\sigma, b, h_j) = \lim_{T \to \infty} (2T)^{-1} \ln \int e^{-\int_{-T}^{T} [B_2 + A_j] dt} dq. \tag{5.10}$$

Proof. — Let ψ be a bounded function on path space, measurable with respect to the σ-algebra generated by $\phi(x, t)$ for $t \geq 0$, let Θ denote reflection about $t = 0$, and let $\psi \to \psi_T$ denote a time translation. Then

$$\begin{aligned} - \delta E(\sigma, b, h_j) &= \sup_{\psi} \lim_{T \to \infty} T^{-1} \ln \langle \psi, e^{-TH(\sigma, j)} \psi \rangle \\ &= \sup_{\psi} \lim_{T \to \infty} T^{-1} \ln \int \Theta\psi \exp \left[- \int_0^T (B_2 + A_j) dt \right] \psi_T dq \\ &\leq \sup_{\psi} \lim_{T \to \infty} \left\{ T^{-1} \ln \| \psi \|_\infty + T^{-1} \ln \int \exp \left[- \int_0^T (B_2 + A_j) dt \right] dq \right\}. \end{aligned}$$

This completes the proof, since the function $\Omega \equiv 1$ is an allowed choice of ψ.

LEMMA 5.4. — $-\delta E(\sigma, b, h_j)$ is non-negative and convex in σ, h_j.

Proof. — The convexity follows from a Schwarz inequality relative to the inner product defined by the Euclidean weak coupling measure dq, combined with Lemma 5.3 $-\delta E \geq 0$ by the normalization $\langle : \varphi^j : \rangle = 0$.

LEMMA 5.5. — $-\delta E(\sigma, b)$ is convex and monotone increasing, as a function of b.

Proof. — Since $-\delta E \geq 0$ and $-\delta E(\sigma, 0) = 0$, the derivative $d(-\delta E)/db$ is non-negative for $b = 0$. Assuming convexity, the derivative is increasing, and hence is always non-negative, so $-\delta E$ is monotonic increasing. To prove convexity, we use a Schwarz inequality relative to the Osterwalder-Schrader inner product defined by a line $x =$ const. Thus

$$-\delta E(\sigma, b) \leq -\frac{1}{2}(\delta E(\sigma, b') + \delta E(\sigma, b''))$$

where $b = \frac{1}{2}(b' + b'')$. Here $x =$ const. determines the choice of b' and b'', and by varying the constant we obtain an arbitrary division.

Remark. — The lemma applies to any localized perturbation, *e. g.* $B_2 + A_j$, so long as each h_j is the characteristic function of Λ_2.

Proof of Theorem 5.2. — We follow [*17*], which simplifies the original method of [*8*]. It is no loss of generality to assume $b \geq a$. As above, we use a Schwarz inequality relative to the lines $x = \pm a$. For $x \in [-a, a]$ we use the norm of the semigroup (with x and t interchanged). Since dq is rotation invariant, it is invariant under rotation by the angle $\theta = \pi/2$. The vacuum energy for fixed x is bounded by $0(1)[1 + |h(x)|^{d/d-j}]T$, where the second factor contributes for $|h|$ large. The factor $0(1)$ contains the σ-dependence. The bound is linear in T by the linear bound on the vacuum energy in the volume. Integrating over x, gives $0(T)(a + \|h\|_{d/d-j}^{d/d-j})$ and $-\delta E(\sigma, b, h_j) \leq -\delta E(\sigma, b) + 0(1)(a + \|h_j\|_{d/d-j}^{d/d-j})$.

In order to apply Theorem 5.2, we let $T_2 \nearrow \infty$, $b \nearrow \infty$, where $[-T_2, T_2]$ is the time interval in which B_2 is inserted in the path space integral. The $T_2 \nearrow \infty$ limit is taken first. In this limit, the effect of the ϕ^j perturbation is estimated using Theorem 5.2. We require,

LEMMA 5.6. — Let $\Omega \equiv 1$ and let

$$\hat{H} = H + B_2 - \delta E(\sigma, b).$$

Then

$$\lim_{T \to \infty} \frac{\| e^{-(T-a)\hat{H}}\Omega \|}{\| e^{-T\hat{H}}\Omega \|} = 1. \tag{5.11}$$

Proof. — We let $\widehat{H} = \int \lambda dE_\lambda$. By Lemma 5.4, $0 = \inf \text{suppt.}\, d\langle E_\lambda \Omega, \Omega \rangle$ $\equiv \inf \text{suppt.}\, d\mu(\lambda)$.
We consider

$$dv_t(\lambda) = \frac{e^{-t\lambda} d\mu(\lambda)}{\int_0^\infty e^{-t\lambda} d\mu(\lambda)}.$$

Thus

$$\frac{\langle \Omega, e^{-(t-a)\widehat{H}}\Omega \rangle}{\langle \Omega, e^{-t\widehat{H}}\Omega \rangle} = \int_0^\infty e^{a\lambda} dv_t(\lambda).$$

Splitting the integral, $\int_0^\infty = \int_0^\varepsilon + \int_\varepsilon^\infty$, we find the contribution $\int_0^\varepsilon \to 1$, and $\int_\varepsilon^\infty \to 0$. which completes the proof.

By Theorem 5.1,

$$\int e^{-\int_{-t_j}^{t_j} A_j(t)dt} e^{-\int_{-T_2}^{T_2} B_2(t)dt} dq \Big/ \int e^{-\int_{-T_2}^{T_2} B_2(t)dt} dq = \langle \theta_{T_2}, e^{-2t_j H_1} \theta_{T_2} \rangle$$

where

$$H_1 = H + B_2 - \delta E(\sigma, b) + A_j = \widehat{H} + A_j$$

and

$$\theta_{T_2} = \frac{e^{-(T_2 - t_j)\widehat{H}}\Omega}{\| e^{-T_2 \widehat{H}}\Omega \|}.$$

By Theorem 5.2, and Lemma 5.6,

$$\lim_{T_2 \to \infty} \langle \theta_{T_2}, e^{-2t_j H_1} \theta_{T_2} \rangle / \| \theta_{T_2} \|^2 \leq \exp t_j 0(1)\big(a + \| h_j \|_{d/d-j}^{d/d-j}\big).$$

Thus by Lemma 5.3,

$$\ln \| e^{-H_1} \| \leq 0(1)\big(a + \| h_j \|_{d/d-j}^{d/d-j}\big)$$

in the limit $b = \infty$. Thus we have proved

THEOREM 5.7. — Let $H(\sigma)$ be the even + linear $\mathscr{P}(\phi)_2$ Hamiltonian with weak coupling boundary conditions given by the mass perturbation

$$\frac{1}{2}\sigma \int :\phi^2: dx, \ \sigma \leq 0.$$

Then,

$$e^{-(H(\sigma) \pm A_j)t} \leq e^{ct} \tag{5.12}$$

where

$$c = 0(1)\big(a + \| h_j \|_{d/d-j}^{d/d-j}\big). \tag{5.13}$$

REMARK 5.8. — The same argument may be used with the Hamiltonian $H(\sigma, \mu)$, for a $\mathscr{P}(\phi)_2$ = even + linear Hamiltonian with weak coupling boundary conditions given by the perturbation $\frac{1}{2}\sigma\int :\phi^2: dx - \mu\int \phi dx$, with $-\sigma, \mu \geq 0$. For a general $\mathscr{P}(\phi)_2$ theory ($\mathscr{P} \neq$ even + linear), Theorem 5.7 remains valid, but in this case the $b = \infty$ infinite volume limit is defined by convergent subsequences.

REMARK 5.9. — The same argument can be used to obtain semigroup estimates of the type (5.12-13) for infinite volume, weakly coupled $\mathscr{P}(\phi)_2$ models. In this case, the convergence of the Schwinger functions as $T \to \infty$ follows from the cluster expansion, rather than monotonicity.

REMARK 5.10. — We improve on (5.13), in order to allow perturbations A_j and test functions h_j which do not have compact support. Using the bounds (5.12)-(5.13), valid for $b \leq \infty$, we have

$$\begin{aligned} -\delta\widehat{E}(\sigma, b, h_j) &\equiv \ln \| \exp(-\widehat{H} + A_j) \| \\ &\leq 0(1)\left(a + \| h_j \|_{d/d-j}^{d/d-j}\right). \end{aligned}$$

Since $-\delta\widehat{E}$ is convex in h_j and vanishes for $h_j = 0$, the above inequality implies

$$-\delta\widehat{E} \leq 0(1)\left(a^{j/d}\| h_j \|_{d/d-j} + \| h_j \|_{d/d-j}^{d/d-j}\right).$$

We write $h_j = \sum_{n=-\infty}^{\infty} \chi_{[n,n+1]} h_j$ and use convexity again to obtain the bound

$$-\delta\widehat{E} \leq 0(1)\left(\|(1 + |x|)^r h_j \|_{d/d-j} + \| h_j \|_{d/d-j}^{d/d-j}\right), \tag{5.14}$$

for some r. This bound holds uniformly for $b \leq \infty$ and for any h_j with $(1 + |x|)^r h_j \in L_{d/d-j}$.

REFERENCES

[1] O. BRATELLI, Conservation of estimates in quantum field theory. *Comm. Pure and Appl. Math.*, t. **25**, 1972, p. 759-779.

[2] C. CALLAN, Broken scale invariance in scalar field theory. *Phys. Rev.*, **D2**, 1970, p. 1541-1547.

[3] S. COLEMAN, Scaling Anomalies, in *Developments in High Energy Physics*, Academic Press, New York, 1972, p. 280-296.

[4] R. DOBRUSHIN and R. MINLOS, Construction of a one dimensional quantum field via a continuous Markov field. *Funct. Anal. and its Appl.*, t. **7**, 1973, p. 324-325 (English transl.).

[5] J. FRÖHLICH, Schwinger functions and their generating functionals I, II. *Helv. Phys. Acta*, and *Adv. in Mathematics*.

[6] J. GINIBRE, General formulation of Griffiths' inequalities. *Commun. Math. Phys.*, t. **16**, 1970, p. 310-328.
[7] J. GLIMM and A. JAFFE, The $(\lambda\varphi^4)_2$ quantum field theory without cutoffs III. The physical vacuum. *Acta Math.*, t. **125**, 1970, p. 203-261.
[8] J. GLIMM and A. JAFFE, The $\lambda(\varphi^4)_2$ quantum field theory IV. Perturbations of the Hamiltonian. *J. Math. Phys.*, t. **13**, 1972, p. 1558-1584.
[9] J. GLIMM and A. JAFFE, Entropy principle for vertex functions in quantum field models. *Ann. Institut Henri Poincaré*, t. **21**, 1974, p. 1-26.
[10] J. GLIMM and A. JAFFE, Critical point dominance in quantum field models. *Ann. Institut Henri Poincaré*, t. **21**, 1974, p. 27-41.
[11] J. GLIMM and A. JAFFE, φ_2^4 quantum field model in the single-phase region: Differentiability of the mass and bounds on critical exponents, *Phys. Rev.*, **D15**, to appear.
[12] J. GLIMM and A. JAFFE, A remark on the existence of φ_4^4. *Phys. Rev. Lett.*, t. **33**, 1974, p. 440-442.
[13] J. GLIMM and A. JAFFE, Absolute bounds on vertices and couplings. *Ann. Institut Henri Poincaré*, t. **22**, 1975, p. 1-13.
[14] J. GLIMM and A. JAFFE, Two and three body equations in quantum field models. *Commun. Math. Phys.*, to appear.
[15] J. GLIMM, A. JAFFE and T. SPENCER, The Wightman axioms and particle structure in the $P(\varphi)_2$ quantum field model. *Ann. Math.*, t. **100**, 1974, p. 585-632.
[16] J. GLIMM, A. JAFFE and T. SPENCER, The particle structure of the weakly coupled $P(\varphi)_2$ models and other applications of high temperature expansions, in: *Constructive Quantum Field Theory*, G. Velo and A. S. Wightman (eds.), Springer-Verlag, Berlin, 1973.
[17] F. GUERRA, L. ROSEN and B. SIMON, Nelson's Symmetry and the infinite volume behavior of the vacuum in $P(\varphi)_2$. *Commun. Math. Phys.*, t. **27**, 1972, p. 10-22.
[18] F. GUERRA, L. ROSEN and B. SIMON, The $P(\varphi)_2$ Euclidean quantum field theory as classical statistical mechanics. *Ann. Math.*, t. **101**, 1975, p. 111-259.
[19] F. GUERRA, L. ROSEN and B. SIMON, *The pressure is independent of the boundary conditions*, preprint.
[20] J. LEBOWITZ, More inequalities for Ising ferromagnets. *Phys. Rev.*, **B5**, 1972, p. 2538-2541.
[21] J. LEBOWITZ, GHS and other inequalities. *Commun. Math. Phys.*, t. **35**, 1974, p. 87-92.
[22] R. MINLOS and Ya. SINAI, Investigation of the spectra of stochastic operators arising in lattice models of a gas. *Theoretical and Math. Phys.*, t. **2**, 1970, p. 167-176 (English transl.).
[23] K. OSTERWALDER and R. SCHRADER, Axioms for Euclidean Green's functions I, II. *Commun. Math. Phys.*, t. **31**, 1973, p. 83-113 and *Commun. Math. Phys.*
[24] G. PARISI, Field theory approach to second order phase transitions in two and three dimensional systems. *1973 Cargèse lectures.*
[25] Y. PARK, *Lattice approximation of the* $(\varphi^4 - \mu\varphi)_3$ *field theory in a finite volume*, preprint.
[26] E. RIEDEL and F. WEGNER, Tricritical exponents and scaling fields. *Phys. Rev. Lett.*, t. **29**, 1972, p. 349-352.
[27] R. SCHRADER, *Local operator products and field equations in* $P(\varphi)_2$ *theories.*
[28] T. SPENCER, Perturbation of the $P(\varphi)_2$ quantum field Hamiltonian. *J. Math. Phys.*, t. **14**, 1973, p. 823-828.
[29] K. SYMANZIK, Small distance behavior in field theory and power counting. *Commun. Math. Phys.*, t. **18**, 1970, p. 227-246.

(Manuscrit reçu le 2 septembre 1974)

Critical Exponents and Elementary Particles

J. Glimm*
The Rockefeller University, New York, NY 10021, USA

A. Jaffe**
Harvard University, Cambridge, MA 02138, USA

Abstract. Particles are shown to exist for a.e. value of the mass in single phase ϕ^4 lattice and continuum field theories and nearest neighbor Ising models. The particles occur in the form of poles at imaginary (Minkowski) momenta of the Fourier transformed two point function. The new inequality $dm^2/d\sigma \leqq Z$, where $\sigma = m_0^2$ is a bare mass^2 and Z is the strength of the particle pole, is basic to our method. This inequality implies inequalities for critical exponents.

1. Introduction

Euclidean ϕ^4 fields are believed to describe the asymptotic long distance behavior of certain lattice models of statistical mechanics at their critical points (e.g. the Ising model). It was proposed [9] that a construction of ϕ^4 fields could be based on this expectation, and partial results in this direction are given in [6–8, 10, 11, 1, 12]. In this construction, the field ϕ will be nontrivial (i.e. not a free field) only in the case in which the corresponding lattice model critical point is asymptotically nontrivial at long distances.

In order to better distinguish between the trivial and the nontrivial cases, we continue here our investigation [7, 10, 11] of critical exponents (see also § 5). In general, our results have the form

$$\text{canonical exponent} \leqq \text{general exponent}, \tag{1.1}$$

$$\phi\text{-exponent} \leqq \phi^2\text{-exponent}, \tag{1.2}$$

and in particular if the lattice ϕ^2 field (e.g. the Ising model energy-energy correlation) is canonical, then so is the corresponding lattice ϕ-field and also the resulting continuum ϕ-field. The converse to this statement seems to be false, and a counterexample may be found in the ϕ_4^4 lattice field at weak coupling. In this model, there is some evidence that ϕ^2 deviates from canonical by a logarithm.

---

* Supported in part by the National Science Foundation under grant PHY 76-17191
** Supported in part by the National Science Foundation under grant MPS 75-21212

The bounds (1.1) are given in [7] while the bound [10]

$$\eta \leqq \eta_E/2 \tag{1.3}$$

of the form (1.2) is an elementary consequence of Lebowitz' inequality. This inequality compares the anomalous dimension η for ϕ with the anomalous dimension η_E for ϕ^2. One result of this paper is an improvement on the bound [3] $1 \leqq (2-\eta)\nu$, namely

$$1 \leqq (2-\zeta/\nu)\nu \tag{1.4}$$

where ν is the exponent for the mass and ζ the exponent for field strength renormalization. (Note $\zeta/\nu \geqq \eta$ in the case of Euclidean covariance and $\zeta/\nu = \eta$, assuming scale relations.) The stronger upper bounds on η in [11] required stronger hypothesis.

Our proof of (1.4) follows from a stronger inequality for an Ising model, lattice ϕ^4 fields or $d=1,2,3$ continuum ϕ^4 fields. We show that

$$dm^2/d\sigma \leqq Z, \quad \text{assuming} \quad \sigma > \sigma_c \tag{1.5}$$

where Z is the strength of the one particle pole in the truncated two point function, m^{-1} is the correlation length and σ is the bare mass^2 (or inverse temperature). The equality (1.5) improves $dm^2/d\sigma \leqq 1$, established in [7], and integration of (1.5) yields the relation (1.4) for critical exponents. (Recall that $m \searrow 0$ as $\sigma \searrow \sigma_c$ [1, 11].)

The main result of the present paper is that $dm^2/d\sigma \neq 0$ for almost all masses m, so that by (1.5), $Z \neq 0$ for almost all masses. We infer that an elementary particle (a pole in the propagator) exists for almost all masses in the region $\sigma > \sigma_c$, i.e. the single phase region. Also as $\sigma \searrow \sigma_c$,

$$\chi = \int G^{(2)}(x)dx \to \infty, \tag{1.6}$$

and so

$$\Gamma^{(2)}(p=0) \to 0 \quad \text{as} \quad \sigma \searrow \sigma_c.$$

This convergence was assumed in [7] in the derivation of the inequality $\gamma \geqq 1$. We give proofs for continuum fields, and the same methods then carry over to lattice fields and nearest neighbor Ising models on a rectangular lattice. For the Ising models σ (the coefficient of ϕ^2 in the ϕ^4 interaction) is replaced by $-\beta$.

2. Critical Behavior of the Mass and the Field Strength Renormalization

For pure imaginary (Minkowski) momentum ip, p real, we define

$$\chi(p) = \int G(x)e^{-px}dx = G^{\sim}(ip)$$

with $\int \dots dx$ replaced by a summation in the lattice case. We choose ip inside the tube of analytically of $G^{\sim}$. Then

$$\begin{aligned} 0 &\leqq -d\chi(p)/d\sigma = \tfrac{1}{2}\int \langle \phi(x) : \phi(y)^2 : \phi(0)\rangle\, e^{-p(x-y)}e^{-py}dxdy \\ &\leqq \int \langle \phi(x)\phi(y)\rangle \langle \phi(y)\phi(0)\rangle\, e^{-p(x-y)}e^{-py}dxdy = \chi(p)^2 . \end{aligned}$$

Thus

$$0 \leqq -d\Gamma/d\sigma = -d(-\chi(p))^{-1}/d\sigma = -\chi^{-2}d\chi(p)/d\sigma \leqq 1 \tag{2.1}$$

for all such p. Since we have used translation invariance, the proof applies in the infinite volume case, and also in the finite volume periodic case.

To apply this inequality in a finite volume with periodic boundary conditions, we note that the Hamiltonian has discrete spectrum, and that its first excited state occurs in the spectral decomposition of the two point function. Furthermore, the first excited state occurs at zero momentum. Thus we take $p=(p_0, \boldsymbol{p})$, with $\boldsymbol{p}=0$, and $Z^{-1}<\infty$ for all volumes $V<\infty$. Let $m_{p,V}$ be the periodic mass. For $-p_0^2<m^2$, Γ is an analytic function of the independent variables σ and p_0^2 by a standard application of the Paley-Wiener theorem, cf. [11, §2]. The one particle curve $+p_0^2=m^2$ is analytic in any region free of level crossings in the first excited state, and thus is at least piecewise analytic. [15, VII, § 1] can be applied to the Hamiltonian $H(\sigma)$, which is analytic in σ by second order estimates [16]. For example the inequality $H(\mathrm{Re}\,\sigma)^2 \leqq \mathrm{const}\, H(\sigma^-)H(\sigma)+\mathrm{const}$ implies that $H(\sigma)$ is closed, on the domain $\mathscr{D}(H(\mathrm{Re}\,\sigma))$. With the same discrete exceptions at possible level crossings, Γ vanishes along the one particle curve $p^2=-m(\sigma)^2$. From these facts, it follows $d\Gamma/dp^2$ is a multiple of $d\Gamma/d\sigma$ on this one particle curve. In particular, in the $(\sigma, -p^2)$ plane, the vector $(1, dm^2/d\sigma)$ lies along the one particle curve, and so

$$0 = \nabla\Gamma\cdot(dm^2/d\sigma, 1) = [\partial\Gamma/\partial(-p^2)]dm^2/d\sigma + \partial\Gamma/d\sigma \tag{2.2}$$

or

$$0 \leqq Z^{-1} = -\partial\Gamma/\partial p^2|_{p^2=-m^2} = -[\partial\Gamma/\partial\sigma]\,[dm^2/d\sigma]^{-1}.$$

Thus by (2.1),

$$0 \leqq dm^2/d\sigma = -Z\partial\Gamma/\partial\sigma \leqq Z, \tag{2.3}$$

proving (2.3). In the case of level crossings, one sided derivatives $dm^2/d\sigma^\pm$ satisfy (2.3), since Z is semicontinuous at a level crossing.

For later use, we note that

$$dm^2/d\sigma \leqq \mathrm{const}. \tag{2.4}$$

by [7] in the continuum case, or (2.3), since $Z \leqq 1$ for canonical continuum fields. For lattice fields and Ising models, the p-space canonical upper bound [4] on the two point function, and its Herglotz representation [11], show $Z \leqq \mathrm{const}$. Thus by (1.6), the bound (2.4) extends to ϕ^4 lattice fields and Ising models in a finite periodic volume, with a constant independent of the volume.

By passing to the subsequence $V_j \to \infty$, we have $m_{p,V} \to m_{p,\infty}$ but we do not know (yet) that $m_{p,\infty}$ is the mass m_p of the corresponding infinite volume theory. However, by semicontinuity of the spectrum.

$$m_{p,\infty} \leqq m_p.$$

By [5], $m_p = m_d$, where m_d is the mass in a theory with Dirichlet boundary conditions. We define the corresponding critical values $\sigma_{cd} = \sigma_{cp} \leqq \sigma_{c,p,\infty}$. By (2.4) and Ascoli's theorem, we have uniform convergence and uniform Lipschitz continuity of $m^2_{p,V}(\sigma)$ and the limit $m^2_{p,\infty}(\sigma)$. In particular, both m_d and $m_{p\infty}$ go continuously to zero at their critical points [7, 1, 12].

By definition of σ_c, the quantity $dm^2/d\sigma$ cannot vanish in a one-sided neighborhood $(\sigma_c, \sigma_c+\varepsilon]$ of σ_c, and by (1.5) the measure of the set on which $dm^2/d\sigma \neq 0$ is proportional to $m^2(\sigma_c+\varepsilon)$, uniformly in V. It follows that by the diagonal process, we can choose $V_j \to \infty$ so that there is a sequence $\sigma_i \searrow \sigma_{cp\infty}$ with the strict inequality

$$\lim_{V_j \to \infty} dm^2/d\sigma|_{\sigma=\sigma_i} > 0 .$$

For these values of σ_i, we also have Z bounded away from zero by (1.5), which implies $m_p = m_{p\infty}$ for $\sigma = \sigma_i$. Because the m's are monotone in σ, it follows that $\sigma_{cd} = \sigma_{cp} = \sigma_{cp\infty}$.

Extending the above argument, we can work with $\Sigma = \{\sigma : dm^2_{p\infty}/d\sigma > 0\}$, which is a set of positive measure in every interval $(\sigma_c, \sigma_c+\varepsilon)$. On this set $Z \neq 0$ and $m_{p\infty} = m_p$. However since $m_{p\infty}$ is constant on $\sim \Sigma$, while m_p is monotone (and hence monotone on Σ) we must have $m_{p\infty} = m_p$ everywhere. We summarize these results.

Theorem 1. *For $\sigma > \sigma_c$, we have*

$$m_{p\infty} = m_p = m_d$$

and

$$dm^2/d\sigma \leqq Z(\sigma) .$$

Furthermore $Z(\sigma)$ uppersemicontinuous:

$$Z(\sigma) \leqq \limsup_{\sigma' \to \sigma} Z(\sigma') .$$

We now argue that $Z \neq 0$ for all masses m not in a set of measure zero. Consider $m^2(\sigma)$ and $\sigma(m^2)$ as monotone functions. Thus they have bounded variation and are differentiable a.e. with derivatives $m^{2\prime}$, σ' which are positive and L_1. Thus

$$(\sigma')^{-1} = dm^2/d\sigma \neq 0 \text{ a.e.},$$

as a function of m, and by (2.3) $Z \neq 0$ a.e. as a function of m. The region of non uniqueness of mass renormalization corresponds to intervals in which $m^2(\sigma) = \text{const.}$, hence to δ-functions in the measures $d\sigma(m^2)$. These occur for a most countable number of values of m^2.

Remark. The above result completes the proof of the second of three steps of [9] for a possible construction[1] of Φ^4_d fields: that defined by the long distance behavior of the corresponding lattice field at its critical point. The first step was established in [7, 1, 12]. The third step seems to depend on showing scaling behavior as $\sigma \searrow \sigma_c$, i.e. the isolation of leading long distance behavior from next to leading, etc.

3. Divergence of χ at the Critical Point

In this section we show that $\chi_c = \chi(\sigma_c) = \infty$, which is related to the upper bound $\eta \leqq 2$. Again we work in the $(\sigma, -p^2)$ parameter space, but in the $V=\infty$ theory directly. Below or on the one particle curve $-p^2 = m^2$, the quantity $\partial\Gamma/\partial\sigma$ is bounded, while

[1] If the nontrivial part of the critical behavior occurs only in the next to leading terms for all values of the bare charge, then the construction of [9] is inappropriate (cf. § 5)

below this curve or on nonhorizontal portions of this curve, $\partial\Gamma/\partial(p^2)$ is bounded also, and Γ vanishes, at least on non-horizontal portions of the curve. Moreover, by the Herglotz representation for Γ, any bound on $\partial\Gamma/\partial(-p^2)$, with $p^2=-m^2$, extends to $0 \leqq -p^2 \leqq m^2$, since $\partial\Gamma/\partial(-p^2)$ is decreasing in $-p^2$ in this interval. Thus by (2.2) and integration,

$$0 \leqq -\Gamma(p^2=0, \sigma=\sigma_i) \leqq 0(m^2)(dm^2/d\sigma)^{-1}$$

where σ_i is chosen as in § 2, so that $Z \neq 0$, hence $G(p^2=-m^2)=\infty$ and $\Gamma(p^2=-m)=0$. Thus

$$0(1)m^{-2}dm^2/d\sigma \leqq \chi_{\sigma_i}.$$

Assuming χ_{σ_c} finite, we have

$$dm/d\sigma \leqq \text{const.}\, m.$$

Recall that $m(\sigma_c+0)=0$, so that after integration from σ_c to $\sigma>\sigma_c$,

$$0 \leqq m(\sigma) \leqq m(\sigma_c+0)\, e^{\text{const.}(\sigma-\sigma_c)}=0.$$

This is a contradiction, which shows that $\chi_c=\infty$. As a corollary of this result, we observe that σ_c is characterized as the largest value of σ for which the effective potential (see [2]) has zero curvature at the origin:

$$V''|_{\sigma=\sigma_c}=\Gamma^{(2)}(p=0)|_{\sigma=\sigma_c}=0.$$

4. The Ising Limit

We consider the interactions

$$\tfrac{1}{2}\textstyle\sum_{n.n}(\phi_i-\phi_{i'})^2+\sum_i[\lambda(\phi_i^2-1)^2+\tfrac{1}{2}\sigma\phi_i^2] \tag{4.1}$$

and

$$-\textstyle\sum_{n.n}\beta\psi_i\psi_{i'}+\sum_i\lambda_0(\psi_i^2-1)^2 \tag{4.2}$$

where $\sum_{n.n}$ denotes a sum over nearest neighbor pairs. With

$$\psi=(1-(\sigma+2d)/4\lambda)^{-1/2}\phi$$

and

$$\lambda_0=\lambda-(\sigma+2d)/4\lambda, \qquad \beta=1-(\sigma+2d)/4\lambda,$$

these two expressions differ by a constant, and thus define the same lattice field theory. For the purpose of studying the Ising limit (λ or $\lambda_0\to\infty$, β fixed), it is convenient to use the representation (4.2). Let β_{cI} be the critical value of β in the Ising model.

Theorem 2. *As* $\lambda_0\to\infty$, $m(\beta,\lambda_0)\to m_I(\beta)$, *the Ising model mass, for* $\beta<\beta_{c,I}$, *and* $\beta_{c,I} \leqq \lim\limits_{\lambda_0\to\infty}\beta_c(\lambda_0)$.

Proof. As in (2.1), $-d\Gamma/d(-\beta) \leqq 2$ for imaginary momenta below the one particle curve, and as in (2.3), $0 \leqq dm^2/d(-\beta) \leqq 2Z$. These bounds are uniform in $\lambda_0 \leqq \infty$, in the one phase region. Thus they imply uniform convergence (through a subsequence

$\lambda_{0j} \to \infty$) and uniform Lipschitz continuity in β of $m^2(\beta, \lambda_0)$. Let $m^2(\beta, \infty)$ be the limit value of the mass^2, and let $m_I^2(\beta)$ be the Ising model mass^2. We have $m^2(\beta, \infty) \leqq m_I^2(\beta)$ for $\beta \geqq \beta_{c,I}$ by semicontinuity of the spectrum. Next we choose values β_i for which $Z(\beta_i, \lambda_0)$ is bounded away from zero as $\lambda_{0,j} \to \infty$. For these values of β_i, we have $m^2(\beta, \infty) = m_I^2(\beta)$. By monotonicity, $m^2(\beta, \infty) = m_I^2(\beta)$ throughout the single phase region. Convergence of the full sequence follows from uniqueness of the limit, in the single phase region. With convergence of the masses assured for $\beta \leqq \beta_{c,I}$, it follows that $\lim_{\lambda_0 \to \infty} \beta_c(\lambda_0) \geqq \beta_{c,I}$.

The same method applies to variation of the β's which couple between layers. Let $J = (J_1, \ldots, J_d)$ and let

$$H(\sigma) = -\tfrac{1}{2}\Sigma \beta J_j \sigma_i \sigma_{i+e_j} \tag{4.3}$$

where e_j is the unit vector in the jth coordinate direction. Then (4.3) defines a d-dimensional Ising model. For simplicity we consider only the case in which $J_d \to 0$ and set $J_1 = \ldots = J_{d-1} = 1$. Then $\beta_c = \beta_c(J_d)$ is a function of J_d. As above, we have

Theorem 3. *For $J_d \geqq 0$, m is continuous and β_c is semicontinuous as functions of J_d, in the single phase region.*

Remark. Continuity of β_c would follow from a uniform lower bound on $dm^2/d(-\beta)$, for example $\nu < \infty$.

5. Trivial vs. Nontrivial Fields

We give a simple sufficient condition for triviality of the ϕ^4 continuum limit. We parametrize the single phase even ϕ^4 lattice fields by $\lambda(=$bare charge), $m(=$physical mass), $\varepsilon(=$lattice spacing) and $d(=$space time dimension).

Theorem 4. *Suppose that $Z = Z(\lambda, m, \varepsilon, d)$ is continuous and bounded away from zero for $0 \leqq \lambda \leqq \infty$, $0 \leqq m \leqq \delta$, $\varepsilon = 1$, and for some $\delta > 0$, and some $d \geqq 3$. Then the continuum limit of an even single phase d-dimensional ϕ^4 lattice field with arbitrary charge renormalization has a free two point function.*

Proof. The continuum fields are constructed with the arbitrary renormalization $\lambda = \lambda(\varepsilon)$, $0 < m = m(\varepsilon) \leqq$ const, where $\varepsilon_j \to 0$ through a subsequence. Existence of the limit follows from the bounds of [6, 4], which establish compactness of the sequence of Schwinger functions. Properties of the $\varepsilon \to 0$ theory are reduced to properties of the $\varepsilon = 1$ theory by a (canonical) scale transformation. Z is invariant under scale transformations, so that

$$Z(\lambda(\varepsilon), m(\varepsilon), \varepsilon) = Z(\varepsilon^{4-d}\lambda(\varepsilon), \varepsilon\, m(\varepsilon), 1).$$

There is a similar transformation law for the spectral weight of the two point function. The Fourier transformed two point function at zero momentum has the Herglotz representation [11]

$$G^{(2)\sim}(p_0, \mathbf{p} = 0) = Z(\cosh m = \cos p_0)^{-1} + \int_{m+0}^{\infty} (\cosh a - \cos p_0)^{-1} d\varrho(a, \lambda, m, \varepsilon, d)$$

for $\varepsilon = 1$. Continuity of Z at $m = 0$ can be expressed as the following property of the measure $d\varrho$:

$$\lim_{\gamma \to 0} \lim_{m \to 0} \int_{m^2 + 0}^{\gamma} d\varrho(a, \lambda_1(m), 1, d) = 0 \tag{5.1}$$

for any $\lambda = \lambda_1(m)$. We choose $m = \varepsilon m(\varepsilon) \to 0$ and

$$\lambda_1(\varepsilon m(\varepsilon)) = \varepsilon^{4-d} \lambda(\varepsilon).$$

Under scaling, (5.1) transforms into

$$\lim_{\gamma \to 0} \lim_{\varepsilon \to 0} \int_{m(\varepsilon)^2 + 0}^{\varepsilon^{-2}\gamma} d\varrho(a, \lambda(\varepsilon), m(\varepsilon), \varepsilon, d) = 0,$$

which implies that limit two point function is free.

References

1. Baker,G.: Self interacting boson quantum field theory and the thermodynamic limit in d dimensions. J. Math. Phys. **16**, 1324—1346 (1975)
2. Coleman,S.: Secret Symmetry: An introduction to spontaneous symmetry breakdown and guage fields. 1973 Erice Summer School of Physics
3. Fisher,M.: Rigorous Inequalities for critical-point correlation exponents. Phys. Rev. **180**, 594—600 (1969)
4. Fröhlich,J., Simon,B., Spencer,T.: A new method for the analysis of phase transitions and spontaneous breaking of discrete and continuous symmetries. Phys. Rev. Lett. **36**, 804—806 (1976)
5. Fröhlich,J., Simon,B.: Pure states for general $P(\phi)_2$ theories: Construction, regularity and variational equality. Ann. Math., to appear
6. Glimm,J., Jaffe,A.: A remark on the existence of ϕ_4^4. Phys. Rev. Lett. **33**, 440—442 (1974)
7. Glimm,J., Jaffe,A.: The ϕ_2^4 quantum field theory in the single phase region: Differentiability of the mass and bounds on critical exponents. Phys. Rev. D **10**, 536—539 (1974)
8. Glimm,J., Jaffe,A.: Three particle structure of ϕ^4 interactions and the scaling limit. Phys. Rev. D **11**, 2816—2827 (1975)
9. Glimm,J., Jaffe,A.: Critical problems in quantum fields, International Colloquium on Mathematical Methods of Quantum Field Theory, Marseille, June 1975
10. Glimm,J., Jaffe,A.: Critical exponents and renormalization in the ϕ^4 scaling limit, Conference on quantum dynamics: Models and mathematics. Acta Phys. Austriaca Suppl. XVI (1975)
11. Glimm,J., Jaffe,A.: Particles and scaling for lattice fields and Ising models. Commun. math. Phys. **51**, 1—14 (1976)
12. Rosen,J.: Mass renormalization for lattice $\lambda\phi_2^4$ fields. Adv. Math. (to appear)
13. Rosen,J.: The Ising limit of ϕ^4 lattice fields. The Rockefeller University (preprint)
14. Schrader,R.: A possible constructive approach to ϕ_4^4 I, III. Commun. math. Phys. **49**, 131—154 (1976); **50**, 97—102 (1976)
15. Kato,J.: Perturbation theory for linear operators. Berlin-Heidelberg-New York: Springer 1966
16. Glimm,J., Jaffe,A.: A $\lambda\phi^4$ Quantum field theory without cutoffs. I. Phys. Rev. **176**, 1945—1951 (1968) and Singular perturbations of self adjoint operators. Comm. Pure Appl. Math. **22**, 401—414 (1969)

Communicated by R. Haag

Received June 30, 1976

V Particle Structure

Introduction

The papers in this section study the particle spectrum of H, using a number of techniques. The one-particle irreducible functions (vertex functions) are obtained from the Green's functions by a Legendre transform. These functions can be used to study the two or three particle spectrum of H, because of their rapid decay rate. One conjectured property of the single phase ϕ^4 model is the absence of three-particle bound states. This would follow from the conjectured negativity of the six point vertex function, $\Gamma^6 \leq 0$. Other ideas are investigated in a paper on the resummation of one particle lines; these may be useful in a future analysis of the problem of asymptotic completeness.

Bibliography

The entropy principle for vertex functions in quantum field models, Ann. l'Inst. H. Poincaré A21. 1–26 (1974).

Three-particle structure of ϕ^4 interactions and the scaling limit, Phys. Rev. D11, 2816–2827 (1975).

Two and three body equations in quantum field models, Commun. Math. Phys. 44, 293–320 (1975).

Particles and scaling for lattice fields and Ising models, Commun. Math. Phys. 51, 1–13 (1976).

The resummation of one particle lines, Commun. Math. Phys. 67, 267–293 (1979).

The entropy principle for vertex functions in quantum field models

by

James GLIMM (*)
Courant Institute. New York University. New York 10012

and

Arthur JAFFE (**)
Harvard University. Cambridge, Massachusetts 02138

ABSTRACT. — We obtain the (amputated, one particle irreducible) vertex functions for weakly coupled Euclidean $\mathscr{P}(\Phi)_2$ models. The generating function $\Gamma_2 \{ A \}$ for the vertex functions is jointly analytic in A and in the bare coupling constants λ_j. The generating function $G \{ J \}$ for connected Green's functions has a convergent tree-graph expansion in terms of vertex functions and propagators.

CONTENTS

(*) Supported in part by the National Science Foundation under Grant NSF-GP-24003.
(**) Supported in part by the National Science Foundation under Grant NSF-GP-40354X.

1. THE VERTEX FUNCTIONS

1.1. The partition function

The partition function in Euclidean $\mathscr{P}(\Phi)_2$ models,

$$Z\{J\} = \int e^{\hat{\Phi}(J)}dq = \int e^{\int(\Phi(x) - \langle\Phi(x)\rangle)J(x)dx}dq = \int e^{\Phi(J)}dq/e^{\langle\Phi\rangle\int J(x)dx}$$

is the generating function for the Schwinger functions

$$S^{(n)}(x_1, \ldots, x_n) = \int \hat{\Phi}(x_1) \ldots \hat{\Phi}(x_n)dq .$$

of the translated field $\hat{\Phi}(x) = \Phi(x) - \langle \Phi(x) \rangle$.

Here dq is the Euclidean measure on $\mathscr{S}'(R^2)$, associated with the $\mathscr{P}(\Phi)_2$ quantum field. Indicating derivatives by subscripts,

$$S^{(n)}(x_1, \ldots, x_n) = Z_{x_1,\ldots,x_n}\{0\} .$$

The connected parts of the Schwinger functions $G^{(n)}(x_1, \ldots, x_n)$ have the effects of disconnected processes removed. (They have no zero particle intermediate states.) The $G^{(n)}$ functions cluster exponentially, with the exponential decay rate equal to the physical mass m. The generating function

$$G\{J\} = \ln Z\{J\}$$

yields these (connected) Green's functions,

$$G^{(n)}(x_1, \ldots, x_n) = G_{x_1,\ldots,x_n}\{0\} .$$

The vertex functions $\Gamma^{(n)}(x_1, \ldots, x_n)$ are the amputated, one particle irreducible parts of the $G^{(n)}(x_1, \ldots, x_n)$. They have the effects of one particle intermediate states removed, and are believed to cluster with a decay rate $> m$. Thus the vertex functions display the « upper mass gap » in the energy momentum spectrum. Furthermore, the physical charge is defined in terms of a vertex function, so understanding the vertex functions is a step toward understanding charge renormalization.

Construction of the vertex functions is the beginning of Symanzik's program of structure analysis for Green's functions [9]; he proposed construction of n-particle irreducible parts. See also [11]. Related to Symanzik's proposed analysis of Green's functions, is our analysis of the n-particle structure of the energy spectrum [3]. Consider the subspace $\mathscr{H}_E \subset \mathscr{H}$ for the spectral interval [0, E] of the Hamiltonian $H = H^*$. In weak $\mathscr{P}(\Phi)_2$ models, cluster properties yield an explicit construction of $\mathscr{H}_E$. Furthermore, this analysis exhibits n-particle structure, since states of energy $\leq (n + 1)m_0(1 - \varepsilon)$ are spanned by polynomials of degree n in the Euclidean field Φ. This result, for $n = 0, 1$, is used to prove the existence of isolated one particle states [*3*].

In this paper we construct the vertex functions $\Gamma^{(n)}$ and their generating function $\Gamma\{A\}$,

$$\Gamma^{(n)}(x_1, \ldots, x_n) = \Gamma_{x_1,\ldots,x_n}\{A = 0\},$$

for $\mathscr{P}(\Phi)_2$ models with weak coupling. We establish analyticity of $\Gamma\{A\}$ in A, as well as in the bare parameters. In a subsequent paper we establish the Callan-Symanzik equations, giving the change $d\Gamma/d\sigma$ for a mass perturbation $\sigma \int :\Phi^2: d\vec{x}$.

In the definition of the vertex functions, amputation means that in momentum space $G^{(n)}$ is divided by the product

$$\prod_{j=1}^{n} \int G^{(2)}(x, 0) e^{ip_j x} dx$$

of propagators in each momentum variable. In perturbation theory language, amputation removes all mass subdiagrams attached to external legs of diagrams contributing to $G^{(n)}$.

One particle irreducible (1PI) graphs are those which cannot be disconnected by removal of a single line. We use a Legendre transformation (variational or entropy principle) to obtain $\Gamma\{A\}$ from $G\{J\}$. The thermodynamic parameters are identified as follows. J is a chemical potential or magnetic field, and so its conjugate variable has the role of N (= number of particles) or magnetization, M. With $\ln Z = PV$, we see that Γ defined by (1.2.1) is essentially the negative of the Helmholtz free energy $N - PV$. Entropy is the variable conjugate to temperature, and $kT = \beta^{-1}$ is the inverse coefficient of the energy in dq. Below we introduce a coefficient λ_j for each homogeneous contribution to the energy. Conjugate to each λ_j is an « entropy » S_j associated with the jth Legendre transform. The case $j = 2$ is discussed in [*11*] and leads to two particle irreducible graphs. Suppression of the factor $e^{-\langle \mathscr{T}(J) \rangle}$ in Z leads to graphs which are one particle irreducible in a stronger sense than the graphs considered here, in that tadpoles (i. e. one particle reducible parts with no external legs) are also removed, see [*11*].

This method goes back to De Dominicis and Martin in statistical mechanics [*2*] and to Jona-Lasinio in quantum field theory [*7*]. We follow work of Symanzik [*10*]. Our contribution is to obtain bounds on $Z\{J\}$, $G\{J\}$ and $G^{(2)}$ which permit the analysis to be carried out rigorously. We also obtain bounds on the vertex functions $\Gamma^{(n)}$. The n-dependence of these bounds ensures convergence of the expansion of $G\{J\}$ (and also $Z\{J\}$ in terms of « tree-like graphs » with vertex kernels $\Gamma^{(r)}$ and propagators $G^{(2)}$. In this manner we exhibit graphically the 1PI property of the vertex functions.

Let H_p denote the complex Sobolev space $\mathscr{D}((-\Delta+1)^{p/2})$, with norm

$$\| f \|_p^2 = \int_{x \in \mathbb{R}^2} |((-\Delta+1)^{p/2} f)(x)|^2 dx .$$

Also let

$$H_{p,\varepsilon} = \{ f : f \in H_p, \| f \|_p < \varepsilon \} .$$

We define analytic functions and multilinear forms $\mathscr{ML}_n$ in the Appendix.

The partition function $Z\{J\}$ is a limit of finite volume partition functions, at least for $J \in C_0^\infty$. The finite volume partition functions are jointly analytic in $J \in H_{-1}$ and in the bare parameters of the theory in any « positive but weak coupling » region of the bare parameter space (See Chapter 2). By definition, in such a region each coupling constant λ_j, the coefficient of $:\Phi^j:$ in the interaction Hamiltonian, is sufficiently small with respect to the mass m_0 in the free Hamiltonian. We also require $|\lambda_j| \le O(1) |\mathrm{Re}\, \lambda_l|$, and $\mathrm{Re}\, \lambda_l > 0$ where λ_l is the coupling constant of the highest degree term, and of course l is even. In this region, with $\| J \|_{-1}$ small the finite volume partition functions are uniformly bounded. Hence by Proposition A3, the infinite volume limit $Z\{J\}$ is analytic for $\| J \|_{-1}$ small and for λ_j in the positive but weak coupling region. { We remark that this argument does not yield derivatives of $Z\{J\}$ with respect to λ_l at $\lambda_l = 0$, but Dimock [*1*] has shown that the Schwinger functions are C^∞ in the λ's in a region including $\lambda_l = 0^+$. } We establish our basic estimates for $Z\{J\}$ in Chapter 2. Fröhlich [*6*] previously used properties of $Z\{J\}$ to study the Schwinger functions in $\mathscr{P}(\Phi)_2$ models. Feldman [*5*] has results for the finite volume Φ_3^4 model.

We assume throughout this paper that the coupling constants λ_j lie in a weak coupling region and that $J \in H_{-1,\varepsilon}$.

PROPOSITION 1.1.1. — The partition function $Z\{J\}$ is jointly analytic in λ_j and $J \in H_{-1,\varepsilon}$, for $\varepsilon > 0$ sufficiently small.

We postpone the proof to Chapter 2. By Proposition A5 we have.

COROLLARY 1.1.2. — Let $J \in H_{-1,\varepsilon}$, for ε sufficiently small. Then

$$Z_{x_1,\dots,x_n}\{J\} \in \mathscr{ML}_n(H_{-1})$$

with norms bounded by $O(1)L^n n!$

Remark. — Fröhlich obtained a bound $(n!)^{1/2}$ by a more careful analysis [*6*], but with a different norm.

PROPOSITION 1.1.3. — $Z\{J\} \neq 0$ and $Z\{J\}$ is bounded away from zero for $\| J \|_{-1}$ small.

Proof. — For real λ_j and J, the Schwarz inequality yields

$$1 = Z\{0\} = \int dq \leq \left[\int e^{\Phi(J)}dq \int e^{-\Phi(J)}dq \right]^{1/2} = Z\{J\}^{1/2} Z\{-J\}^{1/2}.$$

By Proposition 1.1.1, $Z\{-J\}$ is bounded, so $Z\{J\}^{-1}$ is also bounded. For the general (complex) case, $Z\{0\} = 1$ and $Z\{J\} \neq 0$ by continuity.
Recall $G\{J\} = \ln Z\{J\}$. By Proposition 1.1.3, $G\{J\}$ exists.

PROPOSITION 1.1.4 [*10*]. — For λ_j, J real, $G\{J\}$ is a convex function of J.

Proof. — Let $J = \alpha_1 J_1 + \alpha_2 J_2$ be a convex sum with J_i real. By Hölder's inequality with $p_i = \alpha_i^{-1}$,

$$\begin{aligned} G\{J\} &= \ln Z\{\alpha_1 J_1 + \alpha_2 J_2\} \\ &\leq \ln Z\{J_1\}^{\alpha_1} Z\{J_2\}^{\alpha_2} = \alpha_1 G\{J_1\} + \alpha_2 G\{J_2\}. \end{aligned}$$

By Definition A1, we have

PROPOSITION 1.1.5. — $G\{J\}$ is analytic in H_{-1}, and in the λ_j. The Taylor series coefficients

$$G_{x_1,\ldots,x_n}\{J\} \in \mathcal{ML}_n(H_{-1})$$

have norms $O(1)L^n n!$. Here L is uniformly bounded for λ_j and J as above.

PROPOSITION 1.1.6. — For complex coupling constants in a weak coupling region, or for real coupling with positive physical mass, $G^{(2)}(x, y)$ is the integral kernel of a bounded operator $g^{(2)}$ from H_{-1} to H_1 with a bounded inverse from H_1 to H_{-1}.
We prove this result in Chapter 2.
By analyticity in λ_j, $dG^{(2)}(x, y)/d\lambda_j$ is also bounded, from H_{-1} to H_1.
We define the self energy part Π of $g^{(2)}$ by the resolvent equation,

$$g^{(2)} = c + g^{(2)}\Pi c, \tag{1.1.1}$$

or

$$\Pi = c^{-1} - \Gamma = -\Delta + m_0^2 - \Gamma.$$

Here $c = (-\Delta + m_0^2)^{-1}$ Let $c' = (-\Delta + m_0^2 + 2\lambda_2)^{-1}$.

PROPOSITION 1.1.7. — With λ_j in a weak coupling region, the self energy part has the form

$$\Pi = -2\lambda_2 + \Pi', \tag{1.1.2}$$

where Π' is a bounded transformation from H_1 to H_{-1} with norm $\sum_{j=3}^{l}\sum_{k=1}^{l} O(|\lambda_j\lambda_k|)$. Thus for sufficiently weak coupling,

$$g^{(2)} = c + c\sum_{n=1}^{\infty}(\Pi c)^n \tag{1.1.3}$$
$$= (-\Delta + m_0^2 - \Pi)^{-1}$$
$$= (-\Delta + m_0^2 + 2\lambda_2 - \Pi')^{-1}$$
$$= c' + c'\sum_{n=1}^{\infty}(\Pi' c')^n .$$

Remarks. — We give the proof in Chapter 2. The equation (1.1.4) shows how Π or Π' determines the mass renormalization in the propagator. For a pure quadratic interaction, by (1.1.3), $\Pi' = 0$, and

$$g^{(2)} = (-\Delta + m_0^2 + 2\lambda_2)^{-1} .$$

1.2. The Legendre transformation $G\{J\} \leftrightarrow \Gamma\{A\}$

The generating function $\Gamma\{A\}$ for the vertex functions is related to the generating function $G\{J\}$ for Green's functions by a Legendre transformation. We give two formulations [2] [7] [10]: the first formulation involves a variational principle, while the second formulation relies on solving a nonlinear functional equation. Both formulations are defined and agree for real λ_j, A, J, while the second formulation generalizes to the complex case, see Corollaries 1.2.4-1.2.5. Throughout this section we assume that $A \in H_{1,\delta}$. We obtain $\Gamma\{A\}$ and find that it is analytic in A, for δ sufficiently small.

DEFINITION 1.2.1. — Let A, λ_j be real. Define

$$\Gamma\{A\} = \inf_J\,[-A(J) + G\{J\}], \tag{1.2.1}$$

where the infimum runs over real $J \in H_{-1,\varepsilon}$. We take δ small enough so that the infimum is not attained on the boundary.

Given $J \in H_{-1,\varepsilon}$, the expression

$$A(x) = G_x\{J\} \tag{1.2.2}$$

defines a bounded transformation $J \to A$ from $H_{-1,\varepsilon}$ into H_1. The inverse mapping $A \to J$ from $H_{1,\delta}$ to H_{-1} is obtained by solving (1.2.2) as an equation for J.

PROPOSITION 1.2.2. — Given A, λ_j real, the following conditions are equivalent, for $J \in H_{-1,\varepsilon}$:

(*i*) J minimizes (1.2.1),

(*ii*) J satisfies equation (1.2.2).

Proof. — Since (*ii*) is the variational equation associated with (1.2.1), it follows that (*i*) $\Rightarrow$ (*ii*). We assume (*ii*), and let $M \in H_{-1}$. The function

$$f(\alpha) = -A(J + \alpha M) + G\{J + \alpha M\}$$

is convex (Proposition 1.1.4) and has derivative 0 at $\alpha = 0$, by (*ii*). Since convexity implies f' is monotone, $f' \geq 0$ for $\alpha > 0$ and $f' \leq 0$ for $\alpha < 0$. Since M is arbitrary, J is a minimum, implying (*i*).

A direct proof of the existence and uniqueness of a minimizing J in (1.2.1) can be given. However, we study here the solution to (1.2.2). In both approaches, our proofs require $\| A \|_1$ to be small.

THEOREM 1.2.3. — Consider the complex case and let $\Gamma A \in H_{-1,\varepsilon}$ for ε sufficiently small. Then the equation (1.2.2) has a unique solution $J_A \in H_{-1,2\varepsilon}$. The solution J_A is jointly analytic in A and in λ_j.

Remark. — As before, Γ denotes the linear operator inverse to $g^{(2)}$, the operator with kernel $G^{(2)}(x, y) = G_{x,y}\{0\}$.

Two immediate corollaries are:

COROLLARY 1.2.4. — Let A, λ_j be real, $A \in H_{1,\delta}$. Then (1.2.1) has a unique minimizing function $J = J_A$, and

$$\Gamma\{A\} = -A(J_A) + G\{J_A\}. \tag{1.2.3}$$

COROLLARY 1.2.5. — The generating function $\Gamma\{A\}$ is analytic in $A \in H_{1,\delta}$. In other words, for A, $A + B \in H_{1,\delta}$,

$$\Gamma\{A + B\} = \sum_{n=0}^{\infty} \frac{1}{n!} \Gamma^{(n)}_{B,B,\ldots,B}\{A\}.$$

The Taylor series coefficients $\Gamma^{(n)}_{x_1,\ldots,x_n}\{A\}$ are elements of $\mathcal{ML}_n(H_1)$, with norms $O(L^n n!)$, and are analytic functions of A and the λ_j.

PROPOSITION 1.2.6 [*10*]. — In the real case, $\Gamma\{A\}$ is concave: for real A, $B \in H_{1,\delta}$ and $0 \leq \alpha \leq 1$,

$$\Gamma\{\alpha A + (1 - \alpha)B\} \geq \alpha\Gamma\{A\} + (1 - \alpha)\Gamma\{B\}.$$

Proof. — By Definition 1.2.1,

$$\begin{aligned}\Gamma\{\alpha A + (1-\alpha)B\} &= \inf_J [\alpha(-A(J) + G\{J\}) + (1-\alpha)(-B(J) + G\{J\})] \\ &\geq \alpha \inf_J [-A(J) + G\{J\}] + (1-\alpha) \inf_J [-B(J) + G\{J\}] \\ &= \alpha\Gamma\{A\} + (1-\alpha)\Gamma\{B\}.\end{aligned}$$

Remark. — Corollary 1.2.5 holds, in particular, for $A = 0$. Then $\Gamma^{(n)}(x_1, \ldots, x_n) = \Gamma_{x_1,\ldots,x_n}\{0\}$ and we obtain bounds on the vertex functions for weak $\mathscr{P}(\Phi)_2$ models. We note $\Gamma^{(1)}(x) = \Gamma_x^{(1)}\{0\} = 0$, and $-\Gamma^{(2)}(x, y) = -\Gamma_{x,y}^{(2)}\{0\}$ is the kernel of the linear operator $\Gamma = [g^{(2)}]^{-1}$.

Remark. — By Proposition 1.1.6, the operators $\Gamma : H_1 \to H_{-1}$ and $\Gamma^{-1} : H_{-1} \to H_1$ are both bounded. Thus equation (1.2.2) is equivalent to the equation

$$J = \Gamma A + \Gamma K\{J\}, \tag{1.2.4}$$

where K is the nonlinear operator defined by

$$K\{J\}(x) = -G_x\{J\} + G^{(2)}(x, J).$$

Proof of Theorem 1.2.3. — Explicitly,

$$K\{J\}(x_1) = -\sum_{n=3}^{\infty} \frac{1}{(n-1)!} \int dx_2 \ldots dx_n G^{(n)}(x_1, \ldots, x_n) \cdot J(x_2) \ldots J(x_n). \tag{1.2.5}$$

By Propositions 1.1.5 and 1.1.6, the operator ΓK is a strict contraction on $H_{-1,2\varepsilon}$, for ε sufficiently small. In fact,

$$\begin{aligned} \| \Gamma K\{f\} \|_{-1} &\le O(1) \sum_{n=3}^{\infty} \frac{n!L^n}{(n-1)!} \| f \|_{-1}^{n-1} \\ &\le O(1) \| f \|_{-1}^2 \le \frac{1}{2} \| f \|_{-1}. \end{aligned} \tag{1.2.6}$$

Thus for $\Gamma A \in H_{-1,\varepsilon}$, the operator

$$T_A : f \mapsto \Gamma A + \Gamma K\{f\}$$

maps into itself the sphere in H_{-1} centered at ΓA, and with radius ε. $T_A f$ is analytic in A. Likewise

$$\| T_A^n f \|_{-1} < 2\varepsilon, \tag{1.2.7}$$

and $T_A^n f$ is analytic in A.

Define $h_n(A) = T_A^n f$, with $\Gamma A \in H_{-1,\varepsilon}$. By (1.2.7), the norms $\| h_n(A) \|_{-1}$ are uniformly bounded. Also

$$\| h_{n+1}(A) - h_n(A) \|_{-1} = \| \Gamma K\{h_n(A)\} - \Gamma K\{h_{n-1}(A)\} \|_{-1} \tag{1.2.8}$$

We bound (1.2.8) by using (1.2.5) and Proposition 1.1.5, in a fashion similar to the derivation of (1.2.6). We obtain for $f \in H_{-1,\varepsilon_1}$ (with $\varepsilon_1 \le \varepsilon$ sufficiently small),

$$\| h_{n+1}(A) - h_n(A) \|_{-1} \le \frac{1}{2} \| h_n(A) - h_{n-1}(A) \|_{-1}.$$

By iteration,

$$\| h_n(A) - h_m(A) \|_{-1} \leq \sum_{j=m}^{n-1} 2^{-j+1} \leq O(2^{-m}).$$

By Proposition A3, there exists an analytic function $h : H_{1,\varepsilon} \to H_{-1,2\varepsilon}$ such that

$$\| h_n(A) - h(A) \|_{-1} \to 0.$$

In particular,

$$J_A \equiv h(A) = \lim_{n \to \infty} T_A^n f \in H_{-1,2\varepsilon}$$

exists and is a fixed point of T_A, i. e. a solution to (1.2.4) or (1.2.2). Uniqueness of the solution J_A follows. For two solutions $J_1, J_2 \in H_{-1,2\varepsilon}$, we obtain as above, for ε sufficiently small,

$$\begin{aligned} \| J_1 - J_2 \|_{-1} &= \| \Gamma K \{ J_1 \} - \Gamma K \{ J_2 \} \|_{-1} \\ &\leq O(1) \| K \{ J_1 \} - K \{ J_2 \} \|_{-1} \\ &\leq O(\varepsilon) \| J_1 - J_2 \|_{-1} . \end{aligned}$$

Choosing ε sufficiently small, $O(\varepsilon) < 1$, so $J_1 = J_2$.

The solution J_A is analytic in A, as follows by the analyticity of $h(A)$. Analyticity in λ_j follows similarly.

We now study the inverse Legendre transformation $\Gamma \{ A \} \to G \{ J \}$. Our treatment follows that for $G \{ J \} \to \Gamma \{ A \}$ above.

PROPOSITION 1.2.7. — In the real case, with $\| J \|_{-1}$ small,

$$G \{ J \} = \sup_A [J(A) + \Gamma \{ A \}], \tag{1.2.9}$$

where the supremum runs over $A \in H_{1,\delta}$. A maximizing A always exists and is given by (1.2.2).

Proof. — By Definition 1.2.1, $G \{ J \} \geq \Gamma \{ A \} + J(A)$, with equality for A defined by (1.2.2). This completes the proof.

By Corollary 1.2.5, the expression

$$J(x) = - \Gamma_x \{ A \} \tag{1.2.10}$$

defines a bounded transformation $A \mapsto J$, from $H_{1,\varepsilon}$ to H_{-1}. The inverse mapping $J \mapsto A$ is obtained by solving (1.2.10). From the concavity of $\Gamma \{ A \}$, Proposition 1.2.6, we obtain analogously to Proposition 1.2.2:

PROPOSITION 1.2.8. — Consider the real case. Given J, the following conditions are equivalent, for $A \in H_{1,\varepsilon}$:

(*i*) A maximizes (1.2.9),
(*ii*) A satisfies (1.2.10).

THEOREM 1.2.9. — Consider the complex case and let $\Gamma^{-1} J \in H_{1,\varepsilon}$, with ε sufficiently small. Then the equation (1.2.10) has a unique solution $A_J \in H_{1,2\varepsilon}$. The function A_J is jointly analytic in J and in λ_j, and is the inverse function to J_A of Theorem 1.2.3, i. e., $J_{A_J} = J$.

Proof. — Recall that the operator Γ has the kernel $-\Gamma^{(2)}(x, y)$. We then rewrite (1.2.10) as

$$A = \Gamma^{-1}J + \Gamma^{-1}\mathscr{F}\{A\} \equiv \mathscr{M}_J\{A\}, \tag{1.2.11}$$

where

$$\mathscr{F}\{A\}(x) = \Gamma_x\{A\} - \Gamma A = \sum_{n=3}^{\infty} \frac{1}{(n-1)!}\Gamma^{(n)}(x, A, \ldots, A). \tag{1.2.12}$$

We follow the proof of Theorem 1.2.3 to obtain the existence, uniqueness and analyticity of $A_J = \lim_n \mathscr{M}_J^n \Gamma^{-1}J$. Differentiating (1.2.3) and using (1.2.2) yields (1.2.10). Thus A_J and J_A are inverse functions.

COROLLARY 1.2.10. — In the real case, (1.2.9) has a unique maximizing function $A = A_J$, for $J \in H_{-1,\varepsilon}$ and

$$G\{J\} = J(A_J) + \Gamma\{A_J\}. \tag{1.2.13}$$

Legendre transforms have been considered from a general point of view, with $H_{\pm 1}$ replaced by topological vector spaces X and Y put in duality by a bilinear form, see [*8*]. The formula $J_A = J$ and Propositions 1.2.2 and 1.2.8 are valid in this general context.

1.3. Tree graphs

The Legendre transformation

$$G\{J\} \to \Gamma\{A\} \tag{1.3.1}$$

and its inverse

$$\Gamma\{A\} \to G\{J\} \tag{1.3.2}$$

generate convergent expansions for $G\{J\}$ and $\Gamma\{A\}$ in terms of tree-like graphs. For instance, in the series for $G\{J\}$, each vertex in a tree graph corresponds to a kernel $\Gamma^{(r)}(x_1, \ldots, x_r)$ and each line corresponds to a kernel $G^{(2)}(x, y)$ (propagator). In perturbation theory, such graph expansions are used to define the vertex functions.

We start from the identities

$$G_x\{J\} = + A_J(x) \tag{1.3.3}$$

and

$$\Gamma_x\{A\} = - J_A(x) \tag{1.3.4}$$

obtained from (1.2.2) and (1.2.10), along with Propositions 1.2.2 and

1.2.8. These identities relate $G\{J\}$ and $\Gamma\{A\}$ directly to A_J and J_A, and provide our tree graph expansions. For instance, the power series for (1.3.3),

$$A_J(x) = \sum_{n=2}^{\infty} \frac{1}{(n-1)!} G^{(n)}(x, J, \ldots, J) \tag{1.3.5}$$

has an alternative expression from iterating (1.2.11) to obtain A_J. We identify the coefficients of order $n-1$ in J in these two expressions, and thereby obtain an expression for $G^{(n)}(x_1, \ldots, x_n)$. This expression is a finite sum, and each term is an integral that we represent by a graph. The graphs contributing to $G^{(n)}(x_1, \ldots, x_n)$ have n external lines labelled by $x_1, \ldots, x_n$. The graphs have a « root », the external line labelled by x_1, and branch in one direction away from the root. Each vertex with r legs has the kernel $\Gamma^{(r)}(y_1, \ldots, y_r)$ and each line has the kernel $G^{(2)}(x, y)$, the kernel of Γ^{-1}. For instance, the tree-like graph of Figure 1 contributes to $G^{(5)}$ and corresponds (up to a combinatoric factor) to the integral

$$\int G^{(2)}(x_1, y_1)\, G^{(2)}(x_2, y_2)\, G^{(2)}(y_3, z_1)\, G^{(2)}(x_3, z_2)\, G^{(2)}(x_4, z_3) \; G^{(2)}(x_5, z_4)\, \Gamma^{(3)}(y_1, y_2, y_3)\, \Gamma^{(4)}(z_1, z_2, z_3, z_4)\, dy\, dz\,. \tag{1.3.6}$$

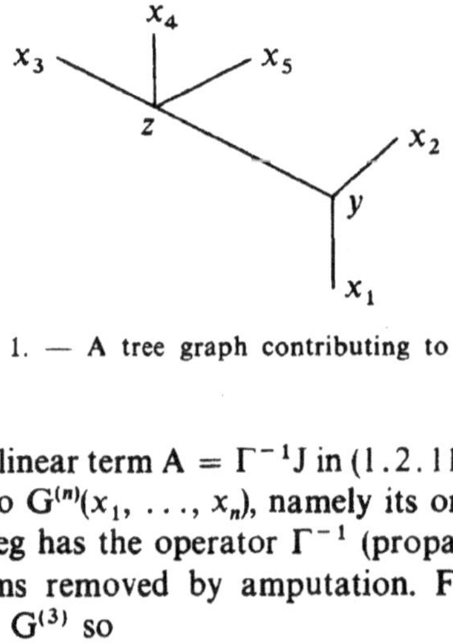

FIG. 1. — A tree graph contributing to $G^{(5)}$.

Retaining only the linear term $A = \Gamma^{-1}J$ in (1.2.11), we obtain the single vertex contribution to $G^{(n)}(x_1, \ldots, x_n)$, namely its one particle irreducible part. Each external leg has the operator Γ^{-1} (propagator) which replaces the mass subdiagrams removed by amputation. For $n = 3$, this is the only contribution to $G^{(3)}$ so

$$G^{(3)}(x_1, x_2, x_3) = \int dy\, G^{(2)}(x_1, y_1)\, G^{(2)}(x_2, y_2)\, G^{(2)}(x_3, y_3) \; \Gamma^{(3)}(y_1, y_2, y_3)\,. \tag{1.3.7}$$

Substituting at least one nonlinear term from $\mathcal{F}\{A\}$ in (1.2.12) into (1.2.11), we obtain the one particle reducible part of $G^{(n)}(x_1, \ldots, x_n)$,

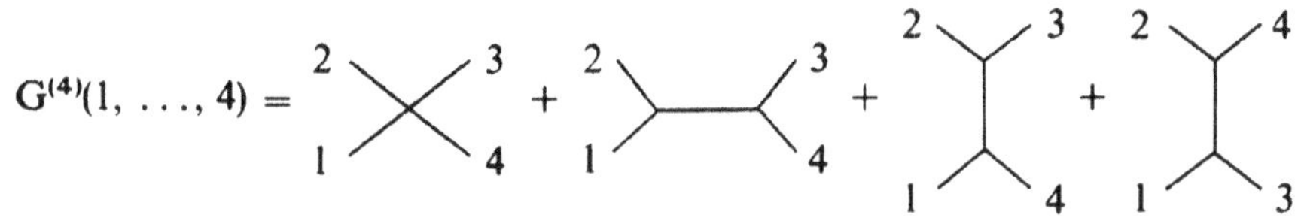

FIG. 2. — In the tree graph expansion of $G^{(n)}$, vertices have kernels $\Gamma^{(r)}$ and lines are propagators $G^{(2)}(x, y)$.

represented by tree graphs with more than one vertex. For example, $G^{(4)}(x_1, \ldots, x_4)$ has the expansion given in Figure 2. Here we abbreviate x_j by j.

For fixed n, the number of tree graphs contributing to $G^{(n)}$ is finite. This number diverges as $n \to \infty$, but nevertheless the power series expansion of $G\{J\}$ in powers of J is convergent, since $G_x\{J\} = A_J(x)$ and A_J is analytic in J. Thus we have proved,

PROPOSITION 1.3.1. — $G\{J\}$ has a convergent tree graph expansion in terms of vertices $\Gamma^{(r)}$ and propagators $G^{(2)}$.

Likewise, we use (1.3.4) and (1.2.4) to generate a tree graph expansion for $\Gamma\{A\}$. In this case, the vertices have kernels equal to the amputated Green's functions,

$$G_a^{(r)}(x_1, \ldots, x_r) = \left(\prod_{i=1}^{r} \Gamma(i)\right) G^{(r)}(x_1, \ldots, x_r),$$

where $\Gamma(i)$ denotes the inverse propagator acting on the variable x_i. Also $r \geq 3$. The lines have kernels given by the propagator $G^{(2)}(x, y)$, as before. In addition, because of the -1 in (1.2.5), there is a factor -1 for each vertex as well as an overall factor -1 from (1.3.4). The sign of a term is therefore $+1$ for an odd number of vertices and -1 for an even number.

Retaining the linear factor $J = \Gamma A$ in (1.2.4), we obtain the single vertex contribution $G_a^{(n)}(x_1, \ldots, x_n)$ to $\Gamma^{(n)}(x_1, \ldots, x_n)$. For $n = 3$, this is the only contribution, and $\Gamma^{(3)}(x_1, x_2, x_3) = G_a^{(3)}(x_1, x_2, x_3)$, which is another form of (1.3.7).

Substituting at least one nonlinear term from (1.2.5) into (1.2.4) we obtain the remaining contributions to $\Gamma^{(n)}$, represented by tree graphs as before. For example, $\Gamma^{(4)}(x_1, \ldots, x_4)$ has the expansion given in Figure 3.

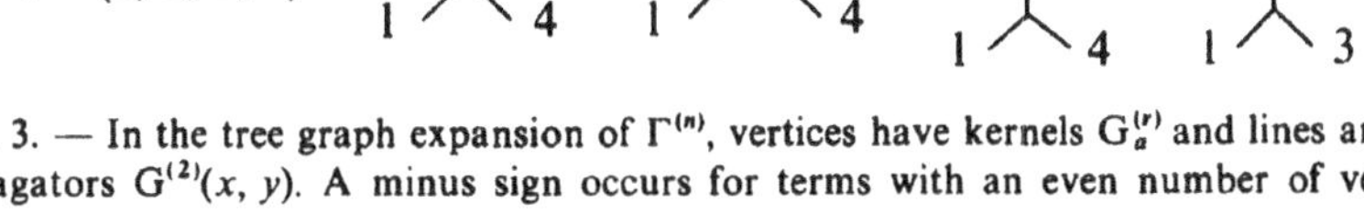

FIG. 3. — In the tree graph expansion of $\Gamma^{(n)}$, vertices have kernels $G_a^{(r)}$ and lines are propagators $G^{(2)}(x, y)$. A minus sign occurs for terms with an even number of vertices.

PROPOSITION 1.3.2. — The analytic function $\Gamma\{A\}$ has a convergent tree graph expansion in terms of vertices $G_a^{(r)}$, of propagators $G^{(2)}$, and with a minus sign for each graph having an even number of vertices.

2. ESTIMATES

In this chapter we establish the estimates stated in Chapter 1. One new feature of our estimate on $Z\{J\}$ is the global bound for $J \in H_{-1}$, as opposed to the local bounds for $J \in L_2$, suppt. J compact, established previously. It is clear that if $\langle \Phi \rangle \neq 0$,

$$\exp \langle \Phi(J) \rangle = \exp\left[\langle \Phi(\cdot) \rangle \int J(x)\, dx \right] \tag{2.1}$$

is bounded in L_1, but not in H_{-1}. This is the reason we substract $\langle \Phi \rangle$ from Φ in the definition of Z. It ensures that no contribution in ln Z is linear in J, and it eliminates the parts of Z which are unbounded in H_{-1}.

In order to estimate $\Phi - \langle \Phi \rangle$, it is convenient to introduce a new symmetry by a tensor product (doubling) procedure, introduced by Ginibre in statistical mechanics. Let $\mathcal{S}'(R^2)^\sim$, $dq^\sim$ be an isomorphic copy of $\mathcal{S}'(R^2)$, dq and let

$$\hat{\mathcal{S}}' = \mathcal{S}'(R^2) \otimes \mathcal{S}'(R^2)^\sim$$
$$d\hat{q} = dq \otimes dq^\sim .$$

For any function A on $\mathcal{S}'(R^2)$, let $A^\sim$ be its isomorphic image, a function on $\mathcal{S}'(R^2)^\sim$, and define an odd function $A_0 = A - A^\sim$ or in tensor product notation

$$A_0 = A \otimes I - I \otimes A^\sim .$$

Then

$$\langle AB \rangle - \langle A \rangle \langle B \rangle = \frac{1}{2} \langle A_0 B_0 \rangle , \tag{2.2}$$

which gives a natural way to perform the subtraction of $\langle \Phi \rangle$ as it occurs in the expansion.

Before getting into the detailed proof of Proposition 1.1.1, we give the proof of Propositions 1.1.6 and 1.1.7. In the case of real coupling constants, let

$$G^{(2)}(p)^\sim = \int_0^\infty \frac{d\rho(a)}{p^2 + a} \tag{2.3}$$

be the spectral representation for the Fourier transform of the connected two point Euclidean Green's function (We assume here Lorentz invariance together with the bound $\int (1 + a)^{-1} d\rho(a) < \infty$). The positivity condition ensures $d\rho(a)$ is a positive measure.

THEOREM 2.1 (*a*). — For real coupling constants $G^{(2)} : H_{-1} \to H_1$ is continuous if and only if

$$\int_0^\infty d\rho(a) + \int_0^\infty a^{-1} d\rho(a) < \infty . \tag{2.4}$$

(*b*) $\Gamma = G^{(2)^{-1}} : H_1 \to H_{-1}$ is continuous.

Proof. — (*a*) Note that $G^{(2)\sim}$ is the Fourier transform of $G^{(2)}(x, 0)$. Then $G^{(2)} : H_{-1} \to H_1$ is a bounded operator, if and only if $(1 + p^2)G^{(2)\sim}(p)$ is a bounded function of p. Assume $(1 + p^2)G^{(2)\sim}(p)$ is bounded. Since $d\rho(a)$ is positive,

$$\int (a^{-1} + 1) d\rho(a) = G^{(2)\sim}(0) + \sup p^2 G^{(2)\sim}(p) < \infty .$$

Conversely, assuming (2.4), we have

$$\sup_p (1 + p^2)G^{(2)\sim}(p) \leq \sup \int (p^2 + a)^{-1} d\rho(a) + \sup \int p^2(p^2 + a^2)^{-1} d\rho(a)$$
$$= \int a^{-1} d\rho(a) + \int d\rho(a) < \infty .$$

(*b*) $\Gamma : H_1 \to H_{-1}$ is a bounded operator if and only if $[G^{(2)\sim}(p)(p^2 + 1)]^{-1}$ is a bounded function of p. There is an interval $[0, \beta]$, such that $\int_0^\beta d\rho(a)$ is nonzero. Then

$$\inf_p (p^2 + 1)G^{(2)\sim}(p) \geq \inf_p \int_0^\beta d\rho(a)(p^2 + 1)(p^2 + a)^{-1}$$
$$\geq \min \{ 1, \beta^{-1} \} \int_0^\beta d\rho(a) > 0 ,$$

and Γ is bounded.

Remarks. — 1. The hypotheses hold in weakly coupled $\mathscr{P}(\Phi)_2$ models. Hence both $G^{(2)} : H_{-1} \to H_1$ and its inverse $\Gamma : H_1 \to H_{-1}$ are bounded operators. In fact, all $\mathscr{P}(\Phi)_2$ quantum field models obey the canonical commutation relations (before field strength renormalization) so for these models

$$\int d\rho(a) = 1 .$$

Weakly coupled $\mathscr{P}(\Phi)_2$ models have isolated one particle states, so $\int a^{-1} d\rho(a) < \infty$ and Proposition 1.1.6 follows for λ_j real.

2. More generally, (*a*) holds for $\mathscr{P}(\Phi)_2$ models except at the critical points (points of zero mass). It should also hold for Φ_3^4, or for other models with a strictly positive mass and a finite field strength renormalization

constant $Z^{-1} = \int d\rho(a)$. Furthermore, (*b*) should hold in any Wightman field theory.

Proof of Proposition 1.1.7. — Let $c = (-\Delta + 1)^{-1}$ have the kernel $C(x - y)$. We integrate by parts on q-space to obtain

$$S^{(2)}(x - y) = C(x - y) - \int dx' C(x - x') \int V'(\Phi(x'))\Phi(y) dq,$$

where V' is the derivative of the interaction polynomial V. Similarly,

$$G^{(2)}(\bar{f}, f) - \| f \|_{-1}^2 = - \langle V'(c\bar{f})\Phi(f) \rangle_T,$$

where T denotes the connected (truncated) part,

$$\langle V'\Phi \rangle_T = \langle V'\Phi \rangle - \langle V' \rangle \langle \Phi \rangle.$$

Let $K(x, y) = \langle V'(\Phi(x))\Phi(y) \rangle_T$. By translation invariance of the measure dq, $K(x, y) = \zeta(x - y)$. Hence the operator k with kernel $K(x, y)$ commutes with c, and k is bounded on H_p if and only if k is bounded on H_0, as we also see by

$$(2.5) \quad | K(f, g) | = | \langle f, kg \rangle | = | \langle c^{p/2} f, k c^{-p/2} g \rangle | \leq \| k \| \; \| f \|_{-p} \| g \|_p .$$

Here $\| k \|$ denotes the operator norm on H_0 (hence on H_p).

We bound $\| k \|$ using estimates from the cluster expansion [3]. Locally, $K(x, y)$ is L_2 and hence L_1. It follows that $\zeta(x)$ is locally L_1. As a consequence of the spectral condition and the mass gap for the relativistic weak coupling $\mathcal{P}(\Phi)_2$ model with real coupling constants [3], $\zeta(x)$ is analytic for $x \neq 0$ and for $|x|$ bounded away from zero,

$$(2.6) \qquad | \zeta(x) | \leq O(\sup_j | \lambda_j |) e^{-m|x|}.$$

Thus $\zeta(x)$ is L_1 and

$$(2.6') \qquad \| k \| \leq \| \zeta \|_{L_1} < O(\sup_j | \lambda_j |).$$

It follows by (2.5) that k is a bounded operator on H_1, and that

$$(2.7) \qquad \Pi = - \Gamma k$$

is a bounded operator from H_1 to H_{-1}.

We next isolate the first order part of Π, by writing

$$V'(\Phi(x)) = \lambda_1 + 2\lambda_2 \Phi(x) + W(x),$$

with

$$W(x) = \sum_{n=3}^{l} n\lambda_n : \Phi(x)^{n-1} : .$$

By definition,

$$K(x, y) = 2\lambda_2 G^{(2)}(x, y) + K_1(x, y),$$

where $K_1(x, y) = \langle W(x)\Phi(y) \rangle_T$ is the integral kernel of the operator k_1. Thus

$$\Pi = -2\lambda_2 - \Gamma k_1, \tag{2.8}$$

and as above, Γk_1 is bounded from H_1 to H_{-1}. Since deg $W \geq 2$, integrating by parts produces at least one more V' vertex in the evaluation of $\langle W\Phi \rangle_T$. This proves that $\| k_1 \|$ is at least second order in the λ_j, and at least first order in some λ_j, $j \geq 3$. The remaining assertions immediately follow for real coupling.

For complex coupling, we use the cluster expansion [*4*] to replace (2.6) by the estimate

$$\int_{|x| \leq 1} |\zeta(x + y)|\, dx \leq O(1)e^{-(m-\varepsilon)|y|}.$$

From (2.6′) we deduce Proposition 1.1.6 in the case of complex coupling constants, and then the remainder of the proof follows as before.

We begin the proof of Proposition 1.1.1 with two preliminary estimates. We use the notation of [*4*].

LEMMA 2.2. — The operator $(-\Delta + 1)^{1/2}$ maps $L_2(R^2)$ boundedly into $L_p(R^2)$ for all $p \in [2, \infty)$.

Proof. — This follows by Hausdorff-Young and Hölder inequalities. We consider the operators

$$\partial^b C_\gamma = \partial^b(-\Delta_\gamma + m_0^2)^{-1} = (-\Delta_\gamma + m_0^2)^{-1} - (\Delta_{\gamma\cup b} + m_0^2)^{-1}$$

and

$$\partial^\Gamma C_\gamma = \partial^{b_1} \ldots \partial^{b_m} C_\gamma$$

where b and b_i are lattice segments and $\Gamma = \{ b_1, \ldots, b_m \}$. Let

$$d(\alpha, \beta, \Gamma) = \sup_{\substack{i=\alpha,\beta \\ b\in\Gamma}} \text{dist}\,(\Delta_i, b).$$

Recall that Δ_i, Δ_α and Δ_β are lattice squares, subsets of R^2, while Δ_γ, $\Delta_{\gamma\cup b}$, are Laplace operators with zero Dirichlet boundary data on lattice curves γ, $\gamma \cup b$, ...

LEMMA 2.3. — Let q be given, $1 \leq q < \infty$, and let m_0 be sufficiently large. The operator

$$(-\Delta + 1)^{1/2}\partial^\Gamma C_\gamma(-\Delta + 1)^{1/2} : L_2(\Delta_\alpha) \to L_2(\Delta_\beta)$$

has a norm bounded by

$$m_0^{O(1)-|\Gamma|/2q} K(\Gamma) \exp\,[-m_0 d(\alpha, \beta, \Gamma)/2]$$

where the constant $K(\Gamma)$ is independent of α, β, γ, Γ, m_0. Furthermore, the combinatoric estimate of [4], Proposition 8.2 holds.

The proof of the combinatoric estimates follows as in [4]. The analytic estimate is based on

LEMMA 2.4. — Let $(\Delta_\gamma - m_0^2)u = 0$ on some subdomain $\mathscr{D}$, and let $u \in L_2$. Then $\nabla u \in L_2$ on any interior subdomain $\mathscr{E}$ which is bounded and bounded away from $\partial\mathscr{D}$. Also

$$\| \nabla u \|_{L_2(\mathscr{E})} \leq \text{const.} \| u \|_{L_2(\mathscr{D})} .$$

Proof. — Away from γ, standard regularity estimates for elliptic operators apply. Thus $u \in C^\infty$ and the Sobolev norms are dominated by $\| u \|_{L_2(\mathscr{D})}$. Thus we consider u up to γ, which consists of lattice line segments. Along the interior of a line segment, u is regular by a reflection principle. For instance, consider the data u given on a semicircle. Extend the data to the opposite semicircle by (-1) times the reflection about the diameter. This determines a harmonic function, zero along the diameter, and hence regular up to the diameter. In the case of an interior corner, a similar reflection argument gives regularity up to γ.

It remains to consider u in a neighborhood of an exterior corner or endpoint ξ of γ. Let C be a circle about ξ, and let $v = \zeta u$, with ζ a smooth function, defined on the interior of C, equal to zero near ξ and equal to one near C. Then v, ∇v and $f = \Delta_\gamma v$ are bounded by const. $\| u \|_{L_2(\mathscr{D})}$. However, $\Delta_{\gamma \cup C}(v - u) = f$, so

$$\nabla u = \nabla v - \nabla(\Delta_{\gamma \cup C})^{-1} f .$$

Since $\nabla(\Delta_{\gamma \cup C})^{-1}$ is a bounded operator, ∇u is bounded in L_2 norm on a neighborhood of ξ.

This completes the proof of Lemma 2.4.

Away from Γ, the kernel of $\partial^\Gamma C_\gamma$ is a solution of the equations

$$(-\Delta_{\gamma,x} + m_0^2)\partial^\Gamma C_\gamma(x, y) = 0 = (-\Delta_{\gamma,y} + m_0^2)\partial^\Gamma C_\gamma(x, y) .$$

Thus by Lemma 2.4, away from Γ, L_2 bounds on $\partial^\Gamma C_\gamma$ [4, Proposition 8.1] imply corresponding L_2 bounds on $\nabla_x \nabla_y \partial^\Gamma C_\gamma(x, y)$. For x or y near Γ, we consider several cases. If either x or y is near $\Gamma_0 \subset \Gamma$, but bounded away from $\Gamma \sim \Gamma_0$, then L_2 bounds on

$$\partial^{\Gamma \sim \Gamma_0} C_{\gamma \cup \Gamma_{00}}(x, y) , \qquad \Gamma_{00} \subset \Gamma_0$$

imply L_2 bounds on

$$\nabla_x \nabla_y \partial^{\Gamma \sim \Gamma_0} C_{\gamma \cup \Gamma_{00}}(x, y) , \qquad \Gamma_{00} \subset \Gamma_0 .$$

The latter give (by linear combination and the definition $\partial^\Gamma = \partial^{\Gamma_0}\partial^{\Gamma \sim \Gamma_0}$ above) L_2 bounds on

$$\nabla_x \nabla_y \partial^\Gamma C_\gamma(x, y) ,$$

with the same decay as the bounds on $\partial^\Gamma C_\gamma$ of [4, Proposition 8.1]. Here we assume $\Gamma_0 \neq \Gamma$ and use the distance $d(\alpha, \beta, \Gamma \sim \Gamma_0)$ to give the decay. If $\Gamma_0 = \Gamma$, then the decay comes only from the separation between x and y. For x bounded away from y

$$(-\Delta_y + m_0^2)C_\gamma(x, y) = 0 = (-\Delta_x + m_0^2)C_\gamma(x, y)$$

and as above we find that

$$\nabla_x\nabla_y C_{\gamma\cup\Gamma_{00}}(x, y), \qquad \Gamma_{00} \subset \Gamma_0 = \Gamma$$

and

$$\nabla_x\nabla_y\partial^\Gamma C_\gamma(x, y)$$

have L_2 bounds with the same decay as $\partial^\Gamma C_\gamma$.

Finally for x, y and Γ all near each other, so $|\Gamma| \leq O(1)$, we use the fact that

$$C_\gamma \leq C_\phi = (-\Delta + m_0^2)^{-1}$$

to see that

$$\| \nabla_x\nabla_y C_\gamma \|$$

is bounded, where $\| \cdot \|$ denotes the L_2 operator norm. In this case, we do not obtain the factor $m_0^{-(2/q)|\Gamma|} = m_0^{-O(1)}$. Combining all cases gives the proof of Lemma 2.3.

The cluster expansion of [4] is based on localizations of the interaction occurring in exponents such as $\exp(\Phi(J))$. With $J \in H_{-1}$, we cannot localize by multiplying J by a characteristic function χ_α of a unit lattice square Δ_α. Instead we define

$$J_\alpha = (-\Delta + 1)^{1/2}\chi_\alpha(-\Delta + 1)^{-1/2}J.$$

Thus $J_\alpha \in H_{-1}$ and

$$J = \sum_\alpha J_\alpha, \qquad \|J\|_{-1}^2 = \sum_\alpha \|J_\alpha\|_{-1}^2.$$

We let γ denote a union of lattice segments in R^2, and let Δ_γ be the Laplace operator with Dirichlet boundary conditions on γ. We let $d\Phi_\gamma$ denote the Gaussian measure on $\mathscr{S}'$ with mean zero and covariance $(-\Delta_\gamma + m_0^2)^{-1} = C_\gamma$. We may take the infinite volume limit for the interacting theory with the Dirichlet contour γ and define for $h \in C_0^\infty$,

$$dq_\gamma = e^{-V(h)}d\Phi_\gamma.$$

We note that dq_γ is not normalized, and refers to a finite volume approximation to the infinite volume measure. We let $J \in C_0^\infty$ and define

$$Z\{J, \gamma\} = \int e^{\Phi(J)}dq_\gamma / e^{\int\langle\Phi(x)\rangle J(x)dx}$$

$$Z_0\{J, \gamma\} = Z\{J, \gamma\}\, e^{\int\langle\Phi(x)\rangle J(x)dx} = \int e^{\Phi(J)}dq_\gamma.$$

Our estimates will be uniform for $\| J \|_{-1}$ small. The contour γ does not occur in the expectation $\langle \Phi \rangle$ in the exponent. In case $R^2 \sim \gamma$ has two components—Ext γ and Int γ—the measure dq_γ factors into an interior and an exterior measure. In an obvious notation

$$Z_0 \{ J, \text{Int}\, \gamma \} = \int \exp\left[\int_{\text{Int}\,\gamma} \Phi(x) J(x) dx \right] dq_\gamma ,$$
$$Z \{ J, \gamma \} = Z \{ J, \text{Ext}\, \gamma \} Z_0 \{ J, \text{Int}\, \gamma \} .$$

Observe that the normalization differs $\left(\text{by the factor } \int dq \right)$ from that of § 1.1. The statement of the following proposition is independent of the normalization, since it concerns a ratio of partition functions. In the proof we use the above unnormalized measures.

PROPOSITION 2.5. — Let $R^2 \sim \gamma$ have two components and let J belong to $H_{-1,\varepsilon}$. In a weak coupling region,

$$| Z \{ J, \text{Ext}\, \gamma \} / Z \{ J \} | \leq e^{O(|\text{Int}\,\gamma| + \|J\|^2_{-1})} .$$

Proof. — The normalizing factor $\exp\left[\int \langle \Phi(x) \rangle J(x)\, dx \right]$ cancels between the numerator and denominator and plays no role in the proof. Let $(\text{Int}\,\gamma)^*$ be the set of lattice lines in Int γ. We write

$$(2.9) \quad \left| \frac{Z \{ J, \text{Ext}\, \gamma \}}{Z \{ J \}} \right| = \left| \frac{Z \{ J, \gamma \}}{Z \{ J \} Z_0 \{ J, \text{Int}\, \gamma \}} \right|$$
$$= \left| \frac{Z_0 \{ J, (\text{Int}\,\gamma)^* \}}{Z_0 \{ J, \text{Int}\,\gamma \}} \times \frac{1}{Z_0 \{ J, (\text{Int}\,\gamma)^* \}} \times \frac{Z \{ J, \gamma \}}{Z \{ J \}} \right|$$

and we bound each of these factors by $\exp\left[O(|\text{Int}\,\gamma| + \| J \|^2_{-1}) \right]$. We first bound $| Z \{ J, \gamma \} / Z \{ J \} |$. This bound is proved in the finite volume approximation, uniformly in the volume, using the Kirkwood-Salsburg equations [*4*, Proposition 5.2]. The same argument, applied within Int γ, bounds the ratio $| Z_0 \{ J, (\text{Int}\,\gamma)^* \} / Z_0 \{ J, \text{Int}\,\gamma \} |$. We have only to indicate how the proof of [*4*] is modified to allow estimates depending on $J \in H_{-1}$.

A typical term in the expansion has the form

$$(2.10) \quad \int K(\Phi, J, \gamma') e^{\int (\Phi - \langle \Phi \rangle) J(x) dx} dq_{\gamma'} / Z \{ J \}$$
$$= \int K(\Phi, J, \gamma') e^{\int_{\text{Int}\,\gamma'} \Phi(x) J(x) dx} dq_{\text{Int}\,\gamma'} Z \{ J, \text{Ext}\, \gamma \} / Z \{ J \}$$

where $K(\Phi, J, \gamma')$ denotes some function on path space. The Kirkwood-Salsburg equations for $\rho(\gamma) = Z\{J, \gamma\}/Z\{J\}$ have the form

$$\rho = 1 + \mathscr{K}\rho,$$

see [4, Chapters 3, 6]. These equations have a unique solution $(I - \mathscr{K})^{-1}1 = \rho$ satisfying

$$|\rho(\gamma)| \leq \exp(|\operatorname{Int}\gamma|)$$

if the operator $\mathscr{K}$ is a contraction. Using the Banach space as in [4], we require bounds of order $\exp(-K|\operatorname{Int}\gamma'|)$ on the kernel in (2.10),

$$\int K(\Phi, J, \gamma') e^{\int_{\operatorname{Int}\gamma'} \Phi(x)J(x)dx} dq_{\operatorname{Int}\gamma'} Z_0\{J, (\operatorname{Int}\gamma')^*\}^{-1}. \tag{2.11}$$

By the Schwarz inequality, (2.11) is dominated by

$$\left[\int |K(\Phi, J, \gamma')|^2 d\Phi_{\operatorname{Int}\gamma'}\right]^{1/2} \left[\int dq'_{\operatorname{Int}\gamma'}\right]^{1/4} \times \left[\int \exp\left[4 \operatorname{Re} \int_{\operatorname{Int}\gamma'} \Phi(x)J(x)dx\right] d\Phi_{\operatorname{Int}\gamma'}\right]^{1/4} Z_0\{J, (\operatorname{Int}\gamma')^*\}^{-1}. \tag{2.12}$$

Here the measure dq' has all the parameters λ_j replaced by $4 \operatorname{Re} \lambda_j$. The integral over dq' is independent of J, and by standard estimates is dominated by $\exp(O|\operatorname{Int}\gamma'|)$. The last factor in (2.12) can be calculated explicitly as

$$\exp\left[2 \int_{\operatorname{Int}\gamma'} \operatorname{Re} J(x) C_\gamma(x, y) \operatorname{Re} J(y) dx\, dy\right] \leq \exp\left[2\|J\|_{-1}^2\right].$$

In the first integral, only the J-vertices require special treatment. We use the H_{-1} localization $J = \Sigma J_\alpha$ as above. The case of a contraction $\partial^\Gamma C_\gamma$ joining J_α and J_β contributes to K the numerical factor

$$|\langle J_\alpha \partial^\Gamma C_\gamma J_\beta \rangle| \leq \|\chi_{\Delta_\alpha}(-\Delta + 1)^{1/2} \partial^\Gamma C_\gamma (-\Delta + 1)^{1/2} \chi_{\Delta_\beta}\| \; \|J_\alpha\|_{-1} \|J_\beta\|_{-1}.$$

The sum over α, β of these factors is bounded by Lemma 2.3. In the case of a contraction $\partial^\Gamma C_\gamma$ joining J_α to a vertex from dq, the vertex is multiplied by the function $\partial^\Gamma C_\gamma(x, J_\alpha)$. By Lemmas 2.2 and 2.3, this function is L_p for all $p < \infty$, with a norm bounded by the decay rate specified in Lemma 2.3. Hence for $p \in [2, \infty)$,

$$\|\partial^\Gamma C_\gamma(\cdot, J)\|_{L_p} \leq \text{const.} \|J\|_{-1}.$$

Multiplication of a vertex from dq by an L_p function does not affect the proofs of [4], and so with the above modifications the proof is that of [4].

Finally, to complete the bound on (2.12), we must show that

$$|Z_0\{J, (\text{Int } \gamma)^*\}|^{-1} \leq e^{O|\text{Int } \gamma|},$$

which then also yields the desired bound on (2.9). We note that

$$Z_0\{J, (\text{Int } \gamma)^*\} = \Pi Z_0\{J, \text{Int } \Delta_\alpha\}$$

where the product ranges over $\Delta_\alpha \subset \text{Int } \gamma$. Thus we must show that

$$|Z_0\{J, \text{Int } \Delta_\alpha\}|^{-1} \leq O(1). \tag{2.13}$$

Using a Duhamel formula and a Schwartz inequality with respect to the free measure, one shows that

$$|Z_0\{J, \text{Int } \Delta_\alpha\} - Z_0\{0, \text{Int } \Delta_\alpha\}| \leq O(\|J\|_{-1}).$$

Hence for small λ_j, (2.13) holds and the proof of the proposition is complete. In the foregoing, we have for notational convenience suppressed the dependence on the multiparameter s which interpolates between the various measures dq_γ. The factor $m_0^{O(1)}$ from Lemma 2.3 is compensated for by factors $O(|\lambda_j|) \ll 1$.

Proof of Proposition 1.1.1. — We obtain a uniform bound on the finite volume approximations to $Z\{J\}$. The desired result then follows from convergence for $J \in C_0^\infty$ and Proposition A3. We proceed by induction, adding one local piece J_α at each step. Let $\{\zeta_\alpha : \alpha \in Z^2\}$ be a C^∞ partition of unity, invariant under lattice translations, and let $J_\alpha = \zeta_\alpha J$. We will show later that $\Sigma \|J_\alpha\|_{-1}^2 \leq \text{const.} \|J\|_{-1}^2$. Let $B \subset Z^2$ be a finite set, let $\alpha \in Z^2 \sim B$, and define

$$J_B = \sum_{\beta \in B} J_\beta$$

$$J(s) = (1 - s)J_B + s(J_\alpha + J_B) = J_B + sJ_\alpha .$$

To state the inductive hypothesis, we define

$$c_{\beta_1\beta_2} = O(1)e^{-(m_0 - \varepsilon)|\beta_1 - \beta_2|}$$

$$C_{B,\beta} = \sum_{\beta_1 \in B} c_{\beta_1,\beta} \leq C \equiv O(1) \sum_{\beta_1 \in Z^2} c_{\beta_1,\beta} .$$

The inductive hypothesis is

$$|Z\{J_B\}| \leq \exp\left[\sum_{\beta \in B} C_{B,\beta} \|J_\beta\|_{-1}^2\right], \tag{2.14}$$

which yields

$$|Z\{J\}| \leq \exp\left[C\Sigma \|J_\beta\|_{-1}^2\right] \leq \exp\left(C' \|J\|_{-1}^2\right).$$

We have, with $\hat{\Phi} = \Phi - \langle \Phi \rangle$,

$$Z\{J_B + J_\alpha\} - Z\{J_B\} = \int_0^1 \frac{d}{ds} Z\{J(s)\}\, ds$$
$$= \int_0^1 ds \int [(\Phi(x) - \langle \Phi \rangle) J_\alpha(x) dx] e^{\int \hat{\Phi}(J(s))} dq$$
$$= \int_0^1 ds [\langle \Phi(J_\alpha) e^{\hat{\Phi}(J(s))} \rangle - \langle \Phi(J_\alpha) \rangle \langle e^{\hat{\Phi}(J(s))} \rangle].$$

Assuming the bound (2.14) on $Z\{J_B\}$,

$$\text{(2.15)} \quad |Z\{J_B + J_\alpha\}| \leq \exp\left[\sum_{\beta \in B} C_{B,\beta} \| J_\beta \|_{-1}^2\right] + \sup_s |\langle \Phi(J_\alpha) e^{\hat{\Phi}(J(s))} \rangle_T|.$$

We show that

$$\text{(2.16)} \quad |\langle \quad \rangle_T| \leq |Z\{J_B\}| \left[(C_{B,\alpha} + c_{\alpha\alpha}) \| J_\alpha \|_{-1}^2 + \sum_{\beta \in B} c_{\beta\alpha} \| J_\beta \|_{-1}^2\right].$$

Substituting in (2.15) yields

$$|Z\{J_B + J_\alpha\}| \leq \exp\left[\sum_{\beta \in B \cup \alpha\}} C_{B \cup \alpha, \beta} \| J_\beta \|_{-1}^2\right]$$

and completes the induction.

To bound $\langle \quad \rangle_T$, we choose

$$A = \Phi(J_\alpha), \qquad B = e^{\hat{\Phi}(J(s))}$$

in (2.2). We apply the cluster expansion to $\langle A_0 B_0 \rangle$, and each expansion term is even under the symmetry $\Phi \leftrightarrow \Phi^{\sim}$, while A_0 and B_0 are individually odd. Using the fact that

$$B_0 = (e^{\hat{\Phi}(J(s))} - I)_0 = (e^{\hat{\Phi}(J(s))} - I) \otimes I - I \otimes (e^{\hat{\Phi}(J(s))} - I),$$

we see that every term in $\langle A_0 B_0 \rangle$ is at least quadratic in J:

$$(\text{linear in } J_\alpha) \times (\text{linear in } J(s)).$$

In the cluster expansion we choose contours γ which contain suppt J_α. Then the region exterior to the contour contributes the factor

$$O(1) |Z\{J_B, \text{Ext}\, \gamma\}| \leq e^{O|\text{Int}\, \gamma| + \|J_B\|_{-1}^2} |Z\{J_B\}|,$$

by Proposition 2.5. The region Int γ contributes $O(1) \exp[-K |\text{Int}\, \gamma|]$ as in [*4*, Proposition 5.2]. We choose K sufficiently large, and the remaining estimates follow as in [*4*].

To complete the proof, we only need to show that

$$\Sigma \| J_\alpha \|_{-1}^2 \leq \text{const.} \| J \|_{-1}^2.$$

In fact

$$\Sigma \| J_\alpha \|_{-1}^2 = \Sigma \| \zeta_\alpha J \|_{-1}^2 = \| J \|_{-1}^2 + \Sigma \langle \zeta_\alpha J, [(-\Delta + 1)^{-1}, \zeta_\alpha] J \rangle .$$

The second term above is dominated by

$$\begin{aligned} (\Sigma \| J_\alpha \|_{-1}^2)^{1/2} (\Sigma \| (-\Delta + 1)^{-1/2} [-\Delta, \zeta_\alpha] (-\Delta + 1)^{-1} J \|_0^2)^{1/2} \\ \leq (\Sigma \| J_\alpha \|_{-1}^2)^{1/2} \| \Sigma - \Delta + 1)^{-1/2} [-\Delta, \zeta_\alpha] (-\Delta + 1)^{-1} \\ [-\Delta, \zeta_\alpha] (-\Delta + 1)^{-1/2} \|^{1/2} \| J \|_{-1} \\ \leq \text{const.} \| J \|_{-1} (\Sigma \| J_\alpha \|_{-1}^2)^{1/2} , \end{aligned}$$

from which the desired bound follows.

APPENDIX

ANALYTIC FUNCTIONS

We consider functions f which map a subset Ω of one Banach space E into a Banach space F. The case $E = C^n$ is the case of vector valued analytic functions. Consider two conditions:

(i) Ω is open in the norm topology.

(ii) Suppose that for each finite sequence $A_j \in E$, $z_j \in C$ such that $z_1A_1 + \ldots + z_nA_n \in \Omega$, the function $f(z_1A_1 + \ldots + z_nA_n)$ is analytic in $z = (z_1, \ldots, z_n)$.

Definition A1. — If (i), (ii) are satisfied, the function $f : \Omega \to F$ is *analytic* in Ω. If $A \in \Omega$, we say that $f(\cdot)$ is analytic at A. If $\Omega = E$, then f is *entire*.

Definition A2. — Let $\mathcal{ML}_n(E, F)$ be the space of bounded multilinear forms of n arguments in E, with values in F. Let $\mathcal{ML}_n(E, C) = \mathcal{ML}_n(E)$.

By Vitali's theorem and a 3ε-argument we obtain

Proposition A3. — Let $f_n : \Omega \to F$ be a sequence of functions analytic in Ω, with $\| f_n \|_F$ uniformly bounded on Ω. If $f_n \to f$ pointwise on a dense subset of Ω, then f is analytic in Ω. Also both f_n and the derivatives of $f_n(\Sigma z_i B_i)$ converge pointwise in F norm to f or its derivatives.

Proposition A4. — If $f : \Omega \to F$ is analytic at $A \in \Omega$, then the derivative

$$\frac{d}{dz_1} \cdots \frac{d}{dz_n} f\left(A + \sum_{i=1}^{n} z_i B_i\right)\bigg|_{z=0} = h_A^{(n)}(B_1, \ldots, B_n)$$

is an element of $\mathcal{ML}_n(E, F)$ with norm less than $\alpha\beta^n n!$, for constants $\alpha = \alpha(A)$ and $\beta = \beta(A)$.

Proof. — Consider $n = 1$. By analyticity

$$\| f(A + z_1(B + C)) - f(A) - z_1 h_A^{(1)}(B + C) \|_F \leq O(|z_1|^2).$$

Also for $n = 2$,

$$\| f(A + z_1B + z_2C) - f(A) - z_1 h_A^{(1)}(B) - z_2 h_A^{(1)}(C) \|_F \leq O(|z_1|^2 + |z_1 z_2| + |z_2|^2).$$

Setting $z_1 = z_2$, we have

$$h_A^{(1)}(B + C) = h_A^{(1)}(B) + h_A^{(1)}(C).$$

The multilinearity of $h_A^{(n)}$ for $n > 1$ follows similarly by induction.

We choose $\| B_i \|_E < 1$. Then by the Cauchy integral formula (with integration over a product of circles of radius $O(n^{-1})$),

$$\| h_A^{(n)}(B_1, \ldots, B_n) \|_F \leq \alpha(A)(\beta n)^n.$$

By multilinearity, for arbitrary $B_i \in E$,

$$\| h_A^{(n)}(B_1, \ldots, B_n) \|_F \leq \alpha(A)(\beta n)^n \prod_{i=1}^{n} \| B_i \|_E,$$

which completes the proof.

PROPOSITION A5. — If f is an entire function, and $\varepsilon > 0$,

$$\| h_A^{(n)}(B_1, \ldots, B_n) \|_F \leq \alpha(A)\varepsilon^n n! \prod_{i=1}^{n} \| B_i \|_E .$$

This proof follows the usual proof for $E = C^n$.

REFERENCES

[1] J. DIMOCK, Asymptotic perturbation expansion in the $\mathscr{P}(\Phi)_2$ quantum field theory. *Commun. Math. Phys.*, t. **35**, 1974, p. 347-356.

[2] C. DE DOMINICIS and P. MARTIN, Stationary entropy principle and renormalization in normal and superfluid systems I, II. *Jour. Math. Phys.*, t. **5**, 1964, p. 14-30, p. 31-59.

[3] J. GLIMM, A. JAFFE and T. SPENCER, The Wightman axioms and particle structure in the $\mathscr{P}(\Phi)_2$ quantum field model. *Ann. Math.*, to appear.

[4] J. GLIMM, A. JAFFE and T. SPENCER, The particle structure of the weakly coupled $\mathscr{P}(\Phi)_2$ models and other applications of high temperature expansions. Part II. The cluster expansion. In: *Constructive quantum field theory*, Ed. by G. Velo and A. Wightman, Lecture notes in physics, Vol. 25, Springer Verlag, Berlin, 1973.

[5] J. FELDMAN, *The $\lambda\Phi_3^4$ field theory in a finite volume.*

[6] J. FRÖHLICH, Schwinger functions and their generating functionals. *I. Helv. Phys. Acta.*

[7] G. JONA-LASINIO, Relativistic field theories with symmetry breaking solutions. *Nuovo Cimento* (L), t. **34**, 1964, p. 1790-1795.

[8] J. MOREAU, Sous-différentiabilité. In : *Colloquium on convexity*, ed. by Fenchel. Kobenhavns Universitet Matematisk Institut, Copenhagen, 1967.

[9] K. SYMANZIK, On the many-particle structure of Green's functions in quantum field theory. *J. Math. Phys.*, t. **1**, 1960, p. 249-273.

[10] K. SYMANZIK, Renormalizable models with simple symmetry breaking. I. Symmetry breaking by a source term. *Commun. Math. Phys.*, t. **16**, 1970, p. 48-80.

[11] A. N. VASILEV and A. K. KAZANSKII, Legendre transforms of the generating functionals in quantum field theory. *Teoret. i Mat. Fizika*, t. **12**, 1972, p. 352-369.

(Manuscrit reçu le 6 février 1974)

Three-particle structure of ϕ^4 interactions and the scaling limit

James Glimm*
Rockefeller University, New York, New York 10021

Arthur Jaffe†
Harvard University, Cambridge, Massachusetts 02138
(Received 13 January 1975)

We show that the inverse propagator $\Gamma(x)$ and the proper self-energy part $\Pi(x)$ are positive for all $x \neq 0$. These results combined with some postulates on the sign or the decay of the vertex functions, in particular, on the decay rate for $\Gamma(x)$, yield the absence of three-particle bound states and the existence of the scaling (Gell-Mann, Low, Wilson) limit for the ϕ^4 interaction.

I. INTRODUCTION

Weak interactions or repulsive interactions are often more tractable than general nonlinear interactions. In the study of particles in quantum field models, the most detailed information is known in these two cases. For weak interactions, the cluster expansions provide extensive information.[1,2] For even ϕ^4 interactions, i.e., ϕ^4 interactions without symmetry breaking, correlation inequalities exclude the even-bound-state spectrum;[3] for earlier partial results see Refs. 2 and 4. In both methods, the proofs rely on decay estimates for the Schwinger functions or Green's functions to establish spectral properties of the mass operator $M=(H^2-P^2)^{1/2}$.

In this paper we call attention to the importance of correlation inequalities or decay estimates on the inverse propagator and the Euclidean vertex functions $\Gamma^{(n)}(x_1, \ldots, x_n)$ (one-particle irreducible Euclidean Green's functions). On the one hand, assuming certain decay estimates, we prove in Secs. V A and V B the absence of odd (three-particle) bound states. By "absence of odd bound states" we just mean the absence of spectrum for the mass operator $M \restriction \mathcal{H}_0$ in the spectral interval $(m, 3m)$. Here m is the physical mass and $\restriction \mathcal{H}_0$ denotes the restriction of M to the subspace $\mathcal{H}_0$ of vectors which are odd under the $\phi \to -\phi$ symmetry. In addition, we find that $\Gamma^{(6)} \leq 0$, or a decay estimate for $\Gamma^{(6)}$, yields a $3m$ exponential decay rate for the inverse propagator $\Gamma(x)$,

$$\Gamma(x) \leq \text{const} \times e^{-3m|x|} .$$

Furthermore, information on the spectral threshold in the inverse propagator, e.g.,

$$\Gamma(x) \leq \text{const} \times e^{-(m+\epsilon)|x|} ,$$

is sufficient to prove the existence of the scaling limit of Gell-Mann, Low, and Wilson. In this limit for $\lambda\phi_2^4$, λ (the bare coupling) $\to\infty$ while m (the physical mass) is held fixed. See Sec. V C. According to renormalization-group ideas, the control of this limit in two dimensions would be comparable (as far as renormalization difficulties go) to an existence proof for ϕ_4^4.

The remainder of this paper is concerned with the reduction of the problem to the decay estimates which we postulate in Sec. V. In Sec. II, we relate the propagator G and inverse propagator Γ to standard representations for Herglotz functions, and we show that the kernel of Γ (and of the proper self-energy part Π) satisfies

$$\Pi(x)=\Gamma(x) \geq 0, \quad x \neq 0 . \tag{1.1}$$

In Sec. III, we derive a functional-integration by parts formula,

$$\langle \phi(x)(1-P_0)A\rangle = \int G(x-y)\left[\left\langle \frac{\delta A}{\delta\phi(y)}\right\rangle - \langle \mathcal{V}'(y)(1-P_0-P_1)A\rangle\right] dy . \tag{1.2}$$

Here $\mathcal{V}$ is the interaction polynomial, G is the propagator, and P_0, P_1 are the projections onto the Euclidean vacuum Ω and the states $(1-P_0)\phi\Omega$, respectively. (See Sec. III for details.) In Sec. IV, we apply these ideas to derive the representation

$$\Gamma(x)=\langle \mathcal{V}'(x)(1-P_0-P_1)\mathcal{V}'(0)\rangle , \quad |x| \neq 0 . \tag{1.3}$$

The bounds of Sec. V use this representation.

II. INTEGRAL REPRESENTATIONS FOR (THE PROPAGATOR AND) THE INVERSE PROPAGATOR

A. The propagator as a Herglotz function

Let $\phi(x)$ be a real, scalar field satisfying the Osterwalder-Schrader axioms.[5] [An Osterwalder-Schrader theory is the analytic continutation to imaginary time of a relativistic, Wightman theory. The Osterwalder-Schrader axioms give conditions on the imaginary-time (Euclidean) Schwinger functions which ensure the existence of an analytic continuation to a real-time theory satisfying the Wightman axioms. See Refs. 5 and 10 for details.] The two-point Green's function (propagator) is defined by

$$G(x-y)=\langle\phi(x)\phi(y)\rangle-\langle\phi(x)\rangle^2 . \quad (2.1)$$

By analytic continuation of the (real time) vacuum expectation value, it follows that for $x\neq 0$

$$G(x)=\int_{m^2}^{\infty} d\rho(a)C_a(x) , \quad (2.2)$$

where $C_a=(-\Delta+a)^{-1}$ and where $d\rho(a)$ is a positive, tempered measure.

In this section we assume an additional bound

$$\int_{m^2}^{\infty}\frac{d\rho(a)}{1+a^2}<\infty , \quad (2.3)$$

which does not follow from the axioms. The bound (2.3) ensures that the representation (2.2) determines a canonical extension of $G(x)$ to a tempered distribution. The assumption (2.3) is reasonable, since for all renormalizable interactions, perturbation theory suggests $\int(1+a)^{-1}d\rho<\infty$, which ensures (2.3). Furthermore, in canonical theories, $\int d\rho=1$. In fact, (2.3) is equivalent to the assumption that $G(x)$ has a Fourier transform $\tilde{G}(p)$ which as a function of $z=-p^2$ is a Herglotz function of class $P(-\infty, m^2)$, in case $m>0$.

By definition, a Herglotz function is an analytic function defined on the upper half plane and having positive imaginary part. A Herglotz function of class $P(a, b)$ is a Herglotz function which can be continued analytically into the lower half plane across the real interval (a, b) by the formula $f(z^*)=f^*(z)$. A Herglotz function $f(z)$ of class $P(-\infty, m^2)$ has a unique, canonical representation

$$f(-z)=-\alpha z+\beta+\int_{m^2}^{\infty}\left(\frac{1}{a+z}-\frac{a}{a^2+1}\right)d\mu(a) . \quad (2.4)$$

where $\alpha\geq 0$, β is real, and $(1+a^2)^{-1}d\mu(a)$ is a positive, finite measure supported on the interval $[m^2, \infty)$. The relation (2.4) is sometimes called a once-subtracted dispersion relation. An analysis of $\tilde{G}(p)$ in terms of representations such as (2.4) has also been given by Symanzik.[6] See Donoghue, Ref. 7, pp. 18–26, for general properties of Herglotz functions, including a proof of (2.4).

Proposition 2.1. Let $F(x)$ be a Euclidean invariant distribution with Fourier-Laplace transform $\tilde{F}(p)$ which as a function of $z=-p^2$ belongs to $P(-\infty, m^2)$. Then

$$F(x)=\int_{m^2}^{\infty} C_a(x)d\mu(a)>0 \text{ for } x\neq 0 , \quad (2.5)$$

where $(1+a^2)^{-1}d\mu(a)$ is a positive, finite measure. Furthermore, $\tilde{F}(p)=f(-p^2)$, where $f(-p^2)$ satisfies (2.4).

Proof. By general properties of Herglotz functions, $\tilde{F}(p)=f(-p^2)$, as described above. Taking the Fourier transform of (2.4) yields

$$F(x)=(\alpha\Delta+\beta)\delta(x)+\int_{m^2}^{\infty} C_a(x)\frac{d\mu(a)}{1+a^2} + \Delta\int_{m^2}^{\infty} C_a(x)\frac{a\, d\mu(a)}{1+a^2} , \quad (2.6)$$

so only the last two terms contribute for $x\neq 0$. For $a>0$ and $x\neq 0$, $C_a(x)$ is positive, real analytic, and exponentially decreasing in $\sqrt{a}|x|$, and Δ may be taken under the integral sign. Since $\Delta C_a(x)=aC_a(x)$ for $x\neq 0$, we have (2.5) for $x\neq 0$, assuming $m>0$. If $m=0$, we have

$$F_\epsilon(x)=(\alpha\Delta+\beta)\delta(x)+\int_{\epsilon^2}^{\infty}[C_a(x)+\Delta C_a(x)a]\frac{d\mu(a)}{1+a^2} = \int_{\epsilon^2}^{\infty} C_a(x)d\mu(a), \quad x\neq 0 , \quad (2.7)$$

and $0<F_\epsilon(x)\nearrow F(x)$ in $\mathcal{S}'$ as $\epsilon\searrow 0$ for $x\neq 0$. By positivity $F_\epsilon(x)\to F(x)$ pointwise and (2.5) is valid. We also obtain a regularity condition on $d\mu(a)$ at $a=0$ in $d=2$ dimensions. For $d>2$, we modify the reasoning to allow for a δ function in $d\mu$ at $a=0$.

B. Positivity of $\Gamma(x)$, the inverse propagator

Assuming (2.2) and (2.3), $G(x)$ is a tempered distribution. Thus, $f\to G*f$ defines a linear operator on L_2 with domain $\mathcal{S}$. Define the inverse propagator Γ by

$$\Gamma\equiv -G^{-1} ,$$

which acts as convolution by $\Gamma(x)$. By definition, $\tilde{\Gamma}(p)=-1/\tilde{G}(p)$, and by positivity of the Euclidean Hilbert-space inner product, $\tilde{G}(p)>0$. Thus, $\tilde{\Gamma}(p)<0$ and Γ is a negative operator.

Theorem 2.2 Assume (2.2) and (2.3). Then $\Gamma(x)$ satisfies the conditions of Proposition 2.1. In particular,

$$\Gamma(x)=\int_{\overline{m}^2}^{\infty} d\nu(a)C_a(x), \quad x\neq 0, \quad (2.8)$$

where $(1+a^2)^{-1}d\nu(a)$ is a positive, finite measure. If in addition $\int(1+a)^{-1}d\rho<\infty$, then $\overline{m}\geq m$; thus for $x\neq 0$, $\Gamma(x)$ is real analytic, positive, monotonic decreasing in $|x|$, and has an exponential decay rate $\overline{m}\geq m$.

Remark. If the spectral measure $d\rho(a)$ for $\tilde{G}(p)$ has no mass in the (nonempty) interval (m^2, M^2), then $m<\overline{m}\leq M$. The Castillejo-Dalitz-Dyson (CDD) radius is the radius of analyticity of the inverse propagator. It equals $\overline{m}$, and $\overline{m}<M$ is the case of CDD poles in $\tilde{\Gamma}(p)$. See Ref. 8,

for example.

Proof. The function $f(z)\neq 0$ is a Herglotz function if and only if $-1/f(z)$ is a Herglotz function. If $f(z)\in P(-\infty, m^2)$ and $\int(1+a)^{-1}d\mu(a)<\infty$, then assuming (2.2),

$$f(-z)=\int_{m^2}^{\infty}\frac{1}{a+z}\,d\mu(a)\,.$$

Then $-1/f(z)\in P(-\infty, m^2)$.

Theorem 2.3. Assume (2.2) and $\int d\rho(a)=1$. Then

$$\tilde{\Gamma}(p)=-(p^2+\beta)+\int_{\bar{m}^2}^{\infty}\left(\frac{1}{p^2+a}-\frac{a}{a^2+1}\right)d\nu(a)\,, \tag{2.9}$$

where $(1+a^2)^{-1}d\nu(a)$ is a positive, finite measure and $\bar{m}\geq m$.

Proof. Since $\int d\rho(a)=1$, it follows that $p^2\tilde{G}(p)\to 1$ as $p^2\to\infty$, and hence $p^{-2}\tilde{\Gamma}(p)\to -1$. By (2.4),

$$\tilde{\Gamma}(p)=-(\alpha p^2+\beta)+\int_{\bar{m}^2}^{\infty}\left(\frac{1}{p^2+a}-\frac{a}{a^2+1}\right)d\nu(a)\,,$$

with $(1+a^2)^{-1}d\nu(a)$ a positive, finite measure. By the dominated convergence theorem, as $p^2\to\infty$,

$$p^{-2}\int\left(\frac{1}{p^2+a}-\frac{a}{a^2+1}\right)d\nu(a)$$

$$=p^{-2}\int\frac{(1-p^2a)}{(p^2+a)(a^2+1)}\,d\nu(a)\to 0\,.$$

Thus $p^{-2}\tilde{\Gamma}(p)\to-\alpha$ and $\alpha=1$. As above, $\bar{m}\geq m$.

We now assume a further regularity requirement on $d\rho(a)$, namely the existence of the first moment,

$$\int a\,d\rho(a)<\infty\,. \tag{2.10}$$

The bound (2.10) is known to hold for $\mathcal{P}(\phi)_2$ models. [This is, however, the only simple example with finite mass renormalization, and hence the only example where we expect (2.10).] In the case of (2.10), we may assume the canonical normalization of ϕ,

$$\int d\rho(a)=1\,. \tag{2.11}$$

Theorem 2.4. Assume (2.2), (2.10), and (2.11). Then

$$\tilde{\Gamma}(p)=-\left(p^2+\int a\,d\rho\right)+\int_{\bar{m}^2}^{\infty}\frac{d\nu(a)}{p^2+a}\,, \tag{2.12}$$

where $(1+a)^{-1}d\nu(a)$ is a positive, finite measure and $\bar{m}\geq m$.

Proof. We have

$$\tilde{\Gamma}(p)=-(p^2+m^2)\left[\int\left(\frac{p^2+m^2}{p^2+a}\right)d\rho(a)\right]^{-1}$$

$$=-(p^2+m^2)\left[1-\int\left(\frac{a-m^2}{p^2+a}\right)d\rho(a)\right]^{-1}$$

$$=-p^2-m^2-\int(a-m^2)d\rho+o(1)$$

$$=-p^2-\int a\,d\rho+o(1).$$

For a positive measure $d\mu(a)$ with $\int(1+a^2)^{-1}d\mu(a)<\infty$,

$$\int_{\bar{m}^2}^{\infty}\left(\frac{1}{p^2+a}-\frac{a}{a^2+1}\right)d\mu(a)=\text{const}+o(1)$$

if and only if $\int(1+a)^{-1}d\mu(a)<\infty$. Combining these two facts with (2.9) yields the theorem.

Remark. Assuming (2.2) and (2.10), but not (2.11),

$$\tilde{\Gamma}(p)=-\left(\int d\rho\right)^{-1}\left(p^2+\frac{\int a\,d\rho}{\int d\rho}\right)+\int_{\bar{m}^2}^{\infty}\frac{d\nu(a)}{p^2+a}\,. \tag{2.13}$$

C. Field-Strength renormalization

In this section we assume the existence of isolated one-particle spectrum in the propagator. We assume

$$\tilde{G}(p)=\int\frac{d\rho(a)}{p^2+a}=\frac{Z}{p^2+m^2}+\int_{M^2}^{\infty}\frac{d\sigma(a)}{p^2+a}\,, \tag{2.14}$$

where $m>0$ and where $d\sigma(a)$ is supported in $[M,\infty)$, with $m<M$. We also assume $\int d\rho<\infty$, so it is no loss of generality to assume a canonical theory, $\int d\rho=1$. A theory satisfying these assumptions is called a canonical, isolated-particle (CIP) theory.

Theorem 2.4. For a CIP theory,

$$Z^{-1}=1+\int_{\bar{m}^2}^{\infty}\frac{d\nu(a)}{(a-m^2)^2}\,. \tag{2.15}$$

Proof. Since $\tilde{\Gamma}(p)$ is analytic as a function of p^2 for $|p^2|<\bar{m}^2$, and since $m<\bar{m}$, it follows that $\tilde{\Gamma}(p)$ is analytic at $p^2=-m^2$, the pole of $\tilde{G}(p)$. Thus,

$$Z^{-1}=-\left(\frac{d\tilde{\Gamma}}{dp^2}\right)_{p^2=-m^2}\,,$$

and differentiating (2.9) yields (2.15).

Definition 2.5. The renormalized propagator $\tilde{G}_{\text{ren}}(p)$ for a CIP theory is

$$\tilde{G}_{\text{ren}}(p)=Z^{-1}\tilde{G}(p)=\int\frac{d\rho_{\text{ren}}(a)}{p^2+a}\,, \tag{2.16}$$

where

$$d\rho_{\rm ren}(a) = Z^{-1} d\rho(a)$$

$$= \left[1 + \int_{\bar{m}^2}^{\infty} \frac{d\nu(b)}{(b-m^2)^2}\right] d\rho(a) . \qquad (2.17)$$

Theorem 2.6. For a CIP theory,

$$\frac{1}{p^2+m^2} \leq \tilde{G}_{\rm ren}(p) \leq \left(\frac{\bar{m}^2}{\bar{m}^2 - m^2}\right)^2 m^{-2} , \qquad (2.18)$$

when p is real.

Comment. For $\bar{m}/m - 1 > \epsilon$, $m > \epsilon$, the bound (2.18) on $\tilde{G}_{\rm ren}(p)$ is independent of all parameters. Because $M \geq \bar{m}$, the bound $\bar{m}/m \geq 1+\epsilon$ also excludes level crossings in the state for the particle of lowest mass. For a canonical $\mathcal{P}(\phi)$ model without bound states or CDD zeros in $\tilde{G}(p)$, we have $\bar{m}/m \geq 2$. As a converse to Theorem 2.6 we have Theorem 2.7.

Theorem 2.7. For a CIP theory, $\bar{m}/m$ is bounded away from 1 if $M^2\tilde{G}_{\rm ren}(p=0)$ is bounded and if M/m is bounded away from one.

Proof of Theorem 2.6. The lower bound in (2.18) follows by definition, so we need only prove the upper bound. To begin, we note that

$$\int_{\bar{m}^2}^{\infty} \frac{d\nu(a)}{(a-m^2)^2} \leq \left(\frac{\bar{m}^2}{\bar{m}^2-m^2}\right)^2 \int_{\bar{m}^2}^{\infty} \frac{d\nu(a)}{a^2} , \qquad (2.19)$$

since $a(a-m^2)^{-1}$ is decreasing in a for $a > m^2$.

We have two expressions for $d\tilde{\Gamma}(p)/dp^2$: One is obtained from (2.9), and the other is obtained from (2.14) and the relation $\tilde{\Gamma}' = \tilde{G}'G^{-2}$. Equating these we obtain

$$1 + \int_{\bar{m}^2}^{\infty} \frac{d\nu(a)}{(p^2+a)^2} = \int_{m^2}^{\infty} \frac{d\rho(a)}{(p^2+a)^2} \left(\int_{m^2}^{\infty} \frac{d\rho(a)}{p^2+a}\right)^{-2} . \qquad (2.20)$$

Using (2.20) with $p^2=0$, we infer from (2.17)–(2.19)

$$\tilde{G}_{\rm ren}(p) \leq \left(\frac{\bar{m}^2}{\bar{m}^2-m^2}\right)^2 \left(1 + \int_{\bar{m}^2}^{\infty} \frac{d\nu(a)}{a^2}\right) \int_{m^2}^{\infty} \frac{d\rho(a)}{p^2+a}$$

$$\leq \left(\frac{\bar{m}^2}{\bar{m}^2-m^2}\right)^2 \int_{m^2}^{\infty} \frac{d\rho(a)}{a^2} \left(\int_{m^2}^{\infty} \frac{d\rho(a)}{a}\right)^{-1}$$

$$\leq \left(\frac{\bar{m}^2}{\bar{m}^2-m^2}\right)^2 m^{-2} ,$$

completing the proof.

Proof of Theorem 2.7. If $\bar{m} < M$, we refer to the definition [Ref. 8, formula (6.5)],

$$(\bar{m}/m)^2 - 1 = \left[m^2 \int_{M^2}^{\infty} \frac{d\rho_{\rm ren}(a)}{a-\bar{m}^2}\right]^{-1}$$

$$= [(M/m)^2 - 1]\left[(M^2-m^2)\int_{M^2}^{\infty} \frac{d\rho_{\rm ren}(a)}{a-\bar{m}^2}\right]^{-1} . \qquad (2.21)$$

Also, using (2.21) and the monotonicity of $a(a-\bar{m}^2)^{-1}$,

$$(M^2-m^2)\int_{M^2}^{\infty} \frac{d\rho_{\rm ren}(a)}{a-\bar{m}^2} = (M^2-\bar{m}^2)\int_{M^2}^{\infty} \frac{d\rho_{\rm ren}(a)}{a-\bar{m}^2} + 1$$

$$\leq M^2 \int_{M^2}^{\infty} \frac{d\rho_{\rm ren}(a)}{a} + 1$$

$$\leq M^2 \tilde{G}_{\rm ren}(0) + 1 .$$

So by (2.21),

$$(\bar{m}/m)^2 - 1 \geq [(M/m)^2 - 1][M^2\tilde{G}_{\rm ren}(0)+1]^{-1} ,$$

completing the proof.

III. INTEGRATION BY PARTS IN THE PHYSICAL MEASURE

We consider integration by parts in Euclidean $\mathcal{P}(\phi)$ boson models. The Euclidean Hilbert space is spanned by states $\phi(x_1)\cdots\phi(x_n)\Omega$, $n=0,1,\ldots$, where Ω is the Euclidean vacuum. See Refs. 5 and 10 for more details. Function-space integration by parts yields the formula

$$\langle\phi(x)(1-P_0)A\rangle$$

$$= \int C(x-y)\left[\left\langle\frac{\delta A}{\delta\phi(y)}\right\rangle - \langle\mathcal{V}'(y)(1-P_0)A\rangle\right] dy , \qquad (3.1)$$

where P_0 is the projection onto Ω and

$$C = (-\Delta + m_0^2)^{-1}, \quad \mathcal{V}'(y) = :\mathcal{P}'(\phi(y)): .$$

In following the point of view that we should express our formulas in terms of physical (renormalized) parameters, we wish to replace C in (3.1) by the Green's function G. Let P_1 denote the projection onto the span of $(1-P_0)\phi\Omega$. Explicitly,

$$P_1\theta = -\int (1-P_0)\phi(x)\Omega\Gamma(x-y)\langle\phi(y)\Omega, 1-P_0)\theta\rangle \, dx\, dy .$$

Theorem 3.1. In each pure phase of a noncritical $\mathcal{P}(\phi)_2$ model,

$$\langle\phi(x)(1-P_0)A\rangle$$

$$= \int G(x-y)\left[\left\langle\frac{\delta A}{\delta\phi(y)}\right\rangle - \langle\mathcal{V}'(y)(1-P_0-P_1)A\rangle\right] dy . \qquad (3.2)$$

Remark. We interpret subtraction of the one-particle contribution $\langle\mathcal{V}'(y)P_1A\rangle$ in (3.2) as providing the mass renormalization of (3.1). In this way we replace C (characterized by the bare mass m_0) by G (characterized by the physical mass m).

Proof. Using (3.1) with $A = \phi(y)$ gives

$$C(x-y) = G(x-y) + \langle\phi(x)(1-P_0)(C\mathcal{V}')(y)\rangle . \qquad (3.3)$$

The validity of (3.1) in each pure phase is proved in Ref. 8. Substituting (3.3) into (3.1) gives

$$\langle\phi(x)(1-P_0)A\rangle = \int G(x-y)\left[\left\langle\frac{\delta A}{\delta\phi(y)}\right\rangle - \langle \mathcal{V}'(y)(1-P_0)A\rangle\right]dy$$
$$+\int \langle\phi(x)(1-P_0)(C\mathcal{V}')(y)\rangle\left[\left\langle\frac{\delta A}{\delta\phi(y)}\right\rangle - \langle \mathcal{V}'(y)(1-P_0)A\rangle\right]dy$$
$$=\int G(x-y)\left[\left\langle\frac{\delta A}{\delta\phi(y)}\right\rangle - \langle \mathcal{V}'(y)(1-P_0)A\rangle\right]dy + \int \langle\phi(x)(1-P_0)\mathcal{V}'(y)\rangle\langle\phi(y)(1-P_0)A\rangle dy \ . \tag{3.4}$$

By Euclidean invariance, $\langle\phi(x)\mathcal{V}'(y)\rangle = \langle\mathcal{V}'(x)\phi(y)\rangle$, and by translation invariance the last term in (3.4) is

$$-\int G(x-y)\langle\mathcal{V}'(y)(1-P_0)\phi(z)\rangle\Gamma(z-t)\langle\phi(t)(1-P_0)A\rangle\, dy\, dz\, dt = \int G(x-y)\langle\mathcal{V}'(y)P_1A\rangle dy \ ,$$

from which (3.2) follows.

Corollary 3.2. In each pure phase of a noncritical $\mathcal{P}(\phi)_2$ model,

$$\Gamma_x\langle\phi(x)(1-P_0)A\rangle = \int \Gamma(x-y)\langle\phi(y)(1-P_0)A\rangle dy$$
$$= -\left\langle\frac{\delta A}{\delta\phi(x)}\right\rangle + \langle\mathcal{V}'(x)(1-P_0-P_1)A\rangle \ . \tag{3.5}$$

Remark. In the case of an even theory, P_0 does not occur in the formulas above.

IV. THE PROPER SELF-ENERGY PART

The proper self-energy operator Π is defined by the equation

$$\Pi = C^{-1} + \Gamma \ , \tag{4.1}$$

where $C = (-\Delta + m_0^2)^{-1}$. Let $\Pi(x-y)$ be the kernel of Π. Since C^{-1} has the local kernel $(-\Delta + m_0^2) \times \delta(x-y)$, by Theorem 2.2 we have Theorem 4.1.

Theorem 4.1. Assuming (2.2)–(2.3),

$$\Pi(x) = \Gamma(x) \geq 0, \quad \text{for } x \neq 0 \ . \tag{4.2}$$

Using the integration by parts formula of Sec. III, we obtain a representation for $\Pi(x)$.

Theorem 4.2. For a noncritical $\mathcal{P}(\phi)_2$ model,

$$\Pi(x-y) = -\langle\mathcal{V}''(x)\rangle\delta(x-y) + \langle\mathcal{V}'(x)(1-P_0-P_1)\mathcal{V}'(y)\rangle \ . \tag{4.3}$$

Proof. Applying the operators Γ and C^{-1} to (3.3) and using (4.1) gives

$$\Gamma(x-y) + C^{-1}(x-y) = \int \Gamma(x-z)\langle\phi(z)(1-P_0)\mathcal{V}'(y)\rangle dz$$
$$= \Pi(x-y) \ . \tag{4.4}$$

Thus by applying (3.5) to (4.4) we obtain (4.3).

Theorem 4.3. For a noncritical $\mathcal{P}(\phi)_2$ model with the canonical normalization (2.10),

$$m^2 = m_0^2 + \langle\mathcal{V}''(x)\rangle - \int_{m^2}^{\infty}(a-m^2)d\rho(a) \ . \tag{4.5}$$

Remark. If $\mathcal{V}$ is Wick ordered in the physical vacuum, then $\langle\mathcal{V}''(x)\rangle = 0$ and

$$m^2 - m_0^2 = -\int_{m^2}^{\infty}(a-m^2)d\rho(a) \leq 0 \ . \tag{4.6}$$

This is the old result of Lehmann that $m \leq m_0$ (see Ref. 9).

Proof. Let $h(p)$ be the Fourier transform of $\langle\mathcal{V}'(x)(1-P_0-P_1)\mathcal{V}'(y)\rangle$. We claim $|h(p)| \leq O(|p|^{-2(1-\epsilon)})$ as $|p| \to \infty$. This is the case for the Fourier transform of $\langle\mathcal{V}'(x)(1-P_0)\mathcal{V}'(y)\rangle$, by the local regularity estimates of Ref. 8. Let $g(p)$ be the Fourier transform of $\langle\mathcal{V}'(x)(1-P_0)\phi(y)\rangle$. Then the Fourier transform of $-\langle\mathcal{V}'(x)P_1\mathcal{V}'(y)\rangle$ is $g^2\tilde{\Gamma}$. For $\mathcal{P}(\phi)_2$, $|g(p)| \leq O(|p|^{-2+\epsilon})$ as $|p| \to \infty$, by a similar argument. Since $|\tilde{G}(p)| \geq O(|p|^{-2})$ by (2.2), $|g^2\tilde{\Gamma}| \leq O(|p|^{-2+2\epsilon})$, showing $|h| \leq O(|p|^{-2(1-\epsilon)})$.

Taking the Fourier transform of (4.3),

$$\tilde{\Gamma}(p) = -(p^2 + m_0^2) + \tilde{\Pi}(p)$$
$$= -[p^2 + m_0^2 + \langle\mathcal{V}''(x)\rangle] + O(|p|^{-2(1-\epsilon)}) \ .$$

Comparing with (2.12) yields

$$\int a\, d\rho(a) = m_0^2 + \langle\mathcal{V}''(x)\rangle \ , \tag{4.7}$$

to complete the proof of (4.5).

Corollary 4.4. For a $\mathcal{P}(\phi)_2$ model,

$$\Gamma(x) + C_\alpha^{-1}(x) = \langle\mathcal{V}'(x)(1-P_0-P_1)\mathcal{V}'(0)\rangle \ , \tag{4.8}$$

where $\alpha = \int a\, d\rho$ and $\tilde{C}_\alpha(p) = (p^2+\alpha)^{-1}$. The Fourier transform of (4.8) is $O(|p|^{-2+\epsilon})$ as $|p| \to \infty$, and (4.8) is positive for all x.

V. ON THE ABSENCE OF THREE-PARTICLE BOUND STATES AND THE SCALING LIMIT FOR ϕ_2^4

A. $\Gamma^{(6)}$ and three-particle bound states

In an even $\mathcal{P}(\phi)$ model, the Hilbert space $\mathcal{H}$ decomposes into even and odd subspaces $\mathcal{H}_e, \mathcal{H}_o$ which are invariant under the symmetry $\phi \to -\phi$,

$$\mathcal{H} = \mathcal{H}_e \oplus \mathcal{H}_o \ . \tag{5.1}$$

The physical mass m is

$$m = \inf \sigma(M \restriction \mathcal{H}_o), \tag{5.2}$$

where σ denotes the spectrum. The eigenfunctions of the mass operator M are the vacuum, the one-particle states, and bound states. We expect even-bound states only in the spectral interval $(m, 2m)$ and odd-bound states only in the spectral interval $(m, 3m)$.

In the even ϕ^4 model,

$$(0, 2m) \cap \sigma(M \restriction \mathcal{H}_e) = \emptyset, \tag{5.3}$$

i.e., two-particle (even) bound states do not occur. This result was established by Spencer[3]; it extends the methods of Feldman[4] and weak-coupling results.[2] We consider here the odd subspace $\mathcal{H}_o$ in the even ϕ^4 model. We investigate the consequences of certain (as yet unproved) inequalities for vertex functions.

Let Γ be the inverse propagator, and let $\Gamma^{(n)}$ denote the *unamputated*, one-particle irreducible part of the connected n-point Green's function $G^{(n)}$. For instance,

$$\Gamma^{(4)}(x_1, \ldots, x_4) = G^{(4)}(x_1, \ldots, x_4)$$

and

$$\Gamma^{(6)}(x_1, \ldots, x_6) = G^{(6)}(x_1, \ldots, x_6) + \sum \int G^{(4)}(x_{i_1}, x_{i_2}, x_{i_3}, z)\Gamma(z - z') \times G^{(4)}(z', x_{i_4}, x_{i_5}, x_{i_6})dz\,dz', \tag{5.4}$$

where the sum ranges over the ten choices of two three-element subsets $\{i_1, i_2, i_3\}, \{i_4, i_5, i_6\}$ of $\{1, \ldots, 6\}$.

For a ϕ^4 interaction, the lowest nonzero order of perturbation theory predicts that (at least for weak coupling)

$$\Gamma^{(4n+2)}(x_1, \ldots, x_{4n+2}) \le 0, \quad n = 1, 2, \ldots. \tag{5.5}$$

Here we assume a weaker form of (5.5) in the case $n = 1$, namely

$$\Gamma^{(6)}(x_1, x_2, x_3, y_1, y_2, y_3) \le O(1)e^{-3md}, \tag{5.6}$$

where

$$d = \text{dist}(\{x_1, x_2, x_3\}, \{y_1, y_2, y_3\}).$$

Let $\mathcal{E}_n^+$ be the subspace of the Euclidean Hilbert space $\mathcal{E}$ which is spanned by vectors of the form $\phi(f_1) \cdots \phi(f_n)\Omega$, with the f_i's supported at positive times. Let E be the canonical projection of $\mathcal{E}^+ = \{\text{span}\mathcal{E}_n^+, n = 0, 1, \cdots\}$ onto $\mathcal{H}$, and let $E\mathcal{E}_n^+ = \mathcal{H}_n$. See Ref. 5.

Theorem 5.1. Assume (5.6) in an even ϕ_2^4 model. Then

$$(m, 3m) \cap \sigma(M \restriction \mathcal{H}_1) = \emptyset, \tag{5.7}$$

$$(m, 3m) \cap \sigma(M \restriction \mathcal{H}_3) = \emptyset. \tag{5.8}$$

Here σ denotes the spectrum (of the mass operator).

Remark. We can interpret these statements as the absence of bound states in the two-point function (5.7) or the absence of bound states in the six-point function (5.8).

Theorem 5.2. Assume (5.6) in an even ϕ_2^4 model. Then $M \ge 3m$ where M is defined in (2.14) and

$$0 \le \Gamma(x) \le O(1)e^{-3m|x|}, \quad |x| \to \infty. \tag{5.9}$$

Thus $\tilde{G}(p)$ has no CDD zeros for $|p| < 3m$, and $\bar{m} \ge 3m$.

Theorem 5.3. Assume (5.6) in a weak-coupling $\lambda\phi_2^4$ model. Then for λ sufficiently small,

$$(m, 3m) \cap \sigma(M \restriction \mathcal{H}_o) = \emptyset.$$

Remark. Theorem 5.3 is a corollary of Theorem 5.1 and the n-particle cluster expansion.[1] In particular, the states $\mathcal{H}_1 \oplus \mathcal{H}_3$ span $\mathcal{H}_o$ up to energy $4m - \epsilon$ for $\lambda \le' \lambda(\epsilon)$. By the Haag-Ruelle scattering theory, we then have $\sigma(M \restriction \mathcal{H}_o) = \{m\} \cup [3m, \infty)$.

Let $P_1^{\mathcal{H}}$ denote the projection of $\mathcal{H}$ onto the one-particle eigenspace of the mass operator M. We require a relation between $P_1^{\mathcal{H}}$ and the Euclidean projection P_1 of $\mathcal{E}$ onto $\mathcal{E}_1 = \text{span}(1 - P_0)\phi\Omega$. Let T_t denote Euclidean time translation and let Θ denote Euclidean time inversion. We show that on $\mathcal{E}^+ \times \mathcal{E}^+$ the operator $A_t \equiv \Theta(E^* P_1^{\mathcal{H}} E - P_1)T_t$ has an exponential decay rate $\bar{m}$, the decay rate of $\Gamma(x)$, i.e., the CDD radius.

Theorem 5.4. Consider a CIP theory. Let $\chi \in \mathcal{E}^+$ be supported in the Euclidean time interval $[0, T]$. Assume $E\chi \in C^\infty(H)$ and for $t \ge 0$ let

$$F(t) \equiv \langle \chi, A_t \chi \rangle_{\mathcal{E}} = \langle E\chi, P_1^{\mathcal{H}} e^{-tH} E\chi \rangle_{\mathcal{H}} - \langle \Theta\chi, P_1 T_t \chi \rangle_{\mathcal{E}}.$$

Then there exists $r < \infty$ such that for any $a < \bar{m}$,

$$\left(1 - \frac{d^2}{dt^2}\right)^{-r} e^{at}\theta(t)F(t) \in L_2, \tag{5.10}$$

where $\theta(t)$ is the Heaviside step function.

Proof of Theorem 5.2. We use Theorem 4.2 and Eq. (4.2). Thus for $x \ne y$,

$$0 \le \Gamma(x - y) = 16\lambda^2 \langle :xxx: (1 - P_1): yyy: \rangle = 16\lambda^2 \lim_{\substack{x_i \to x \\ y_i \to y}} \langle x_1 x_2 x_3 (1 - P_1) y_1 y_2 y_3 \rangle. \tag{5.11}$$

Here $\langle : xy:z \rangle$ denotes $\langle : \phi(x)\phi(y) : \phi(z) \rangle$, etc., and the Wick ordering is omitted since

$$(1 - P_1) : \phi(y_1)\phi(y_2)\phi(y_3) : \Omega = (1 - P_1)\phi(y_1)\phi(y_2)\phi(y_3)\Omega.$$

We expand (5.11) in terms of connected Green's functions, and we expand the term $G^{(6)}$ using (5.4). Thus for $x-y\neq 0$,

$$0\leq (4\lambda)^{-2}\Gamma(x-y)$$
$$=6G(x-y)^3+9G(x-y)G^{(4)}(xxyy)$$
$$-9\int G^{(4)}(xxyz)\Gamma(z-z')G^{(4)}(z'xyy)dz\,dz'$$
$$+\Gamma^{(6)}(xxxyyy)\,. \qquad (5.12)$$

Diagrammatically, we write for $x\neq y$ the expansion in Fig. 1, where the $x-y$ channel is drawn from left to right. We show below that each term above is bounded above by $O(1)\exp(-3m|x-y|)$, and thus (5.9) follows. Thus, by the definition of $\overline{m}$, it follows that $\overline{m}\geq 3m$, and $\tilde{G}(p)$ has no CDD zeros for $|p|<3m$. Since $M\geq\overline{m}$, it follows that $M\geq 3m$, to complete the proof.

To establish the required estimates, we use

$$0\leq G(x)\leq O(1)|x|^{-1/2}e^{-m|x|}\,,\quad |x|\to\infty\,, \qquad (5.13)$$

which follows from the corresponding bound on $C(x)$ and $\int d\rho=1$. It follows that $G(x-y)^3$ has an exponential decay rate $3m$. Furthermore, by the Lebowitz inequality $G^{(4)}(xxyy)\leq 0$ (see Ref. 10), so the term linear in $G^{(4)}$ is bounded above by zero. The bound on $\Gamma^{(6)}$ follows by the assumption (5.6). Thus we obtain the inequality of Fig. 2.

In order to bound the integral in Fig. 2, we use the decay properties of $G^{(4)}$. In fact, by the second Griffiths inequality and the Lebowitz inequality (see Ref. 10)

$$0\leq -G^{(4)}(xyzt)\leq G(x-z)G(y-t)+G(x-t)G(y-z)\,. \qquad (5.14)$$

Furthermore, by Corollary 4.4,

$$-9\int G^{(4)}(xxyz)\Gamma(z-z')G^{(4)}(z'xyy)dz\,dz'$$
$$\leq 9\int G^{(4)}(xxyz)C_\alpha^{-1}(z-z')G^{(4)}(z'xyy)dz\,dz'$$
$$=9\int\langle xxyz\rangle_T C_{m_0^2}^{-1}(z-z')\langle z'xyy\rangle_T dz\,dz'$$
$$+9(\alpha-m_0^2)\int\langle xxyz\rangle_T\langle zxyy\rangle_T dz\,. \qquad (5.15)$$

We use (5.12) and (5.14) to bound the last term in (5.15). Namely

$$\int\langle xxyz\rangle_T\langle zxyy\rangle_T dz\leq 4\langle xy\rangle^2\int\langle xz\rangle\langle zy\rangle\,dz$$
$$\leq O(1)e^{-3m|x-y|}\,. \qquad (5.16)$$

Thus,

$$0\leq\Gamma(x-y)$$
$$\leq O(1)e^{-3m|x-y|}$$
$$+144\lambda^2\int\langle xxyz\rangle_T C^{-1}(z-z')\langle z'xyy\rangle_T dz\,dz'\,.$$

To bound the integral above, we use the integration by parts formula (3.1) to write

$$\int C^{-1}(z-z')\langle z'xyy\rangle_T dz'=-\langle\mathcal{V}'(z)\phi(x)\phi(y)\phi(y)\rangle_T$$
$$\equiv-\langle\overline{z}xyy\rangle_T\,,$$

where we let $\overline{z}$ denote $\mathcal{V}'(z)$. Thus,

$$0\leq\Gamma(x-y)$$
$$\leq O(1)e^{-3m|x-y|}-144\lambda^2\int\langle xxyz\rangle_T\langle\overline{z}xyy\rangle_T dz\,. \qquad (5.17)$$

In order to bound $|\langle\overline{z}xyy\rangle_T|$, we use Hamiltonian estimates. We introduce Euclidean coordinates $x=(x_0,x_1)$ in R^2 so that $x_1=y_1$ and $x_0<y_0$. We consider three cases: $z_0\leq x_0+1$, $x_0+1\leq z_0\leq y_0-1$, $y_0-1\leq z_0$. If $z_0\leq x_0+1$, we write

$$\langle\overline{z}xyy\rangle_T=\langle\overline{z}x(1-P_0)yy\rangle-2\langle\overline{z}y\rangle\langle xy\rangle\,.$$

Since the mass spectrum on $\mathcal{H}_e$ starts at $2m$, it follows that

$$|\langle\overline{z}xyy\rangle_T|\leq O(1)e^{-2m|x-y|}\,.$$

The absence of singularities at the coincidence $\overline{z}=x\neq y$ follows as in Ref. 8 from the fact that the local behavior is dominated by the low-order perturbation contribution. Thus, in this case by (5.14),

$$\left|\int_{z_0\leq x_0+1}\langle xxyz\rangle_T\langle\overline{z}xyy\rangle_T dz\right|\leq O(1)e^{-3m|x-y|}\int\langle xz\rangle dz$$
$$\leq O(1)e^{-3m|x-y|}\,.$$

If $y_0-1\leq z_0$, then we use

$$|\langle\overline{z}xyy\rangle_T|\leq O(1)e^{-m|x-y|}\,,$$

and by (5.14),

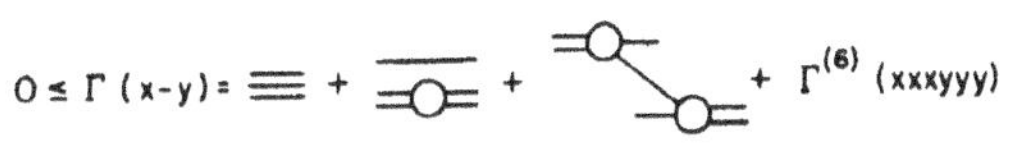

FIG. 1. Diagram expansion for Γ in a ϕ^4 theory according to Eq. (5.12).

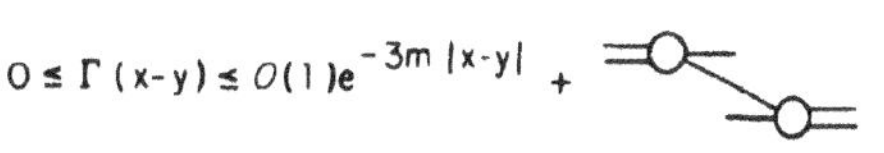

FIG. 2. Bound on $\Gamma(x-y)$.

$$\left|\int_{z_0\le y_0-1}\langle xxyz\rangle_T\langle\bar{z}xyy\rangle_T dz\right|$$

$$\le O(1)\int_{z_0\le y_0-1} e^{-m|x-z|-2m|z-y|}dz$$

$$\le O(1)e^{-3m|x-y|} .$$

Finally, if $x_0+1\le z_0\le y_0-1$, we have

$$|\langle x\bar{z}yy\rangle_T|\le O(1)e^{-m|x_0-z_0|-2m|z_0-y_0|} ,$$

and

$$\left|\int_{x_0+1\le z_0\le y_0-1}\langle xxyz\rangle_T\langle x\bar{z}yy\rangle_T dz\right|$$

$$\le O(1)\int\langle xz\rangle\langle xy\rangle e^{-m|x_0-z_0|}e^{-2m|z_0-y_0|}dz$$

$$\le O(1)e^{-3m|x-y|} .$$

Thus, we obtain from (5.17)

$$0\le\Gamma(x-y)\le O(1)e^{-3m|x-y|} \tag{5.18}$$

to complete the proof.

Proof of Theorem 5.1. We assume Theorem 5.4, and choose $\chi=\phi(f)^3\Omega$, where $f\in C_0^\infty$, $f\ge 0$, and support of f is contained in the strip $0<x_0\le T$. Linear combinations of such vectors (with different f, T) span $\mathcal{E}_3^+$. Thus, it is sufficient to show for $t\ge 0$ that

$$0\le W(t)\equiv\langle E\chi, e^{-tH}(1-P_1^{\mathcal{K}})E\chi\rangle_{\mathcal{K}}$$

$$\le O(1)e^{-at} , \tag{5.19}$$

for any $a<3m$. We establish (5.19) for $a<\bar{m}$, and appeal to Theorem 5.2 for the fact that $\bar{m}=3m$. The positivity of $W(t)$ follows by the positivity of the inner product in $\mathcal{K}$ and the spectral theorem. Thus, we need only establish the upper bound in (5.19).

We begin by repeating the proof of Theorem 5.2 with $\langle\Theta\chi,(1-P_1)T_t\chi\rangle_{\mathcal{E}}=\langle\phi(\vartheta f)^3(1-P_1)\phi(f_t)^3\rangle_{\mathcal{E}}$ replacing $\langle xxx(1-P_1)yyy\rangle_{\mathcal{E}}$. Here $f_t(x_0,x_1)=f(x_0-t,x_1)$. In this manner we obtain the upper bound

$$\langle\phi(\vartheta f)^3(1-P_1)\phi(f_t)^3\rangle_{\mathcal{E}}\le O(1)e^{-3m|t|} . \tag{5.20}$$

With $\chi=\phi(f)^3\Omega$ and $t\ge 0$, we see that the function $F(t)$ of Theorem 5.4 equals

$$F(t)=\langle\phi(\vartheta f)^3(1-P_1)\phi(f_t)^3\rangle_{\mathcal{E}}$$

$$-\langle E\phi(f)^3\Omega, e^{-tH}(1-P_1^{\mathcal{K}})E\phi(f)^3\Omega\rangle_{\mathcal{K}} .$$

Thus, by (5.20),

$$0\le\theta(t)W(t)=\theta(t)\langle\phi(\vartheta f)^3(1-P_1)\phi(f_t)^3\rangle_{\mathcal{E}}-\theta(t)F(t)$$

$$\le O(1)\theta(t)e^{-3mt}-\theta(t)F(t) , \tag{5.21}$$

where $F(t)$ satisfies (5.10), i.e., $\theta(t)F(t)$ belongs to a Sobolev-Paley-Wiener space.

Suppose the vector $(1-P_1^{\mathcal{K}})E\phi(f)^3\Omega$ has energy spectrum which intersects the interval $[0,\bar{m})$. Then by the spectral theorem, there exists $\epsilon>0$ such that for $a<\bar{m}$, but $\bar{m}-a$ sufficiently small,

$$e^{at}\theta(t)W(t)\ge\text{const}\times\theta(t)e^{\epsilon t} .$$

Since $(1-d^2/dt^2)^{-r}$ has a positive kernel, this lower bound contradicts (5.21) and (5.10) and completes the proof of Theorem 5.1.

Proof of Theorem 5.4. We study the functions

$$F_{\mathcal{K}}(t)=\theta(t)\langle E\chi, e^{-tH}P_1^{\mathcal{K}}E\chi\rangle_{\mathcal{K}}$$

and

$$F_{\mathcal{E}}(t)=\langle\Theta\chi,P_1T_t\chi\rangle_{\mathcal{E}} .$$

Furthermore, let

$$F_1(t)=F_{\mathcal{K}}(t)-F_{\mathcal{E}}(t) .$$

We note that $F_1(t)$ is continuous, except for a jump discontinuity at $t=0$, and by the spectral theorem

$$|F_1(t)|\le O(1)e^{-m|t|} .$$

We show below that F_1 can also be written

$$F_1(t)=F_A(t)+F_B(t) ,$$

where (i) F_A, F_B are each continuous, except for a jump discontinuity at $t=0$, (ii) $F_B(t)$ is supported in $t\le 0$, (iii) the Fourier transform $\tilde{F}_A(p_0)$ is an analytic function of p_0 in the strip $|\mathrm{Im}\,p_0|<\bar{m}$, and (iv) in each strip $|\mathrm{Im}\,p_0|\le a<\bar{m}$,

$$|\tilde{F}_A(p_0)|\le O(1)(1+|\mathrm{Re}\,p_0|)^s ,$$

where $O(1)$, but not s, depends on a. The following Paley-Wiener estimate then establishes the theorem: We let $1=\varphi_1(t)+\varphi_2(t)+\varphi_3(t)$ where each φ_j is C^∞, where $\varphi_j(t)$ is supported on the intervals $[-\infty,-1]$, $[-2,2]$, $[1,\infty]$ for $j=1,2,3$, respectively. We claim

$$\left(1-\frac{d^2}{dt^2}\right)^{-r}e^{at}\theta(t)\varphi_j(t)F_A(t)\in L_2 ,$$

for $j=1,2,3$. But by (i) and (ii),

$$\theta(t)F(t)=\theta(t)F_1(t)$$

$$=\theta(t)F_A(t)$$

$$=\theta(t)\sum_{j=1}^{3}\varphi_j(t)F_A(t) ,$$

so (5.10) follows.

Since $\varphi_1(t)\theta(t)=0$, the estimate for $j=1$ is trivial. For $j=2$, φ_j has compact support. Thus by (i),

$$\left\|\left(1-\frac{d^2}{dt^2}\right)^{-r}e^{at}\theta\varphi_2F_A\right\|_2\le\text{const}\times\|\varphi_2F_A\|_2\le\text{const.}$$

We thus are reduced to consider $\theta\varphi_3F_A=\varphi_3F_A$.

$$\left\| \left(1-\frac{d^2}{dt^2}\right)^{-r} e^{at}\varphi_3 F_A \right\|_2$$
$$\le \sum_{n=0}^{\infty} \frac{a^n}{n!} \left\| \left(1-\frac{d^2}{dt^2}\right)^{-r} t^n \varphi_3 F_A \right\|_2$$
$$= \sum_{n=0}^{\infty} \frac{a^n}{n!} \| (1+p_0^2)^{-r} (\varphi_3 F_A)^{\sim(n)}(p_0) \|_2$$
$$\le O(1) \sum_{n=0}^{\infty} \left(\frac{a}{a+\epsilon}\right)^n \| (1+p_0^2)^{-1} \|_2 < \infty ,$$

where the superscript tilde denotes Fourier transformation. Here we use

$$|(1+p_0^2)^{-r+1} (\varphi_3 F)^{\sim(n)}(p_0)| \le \text{const} \times n! \, (a+\epsilon)^{-n} .$$

To establish this inequality we appeal to

$$|(\varphi_3 F_A)^{\sim(n)}(p_0)|$$
$$= \left| \int \tilde{\varphi}_3(p_0 - q) \tilde{F}_A^{(n)}(q) dq \right|$$
$$\le \|(1+r^2)^{-s} \tilde{F}_A^{(n)}(r)\|_\infty \int (1+q^2)^s |\tilde{\varphi}_3(p_0-q)| dq$$
$$\le O(1)(1+p_0^2)^s \|(1+r^2)^{-s} \tilde{F}_A^{(n)}(r)\|_\infty$$
$$\le O(1)(1+p_0^2)^s n! \, (a+\epsilon)^{-n} .$$

The bound on $\tilde{F}_A^{(n)}(r)$ follows by (iii), (iv), and the Cauchy integral formula. We choose $r = s+1$ to complete the proof.

Thus we are reduced to establishing the representation $F_1 = F_A + F_B$, with the properties claimed above. We introduce some notation: Let $|p_1\rangle$ denote the one-particle state in $\mathcal{H}$ with momentum p_1, and normalized so that $\langle p|p'\rangle_{\mathcal{H}} = \delta(p-p')$. By Lorentz covariance

$$H|p_1\rangle = \mu(p_1)|p_1\rangle = (p_1^2+m^2)^{1/2}|p_1\rangle .$$

Since $E\chi \in C^\infty(H)$, the inner product $\langle E\chi|p_1\rangle_{\mathcal{H}}$ is a rapidly decreasing function of p_1.

Using the spectral theorem, we can compute $\tilde{F}_{\mathcal{H}}(p_0)$, namely

$$\tilde{F}_{\mathcal{H}}(p_0) = -i \int \frac{dp_1}{p_0 - i\mu(p_1)} |\langle E\chi|p_1\rangle_{\mathcal{H}}|^2 .$$

Next we study $\tilde{F}_{\mathcal{S}}(p_0)$. Define

$$G(z) \equiv \langle \Theta\chi, \phi(z) \rangle_{\mathcal{S}} .$$

By the definition of P_1,

$$F_{\mathcal{S}}(t) = -\int G(z)\Gamma(z-z')\bar{G}(-z_0'-t, -z_1')dz\,dz'$$

and

$$\tilde{F}_{\mathcal{S}}(p_0) = -\int \tilde{G}(p)\tilde{G}(-p)^- \tilde{\Gamma}(p) dp_1 .$$

Let us assume the representation

$$\tilde{G}(p) = \frac{G_1(p_1)}{p_0 - i\mu(p_1)} + \frac{G_2(p_1)}{p_0 + i\mu(p_1)} + G_3(p_0, p_1) , \tag{5.22}$$

where

$$G_1(p_1) = -i\langle E\chi|p_1\rangle_{\mathcal{H}} \left(\frac{Z}{2\mu}\right)^{1/2} ,$$

and where G_3 is analytic for $|\text{Im}\, p_0| < \overline{m}$. Substituting the representation (5.22) into $\tilde{F}_{\mathcal{S}}$ gives nine terms. The $G_1 - G_1$ term is

$$\int \left(\frac{Z}{2\mu}\right) |\langle E\chi|p_1\rangle_{\mathcal{H}}|^2 \frac{1}{(p_0 - i\mu)^2} \tilde{\Gamma}(p) dp_1$$
$$= -i \int \frac{1}{(p_0 - i\mu)} |\langle E\chi|p_1\rangle_{\mathcal{H}}|^2 \left(\frac{iZ\tilde{\Gamma}(p)}{2\mu(p_0 - i\mu)}\right) dp_1$$
$$= \tilde{F}_{\mathcal{H}}(p_0) + R(p_0) ,$$

where

$$R(p_0) = \int dp_1 |\langle E\chi|p_1\rangle_{\mathcal{H}}|^2 \frac{[Z/2\mu - i\int_{M^2}^\infty [(p_0+i\mu)/(p^2+a)]\, d\rho(a)]}{[Z + \int_{M^2}^\infty [(p^2+m^2)/(p^2+a)]\, d\rho(a)]} .$$

The properties of $R(p_0)$ depend on (v) the rapid decrease of $|\langle E\chi|p_1\rangle|$, (vi) the fact that $M \ge \overline{m}$, and (vii) the fact that

$$Z + \int_{M^2}^\infty \left(\frac{p^2+m^2}{p^2+a}\right) d\rho(a)$$

is analytic and nonzero for $|\text{Im}\, p_0| < \overline{m}$ (see Ref. 8, Chapter 6). Thus, $R(p_0)$ is analytic if $|\text{Im}\, p_0| < \overline{m}$. In each strip $|\text{Im}\, p_0| \le a < \overline{m}$, a bound $|R(p_0)| \le O(1)[1 + (\text{Re}\, p_0)^2]^s$ follows from the representation for $R(p_0)$, and hence $R(p_0)$ is a contribution to $\tilde{F}_A(p_0)$.

Next we study the $G_2 - G_2$ contribution to $\tilde{F}_{\mathcal{S}}$. As above, we find the sum of an integral over a pole at $p_0 = -i\mu$ (instead of $p_0 = i\mu$) and a piece $R_2(p_0)$ analytic in $|\text{Im}\, p_0| < \overline{m}$. The inverse Fourier transform of the pole term has the form

$$F_B(t) = \theta(-t)\langle \psi, e^{tH} P_1^{\mathcal{H}} \psi \rangle_{\mathcal{H}} ,$$

for some vector $\psi \in \mathcal{H}$. The term $R_2(p_0)$ is another contribution to $\tilde{F}_A$.

The remaining terms, $G_i - G_j$ with $i \ne j$ or with $i = j = 3$, are analytic for $|\text{Im}\, p_0| < \overline{m}$, by similar calculation. They, too, obey a polynomial bound in $1 + (\text{Re}\, p_0)^2$, and thus contribute to $\tilde{F}_A$.

To complete the proof, we show that G_3 defined by (5.22) has the analytic properties that we assumed. We first study

$$G_+(p) \equiv \int_{z_0 \geq 0} G(z) e^{-ipz} dz .$$

For $z_0 \geq 0$,

$$G(z) = \langle E\chi, e^{-z_0 H}[P_1^{\mathcal{K}} + (1 - P_1^{\mathcal{K}})] E\phi(z_1, z_0 = 0)\Omega\rangle_{\mathcal{K}} .$$

The one-particle contribution gives the G_1 pole term in (5.22). We use

$$\langle p_1 | \phi(z_1, z_0 = 0)\Omega\rangle_{\mathcal{K}} = \left(\frac{Z}{2\mu}\right)^{1/2} e^{-ip_1 z_1} ,$$

and the spectral theorem to compute $G_1(p_1)$. The contribution from $(1 - P_1^{\mathcal{K}})$ is part of $G_3(p)$, which follows by the spectral theorem and $M \geq \overline{m}$. Similarly, we analyze

$$G_-(p) = \int_{z_0 \leq -T} G(z) e^{-ipz} dz ,$$

and find the G_2 pole plus another contribution to G_3. Finally,

$$G_0(p) = \int_{-T}^{0} dz_0 \int dz_1 G(z) e^{-ipz} ,$$

is entire in p_0, since it has compact support in z_0.

B. $\Gamma^{(4n+2)}$ and three-particle bound states

In this section we assume that

$$\Gamma^{(4n+2)}(x_1, \ldots, x_{2n+1}, y_1, \ldots, y_{2n+1}) \leq O(1) e^{-3md} , \tag{5.23}$$

where $d = \text{dist}\{(x_1, \ldots, x_{2n+1}), (y_1, \ldots, y_{2n+1})\}$, and $n = 1, 2, \ldots$.

Theorem 5.5. Assume (5.23) in an even ϕ^4 model. Then

$$(m, 3m) \cap \sigma(M \restriction \mathcal{K}_0) = \emptyset . \tag{5.24}$$

Proof. The basic idea of the proof is the same as the proof of Theorems 5.1 and 5.2 and as the proof of Feldman,[4] although the diagrams that occur and the estimates are slightly more complicated. We merely present the algebraic structure, without giving detailed estimates. As in the proof of (5.8), it is sufficient to establish the Euclidean estimate

$$\langle \hat{\theta}, T_t(1 - P_1)\theta\rangle_{\mathcal{E}} \leq O(1) e^{-3mt} , \tag{5.25}$$

where

$$\theta = \phi(f_1) \cdots \phi(f_{2n+1})\Omega \in \mathcal{E}^+_{2n+1} , \tag{5.26}$$

$f_i \geq 0$, $f_i \in C_0^\infty$, and $\hat{\theta}$ is the time inverted vector. The connection between the Euclidean estimate (5.25) on $\mathcal{E}_{2n+1}$ and the spectral estimate (5.24) on $\mathcal{K}_{2n+1}$ then follows by

$$|\langle \hat{\theta}, T_t(1 - P_1)\theta\rangle_{\mathcal{E}} - \langle E\theta, e^{-tH}(1 - P_1^{\mathcal{K}})E\theta\rangle_{\mathcal{K}}| \leq O(1) e^{-at} , \tag{5.27}$$

for all $a < 3m$, as in the proof of (5.8). Since $\mathcal{K}_0 = \oplus \mathcal{K}_{2n+1}$, the statement (5.24) follows.

To establish (5.25), we expand the inner product into connected functions. Further, expand the $(4n+2)$-point connected part $G^{(4n+2)}$ into $\Gamma^{(4n+2)}$ [bounded above by (5.23)] plus a sum of terms. Each term in the expansion of (5.25) is represented by a sum of (possibly disconnected) tree graphs. These graphs have $2(2n+1)$ external legs; of them $2n+1$ are labeled $x_1, \ldots, x_{2n+1}$ and called initial legs, while $2n+1$ are labeled $y_1, \ldots, y_{2n+1}$ and called final legs. The vertices of the graphs are connected Green's functions $G^{(r)}(x_1, \ldots, x_r)$, with $4 \leq r \leq 4n+2$, while the lines of the graphs are either G (external lines) or Γ (internal lines). In addition there are combinatoric factors and a factor (-1) for each Γ (internal line). The subtraction $-\langle \hat{\theta}, T_t P_1 \theta\rangle_{\mathcal{E}}$ removes graphs that are one-particle reducible in the channel $(\{x_1, \ldots, x_{2n+1}\}, \{y_1, \ldots, y_{2n+1}\})$. Thus, each graph is formally one-particle irreducible in the $x - y$ channel, and hence they are formally two-particle irreducible by $\phi \to -\phi$ symmetry.

In order to establish the desired three-particle decay (5.25), we estimate each graph separately, as in the proof of Theorems 5.1 and 5.2. In performing these estimates we make use of the following facts:

(1) Connected graphs with an even number of initial legs also have an even number of final legs, by $\phi \to -\phi$ symmetry. These graphs decay as $O(e^{-2mt})$, since they reflect the absence of a two-particle bound-state spectrum on the even subspace $\mathcal{K}_e$ (see Ref. 3).

(2) In proving (5.25) for a given n, we can assume its validity for $\theta \in \mathcal{E}^+_{2r+1}$, $r < n$, i.e., we argue by induction on n.

As a consequence of (1) and (2), all disconnected graphs which contribute to $\langle \hat{\theta}, T_t(1 - P_1)\theta\rangle_{\mathcal{E}}$ are bounded by $O(1)e^{-3mt}$. Thus,

$$\langle \hat{\theta}, T_t(1 - P_1)\theta\rangle_{\mathcal{E}} = O(1) e^{-3mt} + C ,$$

where C is the sum of (connected) tree graphs generated by the expansion of $G^{(4n+2)}$ into $\Gamma^{(4n+2)}$ + tree graphs, and the graph in Fig. 3 does not occur. Each of these remaining graphs is estimated using the decay properties of the $G^{(r)}$ and integration by parts, as in the proof of Theorem 5.2.

FIG. 3. Connected graph excluded by the one-particle subtraction $(I - P_1)$.

C. The CDD radius and the scaling limit

Conventional ideas of the renormalization group place considerable importance in the scaling limit, namely the limit λ (bare coupling) $\to\infty$ with m (physical mass) fixed. If $d<4$, these theories are the limits ($\lambda=\infty, m_0=\infty$) of superrenormalizable models of mass m. Standard arguments indicate that the infinitely scaled theories (if they exist) are not renormalizable in dimension $d<4$. Thus, the construction of the infinitely scaled ϕ_2^4 models would establish the existence of ϕ_2^4 theories of greater singularity than ϕ_4^4. Furthermore, one expects that the $m=0$ limit of the infinitely scaled theories is scale invariant, and is the Gell-Mann–Low–Wilson fixed point. The $m=0$ limit of the infinitely scaled theories is also a hyperbolic fixed point of the scale transformation group. The infinitely scaled transformations form the unstable manifold associated with this fixed point, while the $m=0$, $\lambda<\infty$ critical theories form the stable manifold.

We find that $\overline{m}/m\geq 1+\epsilon$ is a sufficient condition to ensure the existence of the scaling limit. Combined with Ref. 11 and Ref. 5, the results of this paper are a step toward the construction of a ϕ_4^4 field theory. Indeed, the missing steps are (a) the proof that $\overline{m}/m$ is bounded away from 1, and (b) the adaptation of certain arguments depending on Lorentz invariance to the case of a cutoff field theory. Conventional ideas indicate that this construction for ϕ_d^4, with $d>4$, yields a free field. The nontriviality of this construction for ϕ_4^4 is related to the logarithmic divergence expected for the field-strength renormalization Z^{-1}.

We illustrate the scaling limit in Fig. 4, where we plot the physical mass m for theories with fixed bare mass m_0, as a function of the dimensionless bare coupling $g_0=\lambda m_0^{d-4}$. Scaling maps these curves onto one another, and in the figure we assume that $m\to 0$ continuously as we approach the critical coupling $g_{0,c}$. The infinite scaling (dotted) line lies over $g_{0,c}$, and theories on this line have $\lambda=\infty$, $m_0=\infty$. The infinite scaling line is parameterized by m. In Fig. 5 we draw our schematic picture of the presumed ϕ_2^4 scaling trajectories, as described above.

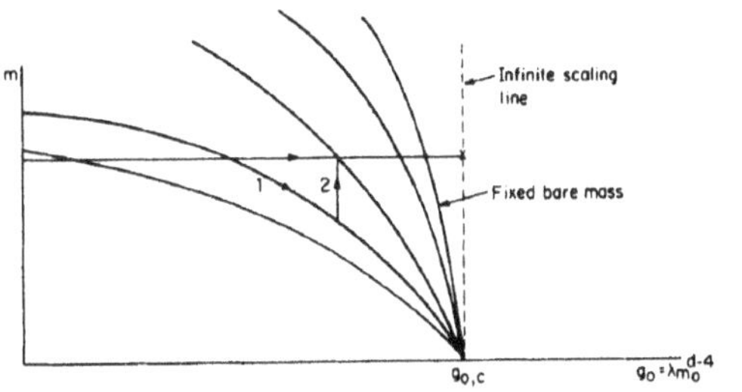

FIG. 4. The scaling limit for ϕ^4. Note movement along the m = const line can be achieved by (1) increasing λ with m_0 fixed, followed by (2) scaling to increase the mass back to m (thus increasing m_0 and λ with λm_0^{d-4} fixed).

Theorem 5.6. Assume $\overline{m}/m$ is bounded away from one for m fixed and $g_0<g_{0,c}$ in an even $\lambda\phi_2^4$ model. Then there exists an infinitely scaled, Osterwalder-Schrader theory with mass m. The Schwinger functions $S_m^{(n)}$ of this model are the limits (through subsequences) of the Schwinger functions $S_{m,\lambda}^{(n)}$ for the model with mass m and coupling λ,

$$S_m^{(n)}=\lim_{\lambda_j\to\infty} S_{m,\lambda_j}^{(n)}. \tag{5.28}$$

Proof. For m fixed and $\overline{m}/m$ bounded away from 1, we apply Theorem 2.6 to bound the two-point Schwinger function in momentum space,

$$(p^2+m^2)^{-1}\leq\tilde{S}_{\rm ren}(m,\lambda,p)=\tilde{G}_{\rm ren}(m,\lambda,p)\leq O(1)$$

with $O(1)$ uniform in λ. Thus,

$$\tilde{S}_{\rm ren}(m,p)=\lim_{\lambda_j\to\infty}\tilde{S}_{\rm ren}(m,\lambda_j,p)$$

exists and is bounded. The $\tilde{S}(m,\lambda,p)$ are Euclidean invariant and Osterwalder-Schrader positive, and hence so is $\tilde{S}_{\rm ren}(m,p)$. Thus,

$$\tilde{S}_{\rm ren}(m,p)=\int_{m^2}^{\infty}\frac{d\rho(a)}{p^2+a},$$

where $d\rho(a)$ is positive and

$$\int_{m^2}^{\infty}\frac{d\rho(a)}{a}<\infty.$$

As in Ref. 11 (see also Ref. 12 and 13,), the n-

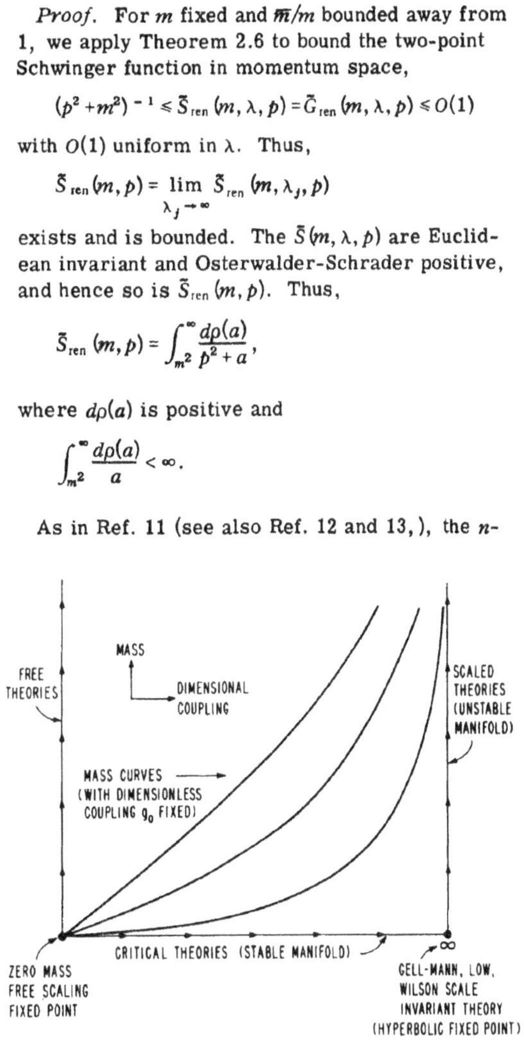

FIG. 5. Trajectories of the scaling group.

point Schwinger functions are bounded in terms of S_{ren}, namely

$$|S_{\mathrm{ren}}^{(n)}(m,\lambda,f_1,\ldots,f_n)| = |Z^{-n/2}S^{(n)}(m,\lambda,f_1,\ldots,f_n)| \leq O(1)\prod_{i=1}^{n} S_{\mathrm{ren}}^{(2)}(m,\lambda,f_i{}^*\otimes f_i)^{1/2}.$$

Hence, passing to subsequences, we obtain limits as $\lambda\to\infty$ for each $S_{\mathrm{ren}}^{(n)}(m,\lambda)$. Since each approximate $S_{\mathrm{ren}}^{(n)}$ satisfies the Osterwalder-Schrader axioms with a mass gap $m>0$, so do the limiting Schwinger functions, as covariance, positivity, symmetry, and absence of spectrum are all preserved under limits.

Remarks 5.7. While the hypotheses of Theorem 5.2 are sufficient to ensure $\mathfrak{M}/m=3$, the weaker hypothesis

$$\Gamma^{(6)}(xxxyyy)\leq O(1)e^{-(m+\epsilon)|x-y|},\quad |x-y|\to\infty, \tag{5.29}$$

uniformly for $g_0<g_{0,c}$ and m fixed, is sufficient to establish in a $\lambda\phi_2^4$ model that

$$\Gamma(x)\leq O(1)e^{-(m+\epsilon)|x|},\quad |x|\to\infty. \tag{5.30}$$

Hence $\mathfrak{M}/m-1\geq\epsilon/m$, and the scaling limit for mass m exists.

Remark 5.8. An alternative construction of an infinitely scaled ϕ_d^4 field theory starts from the d-dimensional Ising model I_d with lattice spacing ϵ, and takes $\epsilon\to 0$ with $m=\xi^{-1}=$ inverse correlation length fixed. Since this construction differs from the previous one ($\lambda\to\infty$) only by a change in the order of the ϵ,λ limits, it is attractive to suppose that both constructions yield the same theory. For $d=2$, a detailed analysis by Wu and co-workers[14] on the closed-form solutions of I_2 suggests that the required bounds on the two-point function may be attainable in order to assert the existence of the infinitely scaled, n-point correlation functions of I_2, using compactness and subsequences as above.

---

*Work supported in part by the National Science Foundation under Grant No. MPS 74-13252.

†Work supported in part by the National Science Foundation under Grant No. MPS 73-05037.

[1] J. Glimm, A. Jaffe, and T. Spencer, Ann. Math. 100, 585 (1974).

[2] J. Glimm, A. Jaffe, and T. Spencer, in *Constructive Quantum Field Theory*, Springer Lecture Notes in Physics, edited by G. Velo and A. Wightman (Springer, Berlin, 1973), Vol. 25.

[3] T. Spencer, Commun. Math. Phys. 39, 77 (1974).

[4] J. Feldman, Can. J. Phys. 52, 1583 (1974).

[5] K. Osterwalder and R. Schrader, Commun. Math. Phys. 31, 83 (1973); *ibid.* (to be published).

[6] K. Symanzik, J. Math. Phys. 1, 249 (1960).

[7] W. Donoghue, Jr., *Monotone Matrix Functions and Analytic Continuation* (Springer, Berlin, 1974).

[8] J. Glimm and A. Jaffe, Commun. Math. Phys. (to be published.

[9] H. Lehmann, Nuovo Cimento 11, 342 (1954).

[10] *Constructive Quantum Field Theory*, Springer Lecture Notes in Physics, edited by G. Velo and A. Wightman (Springer, Berlin, 1973), Vol. 25.

[11] J. Glimm and A. Jaffe, Phys. Rev. Lett. 33, 440 (1974).

[12] J. Glimm and A. Jaffe, Ann. Inst. H. Poincaré 22, 1 (1975).

[13] C. Newman (unpublished).

[14] T. T. Wu, private communication. Also see C. Tracy and B. McCoy, Phys. Rev. Lett. 31, 1500 (1973).

Two and Three Body Equations in Quantum Field Models*

James Glimm**
Rockefeller University, New York, N.Y. 10021, USA

Arthur Jaffe***
Harvard University, Cambridge, Mass. 02138, USA

Received December 27, 1974

Abstract. In each pure phase of a $\mathscr{P}(\phi)_2$ quantum field model, we establish local regularity of the Green's functions and exponential decay for noncritical models. We establish the existence of two-particle and three-particle Bethe-Salpeter kernels in the Euclidean region.

Contents

1. Introduction

The study of particles in weakly coupled $\mathscr{P}(\phi)_2$ quantum field theories was begun in [25, 26]. The cluster expansions developed in these papers resulted in the construction of isolated one-particle states. According to the Haag-Ruelle theory, the existence of n-particle in and out states, and the existence of an isometric S-matrix, follow from the existence of isolated one particle states. The ϕ^4 model, in the single phase region, has been shown to be repulsive in the sense that no even mass spectrum occurs in the two particle bound state interval $(m, 2m)$ [46, 8, 25]. The presence of bound state mass spectrum was indicated for the $\phi^6 - \phi^4$ interaction [25]. Spencer [47] and Spencer and Zirilli [48] have a more detailed analysis of the energy momentum spectrum for weak coupling, which goes up to the threshold $E \leqq 3m - \varepsilon$, and uses the Bethe-Salpeter equation. An early version of their work motivated the present paper.

* We thank the Institut des Hautes Etudes Scientifiques for hospitality.
** Supported in part by the National Science Foundation under Grant MPS 74-13252.
*** Supported in part by the National Science Foundation under Grant MPS 73-05037.

In this paper we study the Bethe-Salpeter kernel in the wider class of noncritical $\mathscr{P}(\phi)_2$ models, and we also study three body kernels. By definition, a quantum field is called critical if its mass is zero; otherwise it is noncritical. We expect particles and a complete set of scattering states for all noncritical $\mathscr{P}(\phi)_2$ theories.

There are at present five (partially overlapping) methods to construct the infinite volume $\mathscr{P}(\phi)_2$ model. The original method [14] of compactness and convergent subsequences uses operator algebra techniques and estimates based in part on Euclidean space time localization and decoupling in the Feyman-Kac formula. This construction applies to the widest class of interactions [14, 42, 43]. The second method is the cluster expansion [25, 26], which applies in the case of weak coupling or large external field [45], and which gives the most complete information about the resulting model.

The third method [39, 28, 9, 44] in based on monotone convergence and upper bounds. The monotonicity is based on Griffiths' inequalities [12, 28], while the upper bound originates in the estimate [15]

$$\pm\phi(h) \leqq C(h)(H+I), \tag{1.1}$$

see also [27, 44, 9, 10]. The passage from this ϕ-bound to estimates on Wightman functions is given in [17]. The same simple proof works for Schwinger functions. It is well known that bounds on Wightman functions imply bounds on Schwinger functions at noncoinciding points, see for example [40, 41]. In [28], a theory of conditional expectations based on the Markov property [49, 38] is developed to handle this point. A variant [24] of the monotonic construction replaces Dirichlet boundary conditions in (1.1) by weak coupling boundary conditions.

The main advantage of the third method is a proof of Lorentz invariance, of correlation inequalities and (for ϕ^4 only) of the Lee Yang theorem. See [8, 21–24, 28, 29, 44, 46] for applications.

In the fourth construction [5, 11], the compactness of method one is formulated in a probability framework. The fifth construction is to obtain the weakly coupled ϕ^4 model by Borel summation of the perturbation series [6]. The convergence of this method is based on new estimates which require cancellation of vacuum graphs in the cluster expansion. This method gives a type of uniqueness which is particularly satisfying in view of the importance of perturbation theory in calculations.

We study the $\mathscr{P}(\phi)_2$ model obtained by one of these procedures. Much of Chapter 2 is an assimilation of results known in the literature for some time. Our starting point is the ϕ^j and π bounds

$$\begin{aligned} &\pm:\phi^j(h): \leqq C(h,j)(H+I)\\ &\quad\pm\pi(h) \leqq C_\pi(h)(H+I)\,. \end{aligned}$$

In Chapter 3 we establish the integration by parts formula for the Schwinger functions $S^{(n)}$ in each pure phase. We use this result in Chapter 4 to analyze the connected parts of the Schwinger functions (Euclidean Green's functions $G^{(n)}$). Let $H_{\pm 1}$ be the Sobolev space defined by the inner product

$$\langle f,g\rangle_{\pm 1} = \langle f, (-\Delta+m_0^2)^{\pm 1} g\rangle_{L_2}\,. \tag{1.2}$$

Let $H_{\pm 1}^{\otimes j}$ be a j-fold tensor product of such spaces. For noncritical $\mathscr{P}(\phi)_2$ theories, we prove that $G^{(n)}(x_1, \ldots, x_n)$ is the kernel of a bounded operator

$$G^{(n)}: H_{-1}^{\otimes j} \to H_1^{\otimes n-j}, \tag{1.3}$$

for $0<j<n$. Here we use the ϕ^j bound, and exponential clustering, which follows from the existence of a mass gap. The stronger tree graph decay rate of the $G^{(n)}$ is known for weak coupling [6]. Since $G^{(n)}$ is translation invariant, it cannot be compact. However, we show that the restriction of $G^{(n)}$ to a subspace of definite total momentum is Hilbert-Schmidt. This is a crucial bound on the $G^{(n)}$.

An immediate corollary of this bound on $G^{(n)}$ is the fact that noncritical Schwinger functions $S^{(n)}$ for the field $\hat{\phi}=\phi-\langle\phi\rangle$ are continuous multilinear functionals on the n-fold product $H_{-1}\times\ldots\times H_{-1}$. This result was established for weak coupling in [20]. A second corollary is the existence of the vertex functions $\Gamma^{(n)}(x_1, \ldots, x_n)$ (direct correlation functions) for noncritical $\mathscr{P}(\phi)_2$ theories. The $\Gamma^{(n)}$ are bounded operators from $H_1^{\otimes j}$ to $H_{-1}^{\otimes n-j}$, $0<j<n$, improving slightly a bound from [20].

In Chapter 6 we show that the inverse propagator $\Gamma(x)$ has an exponential decay rate equal to the CDD radius $\bar{m}$. Here $m<\bar{m}\leq M$ where m is the single particle mass and (m, M) is the upper gap in the spectrum of the two point function.

In Chapter 7 we construct the 2-particle Bethe-Salpeter kernel for non-critical $\mathscr{P}(\phi)_2$ models. We use the compactness of the $G^{(n)}$ on constant momentum subspaces, and apply a Fredholm alternative. The exclusion of the eigenvalue zero in the Fredholm alternative is related to the following property, which we establish in Chapter 5 for a $\mathscr{P}(\phi)_2$ measure: If a polynomial $\mathscr{R}(\phi)$ takes the value zero almost everywhere on $\mathscr{S}'(R^2)$, then $\mathscr{R}=0$, i.e. the coefficients of $\mathscr{R}$ vanish. This property can be rephrased by saying that the Euclidean $\mathscr{P}(\phi)_2$ measure is not supported on any algebraic variety in $\mathscr{S}'(R^2)$. We also require this property for polynomials $\mathscr{R}(\phi)$ with a definite momentum. It is a consequence of integration by parts, or of the existence [11] of Euclidean momentum operators π. We remark that some time ago Bros [3] constructed the two body Bethe-Salpeter kernel in the context of the Wightman axioms, plus additional assumptions. Finally in Chapter 8, we consider three particle equations.

The two body Bethe-Salpeter kernel is a bounded operator

$$K: H_1^{\otimes 2} \to H_{-1}^{\otimes 2} \tag{1.4}$$

satisfying the integral equation

$$R=R_0+R_0KR \tag{1.5}$$

in the case of even $\mathscr{P}(\phi)$ models. Here R and R_0 map $H_{-1}^{\otimes 2}$ into $H_1^{\otimes 2}$ and are defined by their kernels

$$\begin{aligned} R_0(x_1, x_2; y_1, y_2) &= G^{(2)}(x_1, y_1)G^{(2)}(x_2, y_2)+G^{(2)}(x_1, y_2)G^{(2)}(x_2, y_1), \\ R(x_1 x_2; y_1 y_2) &= G^{(4)}(x_1, x_2, y_1, y_2)+R_0(x_1, x_2; y_1, y_2). \end{aligned} \tag{1.6}$$

Thus

$$K=R_0^{-1}-R^{-1}, \tag{1.7}$$

and formally K is a sum of graphs two particle irreducible between initial and final states. The construction of K, however, does not have an immediate significance for the particle structure. In fact the estimates [47] which yield decay properties of $K(x_1, x_2; y_1, y_2)$ depend on the cluster expansion, and are known only for weak coupling. The methods of Lebowitz and Penrose [31, 32] from statistical mechanics given an analytic continuation of decay estimates. The Lebowitz-Penrose results pertain only to the gap between the ground state and the first excited state, and are not strong enough to give the existence of particles. The extension of the Lebowitz-Penrose result [32] to field theory is contained in [29]. It is an interesting speculation that nontrivial decay properties for $K(x_1, x_2, y_1, y_2)$ are a consequence of the existence of a mass gap. However analyticity properties in the coupling should not be a consequence of a positive mass, even in the case $\mathscr{P}$=even, cf. [24, Fig. 2.1].

2. ϕ^j Bounds and Schwinger Functions

Let $\mathfrak{A}$ be the C^*-algebra of the $V=\infty$ limit, acting on a Hilbert space $\mathscr{H}$. By [1], $\mathfrak{A}'=\mathscr{Z}$, the center of $\mathfrak{A}''$. The direct integral decomposition is defined with respect to $\mathscr{Z}$; we write it as

$$\mathscr{H}=\int \mathscr{H}(\xi)d\xi \tag{2.1}$$

so that $\xi \in \Xi$ is a label for the set Ξ of phases. Let $\Omega_\xi \in \mathscr{H}_\xi$ be the vacuum and let H_ξ be the Hamiltonian. For example, in the ϕ_2^4 model we expect a single phase for $\lambda<\lambda_c$, and two phases for $\lambda>\lambda_c$. We expect a continuum of phases in the case of phase transitions with a continuous group of internal symmetries, so this direct integral machinery may not be too elaborate for the use we make of it. See [25, 11] for a further discussion of phases.

We now state our basic ϕ and π bounds. Let

$$1 \leqq j \leqq d = \deg \mathscr{P}.$$

For general polynomials, all constructions [14, 11] assume free boundary conditions. For $\mathscr{P}$=even+linear, we allow either free or weak coupling boundary conditions [24]. We let V denote the volume of interaction; for weak coupling boundary conditions, V is the length of the interval $[-b, b]$ in which the mass perturbation $\sigma\int_{-b}^{b} :\phi^2(x): dx$ is inserted. Note that for $V=\infty$, weak coupling and Dirichlet boundary data define the same theory, for $\mathscr{P}$=even+linear [24].

Theorem 2.1 a (*ϕ^j and π bounds*). *For $V<\infty, j \leqq d$*

$$\pm\phi(h) \leqq \|h\|_1 C(H+I), \tag{2.2a}$$

$$\pm\pi(h) \leqq \|h\|_2 C(H+I), \tag{2.2b}$$

$$\pm :\phi^j:(h) \leqq H + O(\|(|x|^r+1)h\|_{d/d-j} + \|h\|_{d/d-j}^{d/d-j}), \tag{2.2c}$$

for h real and in the appropriate L_p space, and for $r>1+(j|d)$. Here C is a constant, independent of h and V. In the case $j=d$, we also require $\|h\|_\infty$ small.

Theorem 2.1b (*ϕ^j bounds*). *For $j \leqq d$, $V = \infty$ and in a.e. pure phase, the bounds* (2.2a) *and* (2.2c) *are valid. Moreover for free boundary conditions*

$$\pm :\phi^j:(h) \leqq H + O(\|h\|_1 + \|h\|_{d/d-j}^{d/d-j}). \tag{2.2d}$$

Proof of Theorem 2.1a. (2.2a) was proved in [15]. The bound (2.2b) was established in [16, 17] for $V < \infty$, and the same methods can be used directly in the infinite volume limit. Bounds related to (2.2c) were first given in [15] and have been improved independently by Fröhlich [11] and the authors (unpublished). For weak coupling, Dirichlet, or free boundary conditions and $V < \infty$, we use the inequality

$$\pm :\phi^j:(h) \leq H + O(a) + O(1)\|h\|_{d/d-j}^{d/d-j}$$

for support $h \subset [-a, a]$, cf. [24, 51]. Because the vacuum energy of the $:\phi^j:$ perturbation is convex, and is zero for $h = 0$, this yields

$$\pm :\phi^j:(h) \leqq H + O(a)^{j/d}\|h\|_{d/d-j} + O(1)\|h\|_{d/d-j}^{d/d-j}.$$

We write $h = \sum h_n$, suppt $h_n \subset [n, n+1]$, and then by convexity again,

$$\pm :\phi^j:(h) \leqq H + O(1)\sum \|(|n|+1)h_n\|_{d/d-j} + O(1)\|h\|_{d/d-j}^{d/d-j}.$$

Finally,

$$\begin{aligned}\sum \|(|n|+1)h_n\|_{d/d-j} &= \sum (|n|+1)^{-\alpha}\|(|n|+1)^{1+\alpha}h_n\|_{d/d-j} \\ &\leqq \text{const.}\ \|(|x|+1)^{1+\alpha}h\|_{d/d-j}\end{aligned}$$

for $\alpha > j/d$.

Proof of Theorem 2.1b. The transfer of the ϕ bound to the infinite volume limit is given in [15]. The infinite volume limit of the $:\phi^j:$ bounds is established in [24].

There is also an abstract method for transferring an estimate of the form

$$\pm A \leqq H + I \tag{2.3}$$

to the infinite volume limit, and then into almost every phase [2]. This method requires an additional estimate, for example

$$\|A(H+I)^{-1}\| + \|[H, A](H+I)^{-1}\| < \text{const.} \tag{2.4}$$

uniform in the volume V. Such an estimate holds in the case $A = \phi(h)$, $h \in \mathscr{S}(R^1)$, using [15, Lemma 1.1] and (2.2a)–(2.2c). Hence we obtain (2.2a), with $h \in \mathscr{S}$, in the $V = \infty$ limit and in almost every pure phase. The role of the inequality (2.6) is to provide a domain of essential self-adjointness[1] for $A = \phi(h)$. Having established the bound and decomposition to pure phases for $h \in \mathscr{S}$, we pass to $h \in L_1$ using (2.2a). (2.2c) is a consequence of (2.2d), for $V = \infty$. The bound (2.2d) is given in [11].

[1] The general method of using commutation estimates to establish self adjointness was introduced in [13] to prove self adjointness of space-time averaged field operators $\phi(f)$. One abstraction of this method was presented in [18], and another abstraction was presented in [15]. These results established essential self adjointness of an operator or form A, satisfying relative bounds with respect to a positive self adjoint operator H. In our present applications, we use [15: Lemma 1.1, Theorem 1.2] to establish essential self adjointness of $\phi(h)$. Other abstractions and applications have also been given by Nelson [37], McBryan [33—35], Faris und Lavine [7], and McBryan and Park [36].

In the case of weak coupling boundary conditions, the basic method relies on uniform estimates [10, 24] for the analytic generating functionals of the ordinary and generalized Schwinger functions. By compactness, these functionals have a limit point, as $b \to \infty$, and the limit point satisfies the same estimate. However, convergence of the Schwinger functions as $b \to \infty$ [24; see also Theorem 2.2, 2.4 below] implies that the analytic functions actually have a limit as $b \to \infty$. The Schwinger functions are constructed from the generating functionals [10] and yield the infinite volume Hilbert space [10, 40, 41]. On this Hilbert space, the bound (2.2c) holds in the sense of a norm estimate on the perturbed semigroup [24]. Further details are omitted.

Theorem 2.2. *Let $h, \chi \in C_0^\infty$, $\int \chi dx = 1$, $0 \leqq \chi$, and $V < \infty$, and*

$$\delta H_\chi = \int \prod_{i=1}^{j} (\phi * \chi)(x) h(x) dx .$$

Then

$$\pm \delta H_\chi \leqq H + O(1) \tag{2.5}$$

uniformly as $V \to \infty$. Let $\chi_i = \kappa_i \chi(\kappa; x)$, $i = 1, 2$. Then

$$\pm (\delta H_{\chi_1} - \delta H_{\chi_2}) \leqq H + o(1) \tag{2.6}$$

uniformly in V, as $\kappa_1, \kappa_2 \to \infty$. Again for $j = d$, we require $\|h\|_\infty$ small.

Theorem 2.2 follows as an extension of the methods of [15] and [51] for free and Dirichlet boundary conditions, respectively. It is presumably also true for weak coupling boundary conditions, but we do not require this fact.

We need to introduce certain function spaces related to the estimates of Theorem 2.1 Let

$$\begin{cases} \mathscr{L}_{1,p} = \mathscr{L}_1(R^2) \cap \mathscr{L}_p(R^2), & p < \infty \\ \mathscr{L}_{1,\infty,\varepsilon} = \mathscr{L}_1(R^2) \cap \{f : \|f\|_\infty < \varepsilon\} . \end{cases} \tag{2.7}$$

Whenever $p = \infty$, the space $\mathscr{L}_{1,\infty,\varepsilon}$ will be understood, even if the ε is not explicitly included.

Before studying the quantum field on the space $\mathscr{H}(\xi)$, we require more information about its action on $\mathscr{H}$.

Theorem 2.3. *The quantum field on $\mathscr{H}$, specified by a C^* algebra construction, is unitarily equivalent to an Osterwalder-Schrader* [40, 41] *reconstructed field. The Schwinger functions are moments of a unique measure dq on $\mathscr{S}'(R^2)$.*

Remark. Since the covariance of the field under Lorentz transformations is not known, the corresponding imaginary time fields on $\mathscr{S}'(R^2)$ will be time translation covariant, but the Euclidean invariance is not known.

Proof. We verify the axioms (A1)–(A5), B, C of Fröhlich [9] in order to obtain the imaginary time fields and the reconstructed fields. Thus we want to study the generating functions

$$J(f) = \int e^{i\phi(f)} dq \tag{2.8}$$

of the imaginary time measure. For f piecewise constant in t, $J(f)$ has a Hamiltonian definition. If $h_1(x), \ldots, h_n(x)$ are piecewise constant values of $f(\cdot, t)$, then

$$J(if) = \langle \Omega, e^{-(t_2 - t_1)(H + \phi(h_1))} \ldots e^{-(t_n - t_{n-1})(H + \phi(h_n))} \Omega \rangle .$$

For a finite space cutoff, we know that e^{-tH} is positivity preserving and that the imaginary time measure on $\mathscr{S}'(R^2)$ exists. Also $J(f)$ is complex analytic in $f \in \mathscr{S}(R^2)$, for example, and can be bounded using Theorem 2.1.

For any operator H, we define its vacuum energy to be $E(H) = \inf$ spectrum H. Let

$$\delta E(t, f) = E(H + \phi(f(\cdot, t))) - E(H) ,$$

for f real. Then by Theorem 2.1,

$$\begin{aligned} |J(f)| &\leqq \exp - \textstyle\int_{-\infty}^{\infty} \delta E(t, \operatorname{Im} f) dt \\ &\leqq \exp\left(\text{const.}\left(\| f \|_{d/d-1}^{d/d-1} + \textstyle\int \| (|x|^r + 1) f(\cdot, t) \|_{d/d-1} dt\right)\right) \end{aligned} \tag{2.9}$$

with a constant uniform in the space cutoff. By Vitali's theorem, J is complex analytic for $f \in \mathscr{L}$ where $\mathscr{L}$ is defined by the norms in (2.9). Following Fröhlich [10], we conclude that the space cutoff Schwinger n-point function is a bounded multilinear functional on $\mathscr{L} \times \ldots \times \mathscr{L}$. By multilinearity, we can restrict each test function f in the n-point Schwinger function to be supported in a strip

$$\{x, t : |t - i| \leqq 1/2\} ,$$

$i \in Z$. Let n_i be the number of arguments of the Schwinger function restricted to this strip; thus $\sum n_i = n$. With this notation, the Cauchy integral formula combined with the bound (2.9) shows that the n-point Schwinger function is bounded by $O(\prod_i n_i!) K^n$, as a functional on $\prod^n \mathscr{L}$, $K = O(\int \| (|x|^2 + 1) \cdot \|_{d/d-1} dt + \| \cdot \|_{d/d-1}^{d/d-1})$. This bound is uniform in n and in the space cutoff region. We know that the space-cutoff Wightman functions converge [15] (after selection of a subsequence), and by analytic continuation the $V \to \infty$ limit of the Schwinger functions exists also. The $V = \infty$ Schwinger functions have the same $\mathscr{L}$ norm, $O(\prod n_i! K^n)$, which allows summation of the power series. The resulting functional $J(f)$ is complex analytic, satisfies (2.9), and is the limit of the finite volume J's. The axioms of Fröhlich are readily verified, because the estimates and positivity conditions are preserved in the $V = \infty$ limit. We only comment that his axiom CI, which yields sharp time fields, is proved using the second order estimate of [15, Lemma 1.1],

$$\pm \phi(f)^2 \leqq |f|_{\mathscr{S}}^2 (H + I)^2 , \tag{2.10}$$

where $|f|_{\mathscr{S}}$ is a Schwartz space norm on $f \in \mathscr{S}$, independent of $V \leqq \infty$.

From [10], we thus obtain the existence of an imaginary time field theory over $\mathscr{S}'(R^2)$ and an Osterwalder-Schrader reconstructed field theory whose Schwinger functions are just the functional derivatives of $J(f)$. Thus the reconstructed Schwinger functions (and Wightman functions) agree with those of the field theory on $\mathscr{H}$, and so the two field theories are unitarily equivalent [15].

Theorem 2.4. *Let* $1 \leq j_\nu \leq d = \deg \mathscr{P}$, $1 \leq \nu \leq n$. *The generalized Schwinger functions*

$$\mathscr{S}_j^{(n)}(x) = \int \prod_{\nu=1}^n :\phi^{j_\nu}(x_\nu): dq$$

are continuous as multilinear forms on $\prod_{\nu=1}^n \mathscr{L}_{1,d/d-j_\nu}$, *with norms bounded by* $(\prod n_i!)K^n$ *as above. They are the functional derivatives of*

$$Z(h_1, \ldots, h_d) = \int e^{-\sum_{j=1}^d :\phi^j(h_j):} dq \,.$$

Z is bounded and analytic in $h_j \in \mathscr{L}_{1,d/d-j}$.

Proof. The proof is the same as that of Theorem 2.3, once we know that the $V \to \infty$ limit of the generalized Schwinger functions exists. We insert a momentum cutoff κ and use the bounds which result from Theorem 2.2. We conclude that

$$\mathscr{L}_{j,\kappa}^{(n)} = \int \prod_{\nu=1}^n :\phi_\kappa^{j_\nu}(x): dq$$

is bounded uniformly in κ and V, and that

$$\mathscr{S}_{j,\kappa}^{(n)} \to \mathscr{S}_j^{(n)} \tag{2.11}$$

uniformly in V. Since $\mathscr{S}_{j,\kappa}^{(n)}$ can be expressed in terms of ordinary Schwinger functions by undoing the Wick order, the existence of the $V \to \infty$ limit in $\mathscr{S}_{j,\kappa}^{(n)}$ follows. By a 3ε argument, the $V \to \infty$ limit exists in $\mathscr{S}_n^{(n)}$ also, completing the proof.

Remark. The limit (2.11) holds for $V = \infty$ also and it holds for a.e. phase $\xi \in \Xi$ uniformly in ξ. We next show that the estimates of Theorem 2.4 on the generalized Schwinger functions are valid in a.e. phase.

Theorem 2.5. *For the* $\mathscr{P}(\phi)_2$ *field theory acting on* $\mathscr{H}(\xi)$, *the Schwinger functions satisfy the bounds of Theorem 2.4, uniformly in* ξ, *and are the functional derivatives of a complex analytic function* $Z_\xi(h_1, \ldots, h_d)$, *bounded as in Theorem 2.4.*

Proof. For test functions smooth in t, the Schwinger functions with $\phi(x)$, $:\phi_\kappa^j(x):$ and $:\phi_\kappa^j(x): - :\phi_{\kappa'}^j(x):$ vertices are defined, using the estimate of Theorem 2.1 in the argument of [13]. For E a projection in $\mathscr{Z}$, the same statement applies to the Schwinger functions $\mathscr{S}_E$ defined in the vacuum $\Omega_E = E\Omega$. Assuming that h has compact support in t, we bound $\mathscr{S}_E^{(N)}$ by $K^N(\prod_i N_i!)\|E\Omega\|^2$, using the Schwarz inequality:

$$\begin{aligned}|\langle \Omega_E A \Omega_E \rangle| &\leq \|E\Omega\| \langle \Omega_E A^* A \Omega_E \rangle^{1/2} \\ &\leq \ldots \leq \|E\Omega\|^{2-2^n} \langle \Omega (A^*A)^{2^n} \Omega \rangle^{2^{-n-1}} .\end{aligned}$$

Here A is the integral of a time ordered product of $\phi(x), \ldots$ vertices as above and of e^{-tH} operators. We suppose that the test functions for A also have compact support, and we choose the strips in Theorem 2.4 to match the support of the test functions in A. Then $(A^*A)^{2^n}$ is also such a time ordered product, with test functions supported in 2^{n+1} times as many time strips. If A is defined by N functional derivatives of Z, $(A^*A)^n$ has $2^{n+1}N$ such functional derivatives; however, only N_i occur in a given time strip. Thus

$$\langle \Omega (A^*A)^{2^n} \Omega \rangle \leq (K^N \textstyle\prod_i N_i!)^{2^{n+1}}$$

which yields

$$|\langle \Omega_E A \Omega_E \rangle| \leqq \|E\Omega\|^2 K^N \prod_i N_i! .$$

Thus we see that $\mathscr{S}_E^{(n)}/\|E\Omega\|^2$ satisfies the bounds of Theorem 2.4, unformly in E. Passing to the pure phases, we have

$$\mathscr{S}_E^{(n)} = \int_{\Xi_E} \mathscr{S}_\xi^{(n)} d\xi$$
$$\|E\Omega\|^2 = \int_{\Xi_E} d\xi ,$$

which Ξ_E is the subset of Ξ corresponding to $E \in \mathscr{Z}$. It follows that $\mathscr{S}_\xi^{(n)}$ is $L_\infty(\Xi, d\xi)$, as a functional on $\mathscr{L}_{1,d/d-j_1} \times \ldots \times \mathscr{L}_{1,d/d-j_n}$ depending on ξ. These bounds permit summation of the exponential series, yielding Z_ξ, for a.e. ξ. They also permit removal of the ultraviolet cutoff in the $\phi^j(x)$ vertices. This completes the proof.

Results similar to Theorems 2.3–2.5 were obtained independently by Fröhlich [11]. As a reformulation of results of [11], we have the following two theorems:

Theorem 2.6. *For the $\mathscr{P}(\phi)_2$ field theory acting on $\mathscr{H}(\xi)$, a.e. pure phase is unitarily equivalent to an Osterwalder-Schrader reconstructed theory. If dq_ξ denotes the associated path space measure on $\mathscr{S}'(R^2)$, then*

$$Z = \int Z_\xi dq .$$

Theorem 2.7. *Under the unitary equivalence of Theorem 2.3, $\mathscr{Z} = \mathfrak{A}'$ is identified with the algebra $\mathscr{M}_{\text{inv}}$ of time translation invariant bounded measurable function on $\mathscr{S}'(R^2)$.*

Proof. By [11, Lemma 2.17], $\mathscr{M}_{\text{inv}} \subset \mathscr{M}_\infty \cap \mathscr{M}_+ \cap \mathscr{M}_-$, where $\mathscr{M}_\infty$ is L_∞ of the σ-field $\sum_\infty$ of sets measurable at ∞, and $\mathscr{M}_\pm$ is L_∞ of the σ-field $\sum_\pm$ generated by functions $\phi(x, t)$, for $\pm t \geqq 0$. Let G be a positive element of $\mathscr{M}_{\text{inv}}$. Then G is invariant under the time reflection operator θ [11, Proof of Theorem 2.19]. Let E_{inv} be the conditional expectation projection onto $\mathscr{M}_{\text{inv}}$. Then for $f \in \mathscr{M}_+$,

$$0 \leqq \int (\theta G f)^- G f \, dq$$
$$= \int (\theta f)^- f G^2 dq$$

and since this is true for all $G \geqq 0$ in $\mathscr{M}_{\text{inv}}$,

$$0 \leqq E_{\text{inv}}(\theta f)^- f .$$

Hence

$$\int (\theta G f)^- G f \, dq \leqq \|G^2\|_\infty \int E_{\text{inv}}(\theta f)^- f \, dq$$
$$= \|G^2\|_\infty \int (\theta f)^- f \, dq = \|G^2\|_\infty \langle f, f \rangle .$$

It follows that G defines a bounded operator on $\mathscr{H}$. G is seen to commute with $\mathscr{M}_0$, the functions of the time zero field, and G commutes with time translation, on $\mathscr{H}$. Thus $G \in \mathfrak{A}' = \mathscr{Z}$.

We have proved that $\mathscr{M}_{\text{inv}}$ is a subalgebra of $\mathscr{Z}$. Thus we can define a direct integral decomposition of $\mathfrak{A}$ with respect to $\mathscr{M}_{\text{inv}}$. Each component $\mathfrak{A}(\zeta)$ is an Osterwalder-Schrader reconstruction with path space measure dq_ζ equal to a component of dq under direct integral decomposition of dq with respect to $\mathscr{M}_{\text{inv}}$. Thus the measures dq_ζ are ergodic under time translation [11, Theorem 2.16].

Hence they have a cluster property, from which the irreducibility of $\mathfrak{A}(\zeta)$ follows. By general properties of C^*-algebra decompositions, this shows that $\mathscr{Z}=\mathscr{M}_{\text{inv}}$.

3.1. Integration by Parts

The Green's functions $\mathscr{G}^{(n)}$ are defined below as the connected (truncated) parts of the Schwinger functions $\mathscr{S}^{(n)}$, with the truncation defined in the standard fashion. For the case at hand, namely a direct integral of irreducible theories, the truncation is performed in each pure phase ξ, defining $\mathscr{G}_\xi^{(n)}$, and then by definition $\mathscr{G}^{(n)}=\int\mathscr{G}_\xi^{(n)}d\xi$. The ultraviolet behavior of $\mathscr{S}^{(n)}$ and $\mathscr{G}^{(n)}$ are determined by perturbation theory, and these singularities coincide with the local singularities of a massive free theory. In order to establish this fact, we integrate by parts on path space. Successive integration by parts yields a representation of $\mathscr{S}^{(n)}$ or $\mathscr{G}^{(n)}$ as a sum of low order perturbation contributions, plus a remainder which is regular in x. Such an expansion is possible because the $\mathscr{P}(\phi)_2$ model is superrenormalizable.

The formula for integration by parts in the $\mathscr{S}^{(n)}$ is

$$\int :\phi^j:(h)R(\phi)dq=\int dq\int A(x)\left[\frac{\delta R}{\delta\phi(x)}-R(\phi)\mathscr{P}'(\phi(x))\right]dx\,, \tag{3.1.1}$$

where

$$A(x)=\int dyC(x-y)h(y):\phi^{j-1}:(y)\,,$$
$$R(\phi)=\prod_{i=1}^n:\phi^{j_i}:(h_i)\,.$$
$$C=(-\Delta+m_0^2)^{-1}\,.$$

Theorem 3.1.1. *The partial integration formula* (3.1.1) *is valid for the* $V=\infty$ *field theory and in a.e. phase* $\xi\in\Xi$.

Remark. The left side of (3.1.1) is defined by Theorem 2.6 and 2.10, while the right side of (3.1.1) has kernels which do not factor. To deal with this situation, we integrate by parts in each linear factor in $:\phi^j:(h)R(\phi)$. This result produces a sum of many terms, of the form

$$W=\int[\prod_{\nu=1}^r:\mathscr{P}^{(\alpha_\nu)}(\phi(y_\nu)):]w(y)dy\,, \tag{3.1.2}$$

where

$$w(y)=\int v(x,y)\prod_{i=1}^n h_i(x_i)dx\,,$$

and where $v(x,y)$ is a constant times a product of factors $C(x_i-y_j)$ and $C(x_i-x_j)$.

Let $\mu^{-\varepsilon}$ be the operator $(-\Delta+m_0^2)^{-\varepsilon/2}$, and let $\mu_{\otimes j}^{-\varepsilon}$ be the product of $\mu^{-\varepsilon}$ acting on j variables.

Lemma 3.1.2. *In* (3.1.2), $\alpha_\nu\geq 1$ *and* $C(x)\in\mathscr{L}_{d/\alpha_\nu}$.

Proof. By definition $\alpha_\nu\geq 1$, so $d/\alpha_\nu<\infty$. Also $C(x)$ is L_p for all $p<\infty$.

Lemma 3.1.3. *For* $\varepsilon<0$ *and any choice of exponents* l_ν *in* Z^+, *the function*

$$\mu^{-\varepsilon}\prod_{\nu=1}^r C(x-y_\nu)^{l_\nu} \tag{3.1.3}$$

is bounded and continuous. Here $\mu^{-\varepsilon}$ acts on x.

Proof. $C(x-y_\nu)$ has a Fourier transform of magnitude $(p^2+m_0^2)^{-1}$. Thus the $l=\sum l_\nu$ fold convolution is bounded by $O(1)(p^2+m_0^2)^{-1+\varepsilon/2}$, by standard power counting estimates, see e.g. [19]. After multiplication by $\mu^{-\varepsilon}$, the Fourier transform is integrable, and hence (3.1.3) is bounded. Continuity follows similarly, since $|(e^{ipx}-1)|\leqq O(|px|^{\varepsilon/3})$.

Theorem 3.1.4. *For any $\varepsilon>0$, $\mu_{\otimes n}^{-\varepsilon}\mathscr{S}^{(n)}$ is a bounded continuous function, with bounds uniform in $\xi\in\Xi$.*

Theorem 3.1.5. *Let $\mathscr{C}_{\{j\}}=\prod_{1\leqq i<i'\leqq n}(1+C(x_i,x_{i'})^{(j_i+j_{i'})})$. Then $\mathscr{C}_{\{j\}}^{-1}\mathscr{S}^{(n)}$ is a bounded continuous function, with bounds uniform in $\xi\in\Xi$.*

Proof of Theorems 3.1.1, 3.1.4, 3.1.5. The integration by parts formula (3.1.1) is known for $V<\infty$. In that case by Theorem 2.6, the representation (3.1.2) and the Lemmas 3.1.2, 3.1.3 above,

$$\begin{aligned}
|(\mu_{\otimes n}^{-\varepsilon}\mathscr{S}^{(n)})(h)| &= |\mathscr{S}^{(n)}(\mu_{\otimes n}^{-\varepsilon}h)|\\
&\leqq \textstyle\sum_{\text{finite sum}}\int dx\prod\|C(y_\nu-x_i)\|_{\mathscr{L}_{d/\alpha_\nu}}\prod C(x_i-x_j)|\mu_{\otimes n}^{-\varepsilon}h|\\
&\leqq \textstyle\sum_{\text{finite sum}} O(1)\int dx\prod C(x_i-x_j)\mu_{\otimes n}^{-\varepsilon}|h|\\
&\leqq \textstyle\sum_{\text{finite sum}} O(1)\int |h|\mu_{\otimes n}^{-\varepsilon}\prod C(x_i-x_j)\\
&\leqq O(1)\|h\|_{\mathscr{L}_1},
\end{aligned}$$

where $h=\prod_{i=1}^n h_i(x_i)$. Since the dual of $\mathscr{L}^1$ is $\mathscr{L}^\infty$, in a finite volume $\mu_{\otimes n}^{-\varepsilon}\mathscr{S}^{(n)}$ is $\mathscr{L}^\infty$. Since $\mu_{\otimes n}^{-\varepsilon}$ maps $\mathscr{L}^\infty$ into bounded, continuous functions, Theorem 3.1.4 holds for $V<\infty$, with bounds uniform in V. Similarly, Theorem 3.1.5 holds for $V<\infty$ with bounds uniform in V. By limits, the same bounds are valid for $V=\infty$. As in Chapter 2, these estimates decompose into a.e. phase ξ. This establishes the existence of the right side of (3.1.1). The existence of the $V=\infty$ limit for (3.1.1) follows as above.

For a polynomial $Q(\phi)$, let $Q_T=T^{-1}\int_0^T Q(\phi(\cdot+t))dt$. Then Q_T is bounded in L_p, for all $T<\infty$ and $p<\infty$. Let Q_∞ be any w^*-L_p limit point. In the integration by parts formula, if we insert a factor Q_T in both sides, a new term is included due to contraction to Q_T. This term vanishes, however, as $T\to\infty$. Thus the formula remains valid even when the integrands on both sides are multiplied by Q_∞.

Now Q_∞ is time translation invariant, and thus bounded functions of Q_∞ belong to $\mathscr{Z}$, the center of $\mathfrak{A}$. Let $\mathscr{Z}_\infty$ denote the subalgebra of $\mathscr{Z}$, generated by all such Q_∞. Then integration by parts remains valid in the direct integral theory, decomposed with respect to $\mathscr{Z}_\infty$. We must only show that a theory for which $\mathscr{Z}_\infty=\{\text{constants}\}$ is irreducible.

We show reducibility implies $\mathscr{Z}\neq\{\text{constants}\}$. Consider a theory with a multiple vacuum and let P denote the projection onto the space of vacuum vectors. Suppose R is a polynomial such that $R\Omega\perp\Omega$, $P\Omega=\Omega$, and $\|PR\Omega\|=1$. Then

$$1=\lim_{T\to\infty}\frac{1}{T}\int_0^T dt\langle R\Omega, e^{-tH}R\Omega\rangle=\langle R\Omega, R_\infty\Omega\rangle$$

and

$$0=\langle\Omega, e^{-tH}R\Omega\rangle=\langle\Omega, R_\infty\Omega\rangle.$$

Thus const. $\neq R_\infty \in \mathscr{Z}_\infty$, and the proof is complete.

3.2. Green's Functions and Graphs

The integration by parts formula will be used in our study of Green's functions, i.e. connected or truncated Schwinger functions. Here we define truncations for the generalized Green's functions and relate these to the integration by parts formula.

Let $A_1, \ldots, A_n$ be functions of ϕ, e.g. $A_i = :\phi^{j_i}:(h_i)$. Let P be the set of partitions of $\{1, \ldots, n\}$ and for $\varrho \in P$, let $\varrho = \{\sigma_1, \ldots, \sigma_j\}$, $|\varrho| = j$. Each σ_ν is a subset of $\{1, \ldots, n\}$.

Definition 3.2.1. $\langle\prod_i A_i\rangle_T$ is defined recursively by

$$\langle A_1 \ldots A_n\rangle = \sum_{\varrho\in P}\prod_{\sigma\in\varrho}\langle\prod_{i\in\sigma} A_i\rangle_T. \tag{3.2.1}$$

Remark. $\langle\prod A_i\rangle_T$ depends not only on $\prod A_i$, but also on its factorization into the A_i's.

Proposition 3.2.2. *The inversion of* (3.2.1) *is*

$$\langle A_1 \ldots A_n\rangle_T = \sum_{\varrho\in P}(-1)^{|\varrho|}(|\varrho|-1)!\prod_{\sigma\in\varrho}\langle\prod_{i\in\sigma} A_i\rangle. \tag{3.2.2}$$

Proof. This is standard.

Consider the case that each A_i is a monomial in the field. Successive integration by parts produces a sum of terms, and each term is labelled by a graph D. If $\mathscr{D}$ is the set of all such graphs, the identity has the form

$$\langle\prod_i A_i\rangle = \sum_{D\in\mathscr{D}}\langle\prod_\nu B_\nu\rangle \tag{3.2.3}$$

where B_ν are monomials in the field [formed from derivatives of the A_i or of $\mathscr{P}$, as in (3.1.1)] and determined by D. The vertices of the graph are the B_ν, while the lines arise from contractions $C(x-x')$.

Theorem 3.2.2. *With the above definitions of* $\mathscr{D}$ *and* B_ν

$$\langle\prod_{i=1}^n A_i\rangle_T = \sum_{D\in\mathscr{D}}\langle\prod_\nu B_\nu\rangle_T.$$

Proof. It is sufficient to treat the case of one integration by parts, in one linear factor from A_1. We use (3.2.2) to define $\langle\prod A_i\rangle_T$, and integrate by parts on the right side in the factor containing A_1. We group the result into $n+1$ terms. There are $n-1$ terms in which A_1 contracts to an A_i, $1 \leqq i \leqq n$. There is one term with A_1 contracted to a new $:\mathscr{P}(\phi):$ vertex, and if A_1 is not Wick ordered, a contraction of A_1 to itself is also possible. Let B_1' be the function of ϕ which results from the contraction of A_1 to $:\mathscr{P}:$, and let B_1'' be the result of contracting A_1 to itself. Since these last two terms occur for each $\varrho \in P$, they contribute

$$\begin{aligned}&\sum_{\varrho\in P}(-1)^{|\varrho|}(|\varrho|-1)!\Bigg(\prod_{\substack{\sigma\in\varrho\\ 1\notin\sigma}}\langle\prod_{i\in\sigma} A_i\rangle\Bigg)\Bigg\langle\Bigg(\prod_{\substack{i\in\sigma_i\\ i\neq 1}} A_i\Bigg)B_1' + B_1''\Bigg)\Bigg\rangle\\ &\quad=\langle B_1'\prod_{i>1} A_i\rangle_T + \langle B_1''\prod_{i>1} A_i\rangle_T.\end{aligned}$$

Next consider a term with A_1 contracted to A_i, and let B_{1i} be the result of this contraction. Such a term arises only for certain $\varrho \in P$, namely ϱ which do not separate 1 and i. Let P_i be the set of all such partitions. This term is

$$\sum_{\varrho \in P_i} (-1)^{|\varrho|}(|\varrho|-1)! \left(\prod_{\substack{\sigma \in \varrho \\ 1 \notin \sigma}} \langle \prod_{i \in \sigma} A_i \rangle \right) \left\langle B_{1i} \prod_{\substack{v \in \sigma_1 \\ 1 \neq v \neq i}} A_v \right\rangle$$
$$= \langle B_{1i} \prod_{1 \neq v \neq i} A_v \rangle_T .$$

Combining these $n+1$ terms completes the proof of Theorem 3.2.2.

Remark 3.2.3. For any graph D which is not connected, but which has a vacuum component (so that $\deg B_v = 0$, or $B_v = \text{const.}$ for some v), the term $\langle \prod_v B_v \rangle_T$ vanishes.

The truncations defined by (3.2.2) are not linear in the state $\langle\ \rangle$. Thus there is a difference between truncating in the state defined by $\Omega \in \mathscr{H}$ and truncating in a.e. pure phase ξ. The latter is the correct definition, because it leads to Green's functions with nice cluster properties. Using the measure dq_ξ, we define the general Green's functions as

$$G_\xi^{(n)}(x_1, \ldots, x_n) = (\textstyle\int :\phi(x_1)^{j_1}:\ldots:\phi(x_n)^{j_n}: dq_\xi)_T \tag{3.2.4}$$

and

$$G^{(n)}(x_1, \ldots, x_n) = \textstyle\int_\Xi G_\xi^{(n)}(x_1, \ldots, x_n) d\xi . \tag{3.2.5}$$

It follows immediately that $G^{(n)}$ and $G_\xi^{(n)}$ satisfy the bounds of Theorems 3.1.4 and 3.1.5. We improve these bounds in Chapter 4. Using integration by parts, as in Theorem 3.1.1, we no longer need to restrict j_v to be less than $d = \deg \mathscr{P}$. However, in this case, we lose the nice $n!$ bound on the $\mathscr{S}^{(n)}$ from Chapter 2. In principle, the $G^{(n)}$ should have an $n!$ estimate (as derivatives of an analytic function) for factorizing test functions, and for $j_v \leq d$. See [20, 11].

We call *ordinary* the Green's functions

$$G_\xi^{(n)}(x_1, \ldots, x_n) = (\textstyle\int \phi(x_1) \ldots \phi(x_n) dq_\xi)_T , \tag{3.2.6}$$

or their integral, and *generalized* the Green's functions (3.2.4).

4. Regularity of the Green's Functions

There are two aspects of the regularity of the Green's functions: their short distance singularities (ultraviolet behavior) and their long distance decay (infrared behavior). The intermediate range behavior is settled by the general fact that the $\mathscr{S}^{(n)}$ are real analytic at noncoinciding points. The infrared behavior is controlled by the assumption that we are dealing with a noncritical theory. In the case of a physical mass $m > 0$, the $G^{(n)}$ decay exponentially under large separation of points, $|G^{(n)}| \leq e^{-mr}$. We assume that decay occurs in all Euclidean directions, as is the case for a Lorentz invariant theory with $m > 0$. The following theorem bounds the ultraviolet behavior:

Theorem 4.1. *For $n > 2$, the ordinary $\mathscr{P}(\phi)_2$ Green's functions* (3.2.6) *are bounded and continuous for a.e. $\xi \in \Xi$, with bounds uniform in ξ. For $n = 2$,*

$G_\xi^{(2)}(x-x')-C(x-x')$ has this property. For the generalized Green's functions of a $\mathscr{P}(\phi)_2$ theory, $\mu_{\otimes n}^{-\varepsilon}G_\xi^{(n)}$ and $\mathscr{C}_{\{j\}}^{-1}G_\xi^{(n)}$ are bounded and continuous for a.e. $\xi\in\Xi$ (cf. Theorems 3.1.4–3.1.5), with bounds uniform ξ.

Proof. The bounds on $\mu_{\otimes n}^{-\varepsilon}G^{(n)}$ and $\mathscr{C}_{\{j\}}^{-1}G^{(n)}$ follow from Theorems 3.1.4–3.1.5 and the definition of $G^{(n)}$. The bound on the ordinary Green's functions (all $j_\nu=1$) follows from the proof of these theorems, once we observe that there are no $\phi(x_i)-\phi(x_j)$ contractions in $G^{(n)}$, because of the truncations. Here we require $n>2$, or if $n=2$, we exclude the one (zero order) term with such a contraction.

For the long distance behavior we use the comparison function

$$\mathscr{E}_\varepsilon=\prod_{1\leqq i<l\leqq n}e^{\varepsilon|x_i-x_l|}.$$

Theorem 4.2. *For a noncritical $\mathscr{P}(\phi)_2$ model with exponential decay $\geqq m$ in all Euclidean directions, the functions*

$$\mathscr{E}_\varepsilon\mu_{\otimes n}^{-\delta}G^{(n)},\quad \mathscr{E}_\varepsilon\mathscr{C}_{\{j\}}^{-1}G^{(n)}$$

are bounded and continuous for some $\varepsilon>0$ and for any $\delta>0$. For the ordinary Green's functions, the same is true for

$$\mathscr{E}_\varepsilon G^{(n)}.$$

Proof. We modify the proof of Theorems 3.1.4–3.1.5 to obtain the exponential decay. Let d be the diameter of $\{x_1,\ldots,x_n\}$, and let (x_i, x_j) be the pair determining this diameter. We consider two parallel lines separated by distance r which are perpendicular to the line (x_i,x_j) and which separate $\{x_1,\ldots,x_n\}$ into two nonempty clusters $\{I_1,I_2\}$. We assume each x_i has a distance at least r from the lines. We assert such lines always exist with $r\geqq d/3n$, see [30].

For a term labelled by a graph D, let $\varrho(D)$ be the partition of $\{1,\ldots,n\}$ determined by the connected components of D. If ϱ does not refine $\{I_1,I_2\}$, then G contains a sequence of contractions of length r. In this case, the term is dominated by $O(e^{-m_0r})$, so the corresponding contributions to $G^{(n)}$ have exponential decay. Thus we consider partitions ϱ which refine $\{I_1,I_2\}$. In each integration dz over a generalized Green's function in a connected component, we restrict dz to lie on the same side of the parallel lines as the x_i variables of that component. The contribution from the complementary z region is again exponentially small. However, the contribution with z restricted in this fashion is also exponentially small, as one sees by using the Schwarz inequality and the mass $m>0$ in a standard fashion.

We now obtain expressions for the constant momentum subspace of $H_{-1}^{\otimes j}$ and for the restriction of $G^{(n)}$ to this subspace. In $L_2(R^{2j})=L_2(R^2)^{\otimes j}$, the functions $f(x_1,\ldots,x_n)$ of zero total momentum are invariant under the transformation $x_i\to x_i+a$. We define a norm on such functions

$$|f|=[\int|f(x_1,\ldots,x_n)|^2dx_2\ldots dx_n]^{1/2}. \tag{4.1}$$

namely the L_2 norm in $x_2,\ldots,x_n$, and we let $\mathscr{H}(0)$ denote the zero momentum Hilbert space. Functions of total momentum P have the form

$$f_P=e^{iP\Sigma x_j/n}f\in\mathscr{H}(P),\quad f\in\mathscr{H}(0),$$

and $|f_P|=|f|$. We also write $\mathscr{H}^{(n)}(P)$ to indicate the number of particles.

The norm in $H_{\pm 1}^{\otimes j}$ is determined by $\mu_{\otimes j}^{\pm 1}$ and the mappings

$$H_1^{\otimes j} \underset{\mu \otimes j}{\longrightarrow} L_2(R^2)^{\otimes j} \underset{\mu \otimes j}{\longrightarrow} H_{-1}^{\otimes j}$$

are unitary. We define the momentum P vectors $\mathscr{H}_{\pm}(P)$ in $H_{-1}^{\otimes j}$ by

$$\mathscr{H}_{\pm}(P) = \mu_{\otimes j}^{\pm 1} \mathscr{H}(P), \tag{4.2}$$

with the norm $|\mu_{\otimes j}^{\pm 1} f_P| = |f_P|_{\pm}$.

We consider $G^{(n)}(x_1, \ldots, x_n)$ as the kernel of an integral operator $G^{(n)}$ from $H_{-1}^{\otimes j}$ to $H_1^{\otimes n-j}$. This is unitarily equivalent to the operator

$$\mu_{\otimes n} G^{(n)} \equiv \mu_{\otimes n-j} G^{(n)} \mu_{\otimes j}$$

as a transformation from $L_2^{\otimes j}$ to $L_2^{\otimes n-j}$. The unitary equivalence commutes with translations.

If we extend $G^{(n)}$ to a map from $\mathscr{H}_{-}^{(j)}(P)$ to $\mathscr{H}_{+}^{(nn-j)}(P)$, this is unitarily equivalent to $\mu_{\otimes n} G^{(n)}$ as a map from $\mathscr{H}^{(j)}(P)$ to $\mathscr{H}^{(n-j)}(P)$.

Theorem 4.3. (a) *For a noncritical $\mathscr{P}(\phi)_2$ model, and $0<j<n$, the ordinary Green's functions $G^{(n)}$ is a bounded operator from $H_{-1}^{\otimes j}$ to $H_1^{\otimes n-j}$.*

(b) *If in addition $n>2$, then $G^{(n)}$ defines a Hilbert-Schmidt map, $\mu_{\otimes n} G^{(n)}$, from $\mathscr{H}^{(j)}(P)$ to $\mathscr{H}^{(n-j)}(P)$.*

Remark. In essence, this theorem states that after differentiating each variable of $G^{(n)}$ once, the result $\mu_{\otimes n} G^{(n)}$ is an L_2 function of the difference variables. The proof also shows that $\mu_{\otimes n} G^{(n)}$ is continuous in P, and vanishes in Hilbert-Schmidt norm as $P \to \infty$.

Theorem 4.4. *For a noncritical $\mathscr{P}(\phi)_2$ model, and $0<j<n$, the generalized $G^{(n)}$ is a bounded operator from $H_{-1+\varepsilon}^{\otimes j}$ to $H_{1-\varepsilon}^{\otimes n-j}$, for $\varepsilon>0$.*

We prove Theorem 4.3; the proof of Theorem 4.4 is similar. We integrate by parts in each linear factor $\phi(x_i)$ in $G^{(n)}$. Because of truncation (see Theorem 3.2.2) there are no $\phi(x_i) - \phi(x_k)$ contractions. If $n \leqq \deg \mathscr{P}$, there is one factor proportional to

$$\int \prod_{i=1}^{n} C(x_i - z) dz = F(x_1, \ldots, x_n) \tag{4.2}$$

and this term fulfills the conclusions of the theorem. In fact, the Fourier transform of $\mu_{\otimes n} F$ is proportional to

$$\delta\left(\sum_{i=1}^{n} p_i\right) \prod_{i=1}^{n} \mu(p_i)^{-1},$$

which is the kernel of a bounded operator from $L_2^{\otimes j}$ to $L_2^{\otimes n-j}$, verifying the conclusion of (a). Furthermore,

$$\begin{aligned} |\mu_{\otimes n} F|^2 &= \text{const. } \left\| \mu\left(\sum_{i=2}^{n} p_i\right)^{-1} \prod_{i=2}^{n} \mu(p_i) \right\|_{L_2}^2 \\ &\leqq \text{const.}, \end{aligned}$$

proving that (4.2) also satisfies the conclusion of (b).

The remaining terms are all of the form

$$T = \int \prod C(x_i - y_{l_i}) G^{(r)}(y_1, \ldots, y_r) dy,$$

where $G^{(r)}$ is a generalized Green's function. By Theorem 4.2, $\mu_{\otimes n}T$ is continuous, and exponentially decreasing in its difference variables. The theorem follows from these facts.

5. The Support of dq

Following Fröhlich [11], we introduce the Euclidean π operators which generate the translation group $\phi \to \phi + f$.

$$e^{i\pi(f)}: F(\phi) \to F(\phi + f)\left(\frac{d(q+f)}{dq}\right)^{1/2}. \tag{5.1}$$

The Jacobian in (5.1) is

$$\frac{d(q+f)}{dq} = \exp - [\delta\mathscr{P}(f) + \phi((-\Delta + m_0^2)f) + \tfrac{1}{2}\langle f, (-\Delta + m_0^2)f\rangle], \tag{5.2}$$

where

$$\delta\mathscr{P}(f) = \int [:\mathscr{P}(\phi + f): - :\mathscr{P}(\phi):]dx.$$

In view of the bounds of Chapter 2, we choose

$$f \in \mathscr{L}_{1,d} \cap (-\Delta + m_0^2)^{-1} \mathscr{L}_{1,d/d-1} \cap H_1, \tag{5.3}$$

where H_1 is a Sobolev space. Then (5.2) belongs to $L_p(\mathscr{S}'(R^2), dq)$ for all $p < \infty$.

Theorem 5.1 [11]. *The Euclidean π operators are self adjoint operators on the Euclidean space $\mathscr{H}_{\mathscr{E}} = L_2(\mathscr{S}'(R^2), dq)$. The generating functional*

$$E(f, g) = \langle \Omega, e^{i\phi(g)} e^{i\pi(f)} \Omega\rangle \tag{5.4}$$

is entire analytic for $g \in L_{1,d/d-1}$ and f in (5.3).

Theorem 5.2. *The Euclidean operators exist as self adjoint operators on $\mathscr{H}_{\mathscr{E}}(\xi) = L_2(\mathscr{S}'(R^2), dq_\xi)$. The generating functional E_ξ, defined as in* (5.4), *characterizes the Weyl form of the commutation relations.* (*It satisfies positive definiteness, continuity and commutation relations.*) *Moreover $E = \int E_\xi d\xi$, and E_ξ satisfies the bounds of Theorem 5.1.*

This theorem is essentially contained in [11]. The key step is to show that (5.2) is the Jacobean of the translated measure on $\mathscr{H}_{\mathscr{E}}(\xi)$.

It is convenient to work with a domain of smooth vectors $\mathscr{D}$ which is dense in $\mathscr{H}_{\mathscr{E}}(\xi)$. Let $\mathscr{D}$ be the linear span of $\Omega \equiv 1$ and the vectors

$$A\Omega, \quad A = A_n = \left(\textstyle\prod_{i=1}^n \phi\right)(g_n), \tag{5.5}$$

where g_n is a symmetric function in $\mathscr{S}(R^2)^{\otimes n}$. Then

$$\begin{aligned} &\phi(f)\mathscr{D} \subset \mathscr{D}, \quad f \in \mathscr{S}(R^2), \\ &U(a)\mathscr{D} \subset \mathscr{D}, \end{aligned} \tag{5.6}$$

where we assume that Ω is invariant under the action of Euclidean translations $U(a)$. $\mathscr{D}$ is not invariant under $\pi(f)$, and so we introduce the domain $\mathscr{D}'$, spanned by the vector Ω and

$$A\Omega,\ A=\prod_{i=1}^{n} :\phi^{\nu_i}(g_n):, \qquad \nu_i \leqq d\,. \tag{5.5'}$$

Then

$$:\phi^{\nu}(f):\mathscr{D}'\subset\mathscr{D}'; \qquad U(a)\mathscr{D}'\subset\mathscr{D}'; \tag{5.6'}$$

for $\nu\leqq d$, and

$$\pi(f)\mathscr{D}'\subset\mathscr{D}',\ f\in\mathscr{S}(R^2)\,. \tag{5.7}$$

The relation (5.7) follows from the canonical commutation relations and the identity

$$\pi(f)\Omega=-\frac{i}{2}\{\mathscr{P}'(f)+\phi((-\Delta+m_0^2)f)\}\Omega\,, \tag{5.8}$$

which is the infinitesimal form of (5.2).

Theorem 5.3. *The L_2 inner product defined by dq_ξ is nondegenerate on $\mathscr{D}\times\mathscr{D}$.*

Remark. The equation $B=0$, where

$$B=\sum_{n=0}^{N} A_n(q) \tag{5.9}$$

defines a polynomial variety in $\mathscr{S}'(R^2)$. The theorem states that dq_ξ is not supported on any polynomial variety. Here A_n is defined by (5.5).

Proof. Assume that for all $C\in\mathscr{D}$, $\int CBdq_\xi=0$. Since $B\in L_2(dq_\xi)$ and since $\mathscr{D}$ is dense in $\mathscr{D}'$ in the $L_2(dq_\xi)$ norm, by the remark following Theorem 2.4, $B\perp\mathscr{D}'$ also. Thus

$$0=\int C_1BC_2dq_\xi=\langle C_1BC_2\rangle\,, \qquad C_i\in\mathscr{D}'\,,$$

and so for all $f_j\in\mathscr{S}(R^2)$, we have

$$\begin{aligned}0&=\langle\Omega[\pi(f_1),[\pi(f_2),[\ldots,[\pi(f_N),B]\ldots]]]\Omega\rangle\\&=N!(-i)^N g_N(f_1\otimes\ldots\otimes f_N)\,.\end{aligned} \tag{5.10}$$

It follows that $g_N\equiv 0$, and by induction $g_n\equiv 0$ for all n. This completes the proof.

We need to improve this result by allowing A_n to lie in a constant momentum subspace. We denote the pure phase vacuum $\int\cdot\,dq_\xi$ by $\langle\cdot\rangle=\langle\cdot\rangle_\xi$, as above, and we assume a nonzero mass gap.

Theorem 5.4. *Assume an exponential decay rate for correlations, uniformly in all Euclidean directions. Given $p\in R^2$, suppose*

$$0=\int[\langle C,U(a)e^{ipa}B\rangle-\langle C\rangle\langle e^{ipa}B\rangle]da \tag{5.11}$$

for all $C\in\mathscr{D}$. Then

$$0=\int g_n(x_1-a,\ldots,x_n-a)e^{ipa}da, \qquad n\geqq 1\,. \tag{5.12}$$

Proof. The integrand in (5.11) is rapidly decreasing in a, as a consequence of the nonzero mass gap and Theorem 4.2. In fact, the expansion into connected Green's functions yields a sum of products, and each product contains at least one factor with variables from both B and C. This factor decays rapidly in $|a|$ by Theorem 4.2 and the fact that $g_n \in \mathscr{S}$, while the other factors are bounded. We then follow the proof of Theorem 5.3.

For application in Chapters 7,8, we want to consider kernels $g_n \in H_{-1}^{\otimes n}$ rather than in $\mathscr{S}$. In this case, A_n given by (5.5) is not in general defined (even for the free field) but requires some subtraction, e.g. Wick ordering would be sufficient. We define the Wick ordering with respect to the measure dq_ξ, i.e. with respect to the function $\Omega \equiv 1$.

In order to analyze higher portions of the energy spectrum, it is also necessary to subtract states associated with the span of $n=1,2,\ldots$ fields, $\phi(f)$, $\phi(f)\phi(g)$, etc. We consider here the cases $n=0,1$. We discuss briefly the $n=2$ subtraction in Chapter 8, when we introduce equations for three particle states.

Let P_j be the projection of $\mathscr{H}_{\mathscr{E}}(\xi)$ onto the closed subspace of degree j polynomials in ϕ which are orthogonal to all degree l polynomials, $l<j$.

Theorem 5.5. *With the hypotheses of Theorem 5.4,*

$$(1-P_0)B=B-\langle B\rangle \tag{5.13}$$

$$P_1 B=-\int\langle B\hat{\phi}(y)\rangle\Gamma^{(2)}(y-z)\hat{\phi}(z)dydz \tag{5.14}$$

where $\hat{\phi}=\phi-\langle\phi\rangle$, *$B$ is defined by* (5.9) *and each* $g_n \in \mathscr{S}(R^{2n})$.

Proof. Since (5.13) is obvious, we need only verify (5.14). By Theorem 4.2, $g(y)=\langle B\hat{\phi}(y)\rangle$ is a rapidly decreasing element of H_1. Since $\Gamma^{(2)}:H_1\to H_{-1}$ [20], $h=\Gamma^{(2)}g\in H_{-1}$, and

$$\langle\hat{\phi}(\bar{h})\hat{\phi}(h)\rangle=G^{(2)}(\bar{h}\otimes h)\leqq O(1)\|h\|_{-1}^2\,, \tag{5.14a}$$

cf. [20]. Thus the right side of (5.14) is in $L_2(\mathscr{S}'(R^2), dq_\xi)$. Finally we note that the right side of (5.14) equals: 0 for $B=1$; $\hat{\phi}(x)$ for $B=\phi(x)$; and 0 if $B\perp$ to 1 and $B\perp\phi(x)$, all x. These properties characterize P_1, hence (5.14).

Corollary 5.6. *Let*

$$B=\int\hat{\phi}(x)\hat{\phi}(y)g(x,y)dxdy\,. \tag{5.15}$$

Then (5.13) *and* (5.14) *admit extensions to* $g\in H_{-1}^{\otimes 2}$, *by continuity.*

Proof. We note

$$\langle B\hat{\phi}(z)\rangle=\int G^{(3)}(x,y,z)g(x,y)dxdy\,.$$

Thus by Theorem 4.3, we see that (5.14) extends by continuity to $g\in H_{-1}^{\otimes 2}$. To establish (5.13) on $H_{-1}^{\otimes 2}$, we note that

$$\|(1-P_0)B\|^2=G^{(4)}(g^-\otimes g)+\langle g, G^{(2)}\otimes G^{(2)}g\rangle_{L_2(R^2)}\,,$$

and each term extends to $H_{-1}^{\otimes 2}$ by continuity, using Theorem 4.3 and (5.14a).

We now assume the decay properties of $\Gamma^{(2)}(x)$ established in Chapter 6, to sharpen our estimates.

Theorem 5.7. *Suppose g is symmetric and is exponentially decreasing in the variable $x+y$, so that for all β, with $|\beta|$ sufficiently small,*

$$e^{\beta\cdot(x+y)}g(x,y)\in H_{-1}^{\otimes 2}\,.$$

Then with B given by (5.15),

$$\Gamma^{(2)} * \langle B\hat{\phi}(\cdot)\rangle \in H_{-1} \tag{5.16}$$

is exponentially decreasing.

Proof. As above, $\langle B\hat{\phi}(\cdot)\rangle$ is exponentially decreasing, and hence an element of $H_{1,\bar{m}}$, as defined in (6.9). By Theorem 6.4, and $\bar{m} \geqq m$, (5.16) is an element of $H_{-1,m}$.

We remark that in the hypothesis of Theorem 5.4, we could write (5.11) as $0 = \int \langle C, (1-P_0)U(a)e^{ipa}B\rangle da$.

Theorem 5.8. *With B, g as above, suppose that*

$$0 = \int \langle C, (1-P_0-P_1)U(a)e^{ipa}B\rangle da \tag{5.17}$$

for all $C \in \mathscr{D}$. *Then*

$$0 = \int g(x_1 - a, x_2 - a)e^{ipa}da\,. \tag{5.18}$$

Proof. We treat $(1-P_0)$ and P_1 separately, since each of these projections has rapid decay in a. We see this by expanding each term as a sum of products of connected Green's functions. Using the exponential decay of g in the variable $x+y$, the (mass gap) exponential decay of the connected Green's functions and Theorem 6.4, the decay in a follows. Two commutators with π then yield (5.18).

Remark 5.9. The decay of the integrand does not depend on the local regularity of $C \in \mathscr{D}$. In particular, we may take $C = B$.

6. Decay of $\Gamma^{(2)}(x)$ and CDD Zeros

We consider an Osterwalder-Schrader theory with the (connected) momentum space two point function

$$\chi(p) = G^{(2)}(P) = \frac{Z}{p^2+m^2} + \int_{a \geqq M^2} \frac{dv(a)}{p^2+a}\,. \tag{6.1}$$

We assume $0 < m \leqq M$. If, in addition, $m < M$ and $Z \neq 0$, the model has an isolated one particle state of mass m, and in that case Z is the field strength renormalization constant. By positivity, $Z \geqq 0$ and $dv \geqq 0$, and we assume $\int dv(a)a^{-1} < \infty$. If the theory is canonical, then in addition

$$Z + \int dv(a) = 1\,. \tag{6.2}$$

The inverse propagator $-\chi(p)^{-1}$ is also the one particle irreducible part of $\chi(p)$ The Fourier transform of $-\chi(p)^{-1}$ is $\Gamma^{(2)}(x)$, and convolution by $\Gamma^{(2)}$ is an isomorphism of H_1 onto H_{-1}, by [20, Theorem 2.1 and Remark 2].

We are concerned here with the exponential decay rate $\bar{m}$ of $\Gamma^{(2)}(x)$, since such decay is related to the spectral properties of the theory. We find that if $m < M$, then

$$m < \bar{m} \leqq M\,, \tag{6.3}$$

and $\bar{m}$ is related to the complex zeros of $\chi(p)$. In case $\bar{m} < M$, then the decay rate $\bar{m}$ is not associated with any physical particle. For instance, if M is the mass of an isolated bound state, we find

$$m < \bar{m} < M\,, \tag{6.4}$$

In case the strict inequality (6.4) holds, the propagator $\chi(p)$ has a CDD zero, see Castillejo, Dalitz, and Dyson [4].

Our results are nontrivial only if $m < M$, i.e. if an upper mass gap exists in the propagator.

Definition 6.1. We define the CDD root $\bar{m}$ as the root, in the interval $(m, M]$, of the equation

$$Z + \int_{a \geqq M^2} \frac{m^2 - \bar{m}^2}{a - \bar{m}^2}\, d\nu(a) = 0\,. \tag{6.5}$$

If no such root exists, or if $m = M$, define $\bar{m} = M$.

Remark 6.2. Assume a canonical theory (6.4). Then for $Z \neq 0$

$$\bar{m} = \left[\frac{\zeta}{\zeta + Z} m^2 + \frac{Z}{\zeta + Z} M^2\right]^{1/2}, \tag{6.6}$$

where $\zeta \in [0, 1 - Z]$ is defined by the equation

$$\zeta = \int \frac{M^2 - \bar{m}^2}{a - \bar{m}^2}\, d\nu(a)\,. \tag{6.7}$$

For example, if $d\nu(a)$ is a point mass at $a = M^2$, then $\bar{m} = [(1 - Z)m^2 + ZM^2]^{1/2}$, with $\zeta = 1 - Z$ and (6.4) holds. CDD zeros are absent if and only if

$$\int \frac{M^2 - m^2}{a - M^2}\, d\nu(a) < Z\,. \tag{6.8}$$

We show below that $\Gamma^{(2)}(x)$ has the decay rate $\bar{m}$. In order to make precise the notion of decay rate we introduce the Sobolev-Paley-Wiener space $H_{p,\lambda}$. For $\lambda = 0$, we let $H_{p,\lambda} = H_p$, the usual Sobolev space. For $\lambda > 0$, we have

Definition 6.3. Let $H_{p,\lambda}$ be those distributions f for which

$$e^{\alpha x}(-\Delta + a)^{p/2} f \in L_2 \tag{6.9}$$

for all α with $|\alpha| < \lambda$. Here $a > \lambda^{1/2}$ and such a is sufficiently large not to affect the exponential decay rate of f.

Theorem 6.4. (*a*) *The convolution operator $\Gamma^{(2)}(x)$ is a continuous transformation from $H_{1,\lambda}$ to $H_{-1,\lambda}$ for any $\lambda \leqq \bar{m}$.*

(b) *If $\bar{m} < M$, then $\Gamma^{(2)}(x)$ is not densely defined, from $H_{1,\lambda}$ to $H_{-1,\lambda}$ for any $\lambda > \bar{m}$.*

Remark. This is the precise sense in which $\Gamma^{(2)}(x)$ decays at the rate $\bar{m}$. Thus the presence of CDD zeros is equivalent to a decay rate $\bar{m}$ not associated with a physical particle.

Lemma 6.5. *Let T be a convolution operator with kernel $T(x)$, and suppose that when a is sufficiently large,*

$$F(k) = (a + k^2)^{(q-p)/2}\, \tilde{T}(k)$$

is analytic for $k=\sigma+i\tau$, *with* $|\tau|<\lambda$, *and is bounded in* σ *for each fixed* $|\tau|<\lambda$. *Then the operator*

$$T:H_{p,\lambda}\to H_{q,\lambda}$$

is continuous. (Here $|\tau|$ *is the Euclidean norm.)*

Proof. By Fourier transformation, $H_{p,\lambda}$ contains those functions f such that for $|\alpha|<\lambda$ and k real,

$$e^{\alpha\cdot\nabla}(a+k^2)^{p/2}\tilde{f}(k)\in L_2\,.$$

Thus a convolution operator T is bounded from $H_{p,\lambda}$ to $H_{q,\lambda}$ if for k real,

$$F_\alpha\equiv e^{\alpha\cdot\nabla}(a+k^2)^{(q-p)/2}\tilde{T}(k)e^{-\alpha\cdot\nabla} \tag{6.10}$$

is a bounded operator on L_2 for all real α, $|\alpha|<\lambda$. In case the series converges, F_α is given by the Baker-Campbell-Hausdorff formula as a sum of multiplication operators on L_2,

$$F_\alpha=\sum_{n=0}^{\infty}\frac{1}{n!}\left[(\alpha\cdot\nabla)^n(a+k^2)^{(q-p)/2}\tilde{T}(k)\right].$$

Thus the operator norm of F_α on L_2 is bounded by

$$\|F_\alpha\|\leqq\sum_{n=0}^{\infty}\frac{1}{n!}\|(\alpha\cdot\nabla)^n(a+k^2)^{(q-p)/2}\tilde{T}_{(k)}\|_{L_\infty}\,. \tag{6.11}$$

We estimate (6.11), using the Cauchy integral formula and the assumption of boundedness on $(a+k^2)^{(q-p)/2}\tilde{T}(k)$. By a rotation of coordinates, we may suppose that α is parallel to the x_1 coordinate axis. Thus we estimate the derivatives of $(a+k^2)^{(q-p)/2}\tilde{T}(k)$, for $k=\sigma$ real, by integrating over a disc $|\sigma_1-z_1|\leqq\beta$, centered at σ, and with $|\alpha|<\beta<\lambda$. Then

$$\left|\left(\frac{\partial}{\partial k_1}\right)^n(a+k^2)^{(q-p)/2}\tilde{T}(k)\right|\leqq A\beta^{-n},$$

where $A=A(\beta)$ is independent of n and σ. We conclude that

$$\|F_\alpha\|\leqq A\sum_{n=0}^{\infty}|\alpha|^n\beta^{-n}<\infty\,.$$

Proof of Theorem 6.4. We note

$$-\tilde{\Gamma}^{(2)}(k)(k^2+m^2)^{-1}=\left[Z+\int_{a\geqq M^2}\left(\frac{k^2+m^2}{k^2+a}\right)dv(a)\right]^{-1}, \tag{6.12}$$

and we first establish (*b*). If $\Gamma^{(2)}(x)$ has a decay rate β, the Fourier transform $\tilde{\Gamma}^{(2)}(k)$ has an analytic continuation defined for $|\tau|<\beta$. In particular, for $\sigma=0$,

$$0\neq Z+\int_{a\geqq M^2}\frac{m^2-\tau^2}{a-\tau^2}\,dv(a)\,. \tag{6.13}$$

The integral in (6.13) decreases from 0 to

$$\int\frac{m^2-M^2}{a-M^2}\,dv(a)<0$$

as τ^2 increases from m^2 to M^2. If $\bar{m}<M$, then $\bar{m}$ is the first zero of (6.13). Thus we have $\beta \leqq \bar{m}$, proving (b).

To establish (a), we note that for $A \leqq B$, $B+x>0$, the function $(A+x)(B+x)^{-1}$ is monotonic increasing in x. Thus for $k^2=(\sigma+i\tau)^2$,

$$\operatorname{Re}\left(\frac{k^2+m^2}{k^2+a}\right)=\left.\frac{(\sigma_1^2-\tau^2+m^2)(\sigma_1^2-\tau^2+a)+4\sigma_2^2\tau^2}{(\sigma_1^2-\tau^2+a)^2+4\sigma_2^2\tau^2}\right|_{\sigma_1^2=\sigma_2^2=\sigma^2}$$
$$\geqq \frac{(\sigma_1^2-\tau^2+m^2)(\sigma_1^2-\tau^2+a)}{(\sigma_1^2-\tau^2+a)^2}=\frac{\sigma_1^2-\tau^2+m^2}{\sigma_1^2-\tau^2+a}$$
$$\geqq \frac{-\tau^2+m^2}{-\tau^2+a}. \tag{6.14}$$

For $|\tau|<\bar{m}$, (6.13) is strictly positive. Thus by (6.14),

$$0<Z+\int_{a\geqq M^2}\frac{m^2-\tau^2}{a-\tau^2}\,d\nu(a) \leqq \operatorname{Re}\left[Z+\int\frac{k^2+m^2}{k^2+a}\,d\nu(a)\right].$$

We conclude that the right side of (6.12) is nonsingular and bounded away from zero uniformly in σ, for fixed $|\tau|<\bar{m}$. Furthermore in (6.12).

$$\int_{a\geqq M^2}\frac{k^2+m^2}{k^2+a}\,d\nu(a)$$

is analytic in k^2 for $\tau^2<\bar{m}^2$, since $\operatorname{Re}(k^2+a)=\sigma^2-\tau^2+a\geqq\sigma^2+M^2-\bar{m}^2>0$. Thus $(k^2+m^2)^{-1}\tilde{\Gamma}(k)$ is analytic $|\tau|<\bar{m}$ and bounded uniformly in σ for fixed $|\tau|<\bar{m}$. We conclude that Lemma 6.5 applies to $\Gamma^{(2)}(x)$ in the case $p=-q=1$, $\lambda\leqq\bar{m}$, and the proof is complete.

7. The Two Particle Bethe-Salpeter Kernel

In this chapter, we prove the existence of the Bethe-Salpeter kernel K_ξ in the Euclidean region, for almost every pure phase ξ. Let R and R_0 be defined as in (1.6) so that

$$R_0(x_1,x_2,y_1,y_2)=G_\xi^{(2)}(x_1,y_1)G_\xi^{(2)}(x_2,y_2)+G_\xi^{(2)}(x_1,y_2)G_\xi^{(2)}(x_2,y_1) \tag{7.1}$$

and

$$\begin{aligned}R(x_1,x_2,y_1,y_2)&=R_0(x_1,x_2,y_1,y_2)+G_\xi^{(4)}(x_1,x_2,y_1,y_2)\\&\quad\int G_\xi^{(3)}(x_1,x_2,z)\Gamma_\xi^{(2)}(z,z')G_\xi^{(3)}(z',y_1,y_2)dzdz'\\&=\langle\phi(x_1)\phi(x_2)(I-P_0-P_1)\phi(y_1)\phi(y_2)\rangle\end{aligned} \tag{7.2}$$

Here $\Gamma_\xi^{(2)}=-G_\xi^{(2)-1}$, so that R is one particle irreducible between the initial variables $x=(x_1,x_2)$ and the final variables $y=(y_1,y_2)$. Then K_ξ is defined by

$$R=R_0+R_0K_\xi R \tag{7.3}$$

or

$$K_\xi=R_0^{-1}-R^{-1}. \tag{7.4}$$

Theorem 7.1. *Consider a $\mathscr{P}(\phi)_2$ model and suppose that* (a) *the $V=\infty$ limit is Lorentz invariant and* (b) *the mass $m=m_\xi$ in the pure phase ξ is strictly positive. Then K_ξ defined by* (7.3)–(7.4) *is a bounded operator from* Sym $H_1^{\otimes 2}$ *to* $H_{-1}^{\otimes 2}$.

Remark. We expect that the hypotheses are valid for all noncritical $\lambda\mathscr{P}(\phi)_2-\mu\phi$ models. They have been verified for weak coupling (i.e. $\lambda \ll m_0^2$ or $\mu \gg m_0^2$) [25, 26, 45] and for even $\mathscr{P}(\phi)_2$ models in the single phase region or $\lambda\phi^4-\mu\phi$ with $\mu \neq 0$ [39, 28, 9, 44, 29, 24].

The kernel K_ξ is formally two particle irreducible between initial and final states, so we expect a configuration exponential decay rate $3m$, or $4m$ for an even theory, assuming the absence of bound states and CDD zeros.

Proof. We know that $R_0^{-1}: H_1^{\otimes 2} \to H_{-1}^{\otimes 2}$ is bounded, from the spectral representation for the two point function [20]. To establish the same result for R^{-1}, we work in subspaces $\mathscr{H}_\pm(P)$ of $H_{\pm 1}^{\otimes 2}$ with definite total momentum, see (4.2). By Theorem 4.3 (b), $\mu_{\otimes n}G^{(4)}$ is Hilbert-Schmidt on $\mathscr{H}(P)$, so $G^{(4)}$ is Hilbert-Schmidt from $\mathscr{H}_-(P)$ to $\mathscr{H}_+(P)$. Since $\Gamma^{(2)}$ is an isomorphism of H_1 onto H_{-1} [20], the last term in (7.2) is also Hilbert-Schmidt from $\mathscr{H}_-(P)$ to $\mathscr{H}_+(P)$, with norm 0 as $|P|\to\infty$. By the Fredholm alternative, R^{-1} is defined if R has no null vectors on $\mathscr{H}_-(P)$.

For simplicity, we suppose $\langle\phi\rangle_\xi=0$. In other words, we replace ϕ by $\hat{\phi}$ of Chapter 5. With this assumption, $G^{(2)}$ and $G^{(3)}$ are also Schwinger functions, and (7.2) becomes

$$\begin{aligned} R(x,y) = {} & S_\xi^{(4)}(x,y) - S_\xi^{(2)}(x_1,x_2)S_\xi^{(2)}(y_1,y_2) \\ & - S_\xi^{(3)}(x,z)[G^{(2)}]^{-1}(z,z')S_\xi^{(3)}(z',y)dzdz' . \end{aligned} \tag{7.5}$$

The above subtractions remove intermediate vacuum and one particle states, and are related to the projections P_0 and P_1 of Chapter 5. Let

$$\chi(h) = \int \phi(x_1)\phi(x_2)h(x_1,x_2)dx_1dx_2 . \tag{7.6}$$

Then

$$R(g\otimes h) = \langle \chi(g)(1-P_0-P_1)\chi(h)\rangle_\xi . \tag{7.7}$$

Furthermore, representing $(1-P_0-P_1)\chi$ as a polynomial of degree 2, we find

$$\begin{aligned} (1-P_0-P_1)\chi(h) = {} & \chi(h) - \langle\chi(h)\rangle_\xi \\ & + \int \langle\chi(h)\phi(z)\rangle_\xi \Gamma^{(2)}(z,z')\phi(z')dzdz' , \end{aligned} \tag{7.8}$$

and

$$\langle\chi(h)\phi(z)\rangle_\xi = \int G^{(3)}(x_1,x_2,z)h(x_1,x_2)dx . \tag{7.9}$$

We note

$$\langle|\chi(h)-\langle\chi\rangle|^2\rangle_\xi = \int G^{(4)} - R_0)h^-(x)h(y)dxdy . \tag{7.10}$$

We now consider R acting from $\mathscr{H}_-(P)$ to $\mathscr{H}(P)$, and we suppose that R has a nonzero null space. For simplicity, we treat the case $P=0$. We suppose that there is a nonzero even function f in $L_2(R^2)$ such that

$$0 = \int (\mu_{\otimes 4}R)(x) f(x_1-x_2)^- f(x_3-x_4)dx_2dx_3dx_4 .$$

We multiply this equation by

$$\int g(x_1+x_2-w)^- g(x_3+x_4+w)dw = g^- * g(x_1+x_2+x_3+x_4)$$

and integrate over x_1. For $g \in C_0^\infty$, we obtain

$$0 = \int (\mu_{\otimes 4} R)(x) h(x_1, x_2)^- h(x_3+w, x_4+w) dx dw,$$

where

$$h(y, z) = f(y-z) g(y+z).$$

In terms of (7.6)–7.9),

$$0 = \int dw \langle \chi(h), (I - P_0 - P_1)\chi(h_w)\rangle_\xi, \tag{7.11}$$

where $h_w(y, z) = h(y+w, z+w)$. By Theorem 5.8 and Remark 5.9, with $X(h) = B = C$, we have

$$0 = \int h(y+a, z+a) da = \int f(y-z) g(y+z+2a) da = \text{const.}\ f(y-z).$$

Thus $\chi(h) = 0$ and so R has no null vectors (except zero) in the space of definite momentum.

It follows that $R \restriction \mathscr{H}_-(P)$ has a bounded inverse. Since the inverse depends continuously on P, and remains bounded uniformly as $P \to \infty$, R, defined on $\mathscr{H}_{-1}$ has a bounded inverse. This completes the proof of Theorem 7.1.

8. Three Particle Equations

The three particle scattering equation is defined in analogy to Chapter 7, after subtraction of the two body rescattering processes. We give two procedures for these subtractions, one based on the two-body Bethe-Salpeter kernel, the other based on Faddeev-type equations for two-body rescattering processes. We expect that a quantitative analysis of the three-body equation will yield insight into such problems as: Do three-particle bound states occur? Is S unitary up to the four-particle threshold?

Let $\alpha = 12, 13, 23$ labeled pairs of initial or final particles. A two particle process occurs in a channel labeled by a pair α, β. The free Green's function for a two particle channel is

$$R_0^{\alpha\beta} = G^{(2)}(x_i, y_j)[G^{(2)}(x_{\alpha_1}, y_{\beta_1})G^{(2)}(x_{\alpha_2}, y_{\beta_2}) + G^{(2)}(x_{\alpha_1}, y_{\beta_2})G^{(2)}(x_{\alpha_2}, y_{\beta_1})] \tag{8.1}$$

where $i \notin \alpha$, $j \notin \beta$. The total R_0 is

$$R_0 = \sum_\beta R_0^{\alpha\beta} = \sum_\pi G^{(2)}(x_1 - y_i) G^{(2)}(x_2 - y_j) G^{(2)}(x_3 - y_k), \tag{8.2}$$

where $\sum_\pi$ ranges over the six permutations (i, j, k) of $(1, 2, 3)$.

We also define the two particle scattering kernels

$$R_{\alpha\beta} = G^{(2)}(x_i, y_j) G^{(4)}(x_\alpha, y_\beta), \tag{8.3}$$

where $i \notin \alpha, j \notin \beta$. Graphically, $R_{\alpha\beta}$ represents scattering processes with initial state (i, α), final state (j, β), and two particle scattering in the α, β channel. See Fig. 8.1.

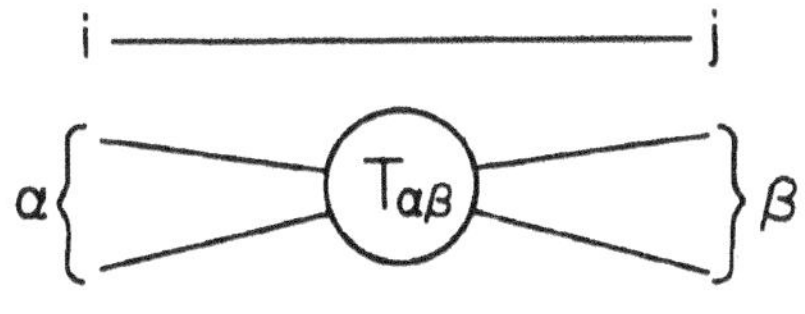

Fig. 8.1. A two particle process

Proposition 8.1. *The operators* $R_0^{\alpha\beta}$, R_0, *and* $R_{\alpha\beta}$ *are bounded transformations* $H_{-1}^{\otimes 3}$ *to* $H_1^{\otimes 3}$. *The norm of* $R_{\alpha\beta}$ *is* $O(\lambda)$ *as* $\lambda \equiv \sup|\lambda_j| \to 0$.

Proof. This follows immediately by the bounds of [20].

Definition 8.2. The two-body Bethe-Salpeter kernel $K_{\alpha\alpha}$ for the three body problem is

$$K_{\alpha\alpha} = R_0^{\alpha\alpha})^{-1} - (R_0^{\alpha\alpha} + R_{\alpha\alpha})^{-1}. \tag{8.4}$$

$K_{\alpha\beta}$ is defined by a permutation of the y variables of $K_{\alpha\alpha}$. The two body amplitude $R_{2\text{-body}}$ satisfies

$$R_{2\text{-body}} = R_0 + \tfrac{1}{9} R_0 \textstyle\sum_{\alpha,\beta} K_{\alpha\beta} R_{2\text{-body}}. \tag{8.5}$$

We remark that $K_{\alpha\beta}$ has the form

$$K_{\alpha\beta} = K \otimes (-\Gamma), \tag{8.6}$$

where K is the Bethe-Salpeter kernel (7.4) for the channel $(\alpha\beta)$ and $-\Gamma$ is the inverse propagator for the remaining particle.

Theorem 8.3. *Under the hypotheses of Theorem 7.1,* $K_{\alpha\beta}$ *and*

$$R_{2\text{-body}}^{-1} = R_0^{-1} - \tfrac{1}{9} \textstyle\sum_{\alpha,\beta} K_{\alpha\beta}$$

are bounded maps from Sym $H_1^{\otimes 3}$ *to* $H_{-1}^{\otimes 3}$, *with* $\|K_{\alpha\beta}\| \leqq O(\lambda)$ *as* $\lambda \to 0$. *Also for* λ *sufficiently small,* $R_{2\text{-body}}$ *is a bounded map from* Sym $H_{-1}^{\otimes 3}$ *to* $H_1^{\otimes 3}$.

Proof. The bound on $K_{\alpha\beta}$ follows by Theorem 7.1 which bounds K of (7.4) and the properties of Γ established in [20]. Since R_0^{-1} is bounded from Sym $H_1^{\otimes 3}$ to $H_{-1}^{\otimes 3}$, the existence and boundedness of $R_{2\text{-body}}^{-1}$ follows. We note by (8.5),

$$(I - \tfrac{1}{9} R_0 \textstyle\sum K_{\alpha\beta}) R_{2\text{-body}} = R_0,$$

and, by the above, the norm of $R_0 K_{\alpha\beta}$ from $H_1^{\otimes 3}$ to $H_1^{\otimes 3}$ is $O(\lambda)$. Thus for λ sufficiently small,

$$(I - \tfrac{1}{9} R_0 \textstyle\sum K_{\alpha\beta})^{-1} : H_1^{\otimes 3} \to H_1^{\otimes 3}$$

is bounded. It follows that

$$R_{2\text{-body}} : H_{-1}^{\otimes 3} \to H_1^{\otimes 3}$$

exists and is bounded.

Theorem 8.4. *Consider* $R_{2\text{-body}}^{-1}$ *restricted to a constant momentum subspace:*

$$R_{2\text{-body}}^{-1}(P) : \mathrm{Sym}\,\mathscr{H}_+(P) \to \mathscr{H}_-(P).$$

Then for $|P|$ sufficiently large, $R_{2\text{-body}}(P)$ is bounded from Sym $\mathscr{H}_-(P)$ *to* $\mathscr{H}_+(P)$. *(Here the minimum $|P|$ may depend on λ.) Also as $|P|\to\infty$,*

$$\|R_{2\text{-body}}(P)-R_0(P)\|_{-,+}\to 0\,.$$

Proof. Each of the operators $R_0, R_0^{\alpha\beta}, R_{\alpha\beta}, K_{\alpha\beta}$ is translation invariant, and thus acts on constant momentum subspaces. By Theorem 4.3 (b), $R_{\alpha\beta}(R_0^{\alpha\beta})^{-1}$ is a bounded operator on Sym $\mathscr{H}_+(P)$, with norm $o(1)$ as $|P|\to\infty$. We choose $|P|$ sufficiently large so that

$$R_0K_{\alpha\beta}=-R_0(R_0^{\alpha\beta})^{-1}\textstyle\sum_{n=1}^{\infty}(-R_{\alpha\beta}(R_0^{\alpha\beta})^{-1})^n$$

has norm on sym $\mathscr{H}_+(P)$ less than 1. Thus $\|\sum_{\alpha\beta}\frac{1}{9}R_0K_{\alpha\beta}\|<1$ on sym $\mathscr{H}_+(P)$ and

$$R_{2\text{-body}}^{-1}=R_0^{-1}(1-\tfrac{1}{9}\textstyle\sum_{\alpha,\beta}R_0K_{\alpha\beta})$$

is invertible. The inverse $R_{2\text{-body}}(P)$ is bounded from Sym$\mathscr{H}_-(P)$ to $\mathscr{H}_+(P)$, since $R_0(P)$ is. Also $(R_0K_{\alpha\beta})(P)$ has norm $o(1)$ as $|P|\to\infty$, so $\|R_{2\text{-body}}(P)-R_0(P)\|_{-,-}$ is $o(1)$. Here $\|\cdot\|_{-+}$ denotes the norm from sym$\mathscr{H}_-(P)$ to $\mathscr{H}_+(P)$.

We remark that instead of using the direct definition (8.5) of $R_{2\text{-body}}$, we could also use Faddeev type equations to separate the two body forces. Let $R(\alpha)$ denote the sum of 2-body rescattering diagrams in which the initial two-body scattering involves the pair α. Then the Feddeev equations are

$$\begin{aligned}R(\alpha)&=R_0^{\alpha\alpha}+R_{\alpha\alpha}\\&\quad+\textstyle\sum_{\beta\neq\alpha}R_{\alpha\beta}(R_0^{\beta\beta}+R_{\beta\beta})^{-1}R(\beta)\\&=R_0^{\alpha\alpha}+R_{\alpha\alpha}\\&\quad\textstyle\sum_{\beta\neq\alpha}R_{\alpha\beta}[(R_0^{\beta\beta})^{-1}-K_{\beta\beta}]R(\beta)\,.\end{aligned}\tag{8.7}$$

As above, we find

Theorem 8.5. *For λ sufficiently small, the Eq.* (8.7) *have a unique solution $R(\alpha):H_{-1}^{\otimes 3}\to H_1^{\otimes 3}$. For constant total momentum P, the equations* (8.7) *have a unique solution $R(\alpha, P):\mathscr{H}_-(P)\to\mathscr{H}_+(P)$, for $|P|$ sufficiently large (depending on λ).*

For small λ, we define

$$\bar{R}_{2\text{-body}}=\textstyle\sum_\alpha(R(\alpha)-R_0^{\alpha\alpha})+R_0\,.\tag{8.8}$$

We expect that $R_{2\text{-body}}=\bar{R}_{2\text{-body}}$.

We now define the three body Bethe-Salpeter kernel $K^{(3)}$, by the equations

$$R^{(3)}(x_1,x_2,x_3,y_1,y_2,y_3)=\langle\textstyle\prod_{i=1}^{3}\phi(x_i)(I-\sum_{j=0}^{2}P_j)\prod_{k=1}^{3}\phi(x_6)\rangle\,,\tag{8.9}$$

and

$$R^{(3)^{-1}}=R_{2\text{-body}}^{-1}-K^{(3)}\,,\tag{8.10}$$

or

$$R^{(3)}=R_{2\text{-body}}+R_{2\text{-body}}K^{(3)}R^{(3)}\,.\tag{8.11}$$

Theorem 8.6. *Under the hypothesis of Theorem 7.1, $R^{(3)^{-1}}$ and $K^{(3)}$ exist as bounded operators from* Sym $H_1^{\otimes 3}$ *to $H_{-1}^{\otimes 3}$, and satisfy* (8.10). *For weak coupling,* (8.11) *also holds.*

Remark. N-body equations, as well as a further study of the three body equations, will be contained in a subsequent article. By definition $-K^{(n)}$ is the connected part of $R^{(n)^{-1}}$.

References

1. Araki, H.: On the algebra of all local observables. Prog. Theor. Phys. **32**, 844—854 (1964)
2. Bratteli, O.: Conservation of estimates in quantum field theory. Comm. Pure and Appl. Math. **25**, 759—779 (1972)
3. Bros, J.: Some analyticity properties implied by the two particle structure of Green's functions in general quantum field theory. In: Gilbert, R., Newton, R. (Eds.): Analytic methods in mathematical physics. New York: Gordon and Breach 1970
4. Castillejo, L., Dalitz, R., Dyson, F.: Low's scattering equation for the charged and neutral scalar theories. Phys. Rev. **101**, 453—458 (1956)
5. Dobrushyn, R., Minlos, R.: Construction of a one dimensional quantum field via a continuous Markov field. Funct. Anal. and its Appl. **7**, 324—325 (1973) (English transl.)
6. Eckmann, J.-P. Magnen, J., Seneor, R.: Decay properties and Borel summability for the Schwinger functions in $P(\phi)_2$ theories. Commun. math. Phys. **39**, 251—271 (1975)
7. Faris, W., Lavine, R.: Commutators and self-adjointness of Hamiltonian operators. Commun. math. Phys. **35**, 39—48 (1974)
8. Feldman, J.: On the absence of bound states in the ϕ^4 quantum field model without symmetry breaking. Canad. J. Phys. **52**, 1583—1587 (1974)
9. Fröhlich, J.: Verification of axioms for Euclidean and relativistic fields and Haag's theorem in a class of $P(\phi)_2$ models. Ann. Inst. H. Poincaré (to appear)
10. Fröhlich, J.: Schwinger functions and their generating functionals. I. Helv. Phys. Acta (to appear)
11. Fröhlich, J.: Schwinger functions and their generating functionals. II. Adv. Math. (to appear)
12. Ginibre, J.: General formulation of Griffiths' inequalities. Comm. Math. Phys. **16**, 310—328 (1970)
13. Glimm, J., Jaffe, A.: A $\lambda\phi^4$ quantum field theory without cutoffs II, the field operators and the approximate vacuum. Ann. of Math. **91**, 362—401 (1970)
14. Glimm, J., Jaffe, A.: The $(\lambda\phi^4)_2$ quantum field theory without cutoffs. III. The physical vacuum. Acta Math. **125**, 203—261 (1970)
15. Glimm, J., Jaffe, A.: The $(\lambda\phi^4)_2$ quantum field theory without cutoffs. IV. Perturbations of the Hamiltonian. J. Math. Phys. **13**, 1558—1584 (1972)
16. Glimm, J., Jaffe, A.: The energy momentum spectrum and vacuum espectation values in quantum field theory. J. Math. Phys. **11**, 3335—3338 (1970)
17. Glimm, J., Jaffe, A.: The energy momentum spectrum and vacuum expectation values in quantum field theory. II. Commun. math. Phys. **22**, 1—22 (1971)
18. Glimm, J., Jaffe, A.: Boson quantum field models. In: Streater, R. (Ed.): Mathematics of contemporary physics. New York: Academic Press 1972
19. Glimm, J., Jaffe, A.: Positivity of the ϕ_3^4 Hamiltonian. Fortschritte der Physik **21**, 327—376 (1973)
20. Glimm, J., Jaffe, A.: Entropy principle for vertex functions in quantum field models. Ann. Inst. H. Poincaré **21**, 1—26 (1974)
21. Glimm, J., Jaffe, A.: ϕ_2^4 quantum field model in the single phase region: Differentiability of the mass and bounds on critical exponents. Phys. Rev. D **10**, 536—539 (1974)
22. Glimm, J., Jaffe, A.: A remark on the existence of ϕ_4^4. Phys. Rev. Lett. **33**, 440—442 (1974)
23. Glimm, J., Jaffe, A.: Absolute bounds on vertices and couplings. Ann. l'Inst. H. Poincaré **22** (to appear)
24. Glimm, J., Jaffe, A.: On the approach to the critical point. Ann. l'Inst. H. Poincaré **22** (to appear)
25. Glimm, J., Jaffe, A., Spencer, T.: The particle structure of the weakly coupled $P(\phi)_2$ model and other applications of high temperature expansions. In: Velo, G., Wightman, A. (Eds.): Constructive quantum field theory. Lecture notes in Physics 25. Berlin Heidelberg New York: Springer 1973
26. Glimm, J., Jaffe, A., Spencer, T.: The Wightman axioms and particle structure in the $P(\phi)_2$ quantum field model. Ann. Math. **100**, 585—632 (1974)
27. Guerra, F., Rosen, L., Simon, B.: Nelson's symmetry and the infinite volume behavior of the vacuum in $P(\phi)_2$. Commun. math. Phys. **27**, 10—22 (1972)

28. Guerra, F., Rosen, L., Simon, B.: The $P(\phi)_2$ Euclidean quantum field theory as classical statistical mechanics. Ann. Math. **101**, 111—259 (1975)
29. Guerra, F., Rosen, L., Simon, B.: Correlation inequalities and the mass gap. III. Mass gap for a class of strongly coupled theories with nonzero external field. Commun. math. Phys., to appear
30. Jost, R.: The general theory of quantized fields. American Mathematical Society, Providence, 1965
31. Lebowitz, J., Penrose, O.: Analytic and clustering properties of the thermodynamic functions and distribution functions for classical lattice and continuum systems **11**, 99—124 (1968)
32. Lebowitz, J., Penrose, O.: Decay of correlations. Phys. Rev. Lett. **31**, 749—752 (1973)
33. McBryan, O.: Local generators for the Lorentz group in the $P(\phi)_2$ model. Nuovo Cimento, **18** A, 654—662 (1973)
34. McBryan, O.: Vector currents in the Yukawa_2 quantum field theory (preprint)
35. McBryan, O.: Self adjointness of relatively bounded quadratic forms and operators. J. Funct. Anal. (to appear)
36. McBryan, O., Park, Y.: Lorentz covariance of the Yukawa_2 quantum field theory. J. Math. Phys. **16**, 105—110 (1975)
37. Nelson, E.: Time ordered operator products of sharp time quadratic forms. J. Funct. Anal. **11**, 211—219 (1972)
38. Nelson, E.: The construction of quantum fields from Markov fields. J. Funct. Anal. **12**, 97—112 (1973)
39. Nelson, E.: Probability theory and Euclidean field theory. In: Velo, G., Wightman, A. (Eds.): Constructive quantum field theory. Lecture notes in Physics 25. Berlin Heidelberg New York: Springer 1973
40. Osterwalder, K., Schrader, R.: Axioms for Euclidean Green's functions. Commun. math. Phys. **31**, 83—112 (1974)
41. Osterwalder, K., Schrader, R.: Axioms for Euclidean Green's functions. II. Commun. math. Phys. **42**, 281—305 (1975)
42. Rosen, L.: A $\lambda\phi^{2n}$ field theory without cutoffs. Commun. math. Phys. **16**, 157—183 (1970)
43. Schrader, R.: Yukawa quantum field theory in two space-time dimensions without cutoff. Ann. Phys. **70**, 412—457 (1972)
44. Simon, B.: The $P(\phi)_2$ Euclidean quantum field theory. Princeton: Princeton University Press 1974
45. Spencer, T.: The mass gap for the $P(\phi)_2$ quantum field model with a strong external field. Commun. math. Phys. **39**, 63—76 (1974)
46. Spencer, T.: The absence of even bound states in ϕ_2^4. Commun. math. Phys. **39**, 77—79 (1974)
47. Spencer, T.: The decay of the Bethe Salpeter kernel in $P(\phi)_2$ quantum field models. Commun. math. Phys. (to appear)
48. Spencer, T., Zirilli, F.: Private communication
49. Symanzik, K.: A modified model for Euclidean quantum field theory. N.Y.U. (preprint) 1964
50. Tucciarone, A.: A relativistic treatment of the three body problem. Nuovo Cimento **41** A, 204—221 (1966)
51. Glimm, J., Jaffe, A.: ϕ^j bounds in $P(\phi)_2$ quantum field models. Proceedings of the International Colloquium on Mathematical Methods in Quantum Field Theory, Marseille, 1975

Communicated by A. S. Wightman

Particles and Scaling for Lattice Fields and Ising Models

James Glimm*
Rockefeller University, New York, N.Y. 10021, USA

Arthur Jaffe**
Harvard University, Cambridge, Mass. 02138, USA

Abstract. The conjectured inequality $\Gamma^{(6)} \leqq 0$ leads to the existence of φ_d^4 fields and the scaling (continuum) limit for d-dimensional Ising models. Assuming $\Gamma^{(6)} \leqq 0$ and Lorentz covariance of this construction, we show that for $d \geqq 6$ these φ_d^4 fields are free fields unless the field strength renormalization Z^{-1} diverges. Let λ be the bare charge and ε the lattice spacing. Under the same assumptions ($\Gamma^{(6)} \leqq 0$, Lorentz covariance and $d \geqq 6$) we show that if $\lambda \varepsilon^{4-d}$ is bounded as $\varepsilon \to 0$, then Z^{-1} is bounded and the limit field is free.

1. Introduction

Even φ^4 fields in the single phase region describe particles which interact through repulsive forces. The evidence in support of this statement includes the absence of even bound states [24, 3, 34] the canonical lower bound on critical exponents [16], the arguments of [19] concerning absence of three-particle bound states and *CDD* zeros (which depend on the conjectured inequality $\Gamma^{(6)} \leqq 0$) and the scaling and numerical arguments of [22, 29] in favor of this $\Gamma^{(6)}$ inequality for $d \leqq 3$.

The purpose of this note is to extend some of these ideas to lattice φ^4 field theories and Ising models. There are two reasons for studying lattice (as opposed to continuum) field theories. The first reason is that all dimensions, including $d \geqq 4$, are possible, and that the results obtained may be related to the existence [17, 16, 1, 30, 21] and triviality/nontriviality problems in these dimensions, see also [32, 33]. The second reason is that the lattice φ^4 field theory is an intermediate link between the continuum φ^4 field theory and the Ising model, and may serve to clarify the relation between Ising model and field theory critical behavior, see [20–22, 26–27, 31]. Beyond these two reasons, we note that lattice field theories are at the outset easier because of the absence of ultraviolet problems but ultimately more difficult because of the absence of the Euclidean rotation symmetry.

* Supported in part by the National Science Foundation under Grant MPS 74-13252
** Supported in part by the National Science Foundation under Grant MPS 75-21212

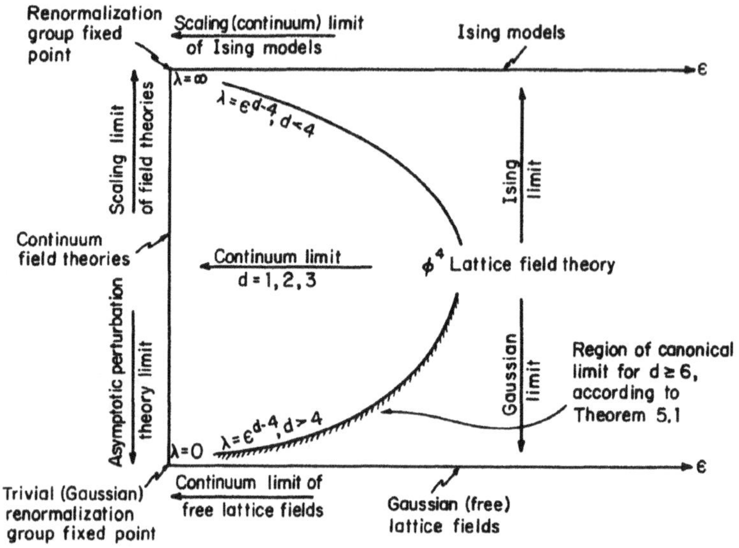

Fig. 1. The $\lambda\varphi^4 + \sigma\varphi^2$ parameter plane. ε = lattice spacing; $\sigma = \sigma(\lambda, \varepsilon, m)$ chosen to give mass m in single phase region; m held fixed (e.g. $m = 1$). The λ axis represents (continuum) field theory. The ε axis at $\lambda = \infty$ represents Ising models [30]. The ε axis at $\lambda = 0$ represents Gaussian lattice fields. The curve $\lambda = \varepsilon^{d-4}$ represents constant dimensionless bare charge. Charge renormalization for $d \geqq 4$ may lead to a reparametrization of the λ axis

As a compensation for the loss of Euclidean invariance and of the resulting Lehmann spectral formula, we use the momentum space canonical upper bound [9] on the two point function, as well as the $\Gamma^{(6)} \leqq 0$ inequality as used in [19].

Our first main result assumes only $\Gamma^{(6)} \leqq 0$. The conclusion is the existence of a continuum φ_d^4 field theory with an arbitrary choice of charge renormalization. Included is the existence of the scaling (continuum) limit for Ising models. The second main result assumes in addition Lorentz invariance of the continuum theory, as well as the (finite) charge renormalization, $\lambda \leqq \varepsilon^{d-4}$, ε being the lattice spacing. Note that $\varepsilon^{d-4} \to 0$ in the continuum limit $\varepsilon \to 0$, for $d \geqq 5$. In this case, the $\varepsilon = 0$ theory is canonical for $d \geqq 6$ (§ 5).

The limits considered here are summarized in Figure 1 above. Coincidence of the two fixed points seems to correspond to the picture in which nontrivial φ^4 field theories cannot be constructed as a limit of lattice cutoff approximations. Similarly, it would seem that the Ising model would be governed by the trivial fixed point only in the case for which a nontrivial fixed point does not occur in Figure 1.

A renormalizability curve $\lambda = O(\varepsilon^{d-4})$ of constant dimensionless bare charge is indicated on this figure. A limit constructed along any such curve is renormalizable. Those limits constructed below this family of curves are superrenormalizable (e.g. the continuum limit with constant λ, $d = 1, 2, 3$). Theories constructed above these curves are nonrenormalizable (i.e. $\lambda\varepsilon^{4-d} \to \infty$ as $\varepsilon \to 0$). The charge λ has dimension (length)$^{d-4}$, so that $\lambda\varepsilon^{4-d}$ is a dimensionless bare charge.

2. The Wave Function Renormalization and Proper Self Energy for Noncritical Lattice Theories

In this chapter, we consider a $P(\varphi)$ lattice field or Ising model. We suppose that the infinite volume limit is defined and is invariant under lattice translations, rotations and reflections. We also assume *FKG* inequalities [6], a positive mass $m>0$ (i.e. an exponential decay rate for correlations in lattice directions) and a single phase. These restrictions should be viewed as restrictions on the boundary conditions and/or coupling constants. In particular they are satisfied for φ^4 interactions and the Ising model approaching the critical point from a single phase direction. Let

$$G^{(2)}(x)=\langle\varphi(x)\varphi(0)\rangle-\langle\varphi(x)\rangle\langle\varphi(0)\rangle\geqq 0 \tag{2.1}$$

be the truncated two point function. The Fourier transform

$$G^{(2)\frown}(p)=\sum e^{-ipx}G^{(2)}(x) \tag{2.2}$$

is nonnegative because $G^{(2)}(x)$ is positive definite. Using invariance of $G^{(2)}(x)$ under reflections and rotations, $G^{(2)\frown}(p)$ is a symmetric function of the variable $\{\cos p_\nu\}_{\nu=1}^d$. Using the selfadjoint transfer matrix, one shows that

$$G^{(2)}(x_1,\ldots,x_d)\leqq G^{(2)}(0,\ldots 0,x_\nu,\ldots 0) \tag{2.3}$$

for any ν [30]. In particular $G^{(2)}$ has an exponential decay rate in all directions and $G^{(2)\frown}$ is analytic for p in a tube τ. By Hartog's theorem, τ is convex and also using the $p_\nu\to -p_\nu$ symmetry, we see that

$$\left\{p\colon \sum_{\nu=1}^{d}|\operatorname{Im}p_\nu|\leqq m\right\}\subset\tau\subset\left\{p\colon \sup_\nu|\operatorname{Im}p_\nu|\leqq m\right\}. \tag{2.4}$$

The use of Hartog's theorem can be avoided by a Schwarz inequality. For $p_1, p_2\in\tau$,

$$\left|\sum G^{(2)}(x)e^{-i(p_1+p_2)x/2}\right|\leqq\left(\sum G^{(2)}(x)e^{\operatorname{Im}p_1 x}\right)^{\frac12}\left(\sum G^{(2)}(x)e^{\operatorname{Im}p_2 x}\right)^{\frac12}.$$

Thus not only is τ convex, but its border coincides with the singularities of $G^{(2)\frown}(p)$, for $\operatorname{Re}p=0$. We have proved the following:

Proposition 2.1. *For a lattice field or Ising model as above, $G^{(2)\frown}$ is analytic in a convex tube τ, satisfying* (2.4).

We distinguish a single lattice direction x, and write $\boldsymbol{x}=\{x_2,\ldots,x_d\}$, $\boldsymbol{p}=\{p_2,\ldots,p_d\}$. The existence of a self adjoint transfer matrix means that $G^{(2)\frown}$ has an analytic continuation for p_1 in the half plane $p_1^2\neq$ neg. real, for $\boldsymbol{p}$ real.

Proposition 2.2. *Consider a lattice field or Ising model as above with $\boldsymbol{p}$ real. Then $G^{(2)\frown}(p)$ is Herglotz as a function of the variable $1-\cos p_1$. For some positive measure $d\varrho(a,\boldsymbol{p})$ depending on $\boldsymbol{p}$,*

$$G^{(2)\frown}(p)=\int_{-1+\cosh m}^{\infty}\frac{d\varrho(a,\boldsymbol{p})}{1-\cos p_1+a}, \tag{2.5}$$

and

$$0<\int\frac{d\varrho}{a}<\infty\,. \tag{2.6}$$

Proof. An explicit calculation shows that the Fourier transform of $e^{-\mu|n|}$ is

$$\Sigma_n e^{-\mu|n|+ipn}=\sinh\mu/(\cosh\mu-\cos p)\,.$$

Substitution of this formula into the spectral representation for the transfer matrix establishes (2.5) after the change of variable $a=-1+\cosh\mu$, while the upper bound in (2.6) follows from

$$\sum_x G^{(2)}(x)e^{-ipx}=G^{(2)\sim}(0,p)=\int\frac{d\varrho(a,p)}{a}\leqq\sum_x G^{(2)}(x)<\infty\,. \tag{2.7}$$

From (2.5) we see that $G^{(2)\sim}(p)$ is monotone decreasing in each argument p_ν for $0\leqq p_\nu\leqq\pi$. We show that $G^{(2)\sim}(\pi,\ldots,\pi)\neq 0$ and then it follows from (2.5) and monotonicity that $d\varrho(a,p)$ is nonzero for any real p. If $G^{(2)\sim}(\pi,\ldots,\pi)=0$, then $d\varrho(a,\pi)=0$ by (2.5), and so $G^{(2)\sim}(p_1,\pi,\ldots,\pi)=0$ for all p_1. Interchanging first and second coordinates, $G^{(2)\sim}(\pi,p_2,\pi,\ldots,\pi)=0$, and so $G^{(2)\sim}(p_1,p_2,\pi,\ldots,\pi)=0$ for arbitrary p_1 and p_2. Continuing in this fashion, we conclude that $G^{(2)\sim}$ and $G^{(2)}$ both vanish, which is a contradiction.

Definition. $\Gamma^{(2)}(x)$ is the convolution inverse to $-G^{(2)}$, so that

$$\Gamma^{(2)\sim}(p)=-1/G^{(2)\sim}(p)\,.$$

The next result is elementary, cf. [19].

Proposition 2.3. *Consider a lattice field theory or Ising model as above.* $\Gamma^{(2)\sim}$ *is analytic in a convex tube* τ_Γ *satisfying*

$$\{p:\textstyle\sum|\mathrm{Im}\,p_\nu|\leqq m\}\subset\tau_\Gamma\,. \tag{2.8}$$

For p *real,* $\Gamma^{(2)\sim}$ *is Herglotz as a function of* $1-\cos p_1$*, and has a unique representation*

$$\Gamma^{(2)\sim}(p)=-\alpha(p)(1-\cos p_1)$$
$$+\beta(p)+\int_{-1+\cosh m}^{\infty}\left[\frac{1}{1-\cos p_1+a}-\frac{a}{a^2+1}\right]d\nu(a,p) \tag{2.9}$$

with $\alpha\geqq 0$*,* β *real,* $d\nu$ *positive and*

$$\int d\nu/(1+a^2)<\infty\,.$$

Remark. For $d\geqq 3$, there is a constant K depending only on d and $G^{(2)}(0)$ such that

$$\alpha+|\beta|+\int(1+a^2)^{-1}d\nu\leqq K\,. \tag{2.10}$$

Proof. By [9],

$$G^{(2)\sim}(p)\leqq\mathrm{const.}\,|p|^{-2}\,. \tag{2.11}$$

By the Herglotz property, used successively in each coordinate, as in the proof of Proposition 2.2,

$$G^{(2)\sim}(p) \leqq \text{const.} \prod_{\nu=1}^{d} |p_\nu|^{-2} G^{(2)\sim}(\pi, \dots, \pi).$$

Thus for any $r>0$ and for $d>3$,

$$\begin{aligned} G^{(2)}(0) &= (2\pi)^{-d} \int G^{(2)\sim}(p)dp \\ &\leqq \text{const.} \left(\int_{\substack{\text{Inf}|p_\nu| \leqq r \\ \nu}} |p|^{-2} dp + r^{-2d} G^{(2)\sim}(\pi, \dots, \pi) \right) \\ &\leqq \text{const.} (r + r^{-2d} G^{(2)\sim}(\pi, \dots, \pi)). \end{aligned}$$

Here the optimal choice of r is

$$r = O(G^{(2)\sim}(\pi, \dots, \pi)^{1/(2d+1)}$$

and so

$$G^{(2)}(0)^{2d+1} \leqq \text{const.} G^{(2)\sim}(\pi, \dots, \pi) \leqq \text{const.} G^{(2)\sim}(p).$$

A uniform upper bound on (2.9) follows. For $d=3$, the exponents must be changed slightly to dominate $r \ln r$ in place of r, but a bound on (2.9) in still valid. Linear combinations of $\cos p_1 = 0, -1$ show that β and $\alpha + \int (a^2+1)^{-1} d\nu(a)$ are each bounded. Since α and $d\nu$ are each nonnegative, the individual terms α and $\int (a^2+1)^{-1} d\nu$ are also bounded, proving (2.10).

We now write $x=(x', x'')$, $p=(p', p'')$ for some decomposition

$$x' = (x_1, \dots, x_j), \qquad x'' = (x_{j+1}, \dots, x_d)$$

of the coordinates, and consider the partial Fourier transform

$$\Gamma^{(2)\sim}(x', p'') \equiv \sum_{x''} e^{-ip''x''} \Gamma^{(2)}(x', x'').$$

Proposition 2.4. *Consider a lattice field or Ising model as above. For $x_1 \neq 0$, $x_2 = \dots = x_j = 0$, p'' real, $\Gamma^{(2)\sim}(x', p'')$ is positive and has an exponential decay rate $\bar{m} \geqq m$. (Note $j \neq 0$, by assumption.)*

Proof. First we consider the case $x' = x_1$ and $p'' = \mathbf{p}$. The Herglotz representation (2.9) in x_1-space has three terms. The first is nonnegative for $x_1 \neq 0$. The second, proportional to β, vanishes for $x_1 \neq 0$. The third is positive for all x_1, so the proof is complete in this case. The general case follows from the case just considered, because integration with respect to $p_2, \dots, p_j$ is equivalent to evaluation at $x_2 = \dots = x_j = 0$.

Definition. If the measure $d\varrho(a, \mathbf{p}=0)$ has a δ-function at $a = -1 + \cosh m$, then $Z>0$ is the strength of this δ-function. Otherwise by definition $Z=0$.

Proposition 2.5. *For a pure phase lattice $P(\varphi)$ field theory or Ising model with $m>0$ and $Z>0$,*

$$\begin{aligned} Z^{-1} &= -(\partial/\partial(1-\cos p_1)) \Gamma^{(2)\sim}(p_1, \mathbf{p})|_{\cos p = \cosh m,\, \mathbf{p}=0} \\ &= \alpha(0) + \int_0^\infty (a - \cosh m + 1)^{-2} d\nu(a, 0) \end{aligned} \tag{2.12}$$

Proof. We write

$$G^{(2)\sim}(p_1,0)=Z(\cosh m-\cos p_1)^{-1}+\int\frac{d\hat{\varrho}}{1-\cos p_1+a}$$

where $\int_{-1+\cosh m}^{-1+\cosh m+\varepsilon} d\hat{\varrho}\to 0$ as $\varepsilon\to 0$. Thus as $\cos p_1\to\cosh m$,

$$Z(\cosh m-\cos p_1)^{-1}\leqq G^{2\sim}(p_1,0)\leqq(Z+\varepsilon)(\cosh m-\cos p_1)^{-1} \tag{2.12a}$$

and the derivative of $\Gamma^{(2)}$ can be evaluated as Z^{-1} from this inequality.

Remark. We have not assumed an upper mass gap, in contrast to [19].

Definition. C_{L,m_0} is the free (Gaussian) lattice covariance,

$$C_{L,m_0}=\left[-1+\cosh m_0+\sum_{\nu=1}^{d}(1-\cos p_\nu)\right]^{-1} \tag{2.13a}$$

and

$$\Pi^\sim(p)=\Gamma^{(2)\sim}(p)+C_{L,m_0}^{-1}. \tag{2.13b}$$

Proposition 2.6. *Consider a lattice field theory with the hypotheses of § 2. Let P_0, P_1 be projections onto the vacuum $\Omega(\equiv 1)$ and the span of $(1-P_0)\varphi\Omega$ respectively. Then*

$$\Pi(x)=-\langle P''\rangle\delta_0(x)+\langle P'(x),(1-P_0-P_1)P'(0)\rangle \tag{2.14}$$

if the Wick ordering covariance for P is C_{L,m_0}.

Proof. As in [19]. The Ising model is excluded because the δ-functions in the single spin measure prevent integration by parts.

Proposition 2.7. *Consider a lattice field theory as above.*

$$\Gamma^{(2)}(x_1,\ldots,x_j,p_{j+1},\ldots,p_d)|_{x_1=0=\ldots=x_j}\geqq 1-\cosh m_0-\langle P''\rangle-2d$$

Proof. We use the integration by parts representation (2.13–14), and omit

$$\sum_{x''}e^{-ip''x''}\langle P'(x'=0,x'')(1-P_0-P_1)P'(0)\rangle.$$

Here $x'=x_1,\ldots,x_j$ and $x''=x_{j+1},\ldots,x_d$. The omitted quantity is positive because

$$x''\mapsto\langle P'(x'=0,x'')(1-P_0-P_1)P'(0)\rangle$$

is positive definite and hence the Fourier transform of a positive measure.

3. Three Particle Decay for $\mathcal{P}(\varphi)=\varphi^4$

In this section we consider $\lambda\varphi^4+\sigma\varphi^2$ lattice field theories with $\sigma>\sigma_c$. Then all hypotheses of §2 are satisfied. Ising models are excluded because we integrate by parts.

Proposition 3.1. *Assume* $\Gamma^{(6)}(xxxyyy)\leqq 0$ *for* $x-y$ *large. Then the CDD radius* $\bar{m}$ *satisfies* $\bar{m}\geqq 3m$. *The analyticity tube* τ_Γ *for* $\Gamma^{(6)}$ *satisfies*

$$\{p:\Sigma|\mathrm{Im}\,p_\nu|\leqq 3m\}\subset\tau_\Gamma$$

Proof. As in [19]. See also Proposition 3.2 below.

This result leads to the existence of φ_d^4 field theories and the scaling limit for Ising models, see §4. In order to establish the triviality of some of these field theories, we need to assume canonical upper bounds for the two point decay rate (as would follow from the Lehmann spectral formula in the continuum case). Explicitly, we assume

$$0\leqq G^{(2)}(x)\leqq\begin{cases}K_1|x|^{-d+2}e^{-m|x|}, & |x|\leqq m^{-1}\\ K_1 m^{d-2}e^{-m|x|}, & |x|\geqq m^{-1}\end{cases}\tag{3.1}$$

This bound is always satisfied for fixed σ, λ, but eventually we will consider the postulate that K_1 is independent of σ and/or λ, which is not known.

Proposition 3.2. *Assume* (3.1) *and* $\Gamma^{(6)}(xxxyyy)\leqq 0$ *for a* $\lambda\varphi^4+\sigma\varphi^2$ *lattice field theory with* $\sigma>\sigma_c$. *Then for*

$$K=(K_1^3+K_1^4)K_2(|\cosh m_0+\langle P''\rangle|+1)$$

where K_2 *is a universal constant and* K_1 *is defined by* (3.1),

$$\Pi(x_1,p=0)\leqq\begin{cases}\lambda^2K|x_1|^{-2d+7}e^{-3m|x_1|} & \text{for}\quad |x_1|\leqq m^{-1}\\ \lambda^2Km^{2d-7}e^{-3m|x_1|} & \text{for}\quad |x_1|\geqq m^{-1},\end{cases}\tag{3.2}$$

Proof. As in [19], we expand Π in terms of connected and one particle irreducible subdiagrams. Negative terms, including $\Gamma^{(6)}$ can be omitted, yielding the bound

$$\Pi(x_1,p=0)\leqq\sum_x\lambda^2\left\{6G^{(2)}(x)^3-36\sum_{z,z'}G^{(4)}(xx0z)\Gamma^{(2)}(z-z')G^{(4)}(z'00x)\right\}.\tag{3.3}$$

By Lebowitz' inequality, $G^4\leqq 0$. Let p_j be the momentum variable conjugate to $z_j-z'_j$. By Proposition 2.4, applied with $p''=(p_2,\dots,p_d)=0$, the terms with $z_1-z'_1\neq 0$ may be omitted. Having done so, we apply Proposition 2.4 again with $z_1-z'_1=0$, $p''=(p_3,\dots,p_d)=0$; we omit terms with $z_2\neq z'_2$. Continuing in this fashion we have

$$\Pi(x_1,p=0)\leqq\sum_x\lambda^2\left\{6G^{(2)}(x)^3+36(2d+1+\langle P''\rangle)\sum_x G^{(4)}(xx0z)G^{(4)}(z00x)\right\},$$

where Proposition 2.7 provides the lower bound on $\Gamma^{(2)}(0)$. Lebowitz' inequality bounds the $G^{(4)}$ factors above, so that

$$\begin{aligned}\Pi(x_1,p=0)\leqq\sum_x\lambda^2\{&6G^{(2)}(x)^3\\ &+144(2d-1+\langle P''\rangle)G^{(2)}(x)^2(G^{(2)}*G^{(2)})(x)\}.\end{aligned}\tag{3.4}$$

Substitution of (3.1) yields (3.2).

4. Existence of Continuum φ_d^4 Fields and the Ising Scaling Limit

The original existence, regularity and Lorentz covariance theorems for φ_2^4 fields were given in [10–12, 15, 2]. A survey of this work, including the contributions of Nelson, Segal and others may be found in [13, 14]. The remaining two Wigthman axioms — uniqueness of the vacuum and its invariance under Lorentz rotations — were verified in [23] for weak coupling, and in [8] for arbitrary coupling. The completion of the φ_3^4 existence program has recently been achieved, see [4, 28] for weak coupling and [5, 7] for arbitrary coupling. The original nonuniqueness proof for relativistic fields was given in [25].

In this section we consider the continuum limit of lattice φ_d^4 fields.

Theorem 4.1. *Assume $\Gamma^{(6)}(xxxyyy)\leqq 0$ for all $\sigma>\sigma_c(\lambda)$ and all λ for $\lambda\varphi^4+\sigma\varphi^2$ lattice fields. Choose a mass $m>0$, and for each choice of lattice spacing $\varepsilon>0$, choose a bare charge $\lambda(\varepsilon)$. Then the limit $\varepsilon\to 0$ defines (through convergent subsequences) a continuum field theory limit of the $\lambda(\varepsilon)\varphi^2+\sigma\varphi^2$ ε-lattice theories. This limit has mass $m>0$ and satisfies the Wightman-Osterwalder-Schrader axioms with the possible exception of Lorentz invariance.*

Proof. By [1, 30], there is a $\sigma=\sigma(\varepsilon)>\sigma_c(\lambda(\varepsilon))$ which gives mass $m>0$ to the ε-lattice field theory. By [17], existence of the limit (through subsequences) follows from a uniform bound on the two point function. As in [19], the bound on the *CDD* radius gives the required bound on the wave function renormalized two point function. Because of the wave function renormalization, the limit theory has a mass m. This completes the proof.

A portion of this proof extends to Ising models.

Proposition 4.2. *Assume $\bar{m}$ (the CDD radius) $\geqq(1+\delta)m$, $\delta>0$, for $T>T_c$ in the d-dimensional Ising model. Then the scaling limit exists and defines a field theory in the sense specified in Theorem 4.1.*

Theorem 4.3. *Assume $\Gamma^{(6)}(xxxyyy)\leqq 0$ for $\lambda\varphi^4+\sigma\varphi^2$ fields with mass m fixed, positive and λ sufficiently large ($\sigma>\sigma_c$). Then the Ising model scale limit exists.*

Proof. By [30], the Schwinger functions of the lattice field theory converge, as $\lambda\to\infty$ to the Schwinger functions of the Ising model. Thus $G^{(2)\sim}(p)$ and $\Gamma^{(2)\sim}(p)$ converge for real p. In fact from (2.3), (2.7) and $m>0$ we have a uniform bound on $\int d\varrho(a,p)/a$ which implies pointwise convergence of $G^{(2)\sim}(p)$ for real p. As in the proof of Proposition 2.2, $G^{(2)\sim}(p)$ is bounded away from zero, so $\Gamma^{(2)\sim}(p)$ converges pointwise and is bounded also. However the remark following Proposition 2.3 yields bounds on α, $|\beta|$ and $\int dv/(1+a^2)$, which by Propositions 2.3 and 3.1 imply that $\Gamma^{(2)\sim}(p_1,0)$ is analytic up to $|\mathrm{Im}\,p_1|<3m$, and bounded uniformly in λ. Hence the same is true for the $\Gamma^{(2)\sim}$ of the Ising model. Thus $\bar{m}\geqq 3m$. This statement is uniform in m, and existence of the scaling limit follows from Proposition 4.2.

5. Critical φ^4 Lattice Theories, $d\geqq 6$

Theorem 5.1. *Consider a $\lambda\varphi^4+\sigma\varphi^2$ lattice field theory with $\sigma>\sigma_c(\lambda)$ and $\Gamma^6(xxxyyy)\leqq 0$. Then $Z\neq 0$. For $d\geqq 6$, the bound on Z^{-1} is independent of $\sigma>\sigma_c(\lambda)$ and λ for λ and m bounded.*

Proof. $Z>0$ because $\Gamma^{(6)}\leqq 0$ implies an upper mass gap. To establish the uniform bound on Z^{-1} for $d\geqq 6$, we use (2.12). By the remark following Proposition 2.3, it is sufficient to estimate the small a behavior of the dv integral, which we do using (2.11), [9] and (3.4). From (2,12),

$$Z^{-1}=\text{const.}+\int_{a=1-\cosh m}^{\infty}\frac{dv}{(a+1-\cosh m)^2}\leqq\text{const.}+\text{const.}\int a^{-2}dv$$
$$\leqq\text{const.}\left(1-\left(\frac{\partial}{\partial p_1}\right)^2\Gamma(p)|_{p=0}\right).$$

(Note in general $\partial/\partial p_1^2)\neq(\partial/\partial p_1)^2$). By (2.13)–(2.14),

$$Z^{-1}\leqq\text{const.}\left(1-\left(\frac{\partial}{\partial p_1}\right)^2\Pi\Gamma(p)|_{p=0}\right)$$

and by (3.4)

$$\begin{aligned}Z^{-1}&\leqq\text{const.}\left(1+\sum_x x_1^2\Pi(x)\right)\\&\leqq\text{const.}\left[1+\sum_x x_1^2(G^{(2)}(x)^3+G^{(2)}(x)^2G^{(2)}*G^{(2)}(x))\right]\qquad(5.1)\\&\leqq\text{const.}\left[1-\left(\frac{\partial}{\partial p_1}\right)^2(G^{(2)\sim}*G^{(2)\sim}*G^{(2)\sim}+G^{(2)\sim}*G^{(2)\sim}*(G^{(2)\sim})^2)\right]\Bigg|_{p=0}.\end{aligned}$$

The point to these manipulations is to replace a derivative of Γ by a derivative of the upper bound on Γ. By the Herglotz property, we are again able to transfer derivatives to the upper bound (2.11) as follows. For simplicity, we consider only the final (and worst) term above. It is

$$-\left(\frac{\partial}{\partial p_1}\right)^2\int(\cos(p-k)^2+a)^{-1}(\cos(k-l)^2+b)^{-1}(\cos(l-q)^2+c)^{-1}(\cos(l-q)^2+e)^{-1}$$
$$\times\, d\varrho(a)d\varrho(b)d\varrho(c)d\varrho(e)dkdldq|_{p=0}.$$

Without loss of generality, we restrict all $d\varrho(a)\ldots d\varrho(e)$ integrations to the interval $[0,1]$. By (2.11), it follows that

$$\int_0^1 d\varrho(a,0)\leqq\text{const.}$$

and so we may replace the $d\varrho$ integrations by $\sup_{a,b,c,e}$. This supremum is evaluated as $0=a=b=c=e$. Thus we have accomplished our goal. The derivative is placed on the upper bound, in this case the zero mass free lattice field. It is convenient to return to x-space, where we find

$$Z^{-1}\leqq\text{const.}\left[1+\sum_x x_1^2\left(x^{-3d+6}+\sum_y(x-y)^{-d+2}y^{-d+2}\right)x^{-2d+4}\right]$$

The theorem follows from this bound.

Theorem 5.2. *Under the hypothesis of Theorem 5.1, and the construction of Theorem 4.1, the two point function is free ($\Pi=0$) at momentum $\mathbf{p}=0$, in the limit $\varepsilon\to 0$.*

Remark. If $\lambda=\lambda(\varepsilon)$ is the bare charge ($\varepsilon=$lattice spacing) then $\lambda(\varepsilon)\leqq\varepsilon^{d-4}$ in the above case.

Corollary 5.3. *Suppose in addition that the limit $\varepsilon\to 0$ above is Euclidean invariant. Then it is a free scalar field of mass m.*

Proof. The two point function is free by the theorem, and the theory as a whole is free by [35, p. 163].

Proof of Theorem 5.2. Let $\Gamma^{\sim}_{\lambda,m,\varepsilon}(p)$ be the inverse propagator for the theory with bare charge λ, mass m and lattice spacing ε. By scaling,

$$\Gamma_{s^{d-4}\lambda,s^{-1}m,s\varepsilon}(p)=s^{-2}\Gamma_{\lambda,m,\varepsilon}(sp)\,. \tag{5.2}$$

Here we choose $\varepsilon=1$, $s\to 0$, and

$$s^{-1}m=\text{const.},\quad \text{i.e.}\quad m=\text{const.}\, s\to 0$$

$$s^{d-4}\lambda=\lambda(s)\leqq O(s^{d-4})$$

so that the λ on the right side of (5.2) is bounded, but possibly s-dependent as $s\to 0$. Now $\Gamma_{\lambda,m,\varepsilon}(p_1,0)$ has a zero at $\cos p=\cosh m$ and as we have just seen, bounded first and second derivatives with respect to p_1 evaluated at $p=0$, $p_1=0(m)$. Furthermore the first derivative vanishes at $p_1=0$ because of the $p_1\to -p_1$ symmetry. The same method bounds one more p_1 derivative, and

$$s^{-2}\Gamma_{\lambda,m,\varepsilon}(sp)|_{p=0}=s^{-2}(Z^{-1}(\cosh m-\cos sp_1)+O(s^3p_1^3))\to Z^{-1}(p_1^2+\text{const.})$$

as $s\to 0$. This completes the proof.

Remark. The vanishing of Γ on the mass shell gives a cancellation which formally allows $d\geqq 5$ in place of $d\geqq 6$ above.

6. Upper Bounds on Critical Exponents

We give upper bounds on certain critical exponents in a single phase $\varphi^4+\sigma\varphi^2$ (lattice) quantum field model. In particular we study the anomalous dimension η, the exponent ζ for the field strength renormalization constant Z and the exponent ν for the physical mass. We choose $\sigma>\sigma_c$ (the single phase region) and consider the limit $\sigma\searrow\sigma_c$.

In this section we assume $\Gamma^{(6)}\leqq 0$, as well as the x-space upper bound

$$G^{(2)}(x,\sigma)\leqq K(\min\{|x|,m^{-1}\})^{-(d-2+\eta)}e^{-m|x|} \tag{6.1}$$

with K uniform in the allowed x, σ. We expect (6.1) to hold for lattice field theories. We then show $\eta=0=\zeta$ for $d\geqq 5$, complementing the results of §5.

Proposition 6.1. *For a pure phase $\lambda\varphi^4+\sigma\varphi^2$ (lattice) field theory with $m>0$, $Z>0$ and satisfying* (6.1), *we have*

$$0\leqq\eta\leqq\zeta/\nu\,. \tag{6.2}$$

If in addition $\Gamma^{(6)}\leqq 0$, then

$$\zeta/\nu\leqq 2\,. \tag{6.3}$$

Here ζ is the exponent for the field strength renormalization constant and ν is the exponent for the mass. Thus $Z \sim m^{\zeta/\nu}$ as $m \to 0$.

Proof. Using (2.12a) evaluated at $p_1 = 0$ gives

$$m^{-2} \leqq \text{const. } Z^{-1} G^{2\sim}(0, \sigma).$$

We bound $G^{2\sim}(0, \sigma)$ by integrating (6.1) to obtain

$$m^{-2} \leqq \text{const. } Z^{-1} m^{\eta - 2}.$$

Thus

$$m^{-\eta} Z \leqq \text{const.},$$

or

$$-\eta + \zeta/\nu \geqq 0.$$

Since the inequality $0 \leqq \eta$ is known we have established (6.2).

Assuming $\Gamma^{(6)} \leqq 0$, we may use (5.1). Then

$$m^{-\zeta/\nu} \sim Z^{-1} \leqq \text{const.} (1 + \sum x_1^2 \Pi(x)).$$

We show $\Sigma \Pi(x) \leqq$ const., from which we infer

$$1 + \sum x_1^2 \Pi(x) \leqq O(m^{-2}),$$

proving $\zeta/\nu \leqq 2$. The summability of $\Pi(x)$ follows from

$$\Sigma \Pi(x) = \Pi^\sim(0) = O(1) + \Gamma^\sim(0). \tag{6.4}$$

Consider σ in the interval (σ_c, σ_1), for $\sigma_1 > \sigma_c$. Since $\Gamma^\sim(0) = -G^{2\sim}(0)^{-1} \leqq 0$, we need only bound $\Gamma^\sim(0)$ from below. However $G^2(x, \sigma)$ is decreasing in σ (a consequence of the second Griffiths inequality). Thus for $\sigma > \sigma_c$

$$\sum_x G^2(x, \sigma) = -\Gamma^\sim(0, \sigma)^{-1}$$

is decreasing in σ and $\Gamma^\sim(0, \sigma)$ is also decreasing in σ. Thus $\Gamma^\sim(0, \sigma_c) \geqq \Gamma^\sim(0, \sigma_1)$.

Theorem 6.2. *Assume $\Gamma^{(6)} \leqq 0$ and (6.1) in a $\lambda\varphi^4 + \sigma\varphi^2$ lattice theory. Then*

$$\eta = \zeta/\nu = 0 \quad \textit{for} \quad d \geqq 5 \tag{6.5}$$

and for $d \leqq 4$

$$0 \leqq \eta \leqq 2(1 - d/5) = \begin{cases} .4 & d = 4 \\ .8 & d = 3. \\ 1.2 & d = 2 \end{cases} \tag{6.6}$$

Proof. We use the bound (5.1) for

$$\begin{aligned} m^{-\zeta/\nu} \sim Z^{-1} &\leqq \text{const.} \left(1 + \sum_x x_1^2 \Pi(x)\right) \\ &\leqq \text{const.} \left(1 + \sum_x x_1^2 [G(x)^3 + G(x)^2 (G * G)(x)]\right). \end{aligned} \tag{6.7}$$

Substituting (6.1) we find the G^3 term in (6.7) sums to

$$O(m^{-8-2d-3\eta}), \tag{6.8}$$

while the $G^2(G*G)$ term sums to

$$O(m^{-(10-2d-4\eta)}). \tag{6.9}$$

By Proposition 6.1, $2-\eta \geqq 0$ and

$$(10-2d-4\eta)-(8-2d-3\eta) \geqq 0.$$

Thus the $G^2(G*G)$ term in (6.7) dominates and by (6.2), (6.3), (6.7),

$$0 \leqq \eta \leqq \zeta/\nu \leqq 10-2d-4\eta. \tag{6.10}$$

For $d \geqq 5$,

$$0 \leqq \eta \leqq \zeta/\nu \leqq 10-2d \leqq 0,$$

proving (6.5). For $d \leqq 4$, (6.10) gives $5\eta \leqq 10-2d$, which is (6.6).

Remark. If the upper bound on $\Pi(x)$ in (3.3) could be improved to show that the $G^{(4)}\Gamma G^{(4)}$ term has the same long distance behavior as the $(G^{(2)})^3$ term, then the above bounds on η are improved to $\eta \leqq \zeta/\nu \leqq 8-2d-3\eta$. In this case $\eta=\zeta=0$ for $d \geqq 4$ (as expected), and for $d \leqq 4$, $\eta \leqq (2-d/2)$.

Acknowledgement. We thank T. Spencer and K. Symanzik for helpful conversations.

References

1. Baker, G.: Self-interacting Boson quantum field theory and the thermodynamic limit in d dimensions. J. Math. Phys. **16**, 1324—1346 (1975)
2. Cannon, J., Jaffe, A.: Lorentz covariance of the $\lambda(\varphi^4)_2$ quantum field theory. Commun. math. Phys. **17**, 261—321 (1970)
3. Feldman, J.: On the absence of bound states for $\lambda\varphi_2^4$ quantum field models without symmetry breaking. Canad. J. Phys. **52**, 1583—1587 (1974)
4. Feldman, J., Osterwalder, K.: The Wightman axioms and the mass gap for weakly coupled Φ_3^4 quantum field theories. Ann. Phys., to appear
5. Feldman, J., Osterwalder, K.: The construction of $\lambda(\varphi^4)_3$ quantum field models, presented at the International Colloquium on Mathematical Methods in Quantum Field Theory, Marseille, June 1975
6. Fortuin, C., Kastelyn, P., Ginibre, J.: Correlation inequalities on some partially ordered sets. Commun. math. Phys. **22**, 89—103 (1971)
7. Fröhlich, J.: To appear
8. Fröhlich, J., Simon, B.: Pure states for general $P(\varphi)_2$ theories: construction, regularity, and variational equality, preprint, 1976
9. Fröhlich, J., Simon, B., Spencer, T.: Infrared bounds, phase transitions and continuous symmetry breaking. Commun. math. Phys. **50**, 79—85 (1976)
10. Glimm, J., Jaffe, A.: A $\lambda\varphi^4$ field theory without cutoffs I. Phys. Rev. **176**, 1945—1951 (1968)
11. Glimm, J., Jaffe, A.: The $\lambda(\varphi^4)_2$ quantum field theory without cutoffs. II. The field operators and the approximate Hamiltonian. Ann. Math. **91**, 362—401 (1970)
12. Glimm, J., Jaffe, A.: The $\lambda(\varphi^4)_2$ quantum field theory without cutoffs. III. The physical vacuum. Acta Math. **125**, 203—267 (1970)
13. Glimm, J., Jaffe, A.: Quantum field models. In: Statistical mechanics (C. DeWitt, R. Stora, eds.), 1970 Les Houches Lectures. New York: Gordon and Breach Science Publishers 1971

14. Glimm, J., Jaffe, A.: Boson quantum field models. In: Mathematics of contemporary physics. London Mathematical Society (R. Streater ed.). New York-London: Academic Press 1972
15. Glimm, J., Jaffe, A.: The $\lambda\varphi_2^4$ quantum field theory without cutoffs. IV. Pérturbations of the Hamiltonian. J. Math. Phys. **13**, 1568—1584 (1972)
16. Glimm, J., Jaffe, A.: The φ_2^4 quantum field model in the single phase region: Differentiability of the mass and bounds on critical exponents. Phys. Rev. D **10**, 536—539 (1974)
17. Glimm, J., Jaffe, A.: A remark on the existence of φ_4^4. Phys. Rev. Letters **33**, 440—442 (1974)
18. Glimm, J., Jaffe, A.: On the approach to the critical point. Ann. de l'Inst. Henri Poincaré **22**, 13—26 (1975)
19. Glimm, J., Jaffe, A.: Three particle structure of φ^4 interactions and the infinite scaling limit. Phys. Rev. D **11**, 2816—2827 (1975)
20. Glimm, J., Jaffe, A.: Particles and bound states and progress toward unitarity and scaling. Presented at the International Symposium on Mathematical Problems in Theoretical Physics. Kyoto, January 23—29, 1975
21. Glimm, J., Jaffe, A.: Critical problems in quantum fields, presented at the International Colloquium on Mathematical Methods of Quantum Field Theory. Marseille, June 1975
22. Glimm, J., Jaffe, A.: Critical exponents and renormalization in the φ^4 scaling limit, presented at the Conference on Quantum Dynamics: Models and Mathematics. Bielefeld, September 1975
23. Glimm, J., Jaffe, A., Spencer, T.: The Wightman axioms and particle structure in the $P(\varphi)_2$ quantum field model. Ann. Math. **100**, 585—632 (1974)
24. Glimm, J., Jaffe, A., Spencer, T.: The particle structure of the weakly coupled $P(\varphi)_2$ model and other applications of high temperature expansions, Part I: Physics of quantum field models; Part II: The cluster expansion. In: Constructive quantum field theory (G. Velo, A. S. Wightman eds.), 1973 Erice Lectures, Lecture Notes in Physics, Vol. 25. Berlin-Heidelberg-New York: Springer 1973
25. Glimm, J., Jaffe, A., Spencer, T.: Phase transitions for φ_2^4 quantum fields. Commun. math. Phys. **45**, 293—320 (1975)
26. Isaacson, D.: The critical behavior of the anharmonic oscillator. Commun. math. Phys., to appear
27. McBryan, O., Rosen, J.: Existence of the critical point in φ^4 field theory. Commun. math. Phys., to appear
28. Magnen, J., Sénéor, R.: The infinite volume limit of the φ_3^4 model. Ann. l'Inst. H. Poincaré, to appear
29. Marcheson, D.: Private communication
30. Rosen, J.: Mass Renormalization for the $\lambda\varphi^4$ Euclidean lattice field theory, preprint, 1975
31. Rosen, J.: The Ising model limit of φ^4 lattice fields, preprint, 1976
32. Schrader, R.: A possible constructive approach to φ_4^4. Commun. math. Phys. **49**, 131—154 (1976)
33. Schrader, R.: A possible constructive approach to φ_4^4 II, preprint, 1976
34. Spencer, T.: The absence of even bound states for $\lambda(\varphi^4)_2$. Commun. math. Phys. **39**, 77—79 (1974)
35. Streater, R., Wightman, A. S.: PCT, spin and statistics and all that. New York: Benjamin 1964

Communicated by K. Hepp

Received May 11, 1976

The Resummation of One Particle Lines*

James Glimm[1] and Arthur Jaffe[2]

[1] The Rockefeller University, New York, New York 10021, USA
[2] Harvard University, Cambridge, Massachusetts 02138, USA

Abstract. We propose a partial resummation for a weak coupling cluster expansion. The resummation gives one particle lines with in/out field propagators. We give a Bethe-Salpeter equation in which one particle subtractions are defined using physical one particle states. By these methods, we show that $P(\phi)_2$ quantum fields in the weak coupling region have only isolated bound state spectrum below the $2m$ threshold. Here P is not restricted to be even.

1. Introduction

We present a new method for the study of mass spectrum, asymptotic completeness and related questions. The method uses exact subtractions of the physical one particle states; technically, it is based on the n-particle cluster expansion [8], and a (new)partial resummation which identifies the physical one particle lines in closed form.

Previous work [3, 4, 11, 12] on spectral properties of $P(\phi)_2$ quantum fields was based on the Euclidean program of defining n-particle irreducible amplitudes [13]. In this program, the vacuum subtractions are performed exactly, while one and higher particle subtractions are replaced by an orthogonalization of the polynomials in the Euclidean field ϕ. Thus in this program, a j-particle irreducible n-point function is defined by projecting $\phi(x_1)\phi(x_2)\ldots\phi(x_n)$ onto the orthogonal complement, in the Euclidean Hilbert space $\mathscr{E}\equiv L_2(\mathscr{S}',d\mu)$, of the subspace spanned by the vectors

$$\phi(y_1)\ldots\phi(y_l), l\leqq j, y_\nu\in\mathbb{R}^2. \tag{1.1}$$

While the resulting subtraction agrees with the subtraction of (physical) j-particle intermediate states for $j=0$, it does not agree in general, and in particular if bound

* Supported in part by the National Science Foundation under Grants PHY 78-08066 and PHY 77-18762. Both authors thank the I.H.E.S., Bures-sur-Yvette, and A. J. thanks C.E.N., Saclay, for their hospitality.

states occur. For $j \geqq 1$ spurious zeros and poles (CDD poles) can be introduced by these subtractions.

The Euclidean program has been successful when $j=0$ (the vacuum) was the only subtraction required. Within this limit, particles and bound states below the continuum threshold have been analyzed, and for even theories asymptotic completeness up to energies below the four particle threshold is known. Various extensions of this work have been given [1, 2, 7, 9]. This work makes it appears likely that the j particle Euclidean subtractions can be incorporated into the spectral analysis program.

We study here the simplest problem requiring a one particle subtraction: the analysis of the four point function up to energies below the three particle threshold, for a general (not necessarily even) interaction at weak coupling. These methods appear useful in the study of higher spectral properties in field theory, and presumably they have an analog in statistical mechanics.

In § 2, we define normal ordering: $\phi(x_1,0)\phi(x_2,0)$: of the square of the time $t=0$ field by subtraction of the physical one particle state. In other words, if $|\Omega\rangle$ is the vacuum state in the physical Hilbert space $\mathscr{H}$, then

$$:\phi^2:|\Omega\rangle \perp |\text{one particle states}\rangle . \tag{1.2}$$

The central object of the study in this paper is then

$$S_{0,T}=\langle\Omega| :\phi(x_1,0)\phi(x_2,0): :\phi(y_1,T)\phi(y_2,T):|\Omega\rangle , \tag{1.3}$$

where $\Omega\equiv 1$ is the Euclidean vacuum state, and the inner product is in the Euclidean Hilbert space $\mathscr{E}$. We let $\langle\cdot\rangle$ now denote an expectation in the Euclidean vacuum Ω. Let

$$\begin{aligned} G_{0,T}=&\langle\phi(x_1,0)\phi(y_1,T)\rangle\langle\phi(x_2,0)\phi(y_2,T)\rangle \\ &+\langle\phi(x_1,0)\phi(y_2,T)\rangle\langle\phi(x_2,0)\phi(y_1,T)\rangle . \end{aligned} \tag{1.4}$$

We replace the Euclidean Bethe – Salpeter equation of the form (in momentum space)

$$S^{\sim}=G^{\sim}+G^{\sim}K^{\sim}S^{\sim}, \tag{1.5}$$

by an equation of the form

$$S_{0,T}=A_{0,T}+\sum_{0\leqq j\leqq T} B_{0,T}S_{j,T} . \tag{1.6}$$

By summing over T, this leads to an equation

$$\sum_{0\leqq T} S_{0,T}=\sum_{0\leqq T} A_{0,T}+\Big(\sum_{0\leqq j} B_{0,j}\Big)\sum_{0\leqq T} S_{0,T} \tag{1.7}$$

with the solution

$$\sum_{0\leqq T} S_{0,T}=\Big(1-\sum_{0\leqq j} B_{0,j}\Big)^{-1}\sum_{0\leqq T} A_{0,T}, \tag{1.8}$$

since $S_{0,T}=S_{j,T+j}$, etc. We see that $\sum_{0\leqq j} B_{0,j}$ plays the role of $G^{\sim}K^{\sim}$ in the Bethe-Salpeter equation. As with $G^{\sim}K^{\sim}$, we achieve a factorization of $\sum_{0\leqq j} B_{0,j}$. The first factor contains the free field behavior of the four point function, and has a square root singularity with a two-sheeted analytic continuation. The first factor arises from two physical one particle lines, up to the time of their first interaction. The second ("amputated") factor is analytic up to energies less than $3(m-\varepsilon)$, and contains the effects of interaction between the one particle lines. Moreover $\sum_{0\leqq j} B_{0,j}$ is compact on a suitable space of functions, analytic in p. The spectral properties of the field theory, for $E\leqq 3(m-\varepsilon)$, follow from these facts.

2. One Particle Subtractions

In this section we review properties of one particle subtractions and establish notation. To simplify the formulas, we assume

$$\langle\Omega|\phi(x,t)|\Omega\rangle=0\,. \tag{2.1}$$

In other words, for a general theory, we make the substitution $\phi\to\hat{\phi}\equiv\phi-\langle\phi\rangle$. Let $|p\rangle$ be the one particle state of momentum p. with the normalization

$$\langle p|p'\rangle=(2\mu(p))^{-1}\delta(p-p')\,, \tag{2.2}$$

where $\mu(p)=(p^2+m^2)$ and m is the one particle mass. Then with

$$\begin{aligned}|x\rangle&=(2\pi)^{-1/2}\int e^{-ipx}|p\rangle dp\\ |f\rangle&=\int f(x)|x\rangle dx\,,\end{aligned} \tag{2.3}$$

we have

$$\langle x|y\rangle=(2\mu)^{-1}(x-y)\,;\langle f|g\rangle=\langle f,(2\mu)^{-1}g\rangle_{L_2}\,, \tag{2.4}$$

where $\mu=(-d^2/dx^2+m^2)^{1/2}$. The field strength renormalization constant Z is defined by

$$\langle y|Z^{-1/2}\phi(x,0)|\Omega\rangle=(2\mu)^{-1}(x-y),\qquad\text{or equivalently} \tag{2.5}$$

$$Z^{1/2}=(2\pi)^{1/2}e^{ipx}2\mu(p)\langle p|\phi(x)|\Omega\rangle\,. \tag{2.6}$$

By Lorentz invariance, Z is independent of x, y, p.

Let $M_0=|\Omega\rangle\langle\Omega|$ and let

$$M_1-M_0=\int|p\rangle\langle p|2\mu(p)dp \tag{2.7}$$

be the orthogonal projection onto one particle states. In general, M_a is the projection onto states of mass $M\equiv(H^2-P^2)^{1/2}\leqq am$. Then $M_0\phi\Omega=0$,

$$M_1\phi(f,0)|\Omega\rangle=Z^{1/2}|f\rangle\,, \tag{2.8}$$

and

$$\begin{aligned}e^{-tH}M_1\phi(f,0)|\Omega\rangle&=Z^{1/2}e^{-tH}|f\rangle\\ &=Z^{1/2}|e^{-t\mu}f\rangle=M_1\phi(e^{-t\mu}f,0)|\Omega\rangle\,.\end{aligned} \tag{2.9}$$

In weakly coupled $P(\phi)_2$ theories, as considered in this paper, $I-M_1$ projects onto states with energy greater than $2(m_0-\varepsilon)$, where m_0 is the bare mass. Thus

$$\|e^{-TH}\{e^{-H}\phi(f,0)|\Omega\rangle-\phi(e^{-\mu}f,0)|\Omega\rangle\}\| \leqq O(1)e^{-2(m_0-\varepsilon)T}. \tag{2.10}$$

We use this formula in § 5 to identify resummed one particle lines.

We define the (physical) normal ordering, so that $:\phi^2:|\Omega\rangle \perp M_1$. Let

$$\alpha_0(x-y)=\langle\Omega|\phi(x,0)\phi(y,0)|\Omega\rangle \tag{2.11}$$

$$\alpha_1(z-x,x-y)=Z^{-1/2}(2\mu_z)\langle z|\phi(x,0)\phi(y,0)|\Omega\rangle. \tag{2.12}$$

Then

$$:\phi(x,0)\phi(y,0):=\phi(x,0)\phi(y,0)-\alpha_0(x-y) -\int\alpha_1(z-x,x-y)\phi(z,0)dz. \tag{2.13}$$

We require for later use regularity properties of α_0 and α_1. Note α_0, as a truncated Schwinger function, is C^∞ except at $x-y=0$, where it has a *ln* singularity, and α_0 and all derivatives are exponentially decreasing as $x-y\to\infty$. The estimates on α_1 in this section are elementary. They will be improved in § 5. We work mainly in a space of fixed total momentum p, and so we seek a bound on

$$\begin{aligned}\tilde{\alpha}_1(p,x-y)&\equiv(2\pi)^{-1/2}\int e^{+ip(x+y-2z)/2}\alpha_1(z-x,x-y)dz\\ &=Z^{-1/2}2\mu(p)\langle p|\phi\left(\frac{x-y}{2},0\right)\phi\left(\frac{y-x}{2},0\right)|\Omega\rangle.\end{aligned} \tag{2.14}$$

Let $\mathscr{H}(p)$ denote the Hilbert space of fixed total momentum p. with inner product $\langle,\rangle_p$. It is convenient to exclude Ω from $\mathscr{H}(0)$, so that

$$\mathscr{H}\oplus\{\lambda\Omega\}=\int_\oplus\mathscr{H}(p)dp. \tag{2.15}$$

In general, vectors in $\mathscr{H}(p)$ are continuum states in $\mathscr{H}$. We can, however, given Φ, $\Psi\in\mathscr{H}-\{\lambda\Omega\}$ write a formula for $\langle\Phi(p),\Psi(p)\rangle_p$, namely

$$\langle\Phi(p),\Psi(p)\rangle_p=\langle\Phi,\Psi\rangle_p=(2\pi)^{-1/2}\int\langle\Phi,e^{-ipa}T_a\Psi\rangle_{\mathscr{H}}da, \tag{2.16}$$

where $T_a\equiv e^{iPa}$ is translation by a.

From the exponential cluster properties of the truncated Schwinger functions, we see that

$$\{\phi(x,0)\phi(y,0)-\alpha_0(x-y)\}\Omega \tag{2.17}$$

has finite norm in $\mathscr{H}(p)$, with a bound uniform in x,y. Optimal estimates give

$$\int f(x-y)\{\phi(x,0)\phi(y,0)-\alpha_0(x-y)\}\Omega d(x-y)=\theta \tag{2.18}$$

a finite norm in $\mathscr{H}(p)$ when f is in the Sobolev space H_{-1}, and in fact

$$\|\theta(p)\|_p^2\leqq\text{const.}\|\mu^{-1}f\|_{L_2}. \tag{2.19}$$

Such bounds are true more generally of vectors

$$\Psi=[\phi(x_1,t_1)\dots\phi(x_n,t_n)-\langle\phi(x_1,t_1)\dots\phi(x_n,t_n)\rangle]\Omega, \tag{2.20}$$

with all $t_j \geq 0$ and no two (x_j, t_j)'s coinciding. Such estimates are proved by integration by parts, cf. [5], to obtain optimal momentum space behavior (i.e. free field singularities at coinciding points in x space). Such vectors Ψ (when smoothed with test functions to become elements of $\mathscr{H}$) are dense in $\mathscr{H}$. Thus the corresponding vectors (2.20) are dense in $\mathscr{H}(p)$ for a.e. p. Furthermore they are continuous in p in the sense that $\langle \psi_1(p), \psi_2(p)\rangle_p$ is continuous in p.

With $f \in \mathscr{S}$, let

$$\theta_t = \phi(e^{t\mu} f, t)\Omega = \int \phi(x,t)(e^{t\mu} f)(x)dx\,\Omega. \tag{2.21}$$

A calculation with the Lehmann spectral formula shows that as $t \to \infty$, the vectors θ_t converge in $\mathscr{H}$ at the rate $O(e^{-t(m-\varepsilon)})$. The same is true, furthermore for $\theta_t(p)$ in $\mathscr{H}(p)$, with convergence uniform in p. The limit is

$$\lim_{t\to\infty} \theta_t = Z^{1/2}\Phi_{ex}(f)\Omega, \tag{2.22}$$

where the time zero asymptotic field Φ_{ex} is given by standard Haag-Ruelle scattering theory. We now use Φ to denote the real time (Minkowski) field, and Φ_{ex} its asymptotic limit. At time zero we identify the Euclidean and Minkowski fields $\Phi(x) \equiv \Phi(x,0) = \phi(x,0) \equiv \phi(x)$.

We claim that after mapping $\Phi_{ex}(f)\Omega$ into $\mathscr{H}(p_0)$ by (2.16), we may take the limit $f(p) \to \delta(p-p_0)$ in the test function. Thus we identify $\tilde{\Phi}_{ex}(p)\Omega\tilde{f}(p)$ as the component of $\Phi_{ex}(f)\Omega$ with momentum p.

To justify this exchange of limits first appeal to the uniformity of the convergence in (2.22) as $t \to \infty$ in each $\mathscr{H}(p)$. It follows that the inner product $\langle \Psi, \Phi_{ex}(f)\Omega\rangle_p$ is a continuous function of p. Take $\tilde{f}(p) = \tilde{f}_1(p)\tilde{f}_2(p)$. By (2.16)

$$\langle \Psi, \Phi_{ex}(f)\Omega\rangle_p = \langle \Psi, \Phi_{ex}(f_1)\Omega\rangle_p \tilde{f}_2(p). \tag{2.23}$$

Because of the continuity of the inner product in p, we can take the limit $\tilde{f}_2(p) \to \delta(p-p_0)$, where $\tilde{f}_1(p_0) = 1$.

Thus

$$\langle \Psi, \tilde{\Phi}_{ex}(p_0)\Omega\rangle_{\mathscr{H}} = \int \langle \Psi, \Phi_{ex}(f)\Omega\rangle_p dp \tag{2.24}$$

and we define $\tilde{\Phi}_{ex}(p_0)\Omega$ as an improper element of $\mathscr{H}$. The vector $\tilde{\Phi}_{ex}(p)\Omega$ has a finite inner product with all vectors Ψ of the form (2.20). Thus also by (2.23),

$$\langle \Psi, \tilde{\Phi}_{ex}(p_0)\Omega\rangle_p = \langle \Psi, \tilde{\Phi}_{ex}(p)\Omega\rangle_{\mathscr{H}}\delta(p-p_0). \tag{2.25}$$

Since the definition, $|p_0\rangle = \tilde{\Phi}_{ex}(p_0)\Omega$, it follows that $\langle \Psi | p_0\rangle$ is also well defined and bounded uniformly in (x_j, t_j) for all $t_j \geq 0$ and no coinciding points, unless $n=2$. In particular, since $\langle p_0 | \Omega\rangle = 0$, it follows that $\tilde{\alpha}_1(p, x-y)$ is bounded as a function of $x-y$. Furthermore, $\tilde{\alpha}_1(p, x-y)$ is Hölder continuous in $x-y$ with exponent less than 1/2.

In terms of the function spaces A_δ of § 6, $\tilde{\alpha}_1(p, \cdot) \in A_\delta$ for $\delta_1 \leq 3/2 - \varepsilon$, $\delta_2 < 0$. To obtain this improvement on the Hölder continuity above, we write

$$\begin{aligned}\langle p | \phi(x_1)\phi(x_2)\Omega\rangle &= e^{\mu(p)}\langle p | e^{-H}\phi(x_1)\phi(x_2)\Omega\rangle \\ &= e^{\mu(p)}\langle p | e^{-H}(I - M_0)\phi(x_1)\phi(x_2)\Omega\rangle,\end{aligned}$$

and the bound the norm of $e^{-H}(I-M_0)\Phi(x_1)\phi(x_2)\Omega$ in $\mathscr{H}(p)$. The improved smoothing comes from e^{-H} and can be exhibited by integration by parts formulas as above. Then (2.19) is replaced by $\|\mu^{-3/2+\varepsilon}f\|_{L_2}$.

3. Span of $:\Phi^n:$

Let

$$:\Phi^j(f,t):\equiv e^{itH}:\Phi(x_1)\dots\Phi(x_j):e^{-itH} \tag{3.1}$$

where $:\Phi(x_1)\dots\Phi(x_j):$ denotes normal ordering defined so that $:\Phi:=\Phi$, $M_0\Phi\Omega=0$, and

$$M_1:\Phi(x_1)\dots\Phi(x_j):\Omega=0, j\geqq 2. \tag{3.2}$$

Theorem 3.1. *Choose n integer* $\geqq 2$, $\varepsilon>0$, *and* $\lambda<\lambda(n,\varepsilon)$. *Then states of the form*

$$M_{n+1-\varepsilon}:\Phi^j(f,t):\Omega, \quad 2\leqq j\leqq n, \tag{3.3}$$

span $(M_{n+1-\varepsilon}-M_1)\mathscr{H}$. *Here f ranges over* $\mathscr{S}(\mathbb{R}^n)$, *and t belongs to an open interval* (a,b).

Remark. Taking $n=2$, the vectors $M_{3-\varepsilon}:\Phi^2:(f,t)\Omega$ span all bound states and two particle states with mass in the interval $(m,(3-\varepsilon)m)$. In other words, all states with mass in this interval occur in the four point function $S_{0,T}$.

Proof. Let E_a denote the projection onto states with energy $\leqq am$. Let $|f,t\rangle_j\equiv :\Phi^j:(f,t)\Omega$. We show that $E_{n+1-\varepsilon}|f,t\rangle_j, j=2,\dots,n$, span $E_{n+1-\varepsilon}(I-M_1)\mathscr{H}$. By Lorentz invariance, the states $M_{n+1-\varepsilon}|f,t\rangle_j$ then span $(M_{n+1-\varepsilon}-M_1)\mathscr{H}$. For states $E_a|f,t\rangle_j$ we note that the total momentum P satisfies

$$P^2\leqq(a^2-1)m^2. \tag{3.4}$$

Secondly we remark that $E_a|f,t\rangle_j$ is an entire function of t. Thus if $\langle\chi|E_a|f,t\rangle_j=0$, for $t\in(a,b)$, it follows by analytic continuation that $\langle\chi|E_a|f,it\rangle_j=0$. Thus it is sufficient to show that $E_{n+1-\varepsilon}|f,it\rangle_j, j=2,3,\dots,n, t\in\mathbb{Z}_+$ span $E_{n+1-\varepsilon}(I-E_1)\mathscr{H}$.

It has already been proved by the cluster expansion [8] that the states

$$E_{n+1-\varepsilon}\{\Omega,\Phi(f,t)\Omega,|f,it\rangle_j,2\leqq j\leqq n,t\in\mathbb{Z}_+\}, \tag{3.5}$$

span $E_{n+1-\varepsilon}\mathscr{H}$. In fact the same method [8], taking (3.4) into account, shows that $\Phi(f,0)\Omega$ can replace $\Phi(f,t)\Omega$ in (3.5). Thus a dense set of states χ in $E_{n+1-\varepsilon}\mathscr{H}$ have the form

$$\chi=E_{n+1-\varepsilon}\{\alpha\Omega+\phi(f)\Omega+\psi\} \tag{3.6}$$

where ψ is in the span of $|f,it\rangle$, $t\in\mathbb{Z}_+$ and $2\leqq j\leqq n$. In (3.6), we restrict f to have Fourier transform $f^\frown(p)$ vanishing unless $p^2\leqq[(n+1-\varepsilon)-1]m^2$ c.f. (3.4).

We take a sequence χ_n of such vectors, converging to a $\chi\in(1-M_1)\mathscr{H}$. Then

$$M_0\chi_n=\alpha_n\Omega\to M_0\chi=0$$

so that $\alpha_n \to 0$. Since the norms $f \to \|\phi(f)\Omega\|$ and $f \to \|M_1 \phi(f)\Omega\|$ are equivalent, and since

$$(M_1 - M_0)\chi_n = M_1 \phi(f_n)\Omega \to M_1 \chi = 0.$$

$\phi(f_n)\Omega \to 0$ also. Thus $\psi_n \to \chi$ and the proof is complete.

4. Integration by Parts

The basic expansion steps depend on integration by parts. Let

$$C_0 = (-\Delta + M_0^2)^{-1}(x, y) = (2\mu)^{-1}[\exp(-|x_0 - y_0|\mu)](\mathbf{x}, \mathbf{y}), \tag{4.1}$$

and let C be a covariance of the class considered in [8]. This class is generated from C_0 by two operations: (a) the block diagonalization

$$C \to \chi_X C \chi_X + \chi_{\mathbb{R}^2 \setminus X} C \chi_{\mathbb{R}^2 \setminus X} \tag{4.2}$$

for X a rectangle in $\mathbb{R}^2$, and (b) the convex sum

$$\{C_j\} \to \sum \beta_j C_j, \sum \beta_j = 1, \beta_j \geqq 0. \tag{4.3}$$

For this class of covariance operators, we define the normalized measure on $\mathscr{S}'(\mathbb{R}^2)$,

$$d\mu_C = \lim_{\Lambda \uparrow \mathbb{R}^2} Z(\Lambda)^{-1} \exp(-\textstyle\int_\Lambda \mathscr{P}(\phi(x))dx) d\mu_C, \tag{4.4}$$

with existence of the limit following from [8].

With $R(\phi)$ a polynomial in ϕ, the formula

$$\frac{DR(\phi)}{D\phi(y)} = \left[\frac{\partial}{\partial\phi(y)} - \mathscr{P}'(\phi(y))\right] R(\phi) \tag{4.5}$$

defines $D/D\phi(y)$ as a densely defined linear operator on $\mathscr{E}$. Integration by parts is the identity

$$\phi(x) = \int_{R^2} C(x, y) \frac{D}{D\phi(y)} dy, \tag{4.6}$$

valid when substituted in the integral

$$\int \phi R(\phi) d\mu_C. \tag{4.7}$$

We use (4.6) to decompose ϕ into a sum of "annihilation" and "creation" operators:

$$\phi(x) = a(x) + a^*(x), \tag{4.8}$$

$$a(x) = \int_{x_0 \leqq y_0} C(x, y) \frac{D}{D\phi(y)} dy; \; a(x)^* = \int_{x_0 \geqq y_0} C(x, y) \frac{D}{D\phi(y)} dy. \tag{4.9}$$

Next consider a time interval (s, t) without contours. This means

$$C(x, y) = C_0(x, y), s \leqq x_0, y_0 \leqq t.$$

Then for $x_0 \leqq z_0 \leqq y_0$

$$C(x,y) = \int_{z_0=\text{const}} C_0(x,z)(2\mu C_0)(z,y)d\mathbf{z},$$

which has the consequence

$$\int_{x_0 \leqq y_0 \leqq z_0} C(x,y)\frac{D}{D\phi(y)}dy = a(x) - (e^{-(z_0-x_0)\mu}a)(z_0,\mathbf{x})$$
$$= a(z_0,\mathbf{x})^* - (e^{-(z_0-x_0)\mu}a^*)(x). \tag{4.10}$$

These identities are used to write $\phi = a + a^*$ and to move a, a^* forward and backward in time. Normally a^* is moved to time $t = -\infty$, i.e. eliminated. Also $a = \phi - a^*$ can be rewritten in terms of ϕ.

Remaining expansion steps concern the insertion and removal of contours, achieved by variation of the parameters β_j in (4.3).

5. The One Particle Line

The main technical tool of this paper is the use of virtual contraction terms to allow the resummation of one particle lines. We explain these ideas here, and for clarity we give a very simple example, namely the evaluation of

$$\int \phi(f,0)R_T(\phi)d\mu = \langle \phi(f,0)R_T\rangle \tag{5.1}$$

with R_T localized at times $t \geqq T$. As an application, we obtain improved bounds on α_1.

The goal of our expansion is to write

$$\langle \phi(f,0)R_T\rangle = \langle \{\alpha + \phi(w,1)\}R_T\rangle + O(e^{-2(m_0-\varepsilon)T}), \tag{5.2}$$

where the right hand side is a partial resummation of the n-particle cluster expansion given in [8]. Having identified the resummation leading to (5.2), we can use (5.2) as an elementary expansion step in the definition of our modified Bethe-Salpeter equation (1.6). In terms of vectors in the physical Hilbert space $\mathscr{H}$, we identify (5.2). Let

$$|\chi\rangle \equiv \phi(f,0)|\Omega\rangle - \langle\Omega|\phi(f,0)|\Omega\rangle|\Omega\rangle - e^H\phi(e^{-\mu}f,0)|\Omega\rangle. \tag{5.3}$$

Then (2.10) says that for any $\theta = R_0|\Omega\rangle$,

$$|\langle \chi, e^{-TH}\theta\rangle| \leqq O(1)e^{-2(m_0-\varepsilon)T}, \tag{5.4}$$

and comparison with (5.2) shows that $w = e^{-\mu}f$, $\alpha = \langle\phi(f)\rangle$.

In order to evaluate (5.1), we use the 1-particle expansion in [8]. Writing $\phi = a + a^*$, we use (4.8), (4.10) to advance the annihilator in time to become a field ϕ at $t = 1$. We use (4.9) to contract a^*. New vertices from the $D/D\phi$ terms are removed by insertion and removal of contours, as in [8]. We divide the expansion terms into two sets as follows: Terms in the first set have $y_0 < T$ (where y_0 is defined in [8, pp. 609–610]). These terms have the form $\langle QR\rangle$, where

$$Q = \sum_j Q_j, \quad Q_j = \alpha(j) + \phi(w(j),t_j), \tag{5.5}$$

where t_j is an integer time, $t_j<T$. The second set of terms has $y_0=+\infty$, and contains contradictions to times $\geqq T$. These are all remainder terms, and the sum of the norms of remainder terms is bounded by $O(1)\|\phi(f)\|\,\|R\|e^{-2(m_0-\varepsilon)T}$ with norms defined in [8]. Thus we write

$$\langle\phi R\rangle=\langle QR\rangle+\text{Remainder}. \tag{5.6}$$

Furthermore, we choose $f(x)$ with $f^{\sim}$ supported in

$$\{p:|p|\leqq\sqrt{3}(m-\varepsilon)\}=\{p:\mu(p)\leqq 2(m-\varepsilon')\}. \tag{5.7}$$

Then the terms $\phi(w(j),t_j)$ in (5.5) can be returned to time $t=1$.

In summary, we have for $n=1$

$$\langle\phi(f,0)R\rangle=\alpha_T\langle R\rangle+\langle\phi(w_T,1)R\rangle+O(e^{-2(m_0-\varepsilon)T}). \tag{5.8}$$

In (5.8), α_T and $\phi(w_T,1)$ are the linear and constant terms in Q. Each term in Q is independent of T. However writing Q as a sum of terms $Q=\sum_i Q_i$, the index set $\{i\}$ depends on T. Thus α_T and w_T also depend on T. In order to cure this defect, we add and subtract certain terms on the right side of (5.8). To identify the missing terms we consider the case $T=\infty$, when (5.8) becomes

$$\langle\phi(f,0)R\rangle=\alpha\langle R\rangle+\langle\phi(w,1)R\rangle. \tag{5.9}$$

For $T=\infty$, there are no error terms, but there are terms in α resulting from contraction of $\phi(f,0)$ to the vacuum in the region $t\geqq T$. We can think of these terms as contractions to the vacuum in the region blocked by R, and we call these terms, virtual contractions of ϕ to R.

The sum of the virtual contraction terms is absolutely convergent and bounded by $O(\exp-2(m-\varepsilon)T)$, because this sum is contained in the expansion of (5.9) to which the estimates of [8] also apply. Thus

$$\alpha pR\rangle-\alpha_T\langle R\rangle=\sum\text{ virtual terms} \tag{5.10}$$

$$\langle\phi(w,1)R\rangle-\langle\phi(w_T)R\rangle=\sum\text{ virtual terms} \tag{5.11}$$

where in (5.10) we sum over all virtual terms which are independent of ϕ, and in (5.11) the sums runs over all terms linear in ϕ. Note that the estimates of [6] apply to operator kernels, such as w_T, $w_{T=\infty}$ and $w_{T=\infty}-w_T$, as well as to expectation values such as $\langle\phi(w_{T=\infty},1)\rangle$. In the norm of [8],

$$\|w_{T=\infty}-w_T\|\leqq O(e^{-2(m-\varepsilon)T}). \tag{5.12}$$

We add and subtract (5.10) and (5.11) to the right side of (5.5). The addition modifies α and w, while the subtraction is absorbed into the error term $O(e^{-2(m_0-\varepsilon)T})$. This yields

$$\langle r(f,0)R\rangle=\alpha\langle R\rangle+\langle\phi(w,1)R\rangle+O(e^{-2(m-\varepsilon)T}), \tag{5.13}$$

where now $\alpha=\alpha_{T-\infty}$ and $w=w_{T=\infty}$ are independent of T.

Now we identify $\alpha=\langle\phi(f,0)\rangle$, and with the convention (2.1), $\alpha=0$. Comparison of (5.2) and (5.4) shows that $w=e^{-\mu}f$.

Thus in (5.13), we have achieved an explicit resummation of the one particle line, with the exact physical mass. The error term is an absolutely convergent series of direct and virtual contraction terms, with two particle decay $\sim e^{-2(m-\varepsilon)T}$.

To conclude the description of the expansion, we correct an error in [8]. In order to return field operator $\phi(f,t)$ to earlier times $s<t$, we must be dealing with translation invariant, i.e., momentum conserving expansion operations. The expansion is based on an integer lattice, artificially introduced in position space. Because of the translation invariance of the original theory, the expansion is term by term invariant under integer (lattice preserving) translations, but not under arbitrary space translations. To recover full translation invariance, we integrate each term over relative lattice positions, generated by the translations

$$\mathbb{Z}\to\mathbb{Z}+\theta,\theta\in(0,1]. \tag{5.14}$$

This integration is done at the end of each expansion step containing no contours, i.e. before moving $\phi(f,t)$ backward in time.

The norms of [8] are not invariant under translation by non-integer θ, as in (5.14), but they are easily replaced by equivalent norms, which are θ-invariant. This is done explicitly in § 6. The expansion remains norm decreasing in these equivalent norms. The extra expansion step,

$$F(\theta,\ldots)\to\int_0^1 F(\theta,\ldots)d\theta$$

is norm preserving, in the new translation invariant, norms. Thus the modified expansion is norm decreasing, and convergent, as before.

The improved estimates on $\tilde{\alpha}_1$ give exponential decrease, in addition to the local regularity established in § 2. In terms of the space A_δ of § 6, we state our result as

Theorem 5.1. *As a function of* $x_{\rm rel}=(x-y)/2$, $\tilde{\alpha}_1(p,x_{\rm rel})\in A_\delta$ *for* $\delta_1<2-\varepsilon$, $\delta_2>0$ *sufficiently small. The norm of* $\tilde{\alpha}_1$ *goes to zero as* $\lambda\to 0$.

We give part of the proof here, and complete the argument in § 9, 10. For $x_{\rm rel}$ in a bounded set, the analysis of § 2 is sufficient. Thus we let $j\leqq x_{\rm rel}\leqq j+1$ and consider j large. As in § 2,

$$\| |(\Phi_{ex}(f)-Z^{-1/2}e^{-jH}(\phi(e^{j\mu}f)))\Omega> \| \leqq O(e^{-(m-\varepsilon)j})$$

by a calculation with the Lehmann spectral formula, and the same bound holds in each space $\mathscr{H}(p)$ of total momentum p. If χ_j represents a smoothed version of the characteristic function of $[j,j+1]$, then we see as in § 2 that for $\chi_j\equiv\chi_j(x_{\rm rel})$,

$$\chi_j\{\tilde{\alpha}_1(p_{\rm tot},x_{\rm rel})-\langle Z^{-1/2}e^{-jH}\phi(e^{j\mu}f)\Omega,\phi^2(x_{\rm rel},p_{\rm tot})\Omega\rangle_p\} \tag{5.15}$$

has an $H_{3/2-\varepsilon}$ norm bounded by $e^{-(m-\varepsilon)j}$, and thus finite norm in A_δ, which is moreover summable in j. By (5.2), we can write

$$\begin{aligned}&\langle\phi(e^{l\mu}f,0),\textstyle\int\phi(x_1,l)\phi(x_2,l)g(x)dx\rangle\\&\quad=\langle\phi(e^{(l-1)\mu}f,0),\textstyle\int\phi(x_1,l-1)\phi(x_2,l-1)g\,dx\rangle\\&\qquad+O(e^{-2(m_0-\varepsilon)l})e^{-\delta j}\|e^{l\mu}f\|\;\|g\|,\end{aligned} \tag{5.16}$$

for $1 \leqq l \leqq j$. The exponent proportional to j will be explained below. The norms in the error term are from [8], and the error term can be bounded by $e^{-\varepsilon' j}\|f\|\,\|g\|$, provided p_{tot}, i.e. the support of f, is restricted to $|p_{\text{tot}}| \leqq \sqrt{3}m_0(1-\varepsilon'')$. We now iterate this identity, and sum over $0 \leqq l \leqq j$ to obtain

$$\begin{aligned}&\langle \phi(e^{ju}f, 0), \int \phi(x_1, j)\,\phi(x_2, j) g(x) dx\rangle \\ &\quad = \langle \phi(f) \int \phi(x_1)\,\phi(x_2) g(x) dx\rangle + O(e^{-\varepsilon' j})\,\|f\|\,\|g\|\,. \end{aligned} \tag{5.17}$$

The first term has exponential decrease in x_{rel}. In fact because of the normalization $\langle\phi\rangle = 0$, the ordinary and truncated three point functions coincide. Thus the first term, at constant momentum p, belongs to A_δ in the variable x_{rel}.

The error terms in (5.16) result from direct and virtual contraction to the observables $\phi(x_1, l-1)$ and $\phi(x_2, l-1)$. However terms with contraction to only one of these two observables can be resummed, and yield zero, again because of the normalization $\langle\phi\rangle = 0$. The terms with direct or virtual contraction to both ϕ's at time $l-1$ have extra decay, determined by $|x_{\text{rel}}|$, i.e., j. The relation of the norms of [8], as in (5.17), to the norms in A_δ is analyzed in §9, 10, and will complete the proof.

6. Function Spaces

In this section we define function spaces suitable for the analysis of our expansion. First we introduce spaces necessary to obtain a translation invariant expansion. Let $\{\Delta_i\}$, $\{\Delta_i'\}$ each be disjoint covers of $\mathbb{R}^n$, and define

$$\|f\| = \sum_i \|f\|_{L_2(\Delta_i)} \tag{6.1}$$

$$\|f\|' = \sum_i \|f\|_{L_2(\Delta_i')}\,. \tag{6.2}$$

We assume that each Δ_i is contained in the union of at most K of the Δ_i''s, and conversely.

Proposition 6.1. *The norms* (6.1) *and* (6.2) *are equivalent, and*

$$\frac{1}{K}\|f\|' \leqq \|f\| \leqq K\|f\|'\,. \tag{6.3}$$

In addition, if $f_a(x) = f(x+a)$, then for all $a \in \mathbb{R}^n$, let

$$|||f||| = \int_\Delta \|f_a\|\, da\,, \tag{6.4}$$

where Δ is a unit n-cube. We assume $(\Delta_i + a) \cap \Delta_j \neq \emptyset$ for at most K j's, for each fixed choice of i and a.

Proposition 6.2. *The norms* (6.1) *and* (6.4) *are equivalent.*

As an application, we change norms in the expansion of [8]. In place of the norms (2.43) of [8], we replace $\|w(\alpha, u, s, y_0, \cdot)\|_{L_2}$ by

$$\int_0^1 \|w(\alpha, u, s, y_0, \cdot + a\theta)\|_{L_2} d\theta, \qquad a = (1, 1, \ldots, 1)\,. \tag{6.5}$$

for a localized kernel w. As described in § 5, the integration over θ arises in the expansion from integration over choices of the position of the space lattice. The step of averaging the expansion over θ preserves the sum of the (averaged) norms of elementary expansion terms.

A second application involves separation of center of mass coordinates. For a function $f(x_1, \ldots, x_j)$ of j variables, we introduce

$$x_{\text{tot}} = \frac{1}{j} \sum_{k=1}^{j} x_k \tag{6.6}$$

$$x_{\text{rel},k} = x_k - x_{\text{tot}}.$$

For $i \in \mathbb{Z}^j$ we define for $I_k = [k, k+1]$,

$$\Delta_i = I_{i_1}(x_{\text{rel},i_1}) \ldots I_{i_{j-1}}(r_{\text{rel},j-1}) I_{i_j(x_{\text{tot}})}. \tag{6.7}$$

With these new Δ's in (6.7), we obtain a new set of equivalent norms. A translation invariant kernel $k(x; y)$, acts on spaces of definite total momentum as follows:

$$k_{\text{rel}}(x_{\text{rel}}; y_{\text{rel}}; p_{\text{tot}}) = (2\pi)^{-1/2} \int e^{ip_{\text{tot}}(y_{\text{tot}} - x_{\text{tot}})} k(x; y) dy_{\text{tot}}. \tag{6.8}$$

In the special case $j = 2$, we deal with kernels for two particle scattering. These are kernels of integral operators from functions of a single relative variable $\frac{1}{2}(y_1 - y_2)$ to functions of a single relative variable $\frac{1}{2}(x_1 - x_2)$. Thus we introduce Hilbert spaces of functions of one variable on which these operators act. We use the space A_δ of symmetric functions with δ_1 fractional L_2 derivatives and with Fourier transforms analytic in a strip $|\text{Im}\, p| < \delta_2 \leqq 4m$, Our choice of the space A_δ is motivated by [12]. Related spaces are considered in [10].

We let $\delta = (\delta_1, \delta_2)$, $|\delta_2| < 4\text{m}$, and let

$$K_\delta = e^{\delta_2 |x|} \mu^{\delta_1}, \tag{6.9}$$

where $\mu = (-d^2/dx^2 + (4m)^2)^{1/2}$. Let us define the inner product

$$\langle f, g \rangle_\delta = \langle K_\delta f, K_\delta g \rangle_{L_2}; \; \|f\|_\delta = \|K_\delta f\|. \tag{6.10}$$

Then A_δ is the Hilbert space C_0^∞ completed in the norm $\| \cdot \|_\delta$. In § 7 we show that K_δ^* may be used in place of K_δ in (6.10). Likewise, since

$$K_\delta^{-1} = K_{-\delta}^*, \tag{6.11}$$

we can define A_δ as the unitary image under $K_{-\delta}^*$ of $L_2(R)$,

$$K_{-\delta}^* : L_2(R) \to A_\delta. \tag{6.12}$$

By (6.11), it follows that the dual space of A_δ is

$$A_\delta^* = A_{-\delta}. \tag{6.13}$$

We say that $\delta > \delta'$ if $\delta > \delta_1'$ and $\delta_2 > \delta_2'$. We state properties of A_δ in three propositions.

Proposition 6.3. *Let $\delta \geqq \delta'$. Then*

$$A_\delta \subset A_{\delta'} \tag{6.14}$$

Let $I(\delta', \delta)$ denote the injection of A_δ into $A_{\delta'}$, for $\delta \geqq \delta'$,

$$I(\delta', \delta): A_{\delta'} \to A_{\delta'}. \tag{6.15}$$

Proposition 6.4. *Let $\delta > \delta'$.*

(i) *$I(\delta', \delta)$ is compact.*

(ii) *If $\delta > \delta' + (\frac{1}{2}, 0)$, then* (6.15) *is Hilbert Schmidt*

(iii) *If $\delta > \delta' + (1, 0)$, then* (6.15) *is trace class.*

We let A^0 denote the Banach space of functions f with norm

$$\|f\| = \sum_n \|f\|_{L_2(n, n+1)} < \infty, \tag{6.16}$$

Proposition 6.5. *Let $\delta_1 \geqq 0$, $\delta_2 > 0$. Then*

$$A_\delta \subset A^0 \subset A_{-\delta}. \tag{6.17}$$

Furthermore the injections $A_\delta \to A^0 \to A_{-\delta}$ are bounded.

Proofs. In the proofs of these three propositions, we refer ahead to § 7. The proof of Propositions 6.3–6.4 involve the study of the operator $K_{\delta'} K_\delta^{-1}$. In fact let $\{f_n\}$ denote a sequence in A_δ. Then

$$K_{\delta'} f_n = (K_{\delta'} K_\delta^{-1})(K_\delta f_n). \tag{6.18}$$

Thus if $\{f_n\}$ is bounded in A_δ, i.e. $\|K_\delta f_n\|_{L_2} \leqq$ const., it is bounded (compact) in $A_{\delta'}$ if $K_{\delta'} K_\delta^{-1}$ is bounded (compact). By Theorem 7.6,

$$K_{\delta'} K_\delta^{-1} = e^{\delta_2' |x|} \mu^{-(\delta_1 - \delta_1')} e^{-\delta_2 |x|} \tag{6.19}$$

is bounded for $\delta \geqq \delta'$ and compact for $\delta > \delta'$. Furthermore, (6.19) is Hilbert-Schmidt under condition (ii) and trace class under condition (iii).

We now proceed to Proposition 6.5. Using Proposition 6.3,

$$A_\delta \subset A_{0, \delta_2} \subset A_{0, -\delta_2} \subset A_{-\delta}.$$

Thus it is sufficient to establish (6.17) for the case $\delta_1 = 0$. Let δ now denote $\delta_2 > 0$, and let

$$M(\delta) = \sum_n e^{-\delta |n|}.$$

Then

$$\|f\|_{-\delta} = \|e^{-\delta |x|} f\|_{L_2} \leqq M(\delta) \sup_n \|f\|_{L_2(n, n+1)} \leqq M(\delta) \|f\|. \tag{6.20}$$

Also

$$\|f\|_{L_2(n, n+1)} \leqq e^{-\delta(|n| - 1)} \|e^{\delta |x|} f\|_{L_2(n, n+1)}.$$

Thus by the Schwarz inequality

$$\|f\| \leqq e^{\delta} M(2\delta)^{1/2} \Big(\sum_n \|e^{\delta|x|} f\|^2_{L_2(n,n+1)}\Big)^{1/2}$$
$$= e^{\delta} M(2\delta)^{1/2} \|f\|_{\delta}.$$

This completes the proof.

We end this section with a remark generalizing the consequences of (6.12) (6.18) for the injection $A_\delta \to A_{\delta'}$.

Proposition 6.6. *The transformation $R: A_\delta \to A_{\delta'}$ is bounded, compact, Hilbert Schmidt or trace class if and only if the operator*

$$K_{\delta'} R K_\delta^{-1}$$

is bounded, compact, Hilbert Schmidt or trace class on $L_2(\mathbb{R})$.

7. L_2 Estimates

In this section let $\mu = (-d^2/dx^2 + a^2)^{1/2}$, and let $k(\alpha, a; x-y)$ denote the kernel of $\mu^{-\alpha}$. Note that $k(\alpha, a; x) = a^{1-\alpha} k(\alpha, 1, ax)$, a scaling relation.

Proposition 7.1. *Let $0 < \alpha < 1$. Then*

$$0 < k(\alpha, a, x) < w(\alpha, a; x) \equiv \text{const.} |x|^{-1+\alpha} e^{-a|x|}. \tag{7.1}$$

If $\alpha > 1$,

$$0 < k(\alpha, a, x) < w(\alpha, a; x) \equiv \text{const.}\, a^{1-\alpha} e^{-a|x|}. \tag{7.2}$$

Proof. Let A be a strictly positive operator with bounded inverse. By the Cauchy integral formula, for $0 < \alpha < 2$,

$$A^{-\alpha/2} = \frac{\sin(\alpha\pi/2)}{\pi} \int_0^\infty (A+b)^{-1} b^{-\alpha/2}\, db. \tag{7.3}$$

Take $A = \mu^2$. The kernel of $(A+b)^{-1}$ is

$$(A+b)^{-1}(x,y) = (2\pi)^{-1/2} \int (p^2 + a^2 + b)^{-1} e^{-ip(x-y)} dp$$
$$= (2\pi)^{1/2} (2m(b))^{-1} e^{-m(b)|x-y|}, \tag{7.4}$$

where $m(b) = (a^2 + b)^{1/2}$. The positivity of k then follows by the positivity of (7.3), (7.4). We now specialize to the case $a = 1$, $\alpha < 1$ from which (7.1) follows by scaling. Inserting (7.4) in (7.3) and dividing the b integration into $\int_0^1 + \int_1^\infty$, we first bound the $\int_0^1$ integral. For $0 \leqq b \leqq 1$, $m(b) \geqq 1 + a/4$. Thus

$$\int_0^1 (A+b)^{-1}(x,0) b^{-\alpha/2} db \leqq \text{const.}\, e^{-|x|} \int_0^1 e^{-b|x|/4} b^{-\alpha/2} db$$
$$\leqq \text{const.}\, e^{-|x|} \min\{|x|^{-1+\alpha/2}, 1\}$$
$$\leqq \text{const.} |x|^{-1+\alpha} e^{-|x|}. \tag{7.5}$$

For the $\int_1^{\infty}$ integral,

$$\int_1^{\infty} m(b)^{-1} e^{-m(b)|x|} b^{-\alpha/2} db$$

$$= 2\int_x^{\infty} \frac{x}{(t^2+x^2)^{1/2}} t^{1-\alpha}|x|^{-2+\alpha} e^{-(t^2+x^2)^{1/2}} dt$$

$$\leqq 2|x|^{-1+\alpha} \int_x^{\infty} t^{-\alpha} e^{-(t^2+x^2)^{1/2}} dt. \tag{7.6}$$

For $t \geqq x$, it follows that for $\varepsilon > 0$ there exists $\beta < \infty$ satisfying

$$(t^2+x^2)^{1/2} \geqq |x| + \varepsilon t - \beta.$$

Thus (7.6) is bounded by

$$\text{const}\,|x|^{-1+\alpha} e^{-|x|}. \tag{7.7}$$

Combining (7.5), (7.7) yields (7.1).

If $\alpha > 1$, we again obtain (7.2) by scaling the $m=1$ bound. In this case we use the n-fold convolution

$$k(\alpha, 1, x) = w * \ldots * w, \qquad w = k(\alpha/n, 1, x) \qquad 0 < \alpha/n < 1, \tag{7.8}$$

and the bound (7.1) to establish $k(\alpha, 1, x) \leqq \text{const}\, e^{-|x|}$. Similarly

Proposition 7.2. *Let* $|x| \geqq \varepsilon > 0$, $-1 < \alpha < \infty$. *Then*

$$\left|\left(\frac{d}{dx}\right)^n k(\alpha, a, x)\right| \leqq C_{n,\varepsilon,a} e^{-a|x|}.$$

Theorem 7.3. *Let* $-1 < \alpha < 1$, $-a < \beta < a$. *Then*

$$\mu^{\alpha} e^{\beta|x|} \mu^{-\alpha} e^{-\beta|x|} \quad \textit{and} \quad e^{\beta|x|} \mu^{\alpha} e^{-\beta|x|} \mu^{-\alpha} \tag{7.9}$$

are bounded, as operators on L_2

Proof. For simplicity consider $\alpha \geqq 0$, $\beta \geqq 0$ in the first operator. The other cases are similar. Let $\zeta \in C_0^{\infty}$, with $\zeta(x) = 1$ in a neighborhood of $x = 0$. We write

$$\mu^{\pm\alpha} = (\mu^{\pm\alpha})_{\zeta} + (\mu^{\pm\alpha})_{1-\zeta}$$

where the subscript ζ or $1-\zeta$ replaces an operator with another whose kernel is multiplied by $\zeta(x-y)$ or $1-\zeta(x-y)$ respectively. Now $(\mu^{\alpha})_{1-\zeta} e^{\beta|x|} \mu^{-\alpha} e^{-\beta|x|}$ is bounded by Propositions 7.1 and 7.2 and

$$e^{\beta|x|} (\mu^{\alpha})_{\zeta} \mu^{-\alpha} e^{-\beta|x|} = I + e^{\beta|x|} (\mu^{\alpha})_{1-\zeta} \mu^{-\alpha} e^{-\beta|x|}$$

is also bounded, for the same reason, as is $e^{\beta|x|} \mu^{-\alpha} e^{-\beta|x|}$. Thus our problem is reduced to bounding

$$T = [(\mu^{\alpha})_{\zeta}, e^{\beta|x|}] e^{-\beta|x|}.$$

Because of the ζ in T, the unboundedness of $e^{\beta|x|}$ plays no role. In fact we choose ζ so that $\sum_j \zeta(x+j)=1$. In momentum space, the kernel of

$$T_j=[(\mu^\alpha)_\zeta, \zeta(x+j)e^{\beta|x|}]e^{-\beta j}$$

is bounded by a multiple of

$$f(p,q)=(\mu(p)^\alpha-\mu(q)^\alpha)(|p-q|+1)^{-2}.$$

Since

$$\sup_p \int |f(p,q)|dq<\infty,$$

T_j is bounded. Moreover $\|T\| \leqq \text{const} \sup_j \|T_j\|$ and the proof is complete.

Corollary 7.4. *The space A_δ is the closure of $C_0^\infty(R)$ in the norm*

$$\|f\|_\delta=\|K_\delta f\|_{L_2} \tag{7.10}$$

or in the norm

$$\|f\|'_\delta=\|K_\delta^* f\|_{L_2}. \tag{7.11}$$

The norms $\|\cdot\|$ and $\|\cdot\|'$ are equivalent.

Proof. We estimate

$$\begin{aligned}\|f\|_\delta&=\|K_\delta f\|=\|K_\delta(K_\delta^*)^{-1}K_\delta^* f\|\\&\leqq \|K_\delta(K_\delta^*)^{-1}\|\,\|f\|'_\delta \leqq \|K_\delta(K_\delta^*)^{-1}\|\,\|K_\delta^* K_\delta^{-1}\|\,\|f\|_\delta.\end{aligned}$$

However

$$K_\delta(K_\delta^*)^{-1}=e^{\delta_2|x|}\mu^{\delta_1}e^{-\delta_2|x|}\mu^{-\delta_1}, \tag{7.12}$$

and

$$K_\delta^* K_\delta^{-1}=\mu^{\delta_1}e^{\delta_2|x|}\mu^{-\delta_1}e^{-\delta_2|x|}. \tag{7.13}$$

Applying the theorem to (7.12)–(7.13) shows that they are bounded, so $\|\cdot\|$ and $\|\cdot\|'$ are equivalent.

Lemma 7.5. *Let $\delta<1$ and let $v(x)$ be bounded, positive and square integrable. Let $T(x,y)$ be the kernel of an integral operator satisfying*

$$|T(x,y)| \leqq v(x)v(y)|x-y|^{-\delta}. \tag{7.14}$$

Then T is compact.

Proof. Let $\chi_n(x,y)$ be the characteristic function of $|x-y|>n^{-1}$, and let $T_n(x,y)=T(x,y)\chi_n(x,y)$. Then T_n is Hilbert Schmidt, and hence compact. Also

$$\|T-T_n\| \leqq \sup_x \int_{|x-y|<n^{-1}} v(x)v(y)|x-y|^{-\delta}dy \leqq O(n^{-1+\delta}).$$

Hence T, as the norm limit of compact operators, is compact.

Theorem 7.6. *Let* $\alpha<\beta$, $\gamma<\varepsilon$, *with* $\alpha, \beta\in(-a,a), \gamma, \varepsilon\in(-1,1)$. *Let*

$$T\equiv e^{+\alpha|x|}\mu^{+\gamma}e^{-\beta|x|}\mu^{-\varepsilon}. \tag{7.15}$$

(i) *T is compact.*
(ii) *If* $\varepsilon-\gamma>\frac{1}{2}$, *T is Hilbert Schmidt.*
(iii) *If* $\varepsilon-\gamma>1$, *T is trace class.*

Proof. We write

$$T=(e^{\alpha|x|}\mu^{-(\varepsilon-\gamma)}e^{-\beta|x|})(e^{\beta|x|}\mu^{\varepsilon}e^{-\beta|x|}\mu^{-\varepsilon}). \tag{7.16}$$

The second factor in (7.16) is bounded by Theorem 7.3. We now show that the first factor is compact, bounded or Hilbert Schmidt. Let S denote this first factor. Then by (7.1), for any $\delta>0$, and assuming $0<\varepsilon-\gamma<1$,

$$\begin{aligned}|S(x,y)|&\leqq \text{const}\, e^{\alpha|x|}|x-y|^{-1+(\varepsilon-\gamma)}e^{-a|x-y|}e^{-\beta|y|}\\ &\leqq \text{const}\, e^{-(\beta-\alpha)|y|}|x-y|^{-1+(\varepsilon-\gamma)}e^{-(a-\alpha)|x-y|}\\ &\leqq \text{const}\, e^{-\delta|x|}e^{-(\beta-\alpha-\delta)|y|}e^{-(a-\alpha-\delta)|x-y|}|x-y|^{-1+\varepsilon-\gamma}\end{aligned} \tag{7.17}$$

By Lemma 7.5, S is compact. Furthermore, if $\varepsilon-\gamma>\frac{1}{2}$, then S is Hilbert Schmidt. If $\varepsilon-\gamma>1$, $0<\delta<\beta-\alpha$,

$$S=(e^{\alpha|x|}\mu^{-(\varepsilon-\gamma)/2}e^{-(\beta-\delta)|x|})(e^{(\beta-\delta)|x|}\mu^{-(\varepsilon-\gamma)/2}e^{-\beta|x|}). \tag{7.18}$$

Each factor in parentheses in (7.18) is Hilbert Schmidt, by (ii). Thus S, as the product of two Hilbert Schmidt operators, is trace class.

Proposition 7.7. *Let* $\delta_2>0$. *The norm*

$$|||f|||_\delta\equiv\max_{\pm}\|e^{\pm\delta_2 x}\mu^{\delta_1}f\|_{L_2} \tag{7.19}$$

is equivalent to the norm $\|f\|_\delta$.

Proof. For $\alpha\geqq 0$, we have the pointwise bound

$$\exp(\pm\alpha x)\leqq \exp\alpha|x|\leqq \exp(\alpha x)+\exp(-\alpha x).$$

Thus

$$|||f|||_\delta\leqq\|f\|_\delta\leqq 2|||f|||_\delta. \tag{7.20}$$

Proposition 7.8. *Let* $|\delta_2|<a$. *Then as an operator from* A_δ *to* A_δ,

$$|||e^{-t\mu}|||_{A_\delta}=\sup_f |||e^{-t\mu}f|||_\delta/|||f|||_\delta\leqq\exp(-t\{a^2-\delta_2^2\}^{1/2}). \tag{7.21}$$

Proof. First we note that the result for $\delta_2<0$ follows from the result for $\delta_2\geqq 0$. In fact if $T:A_\delta\to A_\delta$ is bounded with norm $\|T\|_{A_\delta}$, then $T^*:A_\delta^*\to A_\delta^*$ is bounded with the same norm. By (6.13), $A_\delta^*=A_{-\delta}$, so we take $\delta_2\geqq 0$; let $T=\exp(-t\mu)$. We show that $|||T|||_{A_\delta}$ is bounded as in (7.21). By Proposition 7.7, this shows that T is bounded on A_δ. For convenience, we choose f in the dense set C_0^∞ in A_δ, so $f(p)$ is entire. Then

$$\|e^{\pm\alpha x}\mu^\beta Tf\|_{L_2}=\|\mu(p\pm i\alpha)^\beta\exp(-t\mu(p\pm i\alpha))\tilde{f}(p\pm i\alpha)\|_{L_2}. \tag{7.22}$$

We note that

$$\begin{aligned}\operatorname{Re}\mu(p\pm i\alpha)&=\operatorname{Re}(p^2+a^2-\alpha^2\pm 2ip\alpha)^{1/2}\\ &\geqq(p^2+a^2-\alpha^2)^{1/2}\geqq(a^2-\alpha^2)^{1/2}\,. \end{aligned}\tag{7.23}$$

Thus with $\alpha=\delta_2$, $\beta=\delta_1$,

$$\begin{aligned}|||Tf|||_\delta&\leqq\exp(-t(a^2-\alpha^2)^{1/2})\max_{\pm}\|\mu(p\pm i\alpha)^\beta\tilde f(p\pm i\alpha)\|_{L_2}\\ &=\exp(-t(a^2-\alpha^2)^{1/2})\max_{\pm}\|e^{\pm\alpha x}\mu^\beta f\|_{L_2}\\ &=\exp(-t(a^2-\alpha^2)^{1/2})|||f|||_\delta\,. \end{aligned}\tag{7.24}$$

Proposition 7.9. *Let $T=(e^{-\mu}\otimes e^{-\mu})$. Then acting on A_δ as functions of p_{rel} with p_{tot} fixed,*

$$|||T^t|||_{A_\delta}\leqq\exp(-2t\sqrt{a^2+p_{\mathrm{tot}}^2-\delta_2^2})\tag{7.25}$$

Proof. As above, we assume $\delta_2\geqq 0$. Then

$$\begin{aligned}&\|e^{\pm\alpha x_{\mathrm{rel}}}\mu^\beta Tf\|_{L_2}\\ &=\|\mu(p_{\mathrm{rel}}\pm i\alpha)^\beta\exp\{-t\mu(p_{\mathrm{tot}}+p_{\mathrm{rel}}\pm i\alpha)-t\mu(p_{\mathrm{tot}}-p_{\mathrm{rel}}\pm i\alpha)\}\tilde f(p_{\mathrm{rel}}\pm i\alpha)\|_{L_2}\end{aligned}\tag{7.26}$$

using $p_{\mathrm{tot}}=\frac{1}{2}(p_1+p_2)$, $p_{\mathrm{rel}}=\frac{1}{2}(p_1-p_2)$. Then as in (7.23),

$$\textstyle\sum_\pm\operatorname{Re}\mu(p_{\mathrm{tot}}\pm p_{\mathrm{rel}}\pm i\alpha)\geqq 2(a^2+p_{\mathrm{tot}}^2-\alpha^2)^{1/}$$

Thus

$$\|e^{\pm\alpha x_{\mathrm{rel}}}\mu^\beta Tf\|_{L_2}\leqq e^{-2t(a^2+p_{\mathrm{tot}}^2-\alpha^2)1/2}\|\mu(p_{\mathrm{rel}}\pm i\alpha)^\beta\tilde f\|_{L_2}\,.$$

Corollary 7.10. *For* $\operatorname{Re}E<2(a^2+p_{\mathrm{tot}}^2-\delta_2^2)^{1/2}$,

$$|||(1-e^Ee^{-\mu}\otimes e^{-\mu})^{-1}|||_{A_\delta\to A_\delta}\leqq(1-e^{-[2(a^2+p_{\mathrm{tot}}^2-\delta_2^2)^{1/2}-\operatorname{Re}E]})^{-1}$$

Remark. With T as in Proposition 7.9, the Corollary bounds

$$\tilde T(E)\equiv\sum_{j\geqq 0}T^je^{jE}\tag{7.27}$$

We now show that $\tilde T(E)$ has a two sheeted analytic continuation around a square root branch point at threshold. For simplicity, we take $p_{\mathrm{tot}}=0$ and we do not analyze the singularity at threshold.

Proposition 7.11. *For*

$$\zeta=(4a^2-E^2)^{1/2}\neq 0\,,$$

the operator

$$\tilde T(E(\zeta))=\frac{1}{1-\exp[(4a^2-\zeta^2)^{1/2}-(4a^2+4p_{\mathrm{rel}}^2)^{1/2}]}\tag{7.28}$$

acting from $A_{(\delta_1,\delta_2)}$ to $A_{(\delta_1,-\delta_2)}$, $\delta_2>0$, has a bounded analytic continuation in ζ from $\operatorname{Re}\zeta>2\delta_2$ *(i.e.* $\operatorname{Re}E<2(a^2-\delta_2^2)^{1/2}$*) to the region* $\operatorname{Re}\zeta>-2\delta_2$.

Proof. We see that $\tilde{T}(E(\zeta))$ has poles in $p_{\rm rel}$ at

$$p_{\rm rel} = \pm i\zeta/2. \tag{7.29}$$

As long as the poles do not cross the strip $|\operatorname{Im} p_{\rm rel}| \leqq \delta_2$, i.e. as long as $\operatorname{Re}\zeta > -2\delta_2$, the contour integral defining the inner product

$$\langle f, \tilde{T}(E) f \rangle_{L_2}, \quad f \in A_\delta, \tag{7.30}$$

as written in the momentum space variable $p_{\rm rel}$, can be deformed to avoid the poles (7.29). This procedure defines the analytic continuation of (7.28) to $\operatorname{Re}\zeta > -2\delta_2$. Mapping back to the variable E gives the two sheeted analytic continuation of (7.27) around the square root branch point $E = 2a$. For $\zeta \neq 0$, the contour can be chosen so that (7.28) is bounded along the contours. As $\zeta \to 0$, the poles come together, and the contour, which must go above one pole and below the other, cannot be bounded away from the poles. See also [3, 4, 12].

It is only in the analytic continuation argument that we use the fact that the functions $f \in A_\delta$ are symmetric in x-space. Because of this assumption, (7.30) can be written in momentum space as

$$\langle f, \tilde{T}(E) f \rangle_{L_2} = \int_{-\infty}^{\infty} \tilde{\tilde{f}}(p) (\tilde{T}(E) \tilde{f})(p) dp. \tag{7.31}$$

Thus no absolute values occur in (7.31) and the analytic continuation in E can be accompanied by the deformation of the p integration contour (inside the strip $|\operatorname{Im} p| < \delta_2$) to avoid the poles (7.29) of $\tilde{T}(E(\zeta))$. By the above argument, if $C(\zeta)$ denotes the deformed contour,

$$|\langle f, \tilde{T}(E) f \rangle_{L_2}| \leqq \sup_{C(\zeta)} |\tilde{T}(E(\zeta))| \int_{C(\zeta)} |\tilde{\tilde{f}}(p) \tilde{f}(p)| dp. \tag{7.32}$$

since the sup is dominated by $O(1)$. For $|\operatorname{Im} p|$ bounded away from δ_2, $|\tilde{f}(p)| \leqq \text{const}\, \|f\|_{A_\delta}$. Thus in (7.31),

$$\left| \int_{-L}^{L} \tilde{\tilde{f}}(p) \tilde{f}(p) \tilde{T}(E) dp \right| \leqq \text{const}\, \|f\|_{A_\delta}^2.$$

The tails, $\int_{-\infty}^{-L} + \int_{L}^{\infty}$ are bounded directly, and can be continued analytically to $\operatorname{Re} E \leqq L$ without deformation of contour, as in Proposition 7.9, and Corollary 7.10. Since L is arbitrary, we get the desired analytic continuation.

8. One Particle Lines in a Two Particle Channel

We modify the two particle cluster expansion [8] applied to $S_{0,T}$ in order to define our integral equation. The first modification is inclusion of virtual contraction terms to the $:\phi(T)^2:$ observable, to allow resummation of one particle lines, as explained in § 5. The second modification, which is new to this section, is the inclusion of virtual interaction terms between the independent one particle lines.

The operator

$$e^{-t\mu} \otimes e^{-t\mu} \tag{8.1}$$

generates dynamics in two independent one particle lines, and we want to find this operator by resumming certain terms from the expansion of $:\phi(x_1)\phi(x_2):$. It is easy to find a candidate for this resummation, namely all terms in which there is no contraction between x_1 and x_2 vertices, nor any overlap between the x_1 and x_2 contours. Each term of this nature factors, and thus expresses some type of independent dynamics for x_1 and x_2. However, the sum of these terms does not factor, and does not define independent dynamics for x_1 and x_2. In fact, some terms, expected as part of the $e^{-t\mu}$ dynamics for x_1 do not occur, because the region of space required to insert this term in the expansion is already occupied by other vertices and/or contours belonging to x_2 or to x_1 and x_2 jointly. To compensate, we add and subtract these missing terms. With the addition, the sum of the independent terms also factors, and can be identified with (8.1). When subtracted, the terms are called virtual interaction terms; they contribute along with the direct interaction terms to the Bethe-Salpeter kernel.

In more detail, we restrict the momenta of each x_j test function as in (5.7). Using a smooth momentum space cutoff function, we write the x_j test function as a sum of two test functions. The first has high momentum and the second has purely low momentum. The term with the high momentum has an enhanced decay rate. This fact can be obtained from the convergence proof of [8]. Because of the high decay rate ($\geqq 2m_0 - \varepsilon$ for the single particle) a simplier expansion, not requiring a backward movement in time is possible for the high momentum term. This expansion will be explained later.

The more difficult term involves low momentum. We now explain the expansion for these terms. Let $:\phi^2:_W$ denote normal ordering relative to the bare vacuum. This is the normal ordering used in [8]. By adding and subtracting a constant term and a term linear in ϕ, we can pass from $:\phi^2:$ to $:\phi^2:_W$ or back, as in (2.13). We use this identity to start the expansion with $:\phi^2:_W$, but to yield $:\phi^2:$ terms ($1 < t = i \leqq T$) as a result of the expansion. It is also used to write the time T: $:\phi^2:$ observable as $:\phi^2:_W + \phi + \text{const}$. The linear term $\phi|_{t=0}$ is expanded as in § 5, as will be explained in more detail below.

The $:\phi^2:_W|_{t=0}$ is expanded according to the 2-particle expansion of [8]. Virtual terms are added and subtracted, for interaction with the time T observables and also for interaction between the two one particle lines. For terms without interaction between the two ϕ^2 lines, each ϕ is returned independently to $t=1$. The expansion terminates with any of these terms:

(i) a constant (i.e. ϕ-independent) term,

(ii) a term linear in ϕ, returned to $t=1$,

(iii) two independent ϕ's in the form $:\phi(y_1)\phi(y_2):$, each returned to $t=1$,

(iv) a $:\phi^2:$ at integer time $t=i$, $1 \leqq i < T$ with direct or virtual interaction between the two variables y_1 and y_2

(v) direct or virtual contraction to $:\phi^2:(T)$, the time T observable.

Because of the virtual terms in (v), the terms (i) and (ii) are T independent; because of normal ordering in $:\phi^2:$, they sum to zero. The terms (iii) are exactly the tensor product of the one particle line expansion of § 5, and thus can be identified with the dynamics (8.1). These terms are then $e^{-\mu}\otimes e^{-\mu}S_{1,T}$. The terms (iv) are proportional to $S_{i,T}$, and give $B_{0,i}^{\text{amp}}S_{i,T}$, where $B_{0,i}^{\text{amp}}$ denotes the sum of the proportionality operators for fixed i. Finally the terms (v) have three particle decay, and are bounded by $O(\exp(-3(m_0-\varepsilon)T))$. We denote the terms (v) by $A_{0,T}^{\text{amp}}$.

The expansion of $\phi|_{t=0}$ follows § 5. All terms linear in ϕ are returned to $t=1$. The expansion terminates with terms of type (i), (ii), (iv) or (v), but gives no contribution to the (iii) terms.

In the high momentum term, the two ϕ's in $:\phi^2:_W$ are moved forward in time independently until they interact with each other (direct or virtual interaction or until they contract (direct or virtual contraction) to a time T observable. This produces terms of the type (i), (ii), (iv), (v), as above, but not of type (iii). The (i) and (ii) terms sum to zero, and the (iv) and (v) terms are estimated as below. We note that as long as the two lines do not interact, momentum is conserved within each line. In particular, the high momentum property is retained until interaction.

In both high and low momentum terms, we require an independent average over the lattice displacement $\theta\in(0,1]$ for each of the independent one particle lines. This means that the contours for the distinct lines are inserted along distinct, incompatible lattices.

The estimates on the (iv) and (v) terms of the expansion follow from [8], because all terms, including virtual terms, arise from a finite number of sums, differences or products of convergent expansions. Thus the (iv) and (v) terms are absolutely convergent in the norms of [8]. In view of regularization (see § 9, 10), they are also bounded and absolutely convergent in the operator norms on A_δ, on the space of equal time functions.

In summary, we have

$$S_{0,T}=e^{-\mu}\otimes e^{-\mu}S_{1,T}+\sum_{i=1} B_{0,i}^{\text{amp}}S_{iT}+A_{0,T}^{\text{amp}}. \tag{8.2}$$

Iterating the expansion, we obtain (1.6) with

$$A_{0,T}=\sum_{0\leq j\leq T}(e^{-\mu}\otimes e^{-\mu})^j A_{j,T}^{\text{amp}} \tag{8.3}$$

$$B_{0,T}=\sum_{0\leq j\leq T-1}(e^{-\mu}\otimes e^{-\mu})^j B_{j,T}^{\text{amp}}. \tag{8.4}$$

Thus

$$\tilde{A}(E)=\sum_{0\leq T}A_{0,T}e^{ET}=(I-e^E e^{-\mu}\otimes e^{-\mu})^{-1}\sum_{0\leq T}A_{0,T}^{\text{amp}}e^{ET}=\tilde{T}(E)\tilde{A}^{\text{amp}}(E), \tag{8.5}$$

$$\tilde{B}(E)=\sum_{0\leq T}B_{0,T}e^{ET}=(I-e^E e^{-\mu}\otimes e^{-\mu})^{-1}\sum_{1<T}B_{0,T}^{\text{amp}}e^{ET}=\tilde{T}(E)\tilde{B}^{\text{amp}}(E). \tag{8.6}$$

Thus we write (1.6) as

$$\tilde{S}(E)=\tilde{A}(E)+\tilde{B}(E)\tilde{S}(E). \tag{8.7}$$

9. Decay Estimates in x-Spaces

The two particle expansion of § 8 introduces seven operators which map kernels of const, $\phi(x)$ or $:\phi(x_1)\phi(x_2):$ into kernels of const, ϕ or $:\phi^2:$ at possibly later times.

The simplest of these operators is $e^{-\mu}\otimes e^{-\mu}$. This operator is described in § 7, and it acts on the kernel of $:\phi^2:$ in the mapping

$$:\phi(x_1)\phi(x_2):|_{t=0}\to :\phi(y_1)\phi(y_2):|_{t=1}\,. \tag{9.1}$$

Another simple operation is re-Wick ordering $\delta\alpha_0$. Since integration by parts yields

$$\delta\alpha_0\in A_{\frac{3}{2}-\varepsilon,m-\varepsilon}=A^*_{-\frac{3}{2}+\varepsilon,-m+\varepsilon} \tag{9.2}$$

$\delta\alpha_0$ is a bounded linear functional on all A_δ spaces we consider. Also $\tilde{\alpha}_1(p,x_{\mathrm{rel}})$ for fixed p satisfies

$$\alpha_1(p,\cdot)\in A_{\frac{3}{2}-\varepsilon,\delta_2}=A^*_{-\frac{3}{2}+\varepsilon,-\delta_2},\qquad \delta_2 \gtrsim 0\,. \tag{9.3}$$

and thus is bounded on all A_δ spaces we consider.

Let $\mathscr{K}_j^{(\mathrm{iv})}$ denote the operator which is the sum of all (iv) terms associated with the map

$$:\phi(x_1)\phi(x_2):_W|_{t=0}\to :\phi(y_1)\phi(y_2):|_{t=j}\,. \tag{9.4}$$

The operator $\mathscr{K}^{(\mathrm{iv})}$ has a kernel depending on x_{rel}, y_{rel}, $x_{\mathrm{tot}}-y_{\mathrm{tot}}$.

We similarly define an operator $\mathscr{K}_T^{(\mathrm{v})}$ by the sum of type (v) terms. Then $\mathscr{K}_T^{(\mathrm{v})}$ is determined as a bilinear form on the kernels $f(x_{\mathrm{rel}},p_{\mathrm{rel}})$ and $g(y_{\mathrm{rel}},p_{\mathrm{rel}})$ of the time $t=0$ and time $t=T$ observables.

Namely at time $t=0$, the observable is

$$\int f(x_{\mathrm{rel}},p_{\mathrm{tot}}):\phi^2(x_{\mathrm{rel}},p_{\mathrm{tot}}):_W dx_{\mathrm{rel}}$$

and at Euclidean time $t=T$, it is

$$\begin{aligned}&\int g(y_{\mathrm{rel}},p_{\mathrm{tot}}):\phi^2(y_{\mathrm{rel}},p_{\mathrm{tot}}):dy_{\mathrm{rel}}\\ &\quad=\int g(y_{\mathrm{rel}},p_{\mathrm{tot}}):\phi^2(y_{\mathrm{rel}},p_{\mathrm{tot}}):_W dy_{\mathrm{rel}}\\ &\qquad-\int g(y_{\mathrm{rel}},p_{\mathrm{tot}})\tilde{\alpha}_1(p_{\mathrm{tot}},y_{\mathrm{rel}})dy_{\mathrm{rel}}\tilde{\phi}(p_{\mathrm{tot}})\\ &\qquad+\int g(y_{\mathrm{rel}},p_{\mathrm{tot}})\delta\alpha_0(y_{\mathrm{rel}})dy_{\mathrm{rel}}\,.\end{aligned} \tag{9.5}$$

With $f\in A_\delta$, as a function of x_{rel}, $g\in A_{\delta^1}$, then $\mathscr{K}_T^{(\mathrm{v})}$ is an operator from A_δ to $A^*_{\delta^1}=A_{-\delta^1}$.

Also let $\mathscr{L}_j^{(\mathrm{iv})}$ and $\mathscr{L}_j^{(\mathrm{v})}$ be associated with the map

$$\phi(x)|_{t=0}\to :\phi(y_1)\phi(y_2):|_{t=j} \tag{9.6}$$

and with $\phi(x)$ contractions at time T. Then $\mathscr{L}^{(\mathrm{iv})}$ has a kernel depending on y_{rel}, $x-y_{\mathrm{tot}}$.

For comparison with [8], we restrict $\mathscr{K}$, $\mathscr{L}$ to functions supported in products of unit intervals. The $\mathscr{K}$, $\mathscr{L}$ are bounded from L_2 to L_2. According to § 6, the unit lattice squares in x_1, x_2 space and in y_1, y_2 space can be rotated by $\pi/4$ so that they are alligned with x_{rel}, x_{tot}, y_{rel}, y_{tot}. Let Δ_x, etc. denote the characteristic function of such a square. A minor improvement in the result of [8] includes some x as well as

t exponential decay. With $:\phi^2:$ as the time T observable this yields

$$\|\Delta_x \mathscr{K}_T^{(v)} \Delta_y\|_{L_2 \to L_2^*} \leqq e^{-3(m-\varepsilon)T} e^{-\varepsilon'|x_{tot}-y_{tot}|}$$
$$\leqq [e^{-\varepsilon'|x_{rel}-y_{rel}|} + e^{-\varepsilon'|x_{rel}+y_{rel}|}]. \tag{9.7}$$

The type (iv) terms have an improved decay because the interaction between the two one particle lines give graphs which are connected, and hence decay in $|x_{rel}| + |y_{rel}|$. Thus

$$\|\Delta_x \mathscr{K}_j^{(iv)} \Delta_j\|_{L_2 \to L_2} \leqq e^{-3(m-\varepsilon)j} e^{-\varepsilon'|x_{tot}-y_{tot}|} e^{-\varepsilon'|x_{rel}|-\varepsilon'|y_{rel}|}. \tag{9.8}$$

The graphs in $\mathscr{L}$ are all connected.

$$\|\Delta_x \mathscr{L}_j^{(iv)} \Delta_y\|_{L_2 \to L_2} \leqq e^{-3(m-\varepsilon)j} e^{-\varepsilon'|x-y_{tot}|} e^{-\varepsilon'|y_{rel}|} \tag{9.9}$$

and

$$\|\Delta_x \mathscr{L}_j^{(v)} \Delta_j\|_{L_2 \to L_2^*} \leqq e^{-3(m-\varepsilon)T} e^{-\varepsilon'|x-y_{tot}|} e^{-\varepsilon'|y_{rel}|}. \tag{9.10}$$

Next we consider the operators $\mathscr{K}$, $\mathscr{L}$ at fixed total momentum p, as mappings on the function of relative variables:

$$\begin{aligned} &\mathscr{K}_j^{(iv)} : A_\delta^* = A_{-\delta} \to A_\delta, \quad \mathscr{K}_T^{(v)} : A_\delta \to A_\delta \\ &\mathscr{L}_j^{(iv)} \in A_\delta, \quad \mathscr{L}_T^{(v)} \in A_\delta. \end{aligned} \tag{9.11}$$

For $\delta_1 = 0$ these four norms are bounded as follows:

Theorem 9.1. *For δ_2 sufficiently small and greater than zero, the four operators (9.11) are bounded. The norms are $O(e^{-3(m-\varepsilon)j})$ for $\mathscr{K}_j$ and $\mathscr{L}_j$. Moreover $\mathscr{K}$ and $\mathscr{L}$ are complex analytic in p_{tot}, for $|\mathrm{Im}\, p_{tot}|$ small.*

Proof. For $\mathscr{K}_j^{(iv)}$ we have

$$\|\mathscr{K}_j^{(iv)}\|_{A_\delta \to A_\delta} \leqq \sum e^{\delta_2|x_{rel}| + \delta_2|y_{rel}|} \|\Delta_x \mathscr{K}_j^{(iv)} \Delta_y\|_{L_2 \to L_2}$$
$$\leqq O(\sum e^{-3(m-3)j - |x_{tot}-y_{tot}|(\varepsilon' - |\mathrm{Im}\, p_{tot}|)}) \leqq O(e^{-3(m-\varepsilon)j}).$$

Similarly

$$\|\mathscr{K}_T^{(v)}\|_{A_\delta \to A_\delta} \leqq \sum e^{-\delta_2|x_{rel}| + \delta_2|y_{rel}|} \|\Delta_x \mathscr{K}_T^{(v)} \Delta_y\|_{L_2 \to L_2^*} \leqq O(e^{-3(m-\varepsilon)T}).$$

The remaining bounds on

$$\begin{aligned} &\|\mathscr{L}_j^{(iv)}\|_{A_\delta} \leqq \sum e^{\delta|y_{rel}|} \|\Delta_x \mathscr{L}_j^{(iv)} \Delta_y\|_{L_2 \to L_2} \leqq e^{-3(m-\varepsilon)j} \\ &\|\mathscr{L}_T^{(v)}\|_{A_\delta} \leqq \sum e^{\delta|y_{rel}|} \|\Delta_x \mathscr{L}_T^{(v)} \Delta_y\|_{L_2 \to L_2^*} \leqq e^{-3(m-\varepsilon)T} \end{aligned} \tag{9.12}$$

are similar, to complete the proof.

The proof of (9.12) applies to the error terms in (5.17) and shows their contribution to $\tilde{\alpha}_1$ lies in A_{0,δ_2}, $\delta_2 > 0$.

10. Smoothing and Compactness in x-Space

In this section we exhibit the smoothing property of the operators $\mathscr{K}$ and $\mathscr{L}$ defined in §8 and §9. In particular, this gives the properties of A^{amp} and B^{amp} on the spaces A_δ. Our main technical result, which leads to an analysis of $S_{0,T}$ is the following:

Theorem 10.1. *The connected expansion terms (§8, type* iv*) define bounded operators* $B_{0,j}^{\mathrm{amp}}$ *such that*

$$B_{0,j}^{\mathrm{amp}}: A_{-\delta} \to A_\delta \tag{10.1}$$

with

$$\| B_{0,j}^{\mathrm{amp}} \|_{A_{-\delta} \to A_\delta} \leqq O(\lambda) e^{-3(m-\varepsilon)j} \tag{10.2}$$

when

$$\delta_1 \leqq 2 - \varepsilon, \quad \delta_2 < \varepsilon, \tag{10.3}$$

for some $\varepsilon > 0$ *sufficiently small. The terms of §8, type* v *define the bounded operator* $A_{0,T}^{\mathrm{amp}}$, *where*

$$A_{0,T}^{\mathrm{amp}}: A_{-1+\varepsilon,\delta_2} \to A_{1-\varepsilon,\delta_2} \tag{10.4}$$

with

$$\| A_{0,T}^{\mathrm{amp}} \| \leqq O(1) e^{-3(m-\varepsilon)T}. \tag{10.5}$$

Moreover, $B_{0,j}^{\mathrm{amp}}$ *is compact on the indicated spaces, and trace class if* $\delta_1 \leqq \frac{3}{2} - \varepsilon$.

Remark. It follows that the Laplace transformed operators

$$\begin{aligned} A^{\mathrm{amp}}(E) &= \sum_{0 \leqq T} A_{0,T}^{\mathrm{amp}} e^{ET} \\ B^{\mathrm{amp}}(E) &= \sum_{0 \leqq T} B_{0,T}^{\mathrm{amp}} e^{ET} \end{aligned} \tag{10.6}$$

are complex analytic in E for $\mathrm{Re}\, E \leqq 3(m_0 - \varepsilon)$. This statement is true for $\tilde{A}^{\mathrm{amp}}$ and $\tilde{B}^{\mathrm{amp}}$ as operators on functions of the relative position variable. In particular, they are bounded as operators from A_δ to $A_{\delta'}$ where δ, δ' are given by Theorem 10.1, and $\tilde{B}^{\mathrm{amp}}$ is compact, and has small norm as $\lambda \to 0$.

Proof. The improvement of Theorem 10.1 over the §9 involves a careful examination of the initial (and final) expansion steps. Each term at the end of the expansion has the structure given in §8. The sum of all expansion terms converge in the norms of [8]. We thus inspect individual terms. The $\mathscr{K}_j^{(\mathrm{iv})}$ terms, for example, have the structure

$:\phi^2:(w)$  $:\phi^2:(v)$ (10.7)

where w is the kernel of $:\phi^2:(t=0)$ and v is the kernel of $:\phi^2:(t=j)$. The initial expansion step inserts a contour around the $:\phi^2:(t=0)$ vertices; this contour may contain other vertices and partial contours. Likewise the expansion ends by removing the $:\phi^l:(t<j)$ vertices by a contour. In [8] estimates on the inside vertices $R(\phi)$ and function space integration was obtained by Schwarz inequality in path space

$$\left|\int_{\text{inside}} e^{-P(\phi)}R(\phi)d\phi_{\text{inside}}\right| \leqq |(\int R^2 d\phi \int e^{-2P(\phi)}d\phi). \tag{10.8}$$

However, optimal bounds on the Schwinger functions, e.g. $\int e^{-P(\phi)}R(\phi)d\phi_{\text{inside}}$ improve on this estimate, see [5]. These optimal bounds are obtained by integration by parts of each field in $R(\phi)$, before applying the Schwarz inequality.

In expressions such as for $\mathscr{K}^{(\text{iv})}$ in (10.7), this integration by parts produces "initial" and "final" vertices which contract with the $:\phi^2:(w)$ or $:\phi^2:(v)$. An analysis of these diagrams shows that their continuity in v and w lies not only in L_2, but in the appropriate Sobolev space for Theorem 10.1. For example, the dominant diagrams in $\mathscr{K}^{(\text{iv})}$ have the form

(10.8)

with a single ϕ^4 vertex. Their momentum space decay is estimated for fixed p_{tot} by $O(\lambda)\|\mu^{-2}w\|_{L_2}\|\mu^{-2}v\|_{L_2}$ in the variable p_{rel}. Here each line carries a factor μ^{-1}, and the integration over the time of the intermediate vertex gives an additional factor μ^{-1} which can be divided between w and v. More generally, a large $\mathscr{K}^{(\text{iv})}$ diagram can be estimated by isolating a vertex on either end, and using and L_2 operator norm on the intermediate diagram:

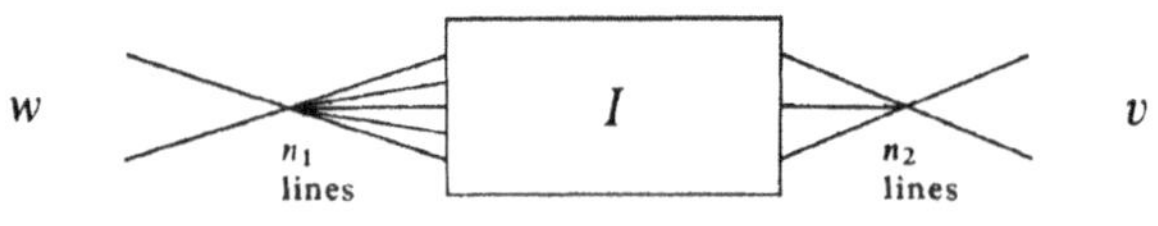

Then I as an operator from $L_2(R^{n_2})$ to $L_2(R^{n_1})$ has an operator norm $\|I\|_{n_1,n_2}$ which can be estimated by the cluster expansion and Schwarz inequality as mentioned above. The single vertex diagrams have Hilbert-Schmidt norms $\leqq O(1)\|\mu^{-2}w\|\ \|\mu^{-2}v\|$ as above. The reduction to these norms is systematized in [6, §6].

Other typical terms which arise involve α_1. (The analysis of such terms completes the proof of Theorem 5.1.) These terms will have a dominant diagram

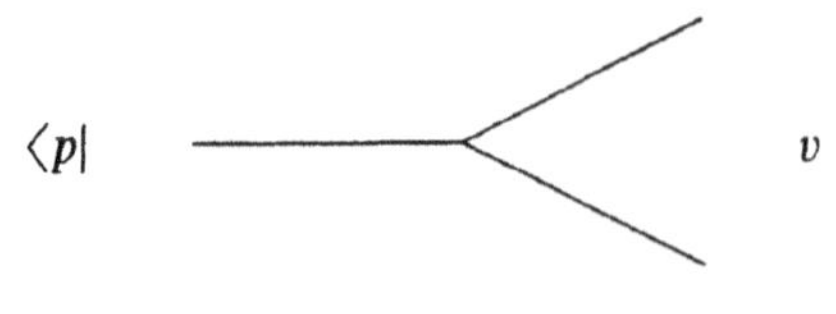

which as above is estimated by

$O(1)\|\mu^{-2}v\|$.

The dominant diagram for $\tilde{A}^{\mathrm{amp}}(E)$ is

arising from the $T=0$ contribution. Here we obtain the estimate $O(1)\|\mu^{-1}v\|_{L_2}\|\mu^{-1}w\|_{L_2}$ arising from a factor μ^{-1} on each line.

We now return to the Eqs. (1.6), (8.3)–(8.7), (10.6)–(10.7). In Laplace transform space the solution for $\tilde{S}(E)$ is

$$\tilde{S}(E)=[1-\tilde{B}(E)]^{-1}\tilde{A}(E). \tag{10.9}$$

To analyze bound states, it is sufficient to regard $\tilde{S}(E)$ as a bilinear form between a dense set of states. We use states $\theta_p=\int f(x_{\mathrm{rel}},p_{\mathrm{tot}}):\phi(x_1)\phi(x_2):dx_{\mathrm{rel}}\Omega$, dense in $\mathscr{H}(p)$ by §3. Thus we study the bilinear form $\langle g,\tilde{S}(E)f\rangle$, or equivalently, we realize (10.9) as an operator equation from $A_{-1+\varepsilon,\delta_2}$ to $A_{1-\varepsilon,\delta_2}=A^*_{-1+\varepsilon,\delta_2}$.

By Proposition 7.11 and Theorem 10.1,

$$\tilde{A}(E):A_{-1+\varepsilon,\delta_2}\to A_{1-\varepsilon,-\delta_2}, \tag{10.10}$$

for $\mathrm{Re}\,E\leqq 3(m_0-\varepsilon)$, with an analytic continuation onto the second sheet, for E not at threshold. Similarly

$$\tilde{B}(E):A_{-2+\varepsilon,-\delta_2}\to A_{2-\varepsilon,-\delta_2} \tag{10.11}$$

is bounded, and

$$\tilde{B}(E):A_{-\frac{3}{2}+\varepsilon,-\delta_2}\to A_{\frac{3}{2}-\varepsilon,-\delta_2} \tag{10.12}$$

is trace class, with norm (away from threshold)

$$\|\tilde{B}(E)\|_{A_\delta\to A_\delta}\leqq O(\lambda)<1\,,\quad \text{for}\quad \delta_1>0\,,\quad \delta_2<0\,. \tag{10.13}$$

Thus $(I-\tilde{B}(E))^{-1}$ is bounded on the same spaces, and

$$\tilde{S}(E):A_{-1+\varepsilon,\delta_2}\to A_{1-\varepsilon,-\delta_2}$$

is bounded.

Near threshold, since $\tilde{B}(E)$ is compact, $(I-\tilde{B}(E))^{-1}$ and $\tilde{S}(E)$ are meromorphic, and the only possible point of accumulation of poles is at threshold.

The asserted properties of bound states follow. See [3, 4, 12] for a more complete analysis of a similar situation.

References

1. Bros, J., Lasalle, M.: Analyticity properties and many body structure in general quantum field theory. III. Commun. Math. Phys. **54**, 33–62 (1977)
2. Combescure, M., Dunlop, F.: N-particle irreducible functions in Euclidean quantum field theory. Ann. Phys. (N.Y.) to appear

3. Dimock, J., Eckmann, J.P.: On the bound state in weakly coupled $\lambda(\phi^6-\phi^4)_2$. Commun. Math. Phys. **51**, 41–54 (1976)
4. Dimock, J., Eckmann, J.P.: Spectral properties and bound state scattering for weakly coupled $P(\phi)_2$ models. Ann. Phys. **103**, 289–314 (1977)
5. Glimm, J., Jaffe, A.: Two and three body equations. Commun. Math. Phys. **44**, 293–320 (1975)
6. Glimm, J., Jaffe, A.: Positivity of the ϕ_3^4 Hamiltonian. Fortschr. Physik **21**, 327–376 (1973)
7. Glimm, J., Jaffe, A.: In: International symposium on mathematical problems in theoretical physics, Araki, H. (ed.). Lecture notes in physics, Vol. 39. Berlin, Heidelberg, New York: Springer 1975
8. Glimm, J., Jaffe, A., Spencer, T.: The Wightman axioms and particle structure in the $P(\phi)_2$ quantum field model. Ann. Math. **100**, 585–632 (1974)
9. Koch, H.: Private communication
10. Gelfand, I., Shilov, G.: Generalized functions, Vol. II, III. New York: Academic Press 1967
11. Spencer, T.: The decay of the Bethe-Salpeter kernel in $P(\phi)_2$ quantum field. Commun. Math. Phys. **44**, 143–164 (1975)
12. Spencer, T., Zirilli, F.: Scattering states and bound states in $\lambda P(\phi)_2$. Commun. Math. Phys. **49**, 1–16 (1976)
13. Symanzik, K.: On the many particle structure of Green's functions in quantum field theory. J. Math. Phys. **1**, 249–273 (1960)

Communicated by A. Jaffe

Received January 9, 1979

VI Bounds on Coupling Constants

Introduction

The ϕ^4 interaction in the single phase region is repulsive, as can be demonstrated either using perturbation theory for weak coupling or using correlation inequalities for all couplings. Furthermore a remarkably simple argument of the authors based on Griffiths' and Lebowitz' inequalities shows that the dimensional physical coupling constant g, defined in terms of the expectation of four fields, is bounded. Thus $0 \leq g \leq$ const. The universal constant depends only on the dimension d of space-time. In other words the nonlinearity in the ϕ^4 equation of motion forces ϕ itself to be small. Using K. Symanzik's Brownian motion representation of correlations, improved correlation inequalities have been derived by Aizenman, Brydges, Fröhlich and Spencer which show that the constant vanishes for $d \leq 5$.

Bibliography

Absolute bounds on vertices and couplings, Ann. l'Inst. H. Poincaré A22, 97–107 (1975).

The coupling constant in a ϕ^4 quantum field theory, in Recent Developments in Gauge Theories, G.'t Hooft et al. Editors, Plenum, New York, 1980.

Absolute bounds on vertices and couplings

by

James GLIMM (1)
Rockefeller University, New York, New York 10021

and

Arthur JAFFE (2)
Harvard University Cambridge, Mass. 02138

ABSTRACT. — We prove absolute upper and/or lower bounds on φ^4 and φ^6 dimensionless vertices and physical coupling constants.

1. DEFINITIONS AND ASSUMPTIONS

We derive absolute bounds on φ^4 and φ^6 dimensionless coupling constants g. For instance, in a pure φ^4 model we prove

$$0 \leq g \leq \text{const.}, \tag{1.1}$$

in the single phase region (no symmetry breaking). We also obtain bounds on associated vertex functions and on connected parts. Our general methods presumably have other consequences for n-particle amplitudes, $n \geq 8$, which we do not pursue here.

DEFINITIONS. — Let

$$\langle 1 \ldots n \rangle \equiv \langle \Phi(x_1) \ldots \Phi(x_n) \rangle \tag{1.2}$$

(1) Supported in part by the National Science Foundation under Grant MPS 74-13252.

(2) Supported in part by the National Science Foundation under Grant MPS 73-05037.

denote the n-point (Euclidean) Schwinger function, and let $\langle 1 \ldots n \rangle_{\mathrm{T}}$ denote its connected part. We let

$$(1.3)\quad \mathrm{S}_{\mathrm{T}}^{(n)}(p) = \mathrm{S}_{\mathrm{T}}^{(n)}(p_1, \ldots, p_{n-1}) = \int \langle 1 \ldots n \rangle_{\mathrm{T}} \exp\left[i \sum_{j=1}^{n=1} p_j x_j\right] dx_1 \ldots dx_{n-1} \Big|_{x_n=0}$$

denote the Fourier transform of $\langle 1 \ldots n \rangle_{\mathrm{T}}$. We encounter $\mathrm{S}_{\mathrm{T}}^{(2)}(p)$ often, and denote it $\chi(p)$. We use the standard spectral representation

$$(1.4)\qquad \chi(p) = \mathrm{S}_{\mathrm{T}}^{(2)}(p) = \int \frac{d\rho(a)}{p^2 + a},$$

where $d\rho(a)$ is the spectral measure. Furthermore, we let

$$(1.5)\qquad \Gamma^{(n)}(p) = \Gamma^{(n)}(p_1, \ldots, p_{n-1})$$

denote the n-point Euclidean vertex function (thus $\Gamma^{(n)}$ is the amputated, one particle irreducible part of $\mathrm{S}_{\mathrm{T}}^{(n)}(p)$). In an even theory, we have for example,

$$(1.6)\qquad \Gamma^{(4)}(p) = \mathrm{S}_{\mathrm{T}}^{(4)}(p) \prod_{i=1}^{4} \chi(p_i)^{-1},$$

where $p_4 = -(p_1 + p_2 + p_3)$. Also

$$(1.7)\quad \Gamma^{(6)}(p) = \left[\mathrm{S}_{\mathrm{T}}^{(6)}(p) - \sum_{\mathrm{X}} \mathrm{S}_{\mathrm{T}}^{(4)}(\mathrm{X})\chi(\mathrm{P}_{\mathrm{X}})^{-1}\mathrm{S}_{\mathrm{T}}^{(4)}(\sim \mathrm{X})\right] \prod_{i=1}^{6} \chi(p_i)^{-1},$$

where $p_1 + \ldots + p_6 = 0$, where $(\mathrm{X}, \sim \mathrm{X})$ is a partition of $(p_1, \ldots, p_6)$ into two subsets of three elements, and where P_{X} is the sum of the momenta in X.

We let $d \leq 4$ denote the space-time dimension and let m denote the mass gap. Note that $\Gamma^{(n)}$ has the dimension of mass to the power $d - \frac{1}{2}n(d-2)$. We define the dimensionless amplitude $g^{(n)}(p)$ by

$$(1.8)\qquad g^{(n)}(p) = -m^{-d-n+\frac{1}{2}nd}\Gamma^{(n)}(p),$$

and the associated coupling constant $g^{(n)}$ by

$$(1.9)\qquad g^{(n)} = g^{(n)}(0).$$

For example,

$$(1.10)\qquad g^{(4)} = -m^{d-4}\chi^{-4} \int \langle 1234 \rangle_{\mathrm{T}} dx_1 dx_2 dx_3,$$

where $\chi \equiv \chi(0)$.

ASSUMPTIONS. — We make some or all of the following assumptions. In the case of weakly coupled $\mathscr{P}(\varphi)_2$, assumptions (*a*)-(*f*) have been proved, while for single phase, even φ_2^4 models (*h*)-(*i*) have been proved. We expect (*a*)-(*f*) to hold for single phase φ_3^4, φ_3^6, and φ_4^4 models, while (*h*)-(*i*) are expected for all single phase even φ_d^4 models, $d \leq 4$. The evidence for the validity of (1.4) in φ_4^4 comes from perturbation theory. The proof of (*i*) in even, single phase φ_2^4 has been established by Cartier (private communication).

(*a*) The Wightman axioms for a scalar boson field, or the Osterwalder-Schrader axioms for the corresponding Euclidean theory. The two point function $\chi(p)$ satisfies (1.4). In particular we are assuming that $\int \frac{d\rho(a)}{a}$ is finite.

(*b*) A positive mass gap, $m > 0$.

(*c*) $\langle \varphi \rangle = 0$.

(*d*) The second Griffiths inequality,

$$\langle A_1 A_2 \rangle \geq \langle A_1 \rangle \langle A_2 \rangle,$$

where (A_1, A_2) is a partition of $(1, \ldots, n)$. For instance,

$$(1.11) \quad -\langle 1234 \rangle_T = -\{\langle 1234 \rangle - \langle 14 \rangle\langle 23 \rangle - \langle 12 \rangle\langle 34 \rangle - \langle 13 \rangle\langle 24 \rangle\} \leq \langle 12 \rangle\langle 34 \rangle + \langle 13 \rangle\langle 24 \rangle.$$

(*f*) The measure $d\rho(a)$ contains a delta function at $a = m^2$, of unit strength (This is a renormalization hypothesis).

(*h*) In the case of a φ^4 interaction, it is appropriate to assume the Lebowitz inequality $\langle 1234 \rangle_T \leq 0$, or the set of Lebowitz inequalities,

$$(1.12) \qquad \langle 12U \rangle \leq \langle 12 \rangle \langle U \rangle + \sum_W \langle W, 1 \rangle \langle \sim W, 2 \rangle$$

where W is an (odd) subset of U.

(*i*) In the case of a pure φ^4 model, we assume

$$(1.13) \qquad \langle 123456 \rangle_T \geq 0.$$

2. BOUNDS ON QUARTIC COUPLINGS

THEOREM 2.1.

(*i*) Assume (*a*)-(*d*) above. Then

$$(2.1) \qquad g^{(4)} \leq \text{const.}\, \chi^{-2} m^{-4}.$$

(*ii*) Assume (*f*) also, then $\chi \geq m^{-2}$, so

$$g^{(4)} \leq \text{const.} \tag{2.2}$$

(*iii*) Assume (*h*) also, then

$$0 \leq g^{(4)} \leq \text{const.}\, \chi^{-2} m^{-4} \leq \text{const.} \tag{2.3}$$

Remark 1. — The constants are pure numbers, independent of all parameters. They can be determined explicitly by the proof.

Remark 2. — The upper bound on $g^{(4)}$ is related to the picture of « critical point dominance » discussed in [*1*]. According to this picture, for $d < 4$, $g^{(4)}$ assumes its maximum value at the critical point, *i. e.* at the onset of phase transitions. According to standard ideas, $m = 0$ at the critical point.

Remark 3. — For $d < 4$, the limit λ_0 (= bare φ^4 coupling constant) $\to \infty$ with m fixed plays a role in the application of the Callan-Symanzik equations to the study of critical phenomena [*2*]. We use Theorem 2.1 below, combined with bounds from [*3*], to show that either this limit exists, or that in this limit $g^{(4)} \to 0$ (see Corollary 2.4). If $g^{(4)}$ is monotone in the dimensionless charge $\lambda_0 m_0^{d-4}$, then the alternative $g^{(4)} \to 0$ is excluded. The possibility that $g^{(4)}$ is monotone is discussed in [*1*], and is a consequence of conventional assumptions made in the study of the Callan-Symanzik equations.

Remark 4. — Other bounds on quartic couplings have been established by Lukaszuk and Martin [*4*], see also Healy [*5*]. Their bounds require different assumptions and methods, and yield different conclusions.

In the case of a pure φ^4 model we also obtain bounds on $g^{(4)}(p)$ for $p \neq 0$.

THEOREM 2.2. — Assume (*a*)-(*h*) above. There exists $\delta > 0$, such that if $|p_i| < \delta m$, then $g^{(4)}(p)$ is analytic in p and bounded uniformly by

$$|g^{(4)}(p)| \leq \text{const.}\, \chi^{-2} m^{-4} \leq \text{const.} \tag{2.4}$$

Proof of Theorem 2.1. — By the positivity of $d\rho$ (Wightman positivity condition)

$$\langle 12 \rangle \geq 0, \tag{2.5}$$

so the right side of (1.11) is positive. By symmetry we then have

$$\begin{aligned} - \langle 1234 \rangle_T \leq (\langle 12 \rangle \langle 34 \rangle + \langle 13 \rangle \langle 24 \rangle)^{\frac{1}{2}} (\langle 14 \rangle \langle 23 \rangle \\ + \langle 12 \rangle \langle 34 \rangle)^{\frac{1}{2}} (\langle 13 \rangle \langle 24 \rangle + \langle 14 \rangle \langle 23 \rangle)^{\frac{1}{2}} \\ = \Big(\sum_{i=1}^{8} A_i \Big)^{\frac{1}{2}} \leq 2 \sup A_i^{\frac{1}{2}}. \end{aligned} \tag{2.6}$$

Here each A_i is either

$$\langle 12\rangle\langle 13\rangle\langle 14\rangle\langle 23\rangle\langle 24\rangle\langle 34\rangle = \prod_{1\le i<j\le 4} (ij)$$

or else a permutation of

$$\langle 12\rangle^2\langle 34\rangle^2\langle 13\rangle\langle 24\rangle.$$

We bound each A_i using Proposition A2.1, of Appendix 2,

$$\langle x0\rangle \le O(1)\,|x|^{-d_e - m(1-\varepsilon)|x|}\chi.$$

An elementary estimate shows that each $A_i^{\frac{1}{2}}$ is integrable over $dx_1dx_2dx_3$. By homogeneity, and (A2.1),

$$\int A_i^{\frac{1}{2}}dx_1dx_2dx_3 \le O(1)m^{-d}\chi^2,$$

from which we conclude

(2.7) $$-\,S_T^{(4)}(0) \le O(1)m^{-d}\chi^2.$$

By (1.8), with $n = 4$, and (1.6) we have (2.1). The theorem then follows.

Proof of Theorem 2.2. — Assuming (*h*) and (2.6),

(2.8) $$|\langle 1234\rangle_T| \le 2\sup_i A_i^{\frac{1}{2}}.$$

We bound the power series coefficients of $S_T^{(4)}(p)$, expanding about $p = 0$. The derivatives $\partial/\partial p_j$ become multiplication by ix_j in the Fourier transform, and are dominated by the exponential decrease of (2.8), estimated by (A2.1). Thus the power series in momenta converges for $|p_i| < m/3$, and establishes the analyticity and boundedness of $S_T^{(4)}(p)$,

(2.9) $$|S_T^{(4)}(p)| \le O(1)m^{-d}\chi^2.$$

Finally, the analyticity and boundedness of $\chi(p)^{-1}$ follows from (1.4), for $|p| < m$, yielding

$$|\chi(p)|^{-1} \le 2\chi(0)^{-1}.$$

Thus $g^{(4)}(p)$ is analytic in p and

$$|g^{(4)}(p)| \le O(1)m^{-d}\chi^{-2} \le O(1)m^{-d+4},$$

with the constant independent of p, m, λ for $|p_i| \le \delta m$. This completes the proof. We have also proved.

COROLLARY 2.3. — Assuming (*a*)-(*h*), $S_T^{(4)}(p)$ is analytic for $|p_i| \le \delta m$ and bounded uniformly by (2.9).

We now let $S_j^{(n)}$ be the n-point Schwinger function for a model satisfying (*a*)-(*h*) and labelled by the index j (for instance, $S^{(n)}_{\lambda_0,m}$ may be the φ^4_2 model, labelled by the bare coupling λ_0 and mass m).

COROLLARY 2.4. — Assume m_j is bounded away from zero. Then either there exists a convergent subsequence of Schwinger functions

$$S^{(n)} = \lim_i S^{(n)}_j, \tag{2.10}$$

with $S^{(n)}$ satisfying (*a*)-(*h*), or else $g^{(4)}_T(p) \to 0$.

Proof. — Assume for some p, with $|p_i| \leq \delta m$, there is a subsequence of $g_r(p)$ bounded away from zero. Then by (2.4)

$$\chi_r^2 \leq \text{const. } m_r^{-4}(g_r^{(4)}(p))^{-1} \leq \text{const.}$$

This bound (uniform in r) on

$$\int \langle 12 \rangle_{T,r} dx_1 = \int \langle 12 \rangle_r dx_1 \leq \text{const.}$$

proves a uniform bound

$$|\langle fg \rangle| \leq |f|_{\mathscr{S}} |g|_{\mathscr{S}} \tag{2.11}$$

for a suitable Schwartz space norm $|\,.\,|_{\mathscr{S}}$. Thus by Theorem 1 of [3], a convergent subsequence of Schwinger functions $S^{(n)}_{j_i}$ exists, and $S^{(n)}$ is a tempered distribution. The properties (*a*)-(*h*) follow from the corresponding properties in the approximate theories.

Conversely, if $g_j(p) \to 0$ for each p with $|p_i| < \delta m$, the uniform bound (2.4) ensures that the family of analytic functions $g_j(p)$ converges to the analytic function zero.

3. BOUNDS ON $S^{(6)}_T(p)$

In this section we prove preliminary bounds on S^6_T.

THEOREM 3.1.

(*i*) Assume (*a*)-(*h*). Then there exists $\delta > 0$ such that if $|p_i| \leq m\delta$, then $S^{(6)}_T(p)$ is analytic and bounded by

$$m^{2d}\chi^{-3} |S^{(6)}_T(p)| \leq \text{const.} \tag{3.1}$$

(*ii*) Assume (*a*)-(*i*). Then

$$0 \leq m^{2d}\chi^{-3} S^{(6)}_T(0) \leq \text{const.} \tag{3.2}$$

The constants are pure numbers, independent of the parameters.

Proof. — We consider $p = 0$. The extension to $p \neq 0$ follows the proof of Theorem 2.2. Using (f), we have

$$\langle 123456 \rangle - \langle 12 \rangle \langle 3456 \rangle \geq 0. \tag{3.3}$$

The definition of the six point connected part is

$$\langle 123456 \rangle_T = \langle 123456 \rangle - \sum_Y \langle Y_1 \rangle \langle Y_2 \rangle + 2 \sum_Z \langle Z_1 \rangle \langle Z_2 \rangle \langle Z_3 \rangle. \tag{3.4}$$

Here $Y = (Y_1, Y_2)$ ranges over the fifteen partitions of $(1, 2, \ldots, 6)$ into two and four element subsets, while $Z = (Z_1, Z_2, Z_3)$ ranges over the fifteen partitions of $(1, \ldots, 6)$ into three pairs. Thus

$$- \langle 123456 \rangle_T \leq \sum_Y{}' \langle Y_1 \rangle \langle Y_2 \rangle - 2 \sum_Z \langle Z_1 \rangle \langle Z_2 \rangle \langle Z_3 \rangle. \tag{3.5}$$

where Σ' omits the partition $\langle 12 \rangle \langle 3456 \rangle$. We use ($h$) and the definition of $\langle 1234 \rangle_T$ to obtain

$$- \langle 123456 \rangle_T \leq \sum_Z{}' \langle Z_1 \rangle \langle Z_2 \rangle \langle Z_3 \rangle, \tag{3.6}$$

where Σ' omits the three partitions which contain the factor $\langle Z_1 \rangle = \langle 12 \rangle$. By symmetry,

$$- \langle 123456 \rangle_T \leq 12 \sup \prod_{j=1}^{15} \prod_{i=1}^{3} \langle Z_i^{(j)} \rangle^{1/15}, \tag{3.7}$$

where $j = 1, 2, \ldots, 15$ labels the two element subsets of $(1, \ldots, 6)$, and $(Z_1^{(j)}, Z_2^{(j)}, Z_3^{(j)})$ one of the twelve partitions of $(1, \ldots, 6)$ into three pairs, no one of which is j.

We now follow the proof of Theorem 2.1 to obtain

$$- S_T^{(6)}(0) \leq \text{const.}\, m^{-2d} \chi^3.$$

proving one side of (3.1).

We next use (1.12) (see [3]) with $U = (3456)$ to obtain

$$\langle 123456 \rangle \leq \langle 12 \rangle \langle 3456 \rangle + \sum_W \langle W, 1 \rangle \langle \sim W, 2 \rangle,$$

where W ranges over the eight odd subsets of U. Thus by (3.4),

$$\langle 123456 \rangle_T \leq - \sum_Y{}'' \langle Y_1 \rangle \langle Y_2 \rangle + 2 \sum_Z \langle Z_1 \rangle \langle Z_2 \rangle \langle Z_3 \rangle$$

where Σ'' ranges over the six partitions containing both 1 and 2 in the four element subset. By the definition of $\langle 1234 \rangle_T$,

$$\langle 123456 \rangle_T \leq - \sum_{Y}'' \langle Y_1 \rangle_T \langle Y_2 \rangle_T + \sum_{Z}' \langle Z_1 \rangle \langle Z_2 \rangle \langle Z_3 \rangle .$$

By (1.11) we note that

$$\langle 123456 \rangle_T \leq 2 \sum_{Z}' \langle Z_1 \rangle \langle Z_2 \rangle \langle Z_3 \rangle ,$$

which is just a bound by twice the right hand side of (3.6). We now proceed as above to obtain the bound (3.2) for $p = 0$.

Finally, we remark that assuming (*i*) yields $0 \leq S_T^{(6)}(0)$, to complete the proof.

4. BOUNDS ON $g^{(6)}$

We use the bounds of Chapter 3 to establish bounds on the six point vertex function in the φ^4 model.

THEOREM 4.1. — Assuming (*a*)-(*h*), for $|p_i| \leq m\delta$, $g^{(6)}(p)$ is analytic and bounded by

$$|g^{(6)}(p)| \leq \text{const.}$$

where δ is the smaller of the constants given by Theorems 2.2 and 3.1.

Proof. — By Theorem 3.1, $m^{2d}\chi^{-3}S_T^{(6)}(p)$ is analytic and bounded. By Theorem 2.2 $g^{(4)}(p)$ is analytic and bounded. Note

$$g^{(4)}(p) = m^{d-4} \prod_{i=1}^{4} \chi(p_i)^{-1} S_T^{(4)}(p).$$

The desired bound follows by

$$g^{(6)}(p) = - m^{2d-6} \Gamma^{(6)}(p)$$

and the representation (1.7).

APPENDIX 1

THE FREE EUCLIDEAN PROPAGATOR

We establish the elementary bounds on

$$c(m^2 ; x) = \frac{1}{\pi} \int \frac{e^{-ipx} dp}{p^2 + m^2} = m^{d-2} c(1 ; mx)$$

used below. By Euclidean invariance, we may evaluate c for $x = x_t = (t, \vec{0}), t > 0$, and

$$f(t) = c(1 ; x_t) = \int \frac{e^{-t\mu}}{\mu} \, d\vec{p}:$$

where $\mu = \mu(\vec{p}) = (\vec{p}^2 + 1)^{\frac{1}{2}}$.

PROPOSITION A1.1. — There is a constant O(1) depending only on d, such that

$$f(t) \leq \begin{cases} O(1) t^{(1-d)/2} e^{-t} & t \geq \frac{1}{2} \\ O(1) t^{-(d-2)} & t \leq \frac{1}{2}, \ d \geq 3 \\ O(1) \mid \ln t \mid & t \leq \frac{1}{2}, \ d = 2 \end{cases}$$

REMARK. — The long and short range exponents of t differ.

Proof. — Let $\mu_t(\vec{p}) = (\vec{p}^2 + t^2)^{\frac{1}{2}}$, so

$$f(t) = t^{-(d-2)} e^{-t} I(t)$$

where

$$I(t) = \int \mu_1^{-1} \exp \left[- (\mu_t - t) \right] d\vec{p}.$$

For $t \geq \frac{1}{2}$, and $t \geq \mid \vec{p} \mid$, we use

$$\mu_t - t \leq \text{const. } \vec{p}^2 t^{-1}$$

to establish

$$I(t) \leq O(t^{(d-3)/2}).$$

Furthermore, for $\frac{1}{2} \leq t \leq \mid \vec{p} \mid$, we use $\mu_t - t \geq \text{const.} \mid \vec{p} \mid$ to give

$$I(t) \leq O(t^{-1}).$$

Thus for $t \geq \frac{1}{2}$, we have established the proposition.

For $t \leq \frac{1}{2}$, and $d \geq 3$, we have

$$I(t) \leq \int \mid \vec{p} \mid^{-1} \exp \left[- \mid \vec{p} \mid + \frac{1}{2} \right] d\vec{p} \leq O(1).$$

To bound the case $d = 2$, with $t \leq \frac{1}{2}$, we use

$$I(t) \leq \text{const.} \int_0^\infty e^{-u} \frac{du}{(t^2 + u^2)^{\frac{1}{2}}} \leq O(\mid \ln t \mid).$$

APPENDIX 2

THE EUCLIDEAN PROPAGATOR

The Euclidean propagator $\langle xy \rangle$ has the spectral representation given by (1.4),

$$\langle xy \rangle = \langle \Phi(x)\Phi(y) \rangle = \int d\rho(a)c(a\,;\,x-y).$$

PROPOSITION A2.1. — Assume (*a*)-(*c*) above, and let $\varepsilon > 0$. Then

$$\langle x0 \rangle \leq O(1)\,|\,x\,|^{-d}e^{-m(1-\varepsilon)|x|}\chi, \tag{A2.1}$$

where O(1) depends only on the dimension d and on ε.

PROOF. — We use proposition A1.1. Let $d \geq 3$. Then

$$\begin{aligned}\langle x0 \rangle &= \int d\rho(a)a^{(d-2)/2}c(1\,;\,\sqrt{a}x).\\ &\leq O(1)\int \frac{d\rho(a)}{a}\,a\,|\,x\,|^{-d+2}e^{-\sqrt{a}|x|(1-\varepsilon/2)}\\ &\leq O(1)e^{-m(1-\varepsilon)|x|}\left(\int \frac{d\rho(a)}{a}\right)|\,x\,|^{-d+2}\left(\sup_a ae^{-\sqrt{a}|x|\varepsilon/2}\right).\end{aligned}$$

The sup occurs for $a = 4(\varepsilon x)^{-2}$. Thus (A2.1) follows. For $d = 2$,

$$\begin{aligned}\langle x0 \rangle &= \int d\rho(a)c(1\,;\,\sqrt{a}x)\\ &\leq O(1)\int_{\mathrm{I}} \frac{d\rho(a)}{a}\,a\,|\ln\sqrt{a}x\,| + O(1)\int_{\mathrm{II}} \frac{d\rho(a)}{a}\,ae^{-\sqrt{a}|x|},\end{aligned}$$

where

$$\mathrm{I} = \{ a : m \leq \sqrt{a} \leq |\,2x\,|^{-1} \} \quad \text{and} \quad \mathrm{II} = \{ a : \sqrt{a} \geq |\,2x\,|^{-1} \}.$$

The integral over II is bounded as above, by (A2.1). The integral over I is bounded using

$$\sup_{a\in\mathrm{I}} a\,|\ln\sqrt{a}\,|\,x\,|\,| \leq |\,x\,|^{-2} \sup_{0\leq u\leq 1} u^2\,|\ln u\,| \leq O(1)\,|\,x\,|^{-2}.$$

This yields the bound $O(1)\,|\,x\,|^{-2}\chi$. Since $|\,mx\,| \leq \frac{1}{2}$, for $a \in \mathrm{I}$, the integral over I is bounded by (A2.1).

REMARK. — The increase in the bound on the short range singularities (from $x^{(2-d)}$ to x^{-d}) is associated with an allowed divergence of the field strength renormalization integral

$$Z^{-1} = \int d\rho(a).$$

REFERENCES

[1] J. GLIMM and A. JAFFE, Critical point dominance in quantum field models. *Ann. Institut Henri Poincaré*, t. **21**, 1974, p. 27-41.

[2] G. PARISI, Field theory approach to second order phase transitions in two and three dimensional systems, *1973 Cargèse lectures*.

[3] J. GLIMM and A. JAFFE, Remark on the existence of φ^4. *Phys. Rev. Lett.*, t. **33**, 1974, p. 440-442.

[4] L. LUKASZUK and A. MARTIN, Absolute upper bounds for $\pi\pi$ scattering. *Nuovo Cimento*, t. **52** *A*, 1967, p. 122-145.

[5] J. HEALY, New rigorous bounds on coupling constants in field theory. *Phys. Rev.*, **D8**, 1973, p. 1904-1914.

(Manuscrit reçu le 9 août 1974)

THE COUPLING CONSTANT IN A ϕ^4 FIELD THEORY

James Glimm*[1] and Arthur Jaffe+[2]

*The Rockefeller University, New York, NY 10021

+Harvard University, Cambridge, MA 02138

Abstract

Under simple hypothesis, the following statements are proved in a ϕ^4 field theory: (1) The coupling constant is continuous in the parameters, e.g. as the scaling limit is approached or as ultraviolet cutoffs are removed. (2) Zero coupling corresponds to a free field or a generalized free field. (3) Different types of amputation in the definition of the coupling constant differ only by a finite nonzero factor.

* * * * *

In this note, we show by elementary methods that the ϕ^4 coupling constant is a continuous function of the parameters under assumptions which should be appropriate for the critical point scaling limit or removal of the ultraviolet cutoff. In particular the limit theory is nontrivial (not a generalized free field) if and only if the coupling constants of the approximating theories are bounded away from zero. Our analysis applies also to the Ising model, at its critical point.

The motivation for this paper is the statement of Baker [1] that high temperature series expansions for the d=3 Ising model

---

1. Supported in part by the National Science Foundation under Grant PHY 78-08066.

2. Supported in part by the National Science Foundation under Grant PHY 79-16812.

give small but significant violations of hyperscaling: $d\nu + \gamma \neq 2\Delta$. If these numbers are significant, we agree with Baker that the critical point defines a trivial field theory.

We work with a coupling constant defined off mass shell (at p=0), in order to be able to perform the analysis. Particle amputation multiplies an external leg of a n-point function by

$$m^2 - p^2 \Big|_{p=0} = m^{+2}$$

while propagator amputation multiplies by

$$\left(Z^{-1} G^{(2)\sim}(p)\right)^{-1} \Bigg|_{p=0} = Z \Big/ \int G^{(2)}(x-y)\, d(x-y)$$

where

$$G^{(2)} = \langle\phi(x)\,\phi(y)\rangle - \langle\phi\rangle^2$$

is the truncated two point function of the ϕ^4 field theory. Also let $\chi = \int G^{(2)}\, d(x-y)$. The the coupling constant is

$$0 \leq g_\ell = -Z^{-2} m^{d+4} \int G^{(4)}(x_1, \ldots, x_4)\, dx_1 \ldots dx_3 \left(\frac{Z}{m^2 \chi}\right)^\ell \tag{1}$$

In g_ℓ, ℓ legs have propagator amputation and $4 - \ell$ legs have particle amputation. The bound

$$Z/(m^2\chi) \leq 1 \tag{2}$$

follows from the definition of field strength renormalization, so that

$$g \equiv g_0 \geq g_1 \geq \cdots \geq g_4 \quad . \tag{3}$$

It was first shown in [2] that

$$g_2 = -m^d \int G^{(4)}/\chi^2 \leq \text{const.} \tag{4}$$

Since g_2 has the critical exponents,

(5) $$g_2 \sim \tau^{-2\Delta+\gamma+d\nu} \quad ,$$

this leads to the inequality

(6) $$-2\Delta + \gamma + d\nu \geq 0 \quad .$$

The subsequent consideration of (4)-(6) by other authors is summarized in [1].

In this paper we work only with g_0 . The finiteness of

(7) $$g_\ell / g_{\ell+1} = Z^{-1} m^2 \chi$$

follows from conventional scaling ideas, which we now assume. In particular we assume

$$0 \leq Z^{-1} m^{-d+2} G^{(2)}(x/m) \leq \begin{cases} O(1)\ x^{-(d-1)/2} e^{-|x|} , & |x| \geq 1 \\ O(1)\ x^{-d+2-\eta} , & |x| \leq 1 \end{cases}$$

We take $\eta < 2$ and the constants O(1) above to be universal constants. Then (7) is bounded uniformly in all parameters. This gives us

Theorem 1. Assume (8) with universal constants. Then with the hypothesis of [2], g and all g_ℓ are bounded, by a universal constant. Either the g_ℓ are all nonzero or they are all zero. In the latter case, and only then, the theory is a generalized free field.

The final statement follows from [3].

Now we turn to the continuity of g under a limit process. We assume that m is bounded away from zero in the limit process. This allows a (positive mass) scaling limit at the critical point, and removal of ultraviolet cutoffs, for noncritical couplings. By scaling, we assume m is fixed at 1.

To illustrate the idea, we see that $g_\ell / g_{\ell+1}$ is convergent by the Lebesgue bounded convergence theorem. Thus is is sufficient to consider g_0 .

Theorem 2. Assume (8) with universal constants and the hypothesis of [2]. The each g_ℓ is continuous in a limit process for which the mass m is fixed and not zero, and for which the two and four point functions $G^{(2)}(x)$ and $G^{(4)}(x_1,\ldots,x_4)$ are pointwise convergent a. e.

Proof. By Lebowitz' and Griffiths' inequalities,

$$0 \le -G^{(4)}(x_1,\dots,x_4) \le G^{(2)}(x_{i_1} - x_{i_2})G^{(2)}(x_{i_3} - x_{i_4})$$

for any permutation $\{i_1,\dots,i_4\}$ of $\{1,\dots,4\}$. After a permutation and translation of variables, we can suppose that $i_\nu = \nu$, $x_1 = 0$, and $|x_2 - x_3| \le |x_3 - x_4|$. Also we may take $m = 1$. Then with $F(x)$ denoting the right hand member of (8),

$$\int F(x_1 - x_2)\, F(x_3 - x_4)\, dx_2\, dx_3\, dx_4$$

$$\le \int_{|x_2-x_3| \le |x_3-x_4|} F(x_3 - x_4)F(x_2)\, d(x_2 - x_3)\, d(x_3 - x_4)\, dx_2$$

$$\le \int x^d\, F(x)\, dx \int F(x)\, dx < \infty \quad .$$

Thus continuity follows from the Lebesgue bounded convergence theorem.

References

1. G. Baker, Phys. Rev. B, 15 p.1553 (1977).

2. J. Glimm and A. Jaffe, Ann. Inst. H. Poincaré 22 p.97 (1975).

3. C. Newman, Commun. Math. Phys. 41 p.1 (1975).

VII Confinement and Instantons

Introduction

These papers are two in a series of articles in which the authors investigate the physics of instanton gases. The abelian four dimensional lattice gauge theory yields a Coulomb dipole gas. The second paper discusses confinement in terms of vortex configurations and their relation to the center of the gauge group.

Bibliography

Instantons in a U(1) lattice gauge theory, Commun. Math. Phys. 56, 195–212 (1977).

Charges, vortices, and confinement, Nuclear Physics B149, 49–60 (1979).

Instantons in a $U(1)$ Lattice Gauge Theory: A Coulomb Dipole Gas

James Glimm*
Rockefeller University, New York, NY 10021, USA

Arthur Jaffe**
Harvard University, Cambridge, MA 02138, USA

Abstract. We study the decomposition $A = A_I + A_{SW}$ of a $U(1)$ lattice gauge field into instanton and spin wave parts. The action also decomposes, $\mathcal{A} = \mathcal{A}_I + \mathcal{A}_{SW} + R$. Here $\mathcal{A}_I$ is a Coulomb dipole gas, $\mathcal{A}_{SW}$ is a zero mass free field, and R is a higher order remainder. We study $\mathcal{A}_I$ in detail, for $d \geqq 4$, in the dilute gas case (which corresponds to the low temperature limit of the gauge field theory). We establish the leading behavior of the free energy: $f \sim \varepsilon^{-d} a\zeta$. Here ε is the lattice spacing, a is a geometrical constant and ζ is an activity defined in terms of a small number of instanton configurations. Our methods suggest the absence of screening in the dilute dipole gas, $d \geqq 4$, in contrast to Debye screening for the dilute monopole gas.

1. Introduction

It has been proposed by Gell-Mann and others that a gauge field coupled to a fermion (quark) field may describe the internal structure of protons, neutrons, etc., as quark triplets, etc. In order to account for the strong binding of individual quarks, as well as the observed property of asymptotic freedom of the physical particles, a qualitative understanding of the phase transitions and critical points of the gauge-quark system appears necessary [11]. The important properties of the critical points are inferred from renormalization group and semiclassical methods.

In this paper we study some steps in this program on a rigorous mathematical level. See [8] for an introduction to the mathematics of lattice gauge fields. Following Polyakov [9], we formally derive the instanton interaction in terms of a dilute Coulomb gas for a $U(1)$ lattice gauge theory. For $d \geqq 4$, the gas consists of vector charges, and a conservation law restricts the charges to occur only in dipoles. Since the dipole-dipole interaction is not integrable at long distances, standard methods do not apply. Nevertheless, we prove an upper and lower bound on the free

* Supported in part by the National Science Foundation under Grant PHY 76-17191
** Supported in part by the National Science Foundation under Grant PHY 75-21212

energy of the dipole gas. The essential step is to use a cancellation between the forces produced on a test dipole by oppositely oriented dipoles occurring in a block. Our conclusions suggest that higher order corrections to the dipole forces do not generate a mass (finite correlation length at sufficiently low temperature. This is in contrast to the $d=3$ dilute Coulomb gas of scalar charges for which Debye suggested a finite correlation length, see [1].

We study in detail the action

$$\mathscr{A} = -\beta \sum \cos(\varepsilon^2 gF) . \tag{1.1}$$

We first separate $F = F_I + F_{SW}$ into two parts: The spin wave part F_{SW} characterizes the oscillations withing one cosine well, while the instanton part F_I characterizes the transitions between different wells of the cosine function. The splitting gives rise to a magnetic charge density Q with the geometric interpretation

$$Q = \partial S ; \tag{1.2}$$

namely Q is the boundary ($d-3$ dimensions) of a Dirac surface S ($d-2$ dimensions). We call Q magnetic (rather than electric) since S is the geometric dual of $[F]$ (a discretization of F). The divergence of the dual of F is generally called a magnetic charge.

For $d=3$ dimensions, the monopole charge Q occurs as the boundary of the Dirac string S. For $d \geqq 4$, Q satisfies the conservation law

$$\partial Q = 0 . \tag{1.3}$$

An elementary charge configuration for $d=4$ is illustrated in Figure 1. In that case the vector charges (arrows) bound an elementary plaquette in the dual lattice.

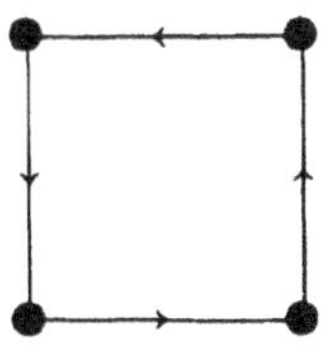

Fig. 1. Elementary dipoles, $d=4$

Because of the conservation law (1.3), the elementary charge configurations in $d \geqq 4$ are dipolar, in fact two orthogonal dipoles for $d=4$.

With the above splitting of F we obtain for the action the splitting

$$\mathscr{A} = \mathscr{A}_I + \mathscr{A}_{SW} + R . \tag{1.4}$$

Each of the three parts in (1.4) is individually gauge invariant. Here $\mathscr{A}_I$ is a Coulomb interaction for the charges Q, while $\mathscr{A}_{SW}$ is the action for a zero mass free field. The remainder R is of order three or more in Q or F_{SW}, and formally vanishes at zero lattice spacing. In the detailed analysis which follows, we neglect R and consider the approximate action $\mathscr{A}_I$ and $\mathscr{A}_{SW}$ composed of independent parts. The free field part $\mathscr{A}_{SW}$ can be solved exactly, while the instanton part $\mathscr{A}_I$ requires a detailed analysis. $\mathscr{A}_I$ also vanishes formally at zero lattice spacing.

In this manner we arrive at the analysis of a Coulomb gas of monopoles ($d=3$) and of dipoles ($d \geqq 4$). The remainder of the paper is devoted to the detailed study of the statistical mechanics of a dipole gas, see also [12]. The basic difficulties arise from the long range nature of the Coulomb action. We develop an expansion method to study the free energy of the dipoles at low density, i.e. $\beta \gg 1$.

Our detailed analysis involves several steps. First we define dipole clusters Q_L (i.e. average or block charges) to replace the elementary dipole excitation Q of Figure 1. We choose the dipole clusters so that the long range interaction between two clusters is the same as the long range interaction between two elementary dipoles, up to a quadrapole error. Here we have the lengths scales

$$\varepsilon \ll R_L \ll r\,; \tag{1.5}$$

ε is the lattice spacing (diameter of the elementary excitation), $R_L = \varepsilon O(\beta^{1/d-2})$ is the diameter of the dipole cluster and r is the inter-cluster separation. The quadrapole-quadrapole interaction is $O(r^{-(d+2)})$, which is integrable at infinity and thus can be treated by standard methods. (The possibility of charge-quadrapole interactions actually requires a second averaging and the replacement of quadrapole by octapole error terms.)

Each dipole cluster has a much reduced self action. The elementary dipoles have self action $O(\beta)$, while the dipole clusters have self action $O(\beta^{-\varepsilon})$. This requires the introduction of an activity

$$\zeta = e^{-O(\beta)} \tag{1.6}$$

which characterizes each averaged dipole cluster. The small activity results in a dilute gas of dipoles in typical configurations, i.e. in an expected density $O(\zeta) \ll 1$.

Using this typical density, we can sum the action up to a statistical length scale $R_{\mathrm{stat}} = \varepsilon \exp(\zeta^{-\delta})$. Thus

$$\varepsilon \ll R_L \ll R_{\mathrm{stat}}\,. \tag{1.7}$$

For $r > R_{\mathrm{stat}}$, we take advantage of the cancellation due to charge symmetry of the dipole gas. For every dipole of positive orientation in Figure 1, there is an elementary dipole with opposite orientation. Thus we can define the total charge δ of a large number ν of dipoles. Configurations which have a large dipole charge are accompanied by a small statistical weight

$$P \leqq e^{-\delta^2/2\nu}\,. \tag{1.8}$$

Here $\nu = \nu_+ + \nu_-$ and $\delta = \nu_+ - \nu_-$ is the deviation from dipole neutrality. The entropy factor (1.8) is just the binomial distribution

$$P = 2^{-\nu}\binom{\nu}{\nu_+}\,. \tag{1.9}$$

We use this small factor to produce conditional convergence on the distance scale $r > R_{\mathrm{stat}}$.

2. The Definition of Instantons and Spin Waves

We consider the simple cubic lattice Z_ε^d with lattice spacing ε. The $U(1)$ gauge field is a Lie algebra valued 1-cochain defined on lattice bonds (i.e. simplices connecting nearest neighbor lattice sites). An element γ of the $U(1)$ group can be written

$$\gamma=\exp(i\varepsilon gA), \qquad -\pi\leqq\varepsilon gA\leqq\pi, \tag{2.1}$$

and we define A to be the gauge field. The 2-cochain defined as the coboundary of A is

$$F=\partial^*A=\varepsilon^{-1}\sum_{i\in\partial p}A(i). \tag{2.2}$$

Thus F is defined on lattice squares (plaquettes p). Since four links bound any plaquette,

$$|\varepsilon^2 gF|\leqq 4\pi. \tag{2.3}$$

We have introduced the factor ε^{-1} in the definition of ∂^* so both A and F formally have limits as $\varepsilon\to 0$. In this limit A is a 1-form, and $F=dA$ is a 2-form. The de Rham cohomology of differential forms on a compact manifold is isomorphic to the simplicial cohomology of a smooth triangulation of the manifold, see [10]. See [7] for the geometry of gauge fields.

The action $\mathscr{A}$ associated with F is defined by

$$\mathscr{A}=-\varepsilon^{d-4}g^{-2}\sum_p\cos(\varepsilon^2 gF)=-\varepsilon^{d-4}g^{-2}\sum_p\cos(\varepsilon^2 g\partial^*A), \tag{2.4}$$

where the sum ranges over plaquettes. We note that

$$\beta=\varepsilon^{d-4}g^{-2} \tag{2.5}$$

is a dimensionless coupling constant which is inversely related to the strength of the interaction. Thus

$$\mathscr{A}=-\beta\sum_p\cos(\varepsilon^2 gF). \tag{2.6}$$

We let $[F]$ denote the discretization of F. By definition

$$\varepsilon^2 g[F]=2\pi j, \quad \text{when} \quad j-\tfrac{1}{2}<(2\pi)^{-1}\varepsilon^2 gF\leqq j+\tfrac{1}{2}, \tag{2.7}$$

and where j is an integer. According to (2.3), the possible values of j are 0, ± 1, ± 2.

We note that $\partial^*F=(\partial^*)^2A=0$, which we can also write $\delta^* * F=0$. Here we introduce the duality map $*$. For the purpose of geometric interpretation we let $*$ be the composition of Hodge duality with the metric. Hodge duality defines an isomorphism between r-cochains on Z_ε^d to $(d-r)$-cochains on the dual lattice $*Z_\varepsilon^d$. The metric defines an isomorphism between $(d-r)$ cochains and $(d-r)$ chains (raising and lowering indices). Thus

$$*: r \text{ cochains} \to d-r \text{ chains}$$
$$r \text{ chains} \to d-r \text{ cochains}$$

and satisfies

$$**=(-1)^{r(d-r)}.$$

Furthermore, under the $*$ isomorphism,

$$*\partial^**=\partial, \quad *\delta^**=\delta,$$

where ∂ is the boundary operator, and δ is its adjoint.

In contrast to the above, $[F]$ does not have vanishing coderivative, and we define the instanton charge density $*Q$ by

$$\partial^*[F]=*Q\,, \tag{2.8}$$

or with $S=[*F]$, $\partial S=\partial[*F]=Q$, cf. (1, 2). Maxwell's equations have the form

$$\partial^* F=0, \quad \delta^* F=-\omega_{d-1}J\,, \tag{2.9}$$

which says that J is the source for F. Similarly Q is the source for $*[F]$. Of course $\partial Q=0$, or equivalently, $\partial^* * Q=0$, which is a conservation law for $d>3$. For $d=4$, the Dirac surface has its boundary Q defined on links, i.e. Q is a 1-cochain. Then $\partial Q=0$ is a Kirkoff law: $\sum_i Q(i)=0$, where the sum runs over $2d=8$ oriented links meeting at a given lattice site. Thus Q is concentrated on closed $d-3$ surfaces (loops, for $d=4$) and it follows that the charge distribution consists of dipoles. For $d=4$, the corresponding quantum mechanical soliton ("boxiton") has also been studied [6]. For $d=3$, however, Q is a 0-cochain, concentrated at points, which corresponds to isolated charges.

This distinction of loop charges (or closed $d-3$ surfaces, $d>3$) vs. point charges ($d=3$) is basic to the study of phase transition in $U(1)$ gauge theories, see also [9]. In particular the $d=4$ loop charge explain the expected occurrence of an ordered phase at sufficiently low temperatures $T>0$ for $d\geqq 4$, and the absence of this phase for $d=3$. The mechanism, that an appropriate boundary first has dimension greater than zero, and that its entropy is thus dominated at low temperature by its energy, is the same as the mechanism governing phase transitions in the Ising model. Thus we expect that the critical dimension for phase transitions in $U(1)$ gauge theories is $d_c=3$, in contrast to the critical dimension $d_c=1$ for the Ising model. The inequality $d_c\geqq 3$ was established in [3].

Now consider the transformation $A(x)\to A(x)+2\pi n_x(\varepsilon g)^{-1}$, with n_x an integer. Under this change of variables,

$$F(p)=(\partial^* A)(p)\to F(p)+2\pi(\varepsilon g)^{-1}(\partial^* n)(p) \tag{2.10}$$

and

$$[F]\to[F]+2\pi\sum_{\partial p} n_x\,.$$

Since $\partial^*\partial^*=0$, Q is unaffected, as is the action $\mathscr{A}$. It follows that S, regarded as a $d-2$ surface, can be deformed by such a change of variables, to an arbitrary $d-2$ surface with the same boundary Q. Unfortunately, there does not seem to be a natural determination of a unique $*[F]$.

The lattice Laplace operator is

$$\Delta=-\delta^*\partial^*-\partial^*\delta^*\,, \tag{2.11}$$

and its dual is

$$\Delta=-\delta\partial-\partial\delta\,.$$

Because the topology of R^d is trivial, we can form Δ^{-1}. The same is true for a finite lattice with Dirichlet boundary conditions or with periodic boundary conditions.

(In the latter case, the topology is trivial except in dimension one.) Consider

$$F_I = [F] + \delta^* \Delta^{-1} * Q . \tag{2.12}$$

Then

$$\begin{aligned} \partial^* F_I &= \partial^* F + \partial^* \delta^* \Delta^{-1} * Q \\ &= *Q + (\partial^* \delta^* + \delta^* \partial^*) \Delta^{-1} * Q = *Q - *Q = 0 . \end{aligned} \tag{2.13}$$

Since the topology is trivial and f is a 2-cochain, we can define A_I by the equation

$$\partial^* A_I = F_I = [F] + \delta^* \Delta^{-1} * Q . \tag{2.14}$$

This is the vector potential for instantons. Finally we define the spinwave potential $A_{SW} \equiv A - A_I$ and $F_{SW} = F - F_I$.

For the action we now have

$$\begin{aligned} \mathscr{A} &= -\beta \sum \cos(\varepsilon^2 g \partial^* A) \\ &= -\beta \sum \cos(\varepsilon^2 g \partial^* A_I) \cos(\varepsilon^2 g \partial^* A_{SW}) + \beta \sum \sin(\varepsilon^2 g \partial^* A_I) \sin(\varepsilon^2 g \partial^* A_{SW}). \end{aligned} \tag{2.15}$$

Observe that

$$\begin{aligned} \cos(\varepsilon^2 g \partial^* A_I) &= \cos(\varepsilon^2 g \partial^* (-\Delta)^{-1} * Q) \\ \sin(\varepsilon^2 g \partial^* A_I) &= \sin(\varepsilon^2 g \partial^* (-\Delta)^{-1} * Q) . \end{aligned} \tag{2.16}$$

so the Dirac surface S or $*[F]$ does not contribute to the action. Also note

$$\langle \delta^* (-\Delta)^{-1} * Q, \partial^* A_{SW} \rangle = \langle (-\Delta)^{-1} * Q, \partial^* \partial^* A_{SW} \rangle = 0 .$$

Based on the φ^4 analysis [4], we expect that the dominant effect on long range order will be given by the decoupled instanton and spin wave fields, i.e. from the terms in (2.15) which are lowest order in εg (and formally ε, g independent as $\varepsilon \to 0$). Thus following Polyakov [9] we introduce the approximate action

$$\begin{aligned} \mathscr{A}_0 &= \text{const} + \tfrac{1}{2} \sum \varepsilon^d (\delta(-\Delta)^{-1} Q)^2 + \tfrac{1}{2} \sum \varepsilon^d (\partial^* A_{SW})^2 \\ &= \text{const} + \mathscr{A}_I + \mathscr{A}_{SW} . \end{aligned} \tag{2.17}$$

After decoupling, we will have long range order only if it occurs in each of the fields A_I and A_{SW}. Since the approximate action for A_{SW} is Gaussian, it can be explicitly computed that long range order occurs in A_{SW} for $d \geqq 4$, but not for $d \leqq 3$ (cf. [3]). Thus for the remainder of the paper we consider only the instanton field generated by the charges Q. Since

$$\begin{aligned} \langle \delta(-\Delta)^{-1} Q, \delta(-\Delta)^{-1} Q \rangle &= \langle Q, (-\Delta)^{-1} \partial \delta (-\Delta)^{-1} Q \rangle \\ &= \langle Q, (-\Delta)^{-2} (\partial \delta + \delta \partial) Q \rangle = \langle Q, (-\Delta)^{-1} Q \rangle , \end{aligned} \tag{2.18}$$

we see that the instanton action is

$$\begin{aligned} \mathscr{A}_I &= \tfrac{1}{2} \langle Q, (-\Delta)^{-1} Q \rangle = \tfrac{1}{2} \sum \varepsilon^d q_i (-\Delta)^{-1} q_j \\ &\sim \tfrac{1}{2} \omega_{d-1} \sum_{i,j} \varepsilon^{2d} q_i r_{ij}^{-d+2} q_j . \end{aligned} \tag{2.19}$$

Thus the $U(1)$ instantons interact with Coulomb forces in the approximation (2.17). Returning to the exact action (2.15), we expect that the Coulomb forces are the leading long distance parts of the instanton action.

3. The Landau Length Scale and the Activity

In the approximation of § 2, the pure instanton action is

$$\mathscr{A}_I = \tfrac{1}{2}\sum \varepsilon^d q_i(-\Delta)^{-1} q_j \,. \tag{3.1}$$

We recognize q_i as the charge density and $\varepsilon^d q_i$ as a charge. Each q_i takes values which are integer multiples of $2\pi\varepsilon^{-3}g^{-1}$, by virtue of (2.7). We let $\beta \equiv \varepsilon^{d-4}g^{-2}$ define the Landau length scale R_L by the formula

$$\beta = (R_L/\varepsilon)^{d-2} \,. \tag{3.2}$$

R_L characterizes the size of individual contributions to (3.1). In other words,

$$R_L = (\varepsilon^{2d-6}g^{-2})1/^{(d-2)} = \varepsilon\beta^{1/(d-2)} \,, \tag{3.3}$$

and the interactions in (3.1) on a length scale less than R_L are strong, while those on a length scale greater than R_L are weak. In particular, if $R_L \gg \varepsilon$, then the *self interaction* is large compared to the *interaction between distinct charges.* The self interaction of a single dipole loop is of the order

$$\varepsilon^{2d}q^2\varepsilon^{-d+2} = O(\varepsilon^{2d-6}g^{-2}\varepsilon^{-d+2}) = O(\beta) \,. \tag{3.4}$$

It is convenient to modify the definition (3.3) in order to make R_L/ε an integer. We set

$$R_L = \varepsilon[\beta^{1/(d-2)}] \,, \tag{3.3'}$$

where here $[\,\cdot\,]$ denotes the integer part. With this definition, the Landau lattice $Z^d_{R_L}$ is a sublattice of Z^d_ε.

Given a charge configuration Q, we will define a decomposition into a typical part Q_t and an atypical part Q_a,

$$Q = Q_a + Q_t \,, \tag{3.5}$$

see Chapter 4. The main property of Q_t is that it is a union of connected components of Q, and that Q_t has a low density of $d-3$ lattice elements. (Here we consider Q as a $d-3$ chain, and hence as a $d-3$ surface in R^d.) Since ∂ is a local operator, we have

$$\partial Q_t = 0 \,. \tag{3.6}$$

We now show how to replace the charge distribution Q_t by an averaged charge distribution $Q_{t,L}$ on the Landau lattice. Each elementary dipole of Q_t in the dual Z^d_ε lattice will be replaced by a large number of weaker dipoles in the dual Landau lattice. In making this replacement, we define the error term δQ_t by the equation

$$\delta Q_t = Q_t - Q_{t,L} \,. \tag{3.7}$$

The physical interpretation of δQ_t is a quadrapole interaction to compensate for the shorter distance (quadrapolar) forces which differ in Q_t and $Q_{t,L}$. The long range (dipolar) forces arising from Q_t to $Q_{t,L}$ will agree, since we define them to have equal (average) dipole densities. We also define Q' by

$$Q' = Q_a + Q_{t,L} = Q - \delta Q_t . \tag{3.8}$$

Note

$$Q = Q_a + Q_{t,L} + \delta Q_t . \tag{3.9}$$

Since $\mathscr{A}_I(Q)$ is quadratic in Q, for two charge configurations Q_i, Q_j it is convenient to define their mutual action $\mathscr{A}_I(Q_i, Q_j)$ by

$$\mathscr{A}_I(Q_i, Q_j) = \tfrac{1}{2}(\mathscr{A}_I(Q_i + Q_j) - \mathscr{A}_I(Q_i) - \mathscr{A}_I(Q_j)) . \tag{3.10}$$

Proposition 3.1. *For any charge distributions* Q, Q_j,

$$\mathscr{A}_I(Q, Q) = \mathscr{A}_I(Q) , \tag{3.11}$$

$$\mathscr{A}_I\left(Q, \sum_{j=1}^{n} Q_j\right) = \sum_{j=1}^{n} \mathscr{A}_I(Q, Q_j) \tag{3.12}$$

and

$$\mathscr{A}_I\left(\sum_{j=1}^{n} Q_j\right) = \sum_{i,j=1}^{n} \mathscr{A}_I(Q_i, Q_j) . \tag{3.13}$$

Proof. These identities follow by expanding the quadratic function $\mathscr{A}_I$, for example

$$\mathscr{A}_I(Q, Q) = \tfrac{1}{2}(\mathscr{A}_I(2Q) - 2\mathscr{A}_I(Q)) = \tfrac{1}{2}(4-2)\mathscr{A}_I(Q) .$$

Corollary. *For* $Q = Q_a + Q_t = Q_a + Q_{t,L} + \delta Q_t = Q' + \delta Q_t$,

$$\mathscr{A}_I(Q) - \mathscr{A}_I(Q') = \mathscr{A}_I(Q_t) - \mathscr{A}_I(Q_{t,L}) + 2\mathscr{A}_I(Q_a, \delta Q_t) . \tag{3.14}$$

Thus the change in action due to replacing typical ε-dipoles by averaged R_L-dipoles is the sum of a quadrapole remainder term, $2\mathscr{A}_I(Q_a, \delta Q_t)$, and the change due to the typical configurations. We estimate (3.14) in Chapter 4.

We now define $Q_{t,L}$. Using (3.6), we write

$$Q_t = \partial S_t \tag{3.15}$$

for some $d-2$ lattice surface S_t. (Henceforth we assume $d \geqq 4$.) This surface S_t is the Dirac surface for the typical charges. We choose S_t to have minimal $d-2$ area. It follows that if $\{Q_t^{(j)}\}$ is a collection of charge configurations supported in lattice volumes of bounded diameter, then the area of $\{S_t^{(j)}\}$ is also bounded. Define

$$S_a = S - S_t \tag{3.16}$$

so

$$\partial S_t = Q_t , \quad \partial S_a = Q_a . \tag{3.17}$$

We first assume that S_t is a single $d-2$ cell in the dual ε lattice, and we define $S_{t,L}$. In the general case where S_t is the sum of a bounded number of elementary $d-2$ cells, we define $S_{t,L}$ by linearity.

Let S_t be a single $d-2$ cell in $*Z^d_\varepsilon$, and consider the collection of $(R_L/\varepsilon)^{d-2}$ dipoles (i.e. $d-2$ surfaces) in $*Z^d_\varepsilon$ which both (i) lie in a single $d-2$ cell in $Z^d_{R_L}$ parallel to S_t and (ii) are closest to S_t. We define $S_{t,L}$ to be this $d=2$ Landau lattice cell, and we assign a charge strength as follows: Since S_t has strength $2\pi j\varepsilon^{d-2}g^{-1}$, we take the strength of each weak dipole (ε lattice dipole making up $S_{t,L}$) to be

$$2\pi j\varepsilon^{d-2}g^{-1}(\varepsilon/R_L)^{d-2} = 2\pi^j\varepsilon^{d-2}g^{-1}\beta^{-1}\,. \tag{3.18}$$

Thus the dipole charge is scaled down by β^{-1}.

We justify this choice by calculating the long range part of the interaction between two pairs of dipole charges, oriented as in Figure 2.

Fig. 2. Two dipole pairs on a line, separated by distance r, illustrated for $d=4$

Before averaging, the mutual action, up to quadrapole corrections is

$$\omega qq'\left(\frac{d^2}{dr^2}r^{-d+2}\right)\varepsilon^2\,,$$

where q, q' are the charges, ω is a constant and ε is the charge separation within a dipole pair. After averaging, there are $(R_L/\varepsilon)^{d-3}$ charges on each face of the boundary of S_L, and each charge is scaled down by β^{-1}. The dipole charge separation is R_L in place of ε. Thus for $r \gg R_L$, the mutual action is

$$\begin{aligned}&\omega\beta^{-2}qq'\left(\frac{d^2}{dr^2}r^{-d+2}\right)R_L^2(R_L/\varepsilon)^{2d-6}\\&=\omega qq'\left(\frac{d^2}{dr^2}r^{-d+2}\right)\varepsilon^2(R_L/\varepsilon)^{2(d-2)}\beta^{-2}\\&=\omega qq'\left(\frac{d^2}{dr^2}r^{-d+2}\right)\varepsilon^2\,,\end{aligned}$$

up to quadrapole corrections; namely the action is unchanged up to a quadrapole error.

Proposition 3.3. *Let Q_t contain at most a bounded number N of charges. Then the self action $\mathscr{A}(Q_{t,L})$ is of magnitude $O(\beta^{-1/(d-2)})$.*

Proof. The charge strength per elementary charge in $Q_{t,L}$ is of the order $\varepsilon^{d-3}g^{-1}\beta^{-1}$, since $Q_{t,L}=\partial S_{t,L}$, and the dipoles have strength $\varepsilon^{d-2}g^{-1}\beta^{-1}$ as explained above. Summation over relative separation of the charge pairs yields the factor ε^{-d+2}, while for fixed relative separation there are $(R_L/\varepsilon)^{d-3}$ charge pairs. Thus the self action has magnitude

$$(\varepsilon^{d-3}g^{-1}\beta^{-1})^2(R_L/\varepsilon)^{d-3}\varepsilon^{-d+2} = O(\varepsilon/R_L) = O(\beta^{-1/(d-2)})\,. \tag{3.19}$$

Note the bound (3.19) uses the fact that $S_{t,L}$ contains a bounded number N of dipoles, since N^2 terms occur as a multiplicative factor in the bound.

As a consequence of this argument, each small charge cluster Q_t can be replaced by weaker averaged charge clusters $Q_{t,L}$ without changing the long range interaction. Aside from the necessity of introducing a quadrapole interaction (which we denote by $\mathcal{Q}$), it is also necessary to introduce an activity ζ to compensate for the large change in self action. For a single ε lattice dipole, the self action is $c_0\beta$, cf. (3.4). Since the averaged dipole has self action $O(\beta^{-1/(d-2)})$, the activity $\zeta=e^{-O(\beta)}$ is associated with a single charge. Likewise, each small dipole cluster $S_t^{(j)}$ will have an activity $\zeta_j=e^{-O(\beta)}$ to compensate for the large change in self action. We discuss this issue in detail in Chapter 4. In terms of the above equations, the change in self action (yielding the activity) arises from the term $\mathcal{A}_I(Q_t)-\mathcal{A}_I(Q_{t'L})$ in (3.14).

We give a lower bound on $\mathcal{A}_I(Q)$ in terms of the charges in Q which have density $2\pi j\varepsilon^{-3}g^{-1}$, i.e. ε lattice charges. We can apply this bound to Q' defined by (3.8), where Q_a are the ε lattice charges.

Proposition 3.4. *Let $Q=Q_a+Q_t$ where Q_a are ε lattice charges. There is a constant $c_1>0$ such that*

$$\mathcal{A}_I(Q)\geqq c_1\beta\sum_{Q_a} j^2\,, \tag{3.20}$$

where $\sum_{Q_a}$ denotes the sum over all $d-3$ cells with nonvanishing charge Q_a, and where the charge density q_i in Q_a takes on values $2\pi\varepsilon^{-3}g^{-1}j$, for $|j|\leqq 12$.

Proof. Let $B=(-\Delta)^{-1}Q$ and observe that

$$\mathcal{A}_I(Q)=\tfrac{1}{2}\langle\delta B,\delta B\rangle\,.$$

For a given charge q in Q, let $\sum^{(q)}$ denote the sum over all $d-2$ cells of the dual lattice for which q lies on the boundary. By definition

$$\delta=\varepsilon^{-1}\sum{}^{(q)}\,.$$

Note that

$$\delta B=\partial\delta(-\Delta)^{-1}S=S+\partial\delta\Delta^{-1}S$$

and hence

$$\partial\delta B=\partial S=Q\,. \tag{3.21}$$

In other words, δB satisfies a conservation law, $\delta\delta B=0$, and in particular for a charge $q\in Q_a$,

$$\sum{}^{(q)}\delta B=2\pi\varepsilon^{-2}g^{-1}j\,. \tag{3.22}$$

However, each $d-2$ cell occurs at most $2(d-2)$ times in the iterated sum $\sum_{Q_a}\sum^{(q)}$, and so omitting terms from $Q_{t,L}$,

$$\mathcal{A}_I(Q)=\tfrac{1}{2}\langle\delta B,\delta B\rangle\geqq\varepsilon^d 2^{-2}(d-2)\sum_{Q_a}\sum{}^{(q)}(\delta B)^2\,. \tag{3.23}$$

Each sum $\sum^{(q)}$ contains $6=2(d-(d-3))$ terms, and so by the Schwarz inequality

$$\left|\sum{}^{(q)}\delta B\right|^2\leqq 36\sum{}^{(q)}(\delta B)^2\,.$$

Thus by (3.22)–(3.23),

$$\mathscr{A}_I Q') \geqq [\pi^2/36(d-2)]\varepsilon^{d-4} g^{-2} \sum_{Q_a} j^2 = c_1 \beta \sum_{Q_a} j^2 ,$$

where $c_1 = \pi^2/36(d-2)$.

We end this section by remarking that a further change in the charge configuration eliminates the long distance part of the quadrapolar interaction $\mathscr{Q}$. In fact, replacing each array of $(R_L/\varepsilon)^{d-2}$ piecewise constant dipoles in $S_{t,L}$ by a linear combination of 2^d translates, we eliminate the quadrapole interaction up to an octapole correction $\mathscr{O}$. So that all error terms arising from linear contributions to charge minus average charge can be estimated with octapole remainders, we perform this one additional averaging in the definition of $Q_{t,L}$.

4. The Dilute Dipole Gas for $d \geqq 4$: Upper Bound on Z

We study $\mathscr{A}_I$ defined in (2.19) for a dipole (instanton) gas and establish bounds on the partition function

$$Z = \sum_Q e^{\mathscr{A}_I(Q)} . \tag{4.1}$$

We obtain the leading contribution to the free energy

$$f_V = V^{-1} \ln Z , \quad \beta \to \infty . \tag{4.2}$$

To any finite configuration of charges Q, we associate the action

$$\zeta_Q = e^{-\mathscr{A}_I(Q)} \tag{4.3}$$

defined by (2.19). As in (3.4), we see that

$$\mathscr{A}_I(Q) = c_Q \beta , \tag{4.4}$$

for a constant c_Q which can be expressed explicitly in terms of $(-\Delta)^{-1}$ via (2.19). The constant c_Q is invariant under scale transformations, and thus (if we use scale transformations to identify charge configurations Q on different lattice spacings) the constant c_Q is independent of ε.

Our main assertion is that it is sufficient to examine a finite number of Q's to determine f_V to leading order. Let $\mathscr{Q}$ be the set of all finite charge configurations, and let

$$\mathscr{Q}_M = \{Q \in \mathscr{Q} : \operatorname{diam} Q \leqq M\varepsilon\}$$
$$\zeta_0 = \sup_Q \{\zeta_Q : Q \in \mathscr{Q}_M\} . \tag{4.5}$$

Since ζ_Q is translation invariant, the sup above is taken over the finite set $\mathscr{Q}_M/Z^d$. Here M is a constant to be chosen later. Let a_0 denote the multiplicity of the maximum, i.e. the number of Q's for which $\zeta_Q = \zeta_0$. In counting a_0, we identify two charge configurations if they are related one to another by a lattice translation. We state our main result.

Theorem 4.1. *There is a $\gamma>0$ and a choice of M in (4.5) such that*

$$f_V = a_0 \varepsilon^{-d} \zeta_0 (1+O(\beta^{-\gamma})) , \tag{4.6}$$

for $V > V(\beta)$ sufficiently large and for β sufficiently large.

Remark. The choice $V > V(\beta)$ eliminates surface boundary effects in leading order. Fabrizio De Angelis, de Falco and Guerra [2] have shown that $\lim_V f_V$ exists for the lattice gauge theory, and Griffiths [5] show this for $d=3$ magnetic dipoles. For the instanton gas considered here, we believe that $\lim f_V = f$ also exists and that $f=\varepsilon^{-d}\sum \zeta_Q + O(\zeta_0^{1+\theta})$, for $\theta<1$. In this section we prove a lower bound on f_V of the form (4.6).

First let us explain the significance of Proposition 3.4. Let c_0 be the coefficient c_Q in (4.4) for Q consisting of a single ε lattice charge loop (charge with density of order $\varepsilon^{-3}g^{-1}$). Let $Q=Q_a+Q_t$ be a decomposition into ε lattice charges Q_a and disjoint charges Q_t. If Q_a contains $L \geqq c_0/c_1$ charges, then Q does not contribute to the maximum in (4.5). Thus only a bounded number of charges will contribute to the maximum in (4.5), independent of M.

We now specify precisely the typical configurations and the application of the construction of Section 3. Decompose the lattice into hypercubic blocks Δ with side length $2R_L$. Let Q be a charge configuration, and let $N(Q,\Delta)$ be the total number of charges in Δ and in $\bar{\Delta}$, the union of Δ and the 3^d-1 blocks adjacent to Δ. If

$$N(Q,\Delta) \leqq 4\cdot 3^d c_0/c_1 \equiv N , \tag{4.7}$$

then $Q \restriction \Delta$ is called typical; otherwise $Q\restriction \Delta$ is called atypical. The typical charge configurations are more complicated than the bounded charge configurations (4.5). In fact a typical configuration may contain several widely separated clusters.

Proposition 4.2. *Let $\delta>0$ be given and let $Q=\sum Q_i$ be a typical configuration, composed of clusters Q_i having minimal separation $D\varepsilon$. Then for $D \geqq D(N,\delta)$ sufficiently large*

$$|\mathscr{A}(Q_i,Q_j)| \leqq \delta\beta N^{-2} , \tag{4.8}$$

for $i \neq j$.

Proof. From the definition (3.10), we have

$$\mathscr{A}_I(Q_i,Q_j) \sim \tfrac{1}{2}\omega_{d-1} \sum_{\substack{q_i\in Q_i\\ q_j \in Q_j}} \varepsilon^{2d} q_i r_{ij}^{-d+2} q_j .$$

Summing over charges $2\pi j \varepsilon^{-3} g^{-1}$ and $r_{ij} \geqq D\varepsilon$ gives

$$|\mathscr{A}_I(Q_i,Q_j)| \leqq c_{Q_i,Q_j} \beta D^{-d+2} , \tag{4.9}$$

where c_{Q_i,Q_j} is a constant. In fact $c_{Q_i,Q_j} \leqq O(N^2)$. Thus (4.8) holds for $D(N,\delta)$ sufficiently large.

With the choice of $D=D(N,\delta)$ as required by Proposition 4.2, we divide $Q\restriction\bar{\Delta}$ into clusters separated by at least $D\varepsilon$. Each cluster Q_j (as $Q\restriction\bar{\Delta}$ itself) has at most N charges, so each cluster has diameter at most $ND\varepsilon$. Choose $M=ND$ in (4.5). Thus if Q is typical in Δ,

$$Q\restriction \bar{\Delta} = \sum Q_j, \quad Q_j \in \mathscr{Q}_M , \tag{4.10}$$

and (4.8) holds. In case the cluster Q_j lies in two or more of the 3^d blocks of $\bar{\Delta}$, we assign Q_j to the block containing its "lower corner" in some lexicographic order of lattice sites in Z^d. For β sufficiently large ($R_L \gg \varepsilon$) each Q_j associated with Δ satisfies

$$\partial Q_j = 0 . \tag{4.11}$$

Let $J(\Delta)$ index clusters associated with Δ. For typical blocks Δ let

$$Q_\Delta = \sum_{j \in J(\Delta)} Q_j . \tag{4.12}$$

Then Q_Δ is the configuration to which we apply the construction of Chapter 3, giving $Q_{\Delta,L}$, which agrees with Q_Δ up to the octapole remainder. Let

$$\begin{aligned} &Q_t = \sum_\Delta Q_\Delta , \quad Q_{t,L} = \sum_\Delta Q_{\Delta,L} , \\ &\partial Q_\Delta = Q_\Delta - Q_{\Delta,L} . \end{aligned} \tag{4.13}$$

Lemma 4.3. *For β sufficiently large,*

$$\mathscr{A}_I(Q_\Delta) - \mathscr{A}_I(Q_{\Delta,L}) \geqq \sum_{j \in J_\Delta} \mathscr{A}_I(Q_j)(1 - \delta - \beta^{-1}) . \tag{4.14}$$

Moreover we may take $\delta = 0$ if J_Δ contains a single element.

Proof. Using Proposition 3.1,

$$\mathscr{A}_I(Q_\Delta) = \sum_{i,j \in J_\Delta} \mathscr{A}_I(Q_i, Q_j) = \sum_{i \in J_\Delta} \mathscr{A}_I(Q_i) + \sum_{\substack{i \neq j \\ i,j \in J_\Delta}} \mathscr{A}_I(Q_i, Q_j) .$$

By Proposition 4.2, we bound the intercluster action by

$$\sum_{i \neq j} \delta \beta N^{-2} \leqq \delta \beta .$$

Since Q_j belongs to the finite set up to translations, of typical charge clusters,

$$\beta \leqq \text{const} \sum_{j \in J_\Delta} \mathscr{A}_I(Q_j) , \tag{4.15}$$

using the constants of (4.4). Thus with a new δ,

$$\mathscr{A}_I(Q_\Delta) \geqq (1 - \delta) \sum_{j \in J_\Delta} \mathscr{A}_I(Q_j) . \tag{4.16}$$

Clearly we may take $\delta = 0$ if J_Δ contains a single element.

We bound $\mathscr{A}_I(Q_{\Delta,L})$ using Proposition 3.3 Thus

$$|\mathscr{A}_I(Q_{\Delta,L})| \leqq O(1) \beta^{-1/(d-2)} ,$$

and by (4.15)

$$\begin{aligned} |\mathscr{A}_I(Q_{\Delta,L})| &\leqq O(1) \beta^{-1/(d-2)} \beta^{-1} \sum_{j \in J_\Delta} \mathscr{A}_I(Q_j) \\ &\leqq \beta^{-1} \sum_{j \in J_\Delta} \mathscr{A}_I(Q_j) \end{aligned} \tag{4.17}$$

for β large. The desired bound follows from (4.16–(4.17).

Lemma 4.4. *There exists $\gamma>0$ such that for β sufficiently large,*

$$\mathscr{A}_I(Q_t)-\mathscr{A}_I(Q_{t,L})\geqq\sum_{\Delta}\sum_{j\in J_\Delta}\mathscr{A}_I(Q_j)(1-\delta_\Delta-\beta^{-\gamma}) \tag{4.18}$$

where $\delta_\Delta=0$ if only one cluster is associated with Δ.

Proof. We expand using Proposition 3.1,

$$\mathscr{A}_I(Q_t)-\mathscr{A}_I(Q_{t,L})$$
$$=\sum_{\Delta}(\mathscr{A}_I(Q_\Delta)-\mathscr{A}_I(Q_{\Delta,L}))+\sum_{\Delta\neq\Delta'}(\mathscr{A}_I(Q_\Delta,Q_{\Delta'})-\mathscr{A}_I(Q_{\Delta,L},Q_{\Delta',L}))\,. \tag{4.19}$$

The first terms in (4.19) are bounded below by Lemma 4.3. The sum over $\Delta\neq\Delta'$ is split into two parts: the sum over adjacent pairs and the sum over nonadjacent pairs. The sum over adjacent pairs is bounded using Proposition 4.2 and the argument in the proof of Lemma 4.3. For nonadjacent pairs we write the terms as

$$\sum_{\Delta,\Delta'}(\mathscr{A}_I(Q_\Delta-Q_{\Delta,L},Q_{\Delta'})+\mathscr{A}_I(Q_{\Delta,L},Q_{\Delta'}-Q_{\Delta',L}))$$
$$=\sum_{\Delta,\Delta'}\mathscr{A}_I(\delta Q_\Delta,Q_{\Delta'}+Q_{\Delta',L})\,. \tag{4.20}$$

The sum (4.20) contributes to the octapole remainder. Due to the subtractions in the definition of δQ_Δ,

$$|\mathscr{A}_I(\delta Q_\Delta,Q_{\Delta'}+Q_{\Delta',L})|\leqq O(1)\beta D^{-d-1}\,,$$

where

$$D=\varepsilon^{-1}\operatorname{dist}(\Delta,\Delta')\geqq(R_L/\varepsilon)=\beta^{1/(d-2)}\,.$$

This bound follows as in the proof of Proposition 4.2, cf. (4.9). The additional powers of D in the denominator arise from the multiple differences which occur in the definition of δQ_Δ.

The summation over Δ and Δ' is achieved by taking $\operatorname{dist}(\Delta,\Delta')=2\boldsymbol{n}R_L$ where $\boldsymbol{n}$ is a vector with integer components, and summing $\boldsymbol{n}$ over Z^d. Thus, using (4.15) and $\sum D^{-d-1}\leqq O(R_L/\varepsilon)^{-d-1}\leqq O(\beta^{-1-\gamma})$,

$$\sum_{\Delta,\Delta'}|\mathscr{A}_I(\delta Q_\Delta,Q_{\Delta'}+Q_{\Delta',L})|\leqq O(\beta^{-\gamma})\sum_{\Delta}\sum_{j\in J_\Delta}\mathscr{A}_I(Q_j)\,.$$

Lemma 4.5. *There exists $\gamma>0$ such that for $\delta>0$ and β sufficiently large*

$$\mathscr{A}_I(Q)\geqq(c_1-\delta)\beta\sum_{Q_a}j^2+\sum_{\Delta}\sum_{j\in J_\Delta}\mathscr{A}_I(Q_j)(1-\delta_\Delta-\beta^{-\gamma})\,. \tag{4.21}$$

Proof. We write

$$\mathscr{A}_I(Q)=\mathscr{A}_I(Q')+\mathscr{A}_I(Q_t)-\mathscr{A}_I(Q_{t,L})+2\mathscr{A}_I(Q_a,\delta Q_t)\,. \tag{4.22}$$

The action $\mathscr{A}_I(Q')$ is bounded below using Proposition 3.4, with $Q'=Q_a+Q_{t,L}$. Thus

$$\mathscr{A}_I(Q')\geqq c_1\beta\sum_{Q_a}j^2\,.$$

The difference $\mathscr{A}_I(Q_t)-\mathscr{A}_I(Q_{t,L})$ is bounded below by Lemma 4.4. The remaining term $\mathscr{A}_I(Q_a, \delta Q_t)$ contributes to the octapole remainder, and it is bounded by

$$|\mathscr{A}_I(Q_a, \delta Q_t)| \leqq O(\delta\beta) \sum_{Q_a} j^2 .$$

Proof of Theorem 4.1 (upper bound on Z). By Lemma 4.5,

$$e^{-\mathscr{A}_I(Q)} \leqq e^{-(c_1-\delta)\beta \sum_{Q_a} j^2} \prod_{\Delta} \prod_{J_\Delta} e^{-\mathscr{A}_I(Q_j)(1-\delta_\Delta-\beta^{-\gamma})} .$$

We substitute $\zeta_j = e^{-\mathscr{A}_I(Q_j)}$. In order to simplify the sum over configurations, and to make the Q_a, Q_t sums factorize, we introduce some spurious terms. We allow overlapping Q_a, Q_t configurations, and we regard two or more elementary clusters in the same or adjacent blocks as separate events with activity

$$(\zeta_{j_1} \dots \zeta_{j_l})^{1-\delta_\Delta} = \zeta_{j_1 \dots j_l} . \tag{4.23}$$

We then allow arbitrary multiplicities of these $\zeta_{j_1}, \zeta_{j_1 j_2} \dots$ events in any block. This allows us to bound the summation over charge configurations, $\sum_{\{Q\}}$, by independent sums $\sum_{\{Q_a\}} \sum_{\{Q_t\}}$ over atypical and typical charge configurations. Thus

$$\begin{aligned} Z = \sum_{\{Q\}} e^{-\mathscr{A}_I(Q)} &\leqq \left(\sum_{\{Q_a\}} e^{-\frac{1}{2}c_1\beta \Sigma Q_a j^2} \right) \left(\sum_{\{Q_t\}} \prod_{\Delta} \prod_{J_\Delta} \zeta_{j_1} \dots \zeta_{j_1 \dots j_2} \right) \\ &\leqq \left(\sum_{\{Q_a\}} e^{-\frac{1}{2}c_1\beta \Sigma Q_a j^2} \right) \left(1 + \sum_j \left(\frac{R_L}{\varepsilon}\right)^d \zeta_j + \sum_{j_1 j_2} \left(\frac{R_L}{\varepsilon}\right)^{2d} \zeta_{j_1 j_2} + \dots \right)^{VR_L^{-d}} . \end{aligned}$$

Here we remark there are a finite number of terms for the typical summation in a given block Δ. The factors $(R_L/\varepsilon)^d$, $(R_L/\varepsilon)^{2d}$, etc. arise from possible locations of the clusters in Δ. Since $(R_L/\varepsilon)^d = \beta^{d/d-2}$, and $\zeta_j \leqq \zeta_0$, $\zeta_{j_1 j_2} \leqq \zeta_0^{2(1-\delta_\Delta)}, \dots$,

$$\begin{aligned} Z &\leqq \left(\sum_{\{Q_a\}} e^{-\frac{1}{2}c_1\beta \Sigma Q_a j^2} \right) \left(1 + \left(\frac{R_L}{\varepsilon}\right)^d a_0 \zeta_0 (1 + O(\beta^{-\gamma})) \right)^{VR_L^{-d}} \\ &\leqq \left(\sum_{\{Q_a\}} e^{-\frac{1}{2}c_1\beta \Sigma Q_a j^2} \right) e^{a_0\zeta_0(1+O(\beta^{-\gamma}))V\varepsilon^{-d}} . \end{aligned}$$

Here we have used the fact that δ_Δ does not appear for a single cluster, to identify the leading term in the exponent as $a_0\zeta_0$.

The sum over atypical configurations can be carried out explicitly. Since Q_a has at least $4 \cdot 3^d c_0/c_1$ charges on the ε lattice, the activity of each atypical configuration in Δ is at least ζ_0^2. Thus the atypical terms can be included in the error $O(\beta^{-\delta})$.

5. The Lower Bound on Z

In this section we prove the lower bound on $Z = \sum_{\{Q\}} e^{-\mathscr{A}_I}$, which completes the proof of Theorem 4.1. Since each term in the sum over configurations is nonnegative, we achieve a lower bound by omitting certain configurations Q. The remaining configurations, called typical, differ from the typical configurations of Section 4 (and in any case are a set of measure zero in the infinite volume measure).

We introduce an infinite sequence $R_1=\varepsilon$, $R_2=R_L$,

$$\begin{aligned} &R_3=\zeta_0^{-2\delta}R_L \\ &R_4=\exp(\zeta_0^{-\delta})R_L \\ &R_n=2^nR_4, \quad n\geqq 4 \end{aligned} \tag{5.1}$$

of length scales. The first restriction defining a typical configuration Q is that Q must be a sum of translates of the a_0 distinct $Q_j\in\mathcal{Q}_M$ with activity $\zeta_0=\zeta_{Q_j}$, as in (4.5). Furthermore we allow only one such elementary cluster in each block of length R_3; within this block, the cluster must have a distance at least $\zeta_0^{-\delta}R_L=\zeta_0^{\delta}R_3$ from the boundary of the block.

Let

$$\mathscr{A}_I(Q_i,R,R')=\sum\mathscr{A}_I(Q_i,Q_j) \tag{5.2}$$

where the sum runs over all elementary clusters $Q_j\neq Q_i$ in Q at a distance between R and R' from the cluster Q_i. For typical configurations,

$$\mathscr{A}_I(Q_i,0,\zeta_0^{\delta}R_3)=0 \tag{5.3}$$

because the minumum intercluster distance is $2\zeta_0^{\delta}R_3$.

Lemma 5.1. *For β sufficiently large and for a typical configuration Q,*

$$|\mathscr{A}_I(Q_i,0,R_4)|\leqq\tfrac{1}{2}\zeta_0^{\delta}\,. \tag{5.4}$$

Proof. At separation r,

$$|\mathscr{A}_I(Q_i,Q_j)|\leqq O(r/R_L)^{-d}\beta^{-1/(d-2)}\,.$$

At minimum separation $r=2\zeta_0^{\delta}R_3=2\zeta_0^{-\delta}R_L$

$$|\mathscr{A}_I(Q_i,Q_j)|\leqq 2^{-d}\zeta_0^{\delta d}\beta^{-1/(d-2)}\,.$$

For Q_j's separated by one or more intermediate block of size R_3, the discrete sum can be estimated by an integral, giving

$$\begin{aligned} |\mathscr{A}_I(Q_i,0,R_4)| &\leqq O(\zeta_0^{\delta d})\beta^{-1/(d-2)}+O(R_3/R_L)^{-d}\beta^{-1/(d-2)}\ln(R_4/R_L) \\ &\leqq\tfrac{1}{2}\zeta_0^{\delta}\,. \end{aligned}$$

Here we have substituted R_3^{-d} as an upper bound on the density of charge clusters.

For distances $R>R_4$, the bound on $\mathscr{A}_I(Q_i,0,R)$ depends on cancellations between oppositely oriented dipoles. To achieve this cancellation, we impose further restrictions on the typical configurations. The map $q\to -q$ is a symmetry of $\mathscr{A}_I$ which reverses the orientation of the elementary dipole clusters. Thus the number a_0 of these clusters is even, and we can label them as $Q_{\pm j}$, $j=1,\ldots,a_0/2$, with $Q_{-j}=-Q_j$ the orientation reversed image of Q_j. We divide R^d into d-cells (block spins) of length R_n, for each length scale R_n, $n\geqq 5$. In any block spin of length R_n, let v_j be the number of times the elementary dipole cluster Q_j occurs. Our final restriction defining a typical configuration is that

$$|\nu_j-\nu_{-j}|\leqq(R_n/R_3)^{3d/4} \tag{5.5}$$

for each $j=1,\dots,a_0/2$, each R_n-block and each $n=5,6,\dots$.

Lemma 5.2. *For β sufficiently large and for typical configurations,*

$$|\mathscr{A}_I(Q_i,0,\infty)|\leqq\zeta_0^\delta\,. \tag{5.6}$$

Proof. Write

$$\mathscr{A}_I(Q_i,0,\infty)=\mathscr{A}_I(Q_i,0,R_4)+\sum_{n=4}^{\infty}\mathscr{A}_I(Q_i,R_n,R_{n+1})\,.$$

We divide the spherical shell contributing to $\mathscr{A}_I(Q_i,R_n,R_{n+1})$ into $O(R_n/R_{4[n/5]})^d$ blocks of length $R_{4[n/5]}$. In each block

$$|\nu_j-\nu_{-j}|\leqq(R_{4[n/5]}/R_3)^{3d/4}$$

is the deviation from neutrality. The dipole interaction strength for a dipole with separation R is bounded by $O(\beta^{-1/(d-2)})(R/R_L)^{-d}$. Thus

$$\begin{aligned}|\mathscr{A}_I(Q_i,R_n,R_{n+1})|&\geqq\sum_j O(\beta^{-1/(d-2)})(R_n/R_L)^{-d}|\nu_j-\nu_{-j}|(R_n/R_{4[n/5]})^d\\&\leqq O(\beta^{-1/(d-2)})R_L^dR_3^{-3d/4}R_{4[n/5]}^{-d/4}\\&\leqq O(\beta^{-1/(d-2)})2^{-[n/5]d}\zeta_0^{3d/2}e^{-\zeta_0^\delta/4}\,.\end{aligned}$$

Summing over n, and choosing β large,

$$\sum_{n=4}^{\infty}|\mathscr{A}_I(Q_I,R_n,R_{n+1})|\leqq\tfrac{1}{2}\zeta_0^d$$

so that Lemma 5.1 completes the proof.

Proof of Theorem 4.1 (completed). We now let $\sum'$ denote the sum over typical configurations. In a bounded region Λ of volume V, a typical configuration Q is specified by the number $N=N(Q)$ of dipole clusters, the choice of $\pm$ orientation for each cluster, the $(a_0/2)^N$ choices of $|j|$ and the

$$\binom{V/R_3^d}{N}(R_3/\varepsilon)^{dN}(1-O(\zeta_0^\delta)^N \tag{5.7}$$

possible positions of the clusters. Here the factor $(1-O(\zeta_0^\delta)^N$ results from the excluded boundary region for the location of each cluster in its block of length R_3. Thus

$$Z=\sum_Q e^{-\mathscr{A}_I}\geqq\sum_Q{}' e^{-\mathscr{A}_I}\leqq\sum_N\sum_{\{Q:N(Q)=N\}}\sum_\pm e^{-\zeta_0^\delta N}(a_0/2)^N\zeta_0^N\,. \tag{5.8}$$

Substitution of (5.7) and $\sum_\pm=2^N$ in (5.8) would give

$$\begin{aligned}Z&\geqq(1+a_0(R_3/\varepsilon)^d e^{-\zeta_0}\zeta_0(1-O(\zeta_0^\delta))^{V/R_3^d}\\&\geqq[1+a_0\varepsilon^{-d}\zeta_0(1-O(\beta^{-\delta}))]^V\end{aligned} \tag{5.9}$$

and complete the proof. Thus it remains to analyze the restriction placed on $\sum'$ by the restriction to typical configurations.

For a particular R_n-length block, the number of choices of $\pm$ which violate (5.5) is bounded by

$$e^{-O(|v_j - v_{-j}|^2/|v_j + v_{-j}|)} 2^{|v_j + v_{-j}|} \leqq e^{-O(R_n/R_3)^{6d/4}/(R_n/R_3)^d} 2^{|v_j + v_{-j}}$$
$$\leqq e^{-O(2^{n/2}\zeta_0^{\delta} e^{\frac{1}{2}\zeta_0^{-\delta}})} 2^{|v_j + v_{-j}|} .$$

Thus the complementary set occurs for at least

$$[1 - e^{-O(^{n/2}\zeta_0^{\delta} e^{\frac{1}{2}\zeta_0^{-\delta}})}] 2^{|v_j + v_{-j}|}$$
$$\cong \left[1 - \left(\frac{R_3}{R_n}\right)^d e^{-O(2^{n/2}\zeta_0^{\delta} e^{\frac{1}{2}\zeta_0^{-\delta}})}\right]^{(R_n/R_3)^d} 2^{|v_j + v_{-j}|}$$

choices. Multiplying these values, we see that (5.5) is satisfied on all R_n-length blocks at once for

$$\left[1 - \left(\frac{R_3}{R_n}\right)^d e^{-O(2^{n/2}\zeta_0^{\delta} e^{\frac{1}{2}\zeta_0^{-\delta}})}\right]^{V/R_3^d} 2^N$$

choices of $\pm$. Multiplying over n shows that

$$\sum_{\pm} \geqq \left[1 - O(1) \sum_n \left(\frac{R_3}{R_n}\right)^d e^{-O(2^{n/2}\zeta_0^{\delta} e^{\frac{1}{2}\zeta_0^{-\delta}}}\right]^{V/R_3^d} 2^N .$$

Substitution in (5.8) now yields (5.9) and completes the proof.

Acknowledgement. We wish to thank O. McBryan and T. Spencer for helpful discussions.

References

1. Brydges, D.: A rigorous approach to Debye screening in dilute classical Coulomb systems. Commun. math. Phys., to appear
2. Fabrizio De Angelis, G., de Falco, D., Guerra, F.: Lattice gauge models in the strong coupling regime. Preprint
3. Glimm, J., Jaffe, A.: Quark trapping for $U(1)$ lattice gauge fields. Phys. Letters **66** B, 67—69 (1977)
4. Glimm, J., Jaffe, A. Spencer, T.: Convergent expansion about mean field theory, Parts I and II. Ann. Phys. **101**, 610—630; **101**, 631—669 (1976)
5. Griffiths, R. B.: Free energy of interacting magnetic dipoles. Phys. Rev. **176**, 655—659 (1968)
6. Kogut, J., Susskind, L.: Hamiltonian formulation of Wilson's lattice gauge theories. Phys. Rev. D **11**, 395—408 (1975)
7. Mayer, M.: Gauge fields as quantized connection forms. In: Conference on differential geometric methods in mathematical physics (eds. K. Bleuler, A. Reetz). Lecture notes in mathematics, Vol. 570. Berlin-Heidelberg-New York: Springer 1977
8. Osterwalder, K., Seiler, E.: Gauge field theories on the lattice. Ann. Phys. to appear
9. Polyakov, A. M.: Quark confinement and topology of gauge theories. Nucl. Phys. B **120**, 429—458 (1977)
10. Singer, I., Thorpe, J. A.: Lecture notes on elementary topology and geometry. Berlin-Heidelberg-New York: Springer 1976
11. Wilson, K.: Confinement of quarks. Phys. Rev. D **10**, 2245—2259 (1974)
12. Høye, J. S., Stell, G.: Ionic solution theory for non-ideal solvents—potential of mean force between ions. Preprint

Communicated by A. Jaffe

Received May 16, 1977

CHARGES, VORTICES AND CONFINEMENT *

James GLIMM
The Rockefeller University, New York, NY 10021, USA

Arthur JAFFE **
Harvard University, Cambridge, MA 02138, USA

Received 28 August 1978
(Revised 20 November 1978)

We study the confinement phase transition. We point out the relation between the decay rates of the Wilson loop integral, Osterwalder and Seiler's n-ality restriction, and Polyakov and 't Hooft's analysis of the center of the gauge group. Vortices and merons are analyzed as possible mechanisms to produce disorder in the center. We show how to realize the 't Hooft commutation relations, and we estimate the critical coupling for vortices to be in approximate agreement with the critical coupling determined by Callan, Dashen and Gross from the analysis of instantons.

1. Introduction

In SU(n) gauge theories, the area decay of the Wilson functional

$$W_\sigma(C) = \langle \mathrm{Tr}_\sigma \mathrm{P e}^{\oint_C A_\mu \, dx^\mu} \rangle \sim e^{-O|\text{area } C|} \tag{1.1}$$

is taken as an indication of quark confinement. Although the sign of the β-function in the renormalization group may be the underlying cause of the disorder of the gauge field which leads to area decay, there have been efforts to describe the dominant modes of disorder directly, e.g., in terms of point charges (merons) [1,2] and vortices (disorder operators) [3]. We examine these two approaches.

2. Particle and quark representations

To distinguish between confined and observable particles, we call the former quarks and the latter particles. The distinction between these possibilities depends

* Supported in part by the National Science Foundation under Grants PHY-78-08066 and PHY-77-18762.
** John S. Guggenheim Fellow.

both on the representation σ of the gauge group G in (1.1) and on the vacuum functional. We use the Wilson functional as the criterion for confinement.

In various examples, one finds that fractional charges are confined, integer charges are not. We make this statement precise by proposing that there is a subgroup $\mathcal{Z}_0$ of the center $\mathcal{Z}$ of G with the property that

$$W_\sigma \sim e^{-O(\text{area})}, \qquad \text{if } \sigma(\mathcal{Z}_0) \neq I\,, \tag{2.1a}$$

$$W_\sigma \sim e^{-O(\text{length})}, \qquad \text{if } \sigma(\mathcal{Z}_0) = I\,. \tag{2.1b}$$

With charges labeled by the dual group $\hat{\mathcal{Z}}$, $\mathcal{Z}_0^\perp$ is the subgroup of $\hat{\mathcal{Z}}$ which occurs in (2.1b), and labels the "integer" charges, while $\hat{\mathcal{Z}} \setminus \mathcal{Z}_0^\perp)$ gives the representation in (2.1a) and labels fractional charges. The subgroup $\mathcal{Z}_0$ depends on the vacuum functional, and thus labels phase transitions in the transverse modes of the gauge theory.

According to perturbation theory, in the asymptotically free phase, all particles are observable and (2.1) holds with $\mathcal{Z}_0 = I$, $\mathcal{Z}_0^\perp = \hat{\mathcal{Z}}$. In this case all charges are integral. At high temperature, on a lattice, we extend [4] to show that (2.1) holds with $\mathcal{Z}_0 = \mathcal{Z}$, $\mathcal{Z}_0^\perp = 0 \in \hat{\mathcal{Z}}$, at least for G = SU(2), or SU(3). In this case all non-zero charges labeled by $\hat{\mathcal{Z}}$ are fractional. To establish (2.1) in general, we require a further hypothesis: that bound states formed from observable particles are also observable.

For pure gauge theories, in $d = 2$, a transfer matrix argument shows all representations $\sigma \neq$ identity are quark representations (have an area decay). Thus $d = 2$, for pure gauge theories does not follow (2.1), but realizes the exceptional case Σ = identity of theorem 2.1 below.

We begin with an abstract characterization of particle states. On physical grounds, we require that particle states yield antiparticle states and also multiparticle states. Expressed in mathematical language, we postulate that if σ and τ are particle representations, then so are the irreducible representations contained in $\sigma \otimes \tau$. Because $W_{\bar{\sigma}} = W_{\sigma^*}$, σ^* is a particle representation if and only if σ is. This class of representations is determined by the following elementary theorem.

Theorem 2.1. Let G be a connected, compact simple Lie group. Let Σ denote a class of irreducible representations closed under complex conjugation and decomposition of tensor products. Then either Σ equals the identity, or Σ is the set of all representations of G which annihilate some subgroup $\mathcal{Z}_0$ of the center $\mathcal{Z}$ of G, i.e., $\Sigma(\mathcal{Z}_0) = I$.

Proof. cf. Chevalley [6]. Let $K = K(\Sigma)$ denote the kernel of the class Σ. The representative functions separate points on G/K, so by the Stone-Weierstrass theorem the representative functions contain all functions on G/K. Thus Σ is the class of all representations on G/K. For a connected simple Lie group G, and a normal subgroup K, either K = G or else K is discrete. In the latter case K is pointwise invariant under inner automorphisms, and hence contained in $\mathcal{Z}$.

As an illustration, let G = U(1), so that the dual group $\hat{G}$ is the integers. Then Σ above is closed under products and inverses, and thus is a subgroup of $\hat{G}$. This agrees with the known [17,18] analysis of the Abelian Higgs model in two dimensions in which confined (quark) representations are labeled by elements of $\hat{G}/\Sigma$. Apparently the methods of [18] would also show that the elements of Σ label particle representations.

We now introduce an action $\mathcal{L}$. For mathematical definiteness we consider a lattice gauge theory, though our results can also be formulated for the continuum limit. We study the plaquette action

$$\mathcal{L}(\mathrm{P})_\tau = \mathrm{Re}\,\mathrm{Tr}_\tau\, U(\mathrm{P})_\tau\,, \qquad U(\mathrm{P})_\tau = \prod_{\mathrm{b}} \tau(U(\mathrm{b}))\,, \tag{2.2}$$

where $U(\mathrm{b}) \in \mathrm{G}$ is the gauge group variable for a bond b, τ is a representation of G and the product $\prod$ ranges over the ordered bonds $\mathrm{b} \in \partial\mathrm{P}$. The total action is summed over plaquettes, and may also be summed over certain representations,

$$\mathcal{L} = \sum_\tau a_\tau \sum_{\mathrm{P}} \mathcal{L}(\mathrm{P})_\tau\,. \tag{2.3}$$

The action defines an expectation $\langle\,\cdot\,\rangle$ and the Wilson functional

$$W_\sigma(\mathrm{C}) = \langle \mathrm{Tr}_\sigma(\mathrm{P}\exp \int_{\mathrm{C}} A)\rangle\,, \tag{2.4}$$

namely the expectation of the trace of the parallel transport about the closed curve C. Here $\mathrm{P}\exp(\int_{\mathrm{C}} A) = U(\mathrm{C})_\sigma$.

We now investigate this mathematical structure in more detail. Since $\mathrm{e}^{\mathcal{L}}$ is a function on G, the Peter-Weyl theorem tells us it has a Fourier expansion in terms of representations of G. Let $\Sigma_{\mathcal{L}}$ be the class of representations thus obtained. Since $\Sigma_{\mathcal{L}}$ satisfies the hypotheses of theorem 2.1, $\mathrm{K}(\Sigma_{\mathcal{L}})$ equals either G or a subgroup of $\mathcal{Z}$. On physical grounds we eliminate $\mathrm{K} = \mathrm{G}$ as being trivial (i.e., $\mathcal{L} = \mathrm{const.}$). Furthermore, we also eliminate $\mathrm{K} = \mathcal{Z}$ since in this case W_σ would vanish identically for all quark representations, as we show in theorem 2.2 below. In case $\mathrm{G} = \mathrm{SU}(2)$ or SU(3), $\mathcal{Z}$ has no proper subgroups and therefore we require $\mathrm{K}(\Sigma_{\mathcal{L}}) = \mathrm{I}$ in that case. In the case of $\mathrm{G} = \mathrm{SU}(n)$, the center is $\mathcal{Z} = \{z^j, j = 0, 1, ..., n-1\}$, where $z = \exp(2\pi\mathrm{i}/n)I$. For the representation σ, the "n-ality" n_σ is defined by the equation:

$$\sigma(z) = \exp(2\pi\, i n_\sigma/n)I\,. \tag{2.5}$$

Abstractly, $n_\sigma : \hat{\mathrm{G}} \to \hat{\mathcal{Z}}$ is defined by $n_\sigma = \sigma \,|\, z \in \hat{\mathcal{Z}}$. Hence the requirement $n_\sigma \neq 0$ is equivalent to the requirement $\sigma(\mathcal{Z}) \not\equiv I$.

Theorem 2.2. Let $\mathrm{K}(\Sigma_{\mathcal{L}}) = \mathcal{Z}_0 \subset \mathcal{Z}$, $\sigma(\mathcal{Z}_0) \neq \mathrm{I}$. Then $W_\sigma = 0$.

Proof.

$$W_\sigma(\mathrm{C}) = \langle \mathrm{Tr}_\sigma\, U(\mathrm{C})_\sigma\rangle = \int_{\mathrm{G}/\mathcal{Z}_0} \mathrm{e}^{\mathcal{L}}\left(\int \mathrm{Tr}_\sigma\, U(\mathrm{C})_\sigma\right) \mathrm{d}\mu/\mathrm{normalization},$$

where $\mathrm{d}\mu$ is the product of Haar measures $\mathrm{d}U(\mathrm{b})$ on bonds b. Fixing all variables but a single bond b, we integrate a single factor $\sigma(U(\mathrm{b}))$. By the orthogonality of group representations, this integral vanishes for $\sigma \not\equiv I$.

The high temperature result of Osterwalder and Seiler is:

Theorem 2.3. For any choice of representation τ in the Lagrangian of an SU(n) lattice gauge theory, for sufficiently high temperature ($a_\tau << 1$), and for $n_\sigma \neq 0$, it follows that W_σ has the area decay (2.5). In other words, if $\sigma(\mathcal{Z}) \not\equiv I$, then at high temperature σ is a quark representation. Theorem 2.3 establishes the quark criterion (2.1) at high temperature.

The next three theorems establish the particle criterion (2.1) at high temperatures, at least for G = SU(2) and SU(3). Consider the set P of irreducible representations σ such that

$$P = \{\sigma : \sigma \in \Sigma\tau \otimes \tau^*, \text{where } \tau(\mathcal{Z}) = I\} \ . \tag{2.6}$$

At high temperatures, for G = SU(2), SU(3) we show that P is exactly the set of particle representations. This may also hold for SU(n).

Theorem 2.4. For any compact group G, the $\sigma \in P$ are representations of $G/\mathcal{Z}$, and $\oplus_{\sigma\in P}$ is a faithful representation of $G/\mathcal{Z}$.

Proof. Since $\sigma(\mathcal{Z}) = I$, σ is a representation of $G/\mathcal{Z}$. To prove the second statement, let $\gamma \in G$, $\gamma \notin \mathcal{Z}$. Then there exists $\chi \in G$ with $\gamma^{-1}\chi\gamma \neq \chi$ and a function f with $f(\gamma^{-1}\chi\gamma) \neq f(\chi)$. But P results from the decomposition of $U_\gamma: f \to f(\gamma^{-1}\chi\gamma)$ on $L_2(G/\mathcal{Z})$.

Theorem 2.5. For G = SU(2) or SU(3), P is the set of all representations σ for which $\sigma(\mathcal{Z}) = I$. In other words, P is the set of n-ality zero representations.

Proof. This follows from a calculation with Clebsch-Gordon coefficients [8,9,15].

Theorem 2.6. Let dimension $d \geq 3$, G = SU(n), $\sigma \in P$. Then W_σ has length decay at high temperatures. In other words, at high temperature, all the representations P are particle representations.

Proof. Expand $\exp(\mathcal{L}(P))$ to obtain $\tau(U(P))$ for selected plaquettes P. For an oriented bond variable b belonging to exactly two of the selected plaquettes (say $b \in P_1$, $-b \in P_2$) it follows from the orthogonality of group representations that

$$\int dU(b)\, \mathrm{Tr}_{\tau_1} U(P_1)\, \mathrm{Tr}_{\tau_2} U(P_2) = \delta_{\tau_1,\tau_2}\, \mathrm{Tr}_{\tau_1} U(P_1 \cup P_2)\,. \tag{2.7}$$

We choose plaquettes forming a tube which contains the curve C. Then all integrations except those in C can be performed by the convolution formula (2.7). The result is

$$\int dU(b)\, \mathrm{Tr}_\sigma U(C) \prod_{b\in C} \mathrm{Tr}_\tau (U(b)\, \mathrm{Tr}_\tau (U(b)^*)\,. \tag{2.8}$$

For $\sigma \in P$, it follows that $\sigma^* \in P$. For each of these remaining integrals, the orthogonality of group representations selects from (2.8) the σ^* component in $\tau \otimes \tau^*$. This then yields a contribution $a_\tau^{O(\text{length } C)}$ to $W_\sigma(C)$. For a_τ sufficiently small (high temperature) the cluster expansion [4,7] shows that this term dominates the expectation $\langle \cdot \rangle$.

3. Vortices

In this and remaining sections of the paper we consider mechanisms which produce disorder and confinement. Here we are concerned with singular gauge transformations which produce vortex configurations in the gauge field. Since vortex configurations reflect geometrical properties of the gauge group G and space-time R^d, we consider a continuum field A.

The singular gauge transformations considered here are multivalued, and not defined on a manifold $S \subset R^d$. On $R^d \setminus S$ we consider a smooth function $x \to U(x)$ with values in G. Along each path in $R^d \setminus S$ there will be a unique continuation of a single valued branch of $U(x)$. Thus $\Pi_1(R^d \setminus S)$ is the obstruction to a single valued choice of U. Since $\Pi_1(R^d \setminus S)$ is trivial if S has codimension three or more, we suppose that codim S = 2.

With U given, the transformed gauge potential is

$$A' = U^{-1}AU + U^{-1}dU \, . \tag{3.1}$$

We require that A is single values. This is the case for the group G = SU(n) when the ambiguity of $U(x)$ lies in the center $\mathcal{Z}$ of G. In other words, suppose that $x \to U(x)/\mathcal{Z}$ is a single valued gauge transformation defined on $R^d \setminus S$, with gauge group $G/\mathcal{Z}$. In that case, two different values $U_1(x)$, $U_2(x)$ of $U(x)$ give in (3.1),

$$A_1' - A_2' = U^{-1}z^{-1}dzU \tag{3.2}$$

for an element $z \in \mathcal{Z}$. Since $\mathcal{Z}$ is discrete for SU(n), $dz = 0$ and A' is single valued. A is also single valued for $U(1)$.

On $R^d \setminus S$, the gauge transformation (3.1) gives $F' = U^{-1}FU$. On S, F' acquires a Dirac delta contribution. In particular, if μ, ν are orthogonal directions, normal to S, then

$$F'_{\mu,\nu} = U^{-1}F_{\mu\nu}U + 2\pi\alpha\delta_{\mu\nu} \, . \tag{3.3}$$

Here α is a function on S with values in the Lie algebra of G and $\delta_{\mu\nu}$ is the δ-function in the two directions normal to S. This formula follows by considering the line integral of A on a curve C, and using the identity for path ordered integrals.

$$R(\mathrm{C}) = \mathrm{Tr}\ \mathrm{P}e^{\oint_\mathrm{C} A'_\mu \mathrm{d}x^\mu}$$

$$= \mathrm{Tr}\ e^{\int_{\mathrm{Int}\,\mathrm{C}} F_{\mu\nu} \mathrm{d}\sigma^{\mu\nu}} \, , \tag{3.4}$$

where C is an infinitesimal circuit around S (not homotopic to zero). Note that R(C) is just the character of the parallel transport of a fibre around C. Similarly we have (using the assumption that the ambiguity of U lies in $\mathcal{Z}$)

$$R'(\mathrm{C}) = \mathrm{Tr}\ \mathrm{P}e^{\oint_\mathrm{C} A'}$$

$$= \operatorname{Tr} \mathrm{P}e^{\oint_C U^{-1}(A+dUU^{-1})U}$$

$$= \operatorname{Tr} \mathrm{P}e^{\oint_C (A+dUU^{-1})}$$

$$= \operatorname{Tr}(\mathrm{P}e^{\oint_C A}\, e^{\oint_C dUU^{-1}})$$

$$= \operatorname{Tr}(\mathrm{P}e^{\oint_C A}\, e^{\oint_C d \ln U}) = \operatorname{Tr}(z(\mathrm{C})\, \mathrm{P}e^{\oint_C A}) \tag{3.5}$$

If G is simple,

$$R'(\mathrm{C}) = z(\mathrm{C})\, R(\mathrm{C})\,. \tag{3.6}$$

Here $z(\mathrm{C}) \in \mathcal{Z}$ is the element of the center given by the homotopy class of C in $\Pi_1(\mathrm{R}^d \setminus \mathrm{S})$. Rewriting (3.5) in infinitesimal form, and using (3.4) yields (3.3).

If Z is discrete, α in (3.3) is normalized in the sense that $z = e^{2\pi\alpha} \in \mathcal{Z}$ is constant on connected components of S. The singular gauge transformation U introduces vorticity into A. The gauge transformation satisfies the 't Hooft commutation relations [3] with the Wilson function $R(\mathrm{C})$. Then (3.6) is the commutation rule

$$R'(\mathrm{C}) = z^j R(\mathrm{C}) \tag{3.7}$$

where j is the winding number of the curve C, and $z = \exp(2\pi i/n)$ for $\mathrm{G} = \mathrm{SU}(n)$.

Because of the δ function in F', the action corresponding to F' is infinite (quadratically divergent). Thus an ultraviolet cutoff is required. We introduce the cutoff in the form of a lattice, and in the lattice theory we define the effective coupling constant g_{eff} by the lattice length and the renormalization group. Then g_{eff} contains all effects of fluctuations on scales shorter than the lattice spacing, and these fluctuations have been integrated out of the theory.

We introduce S as a surface in the dual lattice of codimension 2. Suppose S is the boundary of an oriented hypersurface Σ,

$$\mathrm{S} = \partial\Sigma\,.$$

This assumption leads to $\partial\mathrm{S} = 0$ and the Bianchi identity for F', and will be justified statistically at high temperature in sect. 7. Given $z \in \mathcal{Z}$, we define a multivalued gauge transformation $U(x)$ on sites of the original lattice, to be constant on each side of Σ, and to change by a factor of z on crossing Σ. The resulting $\gamma(\mathrm{b})$ for a bond b transforms as

$$\gamma(\mathrm{b})' = \begin{cases} \gamma(\mathrm{b})z, & \text{if b crosses } \Sigma \text{ in a + direction,} \\ \gamma(\mathrm{b}), & \text{otherwise,} \end{cases}$$

while the plaquette $U(\mathrm{P})$'s have the transformation law

$$U(\mathrm{P})' = \begin{cases} U(\mathrm{P})z, & \text{if P winds S in a + direction,} \\ U(\mathrm{P}), & \text{otherwise.} \end{cases}$$

It follows that the action is modified by an amount proportional to the $d-2$ dimensional area $|S|$ of S. These singular gauge transformations form an Abelian group; they are Euclidean regularizations of 't Hooft's disorder operators. See also [14].

4. The vortex phase transition

The action of the singular gauge transformation is proportional to $g_{\text{eff}}^{-2}|S|$, while its entropy is proportional to $|S|$. For large g_{eff}, the entropy dominates the action and large sized vortex sheets are favored.

One can see that this leads to an area decay in $W_\sigma(\text{C}) = \langle R_\sigma(\text{C})\rangle$ since the number of independent degrees of freedom in the vortex fluctuation is proportional to the area of C, when we are restricted to vortex sheets S passing through the surface spanned by C. With S unbounded, all such vortex sheets (not only those near C) may pass through this surface once, or a non-zero number (mod $n-1$) of times. Then [3,14]

$$R(\text{C}) = z^{\Sigma\,(\text{S, C winding numbers})}\,,$$

and

$$W_\sigma(\text{C}) = \langle R_\sigma(\text{C})\rangle = \text{e}^{-\text{O(no. S degrees of freedom)}}$$

$$= \text{e}^{-\text{O(area)}}\,.$$

We now argue that the effects of interaction is to lower the critical value of g_{eff}, at least for G = SU(2). For SU(2), singular sets cancel on overlap. Thus the composition of singular gauge transformations with singular sets S_i, $i = 1, 2$ is equivalent to a single singular gauge transformation with singular set

$$\text{S} = \text{S}_1 \Delta \text{S}_2 = \text{S}_1 \cup \text{S}_2 \setminus (\text{S}_1 \cap \text{S}_2)\,.$$

Thus overlap reduces the action, favoring more singular sets S_i at any fixed value of g_{eff}. Overlap does not change the (mod 2) winding number

$$\text{winding number } (\text{C}, \text{S}_1 \Delta \text{S}_2) = \sum_{i=1}^{2} \text{winding number } (\text{C}, \text{S}_i), \qquad (\text{mod } 2).$$

One can similarly show that the interaction of S with small, but arbitrary fluction $U(\text{P})$ is attractive. In fact, if the plaquette P winds S, then

$$\begin{aligned} \mathcal{L}'(\text{P}) - \mathcal{L}(\text{P})|_{U=U(\text{P})} &\equiv -\text{Tr}(zU(\text{P})) + \text{Tr}\, U(\text{P}) \\ &= 2\,\text{Tr}\, U(\text{P}) \\ &\leqslant \mathcal{L}'(\text{P}) - \mathcal{L}(\text{P})|_{U=1} \\ &\equiv -\text{Tr}(z) + \text{Tr}\, I = 2\cdot\text{Tr}\, I\,. \end{aligned}$$

The right-hand side is the incremental action of a singular gauge at P, in the presence of a small fluctuation $U(\mathrm{P})$, and the left-hand side is the incremental action of a singular gauge at P, in the presence of the vacuum at P.

For SU(3), the interaction of singular sets S_i again lowers the critical coupling, and presumably the effect of the interaction of S with small fluctuations also lowers the critical coupling.

5. Merons with O(2) symmetry

The influence of merons on confinement has been analyzed previously, using classical meron configurations with O(4) and O(3) symmetry, the suppressed degrees of freedom being reintroduced statistically. Here we use an O(2) symmetric ansatz, *cf.* Taubes [10], which, because of its lower symmetry, allows for arbitrary meron pair locations in a plane to be treated classically. We define the plane by

$$\rho \equiv (x^2 + y^2)^{1/2} = 0 .$$

In a suitable gauge (for which the volume and surface charge densities vanish), the merons satisfy boundary conditions

$$F_{z,t}(\rho = 0) = \pi i \sum_j \pm\delta(z - z_j)\delta(t - t_j) .$$

For a loop C lying in the place $\rho = 0$, the integrand $A_\mu \, \mathrm{d}x^\mu$ lies in an Abelian subalgebra, and so we can evaluate

$$\oint_{\mathrm{C}} A_\mu \, \mathrm{d}x^\mu = \int_{\mathrm{Int\,C}} F_{zt} \, \mathrm{d}z \, \mathrm{d}t$$

$$= i\pi(n_+ - n_-) ,$$

where n_+ (n_-) is the number of merons (antimerons) within C. Thus

$$R(\mathrm{C}) \equiv \mathrm{e}^{\oint A_\mu \, \mathrm{d}x^\mu} = (-1)^{n_+ - n_-} .$$

This formula leads to

$$\langle R(\mathrm{C})\rangle = \mathrm{e}^{-\mathrm{O}(N)} ,$$

where N is the number of effective meron degrees of freedom (the number of independent choices of -1) in C.

At low densities, where the separation between meron pairs is large relative to the separation within a given pair, the pairing of the merons at given locations is essentially unique. Pairs far away from the boundary, C, do not contribute to N, because there is no plausible rearrangement of pairing. Stated differently, a single meron could not be introduced into the interior, because there is no flexibility to allow a sequence of repairing, which would bring in a partner from outside C.

At high densities, the situation is different. When the distance between distinct pairs is similar to the distance within pairs, the choice of pairings is not uniquely determined by the meron locations. In this sense the meron pairs at high density resemble a fluid. A single meron can consistently be introduced into or removed from the interior of C, because the required sequence of repairings of other merons is allowed. Thus in this case all merons within C contribute to N.

Our dense fluid picture results from the idea that meron pairs do not dissociate, either classically or quantum mechanically [2]. Our picture is distinct from the picture of Callan, Dashen and Gross, in which the merons dissociate into a plasma [1], but was inspired by their picture.

Various evidence supports our picture of non-dissociation. A classical analysis by Gidas [19] of infinitesimal deformations about the two-meron solution indicates that there is no relative group freedom, i.e., an 8-parameter family of gauge-inequivalent solutions. This is consistent with the pairings of classical merons on a line [20]. The classical instability of regularized merons by half instantons was known to Callan, Dashen and Gross, *cf.* also [19,21]. Furthermore, it was shown in [22] that dissociated merons are unstable quantum mechanically.

In summary we have $N = O(\text{length})$ at low density (small g_{eff}) and $N = O(\text{area})$ at high density (large g_{eff}). Thus merons are capable of causing the confinement transition.

The SU(3) merons which have been analyzed to date [16] seem to be totally unrelated to triality and to the center $\mathcal{Z}$ of SU(3) (*cf.* sect. 2).

6. Comparison of critical couplings

With an action

$$A = \tfrac{1}{8} g_{\text{eff}}^{-2} \operatorname{Tr} \int F_{\mu\nu}^2(x)\, dx$$

$$= \tfrac{1}{4} g_{\text{eff}}^{-2} \int F_{\mu\nu}^a(x)\, F_a^{\mu\nu}(x)\, dx \,, \qquad (6.1)$$

Callen, Dashen and Gross [1] have estimated the critical value of g_{eff}^2 at

$$g_{\text{c, meron}}^2 = 5.8 \,. \qquad (6.2)$$

Here g_{eff}^2 is defined in terms of a renormalization group transformation, and is a function of the length scale.

Because interactions lower $g_{\text{c, vortex}}^2$, we get an upper bound on the critical temperature from the Z_2 lattice gauge theory. The latter theory is self-dual, so its critical temperature is known in closed form [11]. With the action

$$A = \tfrac{1}{4}g^{-2}\,\mathrm{Tr}\sum_{p} U_p\,,$$

we have

$$g^2_{\mathrm{c,\,vortex,\,lattice}} \leqslant g^2_{\mathrm{c,\,Z_2,\,lattice}} = 1.14\,. \tag{6.3}$$

To find the equivalent critical coupling $g^2_{\mathrm{c,\,vortex,\,YM}}$ for the Yang-Mills interaction (6.1), we follow the method of Foerster [14] (omitting the factor of 2 for a Higgs field). However in place of $g^2_{\mathrm{c,\,Z_2}}$ from (6.3), we begin this calculation with $g^2_{\mathrm{c,\,vortex,\,lattice}} = 0.85$ (see below). This gives

$$g^2_{\mathrm{c,\,vortex,\,YM}} = \pi^2 g^2_{\mathrm{c,\,vortex,\,lattice}} = 8.4\,, \tag{6.4}$$

in order of magnitude agreement with the meron critical value (6.2). We now discuss the correction to (6.3) due to the effect of the SU(2) fluctuations on the Z_2 critical point.

It is convenient to expand $e^{\mathcal{L}}$ in terms of SU(2) group characters. Let Tr_j be the trace of the spin-j representation of SU(2). Then

$$e^{g^{-2}\mathrm{Tr}_{1/2}} = \sum_{j\ \text{half integer}} \alpha_j(g)\,\mathrm{Tr}_j\,. \tag{6.5}$$

Theorem 6.1. In the expansion (6.5),

$$\alpha_j(g) = (2j+1)\,g^2 I_{2j+1}(2g^{-2})\,, \tag{6.6}$$

where $I_\nu(z)$ is the modified Bessel function of the first kind,

$$I_\nu(z) = \sum_{k=0}^{\infty} \frac{(\frac{1}{2}z)^{\nu+2k}}{k!(\nu+k)!}\,.$$

Proof. The effect of multiplication, $\otimes^n\delta \to \otimes^{n+1}\delta$, where δ is the spin-$\frac{1}{2}$ representation of SU(2), generates a random walk on the Clebsch-Gordon coefficients, with an absorbing barrier at $j = -\frac{1}{2}$. This random walk can be solved explicitly in terms of binomial coefficients, see [12], and yields

$$\alpha_j(g) = \sum_{k=0}^{\infty} g^{-4(k+j)} \frac{2j+1}{k!(k+2j+1)!}\,,$$

to complete the proof.

We define the single-site expectation

$$\langle \mathrm{Tr}_j\, U\rangle \equiv \int \mathrm{Tr}_j\, U e^{\mathcal{L}}\, \mathrm{d}U/\text{normalization} = \alpha_j/\alpha_0\,,$$

and so

$$\lim_{g\to 0} \alpha_j/\alpha_0 = 2j + 1 \,, \tag{6.7}$$

$$\lim_{g\to \infty} \alpha_j/\alpha_0 = \delta_{j,0} \,. \tag{6.8}$$

We use the first term in the series, to determine the factor

$$|1 - \tfrac{1}{4}(2 - \langle \mathrm{Tr}_{1/2}\, U\rangle)| = \tfrac{1}{2} + \tfrac{1}{4}(\alpha_{1/2}/a_0) \,,$$

which we regard as a measure of the change in length of the Z_2 spins due to interaction with the SU(2) fluctuations. This change in spin length can be compensated for by a change in coupling constant renormalization, so that

$$g^2_{\mathrm{c,\,vortex}} \simeq (\tfrac{1}{2} + \alpha_{1/2}/4\alpha_0)\; 1.14 \,,$$

or

$$g^2_{\mathrm{c,\,vortex}} \simeq 0.85 \,.$$

7. The high-temperature limit

At high temperature, we attempt a coherent description of the large-scale fluctuations in the gauge field. Vortices play a role, as do magnetic (dipole) charge loops and topological point charges (i.e., merons and instantons). These fluctuations occur on singular sets of dimension 2, 1, 0, respectively.

For a general configuration γ of an SU(2) lattice gauge field, define the set of plaquettes $X = X_\gamma$ as those with $\mathrm{Tr}\, U(P) < 0$. We wish to approximate γ by a vortex configuration γ_S defined by the singular set S, so that

$$U_S(P) = \mathrm{sgn}\; \mathrm{Tr}\; U(P) \,, \tag{7.1}$$

for most plaquettes P and most configurations γ. The problem is that the dual surface X^* is not in general closed, and not in general the boundary of a hypersurface Σ. However, without loss of generality, we may add or subtract from X^* arbitrary closed 2-surfaces S, since these are boundaries of hypersurfaces Σ. Thus, without loss of generality, we may suppose that X^* is the minimal 2-surface with boundary ∂X^*. Stated differently, minimal surfaces with given boundaries ∂X^* span the degrees of freedom for sgn Tr $U(P)$ which are not already included in the vortex degrees of freedom. Because $U(P) = -1$ is unstable relative to $U(P) = +1$, the action of this degree of freedom is proportional to the area $|X^*|$. However, the entropy of this degree of freedom is given by a length $|\partial X^*|$. Thus entropy never dominates over action for large connected portions of X^*. Hence the minimal surfaces X^* remain small (or at least narrow) deviations in sgn Tr $U(P)$, and the dominant large scale fluctuations in sgn Tr $U(P)$ are vortices.

We remark that ∂X^* are magnetic charge loops, and were analyzed at low density in the Abelian case [13]. Because of the cancellation from opposite sides of the loops, the magnetic charge loops are presumably dipoles, hence have forces which are too short range to influence confinement.

References

[1] C. Callan, R. Dashen and D. Gross, Phys. Rev. D17 (1978) 2717.
[2] J. Glimm and A. Jaffe, Phys. Rev. D18 (1978) 463.
[3] G. 't Hooft, Nucl. Phys. B138 (1978) 1.
[4] K. Osterwalder and E. Seiler, Ann. of Phys. 110 (1978) 440.
[5] A.M. Polyakov, Thermal properties of gauge fields and quark liberation, preprint.
[6] C. Chevalley, The theory of Lie groups (Princeton Press, 1946).
[7] J. Glimm, A. Jaffe and T. Spencer, *in* Constructive field theory, Springer Lecture Notes in Physics, vol. 25 (1974).
[8] S. Coleman, J. Math. Phys. 5 (1964) 1343; 6 (1965) 2014.
[9] W. Miller, Symmetry groups and their applications (Academic press, New York, 1972).
[10] C. Taubes, O(2) symmetric solutions to an SU(2) Yang-Mills theory.
[11] R. Balian, J. Drouffe and C. Itzykson, Phys. Rev. D11 (1975) 2098.
[12] W. Feller, Introduction to probability (Wiley, New York, 1957).
[13] J. Glimm and A. Jaffe, Comm. Math. Phys. 56 (1977) 195.
[14] D. Foerster, On quark confinement in a condensate of non-Abelian vortices; An estimate of the coupling needed for vortices to condense in Yang-Mills theory, Cornell preprints.
[15] G. de B. Robinson, Representation of symmetric group (Univ. of Toronto Press, 1961).
[16] J.Z. Imbrie, O(3) symmetric merons in an SU(3) Yang-Mills theory, Lett. Math. Phys. 2 (1978) 483.
[17] C. Callan, R. Dashen and D. Gross, Phys. Lett. 66B (1977) 345.
[18] R. Israel and C. Nappi, Comm. Math. Phys., to appear.
[19] B. Gidas, Deformations and spectral properties of merons, J. Math. Phys., to appear.
[20] J. Glimm and A. Jaffe, Phys. Lett. 73B (1978) 167.
[21] D. Williams, Mathematical remarks on positivity and self-adjointness of the Euclidean, Yang-Mills stability operator, preprint.
[22] J. Glimm and A. Jaffe, Phys. Rev. Lett. 40 (1978) 277.
[23] M. Lüscher, Absence of spontaneous gauge symmetry breaking in Hamiltonian lattice gauge theories, DESY 77/16.

VIII Reflection Positivity

Introduction

Reflection positivity of a Gaussian measure is shown to be equivalent to monotonicity of the Green's function for the Laplace operator with respect to classical Dirichlet or Neumann boundary conditions.

Bibliography

A note on reflection positivity, Lett. Math. Phys. 3, 377–378 (1979).

A NOTE ON REFLECTION POSITIVITY

JAMES GLIMM[1]
The Rockefeller University, New York, NY 10021, U.S.A.

and

ARTHUR JAFFE[2]
Harvard University, Cambridge, MA 02138, U.S.A.

ABSTRACT. We establish reflection positivity of covariance operators, using properties of Dirichlet or Neumann boundary data.

Osterwalder and Schrader isolated a basic positivity property in classical statistical mechanics which leads to the construction of a self adjoint contraction semigroup $\exp(-tH)$ acting on a Hilbert space $\mathcal{H}$. This yields the standard connection between quantum mechanics and path integrals. Verification of their condition in specific examples ultimately relies on a particular positivity condition for the inverse $(-\Delta_B + I)^{-1}$ of a Laplace operator Δ_B acting on $L_2(R^d)$. Here B denotes classical boundary conditions on the finite union of piecewise smooth hypersurfaces Γ in R^d. We show here that the positivity condition can be derived from monotonicity $C_D \leqslant C_N$ of Dirichlet and of Neumann Laplace operators.

Let Π denote a hyperplane in R^d and let $\theta = \theta_\Pi$ denote reflection in Π. Let $R^d_\pm$ denote the two connected components of $R^d \backslash \Pi$. Note that on $L_2(R^d)$, θ is self adjoint and $\theta^2 = I$. Let $\Pi_\pm$ denote the orthogonal projection onto $L_2(R^d_\pm)$.

DEFINITION 1. *An operator C is θ invariant if* $[\theta, C] = 0$.

DEFINITION 2. *A θ invariant operator C is OS positive with respect to θ if*:

$$\Pi_+ \theta C \Pi_+ \geqslant 0. \tag{1}$$

Remark. If $[\theta, C] = 0$, then (1) is equivalent to $\Pi_- \theta C \Pi_- \geqslant 0$.

THEOREM 1. *Let $C = (-\Delta_B + I)^{-1}$ be θ-invariant and let B denote Dirichlet and/or Neumann data on Γ. Then C is OS positive.*

---

[1] Supported in part by National Science Foundation Grant No. PHY 78-08066.
[2] Supported in part by National Science Foundation Grant No. PHY 77-18762.

Proof. Let $P_\pm = \frac{1}{2}(I \pm \theta)$ be the projections onto the even and odd eigenspaces of θ. Note $(I - \theta)C = 2P_-CP_-$, so the positivity condition (1) can be written:

$$\Pi_+ [C - 2P_-CP_-] \Pi_+ \geqslant 0. \tag{2}$$

We assert first, that $2P_-CP_-$, restricted to $\Pi_\pm$, equals C_D restricted to $\Pi_\pm$. Here $C_D = (-\Delta'_B + I)^{-1}$ where Δ'_B is the Laplace operator with Dirichlet data on Π and B data on $\Gamma\backslash\Pi$. Secondly, we assert $C_D \leqslant C$. Hence (2) is a consequence of these assertions.

The first assertion follows from the observation that $(I - \theta)C(x, y)$ vanishes on Π, and that $(I - \theta)C$ satisfies $(-\Delta + I)(I - \theta)C(x, y) = \delta(x - y)$. The second assertion follows from the characterization of the bilinear form $\Pi_\pm C_D^{-1} \Pi_\pm$ as a restriction of the bilinear form C^{-1} to functions vanishing on Π and supported in $\Pi_\pm$.

We remark that on $\Pi_\pm$, $2P_+CP_+ = (I + \theta)C = C_N \geqslant C$ is the Neumann covariance, and a proof could be based on these facts. Note also that on $\Pi_\pm$, $\theta C = \frac{1}{2}(C_N - C_D)$.

We now allow periodic boundary conditions. Consider R^d replaced by the torus T^d; replace Π by a cross-section

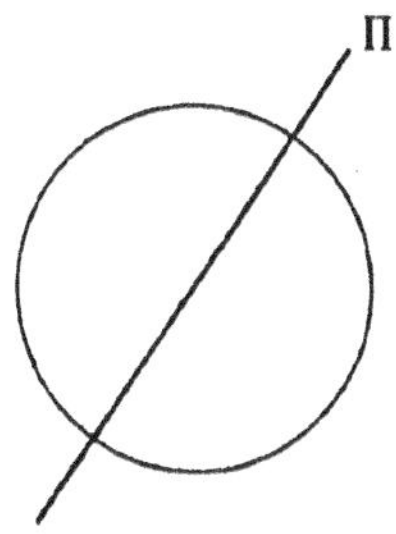

Fig. 1. Hyperplane reflection for a torus.

dividing T^d in two equal parts, cf. Fig. 1. Now $\Pi_\pm$ are the projections onto L_2 functions supported on either side of Π. We allow Dirichlet and/or Neumann data B on Γ as well. With these replacements, Definitions 1 and 2 are made. As above, we show:

THEOREM 2. *Let* $C = (-\Delta_B + I)^{-1}$ *be* θ*-invariant on* $L_2(T^d)$. *Then* C *is OS positive.*

Finally we remark that R^d or T^d can be replaced by a lattice R^d_δ or T^d_δ with spacing δ. In this case we obtain *OS* positivity for classical, θ-invariant boundary conditions on a lattice. Again we use the inequality $C_D \leqslant C$, which can also be established by introducing Dirichlet data with a mass perturbation of C^{-1} on the lattice.

THEOREM 3. *The lattice versions of Theorems* 1 *and* 2 *hold.*

REFERENCE

Osterwalder, K., and Schrader, R., *Commun. Math. Phys.* **31**, 83–112 (1973); **42**, 281–305 (1975).

(Received July 2, 1979)

378

GPSR Compliance

The European Union's (EU) General Product Safety Regulation (GPSR) is a set of rules that requires consumer products to be safe and our obligations to ensure this.

If you have any concerns about our products, you can contact us on ProductSafety@springernature.com

In case Publisher is established outside the EU, the EU authorized representative is:

Springer Nature Customer Service Center GmbH
Europaplatz 3
69115 Heidelberg, Germany

Batch number: 10732714

Printed by Printforce, the Netherlands